航天科技图书出版基金资助出版

导弹与航天发射试验
常用缩略语手册

《导弹与航天发射试验常用缩略语手册》编委会　编

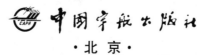

中国宇航出版社
·北京·

版权所有　侵权必究

图书在版编目(CIP)数据

导弹与航天发射试验常用缩略语手册／《导弹与航天发射试验常用缩略语手册》编委会编．--北京：中国宇航出版社，2014.10

ISBN 978-7-5159-0806-9

Ⅰ.①导… Ⅱ.①导… Ⅲ.①导弹发射－发射试验－缩略语－手册②航天器发射－发射试验－缩略语－手册 Ⅳ.①TJ760.6-62②V553-62

中国版本图书馆 CIP 数据核字(2014)第 232169 号

责任编辑	彭晨光		
责任校对	祝延萍	**封面设计**	文道思
出版发行	中国宇航出版社		
社　址	北京市阜成路 8 号	**邮　编**	100830
	(010)68768548		
网　址	www.caphbook.com		
经　销	新华书店		
发行部	(010)68371900	(010)88530478(传真)	
	(010)68768541	(010)68767294(传真)	
零售店	读者服务部	北京宇航文苑	
	(010)68371105	(010)62529336	
承　印	北京画中画印刷有限公司		
版　次	2014 年 10 月第 1 版	2014 年 10 月第 1 次印刷	
规　格	880×1230	**开　本**	1/32
印　张	20.125	**字　数**	825 千字
书　号	ISBN 978-7-5159-0806-9		
定　价	268.00 元		

本书如有印装质量问题，可与发行部联系调换

航天科技图书出版基金简介

航天科技图书出版基金是由中国航天科技集团公司于2007年设立的，旨在鼓励航天科技人员著书立说，不断积累和传承航天科技知识，为航天事业提供知识储备和技术支持，繁荣航天科技图书出版工作，促进航天事业又好又快地发展。基金资助项目由航天科技图书出版基金评审委员会审定，由中国宇航出版社出版。

申请出版基金资助的项目包括航天基础理论著作，航天工程技术著作，航天科技工具书，航天型号管理经验与管理思想集萃，世界航天各学科前沿技术发展译著以及有代表性的科研生产、经营管理译著，向社会公众普及航天知识、宣传航天文化的优秀读物等。出版基金每年评审1～2次，资助10～20项。

欢迎广大作者积极申请航天科技图书出版基金。可以登录中国宇航出版社网站，点击"出版基金"专栏查询详情并下载基金申请表；也可以通过电话、信函索取申报指南和基金申请表。

网址：http://www.caphbook.com
电话：(010) 68767205，68768904

《导弹与航天发射试验常用缩略语手册》
编委会

主　　任：万　全
副 主 任：侯军祥　温庆林　刘占卿
委　　员：刘晓华　肖力田　孙雅度　徐原平　于克俭
　　　　　孙　冲　黄　健　周　旭　宋震宇　王志敏
　　　　　卢　葭　李茂峰　邢科伟　王迎东　吴涧彤
　　　　　王　坚　雷向阳　刘　阳　张　统　曹宗胜
　　　　　徐学文

《导弹与航天发射试验常用缩略语手册》
编辑部

主　　编：李　洁
副 主 编：郭　凯　赵　晨
责任编辑：江红斌　任　飞　杨彩怡　李婷婷　徐元元
　　　　　徐晓蓝　宋紫娇　刘梦梦　王　飞　李玉良
　　　　　任　刚　蒲　婷　饶　雪
校　　审：孙　冲　卢连成　刘　鹰　张凤林　王东峰
　　　　　郝建平
版面设计：黄增光　刘　牧

前　言

进入 21 世纪以来，世界主要国家的导弹与航天技术创新发展愈加迅速，相关技术领域涌现出大量的新事物、新概念、新知识，与此相随的是导弹与航天科技文献中出现了大量用以表达各种概念和术语的缩略语。这些缩略语主要采取单词缩略和词组缩略两种形式，尤以词组缩略数量为最多，表达简便、节省篇幅、使用密度高、缩略范围广，且新词生成快，其广泛使用便利了交流，促进了沟通，也极大地丰富了语言表达。但缩略语的使用也存在着其负面问题，一是导弹与航天领域所涉及的各种技术学科之间交叉融合度较大，在缩略语使用过程中出现了缩略语和被缩略的词组之间并不是一一对应的，同一个缩略语在不同的领域有不同意义的情况；二是缩略语的应用使简明、扼要的语言因其本身失去了表达的清晰度而造成语义模糊，引起理解上的困难；三是科学技术的突飞猛进使大量新缩略语涌现，其出现和发展速度之快令人们措手不及，科研人员需要一段时间和过程才能真正接受并掌握新缩略语。由于目前缺乏一本合适的工具书，科技情报人员和科研人员常常为确定某个释义而耗费大量精力。我们在结合实际工作的基础上，迫切感到有必要编写一本针对导弹与航天发射试验缩略语方面的通用工具书，以此满足导弹与航天技术领域科研人员的现实需求。

《导弹与航天发射试验常用缩略语手册》共收录约 22 000 词条，在词条收录方面着眼于导弹与航天技术领域的科研单位的科技情报研究与科研试验任务需要，侧重收录与国外导弹和航天发射试验及空间飞行任务相关的、具有一定代表性和战略意义的缩略语，主要内容包括：①与空间运载器、航天器（卫星）、飞船、空间站、空间探测相关的学科领域；②导弹的试验、鉴定、发射、应用；③航天发射场及导弹靶场建设发展，设施设备维护等；④工程项目管理；⑤与导弹航天发射试验密切相关的光学、电子、红外、仿真、雷达等学科领域。对入选的词条力求达到译文准确、用词规范、体例科学、编排合理、使用方便。

《导弹与航天发射试验常用缩略语手册》是科技情报研究人员在长期跟踪、收集国外导弹和航天发射试验靶场发展各类文献资料的基础上，结合专业需求并在科研人员协助下完成的一项全面系统的成果。它借鉴和吸收了同类书籍的成功做法，通过横向的比较、补充、修正，提高了本书的专业针对性，学术性较高、先进性、新颖性、实用性较强。我们希望本书不仅能够为导弹与航天技术领域的翻译与科研人员发挥重要的参考作用，而且也能成为其他学科领域人员工作中有价值的工具书。

在本书编写过程中，得到了各级领导和部门的大力关怀与支持，得到了导弹与航天技术领域有关专家的大力指导与帮助，在此，我们特表谢忱，并致敬意！

由于导弹与航天系统是一个综合性系统，其涵盖的学科与专业领域非常广泛，具有一定的复杂性。加之我们水平和掌握的资料有限，书中难免出现不妥和疏漏之处，恳请各界读者不吝赐教，以便我们改进和修订。

《导弹与航天发射试验常用缩略语手册》编委会

2014 年 7 月

目 录

A	……………	1
B	……………	84
C	……………	98
D	……………	141
E	……………	173
F	……………	210
G	……………	233
H	……………	252
I	……………	269
J	……………	296
K	……………	303
L	……………	307
M	……………	343
N	……………	397
O	……………	414
P	……………	437
Q	……………	472
R	……………	477
S	……………	504
T	……………	582
U	……………	608
V	……………	613
W	……………	622
X	……………	629
Y	……………	632
Z	……………	633

A

30WS 30th Weather Squadron 【美国空军】第 30 翼气象中队

45WS 45th Weather Squadron 【美国空军】第 45 翼气象中队

4DLSS 4-Dimensional Lightning Surveillance System 四维雷电监视系统

4DT 4-Dimensional Trajectories 四维轨迹

A Air-Launched 空中发射的

A Forward Acceleration 【代号】前向加速度

A Inertia in Rolling Plane 【代号】滚动平面惯量

A Linear Acceleration 【代号】线性加速度

A Mean Sound Absorption Coefficient 【代号】平均吸声系数

A Power Required to Hover 【代号】空中悬停所需功率

A Slope of Lift Curve 【代号】升力曲线斜度

A Velocity of Sound 【代号】音速；声速

A Spec System Specification 系统说明

A&B Assault and Barrier 突击与屏障

A&CO Acceptance and Checkout 接收与检测

A&CO Assembly and Checkout 装配与检测

A&D Aerospace and Defense 航空航天与防御

A&D Assembly and Disassembly 装配与拆卸

A&DF Assembly and Disassembly Facility 装配/拆卸设备

A&E Aerospace and Equipment 航空航天与设备

A&E Ammunition and Explosives 弹药与爆炸物

A&E Analysis and Evaluation 分析与评估

A&E Architecture and Engineering 体系结构与工程

A&E Azimuth and Elevation 方位角与仰角

A&FM Aerodynamics and Flight Mechanics 空气动力学与飞行力学

A&I Alteration and Inspection 变换与检查

A&I Assembly and Installation 装配与安装

A&L Approach and Landing 进场与着陆

A&MD Air and Missile Defense 防空与导弹防御

A&P Apogee and Perigee 远地点与近地点

A&P Attitude and Pointing 姿态与瞄准

A&PS Administration and Program Support (MSFC Directorate) 【美国马歇尔航天飞行中心管理局】行政与计划保障

A&R Assemble and Repair 装配与修理

A&R Automation and Robotics 自动控制与机器人技术

A

A&RC　Application and Resource Control　应用与资源控制

A&S　Avionics and Software　航空电子设备与软件

A&T　Acceptance and Transfer　接收与传递

A&T　Acquisition and Technology　采办与技术

A&T　Administration and Training　管理与训练

A&T　Assembly and Test　组装与测试

A&TA　Assembly and Test Area　组装与测试区；装配与测试区

A&TP　Assembly and Test Procedure　装配与试验程序；组装与测试程序

A&TQC　Assembly and Test Quality Control　装配与试验质量控制

A&V　Assembly and Verification　装配与验证

A-A MISSILEX　Air-to-Air Missile Exercise　空－空导弹演习

A-D　Analog to Digital　模拟－数字转换

A-RM　Anti-Radio Missile　反无线电导弹

A-S MISSILEX　Air-to-Surface Missile Exercise　空对地导弹演习

A/A　Air to Air　空对空，空空

A/B　Airborne　空运的；机载的；空中的

A/BE　Airborne Equipment　机载设备

A/BF&D　Airborne Fill-and-Drain　空中加注与排放；空中装填与排放

A/BPI　Ascent/Boost-Phase Interceptor　上升/助推段拦截弹

A/BPS　Airborne Power System　机载动力系统

A/BPS　Airborne Propellant System　机载推进系统

A/C　Approach Control　接近管制

A/C　Attitude Control　姿态控制

A/CASP　Air Conditioning Analytical Simulation Program　空气调节分析模拟程序

A/D　Analog to Digital　模拟－数字转换

A/D　Arm/Disarm　打开保险/解除保险

A/D　Assembly and Disassembly　装配与拆卸

A/E　Activity Elements　活性元素

A/EGM　Attack/Effects Guidance Matrix　攻击/效果制导模型

A/F　Air-to-Fuel (Ratio)　空气与燃料比；混合气比

A/FM　Air/Firing Mechanism　空气/击发机构（点火机构）

A/FTP　Acceptance Function Test Procedure　验收功能试验程序

A/G　Air to Ground　空对地

A/G/A　Air-to-Ground-to-Air　空－地－空

A/H　Alter Heading　改变航向

A/L　Air Launch　空中发射

A/L　Air Lock　气闸室；风闸装置；密封（压差隔离，气压过渡）舱

A/L　Approach and Landing　进场和着陆

A/L　Autoland　自动着陆（系统）

A/M　Auto/Manual　自动/手动

A/M　Automatic/Manual　自动（或）手控的

A/MLCMC　Aviation/Missile Life Cycle Management Command　航空/导弹寿命周期管理司令部

A/O　Analog /Output　模拟/输出

A/P　Active/Passive　主动/被动；有源/无源

A/P　Antennas and Propagation　天线

与传播

A/P Automatic Pilot 自动驾驶（仪）

A/PMCU Autopilot Monitor and Control Unit 自动驾驶仪监视与控制装置

A/R Acquisition Radar 搜索雷达；捕获雷达

A/R Autoland Rollout 自动着陆滑行

A/S Airspeed 风速；气流速度；飞行（排气）速度

A/S Arm/Safe 解脱（除）保险／安全；解脱（除）保险与保险

A/S Ascent Stage 火箭上升级

A/S Auxiliary Stage 火箭辅助推进级

A/STAB Auto-Stabilizer 自动稳定器；增稳器（装置）

A/T Action Time 作战时间；作用时间

A/T Angle Tracker 角度跟踪仪（器）

A/T Average Handing Time 平均处理时间

A/T/RM Antenna Transmit/Receive Module 天线发射／接收组件

A1-C and A6-L Early Apollo Suit Prototypes 【美国】早期"阿波罗"登月飞船适配样机

A^{2}ATD Anti-Armor Advanced Technology Demonstration 先进反装甲技术演示

A^{2}C^2 Army Airspace Command and Control 【美国】陆军空域指挥与控制

A^{2}C^{2}S Army Airborne Command and Control System 【美国】陆军空中指挥与控制系统

A^3 Armor/ Anti-Armor 装甲／反装甲

A7-L and A7-LB Actual Apollo Pressure Suits Worn on Missions 【美国】实际任务中的"阿波罗"飞船增压服磨损

AA Absolute Address 绝对地址

AA Absolute Altitude 绝对高度

AA Accelerated Assemblies 加速装配

AA Accelerometer Assembly 加速度计装配

AA Acquisition Activity 捕获；获取

AA Acute Angle 锐角

AA Adaptive Array 自调谐天线阵；自适应阵列

AA Affordability Analysis 费用可承受性分析

AA Angular Accelerometer 角加速度计

AA Antenna Assembly 天线装置

AA Apollo Applications 【美国】"阿波罗"航天应用项目

AA Arrival Angle 入射角；到达角

AA Ascent Abort 上升中止

AA Assembly Area 装配区

AA Attack Assessment 袭击判断；攻击效果评估

AA Auto Acquisition (Radar) 自动捕获；自动搜索（雷达）

AA Auto Analyzer 自动分析仪

AA Automatic Answer 自动应答

AA Automatic Approach 自动进场

AA&D Arming Assembly and Device 解除保险组件与装置

AA/AL Automatic Approach/Autoland 自动进场／自动着陆

AAA Active Acquisition Aid 主动搜索（目标）辅助设备

AAA Airborne Assault Area 空中突击区域

AAA Alternate Assembly Area 备用装配场（区）

AAA Astronaut-Actuated Abort 由航天员操纵的中止

AAA Avionics Air Assembly 航空电子设备气动组件

AAAF Association Aéronautique et Astronautique de France 法国航空航天协会

AAAI American Association for Artificial Intelligence 美国人工智能协会

AAAM Advanced Air-to-Air Missile 先进空－空导弹

AAAMA Air-to-Air Armament Mission Analysis 空－空武器任务分析

AAAMLE Advanced Air-to-Air Missile Launch Envelope 先进空－空导弹发射包线

AAAS American Association for the Advancement of Science 美国科学促进会

AAAS Automated Antenna Alignment System 自动天线校准系统

AAAW Air-Launched Anti-Armour Weapon (UK RAF term) 【英国】空射反装甲武器

AAB Adaptive Angle Bias 自适应角偏差

AABCP Advanced Airborne Command Post 高级空中指挥所

AABM Airborne Anti-Ballistic Missile 机载反弹道导弹

AABNCP Advanced Airborne National Command Post 国家高级空中指挥所

AABS All Attitude Bombing System 全飞行姿态轰炸系统

AAC Acoustical Absorption Coefficient 吸音系数；吸声系数

AAC Acquisition Advice Code 采办建议代码

AAC Adaptive Antenna Control 自适应天线控制

AAC Advanced Adaptive Control 先进自适应控制

AAC Aft Access Closure 后部通道关闭

AAC Alaskan Air Command 【美国】阿拉斯加空军司令部

AAC All-Aspect Capability 【空－空导弹】全向（攻击）能力

AAC Ambient Air Concentration 周围空气浓度

AAC Amplitude Absorption Coefficient 振幅吸收系数

AAC Armament Accuracy Check 武器精度校验

AAC Atomic Absorption Coefficient 原子吸收系数

AAC Automat and Automatic Control 自动装置与自动控制

AAC Automatic Amplitude Control 自动幅度控制

AAC Automatic Answering Control 自动应答控制；自动调用

AAC Automatic Aperture Control 自动孔径控制

AAC Automatic Approach Control 自动进场控制

AACB Aeronautics and Astronautics Coordinating Board 【美国】航空航天协调委员会

AACB Aerospace Activity Coordinating Board 航空航天活动协调委员会

AACC Airborne Alternate Command Center 空中备用指挥中心

AACC Automatic Approach Control Coupler 自动进场控制耦合器

AACE Air-to-Air Combat Environ-

ment 空对空战斗环境
AACLS Aerospace Air Cushion Landing System 航空航天气垫着陆系统
AACP Advanced Airborne Command Post 【美国空军】先进机载指挥所
AACS Active Attitude Control System 主动姿态控制系统
AACS Advanced Automatic Compilation System 先进自动编码系统
AACS Airborne Activity Confinement System 机载活动限制系统
AACS Asynchronous Address Communication System 异步选址通信系统
AACS Attitude and Antenna Control System 姿态与天线控制系统
AACS Attitude and Articulation Control Subsystem 飞行姿态与链接控制子系统
AACT Airborne Atmospheric Compensation and Tracking [Program] 机载大气补偿与跟踪（项目）
AACTS Automated All-Weather Cargo Transfer System 自动全天候货物输送系统
AACTS Automatic Anechoic Chamber Test System 自动无回声舱试验系统
AAD Access to Archival Databases 档案数据库访问
AAD Assigned Altitude Deviation 指定高度偏差
AAD Attitude Anomaly Detector 姿态异常探测器
AAD Average Absolute Deviation 平均绝对偏差
AADC Advanced Avionics Digital Computer 先进航空电子数字计算机
AADC Alaska Aerospace Development Corporation 【美国】阿拉斯加宇航发展公司
AADC Area Air Defense Commander 区域空防指挥官
AADC^2 Army Air Defense Command and Control 【美国】陆军防空指挥与控制
AADCCS Army Air Defense Command and Control System 【美国】陆军防空指挥与控制系统
AADEOS Advanced Air Defense Electro-Optical Sensor 先进防空光学－电子传感器
AADEOS Advanced Air Defense Electro Optical System 先进防空光电系统
AADHS Advanced Avionics Data Handling System 先进航空电子设备数据处理系统
AADPS Algorithmic Architecture Data Processing Subsystem 算法建筑数据处理子系统；算法结构数据处理分系统
AADS Advanced Air Data System 先进大气数据系统
AADS Advanced Air Defense System 高级防空系统
AADS Advanced Airflow Data System 高级气流数据系统
AADS Aero Acoustic Detection System 航空声学探测系统；航空音响探测系统
AADS Airborne Acoustic Detection System 机载声波探测系统
AADS Ascent Air Data System 上升空气数据系统
AAE Abort Advisory Equipment 中止通报设备
AAE Aeronautical and Aerospace Engi-

neering 航空与航天工程（美国伊利诺伊大学系名）

AAE Aerospace Ancillary Equipment 航空航天辅助设备

AAE American Association of Engineers 美国工程师协会

AAE Arms, Ammunition and Explosives 武器、弹药与炸药

AAE Automatic Analysis Equipment 自动分析设备

AAEA Automated Analytical Electrophoresis Apparatus 自动分析电泳装置

AAEC Attitude Axis Emergency Control 【飞行】姿态轴应急控制

AAED Advanced Airborne Expendable Decoy 高级机载一次性使用假目标

AAEL Air-to-Air Eject Launcher 空一空弹射式发射装置

AAES Aerobraking, Aerocapture, and Entry System 空气制动、空气捕获与进入系统

AAESA Assessment of Atmospheric Effects of Stratospheric Aircraft 同温层飞行器大气效应评估

AAF Association Astronautique Francaise 法国宇航协会

AAFB Anderson Air Force Base 【美国】安德森空军基地

AAFB Andrews Air Force Base 【美国】安德鲁斯空军基地

AAFE Advanced Applications Flight Experiment 先进应用飞行试验（美国国家航空航天局旧词）

AAFE Aero-Assist Flight Experiment 【美国航天飞机】空气制动飞行实验

AAFI Audible Automatic Failure Indicator 自动声频故障指示器

AAFIS Advanced Avionics Fault Isolation System 先进航空电子设备故障隔离系统

AAGM Antiaircraft Guided Missile 防空导弹

AAGMC Antiaircraft Artillery and Guided Missile Center 【美国】高射炮与防空导弹中心

AAGMC Antiaircraft Guided Missile Center 防空导弹中心

AAGMS Antiaircraft Guided Missile Site 防空导弹发射阵地

AAGMS Antiaircraft Guided Missile Station 防空导弹站

AAGMS Antiaircraft Guided Missile System 防空导弹系统

AAGW Air-to-Air Guided Weapon 空一空制导武器

AAH Automatic Attitude Hold 自动姿态保持

AAI Angle of Approach Indicator 进场角度指示器

AAI Angle of Attack Indicator 攻角指示器

AAI Azimuth Angle Increment 方位角增量

AAIS Automated Acquisition Information System 自动化采办信息系统

AAL Above Aerodrome Level 高于机场平面

AAL Asynchronous Transfer Mode Adaptation Layer 异步传输模式适配层

AALAAW Advanced Air-Launched Anti-Armor Weapon 先进空射反装甲武器

AALC Advanced Airborne Launch Center 【美国空军】先进空中发射控制中心（使用 EC-135C 飞机控制地下发射井中的"民兵"导弹的发

射阶段）

AALC Autonomous Approach Landing Capability　自主进场着陆能力

AALR Antiaircraft Laser Range-Finder　防空激光测距仪

AALS Advanced Approach and Landing System　先进进场与着陆系统

AALS Aerospace Automatic Landing System　航空航天自动着陆系统

AALT Automatic Azimuth Laying Theodolite　自动方位瞄准经纬仪

AALW Assembled Air Launched Weapon　已装配的空中发射武器

AAM Airborne Antiradiation Missile　机载反辐射导弹

AAM Airborne Armament Maintenance　机载武器维护

AAM Air-to-Air Missile　空－空导弹

AAM Anti-Aircraft Missile　防空导弹

AAM Apogee Adjust Maneuver　远地点适应性操作

AAM Asymmetrical Amplitude Modulation　不对称振幅调制

AAM Auxiliary Aiming Mark　辅助瞄准点/标记

AAM AVUM Avionics Module　【法国】"织女星"运载火箭上面级电子设备舱

AAMDC Army Air and Missile Defense Command　【美国】陆军地域防空与导弹防御司令部

AAMEX Air-to-Air Missile Exercise　空—空导弹演练

AAMGD Advanced Antiradiation Missile Guidance Demonstration　先进反辐射导弹制导演示

AAMGE Air-to-Air Missile Guidance Element　空－空导弹制导元件

AAMP Advanced Architecture Microprocessor　先进体系架构微处理器

AAMREP Air-to-Air Missile Weapon System Report　空－空导弹武器系统报告

AAMRL Armstrong Aerospace Medical Research Laboratory　【美国空军】阿姆斯特朗航空航天医学研究实验室

AANCP Advanced Airborne National Command Post　【美国空军】国家先进空中指挥所

AAO Analog Attitude Output　模拟式姿态输出

AAO Analysis of the Area of Operations　作战地域分析

AAO Astronaut Activities Office　【美国国家航空航天局】航天员飞行管理局

AAO Astronaut Activity Office　【美国】航天员活动办公室

AAP Angle-of-Approach　进场（下滑）角

AAP Angle of Approach Lights　进场灯角；进场下滑角指示灯

AAP Antenna Aperture Package　天线孔径数据包

AAP Apollo Application Program　【美国】"阿波罗"应用项目（后更名为"天空实验室"项目）

AAP Associative Array Processor　相联阵列处理器

AAP Audible Alarm Panel　声频报警信号板

AAPI Attack Assessment Predicted Impact　攻击评估效果预报

AAPP Airborne Auxiliary Power Plant　机载辅助动力装置

AAPS Airborne Angular Positioning Sensor　机载角定位传感器

A

AAQ-6 Airborne Forward-Looking Infrared 机载前视红外装置
AAQ-9 Airborne Infrared Detector 机载红外探测器
AAQS Ambient Air Quality Standards 环境空气质量标准
AAR After Action Report 事后处理报告
AAR After Action Review 事后处理评审
AAR Automated Aerial Refueling 自主空中加注
AAR Automatic Alternative Routing 自动迂回路线
AARGM Advanced Anti-Radiation Guided Missile 先进反辐射制导导弹
AARLOSU Aeronautical and Astronautical Research Laboratory Ohio State University 【美国】俄亥俄州大学航空和航天研究实验室
AARM Advanced Antiradiation Missile 先进反辐射导弹
AARPLS Advanced Airborne Radio Position Location System 先进机载无线电定位系统
AARPS Air Augmented Rocket Propulsion System 空气加力火箭推进系统
AARRS Air Force Aerospace Rescue and Recovery Service 【美国空军】航空航天救援与回收局
AARS Advanced Airborne Radar System 先进机载雷达系统
AARS Attitude Altitude Retention System 姿态高度保持系统
AARS Attitude and Azimuth Reference System 姿态与方位基准系统; 姿态与方位参考系统
AARS Automatic Address Recognition Subsystem 自动地址识别子系统
AARS Automatic Altitude Reporting System 高度自动报知系统
AART Airborne Avionics Research Testbeds 机载航空电子设备研究试验台
AAS Abort Advisory System 中止通报系统
AAS Advanced Antenna System 先进天线系统
AAS Advanced Automatic System 先进自动系统
AAS Advanced Automation System 先进自动化系统
AAS Advanced Automation System 【美国联邦航空总署】先进自动系统
AAS Alerting Automatic System 自动报警系统
AAS American Academy of Sciences 美国科学院
AAS American Astronautical Society 美国宇航协会
AAS Atomic Absorption Spectrometry 原子吸收光谱法
AAS Auto-Alignment Angle Sensor 自动校准角度传感器
AAS Automatic Addressing System 自动寻址系统
AAS Automatic Announcement Subsystem 自动通知子系统
AAS AVUM Avionic Section 【法国】"织女星"火箭上面级航空电子设备段
AASA Advanced Airborne Surveillance Antenna 先进机载监视天线
AASC Aerospace Application Studies Committee 【北约】航空航天（技术）应用研究委员会
AASD Antiaircraft Self-Destroying

(Device)　防空火箭自毁（装置）

AASIR　Advanced Atmospheric Sounder and Imaging Radiometer　先进大气音响与图像辐射仪

AASP　Advanced Acoustic Signal Processor　先进声音信号处理器

AASRS　Advanced Airborne Signal (Intelligence) Reconnaissance System　先进机载信号（情报）侦察系统

AASS　Advanced Acoustic Search Sensors　先进声测传感器

AASS　Advanced Airborne Surveillance Sensors　先进机载监视传感器

AASS　Automatic Abort Sensing System　自动中止传感系统

AASS　Automatic Audio Switching System　自动声频交换系统

AAT　Accelerated Aging Test　加速老化试验

AAT　All Aspect Target　全方位目标

AAT　Architecture Analysis Tool　体系结构分析工具

AAT　Attitude Acquisition Technique　姿态（角）探测技术；姿态捕获技术

AAT　Attitude Angle Transducer　姿态角传感器

AAT-PP　Architecture Analysis Tool-Post Processor　体系结构分析工具-后处理器

AATE　Architectural Assessment Tool – Enhanced　体系结构评估工具（增强型）

AATF　Active Air Target Fuze　空中目标主动式引信

AATPG　Analog Automatic Test Program Generation　模拟式自动测试程序发生器

AATR　Apollo Application Test Requirements　【美国】"阿波罗"计划应用试验要求

AATS　Advanced Automatic Test System　先进自动测试系统

AATS　Advanced Automation Training System　先进自动化培训系统

AATS　Alerting Automatic Telling Status　自动报警状态

AATS　Committee on the Application of Aerospace Technology to Society　【美国国家航空航天局】航空航天技术协会应用委员会

AATSR　Advanced Along Track Scanning Radiometer　【欧洲空间局】先进沿轨扫描辐射计

AATT　Advanced Air Transportation Technology　先进空中运输技术

AATV　Air-to-Air Test Vehicle　空－空试验导弹（英国近距空－空导弹SRAAM的改进型）

AAU　Absolute Alignment Update　绝对对准适时修正

AAU　Angular Accelerator Unit　角加速器组件

AAU　Automatic Addressing Unit　自动寻址单元

AAU　Automatic Answering Unit　自动应答单元

AAUT　Alternate Attitude Update Techniques　备用姿态适时修正技术

AAV　Advanced Aerospace Vehicle　【美国】先进航空航天飞行器（X-31空天飞机的别名）

AAV　Aerospace Audio-Visual　【美国空军】航空航天音频－视频

AAV　Autonomous Aerial Vehicle　自主式空袭兵器

AAV/D　Automatic Alternate Voice/Data　自动语言/数据交替使用

A

AAVCS Airborne Automatic Voice Communication System 机载自动声频通信系统

AAVS Aerospace Audio Visual Service 【美国空军】航空航天声像业务处

AAW Antiaircraft Warfare 反航空兵作战（夺取制空权的作战）

AAW Anti-Air Warfare 防空作战

AAWCS American Airborne Warning and Control System 美国机载告警与控制系统

AAWS Airborne Alert Weapon System (A Battlefield Missile System) 机载警戒武器系统（一种战场导弹系统）

AB Adapter Booster 适配助推器

AB Address Bus 地址总线（母线）

AB Auto-Beacon 自动信标

AB Avionics Bay 电子设备舱

AB/LD Airbrake/Lift Dumpers 减速板/减升板

ABAAMS Assault Breaker Anti-Armor Missile System 突击破甲反装甲导弹系统

ABACS Automated Booster Assembly and Checkout System 助推自动装配与检测系统

ABACUS Advanced Battlefield Computer Simulation 先进战场计算机模拟

ABACUS Air Battle Analysis Center Utility System 空中作战分析中心效用系统

ABACUS Austere Backup Communications Unmanned System 简易备用无人操作通信系统

Abacus Distributed Real-Time Multi-Element Test Environment for HWIL 用于回路硬件的分布式实时多部件测试环境

ABALL Aeroballistic Missile 航空弹道导弹

ABB Automated Beam-Builder 自动桁架制造机（空间制造）

ABC Active Body Control 活动主体控制

ABC Active Boundary Control 主动边界层控制

ABC Advanced Ballistic Computer 先进弹道计算机

ABC Advanced Ballistic Concepts 先进弹道概念；先进弹道方案

ABC Advanced Broadband Communication 先进宽带通信

ABC Advanced Business Communication 【美国】先进商业通信卫星

ABC Aft Bulkhead Carrier 后舱壁托架

ABC Automatic Bandwidth Control 自动带宽控制

ABC Automatic Beam Control 自动波束控制

ABC Automatic Bias Compensation 自动偏压补偿

ABC Automatic Bias Control 自动偏压控制

ABC Automatic Boost Control 自动助推控制

ABCCC Airborne Battlefield Command and Control Capsule 空中战场指挥与控制舱

ABCCC Airborne Battlefield Command and Control Center 空中战场指挥与控制中心

ABCCC Airborne Battlefield Command, Control and Communications 空中战场指挥、控制与通信

ABCCC Airborne Command and Control Center 空中指挥与控制中心

ABCCC Airborne Command, Control and Communications 空中指挥、

控制与通信

ABCCIS Air Base Command and Control Information System 空中基地指挥与控制信息系统

ABCCTC Advanced Base Combat Communications Training Center 先进基地作战通信训练中心

ABCD Advanced Beam Concept Development 【弹道导弹防御组织】先进波束方案研发

ABCD Atomic, Biological and Chemical Defense 原子、生物、化学防护

ABCDC Atomic, Biological, Chemical and Damage Control 原子、生物、化学毁伤控制

ABCS Advanced Beam Control System 先进波束（光束）控制系统

ABCS Airborne Battlefield Countermeasure and Survivability 机载战场对抗措施和生存能力

ABCS Army Battle Command System 【美国】陆军作战指挥系统

ABD Answer-Back Device 应答装置

ABD Applied Ballistics Department 【英国】应用弹道学部

ABE Air-Based Electronics 空基电子设备

ABE Air Burst Effect 空爆效应

ABE Airborne Bombing Evaluation 空中轰炸效果判定

ABEC Analog Backup Engine Control 模拟备份发动机控制

ABERT Auto Bit Error Rate Test 自动比特错误率测试

ABES Air-Breathing Engine System 吸气式发动机系统

ABETS Airborne Beacon Electronic Test Set 机载信标电子测试装置

ABF Annular Blast Fragmentation 【弹头】环形爆炸碎片

ABF Audio Bandpass Filter 声频带通滤波器

ABF Auto Beam Forming 【无源声呐】自动形成波束

ABFM Airborne Field Mill 机载电场仪

ABGD Air Base Ground Defense 空军基地地面防御

ABHRS As-Built Built Hardware Reporting System 实际硬件通报系统

ABI Airborne-Based Interceptor 机载拦截机（导弹）

ABIR Airborne Infrared 机载红外

ABIS Advanced Battlespace Information System 先进作战空间信息系统

ABIS Apollo Bioenvironmental Information System 【美国】"阿波罗"飞船生物环境信息系统

ABIT Automatic Built-In Test 自动内装测试

ABITA Association Belge des Ingenieurs et Techniciens de l'Aeronautuque et de l'Astronautique 【比利时】航空航天工程师与技术员协会

ABITE Antenna Built-In Test Equipment 天线嵌入式测试设备

ABIU Avionics Bay Interface Unit 航空电子设备舱接口单元

ABL Above Base Line 基线以上；高于基线

ABL Airborne Laser 机载激光器

ABL Allegheny Ballistics Laboratory 【美国】阿勒格尼弹道实验室

ABL Allocated Baseline 配置的基线

ABLE Atmospheric Boundary Layer Experiment 大气边界层实验

A

ABLE Automatic Base Line Equipment 自动基线设备
ABLE ACE Airborne Laser Extended Atmospheric Characterization Experiment 机载激光扩展型大气特性实验
ABLEX Airborne Laser Experiment 机载激光实验
ABLV Air-Breathing Launch Vehicle 吸气式运载火箭
ABM Advanced Ballistic Missile 先进弹道导弹
ABM Anti-Ballistic Missile 反弹道导弹
ABM Apogee Boost Maneuver 远地点助推机动飞行
ABM Apogee Boost Missile 远地点助推导弹
ABM Apogee Boost Motor 远地点助推发动机
ABM Army Ballistic Missile 陆军弹道导弹
ABM Asynchronous Balanced Mode 异步平衡模式
ABM Automated Batch Mixing 自动程序组合
ABMA Army Ballistic Missile Agency 【美国】陆军弹道导弹局
ABMD Air Battle Management Demonstration 空战管理演示验证
ABMDA Advanced Ballistic Missile Defense Agency 【美国陆军】先进弹道导弹防御局
ABMDP Advanced Ballistic Missile Defence Program 【美国】先进弹道导弹防御项目
ABMDS Advanced Ballistic Missile Defence System 先进弹道导弹防御系统
ABMEWS Antiballistic Missile Early Warning System 反弹道导弹预警系统
ABMH Anti-Ballistic Missile Hardware 反弹道导弹硬件
ABMM Antiballistic Missile Missile 反弹道导弹（的）导弹
ABMS Apogee Boost Motor Subsystem 远地点助推发动机分系统
ABMT Antiballistic Missile Treaty 反弹道导弹条约
ABNCP Airborne National Command Post 国家空中指挥所
ABNML Abnormal 异常的，反常的；不合标准的
ABO Agent of Biological Origin 【核生化】生物源战剂
ABOSF Automatic Burnout Safety Factor 自动充分燃烧安全系数
ABP Aurora Board of Participants 【欧洲】"欧若拉"项目参与国委员会
ABP Auxiliary Beam Positioning 辅助波束定位
ABPP American Battlefield Protection Program 美国战场防护项目
ABPS Air-Breathing Propulsion System 吸气推进系统
ABR Aeroballistic Rocket 空气弹道火箭；气动弹道火箭
ABRE Air-Breathing Rocket Engine 空气喷气火箭发动机
ABRES Advanced Ballistic Reentry System 【美国】先进弹道再入系统
ABRS Advanced Ballistic Rocket System 先进弹道火箭系统
ABRS Advanced Biological Research System 先进生物研究系统
ABRS Assault Ballistic Rocket System 突袭式弹道火箭系统

ABRV Advanced Ballistic Reentry Vehicle 【美国】先进弹道再入飞行器

ABS American Bureau of Standards 美国标准局

ABS Antenna Base Spring 天线基座弹簧

ABS Antenna Bridge Structure 天线桥式结构

ABS Anti-Burn Shield 抗燃防护物

ABS Anti-Burst System 防爆系统

ABS Antipodal Baseband Signaling 对映基带信号（设备）

ABS Automated Beam Steering 自动化光束控制

ABS Automatic Block Signaling 自动闭塞信号

ABSAA Airborne Sense and Avoid 机载感应与规避

ABT Approximate Burn Time 近似燃烧时间

AC Acknowledge Control 应答控制；（信息）收到控制

AC Aerodynamic Center 气动力中心

AC Altitude Suit 高空飞行（密封）服

AC Analog Computer 模拟计算机

AC Applications Controller 应用控制器

AC Approach Control (Tower) 进场控制（塔台）

AC Atlas-Centaur 【美国】"宇宙神－半人马座"运载火箭

AC Automatic Check 自动检测

AC Automatic Control 自动控制

AC Auxiliary Console 辅助控制台

AC&S Attitude Control and Stabilization 姿态控制与稳定

AC^{2}F Automated Command and Control Facility 自动指挥与控制设施（设备）

ACA Altitude Controller Assembly 高度控制器装配

ACA Astronautics Corporation of America 美国航空航天公司

ACA Asynchronous Communications Adapter 异步通信适配器

ACA Attitude Control Assembly 姿态控制装置

ACALS Airborne Command and Launch Subsystem 机载指挥与发射分系统

ACAMPS Automatic Communication and Message Processing System 【美国】自动通信与信息处理系统

ACAP Advanced Computer for Array Processing 先进阵列处理计算机

ACAP Analysis of Critical Actions Program 临界作用程序分析；关键行动计划分析

ACAP Assessment Correlation Analysis Process 评估相关分析程序

ACARS Automatic Communications and Recording System 自动通信与记录系统

ACAS Airborne Collision Avoidance System 机载防撞系统

ACAS Altocumulus and Altostratus 高积云与高层云

ACAS Analytical Chemistry and Applied Spectroscopy 分析化学与应用光谱学

ACAS Automatic Central Alarm System 中央自动报警系统

ACAS Automatic Collision Avoidance System 自动防撞系统

ACAS Avoidance System 避撞系统

ACAV Automatic Circuit Analyzer and Verifier 自动电路分析器与验证器

ACAVS Advanced Cab and Visual System 先进座舱与目视系统

ACB Adapter Control Block 适配器控制块

ACBA Airborne Communications Bus Architecture 【美国】机载通信总线结构

ACBM Active Common Berthing Mechanism 有效通用型接驳装置

ACBT Automatic Circuit Board Tester 自动电路板测试器

ACBWG Apollo Reentry Communications Blackout Working Group 【美国】"阿波罗"飞船再入通信中断工作组

ACC Active Clearance Control 主动间隙控制（技术）

ACC Active Combustion Control 主动燃烧控制

ACC Aft Cargo Carrier 尾部货物托架

ACC Air Combat Command 【美国空军】空中作战司令部

ACC Airspace Control Center 空域控制中心

ACC Antenna Control Console 天线控制台

ACC Area Control Center 区域控制中心

ACC Area Coordination Center 区域协调中心

ACC Armament Control Computer 弹药控制计算机

ACC Asynchronous Communications Control 异步通信控制

ACC Attitude Control Computer 姿态控制计算机

ACC Automatic Carrier Control 自动载波控制

ACC Automatic Chroma Control 自动色度控制；自动色度调整

ACC Automatic Combustion Control 自动燃烧控制

ACC Automatic Control Console 自动控制台

ACC Auxiliary Crew Compartment 辅助乘员舱

ACC Average Correlation Coefficients 平均相关系数

ACC Avionics Control Computer 航空电子设备控制计算机

ACCAT Advanced Command and Control Architectural Testbed 先进指挥与控制结构试验台

ACCC Area Control Computer Complex 区域管制计算机网

ACCCP Airborne Communications Command and Control Platform 机载通信指挥与控制平台

ACCDS Adapt Communications Correlation Detection System 自适应通信关联探测系统

ACCEL Accelerometer 加速计

ACCESS Assembly Concept for Construction of Erectable Space Structures 装配式空间结构组装方法（美国航天飞机上的舱外装配试验）

ACCESS Automated Control and Checking of Electrical System Support 电气系统保障的自动控制与检测

ACCF Area Communication Control Function 区域通信控制功能

ACCH Associated Control Channel 相关控制信道；随路控制信道

ACCIS Automated Command Control and Information System 自动化指挥控制与信息系统

ACCMS Advanced Checkout, Control, and Maintenance System 先进检查、控制与维修系统

ACCRDS　Area Control Center Radar Display System　区域控制中心雷达显示系统

ACCS　Active Contamination Control System　主动污染控制系统

ACCS　Advanced Checkout and Control System　先进检测与控制系统

ACCS　Advanced Command and Control System　先进指挥与控制系统

ACCS　Aerospace Command and Control System　航空航天指挥与控制系统

ACCS　Air Command and Control System　空中指挥与控制系统

ACCS　Airspace Command/Control System　空域指挥/控制系统

ACCS　Army Command and Control System　陆军指挥与控制系统

ACCS　Attitude Coordinate Converter System　姿态坐标数据换算系统

ACCS　Automated Command and Control System　自动指挥与控制系统

ACCS　Automatic Checkout and Control System　自动检测与控制系统

ACCT　Application of Common Characteristics and Testability　共有特征与可试验性的应用

ACCWT　Alternating Current Continuous Waves Transmitter　交流等幅波发射机

ACD　Active Control Devices　主动控制装置

ACD　Advanced Control Devices　先进控制装置

ACD　Advanced Counterfeit Deterrence　先进防伪

ACD　Architectural Control Document　体系结构控制文件

ACD　Attitude Control Document　姿态控制文件

ACD　Automatic Call Distribution　自动调用分配；自动呼叫分配

ACD　Automatic Chart Display　自动航图显示；自动地图显示器

ACD　Automatic Checkout Device　自动检测装置

ACD　Automatic Closing Device　自动关闭装置

ACD　Automatic Clutter Detection　自动杂波干扰探测

ACD　Automatic Conflict Detection　自动冲突探测

ACDS　Advanced Combat Direction System　先进作战指挥系统

ACDS　Advanced Command Data System　先进指挥数据系统

ACDS　Alarm Communications and Display System　警报传送与显示系统

ACDS　Automatic Comprehensive Display System　自动综合显示系统

ACDT　Advanced Concept Demonstration Technology　先进概念演示验证技术

ACDT　Autocycled Data Test　自动循环式数据测试

ACE　Acceptance Checkout Equipment　验收检测设备

ACE　Actuator Control Electronics　致动控制电子设备

ACE　Advanced Certification Equipment　先进认证设备

ACE　Advanced "Colloid" Experiment　先进"胶体"实验

ACE　Advanced Composition Explorer　先进合成物探测器

ACE　Advanced Compound Engine　先进混合式发动机

ACE　Advanced Controlled Equipment　先进控制设备

A

ACE Advanced Controlled Experiment 先进控制试验

ACE Aerosol, Cloud, Ecosystems 气溶胶—云—生态系统

ACE Aerospace Control Environment 航空航天控制环境

ACE Agile Control Experiment 灵活控制实验（演示可测光学装置在太空的瞄准线稳定性）

ACE Airborne Command Element 机载指挥部件

ACE Altitude Control Electronics 高度控制电子设备

ACE Analysis Control Element 分析控制部件

ACE Antenna Calculating Engine 天线计算引擎

ACE Anti-Radiation Missile (ARM) Countermeasure Evaluator 反辐射导弹对抗鉴定器（装置）

ACE Apollo Checkout Equipment 【美国】"阿波罗"飞船检测设备

ACE Assessment of Combat Effectiveness 战斗效果评估

ACE Atmosphere and Climate Explorer 大气与气候探测器

ACE Atmospheric Control Experimentation 大气控制实验

ACE Attitude Control Electronics 姿态控制电子设备

ACE Automated Commercial Environment 自动商业化环境

ACE Automatic Checkout Equipment 自动检测设备

ACE Automatic Clutter Elimination 自动杂波消除

ACE Automatic Computing Equipment 自动计算装置

ACE Auxiliary Control Element 辅助控制单元

ACE Auxiliary Conversion Equipment 辅助转换设备

ACE Aviation Combat Element 【北大西洋公约组织】航空战斗部件

ACE-SC Acceptance Checkout Equipment-Spacecraft 航天器验收检测设备

ACEA Action Committee for European Aerospace 欧洲航空航天（军事）行动委员会

ACEBP Air-Condition Engine Bleed Pipe 【座舱】空调用发动机引气管

ACEC Ada Compiler Evaluation Capability Ada（语言）编译程序鉴定能力

ACED Aerospace Crew Equipment Development 航空航天机组乘员设备研制

ACEIT Automated Cost Estimating Integrated Tools 自动成本估算集成工具

ACEL Aerospace Crew Equipment Laboratory 【美国】航空航天机组乘员设备实验室

ACES Acceptance and Checkout Evaluation System 验收与检查评估系统

ACES Acceptance Control Equipment System 验收控制设备系统

ACES Acoustic Containerless Experiment System 【美国航天飞机】声学无容器实验系统

ACES Active Control Evaluation for Spacecraft 航天器主动控制评估系统

ACES Advanced Concept Ejection Seat 先进概念性弹射座椅

ACES Advanced Concept Escape System 先进概念性逃逸系统；先进概念性救生系统

ACES Advanced Crew Escape Suit 先进航天员逃生服
ACES Air Collection and Enrichment System 空气收集与浓缩系统
ACES Air Collection Engine System 集气式发动机系统
ACES Atmospheric Combustion Exhaust Simulator 大气完全燃烧模拟器
ACES Atomic Clock Ensemble in Space 太空原子钟组
ACES Automated Command and Control Evaluation System 自动化指挥与控制评估系统
ACES Automatic Checkout and Evaluation Simulator 自动检测与评估模拟器
ACES Automatic Checkout and Evaluation System 自动检测与评估系统
ACES Automatic Checkout Equipment Sequencer 自动检测设备程序
ACES Automatic Code Evaluation System 自动代码评估系统
ACES Automatic Control Evaluation Simulator 自动控制评估模拟器
ACES Automatic Control Evaluation System 自动控制鉴定系统
ACES Automatically Controlled Electrical System 自动控制电气系统
ACET Air-Cushion Equipment Transporter 气垫式装备运输车
ACETEF Air Combat Environment Test and Evaluation Facility 【美国空军】空战环境测试与鉴定设施
AceTR Air Core Enhanced Turbo Rocket 空心增强型涡轮火箭
ACETS Air-Cushion Equipment Transportation System 气垫装备运输系统

ACEU Aerocontrol Electronics Unit 气控电子设备
ACF Acceleration Corrosion Factor 加速腐蚀因子
ACF Access Control Facility 访问控制设施
ACF Area Control Facility 区域控制设施
ACF Authentication Control Function 鉴定控制功能
ACF Auto-Correlation Function 自动相关函数
ACF Axisymmetrical Conical Flow 轴对称锥形流
ACFC Air Cooled Fuel Cooler 气冷式燃料冷却装置
ACFJ Axial-Centrifugal Flow Jet 轴向离心流喷气发动机
ACFS Advanced Concepts Flight Simulator 先进概念飞行模拟器
ACGE Analog Command Generation Equipment 模拟命令生成设备
ACGS Aerospace Cartographic and Geodetic Service 【美国空军】航空航天制图与大地测量处
ACGV Automatically Controlled and Guided Vehicle 自动控制与制导飞行器
ACHEX Aerosol Characterization Experiment 浮质特性实验
ACI Access Control Information 访问控制信息
ACI Acoustic Comfort Index 容许噪声指数
ACI Adjacent Channel Interference 相邻信道干扰
ACI Allocated Configuration Identification 配置构型识别
ACI Allowable Concentration Index 允许的浓度指数

A

ACI　Altitude Command Indicator　高度指挥指示器
ACI　Asynchronous Communication Interface　异步通信接口
ACI　Attitude Controls Indicator　姿态控制指示器
ACI　Automatic Control Instrumentation　自动控制仪表装置
ACIA　Asynchronous Communication Interface Adapter　异步通信接口适配器
ACIA　Asynchronous Communication Interface Alarm　异步通信接口报警器
ACIC　Aeronautical Chart and Information Center　导航图表与信息中心；航空图表与信息中心
ACIL　Automatic Controlled Instrument Landing　自动控制仪器着陆；自动控制盲着陆
ACIM　Availability Centered Inventory Model　可用库存模型
ACINF　Airborne Acoustic Information System　机载声波信息系统
ACINT　Acoustic Intelligence　声频情报
ACIP　Aerodynamic Coefficient Identification Package　气动系数辨识组件
ACIP　Assembly Configuration and Integration Panel　装配配置与综合面板
ACIRU　Attitude Control Inertial Reference Unit　姿态控制惯性基准部件；姿态控制惯性参考装置
ACIS　Advanced Cabin Interphone System　先进舱内对讲系统
ACIS　Arms Control Impact Statement　军备控制影响报告（声明）
ACIS　Automated Control of Industrial Systems　工业系统的自动化控制
ACIS　Avionics Central Information System　航空电子中央信息系统
ACJ　Attitude Control Jet　姿态（角）控制喷流；姿态控制喷射器
ACKTX　Automatic Circuit Exchange　自动电路交换
ACL　Advanced Communication Link　先进通信数据链
ACL　Advanced Computer Laboratory　先进计算机实验室
ACL　Allowable Cargo Load　允许货物负载
ACL　Anti-Collision Light　防撞灯
ACL　Application Control Language　应用控制语言
ACL　Ascent Closed Loop　上升段闭回路
ACL　Attitude Control Logic　姿态控制逻辑（电路）
ACL　Avionics Cooling Loop　航空电子设备冷却回路
ACLE　Automatic Clutter Eliminator　自动干扰抑制器
ACLG　Air Cushion Landing Gear　气垫式起落架
ACLO　Agena Class Lunar Orbiter　【美国】"阿森纳"级月球轨道器
ACLS　Air Cushion Landing System　气垫着陆系统
ACLS　Augmented Contingency Landing Site　加强型应急着陆场
ACLS　Automated Command to Line of Sight　自动化瞄准线(制导导弹)("发射后不管"导弹制导系统)
ACLS　Automated Control and Landing System　自动控制与着陆系统
ACM　Access Control Module　访问控制模块
ACM　Acquisition Control Module

捕获控制模块

ACM Active Countermeasures 有源干扰

ACM Advanced Composite Material 先进复合材料

ACM Advanced Concept Missile 【美国】先进概念导弹

ACM Advanced Cruise Missile 【美国空军】先进巡航导弹

ACM Aerodynamic Configured Missile 气动构型导弹

ACM Air Combat Maneuvering 空战机动

ACM Alarm Control Module 报警控制模块

ACM Allocated Configuration Management 配置构型管理

ACM Attitude Control Motor 姿态控制发动机

ACM Automatic Coding Machine 自动编码机

ACM Auxiliary Control Module 辅助控制模块

ACME Aerodynamic Coefficient Measurement Experiment 气动系数测量实验

ACME Antenna Contour Measuring Equipment 天线外形测量设备

ACME Association of Consulting Management Engineers 【美国】咨询管理工程师协会

ACME Attitude Control and Maneuvering Electronics 姿态控制与机动电子设备

ACMES Attitude Control and Maneuver (ing) Electronic System 姿态（角）控制与机动电子系统

ACMS Advanced Configuration Management System 先进构型管理系统

ACMS Air Combat Maneuvering System 空战机动系统

ACMS Application Control Management System 应用控制管理系统

ACN Airborne Communications Node 机载通信节点

ACN Approval and Clearance Notice 批准放行通知

ACNSS Advanced Communication/Navigation/Surveillance System 先进通信导航监视系统

ACO Acceptance Checkout 验收检测

ACO Adaptive Control Optimization 自适应控制优化

ACO Airspace Control Order 空域控制指令

ACO/MGE Acceptance and Checkout / Maintenance Ground Equipment 地面设备的验收与检测/维护

ACOC Area Communications Operations Center 区域通信操作中心

ACOE Acceptance and Checkout Equipment 验收与检测设备

ACOE Automatic Checkout Equipment 自动检测装置

ACOM Attitude Control Propulsion Motor 姿态控制推进发动机

ACORN Automatic Checkout and Recording Network 自动检测与记录网络

ACOS Arms Control Observation Satellite 军备控制观察卫星

ACOS Automatic Checkout Set 自动检测装置

ACOST Advisory Council on Science and Technology 【英国】科学技术顾问委员会

ACP Acceptance Checkout Procedure 验收检查程序

A

ACP Adaptive Control Process 自适应控制过程

ACP Aerospace Computer Program 航空航天计算机研制项目

ACP Ammunition Control Point 弹药控制点

ACP Analytical Consistency Plan 解析一致性计划

ACP Armament Control Panel 武器控制面板

ACP Astronaut Control Panel 航天员控制操纵台

ACP Attitude Control Processor 姿态控制数据处理机

ACPG Atmospheric Composition Payload Group 【美国空间站】大气成分有效载荷组

ACPL Atmospheric Cloud Physics Laboratory 大气云物理实验室

ACPM Attitude Control Propulsion Motor 姿态控制推进发动机

ACPS Attitude Control Propulsion System (Subsystem) 姿态控制推进系统（分系统）

ACQTRACK Acquisition Tracking 目标探测跟踪；跟踪搜索

ACRBC Acceptance, Checkout, Retest, and Backout Criteria 验收、检测、重新测试与取消标准

ACRC Assure Crew Return Capability 确保乘员返舱能力

ACRE Advanced Cryogenic Rocket Engineering 先进低温火箭工程

ACRF Advanced Computer Research Facility 先进计算机研究设施

ACRIE Automatic Control and Maneuvering Electronic 自动控制与机动电子设备

ACRIM Active Cavity Radiometer Irradiance Monitor 主动空腔辐射计；辐照度监视器

ACRM Advanced Crew Resource Management 先进航天员资源管理

ACRP Advanced Concept Research Project 先进概念研究工程

ACRS Advisory Committee on Reactor Safeguard 【美国】反应堆安全防卫咨询委员会

ACRS Air Cushion Recovery System 气垫回收系统

ACRS Automated Conflict Resolution System 自动冲突消解系统

ACRSIME Active Cavity Radiometer Solar Irradiance Monitor Experiment 主动空腔辐射计太阳辐照度监视实验

ACRV Assured Crew Return Vehicle 确保航天员安全的返回式飞行器

ACRV Astronaut Crew Rescue Vehicle 航天员救援飞行器

ACRV Attitude Controlled ReEntry Vehicle 姿态控制的再入飞行器

ACRV-X Assured Crew Return Vehicle-Experimental 确保乘员安全的返回式飞行器（实验型）

ACS Access Control Server 接入控制服务器

ACS Access Control System 接入控制系统

ACS Active Control System 主动控制系统

ACS Adaptive Control System 自适应控制系统

ACS Advanced Camera for Surveys 先进巡天相机

ACS Advanced Cryptographic System 先进密码系统

ACS AEGIS Combat System 【美国】"宙斯盾"作战系统

ACS Aerial Common Sensor (USA

term) 【美国】航空通用传感器

ACS Aeroflight Control System (Space Shuttle) 【美国航天飞机】大气层飞行控制系统

ACS Air Coating System 空气涂层系统

ACS Airspace Control System 空域控制系统

ACS Alignment Countdown Set 校准倒计时装置

ACS Antenna Checkout Station 天线检测站

ACS Assembly and Command Ship 组装指挥船

ACS Assembly Control System 组装控制系统

ACS Asynchronous Communication Server 异步通信服务器

ACS Atmosphere Control and Supply 大气环境控制与供给

ACS Atmosphere Control System 大气环境控制系统

ACS Attitude Command System 姿态指令系统

ACS Attitude Control and Stabilization 姿态控制与稳定

ACS Attitude Control System 姿态控制系统

ACS Automated Control System 自主控制系统

ACS Automatic Checkout System 自动检测系统

ACS Automatic Coding System 自动编码系统

ACS Automatic Control System 自动控制系统

ACS Auxiliary Communication Shelter 辅助通信掩体

ACS Auxiliary Control System 辅助控制系统

ACS Auxiliary Coolant System 辅助冷却系统

ACS Azimuth Control System 方位（角）控制系统

ACS/Com Attitude Control System/Communications 姿态控制系统与通信

ACSE Access Control and Signaling Equipment 存取控制与信号设备

ACSIS AEGIS Combat System Interface Simulation 【美国】"宙斯盾"作战系统接口模拟

ACSM Apollo Command and Service Module 【美国】"阿波罗"飞船指挥与服务舱

ACSM Assemblies, Components, Spare parts and Materials 组件、部件、备用件与材料

ACSP Advanced Control Signal Processor 先进控制信号处理机

ACSR Associate Committee on Space Research 空间研究联合委员会

ACT Active Cleaning Technology 主动清洁技术

ACT Active Control Technology 主动控制技术

ACT Advanced Circuit Technology 先进电路技术

ACT Advanced Composites Technology 先进复合材料技术

ACT Advanced Concept Technology 先进概念技术

ACT Advanced Control Technology 先进控制技术

ACT Analogical Circuit Technique 模拟电路技术

ACT Area Correlation Tracker 区域相关跟踪仪

ACT Attitude Control Thruster 姿态控制推进器

ACT Automated Control & Distribution of Trainees 受训人员自动控制与分配系统
ACT Automatic Calibration Technique 自动校准技术
ACT Automatic Checkout Techniques 自动检测技术
ACT Automatic Computer Testing 自动计算机测试
ACT Average Cloud Thickness 平均云层厚度
ACT Azimuth Control Torquer 方位（角）控制力矩机构
ACT Azimuth Control Transformer 方位（角）控制变压器
ACT-TO Actual Time and Fuel State at Takeoff 起飞实际时间与燃料状态
ACTCS Active Thermal Control System 主动热控制系统
ACTD Advanced Concept Technology Demonstration 先进概念技术演示
ACTD Advanced Concept Technology Demonstration (Demonstrator) 【美国国防部】先进概念技术验证（机）
ACTD Attitude Control Torque Device 姿态（角）控制扭矩机构
ACTDS Automatically Cued Target Detecting System 自发信号的目标探测（搜索）系统
ACTE Advanced Controls Technology Experiment 先进控制技术实验
ACTE Analytical Communications Test Environment 可分解的通信试验环境
ACTE Automatic Checkout and Test Equipment 自动检查与测试设备
ACTF Attitude Control Test Facility 姿态控制测试设备
ACTIV Advanced Control Technology for Integrated Vehicles 先进一体化飞行器控制技术
ACTIVE Advanced Controls Technology for Integrated Vehicles 综合飞行器先进控制技术
ACTO Automatic Computing Transfer Oscillator 自动计算传递振荡器
ACTS Acoustic Control and Telemetry System 声控制与遥测系统
ACTS Advanced Communication Technology Satellite 【美国】先进通信技术卫星
ACTS Advanced Computational Testing and Simulation 先进的计算测试与仿真
ACTS Advanced Crew Transportation System 【航天器】先进乘员运输系统
ACTS AEGIS Combat Training System 【美国】"宙斯盾"作战培训系统
ACTS Application Control and Teleprocessing System 应用控制与远程（信息）处理系统
ACTS Automatic COBOL Translation System 面向商业的通用语言自动翻译系统
ACTV Advanced Crew Transportation Vehicle 【载人航天器】先进乘员运载工具
ACU Abort Control Unit 中止控制装置
ACU Acceleration Compensation Unit 加速度补偿装置
ACU Annunciator Control Unit 信号器控制装置
ACU Antenna Control Unit 天线控制装置
ACU Antenna Coupling Unit 天线耦合器

ACU	Arm Computer Unit	武器计算机装置
ACU	Arm Control Unit	武器控制部件
ACU	Attitude Control Unit	姿态控制装置
ACU	Audio Control Unit	音频控制装置
ACU	Automatic Calling Unit	自动呼叫装置
ACU	Autopilot Control Unit	自动驾驶控制单元
ACU	Auxiliary Conditioning Unit	辅助调节装置
ACU	Availability Control Unit	可利用性控制装置
ACU	Avionics Control Unit	航空电子设备控制装置
ACUC	Avionic Control Unit Computer	航空电子设备控制装置计算机
ACUO	Avionics Cooling Unit Operator	电子设备与控制系统冷却装置操作员
ACUS	Army Common User System	【美国】陆军通用用户系统
ACV	Automatic Control Valve	自动控制阀
ACVC	Ada Compiler Validation Capability	Ada（语言）编译程序验证能力
ACWC	Advisory Committee on Weather Control	天气控制咨询委员会
ACWEG	Advanced Conventional Weapons Engineering Group	先进常规武器工程部
AD	Actuator Drive	传动装置；传动机构
AD	Adaptor	适配器；接转器
AD	Aerodynamic Damping	气动阻尼
AD	Aerodynamic Disturbance	气动力扰动
AD	Aerospace Defense	航空航天防御
AD	Air Density	空气密度
AD	Apollo Development	【美国】"阿波罗"飞船研制
AD	Architectural Design	体系结构设计
AD	Armament Division	【美国空军】武器系统研究部军械处
AD	ASTIA Documents	AD 报告（美国武装部队技术情报局出版的文献）
AD	Atmospheric Densities	【美国】大气密度测量卫星
AD	Attitude Determination	姿态确定
AD	Automatic Detection	自动探测
AD	Automatic Display	自动显示
AD	Average Depth	平均深度
AD	Average Deviation	平均偏差；平均差
AD	Acceptance Data	验收数据
Ad Int	Advanced Interceptor	先进拦截机（导弹）
AD/C³I	Air Defense/Command, Control, Communications and Intelligence	空防指挥、控制、通信与情报
AD/V	Alternate Data/Voice	数据—语音变换
Ada	a high order computer language being developed by the Department of Defense	【美国】国防部研制的一款高指令计算机语言
ADA	Air Data Assembly	大气数据组件
ADA	Angular Differentiating Accelerometer	角微分加速度计
ADA	Automatic Data Acquisition	自动数据采集（程序）

ADA Azimuth Display of Attenuation 方位（角）的衰减显示

ADA Azimuth Drive Assembly 方位（角）驱动组件

ADAC Alarm Display and Assessment Console 警报显示器与评估控制台

ADAC Attitude Determination and Control 姿态测定与控制

ADAC Automatic Data Acquisition Center 自动数据采集中心

ADAC Automatic Data Analog Computer 自动数据模拟计算机

ADAC Automatic Direct Analog Computer 自动导引（指挥）模拟计算机

ADACC Automatic Data Acquisition and Computer Complex 自动数据采集与计算机综合设备

ADACS Attitude Determination and Control Subsystem 姿态确定与控制分系统

ADACS Attitude Determination and Control System 姿态确定与控制系统

ADACS Automatic Data Acquisition and Control System 自动数据采集与控制系统

ADAI Apollo Documentation Administration Instruction 【美国】"阿波罗"飞船文件管理规程

ADAM Adaptive Digital Avionics Module 自适应数字航空电子设备模块

ADAM Adaptive Dynamic Analysis Maintenance 自适应动力分析维修

ADAM Advanced Data Access Method 先进数据存取方法

ADAM Advanced Database and Modeling 先进数据库与建模

ADAM Advanced Direct-Landing Apollo Mission 【美国】"阿波罗"飞船的先进直接着陆任务

ADAM Automated Design and Manufacturing 自动设计与制造

ADAM Automatic Distance and Angle Measurement 自动测距与测角

ADAM Automatic Distance and Angle Measurement (System) 距离与角度自动测量（系统）

AdaMAT Ada Automated, Static Code, Analysis Tool Ada（语言）自动、静态编码和分析工具

ADAMS Airborne Data Acquisition and Management System 机载数据获取与管理系统

ADAP Aerodynamic Data Analysis Program 气动数据分析项目

ADAPS Armament Delivery Analysis Programming System 武器交付分析程序系统

ADAPS Automatic Data Acquisition and Process System 自动数据采集与处理系统

ADAPT Advanced Design and Production Technologies 先进设计与生产技术

ADAPT Advanced Directed Energy Weapon Active Pointer and Tracker 先进定向能武器主动瞄准与跟踪装置

ADAPT Advanced Directed Energy Weapon Active Precision Tracker 先进定向能武器主动精确跟踪器

ADAPT Automated Data Analysis and Presentation Techniques 自动化数据分析与显示技术

ADAPTS Analogue/Digital/Analogue Process and Test System 模拟/数字/模拟处理与试验系统

ADAR Advanced Design Array Radar

【美国】先进设计阵列雷达
ADAR　Array Radar　相控阵雷达
ADARS　Adaptive Antenna Receiver System　自适应天线接收系统
ADARS　Advanced Defense Avionics Response Strategy　先进防务航空电子设备响应战略
ADAS　Advanced Digital Avionic System　先进数字航空电子设备系统
ADAS　Airborne Data Acquisition System　机载数据采集系统
ADAS　Airborne Data Annotation System　机载数据注释系统
ADAS　Airborne Digital Avionics System　机载数字航空电子设备系统
ADAS　Airborne Dynamic Alignment System　机载动态校准系统
ADAS　Alarm Display and Assessment System　警报显示器与评估系统
ADAS　Alliance Data Automation System　自动数据测取系统
ADAS　Asteroid Data Analysis System　【美国】小行星数据分析系统
ADAS　Automated Weather Observing System Data Acquisition System　自动气象观测系统的数据采集系统
ADAS　Automatic Data Acquisition System　自动数据采集系统
ADAS　Automatic Data Analysis System　自动数据分析系统
ADAS　Auxiliary Data Annotation Set　辅助数据注释装置
ADAT　Automatic Data Accumulation and Transfer　自动数据积累与传输
ADATE　Automatic Digital Assembly Test Equipment　自动数字装配试验设备
ADATOC　Air Defense Artillery Tactical Operations Center　【美国】空防炮兵战术操作中心

ADATS　Automatic Data Accumulator and Transfer System　自动数据积累与传输系统
ADAU　Air Data Acquisition Unit　大气数据采集设备
ADAUPRT　Automated Data Analysis Using Pattern Recognition Techniques　使用模式识别技术的自动数据分析
ADB　Apollo Data Bank　【美国】"阿波罗"飞船数据库
ADBF　Adaptable Beamforming　自适应波束形成
ADBS　Advanced Docking and Berthing System　先进对接停泊系统
ADC　Aerospace Defense Center　航空航天防御中心
ADC　Aerospace Defense Command　【美国空军】航空航天防御司令部
ADC　Aerospace Development Center　航空航天研发中心
ADC　Affiliated Data Center　联合数据中心
ADC　Air Data Computer (Calculator)　大气数据计算机
ADC　Air Data Converter　大气数据交换器
ADC　Air Defense Command　空中防御指令
ADC　Airborne Digital Computer　机载数字计算机
ADC　Airborne Digital Converter　机载数字转换器
ADC　Analog-to-Digital Computer　模拟－数字计算机
ADC　Analog to Digital Conversion　模拟－数字转换
ADC　Analog to Digital Converter　模拟－数字转换器
ADC　Antenna Dish Control　【雷达】天线反射器控制

ADC Area Defense Council 地区防务理事会

ADC Atlantic Data Coverage 大西洋数据涵盖系统

ADC Automatic Digit Control 自动数字控制

ADC Automatic Drift Control 【陀螺】自动漂移控制

ADC Automatic Drive Control 自动驱动控制

ADCA Aerospace Department Chairmen Association 【美国】航空航天部门负责人协会

ADCAP Advanced Capabilities 高性能

ADCAPARM Advanced Capability Antiradiation Missile 先进性能反辐射导弹

ADCAS Automatic Data Collection and Analysis System 自动数据收集与分析系统

ADCC Asynchronous Data Communication Channel 异步数据通信信道

ADCCP Advanced Data Communication Control Procedure 先进数据通信控制规程

ADCE Attitude Determination and Control Electronics 姿态测定与控制电子仪器

ADCLS Advanced Data Collection and Location System 先进数据采集与定位系统

ADCO Attitude Determination and Control Officer (ISS) 【国际空间站】姿态确定与控制官

ADCOM Administrative Command 行政司令部

ADCOM Aerospace Defense Command 【美国空军】航空航天防御司令部

ADCON Advanced Concepts for Terrain Avoidance 【低空】绕飞地面障碍系统先进设计

ADCP Acoustic Data Collection and Processing 声学数据收集与处理

ADCP Advanced (Flight) Control Programmer 先进（飞行）控制编程器

ADCP Advanced Data Communication Protocol 高级数据通信协议

ADCP Advanced Display Core Processor 先进显示核心处理器

ADCP Advanced (Flight) Control Program 先进（飞行）控制项目

ADCS Advanced Defense Communication Satellite 先进国防通信卫星

ADCS Air Data Computer System 大气数据计算机系统

ADCS Attitude Determination and Control Subsystem 姿态确定与控制子系统

ADCSP Anormaly Defense Communications Satellite Program 先进防御通信卫星计划

ADCU Anormaly Detection and Control Unit 异常检测与控制装置

ADD Acoustic Detection Device 声测器；探声装置

ADD Aerospace Digital Development 航空与航天数字（计算机）研发

ADD Agency for Defense Development 【韩国】国防科学研究院

ADD Airstream Direction Detector 气流方向探测器（防失速用）

ADD Allowable Deferred Deficiency 允许推迟处理的缺陷

ADD Arming Decision Detector 解除保险判定探测器

ADD Ascent/Descent Director 上升/下降段指示器

ADDAM　Adaptive Dynamic Decision Aiding Methodology　自适应动态判定辅助方法

ADDAPS　Automatic Digital Dispatching and Process System　自动数字调用与处理系统

ADDAR　Automatic Digital Data Acquisition and Recording　自动数字数据采集与记录

ADDAS　Airborne Digital Data Acquisition System　机载数字数据采集系统

ADDAS　Automatic Digital Data Assembly System　自动数字数据汇编系统

ADDER　Automatic Digital Data Error Recorder　自动数字数据误差记录器

ADDI　Automated Digital Data Interchange　自动化数字数据交换

ADDISS　Advanced Deployable Digital Imagery Support System　先进展开式数字成像支持系统

ADDPB　Automatic Diluter-Demand Pressure Breathing　自动调节氧气浓度的增压供氧

ADDS　Advanced Data Display System　先进数据显示系统

ADDS　Air Data Doppler System　大气数据多普勒系统

ADDS　Air Defense Demonstration System　空防演示验证系统

ADDS　Apollo Documentation Description Standards　【美国】"阿波罗"飞船文件说明标准

ADDS　Application Developmental Data System　应用研制数据系统

ADDS　Applied Digital Data System　应用数字数据系统

ADDS　Army Data Distribution System　【美国】陆军数据分布系统

ADDS　Astro Digital Doppler Speedometer　天文数字多普勒测速仪

ADDS　Atmospheric Distributed Data System　大气分布数据系统

ADDS　Automatic Data Digitizing System　自动数据数字化系统

ADDS　Automatic Data Distribution System　数据自动分配系统

ADDT　Angle Distribution Data Tape　角分布数据带

ADDT　Automatic Dynamic Doppler Tester　自动动态多普勒试验器

ADE　Ada Development Environment　Ada 语言开发环境

ADE　Air Density Explorer　【美国】大气密度探测者卫星

ADE　Antenna Drive Electronics　天线驱动电子设备

ADE　Array Drive Electronics　阵列电子驱动装置

ADE　Automated Data Element　自动数据元

ADE　Automated Data Entry　自动数据输入

ADE　Autotrack Detector Equipment　自动跟踪探测设备

ADECS　Adaptive Engine Control System　自适应发动机控制系统

ADEMS　Advanced Diagnostic Engine Monitoring System　先进发动机诊断监视系统

ADEOS　Advanced Earth Observing Mission　先进地球观测任务

ADEOS　Advanced Earth Observing Satellite　【日本】先进地球观测卫星

ADEOS　Advanced Earth Orbiting Satellite　先进地球轨道卫星

ADEP　Advanced Development Earth

Penetrator 先进研发（性）地面突防导弹
ADEPT Advanced Development Prototype 先进研发（性）样机
ADEMS Alexandria Digital Earth Modeling System 【美国】"亚历山大"数字地球建模系统
ADES Automatic Data Encoding System 自动数据编码系统
ADES Automatic Data Entry System 自动数据进入系统
ADES Automatic Digital Encoding System 自动数字编码系统
ADESS Analog Data Equipment Switching System 模拟数据设备转换系统
ADESS Automatic Data Editing and Switching System 自动数据编辑与转接系统
ADEU Automatic Data Entry Unit 自动数据输入设备
ADEW Airborne Directed Energy Weapons 机载定向能武器
ADEWS Advanced Distributed Electronic Warfare Simulation 先进分布式电子战模拟
ADF Adaptive Digital Filtering 自适应数字过滤
ADF Auxiliary Detonating Fuse 辅助起爆引信
ADF Avionics Development Facility 航空电子设备研发设施
ADFE Automatic Direction Finding Equipment 自动测向设备
ADFS Active Divergence/Flutter Suppression 主动发散-颤震抑制
ADFSC Automatic Data Field System Command 自动数据场系统指令
ADFT Ascent Developmental Flight Test 上升段研发性飞行试验
ADG Accessory Drive Gear 附件传动齿轮；附件传动装置
ADH Advanced Development Hardware 高级开发硬件
ADH Automated Data Handling 自动数据处理
ADHS Analog Data Handling System 模拟数据处理系统
ADHS Automatic Data Handling System 自动数据处理系统
ADI Air Defense Interface 防空界面
ADI Altitude-Direction Indicator 高度-方向指示器
ADI Anti-Detonation Injection 抗爆剂喷射；注入防爆剂
ADI Apollo Document Index 【美国】"阿波罗"飞船文献索引
ADI Attitude Direction Indicator 姿态方位指示器
ADI Automatic Direction Indicator 自动方向指示器
ADIRS Air Data Inertial Reference System 大气数据惯性基准系统
ADIRU Air Data & Inertial Reference Unit 大气数据与惯性基准装置
ADIS A Data Interchange System 一种数据互换系统
ADIS Acquisition and Due-In System 获取与待收系统
ADIS Attitude Director Indicator System 姿态航向指示系统
ADIS Automatic Data Interchange System 自动数据交换系统
ADIS Aviation Data Integration System 【美国国家航空航天局】航空数据综合系统
ADISP Aeronautical Data Interchange System Panel 航空数据交换系统仪表板
ADISP Automated Data Interchange

System Panel 自动数据交换系统仪表板
ADISS Advanced Defense Intelligence Support System 先进防御情报保障系统
ADIT Automatic Detection and Integrated Tracking 自动探测与综合跟踪
ADIZ Air Defense Identification Zone 防空识别区
ADJ Attach, Disconnect, and Jettison 联结、断开与弹射
ADL Aeronautical Data Link 航空数据链
ADL Application Developmental Language 应用开发语言
ADL Armament Development Laboratory 武器研发实验室
ADL Atmospheric Devices Laboratory 大气装置实验室
ADL Automatic Data Link 自动数据链
ADL Automatic Data Logger 自动数据记录仪
ADL Avionics Development Laboratory 电子设备与控制系统研发实验室
ADLC Analogue-Digital Line Converter 模拟—数字线路转换器
ADLIPC Automatic Data Link Plotting System 自动数据传输线路标线系统
ADLP Advanced Data Link Program 高级数据链项目
ADLP Airborne Data Link Protocol 机载数据链协议
ADLP Automated Data Link Protocol 自主数据链协议
ADLS Automatic Drag-Limiting System 自动阻力限制系统
ADLT Advanced Discriminating LADAR Technology 激光雷达先进识别技术
ADM Adaptive Delta Modulator 自适应增量调制器
ADM Advanced Development Model 高级开发（研发）模型
ADM Aid in Decision Making 辅助决策
ADM Air Data Module 大气数据模块
ADM Air-Launched Decoy Missile 空射诱饵导弹
ADM Antenna Drive Mechanism 天线驱动机械装置
ADM Area Defense Missile 地域防御导弹
ADM Atomic Demolition Munitions 原子爆破装置
ADM Attitude Data Multiplexer 姿态数据倍增器
ADM Automated Data Management 自动数据管理
ADM Automated Data Monitoring 自动数据监控
ADMC Actuator Drive and Monitor Computer 致动器驱动与监控计算机
ADMIRE Automatic Diagnostic Maintenance Information Retrieval 自动诊断性维修情报检索
ADMIS Automated Data Management Information System 自动化数据管理信息系统
ADMS Advanced Data Management System 先进数据管理系统
ADMS Air Defense Missile System 【美国】空防导弹系统
ADMS Atmospheric Diffusion Measuring System 大气扩散测量系统
ADMS Automatic Data Monitoring

System 自动数据监控系统
ADMSC Automatic Digital Message Switching Center 数字信息自动交换中心（用于美国海外自动数据网）
ADOC Aerospace Defense Operations Center 空域防御操作中心
ADOCC Air Defense Operations Control Center 空域防御操作控制中心
ADOCS Advanced Digital Optical Control System 先进数字光学控制系统
ADONIS Automatic Digital On-Line Instrumentation System 自动数字联机仪表系统
ADOP Advanced Distributed Onboard Processor 先进分布式箭载（机载、星载）处理器
ADOS Astronautical Defensive Offensive System 航天防御进攻系统
ADP Acceptance Data Package 验收数据包
ADP Acoustic Data Processor 声学数据处理器
ADP Advanced Development Program 先进研发项目
ADP Air Data Probe 大气数据探测器
ADP Automatic Data Processing 自动数据处理
ADP Automatic Deletion Procedure 自动删除程序
ADP Automatic Destruct Program 【导弹在飞行中】自毁程序
ADP Automatic Diagnostic Program 自动诊断程序
ADPA Air Data Probe Assemblies 大气数据探测器组件
ADPC Automatic Data Processing Center 自动数据处理中心
ADPCM Adaptive Differential Pulse Code Modulation 自适应差分脉码调制
ADPCS Advanced Data Processing Control Subsystem 先进数据处理器控制子系统
ADPE Automatic Data Processing Engineering 自动数据处理工程
ADPE Automatic Data Processing Equipment 自动数据处理设备
ADPE Auxiliary Data Processing Equipment 辅助数据处理设备
ADPE/S Automatic Data Processing Equipment and Software 自动数据处理设备与软件
ADPG Atmospheric Dynamic Payload Group 大气动力有效载荷组
ADPIN Automatic Data Processing Intelligence Network 自动数据处理情报网
ADPMI Average Deficiencies per Preventive Maintenance Inspection 每次预防维修检查的平均缺陷数
ADPP Automatic Data Processing Program 自动数据处理程序
ADPS Acceptance Data Package System 接收数据包系统
ADPS ASARS Deployable Processing Station 【美国】先进合成孔径雷达系统部署处理站
ADPS Automatic Data Processing System 自动数据处理系统
ADPS Automatic Display and Plotting System 自动显示与绘图系统
ADPS Auxiliary Data Processing System 辅助数据处理系统
ADPSC Automatic Data Processing Service Center 自动数据处理服务中心
ADPSO Association of Data Processing Service Organizations 【美国、

加拿大】数据处理服务机构协会

ADR Adiabatic Demagnetization Refrigeration 绝热去磁制冷

ADR Advanced Data Recording 先进数据记录

ADR Advanced Data Research 先进数据研究

ADR Analog Data Recognition 模拟数据识别

ADR Analog-Digital Recorder 模拟—数字记录器

ADR Architectural Design Review 体系结构设计评审

ADR Auxiliary Data Record 辅助数据记录器

ADR/HUM Accident Data Recorder and Health Usage Monitor 事故数据记录器与使用状况监控器

ADRA Atlantic Downrange Recovery Area 【美国】大西洋发射中心和沿着试验航向的回收区

ADRC Automatic Digital Rate Changer 自动数字（传输）率变换器

ADRC Automatic Digital Recording and Control 自动数字式记录与控制（系统）

ADRC Automatic Digital Relay Center 自动数字中继中心

ADREP Automatic Data Processing Resource Estimation Procedures 自动数据处理资源评估程序

ADRIFT Archive Data Retrieval Interface Tool 档案数据查询接口工具

ADRIS Automatic Dead Reckoning Instrument System 自动推算定位仪

ADRM Analog-to-Digital Recording System 模拟—数字记录系统

ADRS Advanced Dynamic RF Simulator 先进动态射频模拟器

ADS Advanced Distribution System 先进分布系统

ADS Advanced Docking Simulator 先进轨道对接模拟装置

ADS Aerodynamic Deceleration System 空气动力减速系统

ADS Aerospace Data System 航空航天数据系统

ADS Air Data Sensor 大气数据传感器

ADS Air Data Subsystem 大气数据子系统

ADS Air Data System 大气数据系统

ADS Airspeed and Direction Sensor 风速和风向传感器；空速和航向传感器

ADS Angular Displacement Sensor 角位移传感器

ADS Application Data Structure 应用数据机构

ADS Application Development System 应用开发系统

ADS Asynchronous Data Service 异步数据业务

ADS Attitude Determination Software 姿态确定软件

ADS Attitude Determination System 姿态确定系统

ADS Attitude Determing System 姿态测定系统

ADS Attitude Display System 姿态显示系统

ADS Automatic Dependent Surveillance 自动相关监视

ADS Automatic Destruct System 自动自毁系统

ADSA Advanced Digital Signal Analyzer 先进数字信号分析器

ADSA Air Data Sensor Assembly 大

气数据传感器组件
ADSA Air Derived Separation Assurance 气流分离保证
ADSAM Air-Directed Surface-to-Air Missile 空中定向的地对空导弹
ADSAS Air Derived Separation Assurance System 气流分离保证系统
ADSC Acquisition Data Systems Controller 获取数据系统控制器
ADSCP Advanced Defense Satellite Communication Program 高级国防卫星通信项目
ADSF Automatic Directional Solidification Furnace 【美国航天飞机】自动定向固化（结晶）炉
ADSG Atomic Defense and Space Group 原子防御与航天大队
ADSI Air Force Defense Systems Integrator 【美国】空军防御系统积分仪
ADSI Analog Display Service Interface 模拟显示服务接口
ADSM Air Defense Suppressing Missile 防空压制导弹
ADSO Aerospace Defense Systems Office 航空与航天防御系统署
ADSO Automatic Display Switching Oscilloscope 自动显示转换示波器
ADSP Advanced Digital Signal Processor 先进数字信号处理机
ADSP Automatic Dependent Surveillance Panel 自动相关监视屏
ADSR Attack-Decay-Sustain-Release 起音－衰减－保持－释放
ADSS Aerospace Data System Standards 航空航天数据系统标准
ADSS AETC Decision Support System 【美国空军】教导培训司令部决策支持系统
ADSS Automatic Data Switching System 自动数据转换系统
ADSS Avionics Development Simulation System 电子设备与控制研发模拟系统
ADST Advanced Distributed Simulation Technology 先进分布式模拟技术
ADSU Automatic Dependent Surveillance Unit 自动相关监视装置
ADT Aided Tracking 半自动跟踪；半自动跟踪系统
ADT Air Data Tester 空气数据测试装置
ADT Air Data Transducer 大气数据转换器
ADT Automatic Detection and Tracking 自动探测与跟踪
ADT Autonomous Data Transfer 自动数据转换
ADT&E Advanced Development Test and Evaluation 预研试验与评估
ADTA Air Data Transducer Assembly 大气数据转换器组件
ADTAC Automatic Digital Tracking Analyzer Computer 自动数字跟踪分析计算机
ADTC Armament Development Test Center 【美国空军】装备研制试验中心
ADTI Application Data Transfer Interface 应用数据转换接口
ADTM Antenna Deployment and Trim Mechanism 天线展开与配平机械装置
ADTOC Air Defense Tactical Operations Center 空防战术操作中心
ADTS Automatic Data Test System 自动数据测试系统
ADTU Automatic Digital Test Unit 自动数字式试验装置

ADTU Auxiliary Data Translator Unit 辅助数据译码装置

ADTV Agena Docking Target Vehicle 【美国】"阿森纳"对接目标飞行器

ADU Alignment Display Unit 【惯导】对准显示部件

ADU Avionics Display Unit 电子设备显示装置

ADV Advanced Development Vehicle 先进研制运载器

ADV Ariane Derived Vehicle 【欧洲空间局】"阿里安"衍生型火箭

ADVANCE Airborne Doppler Velocity Altitude Navigation Compass Equipment 机载多普勒速度高度导航罗盘设备

ADVM Auto Data Validity Monitoring 自动数据有效性监测

ADVS Asteroid Deflection Vehicle System 小行星偏转飞行器系统

ADVT-C Advanced Development Verification Test-Contractor 先进开发验证试验－承包人

ADVT-G Advanced Development Verification Test-Government 先进开发验证试验－政府

ADWAR Advanced Directional Warhead 先进定向弹头

ADWCP Automated Digital Weather Communication Program 自动数字气象通信项目

AE Aerospace Engineering 航空航天工程

AE Arianespace 【法国】阿里安航天有限公司

AE Atmosphere Explores 【美国】大气探险者（卫星）

AE Atmospheric Entry 进入大气层

AE/DE Atmospheric Explorer/Dynamics Explorer (Spacecraft) "大气探测者"与"动力探测者"（航天器）

AEA Aft End Assembly 尾部装配；尾部组件

AEA Angular Error Average 角度平均误差

AEA Antenna Elevation Angle 天线仰角

AEA Association of European Astronauts 欧洲航天员协会

AEA Automatic Error Analysis 自动误差分析

AEB Aft Equipment Bay 后部设备舱

AEB Analog Expansion Bus 模拟扩展总线

AEB Avionics Equipment Bay 电子设备与控制系统舱

AEBS Advanced Engine Breathing System 先进发动机通气系统

AEC Advanced Experiment Container 先进实验容器

AEC Aft End Cone 后端锥

AEC Amplifier Electronic Control 放大器电子控制

AEC Automatic Error Correction 误差自动校正

AEC Automatic Exciter Control 激励器自动控制

AECM Active Electronic Counter Measures 有源电子对抗

AECMA Association Européenne des Constructeurs de Matériel Aérospatial 欧洲航空航天器材制造商协会

AECS Advanced Engine Control System 先进发动机控制系统

AECS Advanced Entry Control System 先进进入控制系统

AECS Advanced Environmental Control System 先进环境控制系统

AECS Apollo Environmental Control

AED

System 【美国】"阿波罗"飞船环境调节系统

AED Aerodynamic Equivalent Diameter 气动力当量直径

AED Aerospace Electronical Division 航空航天电子部

AED Astra-Electronics Division 航天电子设备处

AED Automated Engineering Design 自动化工程设计

AEDC Arnold Engineering Development Center 【美国】阿诺德工程发展中心

AEDS Analog Event Distribution System 模拟作用分布系统

AEDS Atmospheric Electric Detection System 大气层电探测系统

AEDS Atmospheric Electron Detection System 大气电子探测系统

AEDT Aviation Environmental Design Tool 航空环境设计工具

AEEP Aerospace Engineering and Engineering Physics 航空航天工程与工程物理学

AEF Aerospace Engineering Facility 航空航天工程设施

AEF Apogee Engine Firing 远地点发动机点火

AEG Apollo Entry Guidance 【美国】"阿波罗"飞船进入导航

Aegis BMD Aegis Ballistic Missile Defense 【美国】"宙斯盾"弹道导弹防御

AEGIS C&D AEGIS Command and Decision 【美国】"宙斯盾"指挥与决策

AEGIS CRC AEGIS Control and Reporting Center 【美国】"宙斯盾"控制与通报中心

AEH Antenna Effective Height 天线有效高度

AEHF Advanced Extremely High Frequency 先进极高频

AEISF Advanced Extend Integration Support Facility 先进可展开型综合保障设施

AEL Advanced Engineering Laboratory 【澳大利亚】先进工程实验室

AEL Aerospace Electronics Laboratory 航空航天电子设备实验室

AEM Acoustical Emission Monitoring 消声监测

AEM Animal Enclosure Module 【美国航天飞机】动物笼舱

AEM Application Explorer Missions 应用探测器（卫星）任务

AEO All Engine Operating 全发动机操作

AEOD Analysis and Evaluation of Operational Data 操作数据的分析与评估

AEOS Advanced Electro-Optical System 先进光电系统

AEOSLI Advanced Electro-Optical System Long-Wave Infrared 先进红外线长波光电系统

AEOSS Advanced Electro-Optical Sensor System 先进光电传感系统

AEOT Advanced Electro-Optical Tracker 先进光电跟踪装置

AEOWS Advanced Electro-Optical Warning Sensor 先进光电告警传感器

AEP Antenna Electronics Package 天线电子仪器组件

AEP AOCS Event Packet 姿态与轨道控制系统事件信息包

AEP Atmospheric Entry Program 进入大气层项目

AEP Avionics Evaluation Program

【美国】电子设备鉴定项目
AEPDS Advanced Electronic Processing and Dissemination System 先进电子处理与分发系统
AEPI Aegis Exoatmospheric Projectile Intercept 【美国】"宙斯盾"轻型外大气层射弹拦截
AEPI Atmospheric Emissions Photometric Imaging Experiment 【美国航天飞机】大气辐射测光成像实验
AEPI Atmospheric Imaging Instrument 大气成像仪
AEPS Advanced Extra-Vehicular Protective System 先进舱外活动保护系统
AEPS Automated Environmental Prediction System 自动环境预测系统
AEPT Advanced Engine/Propulsion Technology 先进发动机/推进技术
AER Antenna Effective Resistance 天线等效电阻
AERCAM Autonomous Extravehicular Robotic Camera 自主舱外遥控照相机
AERIS Automatic Electronic Ranging Information System 自动电子测距信息系统
AERL Aero-Elastic Research Laboratory 气动弹性研究实验室
AERO-A Aeroballistics-Aerodynamics Analysis 航空弹道学—空气动力学分析
AERO-EA Aeroballistics-Experimental Aerodynamics 航空弹道学—实验空气动力学
AEROBEE Aerojet/Bumblebee 冲压式喷气发动机
AERODYN Aerodynamic 空气动力的;气动力的
AEROS Advanced Earth Resources Observation System 先进地球资源观测系统
AEROS Aeronomy Satellite 高层大气物理学卫星;超高层大气研究卫星
AEROSAT Aeronautical Satellite 航空(导航、通信、空中交通管制)卫星
AERS Advanced ESA Resources Satellite 【欧洲空间局】先进资源卫星
AERS Airborne Expendable Rocket System 机载一次性火箭系统
AES Active Electromagnetic System 有源电磁系统
AES Advanced Extravehicular Suit 先进舱外活动航天服
AES Aerospace Electrical Society 【美国】航空航天电气学会
AES Aerospace Electronics System 航空航天电子系统
AES American Engine Society 美国发动机学会
AES Apollo Experiment Support 【美国】"阿波罗"实验保障计划
AES Apollo Extension Systems (Program) 【美国】"阿波罗"扩大应用系统(项目)
AES Army (Tactical Command and Control System) Experimentation Site 【美国】陆军(战术指挥与控制系统)试验场(区)
AES Array Element Study 阵列元研究
AES Artificial Earth Satellite 人造地球卫星
AES Atmosphere Exchange System 大气交换系统
AES Automatic Correction System

A

自动修正系统

AES Auxiliary Encoder System 辅助编码器系统

AESA Active Electronically Scanned Array 有源电子扫描阵；有源电子扫描阵列天线雷达

AESC Aerospace Environment Support Center 航空航天环境保障中心

AESC American Engineering Standards Committee 美国工程标准委员会

AESD Airlock External Stowage Devices 气闸舱外装载装置

AESE Airborne Electrical Support Equipment 机载电气辅助设备

AESG Advanced Electrostatically Suspended Gyroscope 先进静电悬浮陀螺仪

AESOP Airborne Electro-Optical Special Operations Payload 机载光电设备特种作战有效载荷

AESOP An Evolutionary System for On-Line Processing 联机处理中的演变系统

AESWS Advanced Earth Satellite Weapon System 先进地球卫星武器系统

AET Aerosurface End-to-End Test 空气舵端到端测试

AET Automatic Exchange Tester 自动交换试验器

AETF Advanced Engine Test Facility 先进发动机试验设施

AETF Azimuth Error Test Feature 方位误差试验特征

AETS Automatic Engine Trim System 发动机自动调整系统

AEU Antenna Equipment Unit 天线设备单元

AEV Aerothermodynamic Elastic Vehicle 气动热弹性飞行器

AEW Advanced Early Warning 先进早期预警

AEW Airborne Early Warning 机载早期预警

AEW&C Airborne Early Warning and Control 机载早期预警与控制

AEWS Advanced Early Warning System 先进预警系统

AF Attenuation Factor 衰减系数

AF Automatic Following 自动跟踪设备

AF SATCOM Air Force Satellite Communications (System) 【美国】空军卫星通信（系统）

AFA Airframe Assembly 弹体构架组件

AFA Auto Fault Alarm 故障自动报警

AFA Azimuth Follow up Amplifier 方位角跟随放大器；方位角跟踪放大器

AFAM Air Force Acquisition Model 【美国】空军采办模型

AFAOC Air Force Air and Space Operations Center 【美国】空军航天作战中心

AFAR Airborne Fixed Array Radar 机载固定相控阵雷达

AFAS Advanced Field Artillery System 先进战场炮兵系统

AFATDS Advanced Field Artillery Tactical Data System 先进战场炮兵战术数据系统

AFATDS Army Field Artillery Target Direction System 【美国】陆军战场炮兵目标定向系统

AFC Aerodynamic Flight Control 气动飞行控制

AFC Anti-Fading Compensation 抗

衰减补偿

AFC Automatic Flight Control 自动飞行控制

AFC Automatic Flow Control 自动流量控制

AFC Automatic Following Control 自动跟随控制

AFC Automatic Frequency Compensation 自动频率补偿

AFC Automatic Frequency Control 自动频率控制

AFC Automatic Fuel Control 自动燃料控制

AFC^{2}ISRC Air Force Command and Control Intelligence Surveillance Reconnaissance Center 【美国】空军指挥控制信息监察中心

AFC^{2}S Air Force Command and Control System 【美国】空军指挥与控制系统

AFCAS Advanced Flight Control Actuation System 先进飞行控制制动系统

AFCAS Automated Flight Control Augmentation System 自动飞行控制增稳系统

AFCCC Air Force Component Command Center 【美国】空军部队指挥中心

AFCE Automatic Flight Control Equipment 自动飞行控制设备；自动驾驶仪

AFCEE Air Force Center for Environmental Excellence 【美国】空军环境优化中心

AFCOM Air Force Communication Satellite 【美国】空军通信卫星

AFCS Adaptive Flight Control System 自适应飞行控制系统

AFCS Advanced Flight Control System 先进飞行控制系统

AFCS Analog Flight Control System 模拟式飞行控制系统

AFCS Automatic Flight Control System 自动飞行控制系统

AFCS Automatic Fuel Control System 自动燃料控制系统

AFCS Auxiliary Flight Control System 辅助飞行控制系统

AFCS Avionics Flight Control System 航空电子设备飞行控制系统

AFCSC Air Force Cryptological Support Center 【美国】空军密码术支持（保障）中心

AFD Adaptive Flight Display 自适应飞行显示

AFD Aft Flight Deck 【美国航天飞机】飞行翼面后部

AFD Amplitude Frequency Distribution 振幅频率分布

AFD Arming and Fusing Device 解除保险与引信装置

AFD Automatic Fast Demagnetization 自动快速消磁

AFD Automatic Fault Detection 自动故障探测（检测）

AFD Automatic Feeding Device 自动装填装置

AFD Automatic Flaws Detector 自动故障探测器

AFDCP Aft Flight Deck Control Panel 【美国航天飞机】飞行翼面后部控制面板

AFDO Aft Flight Deck Operator 飞行翼面后部操作员

AFDPDB Aft Flight Deck Power Distribution Box 飞行翼面后部配电箱

AFDS Advanced Flight Deck Simulator 先进驾驶舱模拟器

AFDS Automated Flight Data System 自动飞行数据系统

AFDSOC Air Force Defense System Operations Center 【美国】空军防御系统操作中心

AFE Aero-Assist Flight Experiment 【美国航天飞机】空气制动飞行实验

AFE Allowed Failure Effect 容许故障影响

AFERC Air Force Edwards Research Center 【美国】空军爱德华兹研究中心

AFESA Air Force Engineering and Services Agency 【美国】空军工程与勤务局

AFESMC Air Force Eastern Space and Missile Center 【美国】空军东部航天与导弹中心（卡纳维拉尔角）

AFETR Air Force Eastern Test Range 【美国】空军东部试验靶场

AFETRM Air Force Eastern Test Range Manual 【美国】空军东部试验靶场手册

AFEWES Air Force Electronic Warfare Effectiveness Simulator 【美国】空军电子战效能模拟器

AFEX Advanced Furnace for Microgravity Experiment with X-ray Radiography 先进 X 射线微重力实验室

AFF Arming, Fusing and Firing 解除保险、引信待发与点火发射

AFFDL Air Force Flight Dynamics Laboratory 【美国】空军飞行动力学实验室

AFFTC Air Force Flight Test Center 【美国爱德华空军基地】飞行试验中心

AFG Antenna Field Gain 天线场强增益

AFGCS Automatic Flight Guidance and Control System 自动飞行导航与控制系统

AFGIHS Air Force Geographic Information Handling System 【美国】空军地理信息处理系统

AFGL Air Force Geophysics Laboratory 【美国】空军地球物理实验室

AFGS Automatic Flight Guidance System 自动飞行制导系统

AFGWC Air Force Global Weather Center 【美国】空军全球天气（观测）中心

AFI Automatic Fault Isolation 故障自动隔离

AFID Anti-Fratricide Identification Device 防误伤识别装置

AFIRS Autonomous Flight Information Reporting System 自主飞行信息通报系统

AFIRST Advanced Far Infrared Search/Track 先进远红外搜索/跟踪（传感器）

AFIS Airborne Flight Information System 机载飞行信息系统

AFIS Automated Flight Inspection System 自动飞行检查系统

AFISC Air Force Inspection and Safety Center 【美国】空军检查与安全中心

AFIT Automatic Fault Isolation Test 自动故障隔离测试

AFITAE Association Française des Ingénieurs et Techniciens de l'Aéronautique et l'Espace 【法国】航空航天工程师和技术员协会

AFITV AF Instrumented Test Vehicle 【美国】空军测量测试运载器（反卫星目标运载器）

AFIWC Air Force Information Warfare Center 【美国】空军信息战中心

AFL Actual Flight Level 实际飞行高度

AFL Autoland Flight Test 自动着陆飞行测试

AFL Automatic Fault Location 自动故障定位

AFL Automatic Fault Locator 自动故障定位器

AFL Avionics Flying Laboratory 【美国】航空电子设备飞行实验室

AFLIR Advanced Forward Looking Infrared 先进前视红外仪

AFLS Approach Flash Lighting System 进场照明系统

AFM Automatic Fault Finding and Maintenance 自动故障探测和维修

AFM Automatic Flight Management 自动飞行管理

AFMC/OAS Air Force Material Command/Office of Aerospace Studies 【美国】空军装备司令部/航空航天研究办公室

AFMDC Air Force Missile Development Center 【美国】空军导弹研制中心

AFMS Automatic Flight Management System 自动飞行管理系统

AFMSS Air Force Mission Support System 【美国】空军任务支援（保障）系统

AFMTC Air Force Missile Test Center 【美国】空军导弹试验中心

AFOLDS Air Force On-Line Data System 【美国】空军联机数据系统

AFOS Advanced Field Operating System 先进外场操作系统

AFOS Automation of Field Operations and Services 场区操作与维修自动化

AFOSH Air Force Occupational Safety and Health 【美国】空军职业安全与健康

AFOTEC Air Force Operational Test & Evaluation Command 【美国】空军作战试验与论证（鉴定）司令部

AFOTEC Air Force Operational Test and Evaluation Center 【美国】空军作战试验与论证（鉴定）中心

AFP Alternate Flight Plan 【美国国家航空航天局】备份飞行计划

AFPA Automatic Flow Process Analysis 气流自动处理分析

AFPAM Automatic Flight Planning and Monitoring 自动飞行规划与监控

AFPG Apollo Final Phase Guidance 【美国】"阿波罗"飞船末段制导

AFQT Armed Forces Qualification Test 武装部队资格测试

AFR Acceptable Failure Rate 容许故障率

AFR Acceptance Failure Rate 验收故障率

AFR Automatic Field Recognition 自动现场识别

AFR Automatic Frequency Regulation 自动调频

AFRC Area Frequency Response Characteristic 区域频率响应特性

AFRC Automatic Frequency Ratio Controller 自动频率比率控制器

AFRCC Air Force Rescue Control Center 【美国】空军援救控制中心

AFRL Air Force Research Laboratory 【美国】空军研究实验室

AFRPL Air Force Rocket Propulsion Laboratory 【美国】空军火箭推进实验室

AFRS Auxiliary Flight Reference System 辅助飞行基准系统

AFRSI Advanced Flexible Reusable Surface Insulation 【美国航天飞机】先进柔性可重复使用表面隔离材料

AFS Active Flutter Suppression 主动式颤振抑制

AFS Automatic Flight System 自动飞行系统

AFS Avionics Flight Software 航空电子设备飞行软件

AFS Azimuth Follow up System 方位（角）跟随系统

AFSAA Air Force Studies and Analyses Agency 【美国】空军研究与分析局

AFSARC Air Force Systems Acquisition Review Council 【美国】空军系统采办评审委员会

AFSATCOMS Air Force Satellite Communication System 【美国】空军卫星通信系统

AFSC Air Force System Command 【美国】空军武器系统司令部

AFSC Automatic Flight Stabilization and Control 自动飞行稳定与控制

AFSCC Air Force Satellite Control Center 【美国】空军卫星控制中心

AFSCF Air Force Satellite Control Facility 【美国】空军卫星控制设施

AFSCF Air Force Satellite/Spacecraft Control Facility 【美国】空军卫星/航天器控制设施

AFSCN Air Force Satellite Control Network 【美国】空军卫星控制网

AFSCN Air Force Space Control Network 【美国】空军航天控制网

AFSCN Air Force Spacecraft Communication Network 【美国】空军航天器通信网络

AFSFC Air Force Space Forecasting Center 【美国】空军航天预报中心

AFSIG Ascent Flight Systems Integration Group 上升飞行系统综合研究部

AFSMC Air Force Space & Missile Centre 【美国】空军航天与导弹中心

AFSMSC Air Force Space and Missile Systems Center 【美国】空军航天与导弹系统中心

AFSOC Air Force Special Operations Command 【美国】空军特殊操作指挥部

AFSPC Air Force Space Command 【美国】空军航天司令部

AFSPSPM Air Force Study of Performance of Solid Propellant Motors 空军固体燃料火箭发动机性能研究

AFSS Automated Flight Service Station 【美国】自动飞行服务站

AFSS Autonomous Flight Safety System 自主飞行安全系统

AFSSI Air Force System Security Instruction 【美国】空军系统安全指南

AFSSS Air Force Space Surveillance System 【美国】空军航天监视系统

AFSTC Air Force Satellite Test Center 【美国】空军卫星测试中心

AFSTC Air Force Space Technology Center 【美国】空军航天技术中心

AFSTC Air Force Space Test Center 【美国】空军航天测试中心

AFSTS Air Force Space Technology Satellite 【美国】空军航天技术卫星

AFSWC Air Force Special Weapons Center 【美国】空军特种武器研究

中心
AFT Abort Flight Test 中止飞行试验
AFT Adapter Fault Tolerance 适配器容错
AFT Aerodynamic Flight Test 气动飞行试验
AFT Aft Flight Deck 尾部飞行舱面
AFT Atmospheric Flight Test 大气层飞行试验
AFT Automatic Flight Termination 自动飞行中止
AFT Automatic Frequency Tuner 自动调频器
AFTA Acoustic Fatigue Test Article 声疲劳测试装置
AFTA Avionics Fault Tree Analyser 电子设备与控制系统的故障树分析器
AFTAAS Advanced Fast Time Acoustic Analysis System 先进快速声学分析系统
AFTAC Air Force Technical Applications Center 【美国】空军技术应用中心
AFTADS Army Field Artillery Target Data System 【美国】陆军战场炮兵目标数据系统
AFTEC Air Force Test and Evaluation Center 【美国】空军试验与鉴定中心
AFTEF Air Force Test and Evaluation Facility 【美国】空军试验与鉴定所
AFTN Aeronautical Fixed Telecommunications Network 航空固定远程通信网
AFTRCC Aerospace Flight Test Radio Coordinating Council 航空航天飞行试验无线电协调委员会

AFTS Adaptive Flight Training System 自适应飞行训练系统
AFUWT Area Forecast of Upper Winds and Temperatures 高空风和温度区域预报
AFW Auxiliary Feedwater System 辅助供水系统
AFWET Air Force Weapon Effectiveness Test 【美国】空军武器效能测试（系统）
AFWL Air Force Weapon Laboratory 【美国】空军武器实验室
AFWOFS Air Force Weather Observing and Forecasting System 【美国】空军天气观测与预报系统
AFWSMC Air Force Western Space and Missile Center 【美国】空军西部航天与导弹中心
AFWTR Air Force Western Test Range 【美国】空军西部试验靶场
AG Artificial Gravity 人造重力
AG Attitude Gyro 姿态陀螺仪
AG Available Gain 有效增益
AGAA Attitude Gyro Accelerometer Assembly 姿态陀螺加速仪组件
AGAA Automatic Gain Adjusting Amplifier 自动增益调整放大器
AGAGE Advanced Global Atmospheric Gases Experiment 先进全球大气气体试验
AGANI Apollo Guidance and Navigation Information 【美国】"阿波罗"飞船制导与导航信息
AGAP Attitude Gyro Accelerometer Package 姿态陀螺仪加速度计密封装置
AGARD Advisory Group on Aerospace Research and Development 航空航天研究与发展咨询组
AGARD Aerospace Research and De-

velopment 【北大西洋公约组织】航空航天研究与发展

AGATE Advanced General Aviation Transport Experiments 【美国国家航空航天局】先进通用航空运输试验

AGAVE Automatic Gimbaled Antenna Vectoring Equipment 万向（架支撑的）天线自动指向装置；自动定向天线导航设备

AGB Autonomous Guided Bomb 自主式导弹

AGC Apollo Guidance Computer 【美国】"阿波罗"飞船制导计算机

AGC Automatic Gain Control 自动增益控制

AGC Automatic Gauge Control 自动轨距控制；自动计量管理

AGC Automatic Gauge Controller 自动测量控制器

AGC Automatic Generation Control 自动生成控制

AGCA Automatic Ground Controlled Approach 地面控制自动进场

AGCAS Automatic Ground Collision Avoidance System 自动地面避撞系统

AGCCS Air Force Global Command and Control System 【美国】空军全球指挥与控制系统

AGCCS Army Global Command and Control System 【美国】陆军全球指挥与控制系统

AGCI Automatic Ground-Controlled Intercept 自动地面控制的截击

AGCI Automatic Ground-Controlled Interception 地面自动控制拦截

AGCL Automatic Ground-Controlled Landing 自动地面控制着陆/自动引导着陆

AGCS Advanced Guidance and Control System 先进制导与控制系统

AGCS Automatic Ground Checkout System 自动地面检测系统

AGCS Automatic Ground Control Station 地面自动控制站

AGCS Automatic Ground Control System 自动地面控制系统

AGCU Attitude Gyro Coupling Unit 姿态陀螺仪耦合装置

AGCW Autonomously Guided Conventional Weapon 自主制导常规武器

AGD Absolute Gyro Drift 陀螺绝对漂移

AGD Axial-Gear Differential 轴向差动齿轮

AGE Aerospace Ground Equipment 航空航天地面设备

AGE Air/Ground Equipment 空中/地面设备

AGE Apollo Guidance Equipment 【美国】"阿波罗"飞船制导设备

AGE Automatic Ground Equipment 自动地面设备

AGE Automatic Guidance Electronics 自动制导电子设备

AGE Auxiliary Ground Equipment 辅助地面设备

AGE Programmable AGE 程控式航空航天地面设备

AGEI Aerospace Ground Equipment Installation 航空航天地面设备安装

AGEOCP Aerospace Ground Equipment Out of Commission for Parts 因零件问题而不能使用的航空航天地面设备

AGETS Automated Ground Engine Test System 自动化地面发动机试

车系统
AGF Advanced Ground Facility 先进地面设施
AGF Automatic Guided Flight 自动制导飞行
AGGD Apollo Guidance Ground Display 【美国】"阿波罗"飞船制导地面显示器
AGIPA Adaptive Ground Implementing Phased Array 自适应地面部署的相控阵雷达
AGL Above Ground Level 地平面以上
AGL Absolute Ground Level 绝对地面平面
AGM Air-to-Ground Missile 空对地导弹
AGM Attack Guidance Matrix 攻击引导矩阵
AGM Attack Guided Missile 攻击导弹
AGMC Aerospace Guidance and Meteorology Center 【美国】空军航空航天制导与计量中心
AGNIS Apollo Guidance and Navigation Industrial Support 【美国】"阿波罗"项目制导与导航工业保障
AGOES Advance Geosynchronous Observation Environment Satellite 先进地球同步观测环境卫星
AGOSS Automated Ground Operations Scheduling System 自动地面操作调度系统
AGP Avionics Ground Prototype 航空电子设备地面样机
AGPI Automatic Ground Position Indicator 自动地面位置指示器
AGPO Atlas Government Program Office 【美国】"宇宙神"运载火箭政府项目办公室

AGPS Auxiliary Gyro Platform Subsystem 辅助陀螺仪平台子系统
AGRE Active Geophysical Rocket Experiment 实际地球物理火箭实验
AGREE Advanced Ground Receiving Equipment Experiment 先进地面接收设备试验
AGRF Advanced Guidance Research Facility 先进制导研究设施
AGROS Advanced Global Research Observation Satellite 先进全球研究观察卫星
AGRT Automatic Ground Receiver Terminal 自动地面接收机终端
AGS Abort Guidance System 中止制导系统
AGS Active Guidance System 主动式制导系统
AGS Advanced Gimbal System 先进万向架系统
AGS Advanced Guidance Sensor 先进制导传感器
AGS Advanced Guidance System 先进制导系统
AGS Antenna Gain Stabilization 天线增益稳定
AGS Anti-Gravity Suit 抗重力装置
AGS Artificial Gravity Structure 人造重力结构
AGS Ascent Guidance Software 上升制导软件
AGS Ascent Guidance-and-Control System 爬高制导与控制系统
AGS Automatic Gain Stabilization 自动增益稳定
AGSC Analog Gyro Slew Circuits 模拟陀螺回转线路
AGSD Advanced Ground Segment Design 先进地面段设计

AGSE Aerospace Ground Support Equipment 航空航天地面保障设备

AGSL Satellite Launching Ship 【代号】卫星发射船

AGSR All-Weather Ground Surveillance Radar 全天候地面监视雷达

AGSS Attitude Ground Support System 飞行姿态地面支持系统

AGSW Adaptive Gain and Stall Warming Unit 自适应增益与失速告警装置

AGTELIS Automatic Ground Transportable Emitter Location & Identification System 地面移动式发射机自动定位与识别系统

AGV Automated Guided Vehicle 自动制导飞行器；自导运载器

AGVS Air/Ground Voice System 空中／地面声音系统

AGVT Advance Guided Vehicle Technology 先进制导飞行器技术

AH Attitude Hold 姿态保持

AHAM Advanced Heavy Anti-Tank Missile 先进重型反坦克导弹

AHAMS Advanced Heavy Anti-Tank Missile System 先进重型反坦克导弹系统

AHC Attitude Heading Computer 姿态航向计算机

AHCS Advanced Hybrid Computing System 先进混合计算系统

AHD Anti-Handling Device 反拆装置

AHDL Analog Hardware Descriptive Language 模拟硬件说明语言

AHIS Agile Homing Interceptor Simulator 灵敏寻的拦截机（导弹）模拟器（装置）

AHM Avionics Health Management 电子设备和控制系统健康管理

AHMS Advanced Health Management System 先进健康管理系统

AHOB Actual Height of Burst 实际爆炸高度

AHP Aocs Housekeeping Packet 姿态和轨道控制系统辅助信息包

AHR Ablative Heat Rate 烧蚀热率

AHRS Attitude Heading Reference System 姿态航向参照系统

AHS Ablation Heat Shield 烧蚀隔热层；烧蚀防热层

AHS Advanced Habitat Systems 先进适居系统

AHS Automatic Hovering System 自动悬停系统

AHSE Assembly, Handling and Shipping Equipment 装配、搬运与海运设备

AHTP Aerodynamic Heat Test Plan 气动热试验计划

AHTR Auto Horizontal Tail Retrimming after Landing 着陆后水平尾翼自动调整配平

AHWT Ames Hypersonic Wind Tunnel 【美国】艾姆斯高超声速风洞

AI Airspeed Indicator 空气速度指示器

AI Altimeter Indicator 测高指示器

AI Altitude Indicator 高度指示器

AI Attitude Indicator 姿态指示器

AI Autonomous Inspection 自主检验

AIR Artificial Intelligence and Robotics 人工智能与机器人

AIT Assembly, Integration and Test 装配、集成与测试

AI²S Advanced Infrared Imaging Seeker 先进红外成像导引头

AIA Advanced Information Architec-

ture 先进信息结构
AIA Associazione Industrie Aerospaziali 【意大利】航空航天工业协会
AIA Auxiliary Interface Adapter 辅助接口连接器
AIAA American Institute of Aeronautics and Astronautics 美国航空航天协会
AIAA Area of Intense Air Activity 稠密空气层活动区
AIAA-TIS Technical Information Service of American Institute of Aeronautics and Astronautics 美国航空航天学会技术信息服务处
AIAAM Advance-Interrupt (or intercept) Air-to-Air Missile 先进拦截（或拦截）空对空导弹
AIAR American Institute of Aerological Research 美国高空气象研究所
AIC Air Inlet Control 进气控制
AIC Ammunition Identification Code 弹药识别码
AIC Apollo Inter Mediate Chart 【美国】"阿波罗"飞船中等比例尺月球图
AICBM Anti-Intercontinental Ballistic Missile 反洲际弹道导弹
AICS Advanced Imaging Communication System 先进图像数据通信系统
AICS Advanced Interior Communication System 先进内部通信系统
AICS Air Inducting Control System 进气控制系统
AICS Automated Identification and Control System 自动识别与控制系统
AICS Automatic Intake Control System 自动注入控制系统
AICS Automatic Intersection Control System 交叉点自动控制系统
AICT Automatic Integrated Circuit Tester 自动集成电路测试器
AID Agile Interceptor Development 灵敏拦截机（导弹）研制
AID Analog Input Differential 模拟输入差分
AID Attached Inflatable Decelerator 自带充气减速器
AID Automated Inspection of Data 数据自动检查
AID Automatic Internal Diagnosis 【计算机】自动内部诊断
AIDA Artificial Intelligence Discrimination Architecture 人工智能识别结构
AIDAA Associazione Italiana di Aeronautica e Astronautica 意大利航空航天协会
AIDAS Advanced Instrumentation and Data Analysis System 先进仪表测量与数据分析系统
AIDAT Automatic Integrated Dynamic Avionic Tester 自动综合动态航空电子设备测试器
AIDB Artificial Intelligence Data Base 人工智能数据库
AIDE Aerospace Installation Diagnostic Equipment 航空航天设施的（故障）诊断设备
AIDES Automated Image Data Extraction System 自动化图像数据提取系统
AIDPS Automatic Inspection Diagnostic and Prognostic System 自动检测诊断与预报系统
AIDPS Automatic Integrated Diagnostic and Prognostic Subsystem 自动综合故障判断与预报分系统
AIDR Aerospace Internal Data Report

A

【美国空军】航空航天内部数据报告

AIDRS Automated Integrated Data Recording System 自动综合数据记录系统

AIDS Action Information Display System 指令信息显示系统

AIDS Adaptive Intrusion Data System 自适应侵入数据系统

AIDS Advanced Integrated Data System 先进综合数据系统

AIDS Advanced Integrated Display System 先进综合显示系统

AIDS Advanced Interactive Display System 先进交互式显示系统

AIDS Aerospace Intelligence Data System 航空航天情报数据系统

AIDS Airborne Inertial Data System 机载惯性数据系统

AIDS Airborne Integration Data System 机载综合数据系统

AIDS Automatic Infrared Diagnostic System 自动红外（故障）判断系统

AIDS Automatic Integrated Data System 自动综合数据系统

AIDS Automatic Integration Debugging System 自动综合排除故障系统

AIDS Automation Instrument Data Service 自动化仪表数据服务

AIE Airborne Interception Equipment 机载拦截设备

AIED Aeronautical Industry Engineering and Development 航空航天工业工程与研发

AIEWS Advanced Integrated Electronic Warfare System 先进综合电子战系统

AIF Attendance Improvement Formula 维修改进方案

AIFI Automatic Inflight Insertion 飞行中自动插入

AIFSS Automated International Flight Service Stations 国际自动飞行服务站

AIG All Inertial Guidance 全惯性制导

AIG Augmented Inertial Guidance 增强型惯性制导

AIGS All Inertial Guidance System 全惯性制导系统

AII Automated Interchange of Information 信息自动交换

AIL Aerospace Instrumentation Laboratory 航空航天仪表装置实验室

AIL Artificial Intelligence Language 人工智能语言

AIL Avionics Integration Laboratory 电子设备与控制系统综合实验室

AILAS Automatic Instrument Landing Approach System 自动仪表着陆进场系统

AILS Advanced Impact Location System 先进弹着点定位系统

AILS Advanced Integrated Landing System 先进综合着陆系统

AILS Automatic Instrument Landing System 先进仪表着陆系统

AILSA Aerospace Industrial Life Sciences Association 【美国】航空航天工业生命科学协会

AILSS Advanced Integrated Regenerative Life Support System 先进再生式综合生命保障系统

AIM Aeronautical Information Management 航空航天信息管理

AIM Air Intercept Missile 空中拦截导弹

AIM Air-Launched Interceptor Missile

空射拦截导弹

AIMAS Académie Internationale de Medécine Aéronautique et Spatiale 国际航空与航天医学研究院

AIME Advanced Inertial Measurement Equipment 先进惯性测量设备

AIME American Institute of Mechanical Engineers 美国机械工程师协会

AIMS Advanced Inertial Measurement System 先进惯性测量系统

AIMS Advanced Intercontinental Missile System 先进洲际导弹系统

AIMS Altitude Identification and Military System 高度识别与军事系统

AIMS Artes Information Management System 通信系统远景研究信息管理系统

AIMU Advanced Inertial Management Unit 先进惯性测量单元

AINS Advanced Inertial Navigation System 先进惯性导航系统

AINS Area-Inertial Navigation System 区域惯性导航系统

AIOP Analog Input/Output Package 模拟输入/输出组件

AIP Advanced Instrumentation Platform 先进测量平台

AIP Air-Independent Propulsion 不依赖空气的推进技术

AIP Artificial Intelligence Programming 人工智能程序设计

AIP Auto Track Interface Processor 自动跟踪接口处理机

AIP Auto-Igniting Propellant 自燃（火箭）推进剂

AIP Avionics Integration Plan 电子设备与控制系统一体化计划

AIPR Automated Information Processing Resources 自动信息处理资源

AIPS Advanced Integrated Power System 先进综合动力系统

AIR Air Injection Reaction 喷气反力；喷气阻力

AIRAC Aeronautical Information Regulation and Control 航空航天信息规定与控制

AIRAPT International Association for the Advancement of High Pressure Science and Technology 国际高压科学与技术进步协会

AIRBM Anti-Intermediate Range Ballistic Missile 反中程弹道导弹

AIRBOSS Advanced Infrared Ballistic Missile Observation Sensor System 【日本】先进红外弹道导弹观测遥感器系统

AIRBOSS Advanced Infrared Ballistic Missile Optical Sensor System 先进红外弹道导弹光学传感器系统

AIRES Advanced Infrared Entertainment System 先进红外演示系统

AIRES Artificial Intelligence Research in Environmental Sciences 环境科学的人工智能研究

AIRMS Airborne Infrared Measurement System 机载红外测量系统

AIRPAP Air Pressure Analysis Program 大气压力分析程序

AIRS Accident/Incident Reporting System 事故/事件报告系统

AIRS Advanced Inertial Reference System 先进惯性参考系统

AIRS Advanced Infrared Sounder 先进红外探测器

AIRS Aerospace Instrumentation Range Station 航空航天仪表测距站

AIRS Airborne Infrared Surveillance 机载红外监视

A

AIRS Atmospheric Infrared Sounder 【美国】大气红外探测器

AIRS Automated Information Retrieval System 自动信息检索系统

AIRS Autonomous Infrared Sensor 自主式红外传感器

AIRSS Alternative Infrared Satellite System 【美国】替代性红外卫星系统

AIRST Advanced Infrared Search and Track 先进红外搜索与跟踪

AIS Adaptive Interface Suppression 自适应接口抑制

AIS Advanced Identification System 先进识别系统

AIS Advanced Instruction System 先进指令系统

AIS Advanced Instrumentation System 先进仪表系统

AIS Air Injection System 空气喷射系统

AIS Alarm Indication Signal 报警指示信号

AIS American Interplanetary Society 美国行星际学会

AIS Applo Instrumentation Ship 【美国】"阿波罗"飞船测量船

AIS Ascension Island Station 【美国】阿森松岛跟踪站

AIS Attitude Indicating System 姿态指示系统

AIS Automated Identification System 自动识别系统

AIS Automated Information Security 自动化信息安防

AIS Automated Information System 自动化信息系统

AIS Automated Inspection System 自动检查系统

AIS Automated Instrumentation System 自动（化）仪表（测试）系统

AIS Automatic Information Services 自动信息服务

AIS Automatic Interplanetary Station 自动行星际站

AIS Autonomic Information System 自主信息系统

AISF Avionics Integration Support Facilities 电子设备与控制系统综合保障设施

AISOC Advanced Integrated Safety and Optimizing Computer 先进综合安全与优化计算机

AISSP Automated Information System Security Plan 自动信息系统安全计划

AIST Avionics Integration Support Technology 电子设备与控制系统综合支持技术

AIT Active Imaging Testbed 主动（有源）成像试验（测试）台（地）

AIT Assembly, Integration and Testing 装配、组装与测试

AIT Atmospheric Interceptor Technology 大气截击器技术

AIT Auto-Ignition Temperatures 自动点火温度

AIT Automated Identification Technology 自动识别技术

AIT Automated Interactive Target 自动化交互式目标（识别）

AIU Abort Interface Unit 中止接口装置

AIU Alarm Interface Unit 报警接口装置

AIU Antenna Interface Unit 天线接口装置

AIU Avionics Interface Unit 电子设备与控制系统接口装置

AJA Assembly Jig Accessory 装配

夹具附件；装配型架附件
AJD Anti-Jam Display 抗干扰显示器
AJF Anti-Jamming Frequency 抗干扰频率
AJTGPS Anti-Jam Technology for GPS 全球定位系统抗干扰技术
AKM Apogee Kick Motor 远地点起动发动机
AKMCD Apogee Kick Motor Capture Device 远地点起动发动机捕获装置
AL Accuracy Landing 准确着陆
AL Aerodynamics Laboratory 空气动力学实验室
AL Aeronomy Laboratory 高层大气层物理学实验室
AL Aerophysics Laboratory 大气物理学实验室
AL Air Launch 空中发射
AL Airlock 气闸舱；气密室（间）
AL Ames Laboratory 【美国】艾姆斯实验室
AL Approach and Landing 进场与着陆
AL Approach Lighting 进场照明
AL Arrival Locator 到达定位器
AL Assemble/Load 装配与载荷
AL Assigned Level 指定电平
AL Audio Language 声频语言
AL/EMU Airlock/Extravehicular Mobility Unit 气闸/舱外可移动式装置
ALA Autonomous Logistics Analysis 自主后勤分析
ALAAR Air-Launched Air-to-Air Rocket 空射空对空火箭
ALAC Alaskan Air Command 【美国】阿拉斯加空军司令部
ALADNS Automated Location and Data Netting System 自动定位与数据网系统
ALAEE Atmospheric Lyman Alpha Emissions Experiment 【美国航天飞机】大气莱曼粒子发射实验
ALAM Advanced Land-Attack Missile 先进陆攻导弹
ALAM Automated Logistics Assessment Model 自动化后勤保障评估模型
ALAN Advanced Local Area Network 先进局域网
ALARM Air Launched Anti-Radiation Missile 空射反辐射导弹
ALARM Alert, Locate and Report Missile 导弹预警、定位与报告系统
ALARR Air Launch Air Recoverable Rocket 空中发射空中回收的火箭
ALAS Advanced Liquid Axial Stage 先进液体轴向芯级
ALAS Approach Landing Autopilot System (Subsystem) 进场着陆自动驾驶仪系统（分系统）
ALAS Atmosphere Laboratory for Applications and Science 【美国】大气应用与科学实验室
ALAS Automated Logistical Analysis System 自动化后勤保障分析系统
ALASA Airborne Launch Assist Space Access 机载发射辅助空间进入技术
ALASAT Air-Launched Anti-Satellite 空中发射反卫星
ALB-F Air-Land Battle-Future 未来空地一体战
ALBCS Airborne Laser Beam Control System 机载激光束控制系统
ALBD Automatic Load Balancing Device 自动载荷平衡装置

ALBE Air-Land Battlefield Environment 空地一体作战战场环境

ALBM Air-Launched Ballistic Missile 空射弹道导弹

ALBS Air Launched Balloon System 空中发射气球系统

ALC Aft Load Controller 后部荷载控制器

ALC Air Logistics Center 空中后勤保障中心

ALC Asynchronous Link Control 异步数据链控制

ALC Automatic Landing Control 自动着陆控制

ALC Automatic Load Control 自动载荷控制

ALCA Aft Load Control Assembly 后部荷载控制组件

ALCAM Air-Launched Conventional Attack Missile 空射常规攻击导弹

ALCC Airborne Launch Control Center 空中发射控制中心

ALCC Airlift Coordination Center 空运协调中心

ALCE Airlift Coordination Element 空运协调部件（要素）

ALCM Air Launch Cruise Missile 空中发射巡航导弹

ALCM Air Launched Cruise Missile 空射巡航导弹

ALCOR ARPA/Lincoln C-Band Observable Radar 【美国国防部】高级研究计划局研制的林肯型C波段观测雷达

ALCS Automation Launch and Control System 自动化发射与控制系统

ALDCS Active Lift Distribution Control System 升力分配主动控制系统

ALDF Advanced Lightning Direction Finder 高级雷电定向器

ALDS Apollo Launch Data System 【美国】"阿波罗"飞船发射数据系统

ALDT Average Logistics Delay Time 平均后勤延误（滞后）时间

ALE Airborne Laser Experiment 机载激光实验

ALE Atmospheric Lifetime Experiment 大气寿命实验

ALEMS Apollo Lunar Excursion Module Sensor 【美国】"阿波罗"飞船登月舱传感器

ALEP Apollo Lunar Exploration Program 【美国】"阿波罗"月球探测计划

ALERT Acute Launch Emergency Reliability Tip 急性发射紧急可靠性提示

ALERT Automatic Logging Electric Reporting Telemetry 自动记录电子信息报告和遥测

ALEXIS Array of Low Energy X-Ray Imaging Sensors 阵列式低能X射线成像传感器

ALF Auxiliary Landing Field 辅助着陆场

ALFA Automatic Line Fault Analysis 自动线路故障分析

ALFC Automatic Load Frequency Control 自动载荷频率控制

ALFE Advanced Liquid Feed Experiment 先进液体加载实验

ALG Autonomous Landing Guidance 自主着陆制导

ALGS Approach and Landing Guidance System 进场与着陆引导系统

ALH Advanced Liquid Hydrogen 先进液氧

ALHAT Autonomous Landing and

Hazard Avoidance Technology 自主着陆与危险规避技术

ALHTK Air Launch Hit to Kill 空中发射撞击杀伤

ALICE Automated Location of Isolation and Continuity Errors 隔离与连续性误差的自动化定位

ALIMS Automatic Laser Instrumentation Measurement System 自动激光检测系统

ALIRT Adaptive Long Range Infrared Tracker 自适应远程红外跟踪器

ALIRT Advanced Large-Area Infrared Transducer 先进大面积红外传感器

ALIT Automatic Line Insulation Tester 自动线路绝缘测试装置

ALLACM Air-Launched Low-Altitude Cruise Missile 空射低空巡航导弹

ALLS Apollo Lunar Landing System 【美国】"阿波罗"飞船月球着陆系统

ALLS Apollo Lunar Logistics Support 【美国】"阿波罗"飞船登月后勤保障

ALMAGS Apollo LM Abort (or Ascent) Guidance System 【美国】"阿波罗"着陆舱中止（或上升）制导系统

ALMAZ 2nd Generation Soviet Manned Orbital Space Station-Military Salyut 【苏联】"礼炮号"空间站（第二代有人驾驶轨道太空站）

ALMB Air-Launched Missile Ballistics 空射导弹弹道学

ALMV Air-Launched Miniature Vehicle 空射微型飞行器

ALOD Adaptive Locally Optimum Detector 【美国海军】自适应局部最佳探测器

ALOR Advanced Lunar Orbital Rendezvous 先进月球轨道交会（技术）系统

ALORS Advanced Large Object Recovery System 先进大型目标回收系统

ALOS Advanced Land Observing Satellite 先进对地观测卫星

ALOS Advanced Land Observing System 先进对地观测系统

ALP Advanced Lunar Projects 先进登月工程

ALP Air Launch Platform 空中发射平台

ALP Alternative Launch Point 备择发射点

ALP Ascent Load Package 上升载荷部件

ALPO Association of Lunar and Planetary Observers 月球与行星观察者协会

ALPS Accidental Launch Protection System 【导弹核武器】意外发射预防系统

ALPS Advanced Linear Programming System 先进线型程序设计系统

ALPS Advanced Liquid Propulsion System 先进液体推进系统

ALPS Approach and Landing Procedures Simulator 进场与着陆程序模拟器

ALPS Automatic Landing and Position System 自动着陆与定位系统

ALPURCOMS All-Purpose Communication System 多用途通信系统

ALQA Advanced Link Quality Analysis 先进链路质量分析

ALRDI Automatic Laser Ranging and Direction Instrument 自动激光测距与定位仪

ALRR Air-Launched and Recoverable

Rocket 空中发射的可回收火箭

ALS Advanced LANDSAT Sensor 陆地卫星先进遥感器

ALS Advanced Launch System 先进发射系统

ALS Advanced Logistics System 先进后勤保障系统

ALS Alternate Landing Site 备用着陆场

ALS Approach and Landing Simulator 进场与着陆模拟器

ALS Approach Landing System 进场着陆系统

ALS Approach Lighting System 进场照明系统

ALS Area Landing System 区域着陆系统

ALS Automatic Landing System 自动着陆系统

ALS Azimuth Laying Set 方位瞄准装置

ALSA Astronaut Life Support Assembly 航天员生命保障组件

ALSAFECOM All Safety Commands 所有安全指令

ALSARM Advanced Life Support Automated Remote Manipulator 先进生命保障自动化远程控制器

ALSC Automatic Level and Slope Control 自动水平与倾斜控制

ALSCC Apollo Lunar Surface Close-up Camera 【美国】"阿波罗"飞船月球表面特写摄影机

ALSD Apollo Lunar Surface Drill-drill used to bore holes into the lunar surface 【美国】"阿波罗"月球表面钻探机

ALSE Apollo Lunar Sounder Experiment 【美国】"阿波罗"飞船月球探测实验

ALSE Astronaut Life Support Equipment 航天员生命保障设备

ALSEP Apollo Lunar Surface Experiments Package 【美国】"阿波罗"月球表面实验设备组件

ALSMS Automated Life Support Management System 自动生命保障管理系统

ALSOR Air Launched Sounding Rocket 空中发射探空火箭

ALSP Advanced Life Support Pack 先进生命保障包

ALSP Aggregate Level Simulation Protocol 集成模拟协议

ALSRC Apollo Lunar Sample Return Container 【美国】"阿波罗"飞船月球取样容器

ALSS Airlock Support System (Subsystem) 气闸舱保障系统（分系统）

ALSS Apollo Logistic Support System 【美国】"阿波罗"项目后勤保障系统

ALSS Automatic Load Stabilization System 自动载荷稳定系统

ALT Approach and Landing Test 【美国航天飞机】进场与着陆试验

ALT-HOLD Altitude Hold Mode 高度保持模式

ALTB Airborne Laser Test Bed 机载激光试验（测试）台

ALTR Approach and Landing Test Requirement 进场与着陆测试要求

ALTR Approach/Landing Thrust Reverser 进场/着陆反推装置

ALTREC Automatic Life Testing and Recording of Electronic Components 电子组件的自动寿命试验与记录

ALTS Advanced Lunar Transportation System 先进月球（空间）运输系统

ALTV Approach and Landing Test Ve-

hicle 进场与着陆试验飞行器
ALU Advanced Levitation Unit 先进悬浮装置
ALV Air Launched Vehicle 空中发射飞行器（或运载器）
ALV Autonomous Land Vehicle 自主式着陆飞行器
AM Abort Motor 中止发动机
AM Access Module 访问模块
AM Actuator Mechanism 致动机械装置
AM Auxiliary Module 辅助舱
AM Avionics Module 电子设备与控制系统舱
AM/LRAAM Advanced Medium or Long Range Air to Air Missile 先进中距或远距空－空导弹
AM/MDA Airlock Module and Multiple Docking Adapter 【美国】气闸舱与多用途对接舱
AMA Absolute Measurement Accuracy 绝对测量精度
AMA Adaptive Multifunction Antenna 自适应多功能天线
AMA Attitude Measurement Assembly 姿态测量设备
AMARS Automatic Message Address Routing System 自动信息地址线路选择系统
AMARS Automatic Multiple Address Relay System 自动多址中继系统
AMARV Advanced Maneuvering Reentry Vehicle 先进机动再入大气层飞行器
AMAS Advanced Midcourse Active System 先进中程主动（控制）系统
AMB Avionics Maintenance Branch 【美国空军】电子设备与控制系统维修分队
AMBA Adaptive Multibeam Antenna 自适应多波束天线
AMBER Advanced Missile Bomb Ejector Rack 先进导弹弹射架
AMBLER Autonomous Mobile Exploration Robot 自主移动式探险机器人
AMC Advanced Minuteman Computer 【美国】先进"民兵"导弹用计算机
AMC Air Mobility Command 空中移动指挥部
AMC Automatic Mixture Control 自动混合型控制
AMC Auxiliary Maintenance Computer 辅助维护计算机
AMCA Aft Motor Control Assembly 后部发动机控制装配
AMCHIPOT Advanced Mode Control and High Power Optics Techniques 先进模态控制与高能光学技术
AMCL Advanced Material Concepts Laboratory 先进材料方案实验室
AMCM Advanced Missions Cost Model 先进任务成本模型
AMCP Aeronautical Mobile Communications Panel 航空航天移动式通信操纵台
AMCPS Automatic Multiparameter Collection Processing System 多参数自动收集处理系统
AMCS Adaptive Microprogrammed Control System 自适应微程序控制系统
AMCS Advanced Main Coding System 高级主编码系统
AMCS Advanced Missile Control System 先进导弹控制系统
AMCS Alternate Master Control Station 备用主控站
AMCS Attitude Measurement and

Control System (Subsystem) 姿态测量与控制系统（分系统）
AMD Aerospace Medical Division 【美国空军】航空航天医学部
AMD Antimissile Defense 反导弹防御
AMDB Automated Maintenance Data Base 自动维护数据库
AMDI Automatic Miss Distance Indicator 脱靶距离自动指示器
AMDR Automatic Missile Detection Radar 【以色列海军】自动导弹探测雷达
AMDRFM Advanced Monolithic Digital Radio Frequency Memory 先进整体数字射频储存器
AMDS Active Missile Defense System 主动导弹防御系统
AMDS Advanced Missions Docking Subsystem 先进任务停泊分系统
AMDS Area Missile Defense System 地域导弹防御系统
AMDTF Air and Missile Defense Task Force 【美国】空中与导弹防御部队
AMDWS Air and Missile Defense Workstation 防空与导弹防御工作站
AME Angle Measuring Equipment 测角仪
AME Attitude Measurement Electronics 姿态测量电子设备
AME Automatic Monitoring Equipment 自动监控设备
AME AVUM Main Engine 【法国】"织女星"火箭的姿态及游尺上部模块的主发动机
AME/COTAR Angle Measuring Equipment-Correlation Tracking and Ranging 相关跟踪和测距用角度测量装置
AMEC Advanced Multifunction Embedded Computer 先进多功能嵌入式计算机
AMEDAS Automated Meteorological Data Acquisition System 自动气象数据采集系统
AMEL Automated Mars Exploration Laboratory 火星自动探测实验室
AMES Advanced Multiple Environment Simulator 先进多种环境模拟器
AMES Aeronautical Maritime Engineering Satellite 【日本】航空海事工程卫星
AMET Advanced Military Engine Technology 先进军用发动机技术
AMF Abort Motor Facility 中止发动机设施
AMF Apogee Maneuver Firing 远地点机动点火
AMF Apogee Motor Firing 远地点发动机点火
AMFABS Advanced Maintenance-Free Aircraft Battery System 无需维护的先进飞机电池系统
AMFCS Airborne Missile Fire Control System 机载导弹发射控制系统
AMG Angle of Middle Gimbal 中间万向支架转动角
AMG Automatic Magnetic Guidance 自动磁性制导
AMHFS Advanced Miniature High Frequency System 先进小型高频系统
AMI Apogee Motor Igniter 远地点发动机点火器
AMI Avionics Midlife Improvement 电子设备与控制系统中寿改进
AMIC Aerospace Material Informa-

tion Center 【美国】航空航天材料信息中心

AMIR Advance Microwave Imaging Radiometer 先进微波成像辐射仪

AMIR Antimissile Infrared 反导弹红外线

AML Adaptive Maneuvering Logic 自适应机动飞行逻辑

AML Aerospace Medical Laboratory 【美国空军】航空航天医学实验室

AML Astro Meteorological Launcher 天体气象运载火箭

AMLCD Active Matrix Liquid Crystal Display 有源矩阵液晶显示

AMLLV Advanced Multipurpose Large Launch Vehicles 先进大型多用途发射装置；先进大型多用途运载火箭

AMLS Advanced Manned Launch System 【美国】先进载人发射系统

AMM Advanced Manned Mission 先进载人飞行任务

AMM Advanced Multipurpose Missile 先进多用途导弹

AMM Anti-Missile Missile 反导弹导弹

AMMD Advanced Modular Missile Detector 先进组合式弹道探测器

AMMO Army Mobile Missile Operation 陆军机动导弹操作

AMMP Advanced Manned Mission Program 先进载人飞行项目

AMMP Apollo Master Measurement Program 【美国】"阿波罗"飞船主测量项目

AMMRPV Advanced Multimission Remotely Piloted Vehicle 高级多用途遥控车

AMMRS Advanced Multimission Reconnaissance System 先进多任务侦察系统

AMMS Advanced Microwave Moisture Sounder 先进微波湿度探测器

AMMS Automatic Maintenance Management System 自动维护管理系统

AMMTR Antimissile Missile Test Range 反导弹导弹试验靶场

AMOOS Advanced Maneuvering Orbit-to-Orbit Shuttle 高级机动轨道间飞船

AMOR Army Missile Optical Range 【美国】陆军导弹光学靶场

AMOS AF Maui Optical Station 【美国】空军毛伊岛光学观测站

AMOS Automatic Meteorological Observation Station 自动气象观测站

AMOS Automatic Meteorological Oceanographic Survey 海洋气象声学探测

AMP Angular Measurement Precision 角测量精度

AMP Avionics Modernization Program （航空、导弹、航天）电子设备与控制系统现代化项目

AMPG Atlas Mission Planner's Guide 【美国】"宇宙神"运载火箭任务实施手册

AMPM Agency Mission Planning Model 政府部门任务规划模型

AMPMS Advanced Multi-Purpose Missile System 先进多用途导弹系统

AMPOR Airborne Missile Position Optical Recorder 导弹位置机载光学记录器

AMPS Advanced Maneuvering Propulsion System 【美国】先进机动推进系统

AMPS Airborne Multisensor Pod Sys-

tem 机载多传感器吊舱系统
AMPS Atmospheric Magnetospheric Plasma System 大气磁层等离子体系统
AMPS Atmospheric, Magnetospheric and Plasma-in-Space 大气层、磁层和空间等离子区
AMPS Automated Meteorological Profiling System 自动气象计测系统
AMPT Advanced Maneuvering Propulsion Technology 【美国】先进机动推进技术
AMPTE Active Magnetospheric Particle Tracer Explorers 【美国国家航空航天局】主动磁层粒子示踪探测器
AMR Atlantic Missile Range 【美国】大西洋导弹靶场
AMRAAM Advanced Medium Range Air-to-Air Missile 先进中程空对空导弹
AMRFS Advanced Multifunction Radio Frequency System 高级多功能射频系统
AMRI Association of Missile and Rocket Industries 【美国】导弹与火箭工业协会
AMRIR Advanced Medium Resolution Imaging Radiometer 先进中等分辨率成像辐射仪
AMRL Aerospace Medical Research Laboratory 【美国空军】航空航天医学研究实验室
AMRO Atlantic Missile Range Operations 【美国】大西洋导弹靶场操作
AMRPD Applied Manufacturing Research and Process Development 应用制造研究与加工工艺研发
AMRV Astronaut Maneuvering Vehi-
cle 航天员机动运载器
AMS Actuation Mechanism Subsystem 致动机械装置分系统
AMS Advanced Meteoroid Satellite 先进流星体研究卫星
AMS Advanced Meteorological Satellite 【美国】先进气象卫星（由航天局的气象卫星构成）
AMS Advanced Minuteman System 先进"民兵"导弹系统
AMS Aerodynamic Maneuvering System 气动操纵（机动）系统
AMS Alpha Magnetic Spectrometer 阿尔法磁谱仪
AMS American Military Standard 美国军用标准
AMS Apogee Maneuver Stage 远地点机动芯级
AMS Apollo Mission Simulator 【美国】"阿波罗"飞船任务模拟装置
AMS Atmospheric Monitoring Satellite 大气监测卫星
AMS Attitude Maneuvering System 姿态操纵系统
AMS Attitude Measurement Sensor 姿态测量传感器
AMS Automatic Meteorological Station 自动气象站
AMS Automatic Monitoring System 自动监视系统
AMS-H Advanced Missile System-Heavy 先进重型导弹系统
AMSC Advanced Military Spaceflight Capability 先进军事航天飞行能力（提高穿越大气层和低地球轨道载人飞行的能力）
AMSDL Acquisition Management System Data Requirements Control List 采办管理系统数据要求控制清单
AMSEC Analytical Method for Sys-

tem Evaluation and Control 系统评定与控制分析方法
AMSS Advanced Manned Space Simulator 先进载人航天模拟器
AMSS Advanced Mission Sensor System 先进任务传感器系统
AMSS Aeronautical Mobile Satellite Service 航空航天移动卫星服务
AMSS Automatic Meteorological Sensor System 自动气象传感器系统
AMSTE Affordable Moving Surface Target Engagement 可提供的（精确武器对）地面活动目标攻击（系统）
AMSU Advanced Microwave Sounding Unit 先进微波探测装置
AMT Accelerated Mission Testing 加速任务试车
AMT ATCS Mobile Terminal 【美国】陆军战术控制系统移动终端
AMT Automatic Moon Tracking 自动月球跟踪
AMT Automatic Motor Tester 自动发动机测试器
AMTA Antenna Measurement Techniques Association 天线测量技术协会
AMTAS Automatic Model Tuning and Analysis System 自动模态调谐及分析系统
AMTB Attack Management Test Bed 攻击管理试验（测试）台
AMTI Advanced Missile Technology Integration 先进导弹技术一体化
AMTI Air Moving Target Indication 空中移动目标指示（显示）
AMTL Army Materials Technology Laboratory 【美国】陆军材料技术实验室
AMTS Advanced Medium Range Anti-Missile Maintenance Training Simulator 先进中程反导弹维修训练模拟器
AMU Astronaut Maneuvering Unit 航天员机动飞行装置
AMU Atomic Mass Unit 原子质量单位
AMU Avionics Data Multiplexing 电子设备与控制系统数据多路传输
AMUS Avionics Data Multiplexing System 电子设备与控制系统数据多路传输系统
AMVT Advanced Medium Vehicle Technology 先进中型运载器技术
AMWC Alternate Missile Warning Center 备用导弹预警中心
AN Ascending Node 上升节点
AN/TPY Army Navy/Transportable Radar Surveillance 【美国】陆/海军移动式雷达监测
ANAC Automatic Navigation/Attack Controls 自动导航/攻击控制（装置）
ANAE Académie Nationale de l'Air et de l'Espace 【法国】国家航空和空间科学院
ANALYZE Static Code Analyzer 静态密码分析器
ANAO Australian National Aerospace Organization 澳大利亚国家航空航天组织
ANC Active Noise Control 主动噪声控制
ANC Area Navigation Capability 区域导航能力
ANCD Automated Network Control Device 自动网络控制装置
AND Air Force-Navy Aeronautical Design Standard 【美国】空军-海军航空航天设计标准

ANDAS Automatic Navigation and Data Acquisition System 自动导航与数据采集系统

ANDE Atmospheric Neutral Density Experiment 大气中性密度试验

ANDE Atmospheric Neutral Drag Experiment 大气中性拖曳试验

ANF Air Navigation Facility 空中导航设备

ANIK E1 Canadian Telecommunications Satellite's Name 加拿大远程通信卫星

ANITA Analyzing Interferometer for Ambient Air 周围空气的分析干涉计

ANL Argonne National Laboratory 【美国】阿贡国家实验室

ANMD Army National Missile Defense 【美国】陆军国家导弹防御

ANN Artificial Neural Networks 人工神经网络

ANNA Army-Navy-NASA-Air Force 【美国】"安娜"卫星（陆军、海军、国家航空航天局和空军联合研制的测地卫星）

ANRAC Aids to Navigation Radio Control 导航无线电控制辅助装置

ANRS Automatic Navigation Relay Station 自动导航中继站

ANS Autonomous Navigation System 自主导航系统

ANSA Advanced Network System Architecture 先进网络系统体系结构

ANSC Autonomous Navigation System Concept 自主导航系统方案

ANSIM Analogue Simulator 类比模拟器

ANSIR Advanced Navigation System Inertial Reference 先进导航系统惯性参照

ANSSR Aerodynamically Neutral Spin Stabilized Rocket 空气动力学中心线自旋稳定火箭；气动自旋稳定火箭

ANSU Avionics Network Server Unit 电子设备与控制系统网络服务器组件

ANT Antigua Air Station 【美国】安提瓜岛航空站

ANT Automatic Navigation Technology 自动导航技术

ANT Autonomous Navigation Technology 自主导航技术

ANTARES Antenna Tracking Altitude, Azimuth and Range by Electronic Scan 天线通过电子扫描跟踪目标高度、方位角及距离

ANTARES Advanced NASA Technology Architecture for Exploration Studies 用于探索研究的美国国家航空航天局先进技术结构体系

ANTC Advanced Networking Test Center 先进建网测试中心

ANTMS Area Navigation, Test and Management Services 区域导航、测试与管理服务

AO Adaptive Optics 自适应光学

AO Area of Operations 操作区域

AO Ascent Orbit 上升轨道

AO Autonomous Operation 自主运行

AOA Abort Once Around 【美国】亚轨道故障返回（航天飞机发射故障处理模式之一）

AOA Advanced Optical Adjunct 高级光学辅助系统

AOA Analysis of Alternatives 选择性分析（原为成本与运作效能分析）

AOA Angle of Attack 攻击角

AOACO Abort-Once-Around Cutoff 亚轨道故障返回关机

AOAO Advanced Orbiting Astronomical Observatory 先进轨道天文观测台

AOARD Asian Office of Aerospace Research and Development 【美国】航空航天研究与发展亚洲部

AOC Aeronautical Operational Control 航空航天操作控制

AOC Air Operations Center 空中操作中心

AOC Attitude and Orbit Control 姿态与轨道控制

AOCC Air Operations Control Center 空中操作控制中心

AOCE Attitude and Orbit Control Electronics 姿态与轨道控制电子设备

AOCMS Attitude and Orbit Control Measurement System 姿态与轨道控制测量系统

AOCS Attitude and Orbit Control Subsystem 姿态与轨道控制分系统

AOCSOS AOCS On-board Software 星（或箭）载姿态与轨道控制系统软件

AODA Attitude and Orbit Determination Avionics 姿态与轨道测定电子设备与控制系统

AODS All-Ordnance Destruct System 全火工品自毁系统；全弹自毁系统

AOE Area Of Effectiveness 有效范围（区域）

AOEC Aero-Optic Evaluation Center 航空光学鉴定中心

AOFS Active Optical Fuzing System 有源光学引信起爆系统

AOGO Advanced Orbiting Geophysical Observatory 先进轨道飞行地球物理观测站

AOH Apollo Operations Handbook 【美国】"阿波罗"飞船操作手册

AOI Active Optical Imager 主动光学成像仪

AOIB All-Optical Imaging Brassboard 全光学成像模型试验

AOL Atlantic Oceanographic Laboratories 【美国】大西洋海洋实验室

AOL Apollo Orbiting Laboratory 【美国】"阿波罗"飞船轨道实验室

AOLM Apollo Orbiting Laboratory Module 【美国】"阿波罗"飞船轨道实验室舱

AOLO Advanced Orbital Launch Operations 先进轨道发射操作

AOML Atlantic Oceanographic and Meteorological Laboratory 【美国国家海洋气象局】大西洋海洋与气象实验室

AOOS Automatic Orbital Operations System (for satellites and space probes) 在轨飞行自动操纵系统（用于卫星和空间探测器）

AOP Advanced Optical Processor 先进光学处理器

AOP Aerospace Observation Platform 航空航天观测平台

AOP Airborne Optics Platform 机载光学平台

AOP Automatic Operations Panel 自动操作仪表盘

AOR Area of Responsibility 责任区

AORL Apollo Orbital Research Laboratory 【美国】"阿波罗"飞船轨道研究实验室

AORP Autonomous Orbiter Rapid Prototype 自主轨道器快速样机

AOS Acquisition of Signal 信号采集

AOS Advanced Orbit System 高级在轨系统

AOS Airborne Optical Sensor 机载

光学探测器（传感器）
AOSO Advanced Orbiting Solar Observatory 先进轨道太阳观测台
AOSP Advanced On-Board Signal Processor 先进星（弹、机、船）载信号处理器
AOT Actual Operating Time 实际操作时间
AOTF Acoustic-Optic Tunable Filter 声光可调滤波器
AOTSS Advanced Orbital Test Satellite System 先进轨道试验卫星系统
AOTV Aeroassisted-Orbital-Transfer Vehicles 空气制动轨道转移飞行器
AP Absolute Pressure 绝对压力
AP Air Pressure 大气压力
AP Ascent Phase 上升段
AP Atmospheric and Space Physics 大气与空间物理学
AP&AE Attached Payload and Associated Equipment 附属有效载荷及相关设备
AP/DF Approach and Direction Finding Facility 进场与定向设备
AP/TR Approach and Tower Facility 进场与指挥塔设备
APA Abort Programmer Assembly 中止程序装置组件
APA Allowance for Program Adjustment 程序调整的容差
APA(E) Attached Payload Accommodations (Element) 附属有效载荷设备（组件）
APAS Aerodynamic Preliminary Analysis System 空气动力初步分析系统
APAS Androgynous Peripheral Attachment System 异体同构周边式附着系统
APAT Avionics Prototype and Analysis Tool 电子设备与控制系统原型样机与分析工具
APBS Advanced Post Boost System 高级后置助推系统
APC Adapted Payload Carrier 适应有效载荷载体
APC Advanced Propulsion Comparison Study 先进推进剂系统比较研究
APC Aft Power Controller 后部动力控制器
APC All Purpose Capsule 多用途（太空）舱
APC Approach Control 进场控制系统
APC Automatic Pitch Control 自动俯仰操纵
APC Autonomous Payload Controller 自主有效载荷控制器
APCA Aft Power Controller Assembly 后部动力控制器组件
APCC Alternate Processing and Correlation Center 备用处理和相关中心
APCC Atmospheric Pressure and Composition Control 大气压力与组成控制
APCE Antenna Positioned Control Electronics 天线定位装置控制电子设备
APCGF Advanced Protein Crystal Growth Facility 【美国空间站】先进蛋白质晶体生长装置
APCIA Armor Piercing Capped Incendiary Ammunition 带被帽的穿甲燃烧弹
APCIT Armor-Piercing Capped Incendiary with Tracer 带被帽的穿甲燃

烧曳光弹

APCM Advanced Prop-Fan Cruise Missile 先进桨扇式发动机巡航导弹

APCOPPLSRF Analysis and Program for Calculation of Optimum Propellant Performance for Liquid and Solid Rocket Fuel 液体和固体火箭燃料的最佳推进剂性能计算的分析与程序

APCR Apollo Program Control Room 【美国】"阿波罗"飞船程序控制室

APCR Armor Piercing, Composite Rigid 复合硬芯穿甲（弹）

APCS Attitude and Pointing Control System 姿态与航向控制系统

APCS Autonomous Payload Control System 自主有效载荷控制系统

APCS Auto-Pitch Control System 俯仰自动控制系统

APCSI Auto-Pitch Control System Indicator 俯仰自动控制系统指示器

APCU Assembly Power Converter Unit 装配电源转换组件

APDF Apollo Propulsion Development Facility 【美国】"阿波罗"飞船推进装置研制设施

APDP Acquisition Professional Development Program 采办专业人员发展计划

APDS Androgynous Peripheral Docking System 【俄罗斯】异体同构周边式对接系统

APDS Armor-Piercing Discarding Sabot 脱壳穿甲弹

APDS-T Armor-Piercing Discarding Sabot with Tracer 硬芯（脱壳）曳光穿甲弹

APDSMS Advanced Point Defense Surface Missile System 先进的点防御水（地）面导弹系统

APE Absolute Pointing Error 绝对定向误差

APELS Airborne Precision Emitter Location System 机载辐射源精确定位系统

APEOE Furthest Point in Elliptical Orbit from Earth 椭圆轨道远地点

APEOM Furthest Point in Elliptical Orbit from Moon 椭圆轨道近地点

APEX Advanced Phase Conjunction Experiment 先进发展阶段联合实验

APEX Advanced Photovoltaic and Electronics Experiment 【美国空军】先进光电与电子学实验（卫星）

APEX Ariane Passenger Experiments 【法国】"阿里安"火箭搭载实验

APFSDS Armor-Piercing Fin-Stabilized Discarding Sabot 【英国】尾翼稳定脱壳穿甲弹

APFSDS-DU Armor Penetrating, Fin-Stabilized, Discarding Sabot Depleted Uranium 尾翼稳定脱壳贫铀穿甲弹

APFSDS-T Armor-Piercing Fin-Stabilized Discarding Sabot Tracer 尾翼稳定式脱壳（硬芯）曳光穿甲弹

APG Aberdeen Proving Ground 【美国】阿伯丁试验场

APGS Auxiliary Power Generation System 辅助动力发生系统

APHC-T Armor-Piercing High Capacity-Tracer 曳光高能穿甲弹

APHE Armor-Piercing, High Explosive 高爆穿甲（弹药）

APHEBC Armor-Piercing, High Explosive Ballistic Cap 高爆弹道被帽穿甲（弹药）

APHEI Armor-Piercing, High Explosive Incendiary 穿甲高爆燃烧（弹

药）

API Application Program Interface 应用项目接口

API Application Programming Interface 应用编程接口

API Ascent-Phase Intercept 上升段拦截

APIC Advanced Programmable Interrupt Controller 高级可编程序中断控制器

APIC Apollo Parts Information Center 【美国】"阿波罗"飞船零部件信息中心

APICM Armor-Piercing Improved Conventional Munition 改进型常规穿甲弹药

APIF Aerodynamic Propulsive Interactive Force （空）气动（力）推进互作用力

APIRS Attitude and Positioning Inertial Reference System 姿态与定位惯性参照系统

APIS Apogee-Perigee Injection System 远地点－近地点入轨系统

APIT(A) Armor-Piercing Incendiary-Tracer (Ammunition) 曳光穿甲燃烧（弹）

APIXS Alpha Particle Induced X-ray Spectrometer 阿尔法粒子诱导X射线分光计

APLAC Analysis Program Linear Active Circuits 分析程序线性主动回路

APLE Average Power Laser Experiment 平均功率激光实验

APM Advanced Penetration Model 先进突防模型（方式）

APM Advanced Pressurized Module 先进增压舱

APM Approach Path Monitor 进场路径监控

APM Ascent Particle Monitor 上升段粒子监测器

APM Ascent Performance Margin 上升段性能裕度

APM Astronaut Positioning Mechanism 航天员固位机械装置

APM Attached Pressurized Module 搭接的增压舱

APM AVUM Propulsion Module 【法国】"织女星"运载火箭上面级推进舱

APMS Automated Performance Measurement System 自动性能测量系统

APO Apache Point Observatory 【美国】"阿帕奇"瞄准监视台

APOD Aerial Point of Debarkation 空运完成点

APOL Aerospace Program Oriented Language 航空航天程序专用语言

Apollo/LM Apollo/Lunar Module 【美国】"阿波罗"飞船登月舱

APOP Apollo Preflight Operations Procedures 【美国】"阿波罗"起飞前操作程序

APOTA Automatic Positioning of Telemetering Antenna 遥测天线自动定位

APOTV All-Propulsive Orbited Transfer Vehicle 【美国国家航空航天局】全推进轨道转移飞行器

APP Advanced Planetary Probe 先进行星际探测器

APP Armor-Piercing Projectile 穿甲弹

APPECS Adaptive Pattern Perceiving Electronic Computer System 自适应图形识别电子计算机系统

APPF Automated Payload Processing

Facility 自主有效载荷处理设施
APPLE Advanced Propulsion Packaged Liquid Engine 先进推进预装液体燃料（火箭）发动机
APPLE Advanced Propulsion Payload Effects 先进推进有效载荷效应
APPS Auxiliary Payload Power System 辅助有效载荷动力系统；有效载荷辅助电源系统
APRF Aero-Physics/ Propulsion Research Facility 航空物理/推进力研究设施
APRIL Automatically Programmed Remote Indication Logging 自动程序控制的远距指示记录
APRINT Solid Propellant Rocket Interceptor 装有固体推进剂火箭的拦截导弹
APS Absolute Pressure Sensor 绝对压力传感器
APS Acceleration and Pre-Separation 加速度和预分离
APS Accelerometer Parameter Shift 加速度计参数改变
APS Accessory Power Supply 辅助电源
APS Accessory Power System 辅助动力系统；辅助电源系统
APS Accurate Positioning System 精确定位系统
APS Advanced Positioning Sensor 先进定位传感器
APS Afloat Planning System 【美国海军】船上（任务）规划系统
APS Aft Propulsion System 后部推进系统
APS Airborne Power Supply 机载电源
APS Armor Piercing Sabot 脱壳穿甲弹
APS Armor Piercing Shell 穿甲炮弹
APS Ascent Propulsion System 上升推进系统
APS Astronaut Positioning System 航天员定位系统
APS Attitude Propulsion Subsystem 姿态推进分系统
APS Automated Payload Switch 有效载荷自动开关
APS Automatic Phasing System 自动调相系统
APS Automatic Processing System 自动处理系统
APS Autonomous Propulsion System 自主推进系统
APS Auxiliary Power Subsystem 辅助电源分系统
APS Auxiliary Propulsion System 辅助推进系统
APS Axial Propulsion System 轴向推进系统
APSD Armor-Piercing Sabot Discarding 脱壳（硬芯）穿甲（弹）
APSFFF Ames Prototype Supersonic Free Flight Facility 【美国】"艾姆斯"样机超声速自由飞行装置
APSI Advanced Propulsion Subsystem Integration 先进推进系统一体化
APSM Automated Power System Management 动力系统自动管理
APSMT Asia Pacific Satellite Mobile Telecommunication 亚太卫星移动通信系统
APSS Automatic Planetary Space Station 自动行星际空间站
APSSE Armor-Piercing Shell with Secondary Effects 二次效应穿甲弹
APSVDS Armor-Piercing Super-Velocity Discarding Sabot 超高速脱壳穿甲弹

APT Acquisition, Pointing, and Tracking 捕获、瞄准与跟踪

APT Additional Propellant Tank 推进剂附加贮箱

APT Advanced Propulsion Test 先进推进技术试验

APT Apollo Pad Test 【美国】"阿波罗"飞船发射台试验

APT Armor-Piercing with Tracer (ammunition) 曳光穿甲（弹药）

APT Automated Powerplant Test 动力装置自动测试；发动机自动测试

APT Automatic Picture Transmission 自动图像传送

APT Automatic Programmed Tool 自动编程工具

APTAMMO Armor-Piercing and Tracer Ammunition 穿甲与曳光弹药

APTT Apollo Parts Task Trainer 【美国】"阿波罗"飞船零件保障专用训练器

APTU Aerodynamic and Propulsion Test Unit 空气动力与推进试验装置

APTU Auxiliary Power and Thrust Unit 辅助动力与推进装置

APU Auxiliary Power Unit 辅助电源装置

APU Auxiliary Propulsion Unit 辅助推进装置

APUS Auxiliary Power Unit Subsystem 辅助电源装置分系统

APUT Auxiliary Power Unit Test 辅助电源装置测试

APW Advanced Penetrating Warhead 高级侵彻弹头；高级突防弹头

APXS Alpha Particle X-Ray Spectrometer 【美国】"阿尔法"粒子X射线光谱仪

AQ Apollo Qualification 【美国】"阿波罗"飞船鉴定试验

AR Assigned Range 规定距离；给定距离；给定射程

AR&C Automated Rendezvous and Capture 自动交会与截（捕）获

AR&D Automated Rendezvous and Docking 自动交会与对接

AR&D Autonomous Rendezvous and Docking 自主交会对接

AR/IRR Acceptance Review/Integration Readiness Review 验收评审/综合成熟度评审

ARA Aerodynamic Reference Area 气动基准面

ARA Airborne Radar Approach 机载雷达引导进场

ARA Attitude Reference Assembly 姿态基准组件

ARAA Aerodrome Radar Approach Aid 机场雷达进场辅助设备；机场雷达进场导航设备

ARABSAT Arab Satellite 阿拉伯卫星

ARABSAT Arab Satellite Communications Organization 阿拉伯卫星通信组织

ARAC Aerospace Research Application(s) Center 航空航天研究应用中心

ARAD Anti-Radiation Missile 反辐射导弹

ARAE Advanced Radio Astronomy Explorer 先进射电天文探测卫星；先进无线电天文探测器

ARAP Astronaut Rescue Air Pack 航天员救援充气包

ARBC Attitude Reference Bombing Computer 姿态基准轰炸计算机

ARBS Angle Rate Bombing Set 角

速度轰炸装置
ARC　Adjusted Range Correction　调整距离修正量；调整射程修正量
ARC　Advanced Reentry Concepts　先进再入（大气层）概念
ARC　Advanced RISC Computing　高级精简指令集计算机运算
ARC　Aerospace Remote Calculator　航空航天遥控计算机
ARC　Ames Research Center　【美国】艾姆斯研究中心
ARC　Automatic Range Control　自动距离控制；自动射程控制
ARC/SC　Advanced Research Center / Simulation Center　先进研究中心 / 仿真中心
ARCADE　Automatic Radar Control and Data Equipment　自动雷达控制与数据设备
ARCAS　All Purpose for Collection Atmospheric Soundings　收集大气探测数据的通用火箭
ARCAS　Automatic Radar Chain Acquisition System　自动雷达链目标搜索系统
ARCCC　Army Component Command Center　【美国】陆军部队指挥中心
ARCM　Antiradiation Countermeasures　反辐射对抗措施
ARCM　Atlas Roll Control Module　【美国】"宇宙神"运载火箭滚动控制模块
ARCOMSA　Arab League Communications Satellite　阿拉伯联盟通信卫星
ARCON　Automatic Rudder Control　自动方向舵控制
ARCS　Acquisition Radar and Control System　搜索雷达与控制系统
ARCS　Adaptive Reliability Control System　自适应可靠性控制系统
ARCS　Advanced Reconfigurable Computer System　先进可重构计算机系统
ARCS　Aerial Rocket Control System　航空火箭控制系统
ARCS　Aft Reaction Control System (Subsystem)　后部反应控制系统（分系统）
ARCS　Altitude Rate Command System　高度变化率指令系统
ARCS　Automated Reusable Components System　自动化重复使用部件系统
ARCT　Advanced Radar Component Technology　先进雷达部件技术
ARCU　Atlas Remote Control Unit　【美国】"宇宙神"火箭遥控装置
ARD　Atmospheric Reentry Demonstrator　大气再进入演示器
ARD　Average Response Data　平均响应数据
ARDA　Analog Recording Dynamic Analyzer　模拟记录动态分析器
ARDA　Astronautical Research and Development Agency　【美国】航天研究与发展局
ARDC　Automatic Roll Deflection Control　自动横滚偏转控制
ARDCBMD　Air Research and Development Command Ballistic Missile Division　【美国】空军研究与发展司令部弹道导弹部
ARDE　Armament Research & Development Establishment　【印度】武器装备发展研究院
ARDEL　Advanced Radar Detection Laboratory　先进雷达探测实验室
ARDEMS　Artillery Delivered Multipurpose Submunition　火炮发射的

A

多用途子母弹

ARDES Automatic Recorded Data Evaluation System 自动录入数据评估系统

ARDME Automatic Range Detector and Measuring Equipment 自动射程探测器与测量装置

ARDS Advanced Remote Display Station 先进遥控显示台

ARDS Advanced Remote Display System 先进遥控显示系统

ARDS Aerial Rocket Delivery System 航空火箭运载系统

ARDS Automatic Remote Display System 自动远程显示系统

ARDSOC Army Defense System Operations Center 【美国】陆军防御系统操作中心

ARDT Advanced Remote Data Terminal 先进远程数据终端

ARDT Automatic Remote Data Terminal 自动遥测数据终端

ARE Aerothermal Reentry Experiment 气动热再入试验

ARE Allowable Radial Error 允许径向误差

ARE Apollo Reliability Engineering 【美国】"阿波罗"飞船可靠性工程

ARE Automatic Record Evaluation 自动记录评估

ARE Auxiliary Rocket Engine 辅助火箭发动机

ARED Advanced Resistance Exercise Device 先进抗阻力运动设备

AREE Apollo Reliability Engineering Electronics 【美国】"阿波罗"飞船可靠性工程技术电子设备

ARENTS Advanced Research Engineering Nuclear Test Satellite 先进研究工程核试验卫星

ARENTS Advanced Research Enviromental Test Satellite 【美国】先进环境测试研究卫星

AREP Atlas Reliability Enhancement Program 【美国】"宇宙神"运载火箭可靠性强化计划

AREPS Advanced Reconnaissance Electrically Propelled Spacecraft 先进电推进侦察卫星

ARES Advanced Radiation Effects Simulation 先进辐射效应模拟

ARES Advanced Research EMP Simulator 高级研究电磁脉冲模拟器

ARES Advanced Rocket Engine Storable 先进可储存火箭发动机

ARF Aerodynamic Research Facility 气动研究设施

ARF Aerospace Recovery Facility 航空航天回收设施

ARF Assembly and Refurbishment Facility 【美国航天飞机】组装与整修厂房

ARFDS Automatic Reentry Flight Dynamics Simulator 自动返回飞行动力模拟器

ARGMA Army Rocket and Guided Missile Agency 【美国】陆军火箭与导弹局（现为导弹研究发展局）

ARGO Advanced Research Geophysical Observatory 先进地球物理研究观测台；高级研究用地球物理观象卫星

ARGOS Active Response Gravity Offload System 主动反应重力卸载系统

ARGOS Advanced Research and Global Observation Satellite 先进的研究与全球观察卫星

ARGOS Automatic Relay Global Observation System 全球观测自动中

ARGOS National Space-Based Navigation System 【法国】阿戈斯系统（国家天基导航系统）

ARGS Advanced RADEC Ground System （国防支援计划）高级辐射探测能力地面系统

ARGUS Advanced Real-Time Gaming Universal Simulation 通用型高级实时对策仿真（模拟）

ARGUS Automatic Routine Generating and Updating System 自动程序生成与更新系统

ARH Anti-Radiation Homing 主动雷达导引，主动雷达寻的

ARIA Advanced Range Instrumentation Aircraft 先进的靶场测量飞机

ARIA Apollo Range Instrumented Aircraft 【美国】"阿波罗"飞船登月计划中的靶场测量飞机

ARIAA Automatic Radar Identification Analysis and Alarm 自动雷达识别分析与报警

ARID-PCM Adaptive Recursive Interpolated Differential Pulse Code Modulation 自适应递归内插值微分脉冲代码调制

ARIES Active Radio Interferometer for Explosion Surveillance 用于爆炸监视的主动射频干涉仪

ARIES Advanced Radar Information Evaluation System 高级雷达信息鉴定系统

ARIP Automatic Rocket Impact Predictor 火箭弹着点自动预测装置；火箭降落点自动预测器

ARIRS Advanced Range Instrumentation Radar System 先进测距仪表雷达系统

ARIS Advanced Range Instrumentation Ship 先进靶场测量船

ARIS Advanced Range Instrumentation System 先进靶场测量系统

ARIS Altitude and Rate Indicating System 【宇宙飞船】高度与速率指示系统

ARIS Atomic Reactors in Space 空间原子反应堆

ARJ Alpha Rotary Joint 阿尔法旋转接头

ARL Astronautical Research Laboratory 【美国】航天员研究实验室

ARL Authorized Retention Level 核定超期服役数量

ARM Antiradar Missile 反雷达导弹

ARM Anti Radiation Missile 反辐射导弹

ARM Apogee Raising Manoeuvre 远地点上升机动

ARM Apollo Requirements Manual 【美国】"阿波罗"飞船技术要求手册

ARM ARPANET Reference Model 高级研究与计划局网络参考模型

ARM Atmospheric Radiation Measurement 大气辐射测量

ARM Autonomic Response Mechanism 自主响应机构

ARM Availability, Reliability and Maintainability 可用性、可靠性与维修性

ARMAT Antiradar Martel 【英国/法国】"玛特尔"反雷达导弹

ARMAT Antiradiation Missile 【法国】阿玛特巡航导弹；反辐射导弹

ARMMCS Antiradiation Missile Management and Control System 反辐射导弹管理与控制系统

ARMMS Automated Reliability and Maintenance Management System

自动化可靠性与维修管理系统

ARMMS Automatic Reconfigurable Modular Multiprocessor System 自动模块化可重构多处理机系统

ARMS ADPE (Automatic Data Processing Equipment) Resources Management System 自动数据处理设备资源管理系统

ARMS Aerial Radiological Measurement and Survey 空中放射性测量与鉴定（计划）

ARMS Antiradiation Missile Simulator 反辐射导弹模拟器

ARMSS Automated Robotic Maintenance for Space Station 空间站自动机器人维修

ARO All Reflective Optics 全反射光学

AROD Advanced Range and Orbit Determination 先进航（射）程与轨道测定

AROG Automatic Roll-Out Guidance 自动着陆滑跑制导

AROS Airborne Radar Optical System 机载雷达光学系统

ARP Address Resolution Protocol 地址解析协议

ARP Advanced Reentry Program 先进再入研究计划

ARPAETS ARPA Environmental Test Satellite 【美国国防部】高级研究计划局环境测试卫星

ARPCS Atmospheric Revitalization Pressure Control System 大气再生压力控制系统

ARR Advanced Rocket Ramjet 先进火箭冲压喷气发动机

ARR Aerospace Rescue and Recovery 航空航天救援与回收

ARRC Aerospace Rescue and Recovery Center 航空航天救援与回收中心

ARRGP Aerospace Rescue and Recovery Group 【美国空军】航空航天救援与回收大队

ARRMD Affordable Rapid Response Missile Demonstrator 可承担的快速响应导弹验证机

ARROC Army Regional Operations Center 【美国】陆军区域操作中心

ARRS Advanced Rescue & Recovery Service 先进救援与回收服务

ARRS Air Recovery and Rescue Service 空中回收与救援服务

ARRSC Asian Regional Remote Sensing Center 亚洲地区遥感中心

ARRTC Aerospace Rescue and Recovery Training Center 航空航天救援与回收训练中心

ARS Active Ranger System 有源测距仪系统

ARS Active Ranging System 主动测距系统

ARS Advanced Reconnaissance Satellite 先进侦察卫星

ARS Advanced Recovery System 先进回收系统

ARS Advanced Reentry System 先进再入系统

ARS Advanced Rescue System 先进救援系统

ARS Aerospace Recovery System 航空航天回收系统

ARS Aerospace Research Satellite 航空航天研究卫星

ARS Airborne Remote Sensing 机载远程传感（探测）

ARS American Rocket Society 美国火箭学会

ARS Anomaly Resolution System 异

常情况处理系统
ARS　Atmospheric Revitalization System　大气再生系统
ARS　Authorized Return Slip　核准的返回侧滑飞行
ARS　Automatic Recovery System　自动回收系统
ARS-ST　Airborne Receiving System-Surface Terminal　机载接收系统—地面终端
ARS/SL　Automatic Reference System/Sequential Launch　自动基准系统/顺序发射
ARSC　African Remote Sensing Council　非洲遥感委员会
ARSCS　Automated Rear Services Control System　自动化后勤控制系统
ARSFSS　Advanced Reduced Scale Fuel System Simulator　先进缩尺燃油系统模拟器（装置）
ARSSMM　Active-Radar Sea-Skimming Martel Missile　有源雷达掠海飞行玛特尔导弹
ART　Advanced Reusable Technology　先进可重复使用技术
ART　Augmented Reentry Test　增强型再入大气层试验
ARTA　Ariane 5 Research and Technology Accompaniment　【法国】"阿里安"5运载火箭研究与技术附属项目
ARTC　Aerospace Research & Testing Committee　【美国】宇航研究与试验委员会
ARTEMIS　Advanced Relay and Technology Mission　先进中继与技术任务
ARTEMIS　Automated Reporting, Tracking and Evaluation Management Information System　自动报告、跟踪、鉴定管理信息系统
ARTEP　Ariane Technology Experiment Payload　【法国】"阿里安"技术实验有效载荷
ARTEP　Ariane Technology Experiment Platform　【法国】"阿里安"运载火箭技术实验平台
ARTRC　Advanced Range Testing, Reporting And Control　先进靶场试验、报告和控制
ARTRC　Advanced Real Time Range Control (System)　先进实时距离控制（系统）
ARTS　Advanced Radar Targeting System　先进雷达瞄准系统
ARTS　Advanced Radar Terminal System　先进雷达终端系统
ARTS　Automated Radar Tracking System　自动雷达跟踪系统
ARTS　Automated Remote Tracking Station　自动遥控跟踪站
ARTU　Automated Range Tracking Unit　自动距离跟踪装置；自动射程跟踪装置
ARU　Altitude Reference Unit　高度参考单元
ARU　Attitude Reference Unit　姿态基准装置
ARV　Advanced Reentry Vehicle　高级再入飞行器
ARV　Advanced Return Vehicle　先进返回舱
ARV　Aeroballistic Reentry Vehicle　气动弹道再入（大气层）飞行器
ARV　Aerospace Research Vehicle　航空航天研究飞行器
ARV-RSTA　Armed Robotic Vehicle-Reconnaissance, Surveillance and Target Acquisition　侦察、监视和

A

目标捕获无人驾驶武器飞行器

ARW Attitude Reaction Wheel 姿态反应机构

AS Adapter Section 过渡段；适配段

AS Adapter Structure 适配器结构

AS Aft Shroud 后部护罩

AS Apollo-Saturn "阿波罗－土星"火箭

AS Artificial Satellite 人造卫星

AS Atlas Station 【法国】"宇宙神"运载火箭地面站

AS Atmosphere and Space 大气层与大气层外空间

AS&C Aerospace Surveillance and Control 航空航天监视与控制

AS&T Air Pressure Test for Strength and Tightness 强度及气密性气压试验

AS-2xx Apollo-Saturn IB mission 【美国】"阿波罗－土星IB"任务

AS-5xx Apollo-Saturn V mission 【美国】"阿波罗－土星5"任务

ASA Abort Sensor Assembly 中止传感器组件

ASA Accelerometer Sensor Assembly 加速度传感器组件

ASA Adapter Service Area 适配器服务区

ASA Attitude Switch Assembly 姿态切换组件

ASAC Active Satellite Attitude Control 有源卫星姿态控制

ASAC Aerodynamic Surface Assembly and Checkout 气动表面组件与检测

ASALM Advanced Strategic Air-Launched Cruise Missile 先进战略空射巡航导弹

ASALM Advanced Strategic Air-Launched Missile 先进战略空中发射导弹

ASALM Advanced Supersonic Air-Launched Missile 先进超声速空射导弹

ASAM Advanced Surface to Air Missile 先进地空导弹

ASAP Advanced Survival Avionics Program 【美国空军】先进高生存航空电子设备计划

ASAP Aerospace Safety Automation Program 航空航天安全自动化项目

ASARS Advanced Synthetic Aperture Radar System 先进合成孔径雷达系统

ASAS Advanced Solid Axial Stage 先进固体轴向芯级

ASAS Aerodynamic Stability and Augmentation System 气动稳定性与增稳系统

ASAS Aerodynamic Stability Augmentation Subsystem 气动稳定增强分系统

ASAS All Source Analysis System 全源分析系统

ASAS Automated Structural Analysis System 自动结构分析系统

ASAT Anti-Satellite 反卫星

ASAT Anti-Satellite Ballistic Missile Test 反卫星弹道导弹试验

ASAT Antisatellite Satellite 反卫星卫星

ASATS Antisatellite System 反卫星系统

ASATT Anti-Satellite Technology 反卫星技术

ASATW Anti-Satellite Weapon 反卫星武器

ASAW Aerospace Surveillance and

Warning 航空航天监视与报警

ASC Advanced Structural Components 高级结构部件

ASC Aerospace System Center 航空航天系统中心

ASC Army Space Command 【美国】陆军航天司令部

ASCA Advanced Satellite for Cosmology and Astrophysics 用于宇宙学和天体物理学研究的先进卫星

ASCAT Apollo Simulation Checkout and Training 【美国】"阿波罗"飞船模拟检测与训练

ASCB Avionics Standard Communication Bus 电子设备与控制系统的标准通信母线

ASCE Airlock Signal Conditioning Electronics 气闸室信号调节电子设备

ASCENDS Active Sensing of CO_2 Emissions over Nights, Days, and Seasons 全天候、全季节二氧化碳排放主动探测

ASCM Advanced Spaceborne Computer Module 先进天基计算机模块

ASCM Antiship Capable Missile 具有反舰能力的导弹

ASCM Antiship Cruise Missile 反舰巡航导弹

ASCOM Arvi Satellite Communication Project 【印度】阿伟卫星通信计划

ASCP Attitude Set Control Panel 姿态装置控制屏

ASCR Assured Safe Crew Return 确保机组安全返回

ASCS Advanced Satellite Control System 先进卫星控制系统

ASCS Air Supply Control System 供气控制系统

ASCS Altitude Sensing and Control System 高度传感与控制系统

ASCS Altitude Stabilization and Control System 高度稳定与控制系统

ASCS Area Surveillance Control System 区域监视控制系统

ASCS Atmosphere Storage and Control System (Subsystem) 大气存储与控制系统（分系统）

ASCS Atmospheric Storage and Control Section 大气存储与控制单元

ASCS Attitude and Spin Control Subsystem 姿态与自旋控制子系统

ASCS Attitude Sensing and Control System 姿态传感与控制系统

ASCS Attitude Stabilization and Control System 姿态稳定与控制系统

ASCS Automatic Stabilization and Control System 自动稳定与控制系统

ASCU Alarm System Control Unit 警报系统控制装置

ASCU Analog Signal Converter Unit 模拟信号转换装置

ASCU Anti-Skid Control Unit 防滑控制装置

ASD Accelerate-Stop Distance 中断起飞距离；加速停止距离

ASDC Aeronomy and Space Data Center 超高层大气物理学与航天数据中心

ASDC Alternate Space Defense Center 备用空间防御中心

ASDCS Airspace Surveillance Display and Control System 空域监视显示与控制系统

ASDD Air Stream Direction Detector 气流方向探测器

ASDP Advanced Sensor Demonstration Program 先进传感器演示验证项目

ASDS Autonomous Satellite Docking System 自主卫星对接系统

ASE Advanced Space Engine 先进航天发动机，先进空间发动机

ASE Aerospace Support Equipment 航空航天保障设备

ASE Airborne Support Equipment 机载保障设备

ASE Automatic Support Equipment 自动保障设备

ASEA Advanced Solidification Experiment Activity 先进固化实验

ASEDP Army Space Exploitation Demonstration Program 【美国】陆军空间开发利用演示计划

ASEM Augmented Space Environment Monitor 增进型空间环境监测器

ASESS Aerospace Environment Simulation System 航空航天环境模拟系统；宇宙空间模拟装置

ASET Aeronautical Satellite Earth Terminal 航空卫星地面终端；导航卫星地球终端站

ASF Atmospheric Science Facility 大气科学设施

ASFISS Advanced Simulation Facility Interconnection and Setup Subsystem 先进模拟设施联接与装配子系统

ASFTS Auxiliary Systems Functional Test Stand 辅助系统功能试验台

ASGLS Advanced Space Ground Link Subsystem 先进空间地面数据链子系统

ASGPD Attitude Set and Gimbal Position Display 姿态调整与万向支架位置显示器

ASGS All-Weather Guidance System 全天候制导系统

ASI Avionics System Integration 电子设备与控制系统集成

ASIAC Aerospace Structures Information and Analysis Center 【美国空军】航空航天结构信息与分析中心

ASIAS Aviation Safety Information Analysis and Sharing 【美国联邦航空总署】航空安全信息分析与共享

ASIC Application Specific Integrated Circuit 专用集成电路

ASIM Atmospheric Space Interactions Monitor 大气空间相互作用监测器

ASIOE Associated Support Items of Equipment 【美国】装备的相关保障项目

ASIP Arrow System Improvement Program 【美国】"箭"导弹系统改进项目

ASIS Abort Sensing and Implementation System 飞行中止敏感与执行系统

ASL Above Sea Level 高于海平面

ASL Astrosurveillance Sciences Lab 天体监视科学实验室

ASLAMS Automated Ship Location and Attitude Measuring System 自动化舰船定位与姿态测量系统

ASLSS Air-Supply Life-Support System 供气生命保障系统

ASLV Advanced Small Launch Vehicle 先进小型运载器

ASLV Aerospace Structural Material 航空航天结构材料

ASLV Atlas Standard Launch Vehicle 【美国】"宇宙神"标准型运载火箭

ASLV Augmented Satellite Launch Vehicle 【印度】增大推力的卫星运载火箭

ASMDC Army Space and Missile De-

fense Command 【美国】陆军航天与导弹防御司令部
ASMGCS Advanced Surface Movement Guidance and Control Systems 先进地面移动制导与控制系统
ASMS Advanced Strategic Missile System 先进战略导弹系统
ASMS Advanced Synchronous Meteorological Satellite 【美国】先进同步气象卫星
ASMT Air Superiority Missile Technology 空中优势导弹技术
ASO Advanced Solar Observatory 【美国】先进太阳观测台
ASOC Air Support Operations Center 空中保障操作中心
ASOC Atlas V Spaceflight Operation Center 【美国】"宇宙神V"火箭飞行操作中心
ASOS Automated Surface Observation System 地面自动观测系统
ASP Advanced Sensor Platform 先进传感器平台
ASP Advanced Signal Processor 先进信号处理器
ASP Airborne Surveillance Platform 机载监视平台
ASP Airspace Systems Program 【美国国家航空航天局】空域系统项目
ASP Apollo Simple Penetrator-A pointed rod pushed into the lunar surface to determine it's penetration properties 【美国】"阿波罗"飞船简易穿透器—使用1个尖锐杆推进月球表面以测定其穿透性能
ASP Apollo Spacecraft Project 【美国】"阿波罗"飞船项目
ASP Auxiliary Spacecraft Power 航天飞机辅助电源
ASPADOC Alternate Space Defense Operations Center 后备太空防御作战中心
ASPERA Automatic Space Plasma Experiment with Rotating Analyzer 带旋转分析器的自动空间等离子体实验
ASPIRIS Advanced Signal Processing for IR Sensors 红外传感器的先进信号处理
ASPJ Airborne Self-Protection Jammer 机载自动保护干扰机
ASPO Apollo Spacecraft Project Office 【美国】"阿波罗"飞船设计局
ASPOC Active Spacecraft Potential Control 航天器电位动态控制
ASPP Antenna/Solar Panel Positioner 天线与太阳能板定向器
ASPS Adaptable Space Propulsion System 【美国】适用的空间推进系统
ASR Avionics System Review 电子设备与控制系统评审
ASRAAM Advanced Short Range Air-to-Air Missile 先进短程空—空导弹
ASRAAMDP Advanced Short-Range Air-to-Air Missile Development Program 先进短程空—空导弹发展计划
ASRB Advanced Solid Rocket Booster 先进固体火箭助推器
ASRC Alabama Space and Rocket Center 【美国】亚拉巴马航天与火箭中心
ASRDI Aerospace Safety Research and Data Institute 【美国】航空航天安全研究与数据所
ASRF Atmospheric Sciences Research Facility 大气科学研究设施

ASRG Advanced Stirling Radioisotope Generator 先进斯特林放射性同位素发电系统

ASRL Aeroelastic and Structures Research Laboratory 空气动力弹性与结构研究实验室

ASRL Antisubmarine Rocket Launcher 反潜火箭发射器

ASRM Abort Solid Rocket Motor 中止固体火箭发动机

ASRM Advanced Solid Rocket Motor 【美国】先进固体火箭发动机

ASRO Astronomical Roentgen Observatory Satellite 伦琴射线天文观测卫星

ASROC Anti-Submarine Rocket 反潜火箭

ASRWS Apollo Space Radiation Warning System 【美国】"阿波罗"飞船空间辐射报警系统

ASS Advanced Space Station 先进空间站

ASS Advanced Synchronous Satellite 高级同步卫星

ASS Aerospace Surveillance System 航空航天监视系统

ASS Antisatellite Satellite 反卫星卫星

ASS Atmospheric Structure Satellite 大气结构研究卫星

ASS Attitude Sensing System 姿态传感系统；姿态探测系统

ASSCAS Advanced Spacecraft Subsystem Cost Analysis Structure 先进航天器分系统成本分析结构

ASSESS Airborne Science/Shuttle Experiment System Simulation 机载科学与航天飞机实验系统模拟

ASSESS Airborne Science/Spacelab Experiment System Simulation 机载科学与"空间实验室"实验系统模拟

ASSET Aerospace Structure Environmental Test 【美国空军】航空航天结构环境试验

ASSET Aerothermodynamic/Elastic Structural Systems Environments Test 【美国空军】空气热动力－弹性结构系统环境试验

ASSIST Advanced System Support and Information System Technology 先进系统保障与信息系统技术

ASSIST Automated Systems Security Incident Support Team 自动化系统安全事件支援小组

ASSTC Aerospace Simulation and Systems Test Center 空间模拟与系统测试中心

AST Advanced Subsonic Technology 先进亚声速技术

AST Advanced Supersonic Technology 先进超声技术

AST Airborne Surveillance Test Bed 机载监视试验（测试）台

AST Antisatellite Technology 反卫星技术

AST Apollo System Test 【美国】"阿波罗"飞船系统试验

AST Arrow System Test 【美国】"箭"导弹系统试验

ASTA Automatic System Trouble Analysis 自动系统故障分析

ASTAMIDS Airborne Standoff Minefield Detection System 机载远距离雷场探测系统

ASTAP Advanced Statistical Analysis Program 先进统计分析程序

ASTAR Advanced Spacecraft Thermal Analysis Routine 先进航天飞机热分析程序

ASTAR Airborne Search Target Attack Radar 机载搜索目标攻击雷达
ASTAS Aerial Surveillance and Target Acquisition System 空中监视与目标捕获系统
ASTAT Affordable Sensor Technology for Aerial Targeting 可提供的航空目标传感器技术
ASTC Advanced Satellite Tracking Center 先进的卫星跟踪中心
ASTC Army Satellite Tracking Center 【美国】陆军卫星跟踪中心
ASTE Aerospace System Test Environment 航空航天系统试验环境
ASTEI Air-Surface Technology Evaluation and Integration Program 空对地技术评估与综合计划
ASTER Advanced Spaceborne Thermal Emission and Reflection Radiometer 【美国国家航空航天局】先进星载热发射与反射辐射仪
ASTEX Advanced Space Technology Experiment 【美国空军】先进空间技术实验
ASTF Aeropropulsion System Test Facility 航空推进剂系统测试设施
ASTF Aeropropulsion Systems Test Facility 【美国空军】武器系统研究部阿诺德工程技术中心的空气推进系统试验部
ASTIDP Astrobiology Instrument Developement Program 天体仿生学仪器研发计划
ASTMC American Rocket Society Structures and Materials Committee 美国火箭学会结构与材料委员会
ASTOVL Advanced Short Takeoff/Vertical Landing 【美国国家航空航天局】先进短距起飞及垂直着陆
ASTP Advanced Space Technology Program 先进航天技术研究计划
ASTP Advanced Space Transportation Program 先进空间运输计划
ASTP Advanced System and Technology Program(me) 【欧洲空间局】先进系统与技术计划
ASTP Apollo-Soyuz Test Program 【美国】"阿波罗—联盟号"测试项目
ASTRA Application of Space Techniques Relation to Aviation 航天技术在航空领域的应用
ASTRA Astronomical Space Telescope Rescue Agency 空间天文望远镜救援机构
ASTRA Astronomical Space Telescope Research Assembly 空间天文望远镜研究装置
ASTRA Attitude, Steering, Turn, Rate, Azimuth 姿态、操纵、转弯、速度、方位
ASTREX Advanced Space Structures Technology Research Experiment 先进航天体系技术研究实验
ASTRO Advanced Spacecraft Transport Reusable Orbiter 可重复使用的先进航天运输轨道飞行器
ASTRO Aerodynamic Spacecraft Two Stage Reusable Orbiter 可重复使用的气动航天器两级轨道器
ASTROC Automatic Stellar Tracking Recognition and Orientation Computer 自动星体跟踪识别与定向计算机
ASTROM Astronautics Reference Object Model 航天学参照物模型
ASTROS Advanced Star and Target Reference Optical Sensor 先进恒星与目标基准光学传感器
ASTS Avionics System Test Specifica-

ASTV Aero-Assisted Space Transfer Vehicles 空气助推的太空转移飞行器

ASTWG Advanced Spaceport Technology Working Group 先进航天港技术研究部

ASUT Adapter Submit Tester 适配器亚单元测试器

ASV Aerothermodynamic Structural Vehicles 【美国】空气热动力结构（试验）飞行器

ASVT Application Systems Verification and Transfer 【美国】陆地卫星应用系统验证与转让（计划）

AT Action Time 作用时间

AT Apogee Thruster 远地点推进器

AT&L Acquisition, Technology and Logistics 采办、技术与后勤

AT&O Apollo Test and Operations 【美国】"阿波罗"飞船测试与操作

ATA Abort Timing Assembly 中止定时装置

ATA Advanced Test Accelerator 先进测试加速器（装置）

ATA Asynchronous Terminal Adapter 异步通信终端适配器

ATA Automated Target Acquisition 自动目标截获

ATA Automatic Threshold Adjust 自动阈值调整

ATA Automatic Track(ing) Acquisition 自动跟踪截获

ATA Automatic Trouble Analysis 自动故障分析

ATA Avionics Test Article 电子设备与控制系统测试装置

ATAAM Advanced Tactical Air-to-Air Missile 先进战术空空导弹

ATAC Advanced Traceability and Control 先进的跟踪能力与控制

ATAC All Terrain All Climate 全地形全气候

ATACC Advanced Tactical Command Central 先进战术指挥中心

ATACC Advanced Theater Air Command Center 先进战区空中指挥中心

ATACCM Advanced Technology Air-Launched Conventional Cruise Missile 先进技术空射常规巡航导弹

ATACCS Advanced Tactical Air Command and Control System 先进战术空中指挥与控制系统

ATACMS Advanced Tactical Missile System 先进战术导弹系统；先进的战术导弹系统

ATACMS Army Tactical Missile System 陆军战术导弹系统

ATAGS Advanced Technology Anti-G Suit 先进抗重力服

ATAMS Automated Tracking and Monitoring System 自动跟踪与监控系统

ATAO Automatic Terminal Area Operational 自动控制终端区域运行

ATAR Acquisition Tracking and Recognition 截获跟踪与识别

ATARR Advanced Turbine Aerothermal Research Rig 先进涡轮气动热力研究试验台

ATB Abort Test Booster 中止测试助推器

ATBM Anti-Tactical Ballistic Missile 反战术弹道导弹

ATBM Anti-Theater Ballistic Missile 反战区弹道导弹

ATBM Average Time Between Maintenance 平均维修间隔时间

ATBMD Antitactical Ballistic Missile Defense 反战术弹道导弹防御
ATBMS Antiactical Ballistic Missile System 反战术弹道导弹系统
ATC Ablative Thrust Chamber 烧蚀推力燃烧室
ATC Ablative Thrust Control 烧蚀推力控制
ATC Automatic Target Counting 自动目标记数
ATC Automatic Target Cueing 自动目标提示
ATC Automatic Threat Countering 自动威胁对抗
ATC Automatic Tilt Control 自动倾斜控制
ATCCC Advanced Tactical Command and Control Capacity 先进战术指挥与控制能力
ATCCS Army Tactical Command and Control System 【美国】陆军战术指挥与控制系统
ATCE Ablative Thrust Chamber Engine 烧蚀推力室（火箭）发动机
ATCM Advanced Technology Cruise Missile 先进技术巡航导弹
ATCOM Advanced Tactical Combat Model 先进战术作战模型
ATCOS Atmosphere Collection Satellite 大气采集卫星
ATCOS Atmospheric Composition Satellite 【美国空军】大气成分卫星
ATCP Advanced Technology Coorbiting Platform 【日本】先进技术同轨平台
ATCS Active Temperature Control System 有源（主动）温控系统
ATCS Active Thermal Control Subsystem 有源（主动）热控制子系统
ATCS Air Temperature Control System 空气温度控制系统
ATCU Attitude and Translation Control Unit 姿态与平移控制装置
ATD Advanced Technologies Demonstration 先进技术演示验证
ATD Advanced Technology Development 先进技术研发
ATD Automatic Target Detection 自动目标探测
ATDA Agena Target Docking Adapter 【美国】"阿森纳"火箭目标对接器
ATDA Augmented Target Docking Adapter 【美国】改进型目标对接器
ATDC Advanced Technology Development Center 【美国国家航空航天局】先进技术发展中心
ATDE Advanced Technology Demonstrator Engines 先进技术演示验证发动机
ATDL Army Tactical Data Link 【美国】陆军战术数据链
ATDL Atmospheric Turbulence and Diffusion Laboratory 大气湍流和扩散实验室
ATDLS Advanced Tactical Data Link System 先进战术数据链系统
ATDRSS Advanced Tracking and Data Relay Satellite System 先进跟踪与数据中继卫星系统
ATDS Airborne Tactical Data System 机载战术数据系统
ATE Advanced Technology Engine 先进技术发动机
ATE Airborne Test Equipment 机载测试设备
ATE Along Track Error 沿轨道误差
ATE Automatic Test Equipment 自

动测试设备
ATE Avionics Test and Evaluation 电子设备与控制系统测试与评估
ATEC Atlantic Test and Evaluation Center 【美国】大西洋测试与评估中心
ATEGG Advanced Turbine Engine Gas Generator 先进涡轮发动机燃气生成器
ATER Automatic Testing, Evaluation and Reporting 自动测试、鉴定与通报
ATES Advanced Technology Engine Study 先进发动机技术研究
ATEWA Automatic Target Evaluator and Weapon Assignor 自动目标鉴定器与武器指定器
ATF Adaptive Transversal Filter 自适应横向滤波器
ATF Advanced Tactical Fighter 先进战术战斗机
ATF Aeronautical Tracking Facility 宇宙飞行跟踪设施
ATF Astrometric Telescope Facility 【美国】天体测量望远镜设施
ATF Automatic Target Finder 自动目标探测器
ATFE Advanced Thermal Control Flight Experiment 先进热控飞行实验
ATFOS Alignment and Test Facility for Optical System 光学系统校准与试验设备
ATGM Anti-Tank Guided Missile 反坦克导弹
ATH Above the Horizon 地平线；高出地平线
ATHS Airborne Target Handover System 机载目标（跟踪）转交系统
ATI Advanced Terminal Interceptor 高级末端截击导弹
ATIC Advanced Technology Innovation Cell 先进技术创新舱
ATIC Advanced Thin Ionization Calorimeter 先进薄电离量能器
ATILL Advanced Tracking Illuminator Laser 先进跟踪照明激光
ATIM Advanced Technology Insertion Module 先进技术插入模块
ATIMS Automatic Time Interval Measurement System 自动时间间隔测量系统
ATIP Acquisition Tracking and Insertion Program 探测跟踪与导入程序
ATIRCM Advanced Threat Infrared Countermeasures 先进威胁红外对抗
ATIS Automatic Terminal Information Service 自动终端信息服务
ATIS Automatic Terminal Information System 自动终端信息系统
ATJC Annular Turbojet Combustor 环形涡轮喷气发动机燃烧室
ATL Advanced Technology Laboratory 先进技术实验室
ATL Aerospace Test Laboratory 航空航天测试实验室
ATL Applied Technology Laboratory 应用技术实验室
ATLAS Abbreviated Test Language for Avionics Systems 用于电子设备与控制系统的简化测试语言
ATLASS Advanced Technology for Large Area Space Structures 先进大区域空间结构技术
ATLAST Advanced Technology Large Aperture Space Telescope 先进技术大孔径空间望远镜
ATLO Acceptance Test and Launch Operations 验收试验与发射操作
ATLO Assembly, Test and Launch Op-

erations 装配、测试与发射操作
ATM Apollo Telescope Mount (Skylab)【美国"天空实验室"项目】"阿波罗"飞船望远镜装置
ATM Asynchronous Transfer Mode 异步转换模式
ATMDC Apollo Telescope Mount Digital Computer 【美国"天空实验室"项目】"阿波罗"飞船望远镜装置数字计算机
ATMDF Air and Theater Missile Defense Force 【美国】空防与战区导弹防御部队
ATMOS Atmospheric Trace Molecule Spectroscopy 大气痕量分子分光检查
ATMOS Atmospheric Trace Molecules Observed by Spectroscopy 分光观测大气痕量分子
ATMOSE Atmospheric Trace Molecule Spectroscopy Experiment 大气痕量分子分光实验
ATN Advanced TIROS-N 【美国】先进"泰罗斯 N"气象卫星
ATO Abort-to-Orbit 入轨中止
ATO Authority to Operate 授权操作
ATODB Air Tasking Order Data Base 空中任务命令数据库
ATOL Assisted Takeoff & Landing 助推起飞和着陆
ATOL Automatic Take-Off and Landing 自动起飞与着陆
ATOLL Acceptance, Test or Launch Language 验收、试验或发射语言
ATOLL Assembly/Test Oriented Launch Language 用于装配与测试的发射语言
ATOROCKET Assisted Takeoff Rocket 辅助起飞火箭
ATOS Advanced Transport Operating Systems 先进运输操作系统
ATP Acceptance Test Procedure 验收试验规程
ATP Acquisition, Tracking, and Pointing 捕获、跟踪与瞄准
ATP Advanced Technical Payload 先进技术有效载荷
ATP&FC Acquisition, Tracking, Pointing, and Fire Control 捕获、跟踪、瞄准与发射控制
ATPA Alpha Temperature Probe Assembly 阿尔法温度探测器装配
ATPCS Automatic Takeoff Power Control System 自动起飞功率控制系统
ATPFC Acquisition, Tracking & Pointing & Fire Control 捕获、跟踪、瞄准与发射控制
ATR Air Launched Trainer Rocket 空中发射火箭训练机
ATR Air Turbo Rocket 空气涡轮火箭
ATR Apollo Test Requirement 【美国】"阿波罗"飞船试验要求
ATR Attained Turn Rate 可达到的（最大）转弯角速度
ATR Automatic Target Recognition 目标自动识别
ATR Autonomous/Automated Target Recognition 自主/自动目标识别
ATRAN Automatic Terrain Recognition and Navigation (System) 自动地形识别与导航（系统）
ATRD Automatic Target Recognition Device 目标自动识别装置
ATRID Automatic Target Recognition, Identification and Detection 目标自动识别、鉴定与探测
ATRID Automatic Terrain Recognition and Identification Device 自动地形

识别与鉴定装置
ATRJ Advanced Threat Radar Jammer 先进威胁雷达干扰器
ATRR Acceptance Test Readiness Review 验收试验（测试）准备状态评审
ATS Acceptance Test Specification 验收试验（测试）技术说明
ATS Access to Space 进入空间
ATS Acquisition and Tracking System 截获与跟踪系统
ATS Acquisition Target and Search 截获目标与搜索
ATS Advanced Technology Satellite 先进技术卫星
ATS Advanced Technology Spacecraft 先进技术航天器
ATS Advanced Teleprocessing System 先进远程处理系统
ATS Aft Transition Structure 后部过渡结构
ATS Applications Technology Satellites 【美国】应用技术卫星
ATS Attitude Transfer System 姿态转换系统
ATS Automatic Terminal System 自动终端系统
ATSIM Acquisition and Track Simulation 捕获与跟踪模拟
ATSIT Automatic Technique for Selection and Identification 自动目标选择与识别技术
ATSL Ada Technology Support Laboratory Ada语言技术保障实验室
ATSLS Analysis of Thrust Structure Loads and Stresses 推力结构载荷和应力分析
ATSM Advanced Tactical Stand-Off Missile 高级战术远距发射导弹
ATSOCC Applications Technology Satellite Operations Control Center 【美国】应用技术卫星操作控制中心
ATT Algorithm-to-Test Reviews 测试计算评审
ATTD Advanced Technology Transition Demonstration 先进技术转换论证；先进技术过渡演示验证
ATTS Acquisition Target Tracking System 拦截目标跟踪系统
ATTS Automatic Telemetry Tracking System 自动遥测跟踪系统
ATU Airlock Audio Terminal 气闸音频终端
ATV Advanced Technology Validation 先进技术确认
ATV Advanced Test Vehicle 先进型试验飞行器
ATV Advanced Tether Vehicle 先进系留运载器
ATV Aerodynamic Test Vehicle 空气动力试验飞行器
ATV Ariane Transfer Vehicle 【欧洲空间局】"阿里安"运载火箭转运车
ATV Automated Transfer Vehicle 【欧洲空间局】自动转移飞行器
ATV Automated Transport Vehicle 自主运输飞行器
ATV Autonomous Transfer Vehicle 自主转移飞行器
ATVC Ascent Thrust Vector Control 上升段推力矢量控制
ATVC Ascent Thrust Vector Controller 上升段推力矢量控制器
ATVC Automatic Thrust Vector Control 推力矢量自动控制
ATVCC Automated Transfer Vehicle Control Centre 自动转移飞行器控制中心

ATVCD Ascent Thrust Vector Control Driver 上升段推力矢量控制驱动器

ATVS Ada Test and Verification System 软件测试与校验系统

AU Astronomical Units 天文单位

AULS Accidental or Unauthorized Limited Strike 意外的或未授权的有限打击

AUS Advanced Upper Stage 先进上面级

AUSSAT Australian Communications Satellite System 澳大利亚通信卫星系统

AUST COMSAT Australian Communication Satellite 澳大利亚通信卫星

AUTODIN Automatic Digital Network 自动数字网

AUTOGOSS Automated Ground Operations Scheduling System (also AGOSS) 地面操作自动调度系统（亦称 AGOSS）

AUTOLAND Automatic Landing 自动着陆

AUTOSCAN Automatic Satellite Computer Aid to Navigation 卫星计算机辅助自动导航系统

AUTOSCAN Automatic Speed Control and Approach Navigation (System) 自动速度控制与进场导航（系统）

AUV Automatic Underwater Vehicle 自动水下运输车

AUWS Automatic Unmanned Weather Station 无人管理自动气象站

AV Average Value 平均值

AV BAY Avionics Bay 电子设备与控制系统舱

AVC Advanced Vehicle Concept 先进飞行器概念

AVC Automatic Volume Control 容量自动控制

AVD Aerospace Vehicle Detection 航空航天飞行器探测

AVE Aerospace Vehicle Equipment 航空航天飞行器设备

AVE Atmospheric Variability Experiment 大气能变性实验

AVERT Automatic Verification, Evaluation and Readiness Tester 自动证、评价和准备测试器

AVHRR Advanced Very High Resolution Radiometer 先进甚高分辨率辐射计

AVID Aerospace Vehicle Interactive Design 航空航天器交互设计

AVIPSS Automated Virtual Information Production Support System 自动虚拟信息整理保障系统

AVIS Active Vibration Isolation System 有效振动隔离系统；有效防振系统

AVIT Air Vehicle Integration and Test 空中飞行器一体化测试

AVL Automated Vehicle Locations 自动运载器定位

AVL Avionics Verification Laboratory 电子设备与控制系统校验实验室

AVLAN Avionics Local Area Network 电子设备与控制系统本地网络

AVLS Advanced Vertical Launching System 先进垂直发射系统

AVSCOM Aviation Systems Command 【美国陆军】航空系统指挥部

AVSR Avionics Verification Status Room 电子设备与控制系统校验状态间

AVT Advanced Vehicle Technology 先进飞行器技术

A

AVTA Advanced Vehicle Testing Activity 先进飞行器测试工作
AVVI Altimeter Vertical Velocity Indicator 测高仪垂直速度指示器
AVVI Attitude/Vertical Velocity Indicator 【飞行】姿态/垂直速度指示器
AW Assembly Workstand 装配工作台
AW-SRADMS All-Weather Short-Range Air Defense Missile System 全天候近程防空导弹系统
AWACS Airborne Warning and Control System 机载报警与指挥系统
AWAPS Advanced Weather Analysis and Prediction System 先进天气分析与预报系统
AWAS Automated Weather Advisory Station 先进气象通报站
AWCS Automatic Work Control System 自动工作控制系统
AWE Advanced Warfighting Experiment 先进作战实验
AWE Advanced Wave-Effects 高级波动效应
AWESS Automatic Weapon Effect Signature Simulator 自动武器效应特征模拟器
AWIPS Advanced Weather Interactive Processing System 先进气候交互处理系统
AWIPS Automated Weather Interactive Processing System 自动气象交互式处理系统
AWIS Advanced Weather Information System 先进气象信息系统
AWIS All-Weather Identification Sensor 全天候（目标）识别传感器
AWLS All-Weather Landing System 全天候着陆系统

AWO All Weather Operations 全天候操作
AWR Adaptive Waveform Recognition 自适应波形识别
AWS Advanced Warning System 先进报警系统
AWS Aegis Weapon Systems 【美国】"宙斯盾"武器系统
AWS Air Weather Service 全天候服务
AWS Anti-Missile-Weapon System 反导弹武器系统
AWS Arrow Weapons System 【以色列】"箭"式导弹武器系统
AWS Automated Wiring System 自动布线系统
AWSACS All Weather Standoff Attack Control System 全天候敌防区外攻击控制系统
AWSM Air Warfare Simulation Model 空战模拟模型
AWTTP Apollo Wind-Tunnel Testing Program 【美国】"阿波罗"风洞试验项目
AWWIMS Automated Worldwide Warning Indicator Monitoring System 自动化全球报警指示器监控系统
AWYDC All-Weather Yaw Damper Computer 全天候偏航阻尼计算机
AXAF Advanced X-Ray Astrophysics Facility 【美国空间站】先进X射线天文物理设施
AXET Space Plasma Physics X-Ray Telescope 空间等离子体物理X射线望远镜
AXIS Atmospheric X-Ray Image Spectrometer 大气X射线成像光谱仪
AYC Aerodynamic Yaw Coupling 气动力偏航耦合
AYCP Aerodynamic Yaw Coupling

Parameters 气动力偏航耦合参数
AYI Angle of Yaw Indicator 偏航角指示器
AZAR Adjustable Zero, Adjustable Range 可调零点，可调量程
AZAS Adjustable Zero, Adjustable Span 可调零点，可调跨距（跨度）
AZS Automatic Zero Set 自动调零；自动零位调整

B

B Spec　Development Specification　研发说明
B&R　Bomb and Rocket　炸弹与火箭
B&S　Booster and Sustainer　助推器与主发动机；助推发动机和主发动机
B-FACT　Booster Flight-Acceptance Composite　【美国国家航空航天局】助推器飞行验收组合
B/A　Barometric Altimeter　气压测高仪
B/A　Beam Approach　波束引导进场
B/A　Booster Adapter　运载火箭接合装置；助推器适配器
B/A　Buffer Amplifier　缓冲放大器
B/C　Bench Check　台架检测
B/H　Blockhouse　火箭发射掩体
B/L　Baseline　基线
B/O　Booster Orbiter　助推轨道飞行器
B/O　Burnout　燃料烧尽；主动段终点
B^2C^2　Brigade-and-Below Command and Control System　【美国陆军】旅及旅以下的指挥与控制系统
BA　Backup Aerospace Vehicle　备用航空航天运载器
BA　Bank Angle　倾斜角
BA　Beacon Antenna　信标天线
BA　Breakup Altitude　【火箭】级分离高度
BABS　Beam Approach Beacon System　波束控制进场信标系统
BABS　Blind Approach Beacon System　仪表进场信标系统
BAC　Boeing Aerospace Company　【美国】波音航空航天公司
BAC　Booster Assembly Contractor　助推器组装承包商
BACE　Booster Adapter Control Electronics　助推器适配器控制电子设备
BACIMO　Battlefield Atmosphere and Cloud Impacts on Military Operations　战斗空间大气和云对军事行动的影响
BACS　Body Axis Coordinate System　机身轴坐标系统
BACT　Best Available Control Technology　最佳可用控制技术
BAD　Biological Aerosol Detection　生物气溶胶探测
BAD　Blast Attenuator Device　冲击波衰减装置
BAD　Boom Avoidance Distance　防声爆距离
BADCT　Best Available Demonstrated Control Technology　经过验证的现有最佳控制技术
BADD　Battlefield Awareness and Data Dissemination　战场感知与数据传送
BADG　Battle Group Antiair Warfare Display Group　战斗群对空作战显示器组
BADGES　Base Air Defense Ground Environment System　基地防空警戒地面环境系统
BADIC　Biological Analysis Detection Instrumentation and Control　生物

分析检验仪器与控制
BAE Battlefield Area Evaluation 战场范围评估
BAE Beacon Antenna Equipment 信标天线设备
BAF Batiment d'Assemblage Final 【法语】总装测试厂房
BAF/EH Encapsulation Hall of BAF 【法语】最后总装测试厂房内的封装大厅
BAGS Bullpup All-Weather Guidance System 【美国】"小斗犬"导弹全天候制导系统
BAGS Bullpup Automatic Guidance System 【美国】"小斗犬"导弹自动制导系统
BAGSE Boresight Assembly Ground Support Equipment 校靶装配地面保障设备
BAI Backup Aerospace Vehicle Inventory 备用航空航天器库存总数
BAI Barometric Altitude Indicator 气压高度指示器
BAI Battlefield Air Interdiction 战场空中遮断（封锁）
BAI Bearing Altitude Indicator 方位—高度指示器
BAINS Basic Advanced Integrated Navigation System 基本高级综合导航系统
BAIS Blind Approach Instrument System 盲目（按仪表）进场着陆仪表系统
BALD Ballistic Density 弹道空气密度
BALDAS Ballistic Data Acquisition System 弹道数据采集系统
BALFRAM Balanced Force Requirements Attrition Model 力量平衡需求磨损模型

BALID Ballistics Identification 弹道识别
BALLAD Ballistic LORAN Assist Device 【德国/法国】弹道"罗兰"（双曲线远程导弹系统）辅助设备
BALMI Ballistic Missile 弹道导弹
BALS Blind Approach Landing System 盲目（按仪表）进场着陆系统
BALWND Ballistic Wind 弹道风
BAM Ballistic Advanced Missile 高级弹道导弹
BAMB Bending Annular Missile Body 弯曲环形导弹弹体
BAMBI Ballistic Antiballistic-Missile Boost Intercept 弹道反弹道导弹主动段拦截
BAMBI Ballistic Antiballistic-Missile Boost Interceptor 弹道反弹道导弹主动段拦截器
BAMBI Ballistic Missile Bombardment Interceptor 弹道导弹轰炸截击器
BAMBI Ballistic Missile Boost Intercept 弹道导弹主动段拦截
BAMBI Ballistic Missile Booster Interceptor 弹道导弹主动段拦截器
BAMIRAC Ballistic Missile Radiation Analysis Center 弹道导弹辐射分析中心
BAMM Balloon Altitude Mosaic Measurements 高空气球感光嵌镶幕测量技术
BAMS Broad Area Maritime Surveillance 阔域海事监视
BAN Beacon Alphanumerics 【雷达】信标字母数字显示设备
BAN Bionics Adaptive Network 仿生电子自适应网络
BANSHEE Anti-ICBM and Anti-Space

Technique Project Under Study 研究中的反洲际弹道导弹和反太空技术项目
BAP Ballistic Aimpoint 弹道瞄准点
BAP Base Auxiliary Power 地面辅助动力（电源）
BAP Best Adaptive Path 最佳适配路径
BARBICAN Battlefield Automated Radar Bearing Intercept Classification and Analysis 战场自动化雷达定向截获目标分类与分析
BARS Backup Attitude Reference System 备用姿态参考系统
BARS Ballistic Analysis Research System 弹道分析研究系统
BARS Ballistics/Aerodynamics Research System 弹道与空气动力研究系统
BARS Baseline Accounting and Reporting System 基线统计与报告系统
BAS Boresight Adjustment System 校靶调整系统；瞄准点调整系统
BASES Beam Approach Seeker Evaluation System 波束引导进场探寻器鉴定系统
BASICS Battle Area Surveillance and Integrated Communication System 战区监视与综合通信系统
BASIS Battlefield Automatic Secure Identification System 战场自动安全识别系统
BASS Backup Avionics System Software 备份电子设备与控制系统软件
BASS Basic Analog Simulation System 基本模拟仿真系统
BAT Ballistic Aerial Target 弹道飞行目标；弹道式空中目标

BAT BMC^{4}I Advanced Technology 作战管理指挥、控制、通信、计算机与情报的先进技术
BATO Balloon-Assisted Takeoff 气球辅助起飞
BATS Ballistic Aerial Target System 弹道式空中目标系统
BATS Basic Additional Teleprocessing Support 基本辅助远程信息处理支持
BATS Basic Additional Teleprocessing System 基本辅助远程信息处理系统
BATSE Burst and Transient Source Experiment 【γ射线】爆发与瞬时爆发源实验
BATSS Battlefield Targeting Support System 战场目标定位支援系统
BAU Bus Adapter Unit 总线适配器部件
BAV Ballistic Accuracy Verification 弹道准确度确认
BBIM Buoyant Ballistic Inertial Missile 弹性弹道惯性导弹
BBM Body Bending Mode 弹体弯曲状态
BBNAV Blind Bombing/Navigation 盲炸/导航
BBO Booster Burnout 助推器熄火
BBSO Big Bear Solar Observatory 【美国】"大熊"太阳观测台
BBT Booster Burn Time 助推器燃烧时间
BBX Black-Brant X 【美国】"黑雁"X探空火箭
BC Ballistic Coefficient 弹道系数
BC/FC Beam Control/Fire Control 光束控制/发射控制
BC^2 Battlespace Command and Control 作战空间指挥与控制

BCA Basing Concept Analysis 驻扎方案分析
BCA Bias Correction and Acceleration 偏移修正值与加速度
BCA BMDS Capability Assessment 弹道导弹防御系统性能评估
BCAS Base Contracting Automated System 基地签约自动化系统
BCAS Battle Management and C^3 Architecture Simulator 作战管理和指挥、控制与通信体系结构模拟器
BCAS Beacon-Based Collision Avoidance System 信标防撞系统
BCAS BM and C^3 Architecture Simulator 作战管理与指挥、控制、通信体系结构仿真器
BCBL Battle Command Battle Laboratory 作战指挥战斗实验室
BCC Backup Control Center 备用控制中心
BCC Ballistic Control Computer 弹道控制计算机
BCC Beacon Control Console 信标控制台
BCC BETA (Battlefield Exploitation and Target Acquisition) Correlation Center 战场利用与目标搜索相互关系中心
BCCC Ballistic Compressor Computer Code 弹道压缩计算机码
BCCE BM/C^3 Consolidated Capabilities Effort 作战管理/指挥、控制与通信综合能力工作
BCD Baseline Concept Description 基本概念（方案）说明
BCD Bio-Chemical Detector 生物－化学战剂探测器
BCD Burst Control Device 起爆控制装置；爆炸控制装置
BCDSS Battle Command Decision Support System 作战指挥决定支持系统
BCE Backup Control Electronics 备用控制电子设备
BCE Baseline Cost Estimate 基线成本估计；基础成本估算
BCE Battlefield Coordinating Element 战场协调要件（素）
BCE Beam Collimation Error 波束准直误差
BCE Beam Control Element 光束控制单元；光束控制元件
BCE Beamsplitter Control Electronics 分光镜控制电子装置
BCE Bus Control Element 总线控制元件
BCFS Backup Flight Control System 备用飞行控制系统
BCL Bance de Control Lanceur 【法语】运载器测试系统
BCM Ballistic Correction of the Moment 即时弹道修正
BCM Baseline Correlation Matrix 【美国空军】基线相关矩阵
BCM Biological Countermeasures 生物对抗措施
BCO Broad Concept of Operations 扩展的作战概念（方案）
BCOB Booster Cut-Off Backup 助推熄火备份装置
BCRT Bus Controller Remote Terminal 总线控制器遥控终端
BCRV Blunt Conical Reentry Vehicle 钝锥形再入飞行器
BCS Ballistic Computer System 弹道计算机系统
BCS Ballistic Control System 弹道（飞行）控制系统
BCS Baseline Comparison System 基线比较系统

BCS Beam Control System 光束控制系统

BCSC-T BMDS Communication System Complex, Transportable 便携式 BMDS 通信系统成套设备

BCTP Battle Command Training Program 【美国】作战指挥培训项目

BCTS Broadcast Communication Technology Satellite 【日本】广播通信技术卫星

BCU Ballistic Computer Unit 弹道计算机部件

BCU Buffer Control Unit 缓冲控制器

BCV Battle Command Vehicle 【美国】作战指挥车

BCW Biological and Chemical Weapons 生化武器

BDA Battle Damage Assessment 战斗毁伤评估

BDA Beacon Drive Assemblies 信标驱动组件

BDA Blast Danger Area 爆炸危险区域

BDA Bomb Damage Assessment 轰炸损伤评估

BDADS Beacon Data Acquisition and Display System 信标数据采集与显示系统

BDAR Battlefield Damage Assessment and Repair 战场损伤评估与修复

BDAR Bomb Damage Assessment Reconnaissance 轰炸效果判定侦察

BDC Bidirectional Converter 双向转换装置

BDCA Boitier de Destruction Commande'e et Automatique 【法语】指令自毁装置

BDCF Baseline Data Collection Facility 基线数据采用设备

BDCS Battle Damage Control System 作战损伤控制系统

BDI Burst Distances Indicator 爆炸距离指示器

BDIAC Battle-Defense Information Analysis Center 作战防御信息分析中心

BDL Battlefield Demonstration Laser 战场演示验证激光

BDM Ballistic Defense Missile 弹道防御导弹；弹道导弹的防御截击导弹

BDM Booster Deceleration Motors 助推器减速发动机

BDO Burst Detector Optical 【核探测卫星】炸点光学探测器

BDP Baseline Data Package 基线数据包

BDPI Baseline Data Package Integration 综合基线数据包

BDS Bomb Damage Survey 轰炸效果调查

BDS Boost Phase Detection System 助推段探测系统

BDSD Base-Detonating Self-Destroying 带弹底起爆引信和自毁

BDT Birth-to-Death Tracking 【从空间目标与助推器分离至被摧毁】全程跟踪；（对导弹弹头自释放直至将其摧毁的）自始至终跟踪

BDT Bulk Data Transfer 批量数据传送

BDT Burst Delay Timer 爆炸延迟计时器

BDTR Basic Data Transmission Routine 基本数据传输程序

BDX Burst Detector X Ray 【核探测卫星】炸点 X 射线探测器

BDY Burst Detector Y-Sensor 猝发探测器 Y 传感器

BDZ Buffer Demodulator Zero 缓冲解调器归零

BE Beacon Explorers 【美国】信标探险者卫星

BE Booster Engine 助推器发动机

BEAP Banc d'Essais des Propulseurs d'Appoint à Poudre 【法语】固体助推试车台

BEAR Beam Experiment Abroad Rocket 火箭波束实验

BEAST Battle Experiment Area Simulator Tracker 作战试验区仿真器跟踪系统

BECO Before Engine Cutoff 发动机关闭前

BECO Booster Engine Cut Off 助推器发动机关机

BECS Battlefield Electronic CEOI System 战场电子通信电子操作指令系统

BECS Burst Error Control System 爆炸误差控制系统

BED (NASA) Block Error Detector 【美国国家航空航天局】组合误差探测器

BEDAC Burst Error Detection and Correction 爆炸误差检测与修正

BEET Best Estimated Evaluation Trajectory 最佳测定的评估轨迹

BELIER French Sounding Rocket 法国探空火箭

BENS Bounded Error Navigation System 有限误差导航系统

BEO Beyond Earth Orbit 地球轨道以远

BEOP Best Estimate of Orbital Parameters 轨道参数的最佳估计

BEP Beamed Energy Propulsion 束能推进

BEP Best Efficiency Point 最佳有效点

BER Burst Excess Rate 爆炸余率

BES Bang Erection System 【火箭发射】开关控制的起竖系统

BES Booster Exhaust Stream 助推器排气气流

BESC BM/C^3 Element Support Center 作战管理、指挥、控制、通信分队支援中心

BESim Brilliant Eyes Simulator 智能眼模拟器（装置）

BESim/AT Brilliant Eyes Simulator Analysis Tool 智能眼模拟器（装置）分析工具

BESim/RT Brilliant Eyes Simulator Real-Time 实时智能眼模拟器（装置）

BESS Biological Experiment Scientific Satellite 生物实验科学卫星

BESS Biomedical Experiment Scientific Satellite 生物医学实验科学卫星

BEST BM/C^3 Element Support Task 作战管理、指挥、控制、通信分队支援任务

BEST Booster Exhaust Study Test 助推器排气研究测试

BET Best Estimate of Trajectory 【美国国家航空航天局】最佳弹道测算

BET Best Estimated Trajectory 最佳预估轨道

BETA Battlefield Exploitation and Target Acquisition 战场情报利用与目标捕获；战场利用与目标搜索

BETO Booster Engine Thrust Oscillation 助推器发动机推力振荡

BETT Bolt Extrusion Thrust-Terminator 螺栓挤压推力终止装置

BEWS Blast Wave Environment Simulator 冲击波环境模拟装置

BFACS Battlefield Functional Area Command and Control System 战场职能地区指挥与控制系统

BFC Backup Flight Computer 备用飞行计算机

BFC Backup Flight Control (System) 备用飞行控制（系统）

BFCGW Basic Flight Design Gross Weight 基本飞行设计总重

BFCS Backup Flight Control System 备用飞行控制系统

BFCS Ballistic Framing Camera System 弹道成帧摄像机系统

BFCS Batteries of Flight Control System 飞行控制系统蓄电池组

BFDAS Basic Flight Data Acquisition System 基本飞行数据采集系统

BFE Basic Flight Envelope 基本飞行包线

BFM Basic Flight Maneuver 基本飞行机动

BFM Basic Flight Module 主飞行舱

BFM Battle Scale Forecast Model 战斗规模预报模型

BFMDS Base Flight Management Data System 基地飞行管理数据系统

BFS Backup Flight Software 备用飞行软件

BFS Backup Flight System 备用飞行系统

BFT Batch Fabrication Technique 批生产技术

BFTA Bulk Fuel Tank Assembly 燃料贮箱部件；散装燃料贮箱组件

BFTC Boeing Field Test Center 【美国】波音公司外场试验中心

BFTM Ballistic Flight Test Missile 弹道飞行试验导弹

BFTT Battle Force Tactical Training 作战部队战术培训

BFTU Boeing Field Test Unit 【美国】波音公司外场试验装置

BG Blast Gas 爆炸气体

BGG Booster Gas Generator 助推器气体发生器

BGRV Boost Glide Reentry Vehicle 助推滑翔再入（大气层）飞行器

BGS Backup Gimbal Servo 备用常平架伺服装置

BGSS Bus Ground Support System 运载舱地面支援系统

BGV Boost Glide Vehicle 助推滑翔飞行器

BHC Ballistic Height Correction 弹道高度修正量

BHI Burst-Height Indicator 爆炸高度指示器

BHWT Boeing Hypersonic Wind Tunnel 【美国】波音公司高超声速风洞

BI Booster Intergration 助推器对接

BI-MM Base Installation Minuteman 【美国】"民兵"导弹基地设施

BIA Boost, Insertion and Abort 助推、入轨与中止飞行

BIA Booster Inert Assembly 助推器惯性部件

BIA Booster Integrated Assembly 助推器总装；助推器整合装配

BIA Bus Interface Adapter 总线接口适配器

BIATC Built-In Automatic Test and Checkout Equipment 内装自动测试与检查设备；嵌入式自动测试与检测设备

BIC Battlefield Integration Center 战场综合中心

BICES Battlefield Information Collection and Exploitation System 战场

信息收集与开发（应用）系统
BID Built-In Diagnostics 机内诊断（法）
BIDE Blow-In Door Ejector 【发动机喷管】进气门引射器
BIDS Biological Integrated Detection System 生物综合探测系统
BIG Biological Isolation Garment 【美国国家航空航天局】生物隔离服
BIGREV Big Reentry Vehicle 大型再入飞行器
BIGS Booster Inertial Guidance System 助推器（运载火箭）惯性制导系统
BIL Bâtiment d'Intégration Lanceur/LV Integration Building 运载火箭组装厂房
BILL Beacon Illuminator Laser 信标照明激光
BIM Ballistic Intercept Missile 弹道拦截导弹
BIOCORE Biological Cosmic Ray Experiment 生物宇宙射线实验
BIOLABS Biological Measurement of Man-in Space 人在太空进行的生物测量
BIOS Basic Input/Output System 基本输入/输出系统
BIOS Biological Investigation of Space 空间生物研究卫星；【美国国家航天局】太空生物调查
BIOS Biological Satellite 【美国国家航空航天局】生物卫星
BIP Ballistic Improvement Program 弹道改进计划
BIP Bâtiment d'Intégration des Propulseurs/Boosters Integration Building 助推器组装厂房
BIP Bomb Impact Plot 炸弹弹着点标示图
BIPROP Bipropellant 双推进剂
BIRAMIS Bilan Radiatif de l'atmosphere par Microaccelero Metrie Spatiale 【法语】星载微加速度测定地球辐射平衡的卫星
BISA Belgium Institute for Space Aeronomy 比利时太空大气物理学研究所
BISR Bochum Institute of Space Research 【日本】波鸿空间研究所
BIST Built-in-Self-Test 机内自测试
BIT Built-in Test 内部测试
BITC Battle Management Integration Center 作战管理一体化中心
BITE BESC Integration, Test & Evaluation 作战管理、指挥、控制、通信分队的综合、测试与鉴定
BITE Built-in Test Equipment 内部测试设备
BITG Built-in Tracking Generator 嵌入式跟踪发生器
BIU Bay Interface Unit 舱接口单元
BIU Buffer Interface Unit 缓冲器接口组件
BIU Bus Interface Unit 母线接口组件
BIV VEGA Integration Building 【法国】"织女星"运载火箭组装厂房
BKEP Bomb, Kinetic-Energy Penetrator 动能贯穿炸弹
BKEP Boosted Kinetic Energy Penetrator 助推动能穿地弹（跑道破坏弹）
BLA Base de Lancement Ariane 【法语】"阿里安"运载火箭发射场
BLA Bracket and Linkage Assembly 托架与联动装置；支架与联动机构的组装
BLACK BRANT Canadian Sounding

Rockets 加拿大探空火箭
BLADE-GT Blade Life Analysis and Design Evaluation for Gas Turbines 燃气涡轮机叶片寿命分析和设计评估
BLADES Ballistic Missile Defense Long Wave Length Infrared Advanced Exoatmospheric Sensor 弹道导弹防御长波红外型先进外大气层传感器
BLADES BMD Long Wavelength Infrared Advanced Exoatmospheric Sensor 弹道导弹防御用大气层外先进长波红外探测器
BLADT Blast, Dust, Thermal Effects Model 冲击波、尘降、热效应模型
BLAHA Basic Load Ammunition Holding Area 基本装载量弹药储存区
BLAM Ballistically Launched Aerodynamic Missile 按弹道发射的空气动力导弹
BLCCE Ballistic Missile Defense Organization Life Cycle Cost Estimate 弹道导弹防御局寿命周期费用评估
BLDT Balloon Launch Decelerator Test 气球发射减速器试验
BLDTV Balloon-Launch Drop Test Vehicle 气球升空降落试验运载器
BLOW Booster Lift Off Weight 助推器起飞重量
BLP Basic Launch Plan 基本发射计划
BLS Boundary Layer Separation 边界层分离
BLSS Bioregenerative Life Support Systems 生物再生式生命保障系统
BM Ballistic Missile 弹道导弹
BM Battle Management 作战管理
BM ATD Battle Management Advanced Technology Demonstration 作战管理先进技术演示
BM/C^3 Battle Management,Command, Control and Communications 作战管理与指挥、控制和通信
BM/C^3 Battlefield Management/Command, Control and Communications 【美国】战场管理与指挥、控制和通信（系统）
BMAAT Battle Management Architecture Analysis Tool 作战管理体系分析工具
BMAG Body-Mounted Attitude Gyro 导弹姿态陀螺仪
BMAR Ballistic Missile Acquisition Radar 弹道导弹搜索雷达
BMB Base Maintenance Building 基地维修厂房
BMC Ballistic Missile Center 弹道导弹中心
BMC Battle Management Center 作战管理中心
BMC^2 Battle Management, Command and Control 战斗管理、指挥与控制
BMC^{3}I Battle Management Command, Control, Communications and Intelligence 作战管理指挥、控制、通信与情报
BMC^{4}I Ballistic Missile Management Command, Control, Communications, Computers and Intelligence 弹道导弹管理指挥、控制、通信、计算机与情报
BMC^{4}I Battle Management Command, Control, Communications, Computers and Intelligence 作战管理指挥、控制、通信、计算机与情报
BMC^{4}ISR Battle Management C^{4}I Surveillance and Reconnaissance

作战管理指挥、控制、通信、计算机、情报、监视和侦察

BMCS Backup Master Control Station 备用主地面控制站

BMD Ballistic Missile Defense 弹道导弹防御

BMD Ballistic Missile Development 弹道导弹研发（制）

BMDAT Ballistic Missile Defence Advanced Technology 先进弹道导弹防御技术

BMDATC Ballistic Missile Defence Advanced Technology Center 【美国】先进弹道导弹防御技术中心

BMDATP Ballistic Missile Defence Advanced Technology Program 先进弹道导弹防御技术项目

BMDC Ballistic Missile Defence Center 弹道导弹防御中心

BMDCC Ballistic Missile Defense Command/Control Center 弹道导弹防御指挥与控制中心

BMDCM Ballistic Missile Defense Countermeasure 弹道导弹防御对抗措施

BMDCP Ballistic Missile Defense Command Post 弹道导弹防御指挥所

BMDD Ballistic Missile Defense Division 弹道导弹防御处

BMDM Ballistic Missile Defense Monitor 弹道导弹防御监视器

BMDN Ballistic Missile Defense Network 弹道导弹防御网络

BMDO Ballistic Missile Defense Organization 弹道导弹防御组织

BMDOC Ballistic Missile Defense Operations Center 弹道导弹防御作战中心

BMDOICA BMDO Independent Cost Assessment 弹道导弹防御组织的独立成本评估

BMDP Ballistic Missile Defense Program 弹道导弹防御计划

BMDS Ballistic Missile Defense System 弹道导弹防御系统

BMDSAS Ballistic Missile Defense System Architecture Study 弹道导弹防御系统体系结构研究

BMDSCM Ballistic Missile Defense System Command 弹道导弹防御系统司令部

BMDSS Ballistic Missile Defense Support System 弹道导弹防御保障系统

BMDTP Ballistic Missile Defense Test Program 弹道导弹防御试验计划

BME Ballistic Measurement Equipment 弹道测试设备

BMEP Brake Mean Effective Pressure 制动器平均有效压力

BMEW Ballistic Missile Early Warning 弹道导弹预警

BMEWL Ballistic Missile Early Warning Line 弹道导弹预警线

BMEWS Ballistic Missile Early Warning Satellite 【美国】弹道导弹预警卫星

BMEWS Ballistic Missile Early Warning Station 弹道导弹预警站

BMEWS Ballistic Missile Early Warning System 弹道导弹预警系统

BMEWSRCS Ballistic Missile Early Warning System Rearward Communications System 弹道导弹预警系统后向通信系统

BMEWSTP Ballistic Missile Early Warning System Test Procedure 弹道导弹预警系统试验程序

BMFT German D-1 Spacelab Payload

德国 D-1 空间实验室有效载荷

BMI Ballistic Missile Interceptor 弹道导弹截击机

BMIC Battle Management Integration Center 作战管理一体化中心；作战管理综合中心

BMIC Battlespace Management Information Center 作战空间管理信息中心

BML Ballistic Missile Launcher 弹道导弹发射装置

BMLC Basic Mode Link Control 基本型数据链控制

BMLC Basic Multiline Controller 基本多路控制器

BMMC Ballistic Missile Management Complex 弹道导弹管理成套设备；弹道导弹管理综合设施

BMMD Body Mass Measuring Device 物体质量测量装置

BMRA Basic Multirole Avionics 基本多用途电子设备与控制系统

BMRS Ballistic Missile Reentry System 弹道导弹再入系统

BMS Background Measurement Satellite 背景测量卫星

BMT Ballistic Missile Threat 弹道导弹威胁

BMT Ballistic Missile Transport 弹道导弹运输车

BMTD Ballistic Missile Terminal Defense 弹道导弹终端防御

BMTOGW Basic Mission Takeoff Gross Weight 基本（战斗）任务起飞总重（量）

BMTS Ballistic Missile Target System 弹道导弹目标系统

BMWS Ballistic Missile Weapon System 弹道导弹武器系统

BNNT Boron Nitride Nanotubes 氮化硼纳米管

BNS Bombing and Navigation System 轰炸与导航系统

BOA Battlefield Ordnance Awareness 战场军械掌握情报

BOA-MILS Broad Ocean Area-Missile Impact Locating System 远洋区-导弹落点定位系统

BOJ Booster Jettison 助推器抛掉（分离）；助推器抛投

BOMIS Bottom Mounted Instrumentation System 安装在底部的仪表系统

BOMO Bomb or Missile Optics 炸弹或导弹光学设备

BOMOK Boundary Marker, instrument landing system, resumed operation 仪表着陆系统边界指点标恢复工作（代号）

BOOSTERKT Booster Rocket 助推火箭

BOPACE Boeing Plastic Analysis Capability for Engines 波音公司发动机塑性性能分析

BORRG Ballistic Missile Operational Requirements Review Group 弹道导弹操作要求评审小组

BOS Battlefield Operating System 战场操作系统

BOSP Bioastronautics Orbital Space Program 【美国空军】生物航天轨道空间计划；航天生物研究轨道运行计划

BOSS Background Optical Suppression Sensor 背景光学抑制传感器

BOSS Ballistic Offensive Suppression System 弹道攻击压制武器系统

BOSS Basic Object Simulation System 基本对象模拟仿真系统

BOSS Bioastronautical Orbiting Space

Station 航天生物研究轨道运行空间站

BOSS Biological Orbiting Space Station 生物轨道空间站；沿轨道运行的生物学空间站

BOSS BMEWS Operational Simulation System 弹道导弹预警系统工作模拟系统

BOSS Bomb Orbital Strategic System 轨道轰炸战略系统

BOSS Broad Ocean Scoring System 【美国】大面积海域（导弹或航天器）落区确定系统

BOT Burst On Target 击中目标爆炸

BP Biopack 宇宙飞行器中的生物容器或生物舱

BPA Bioshield Power Assembly 【美国国家航空航天局】生物屏蔽动力装置

BPBM Boost Phase Battle Management 助推阶段作战管理

BPC Boost Protective Cover-launch cover used on Apollo CMs 助推覆盖保护层－用于"阿波罗"指令舱的发射保护

BPDMS Basic Point Defense Missile System 基本点防御导弹系统

BPDSMS Basic Point Defense Surface Missile System 基点防御水面导弹系统；基点防御地面导弹系统

BPHIT Brilliant Pebbles Hover Interceptor Test 智能卵石悬停拦截机（导弹）试验

BPI Boost Phase Intercept 助推阶段截击

BPI Boost Phase Interceptor 助推阶段拦截器

BPI/EI Boost Phase Intercept/Exoatmospheric Intercept 助推阶段截击/外大气层截击

BPI-API Boost Phase Intercept-Ascent Phase Intercept 助推阶段截击－上升阶段截击

BPIM Boost Phase Intercept Munitions 助推段拦截弹药

BPJ Booster Package Jettison 助推器组件投射

BPL Boost Phase Leakage 助推段泄漏

BPL Burst Position Locator 爆炸定位器

BPPBS Biennial Planning, Programming and Budgeting System 两年制计划、规划与预算系统

BPRRA Baseline Production Readiness Risk Assessments 基线生产成熟度风险评估

BPS Booster Pressurization System 助推火箭增压系统

BPS Booster Propulsion System 助推器推进系统

BPSS Biopack Subsystem 【宇宙飞行器中】生物容器子系统

BPT Boost Phase Tracking 助推段跟踪

BPT ATD Boost Phase Tracking Advanced Technology Demonstration 助推阶段跟踪先进技术演示

BPTS Boost Phase Tracking System 助推阶段跟踪系统

BPUP Ballistic Protection and Upgrade Package 弹道防护升级包

BPV Biopropellant Valve 生物推进剂阀

BQT Block Qualification Testing 批次鉴定试验

BR Boost Reliability 助推器可靠性

BR Booster Rocket 助推火箭

BR Booster-Regulator 【美国国家航空航天局】助推调节器

BR/RL Bomb Rack/Rocket Launcher 炸弹架/火箭发射器

BRAC Base Realignment and Closure 基地调整与关闭

BRAC Bomb Release Angle Computer 投弹角度计算机

BRAMS Beacon Range Altitude Monitor System 信标距离高度监控系统

BRASS Battlefield Robotic Ammunition Supply System 战场机器人弹药补给系统

BRAVO Business Risk and Value of Operation in Space 空间业务的商业风险与价值

BRAZILSAT Brazilian National Communications Satellite 巴西卫星（巴西国家通信卫星）

BRB Ballistic Recoverable Booster 弹道回收助推器；可回收的弹道式助推器

BRCS Basic Reference Coordinate System 基础参照坐标系统

BRCS Bearing and Range Computer System 方位与距离计算机系统

BRCU Baie de Repartition Charge Unit 有效载荷互联舱

BRCU Booster Remote Control Unit 助推火箭遥控装置

BRET Bistatic Reflected Energy Target 双静态反射能力目标

BRI Bearing and Range Indicator 方位与速度指示器

BRICU Bomb Release Interval Control Unit 投弹间隔控制器

BRL Ballistics Research Laboratory 弹道学研究实验室

BRLTR Ballistic Research Laboratories, Transonic Range 跨声速（范围）弹道研究实验室

BRM Biological Research Module 生物研究舱

BRV Ballistic Reentry Vehicle 再入（大气层）弹道飞行器

BSA Body Surface Area 体表面积

BSD Battlefield Situation Display 作战状态显示

BSDP Booster Stage Discharge Pressure 助推器级排放压力

BSF Bulk Shielding Facility 整体屏蔽设备

BSLSS "Buddy" Secondary Life Support System "伙伴"辅助生命保障系统

BSM Belgian Soyuz Mission 比利时的"联盟号"任务

BSM Booster Separation Motor 助推器分离发动机

BSP BMD Signal Processor 弹道导弹防御系统信号处理器

BSPR Boost/Sustainer Pressure Ratio (Rocket) 助推器/主发动机压力比；助推/主发压比（火箭）

BSRM Booster Solid Rocket Motor 助推器固体燃料火箭发动机

BSS Battlefield Sensor Simulator 战场传感器模拟器

BSS Boeing Satellite Systems 【美国】波音公司人造卫星系统

BSTRKT Boost Rocket 助推火箭

BSTS Boost Surveillance and Tracking System 助推器监测与跟踪系统

BSU Baseband Separation Unit 基带间隔（分离）装置

BT Burn Time (Engine) 发动机燃烧时间

BT ATD Booster Typing Advanced Technology Demonstration 助推器的先进技术演示验证

BTES Ballistic Test Evaluation and

Sealing 弹道试验评估与确认
BTF Ballistic Test Facility 弹道试验设施
BTH Below the Horizon 地平线以下
BTM Booster Tumble Motors 助推器翻转发动机
BTOC Battalion Tactical Operations Center 营战术操作中心
BTP BMEWS (Ballistic Missile Early Warning System) Test Procedure 弹道导弹预警系统试验程序
BTS Biotelemetry System 生物遥测系统
BTS Booster Technology Simulator 助推器技术模拟器
BTTV Ballistic Tactical Target Vehicle 弹道战术目标飞行器
BTV "Battleship" Test Vehicle 【美国】"战舰"试验飞行器
BTV Blast Test Vehicles 爆破试验飞行器
BTWC Biological and Toxin Weapons Convention 传统生物与毒素武器
BUC Backup Fuel Control 备用燃料控制
BUR Backscatter Ultraviolet Radiometer 反向散射紫外辐射仪
BV Booster Validation 助推器验证
BVP Booster Vacuum Pump 助推器真空泵
BVPS Booster Vacuum Pump System 助推器真空泵系统
BVR Beyond Visual Range 超视距
BVRAAM Beyond Visual Range Air-to-Air Missile 超视距空－空导弹
BW Biological Weapons/Biological Warfare 生物武器/生物战
BWBLSV Blended Wing Body Low Speed Vehicle (X-48B) 翼身融合体低速飞机
BZ Beaten Zone 命中地带；着弹地带
BZI Beam Zero Indication 波束零指示

C

C Spec Product Specification 产品说明

C&C Command and Control 指挥与控制

C&C Communications and Control 通信与控制

C&CE Command and Control Electronics 指令与控制电子设备

C&D Command and Decision 指挥与决策

C&D Construction and Demolition 建造与拆除

C&D Control and Display 控制与显示

C&D/A Command and Decision/Auxiliary 指挥与决策/补充（附加）

C&DC Control and Display Console 控制与显示台

C&DH Command and Data Handler 指令与数据处理器；指令与数据处理程序

C&DH Command and Data Handling 指令与数据处理

C&DH Communication and Data Handling 通信与数据处理

C&DS Command and Data Simulator 指令与数据模拟器

C&EA Cause and Effect Analysis 因效分析

C&I Compatibility and Interoperability 兼容性与互操作性

C&M Care and Maintenance 维护保养

C&M Control and Monitoring 控制与监控

C&N Communication and Navigation 通信与导航

C&N Control and Navigation 控制与导航

C&SM Communication and System Management 通信与系统管理

C&ST Computing and Software Technology 计算与软件技术

C&T Communication and Telemetry 通信与遥测

C&T Communication and Tracking 通信与跟踪

C&TSS Communication and Tracking Subsystem 通信与跟踪分系统

C&W Caution and Warning 预防与告警

C&W Control and Warning 控制与告警

C&WS Caution and Warning System 预防与告警系统

C-ISA Centaur Interstage Adapter 【美国】"半人马座"上面级级间段适配器

C-MANPADS Counter-Man Portable Air Defense System 便携式对人空中防御系统

C-RISTA Counter-Reconnaissance, Intelligence, Surveillance and Target Acquisition 反侦察、情报、监视和目标搜索措施

C-to-C Computer-to-Computer 计算机至计算机

C/A Collision Assessment 撞击评估

C/AHRS Compass, Attitude Heading Reference System 【美国】罗盘、

姿态航向参考系统
C/C　Combustion Chamber　燃烧室
C/C　Command Control　指令控制
C/D　Control/Display　控制/显示
C/D　Countdown　倒数计时；逆计数
C/DMD　Configuration/Data Management Division　型号/数据管理部门
C/ELCMC　Communications/Electronics Life Cycle Management Command　通信/电子寿命周期管理指令
C/F　Center Frequency　中心频率
C/L　Closed Loop　闭合回路；闭环
C/NO　Carrier/Noise Spectral Density Ratio　载波/噪声谱密度比
C/O　Checkout　检验；检测；测试
C/O　Cutoff　关机；断开；中止
C/R　Command/Response　指挥/响应
C/R　Commutation Rate　交换率；切换率；转换率
C/S　Course and Speed　航向与速度
C/SCSC　Cost/Schedule Control Systems Criteria　成本/进度控制系统标准
C/W　Carrier Wave　载波
C/W　Caution and Warning　预防与告警
C^2　Command and Control　指挥与控制
C^2/CNI　Command and Control, Communications Navigation and Identification　指挥与控制、通信导航和识别
C^2BFMA　Command and Control Battlefield Function Mission Area　指挥、控制战场职能任务区
C^2BMC　Command and Control, Battle Management, and Communications　指挥、控制，战场管理与通信
C^2C^3S　CECOM Center for Command, Control, and Communications Systems　指挥、控制与通信系统的 CECOM 中心
C^2CAMP　Command and Control Concept and Management Plan　指挥、控制方案与管理计划
C^2CDM　Command and Control Core Data Model　指挥、控制核心数据模型
C^2FMO　Command and Control for Mobile Operations　机动作战的指挥、控制
C^2I　Command, Control, and Intelligence　指挥、控制与情报
C^2IS　Command and Control Information System　指挥、控制信息系统
C^2ISR　Command, Control, Intelligence, Surveillance and Reconnaissance　指挥、控制、情报、监视和侦察
C^2MAA　Command and Control Mission Area Analysis　指挥、控制任务区分析
C^2MADP　Command and Control Mission Area Development Plan　指挥、控制任务区研发计划
C^2MAMP　Command and Control Mission Area Materiel Plan　指挥、控制任务区材料计划
C^2MUVE　Command and Control Multiuser Virtual Environment　指挥和控制多用途虚拟环境
C^2P　Command and Control Planning　指挥、控制规划
C^2P　Command and Control Processor　指挥与控制处理器；指挥与控制处理程序
C^2S　Command and Control Support

指挥、控制保障

C^2S Command and Control Systems 指挥、控制系统

C^2S^2 Command, Control, and Subordinate Systems 指挥、控制与附属系统

C^2SC Command and Control Simulation Center 指挥与控制模拟中心

C^2Sims Command and Control Simulations 指挥与控制模拟

C^2SPR Command and Control System Program Review 指挥、控制系统项目评审

C^2STN Command and Control System Test Network 指挥与控制系统测试网络

C^2TAS Command and Control Timeline Analysis System 指挥与控制计时线分析系统

C^2TED C^2 Theater Exploitation Demonstration 指挥、控制战区开发(利用)演示验证

C^2V Command and Control Vehicle 【美国】指挥与控制车

C^3 Command, Control and Communications 指挥、控制与通信

C^3CM Command, Control, Communications and Countermeasures 指挥、控制、通信与对抗措施

C^3I Command, Control, Communications and Intelligence 指挥、控制、通信与情报

C^3I^2 Command, Control, Communications, Intelligence and Interoperability 指挥、控制、通信、情报与互操作性

C^3IC Coalition Coordination, Communications and Integration Center 联军协调、通信与一体化中心

C^3IEW Command, Control, Communications, Intelligence, and Electronic Warfare 指挥、控制、通信、情报与电子战

C^3IIT C^3I Integration Test C^3I 综合测试

C^3ISR Command, Control, Communications, Intelligence, Surveillance and Reconnaissance 指挥、控制、通信、情报、监视与侦察

$C^3ISR\&SS$ C^3ISR and Space Systems 指挥、控制、通信、情报、监视侦察和空间系统

C^3MP Command, Control, and Communications Master Plan 指挥、控制与通信主计划

C^3N Command, Control, Communication and Vehicle Navigation 指挥、控制与通信和运载器导航

C^3S Command, Control and Communications Segment 【美国】指挥、控制与通信段(国防气象卫星计划)

C^3S Command, Control, and Communications Systems 指挥、控制与通信系统

C^3S Command, Control, and Computer Systems 指挥、控制与计算机系统

C^3TED C^3 Theater Exploitation Demonstration C^3 战区开发(利用)演示验证

C^4 Command, Control, Communications and Computer 指挥、控制、通信与计算机

C^4CM Command, Control, Communication and Computer Counter-Measures 指挥、控制、通信与计算机对抗措施

C^4I Command, Control，Communications, Computer and Intelligence 指挥、控制、通信、计算机与情报

C^4ID Command, Control, Communications, Computers, and Intelligence Dissemination 指挥、控制、通信、计算机与情报通报演示

C^4IEW Command, Control, Communications, Computers, and Intelligence Electronic Warfare 指挥、控制、通信、计算机与情报电子战

C^4IFTW Command, Control, Communications, Computers, and Intelligence for the Warrior "勇士"指挥、控制、通信、计算机与情报（系统）

C^4IM Command, Control, Communications, Computers, and Information Management 指挥、控制、通信、计算机与信息管理

C^4ISR Command, Control, Communications, Computer, Intelligence, Surveillance and Reconnaissance 自动化指挥系统（指挥、控制、通信、计算机、情报、监视与侦察）

C^4KISR Command, Control, Communications, Computer, Kill, Intelligence, Surveillance and Reconnaissance 指挥、控制、通信、计算机、杀伤、情报、监视与侦察

C^4ME Command, Control, Communications, and Computers Mission Environment 指挥、控制、通信与计算机任务环境

C^4S Command, Control, Communication, and Computer Systems 指挥、控制、通信与计算机系统

CA Calibrated Altitude 校准高度；修正表高

CA Collision Avoidance 避撞

CA Compositional Analysis 成分分析；组成分析；结构分析

CA Cone Angle 圆锥角

CA Constant Attenuation 恒定衰减；衰减（减幅）常数

CA Contingency Abort 应急中止

CA Crab Angle 偏流修正角

CA Criticality Analysis 临界（状态）分析

CAA Collision Avoidance Aids 避撞辅助设备

CAA Computer Aids Analysis 计算机辅助分析

CAA Computer Aids Assemble 计算机辅助装配

CAA Controlled Access Area 通行管制区

CAA Crew Access Arm 航天员入舱臂

CAA Critical Assembly Area 关键组装区

CAAIRS Computer Aided Analysis and Information Recover System 计算机辅助分析与信息恢复系统

CAAR Compressed Air Accumulatory Rocket 压缩空气蓄能火箭发动机

CAAS Common Avionics Architecture System 通用电子设备与控制系统体系架构系统

CAAS Computer Aided Approach Spacing 计算机辅助进场（进近）间隔

CAAS Computer Assisted Approach Sequencing 计算机辅助进场（进近）程序

CAATD Crewman's Associate Advanced Technology Demonstration 乘员辅助先进技术演示

CAB Critical Air Blast 临界空中爆炸

CABS Cockpit Airbag System 座舱安全气袋系统；座舱气囊系统

CAC Centralized Approach Control 集中进场控制

CACC Communications and Configuration Console 通信与配置操纵台

CACL Computer Assisted Computer Language 计算机辅助计算机语言

CAD Computer Aided Design 计算机辅助设计

CADA Clear Air Dot Angle 净航向角

CADCI Common Air Defense Communications Interface 通用空防通信接口

CADE Combined Allied Defense Experiment 盟国联合防御实验

CADE Controller/Attitude-Direct Electronics 控制器/姿态指引电子设备

CADI Central Apollo Data Index 中央"阿波罗"飞船数据索引

CADMOS Centre d'Aide au Développement de la Micropesanteur et aux Opérations Spatiales 【法国】微重力操纵空间飞船辅助研发中心

CADP Critical Assignment Development Program 关键性分配与研发项目

CADS Central Attitude Determination System 中央姿态测定系统

CADS Centralized Air Defense System 集中防空系统

CADS Centralized Automatic Dependent Surveillance 集中自动相关性监视

CADS Command and Data Simulator 指令与数据模拟器

CADS Common Airborne Data System 通用机载数据系统

CADSI Communications and Data Systems Integration 通信与数据系统集成

CADSS Combined Analog-Digital Systems Simulator 模拟－数据组合系统模拟装置

CADU Control and Display Unit 控制与显示组件

CAE Computer-Aided Engineering 计算机辅助工程

CAEPE Center d'Achevement Etd'Essais de Propulseurs d'Engins 【法国】导弹推进装置总试验中心

CAESAR CONUS (Continental United States) Attack Engagement Systems Requirements Simulation 美国本土攻击交战系统需求模拟

CAFA Computer Aided Curve Fit and Analysis 计算机辅助曲线拟合与分析

CAFIT Computer Assisted Fault Isolation Test 计算机辅助故障隔离试验

CAFS Checkout and Firing Subsystem 检测与点火分系统

CAGC Clutter Automatic Gain Control 杂乱回波自动增益控制

CAGC Coded Automatic Gain Control 编码自动增益控制（电路）

CAGC Continuous Access Guided Communications 连续存取制导通信

CAGE Commercial Avionics GPS Engine 商业航空电子设备与控制系统GPS发动机

CAGE Computerized Aerospace Ground Equipment 计算机化航空航天地面设备

CAI Close-Controlled Air Interception 近距指挥空中截击

CAI Computer Aided Instruction 计算机辅助指令

CAIB Columbia Accident Investigation Board 【美国】"哥伦比亚号"航

天飞机事故调查委员会

CAIDS Computer Aided Interactive Design System 计算机辅助交互式设计系统

CAIG Cost Analysis Improvement Group 【美国国防部】成本分析改进小组

CAIL CEV Avionics Integration Laboratory 【美国】CEV 电子设备与控制系统集成实验室

CAIRS Computer Assisted Information Retrieval System 计算机辅助信息检索系统

CAIS Common Airborne Instrumentation System 通用机载测量系统

CAIT Computer Aided Instructional Training 计算机辅助指令培训

CAL Completely Assembled for Launch 为发射而总装

CAL Cornell Aeronautical Laboratory 【美国】康奈尔航空航天实验室

CALA Computer Aided Loads Analysis 计算机辅助载荷分析

CALCM Conventional Air Launched Cruise Missile 常规空射巡航导弹

CALCM-ER Conventional Air-Launched Cruise Missile-Enhanced Range 增程常规空射巡航导弹

CALET Calorimetric Electron Telescope 量能型电子望远镜

CALI-DALE Centralized Alarm Interface-Dependent Alarm Equipment 集中告警接口－相关告警设备

CALIPSO Cloud-Aerosol Lidar and Infrared Pathfinder Satellite Observation 云－气溶胶激光雷达与红外引导观测卫星

CALS Computer-Aided Acquisition and Logistics 计算机辅助采办与后勤保障

CALS Computer-Aided Acquisition and Logistics Support 计算机辅助采办与后勤保障

CALS Computer Aided Acquisition and Logistics System 计算机辅助采办与后勤保障系统

CALS Computer Aided Life Cycle Support 计算机辅助寿命周期保障

CALS Computer Aided Logistics Support 计算机辅助后勤保障；计算机辅助后勤支持

CALS Continuous Acquisition and Life-Cycle Support 持续采办与全寿命保障

CaLV Cargo Launch Vehicle 载货运载火箭

CAM Centrifuge Accommodation Module 离心生活舱

CAM Checkout and Automatic Monitoring 检测与自动监控

CAM Collision Avoidance Manoeuvre 避撞机动

CAM Computer Aided Manufacturing 计算机辅助制造

CAM Computer Annunciation Matrix 计算机公告矩阵

CAM Conventional Attack Missile 常规攻击导弹

CAMDEN Cooperative Air and Missile Defense Exercise Network 空防与反导合作训练网络

CAMEO Chemically Active Material Eject in Orbit 【美国】卡梅欧卫星（释放化学活性物质的卫星）

CAMMS Combined Arms Multi-Purpose Missile System 联合武器多用途导弹系统

CAMS Coastal Anti-Missile System 【美国】海岸反导弹系统

CAMS Combat Aviation Management

System 作战飞行管理系统
CAN Certification Analysis Network 认证分析网
CANDOS Communications and Navigation Demonstration on Shuttle 航天飞机通信和导航演示验证
CANE Computer-Assisted Navigation Equipment 计算机辅助导航设备
CAO Counter Air Operation 空防作战
CAOC Combined Air Operations Center 组合式空域操作中心
CAOS Completely Automatic Operational System 全自动操作系统; 全自动作战系统
CAOS Computer-Assisted Oscilloscope System 计算机辅助示波器系统
CAP CCMS Application Programs 指挥控制与监控系统应用项目
CAP Combat Air Patrol 空中作战巡逻
CAP Computer Application Program 计算机应用项目
CAP Configuration and Alarm Panel 配置与报警控制板
CAP Crew Activity Plan 乘员活动计划（航天员的工作和生活安排）
CAPCOM Capsule Communicator 座舱通话装置
CAPE Convective Available Potential Energy 对流性可利用潜能
CaPEE Convection and Precipitation/Electrification Experiment 对流与沉淀/起电实验
CAPPS Checkout, Assembly & Payload Processing Services 检测、装配与有效载荷操作服务
CAPRIS Combat Active/Passive Radar Identification System 战斗主动/被动雷达识别系统
CAPRS Compensation and Performance Review System 补偿与性能评审系统
CAPS Capsule 座舱；航天器密封舱；容器
CAPS Commanders Analysis and Planning Simulation 指挥官分析与规划模拟
CAPS Commanders Analysis and Planning System 指令长分析与计划系统
CAPS Common Attitude Pointing System 普通姿态定向系统
CAPS Corrective Action Problem System 纠正措施问题系统
CAPS Crew Activity Planning System 航天员活动计划系统
CAPS Crew Altitude Protection Suit 乘员高空防护服
CAPSEP Capsule Separation 【空间飞行中】座舱分离
CAPSIM Captive Simulation 捕获模拟
CAPWSK Collision Avoidance, Proximity Warning, Station Keeping (Equipment) 避撞、接近告警与位置保持（设备）
CAR Command Assessment Review 【美国空军】指挥评估评审
CAR Configuration and Acceptance Review 配置与验收评审
CAR Corrective Action Request 纠正措施请求
CARA Collision Avoidance Risk Assessment 避撞风险评估
CARID Customer Acceptance Review Item Disposition 用户验收评定项处理
CARL Compressor Aero Research Lab-

oratory 压缩机航空研究实验室
CARM Counter Anti-Radiation Missile 反反辐射导弹
CARP Computed Air Release Point 经计算的空投点
CARR Customer Acceptance Readiness Review 用户验收准备评审
CARS Collision Avoidance Radar Simulator 避撞雷达模拟器
CARS Combat Arms Regimental Systems 作战武器编组系统
CARS Common Automatic Recovery System (UAV Recovery) 【无人驾驶飞行器回收】通用自动回收系统
CARSRA Computer Aided Redundant System Reliability Analysis 计算机辅助冗余系统可靠性分析
CART Collision Avoidance Radar Trainer 避撞雷达训练器
CART Complete and Ready for Test 完成并准备测试
CARTS Common Automated Radar Terminal System 通用自动化雷达终端系统
CARVER Criticality, Accessibility, Recoverability, Vulnerability, Effect and Recognizability 【攻击目标的】紧要性、可接近性、可恢复性、易损性、（攻击）效果与（攻击目标的）可识别性
CAS Calibrated Airspeed 校正的大气速度
CAS Close Air Support 近距离空中支援
CAS Command Augmentation System 指令增大系统
CAS Commercially Available Software 商业化软件
CAS Compressed Air Storage 压缩空气贮存
CAS Computer-Aided Servicing 计算机辅助检修
CAS Control Actuation Section 【导弹】操纵面偏转段；操纵面偏转部分
CAS Control Augmentation System 控制增稳系统
CAS/M Computer-Aided Servicing/Maintenance 计算机辅助检修/维护
CASA Computer Aided Systems Analysis 计算机辅助系统分析
CASA Cost Analysis Strategy Assessment 成本分析策略评估
CASBAR Collision Avoidance System Using Baseband Refractometer 采用基带折射仪的防撞系统
CASC Combined Acceleration and Speed Control 加速度与速度综合控制
CASE Communications, Analysis, Simulation and Evaluation 通信、分析、模拟与评估
CASE Computer Aided Software Engineering 计算机辅助软件工程
CASE Coordinated Aerospace Supplier Evaluation 航空航天供应商协同鉴定
CASE Crew Accommodations and Support Equipment 航天员生活舱和保障设备；机组配套保障设备
CASES Checkout Atmospheric Science Experiment Set 检测大气科学实验装置
CASES Controls Astrophysics Structures Experiment in Space 空间控制天体物理结构实验
CASIS Center for the Advancement of Science in Space 美国空间科学促进中心

CASM　Close Air Support Missile　近距空中支援导弹

CASOM　Conventionally Armed Stand Off Missile　【美国空军】常规防区外导弹

CASP　CDS Application Support Programs　【美国国家航空航天局】控制与诊断系统应用保障项目

CASPR　Computer Aided Scheduling and Planning Resources　计算机辅助调度与规划资源

CASS　CITE Augmentation Support System　【美国国家航空航天局】货物（有效载荷）综合测试设备的增强型保障系统

CASS　Consolidated Automated Support System　强化型自动保障系统

CAT　Catapult/Cockpit Automation Technology　座舱自动化技术

CaT　Characterization and Transition Reviews　特性与过渡评审

CAT　Computer Aided Test　计算机辅助测试

CATETS　Counter Air Test and Evaluation Test Site　防空作战试验与评价试验场

CATLAS　Centralized Automatic Trouble Locating and Analysis System　集中化自动故障定位和分析系统

CATO　Catapult-Assisted Takeoff　弹射器助推起飞

CATO　Combined Arms Tactical Operations　【美国陆军】联合兵种战术行动

CATO　Common Automated Tactical Operations　通用自动化战术操作（行动）

CATS　Cheap Access To Space　以较低成本进入空间

CATS　Component Acceptance Test System　元器件验收测试系统

CATT　Combined Arms Tactical Trainer　【美国】联合兵种战术训练器

CATT　Computer Assisted Technology Transfer　计算机辅助技术转让

CAU　Command Acquisition Unit　指令采集组件

CAU　Command Activation Unit　指令激活组件

CAUSE　Computer Aided User Oriented System Evaluation　计算机辅助面向用户系统评估

CAV　Cumulative Absolute Velocity　累积绝对速度

CAWSS　Crisis Action Weather Support System　危机反应气象保障系统

CB　Center of Buoyancy　浮力中心

CB　Chemical and Biological　化学与生物

CB　Control Building　控制大楼

CBA　C-Band Transponder Antenna　C 波段应答天线

CBA　Cost Benefit Analysis　成本收益分析

CBC　Common Booster Core (Delta IV First Stage, Boeing)　通用助推芯级（波音公司"德尔它"4 火箭的一子级）

CBD　CINC (commander in chief) BM/C^3 Demonstrator　总司令的弹道导弹/C^3 模拟器

CBE　Cosmic Background Explorer　宇宙背景探测器

CBERS　China-Brazil Earth Resources Satellite　中国—巴西地球资源卫星

CBM　Central Battle Management　中央作战管理

CBM　Common Berthing Mechanism　通用停泊机械装置

CBM Continental Ballistic Missile 洲际弹道导弹

CBM Conventional Ballistic Missile 常规弹道导弹

CBPS Combined Braking/Correction Propulsion System 【俄罗斯 Phobos 探测器】修正－联合推进系统

CBR Chemical, Biological and Radiological 化学、生物与放射性的

CBR Configuration Budget Review 配置预算评审

CBRN Chemical, Biological, Radiological and Nuclear 化学、生物、放射性和核（的）；核生化与辐射（的）

CBS Complex Behavior Simulator 复杂行为模拟器

CBSA Cargo Bay Stowage Assembly 货舱存贮组件

CBSC Chinese Broadcasting Satellite Corporation 中国广播卫星公司

CBT Computer-Based Training 基于计算机的培训

CBU Calibrate Before Use 使用前校准

CBU Clustered Bomb Unit 集束炸弹

CBW Chemical and Biological Weapons 化学生物武器

CBX C-Band Transponder C 波段应答器

CC Change Course 改变航向

CC Channel Controller 频道控制器

CC Closed Cycle 闭循环

CC Combustion Chamber 燃烧室

CC Communications Control 通信控制

CC Crew Compartment 乘员舱

CC Crew(s) Certified 经过认证后的航天员

CC&D Camouflage, Concealment, and Deception 伪装、隐蔽与欺骗

CC&D Common Command and Decision 【美国海军】通用指挥与决策

CC/SOIF Command Center/System Operation and Integration Functions 指挥中心/系统操作与集成功能

CCA Carrier-Controlled Approach 航母控制空域

CCA Circuit Card Assembly 线路卡片装配

CCA Communication Cap Assembly 通信盖装置

CCA Communications Carrier Assembly 通信载波装置

CCA Component Cost Assessment 部件成本评估

CCA Contingency Capabilities Assessment 应急能力评估

CCA Coolant Control Assembly 冷却剂控制装置

CCAAFB Cape Canaveral Auxiliary Air Force Base 【美国】卡纳维拉尔角附属空军基地

CCAPS Computer-Controlled Atlas Pressurization System 计算机控制的"宇宙神"级增压系统

CCATS Command, Communication, and Telemetry System (Subsystem) 指挥、通信与遥测系统（分系统）

CCATS Communication, Command And Telemetry System 通信、指挥和遥测系统

CCB Command Control Block 指挥控制部件，指令控制块

CCB Common Core Booster (Atlas V First Stage, Lockheed/Martin) 通用芯级助推器（洛克希德·马丁公司"宇宙神"5 火箭的一子级）

CCB Component Cooling Building

元器件冷却厂房

CCC Central Computer Complex 中央计算机整套设备

CCC CINC Command Complex 总司令指挥区

CCC Command and Control Center 指令与控制中心

CCC Command Control Communication System 指挥控制通信系统

CCC Communication(s) Control Console 通信控制操纵台

CCC Complex Control Center 综合控制中心

CCC Component Command Center 分队指挥中心

CCC Consolidated Command Center 统一指挥中心

CCC Controller Checkout Console 控制人员检测操纵台

CCCA Close Combat Capabilities Analysis 近距离作战能力分析

CCCI Command, Control, Communications and Intelligence 指挥、控制、通信与情报

CCCS Common Communications Component Set 通用通信设备

CCCU Crew Compartment Cooling Unit 航天器乘员舱冷却装置

CCD Camouflage, Concealment, and Deception 伪装、隐蔽与欺骗

CCD Charge Coupled Devices 电荷耦合器件

CCD Charged Coupled Device 电荷耦合器件

CCD Checkout Command Decoder 检查指令解码器

CCD Configuration Control Document 配置控制文件

CCDC Central Control and Display Console 中央控制与显示台

CCDH Command, Control and Data Handling 指挥、控制和数据处理

CCDP Commercial Crew Development Program 商业乘员开发项目

CCDR Contractor Critical Design Review 承包商关键性设计评审

CCDS Center for the Commercial Development of Space (Program) 【美国国家航空航天局】空间商业开发中心计划

CCE Charge Composition Explorer 【宇宙射线】电荷组成探测器

CCE Combined Cycle Engine 组合循环发动机

CCE Combustion Cycle Engines 燃烧循环发动机

CCE Command Center Element of the SDS (Strategic Defense Command) C^2E 战略防御司令部指挥控制部分的指挥中心

CCE Command Control Equipment 命令控制装置

CCEB Combined Communications-Electronics Board 盟军通信—电子设备委员会

CCEP Commercial COMSEC (Communications Security) Endorsement Program 商业通信安全确认程序

CCEV Command Center Experimental Version 控制中心实验样品

CCF Controller Configuration Facility 控制器配置设施

CCF Converter Compressor Facility 转换压缩设施

CCFF Cape Canaveral Forecast Facility 【美国】卡纳维拉尔角预报设施

CCGE Cold Cathode Gauge Experiment 【美国】"阿波罗"飞船上的冷阴极真空规实验

CCI Crew Command Interface 乘员

指令接口
CCIC Command Control Interface Computer 指令控制接口计算机
CCICap Commercial Crew Integrated Capability 商业乘员综合能力
CCIDES Command and Control Interactive Display Experimentation System 指挥控制交互显示试验系统
CCIL Continuously Computing Impact Line 连续计算弹着线
CCIM Command Computer Input Multiplexer 指挥计算机输入多路调制器
CCIP Continuously Computed Impact Point 连续计算弹着点
CCIS Command and Control Information System 指挥与控制信息系统
CCITSE Common Criteria for Information Technology Security Evaluation 信息技术安全评估通用标准
CCL Closed Circuit Loop 闭合电路回路
CCL Configuration Control Logic 配置控制逻辑
CCLK Centre de Contrôle et de Lancement de Kourou 【法国】库鲁发射基地发射控制中心
CCLMAA Close Combat Light Mission Area Analysis 近距作战轻型任务区分析
CCLS Computer Controlled Launch Set 计算机控制的发射设备
CCLSPR Close Combat Light Systems Program Review 近距作战轻型系统项目评审
CCM Centre de Contrôle de Mission 【法国】任务控制中心
CCM Configuration Control Module 配置控制模块；构型控制模块

CCM Controlled Carrier Modulation 控制载波调制
CCM Crew/Cargo Module 乘员/货物舱
CCMA Crew Correctable Maintenance Action 乘员可修正的维护活动
CCMC Community Coordinated Modeling Center 公共协调建模中心
CCMPS Counter-Countermeasure Parametric Study 反对抗参数研究
CCMS Checkout, Control, and Monitor Subsystem 【美国肯尼迪航天中心发射操作系统】检测、控制与监控分系统
CCMS Close Combat Missile System 近战导弹系统
CCMS Command Control and Monitor System 指挥控制与监控系统
CCMWG Common Cost Methodology Working Group 通用成本研究方法工作小组
CCOC Combustion Chamber Outer Casing 燃烧室外壳
CCOH Corrosive Contaminants, Oxygen, and Humidity 腐蚀性污染物、氧气和湿度
CCP Center Console Panel 中央操纵台控制屏
CCP Charge Capacitance Probe 电荷电容探测器
CCP Closed Cherry Picker 【美国航天飞机】封闭式升降舱
CCP Commercial Change Proposal 商业更改建议
CCP Commercial Crew Program 商业乘员项目
CCP Computer Control Panel 计算机控制操纵台
CCP Configuration Change Point 配置更改点

CCP Configuration Control Panel 配置控制操纵台

CCP Configuration Control Phase 配置控制阶段

CCP Cost Control Program 成本控制项目

CCPAVS Computer Controlled Pressure And Venting System 计算机控制的增压和排气系统

CCPE Cooperative Convective Precipitation Experiment 协同式对流与沉淀实验

CCPS Central Control Processing System 中央控制处理系统

CCR Central Control Room 中央控制室

CCR Configuration Change Request 配置更改请求

CCRF Consolidated Communication Recording Facility 合并通信记录设备

CCRM Continuous Cost-Risk Management 成本—风险连续性评估

CCRP Continuously-Computed Release Point 连续计算的投放点

CCRS Central Command Remoting System 中央指令远程系统

CCS Central Control Section 中央控制段

CCS Central Control Station 中央控制站

CCS Charge Control System 装药控制系统

CCS Combat Control System 作战控制系统

CCS Command and Communication System (Subsystem) 指挥与通信系统（分系统）

CCS Command and Control System 指挥与控制系统

CCS Commercial Communication Satellite 商业通信卫星

CCS Common Core System 通用芯级系统

CCS Communications Control Set 【美国】通信控制设备

CCS Complex Control Set 综合控制装置

CCS Computer Core Segment 计算机核心段

CCS Console Communication System 操纵台通信系统

CCS Contamination Control System 污染控制系统

CCSDS Consultative Committee for Space Data System 空间数据系统协商委员会（1982年成立的国际组织）

CCSI California Commercial Spaceport Inc. 【美国】加利福尼亚商业航天港公司

CCSIL Command and Control Simulation Interface Language 指挥与控制模拟接口语言

CCSMO Cape Canaveral Space Management Office 【美国】卡纳维拉尔角空间管理办公室

CCSS Canada Centre for Space Science 加拿大空间科学中心

CCT Command Control Transmitter 指挥控制发射装置

CCTC Command and Control Technical Center 指挥控制技术中心

CCTDE Compound Cycle Turbine-Diesel Engine 复合式周期涡轮—柴油发动机

CCTF Centaur Cryogenic Tanking Facility 【美国】"半人马座"上面级低温储罐设备

CCU Central Control Unit 中央控制

装置
CCU Communication Control Unit 通信控制装置
CCU Communications Carrier Unit 通信载波装置
CCU Crew Communications Unit 乘员通信装置
CCU Crewman Communications Umbilical 乘员通信脐带
CCV Chamber Coolant Valve 推力室冷却剂阀门
CCV Control Configured Vehicle 控制配置运载器
CCVA Chamber Coolant Valve Actuator 舱冷却剂阀门致动器
CCW Command and Control Warfare 指挥与控制战
CCWS Component Cooling Water System 元器件冷却水系统
CD Coefficient of Drag 牵引系数
CD&SC Central Distribution and Switching Center 中央分布与转换中心
CD&SC Communication Distribution and Switch Center 通信分配与转换中心
CD/SD Command Destruct/Self-Destruct 指令自毁/预编程自毁
CD/V Concept Demonstration/ Validation 方案论证/鉴定阶段
CDA Central Design Activity 【美国空军软件工程中心】中央设计局
CDA Command and Data Acquisition 指挥与数据采集
CDA Core Depot Assessment 核心后方评估
CDA Critical Design Audit 关键设计审查
CDAM Centralized Data Acquisition Module 集中式数据采集模块
CDAM Common Data Acquisition Module 通用数据采集模块
CDB Cast Double Base (Rocket Propellant) 浇铸双基（火箭推进剂）
CDB Central Data Buffer 中央数据缓冲装置
CDB Command Data Base 指令数据库
CDBFR Common Data Buffer 通用数据缓冲装置
CDCE Central Data Conversion Equipment 中央数据转换装置
CDCE Cosmic Dust Collection Experiment 宇宙尘收集实验
CDCF Cosmic Dust Collection Facility 宇宙尘收集设施
CDCR Contractor Drawing Change Request 承包商图纸更改请求
CDCS Central Data Collection System 中央数据收集系统
CDDS Command and Data System 指挥与数据系统
CDDT Countdown Demonstration Test 临射前倒数计时演示试验
CDEC Combat Development and Experimentation Center (or Command) 战斗力生成与实验中心（或指挥部）
CDF Cable Distribution Frame 电缆配线架
CDF Central Data Facility 中央数据设备
CDF Commercial Demonstration Flight 商业演示飞行
CDF Cool-Down Facility 冷却设施
CDF Core Damage Frequency 堆芯损坏频率
CDF Cumulative Distribution Function 累积分配功能
CDF&TDS Circuit Design, Fabrication, and Test Data Systems 回路设计、生产与测试数据系统

CDHC Command and Data Handling Console 指令与数据处理操纵台

CDHF Central Data Handling Facility 中央数据处理设备

CDI Classification, Discrimination, and Identification 分级、分类和认证

CDI Compressed Data Interface 压缩数据接口

CDI Course Deviation Indicator 航线偏离指示器

CDITS Command Destruct Independent Test Sets 指令摧毁独立测试装置

CDL Capability Demonstration Laboratory 能力演示验证实验室（美国航天器的地面测试设施）

CDL Centre de Lancement (Launch Center) 【法国】发射中心

CDL Centre de Lancement (Launch Control Building) 【法国】发射控制中心

CDL Command Definition Language 指令定义语言

CDM Corona Diagnostic Mission 日冕探测任务

CDMC Coupling Display Manual Control-IMU (Inertial Measurement Unit) 惯性测量装置的耦合显示器手控

CDMLS Commutated-Doppler Microwave Landing System 转换式多普勒微波着陆系统

CDMRSF Commercially Developed Microgravity Research Space Facility 【美国国家航空航天局】商业开发空间微重力研究设施

CDMS Command and Data Management Subsystem 指令与数据管理分系统

CDMS Communication and Data Management System 通信与数据管理系统

CDMS Control and Data Management System 控制与数据管理系统

CDO Contingency Deployment Option 应急研发方案

CDOS Customer Data and Operations System 用户数据与操作系统

CDP Command Data Processor 指挥数据处理器

CDP Concept Demonstration Phase 方案演示验证阶段

CDPC Critical Data Processing Center 临界数据处理中心

CDPI Command, Data Processing and Instrumentation 指挥数据处理与仪表

CDPIS Command Data Processing and Instrumentation System 指挥数据处理与仪表系统

CDQR Critical Design and Qualification Review 关键设计与资质评审

CDR Command Destruct Receiver 自毁指令接收机

CDR Controlled Dynamic Range 受控动态范围

CDR Critical Design Review 关键设计评审

CDR AGC Command Destruct Receiver Automatic Gain Control 指令摧毁接收装置的自动增益控制

CDRA Carbon Dioxide Removal Assembly 二氧化碳清除组件

CDRR Component Design Requirements Review 元器件设计要求评审

CDRR Concept Demonstration and Risk Reduction 方案演示验证与风险降低

CDRS Command Destruct Receiver

System 指令自毁接收机系统
CDS　Capabilities Demonstration Satellite　性能示范卫星
CDS　Central Data Subsystems　【美国肯尼迪航天中心发射操作系统】中央数据分系统
CDS　Collision Detector System　撞击探测器系统
CDS　Command Destruct System　指令摧毁系统
CDS　Control Data System　控制数据系统
CDS　Control/Diagnostic System　【美国国家航空航天局】控制/诊断系统
CDS　Coronal Diagnostic Spectrometer　日冕诊断分光计
CDSC　Communications Distribution and Switching Center　通信分配与转换中心
CDSF　Combat Development Support Facility　作战开发保障设施
CDSF　Commercially Developed Space Facility　【美国】商业研发的空间设施
CDSS　Cockpit Door Surveillance System　座舱门监视系统
CDT　Compressed Data Tape　压缩数据带
CDT　Countdown Time　倒计时时间
CDU　Computer Display Unit　计算机显示装置
CDU　Control and Display Unit　控制与显示装置
CDU　Coupling Data Unit　耦合数据装置
CDU　Coupling Display Unit　耦合显示装置
CDUO　Coupling Display Unit-Optics　光学耦合显示装置

CDV　Cabin Depress Valve　座舱减压阀
CDW　Command Data Word　指挥数据命令
CDW　Computer Data Word　计算机数据命令
CE　Capability Enhanced　能力增强
CE　Change Evaluation　更改评估
CE　Combat Enhancement　作战强化
CE　Control Element　控制要素
CE&IS　Combined Elements and Integration Systems　组合要件与集成系统
CE&T　Common Environments & Tools　通用环境与工具
CE/D　Concept Exploration/Definition Phase　概念探索/定义阶段
CEA　Control Electronics Assembly　控制电子设备组件
CEAT　Centre d'Essais Aéronautiques de Toulouse　【法国】图卢兹航空试验中心
CEATM　Cost Effectiveness at the Margin　高效费比
CEC　Combined Environmental Chamber　组合环境燃烧室
CEC　Control Electronics Container　控制电子设备容器
CEC　Control Encoder Coupler　控制编码耦合器
CEC　Cooperative Engagement Capability　协同交战能力
CECLES　Centre Européen pour la Construction des Lanceurs et des Engins Spatiaux　欧盟运载火箭及空间装备发射中心
CED　Concept Exploration and Development　概念探索与研发
CEDAR　Computer-Aided Environmental Design, Analysis and Realiza-

tion 计算机辅助环境设计、分析与实现
CEDAR Coupling, Energetics and Dynamics of Atmospheric Regions 耦合、能量学与大气区动力学
CEDU Comprehensive Engine Diagnostic Unit 发动机综合诊断单元
CEEF Crew Escape Effectiveness Factor 乘员逃逸效力因数
CEEM Cost-Effectiveness Evaluation Model 费用效能评价模型
CEFA Cockpit Emulator for Flight Analysis 用于飞行分析的座舱仿真模拟器
CEGARS Combined Entry Guidance and Attitude Reference System 进入（大气层）综合制导与姿态基准系统
CEIAC Coastal Engineering Information Analysis Center 【美国】海岸工程信息分析中心
CEIT Cargo Equipment Integration Test 货物设备集成测试
CEIT Crew Equipment Integration Test 乘员设备集成测试
CEIT Crew Equipment Interface Test 乘员设备接口测试
CEL Critical Experiment Laboratory 临界试验实验室
CEL Cryogenics Engineering Laboratory 低温工程实验室
CELSA Cost Estimate Logistics Support Analysis 费用估算后勤保障分析
CELSS Closed Ecological Life Support System 密闭生态生命保障系统
CELSS Closed Environmental Life Support System 闭式环境生命保障系统
CELSS Controlled Ecological Life Support System 受控生态生命保障系统
CELSS Controlled Environment and Life Support System 受控环境与生命保障系统
CELV Complementary Expendable Launch Vehicle 【美国】补充性一次使用运载火箭
CEM Central Experiment Module 中央实验舱
CEM Contingency Evaluation Model 应急评估模式
CEM Crew Escape Module 乘员逃逸舱
CEMS Centre d'Etudes de la Météorologie Spatiale 【法国】空间气象研究中心
CEMS Comprehensive Engine Management System 发动机综合管理系统
CENTCOM Central Command 中央司令部
CEO Centre for Earth Observation 地球观测中心
CEO Close Earth Orbit 近地球轨道
CEO Crew Earth Observations 乘员地球观测
CEOI Communications Electronics Operating Instructions 通信—电子设备操作指令
CEOS Committee for Earth Observation Satellites 【欧洲空间局】地球观测卫星委员会
CEP Circular Error Probable 圆概率偏差
CEP Common Engine Program 通用发动机项目
CEP Cooperative Engagement Processor 协同交战处理器

CEP Cylindrical Electrostatic Probe 【美国空间物理探测卫星】圆柱形静电探测器

CEPE Cabin External Payload Equipment 实验舱的外部载荷设备

CEPF Columbus External Payload Facility 【美国航天飞机】外有效载荷设施

CEPMMT Centre Européen de Prévision Météorologiques à Moyen Terme 【法国】欧洲中期气象预报中心

CEPS Command Module Electrical Power System 指挥舱电源系统

CEPSARC Concepts Evaluation Program Schedule and Review Committee 方案评估项目进度与审核委员会

CER Capability Evaluation Review 性能鉴定审查

CERCS Centralized Engine Room Control System 集中发动机室控制系统

CERES Clouds and the Earth's Radiant Energy System 云与地球辐射能系统

CERFACS Centre Européen de Recherche et de Formation Avancée en Calcul Scientifique 【法国】欧盟计算科学发展研究中心

CERS Centre Européen de Recherche Spatiale 【法国】欧洲航天研究中心

CERT Certification 认证；鉴定；确认

CERT Combined Environmental Reliability Test 综合环境可靠性测试

CERT Crew Emergency and Rescue Technology 航天器乘员应急与救援技术

CERV Crew Emergency Rescue Vehicle 乘员应急营救飞行器；乘员组紧急救援车

CERV Crew Emergency Return Vehicle 乘员应急返回飞行器

CERV Crew Escape and Reentry Vehicle 乘员逃逸与再入（大气层）飞行器

CES Control Electronics System 控制电子设备系统

CES Crew Escape System 乘员逃逸系统

CESAR Capsule Escape and Survival Applied Research 太空舱逃逸与救生应用研究

CESC Canadian Space Agency Engineering Support Center 加拿大航天局工程支持中心

CESR Centre d'Etude Spatiale des Rayonnement 【法国】空间辐射研究中心

CET Central Earth Terminals 中央地面终端（欧洲数据中继卫星的中央终端）

CET Combustor Exit Temperature 燃烧室出口温度

CET Crew Escape Technology 乘员逃生技术

CET Critical Experiment Tank 临界实验箱

CETA Crew and Equipment Transfer Assembly 乘员与设备转移装置

CETA Crew and Equipment Transfer Aid 乘员与设备转换辅助装置

CETA Crew Equipment Translation Aid 【美国空间站】乘员设备移动装置

CETEX Contamination by Extraterrestrial Exploration 外空探索造成的污染

CETI Communication with Extra-Terrestrial Intelligence 与外星智能联络

CEU Control Electronics Unit 控制电子设备装置

CEU Cooling Equipment Unit 冷却设备装置

CEV Centre d'Essais en Vol 【法国】航天飞行器试验中心

CEV Crew Exploration Vehicle 【美国国家航空航天局】乘员探索飞行器

CF Centrifugal Force 离心力

CF Conversion Factor 转换因数

CF Critical Field 临界场

CFA Centaur Forward Adapter 【美国】"半人马座"上面级前部适配器

CFA Controlled Firing Area 受控发射区

CFAC Composite Fabrication and Assembly Center 【美国】复合材料制造与装配中心

CFD Computational Fluid Dynamics 计算流体力学

CFDIU Centralized Fault Display Interface Unit 集中式故障显示接口装置

CFDM Constant Factor Delta Modulation 常量因子增量调制

CFDP Composite Flight Data Processing 综合飞行数据处理

CFDS Central Fault Display System 故障中央显示系统

CFE Commercial Equivalent Equipment 商业等效设备

CFE Computational Fluid Effects 计算流体效应

CFES Continuous Flow Electrophoresis System 连续流动电泳系统

CFES Continuous Flow Electrophoresis System (NASA) 【美国国家航空航天局航天飞机上的）连续流电泳系统

CFET Centrifuge-Based Flight Environment Trainer 基于离心机的飞行环境训练器

CFF Columbus Free Flyer 【欧洲】"哥伦布"自由飞行器

CFF Critical Fusion Frequency 临界融合频率

CFFT Critical Flicker Fusion Threshold 临界闪融阈值

CFHF Collision-Free Hash Function 非碰撞散列函数

CFI&I Center for Integration and Interoperability 综合与互通性中心

CFIMS Center for Integrated Mission Support 综合任务保障中心

CFIT Controlled Flight into Terrain 可控飞行撞地

CFJ Centrifugal-Flow Jet 离心射流

CFLR Centaur Forward Load Reactor 【美国】"半人马座"上面级前部载荷反应器

CFLSE Critical Fluid Light Scattering Experiment 【美国国家航空航天局】临界液体光散射实验

CFM Cryogenic Fluid Management 低温流体管理

CFP Conceptual Flight Profile 概念性飞行剖面

CFRP Carbon Fiber Reinforced Plastics 碳纤维增强塑料

CFRP Carbon Fiber Reinforced Polymer 碳纤维增强复合材料

CFS Cryogenic Fluid Storage 低温液体储存

CFSC Cryogenic Fluid Storage Container 低温液体储存器

CFT Captive Flight Trainer 系留飞行训练器

CFT　Cold Flow Test　冷流试验
CFT　Common Facilities Test　通用设施测试
CFTB　Control Flight Test Bed　控制飞行试验台
CFTM　Captive Flight Test Missile　系留飞行试验导弹
CFTS　Captive Firing Test Set　系留点火试验装置
CFTS　Computerized Flight Test System　计算机化飞行试验系统
CFU　Colony Forming Unit　晶团形成装置
CFV　Composite Flight Vehicle　组合飞行器
CG　Center of Gravity　重心
CG　Cloud-to-Ground　云对地
CGAU　Cabin Gas Analysis Unit　座舱气体分析装置
CGC　Command Guidance Computer　指挥制导计算机
CGC　Critical Grid Current　临界栅极电流
CGCD　Crossed Grid Charge Detector　十字栅极电荷探测器
CGF　Crystal Growth Furnace　晶体生长炉
CGG　Contingency Gravity Gradient　应急重力倾斜度
CGL　Controlled Ground Landing　受控着陆
CGLSS　Cloud-to-Ground Lightning Surveillance System　云－地雷电监视系统
CGM　Mace Missile　压缩液态毒气导弹；梅斯毒气导弹
CGP　Central Grounding Point　中心接地点
CGRO　Compton Gamma Ray Observatory　【美国】"康普顿"伽马射线观察卫星
CGS　Centaur Guidance System　【美国】"半人马座"制导系统
CGS　Combined Guidance System　联合制导系统
CGS　Command Guidance System　指令制导系统
CGSE　Common Ground Support Equipment　通用地面保障设备
CGSS　Cryogenic Gas Storage Subsystem　低温气体存贮分系统
CGSS　Cryogenic Gas Storage System　低温气体储存系统
CGV　Critical Grid Voltage　临界栅极电压
CHAALS　Common High-Accuracy Airborne Location System　通用高精度机载定位系统
CHAALS　Communications High Accuracy Airborne Location System　高精度机载定位通信系统
CHAIN　Compartmented High Assurance Information Network　分隔式信息高保证网络
CHALS-X　Communications High Accuracy Location System Exploitable　可利用的高精度定位通信系统
CHAMP　Comet Halley Active Monitoring Program　哈雷彗星动态监测计划
CHAMP　Composite High Altitude Maneuvering PBV　复合高空机动后助推器
CHARM　Composite High Altitude Radiation Model　复合高空辐射模型
CHASE　Coronal Helium Abundance Spacelab Experiment　【欧洲】空间实验室日冕氦丰度实验
CHCS　Cabin Humidity Control Sub-

system 座舱湿度控制分系统
CHeCS Crew Health Care System 机组（乘员）保健系统
CheMin Chemical Mineral Instrument 化学矿物组成探测仪器
CHET Commercial Heavy Equipment Transporter 商业重型设备运输车
CHI Computer-Human Interaction 人机相互作用；人机交互（对话）
CHI Computer/Human Interface 人/机接口
CHIS Center Hydraulic Isolation System 中心液压隔绝系统
CHOP Countermeasures Hands-On Program 干扰依次传递程序
CHS Common Hardware and Software 通用型软硬件
CHX Cabin Heat Exchanger 座舱热交换器
CI Center of Impact 平均弹着点；弹着点散布中心
CI Certification Inspection 认证检查
CI Configuration Inspection 配置检查
CI Configuration Item 配置项
CI Continuous Interlock 连续联运
CI Crew Interface 乘员接口
CI&I Compatibility, Interoperability, and Integration 兼容性、互操作性与集成
CI/LI Corrosion Inhibitor/Lubricity Improver 腐蚀抑制剂/润滑剂
CIA Control Interface Assembly 控制接口组件
CIAC Computer Incident Advisory Capability 计算机事故咨询能力
CIC Combat Information Center 战斗信息中心
CIC Command Interface Control 指挥接口控制
CIC Communications Interface Coordinator 通信接口协调装置
CIC Control and Information Center 控制与信息中心
CIC Crew Interface Coordinator 乘员接口协调员
CICAP Crew Integrated Capability 乘员综合能力
CICAS Computer Integrated Command and Attack System 计算机集成指挥与攻击系统
CICC Cargo Integration Control Center 货物集成控制中心
CICU Computer Interface Conditioning Unit 计算机接口调节装置
CICU Computer Interface Control Unit 计算机接口控制装置
CID Combat Identification 作战识别
CID Computer Interface Device 计算机接口装置
CID Controlled Impact Demonstration 受控撞击演示验证
CIDEX Cometary Ice and Dust Experiment 【美国】彗星冰与尘埃实验
CIDR Configuration Item Design Review 配置项设计评审
CIDR Critical Intermediate Design Review 关键中级设计评审
CIDS Combat Information Demonstration System 作战信息演示系统
CIDS Combine Identified Derivative Sets 综合参数识别装置
CIDS Control, Instrumentation and Diagnostic Systems 控制、测量与诊断系统
CIDS Critical Item Development Specification 关键项（部件）研发（制）说明
CIDSE Consolidated Integrated Development Support Environment 统

一的综合研发保障环境
CIE Communications Interface Equipment 通信接口设备
CIE Computer Interface Electronics 计算机接口电子设备
CIEL Certification and INFOSEC Engineering Laboratory 鉴定与信息保密工程实验室
CIES Community Information Exchange System 共同信息交换系统
CIF Central Instrumentation Facility 中心测量厂房
CIF Central Integration Facility 中心装配厂房
CIFS Computer-Interactive Flight Simulation 计算机交互式飞行模拟
CIG Celestial Inertial Guidance 天文惯性制导；天文惯性导航
CIG Computer Image Generator 计算机图像发生器
CIGSS Common Imagery Ground/Surface System 通用图像地面接口系统
CIGTF Central Inertial Guidance Test Facility 【美国】中央惯性制导试验设施
CII Commonality, Interchangeability and Interoperability (Concept) 【北约】通用性、互换性与相互适应性（概念）
CII Configuration Identification Index 构型识别标记
CIIMS Central Integrated Information Management System 中央集成信息管理系统
CIM Computer Input Multiplexer 计算机输入多路转接器
CIM Computer-Integrated Manufacturing 计算机一体化制造
CIMSS Cooperative Institute for Meteorological Satellite Studies 气象卫星研究合作协会
CIN Center Information Network 中央信息网络
CINC Combined Intelligence Center 盟军情报中心
CINS Cryogenic Inertial Navigating System 低温惯性导航系统
CIOTE Commander's Integrated Open System Technology Evaluator 指挥官集成开放系统技术鉴定器（装置）
CIP Coordinated Instrument Package 协调后的测量数据包
CIP Customer Integration Panel 用户集成操纵台
CIPS Cockpit Instrument Panel Space 座舱仪表板空间
CIR Cargo Integration Review 货物（有效载荷）综合评审
CIR Combustion Integration Rack 燃烧集成机架
CIR Configuration Inspection Report 配置检查报告
CIRHS Critical Items and Residual Hazards List 关键项及剩余危险物清单
CIRIS Completely Integrated Reference Instrumentation System 全集成基准测量仪表系统
CIRRI Cryogenic Infrared Radiance Instrumentation 【美国】低温红外辐射仪
CIRRIS Cryogenic Infrared Radiance Instrument for Shuttle 【美国国防部】航天飞机低温红外线放射设备
CIS Central Integration Site 中央集成区
CIS Communication Interface System 通信接口系统
CIS Communications Interface Shelter

通信接口掩蔽
CIS Component Identification Sheet 元器件识别单
CIS Coronographic Imaging System 日冕仪成像系统
CIS Coupled Impedance Synthesis 耦合阻抗合成
CIS Crew Information System 乘员信息系统
CIS Crickets In Space 空间碎片
CIS Cryogenic Inertial Sensor 低温惯性传感器
CIS Customer Information System 用户信息系统
CISR Climate Impacts of Space Radiation 空间辐射的气候影响
CISS Centaur Integrated Support Structure 【美国】"半人马座"上面级集成式支撑构件
CISS Centaur Integrated Support System 【美国】"半人马座"上面级集成式支撑系统
CISS Center for Information Systems Security 信息系统安全中心
CISS Consolidated Information Storage System 统一信息存储系统
CITE Cargo Integration and Test Equipment 货物综合与测试设备
CITE Cargo Integration Test Equipment 【美国】货物（有效载荷）综合测试设备
CITF CDOS Integration and Test Facility 用户数据和操作系统组装测试设施
CITIS Contractor Integrated Technical Information Service 承包商综合技术信息服务；综合技术信息部门承包商
CIU Communications Interface Unit 通信接口装置

CIU Computer Interface Unit 计算机接口装置
CIU Controller Interface Unit 控制器接口装置
CIU Coupler Interface Unit 耦合器接口装置
CIV Critical Ionization Velocity 临界电离速度
CIVT Cargo Interface Verification Test 货物（有效载荷）接口验证测试
CIWS Close Intercept Weapon System 近程防御系统
CIWS Close-In Weapon System 近战武器系统
CJC Cold Junction Compensator 冷接合补偿器
CKD Component Knocked Down 元器件拆卸
CL Closed Loop 闭合回路；闭环
CL Closed Loop Control Logic 闭环控制逻辑
CL Coefficient of Lift 提升系数
CLA Cushion Launch Area 有减震装置的发射区域
ClAP Climatic Impact Assessment Program 气候影响评估项目
CLASP Composite Launch of Spacecraft Program 航天器组合发射方案
CLASP Computer Language for Aeronautics and Space Programming 航空航天程序设计计算机语言
CLASS Closed Loop Artillery Simulation System 闭环武器模拟系统
CLASS Cross-Chain LORAN Atmospheric Sounding System 交叉链式罗兰大气探测系统
CLASS Large Area Soft X-ray Spectrometer 大区域软X射线光谱仪
CLC Critical Load Cycle 关键荷载

周期

CLCS Checkout & Launch Control System 测试与发射控制系统

CLCSE Center for Life Cycle Software Engineering 寿命周期软件工程中心

CLDST Closed Loop Dynamic Stability Test 闭合回路动态稳定性测试

CLE Command and Launch Equipment 指挥与发射设备

CLEM Cargo Lunar Excursion Module 载货登月舱

CLF Contingency Logistic Flight 应急保障飞行

CLI Command Line Interface 指令线路接口

CLIFS Coordination, Life, Interchangeability, Function and Safety 协调、寿命、互换性、功能与安全性

CLIP Combined Laser Instrumentation Package 组合式激光测量数据包

CLL Critical Load Line 临界负载线

CLM Circum Lunar Mission 绕月飞行任务

CLMC Central Logistics Management Center 中央后勤保障管理中心

CLOC Canadian Logistics Operation Center 加拿大后勤保障操作中心

CLS Capsule Launch System 舱发射系统

CLS (Vertical-) Cell Launch System （垂直一）竖井式（导弹）发射系统

CLS Cislunar Space 地月（轨道）之间的空间

CLS Closed Loop Stimulation 闭环模拟

CLS Command and Launch Station 指挥与发射站

CLS Command and Launch Subsystem 指挥与发射分系统

CLS Commercial Launch Services 商业发射服务

CLS Contingency Landing Site(s) 应急着陆场

CLS Contractor Logistics Support 承包商后勤保障

CLSS Closed Life Support System 闭（环）式生命保障系统

CLT Closed Loop Test 闭环测试

CLT Complete Loss of Thrust 推力全部损耗

CLU Command Launch Unit 指令发射装置

CLUSTER International Solar Terrestrial Physics Program 国际太阳地球物理项目

CLV Crew Launch Vehicle 乘员运载火箭

CLV-P Cargo Launch Vehicle-Piloted 货物运载火箭（有人驾驶的）

CM Cargo Management 货物管理

CM Center of Mass 质量中心

CM Command Module 【载人飞船】指令舱

CM Common Module 通用模块；公共舱

CM Communications Module 通信舱

CM Configuration Management 配置管理

CM Consumables Management 耗用性管理

CM Control Monitor 控制监控

CM Crew Module 乘员舱

CM Crew Module (also Apollo Command Module) 【美国国家航空航天局】乘员舱（亦指"阿波罗"指令舱）

Cm Pitching Moment Coefficient 俯

仰力矩系数
CM&S Communications Maintenance and Storage 通信维护与存贮
CM/LAS Crew Module and Launch Abort System 乘员舱与发射中止系统
CMA Configuration Management Accounting 配置管理统计
CMA&I Countermeasures, Assessments and Integration 对抗措施、评估与综合
CMACS Central Monitor and Control System 中央监视与控制系统
CMAM Commercial Middeck Augmentation Module 【美国航天飞机】中舱商业有效载荷增载舱
CMART Consolidated Missile Asset Reused for Targets 统一导弹设备目标再利用
CMAS Characterization of Microorganisms and Allergens in Spacecraft 航天器内微生物和过敏原的综合表征
CMAT Compatible Materials 兼容性材料
CMC Central Maintenance Computer 中央维护计算机
CMC Ceramic Matrix Composites 陶瓷基复合材料
CMC Checkout and Maintenance Status Console 检查与维修情况控制台
CMC Command Management Center 指令管理中心
CMC Command Module Computer 【美国国家航空航天局】"阿波罗"指令舱计算机
CMC Command Module Computer 指令舱计算机
CMC Crew Module Computer 乘员舱计算机
CMCC Central Mission Control Centre 【欧洲空间局】中央任务控制中心
CMD Command 指挥（权）；指令；信号；目标值
CMD Contamination Monitor Package 污染监控数据包
CMD DCDR Command Decoder 指令解码
CMDES Cruise Missile Defense Expert System 巡航导弹防御专家系统
CMDL (NOAA) Climate Monitoring and Diagnostics Laboratory 【美国海洋气象局】气候监控和诊断实验室
CMDS Cislunar Meteoroid Detection Satellite 月地空间流星探测卫星
CMDS Configuration Management Data System 配置管理数据系统
CME Coronal Mass Ejections 日冕物质抛射
CME Cryogen Management Electronics 低温管理电子设备
CMEST Cruise Missile Engagement Systems Technology 巡航导弹交战系统技术
CMEV Command Message Encoder/Verifier 指令信息编码器/检验器
CMF Command Management Facility 指令管理设施
CMG Control Moment Gyroscope 控制力矩陀螺仪
CMG Control Momentum Gyroscope 【国际空间站】控制力矩陀螺
CMG Cryosphere Monitoring Payload Group 低温层监控有效载荷组
CMGA Control Moment Gyro Assembly 控制力矩陀螺组件

CMIF Core Module Integration Facility 核心模块集成设施
CMIS Core Module Integration Simulator 核心模块集成模拟装置
CML Commercial MPS Lab 商用空间材料加工实验室
CMM Capability Maturity Model 【软件】能力成熟度模型
CMM-SW Capability Maturity Model-Software 能力成熟度模型（软件）
CMMCA Cruise Missile Mission Control Aircraft 巡航导弹任务控制飞机
CMMI Capability Maturity Model Integration 能力成熟度模型集成
CMN Crewman 乘员；航天员
CMOR Canadian Meteor Orbit Radar 加拿大陨星轨道雷达
CMOS Complementary Metal Oxide Semiconductor 互补性金属氧化物半导体
CMP Command Module Pilot 指令舱驾驶员
CMP Command Monitor Panel 指令监控器面板
CMRCS Command Module Reaction Control System 指挥舱反作用控制系统
CMRN Cooperative Meteorological Rocket Network 合作（或协同）气象火箭（探测）网
CMRR Common Mode Rejection Ratio 通用模式弹射比
CMRS Communication Moon Relay System 月球通信中继系统
CMS Command Management System 指挥管理系统
CMS Command Module Simulator 指挥舱模拟器
CMS Contingency Management System (Subsystem) 应急管理系统（分系统）
CMS Corrective Maintenance System 校正性维修系统
CMS-2 Cambridge Monitoring System 【美国IBM公司】剑桥监视系统
CMSI Checkout/Control and Monitor Subsystem Interface 检查/控制与监控分系统接口
CMSO Crew Medical Support Office 乘员医学保障办公室
CMT Crew Maintenance Time 乘员检修时间
CMTF Canadian Mockup and Training Facility 加拿大实体模型与训练设施
CMTI Celestial Moving Target Indicator 天空（或天体）活动目标指示器
CMTM Capsule Mechanical Training Model 【航天器】舱内机械训练模型
CMTM Communications and Telemetry 通信与遥测技术
CMTS Computerized Maintenance Test System 计算机维护测试系统
CMU Command Master Unit 指令控制单元
CMV Common Mode Voltage 通用模式电压
CMWS Common Missile Warning System 通用导弹告警系统
CNC Comissao Nacional de Atividades Espaciais 【巴西】国家空间活动委员会
CNC Critical Noncompliance 关键不符合项
CNE Communications Node Equipment 通信节点设备

CNES　Centre National d'Etudes Spatiales　法国国家空间研究中心

CNG　Compressed Natural Gas　压缩天然气

CNI　Communication Navigation Instrument　通信导航仪表

CNI　Communication, Navigation, Identification　通信、导航、识别（综合系统）

CNIP　C^{2}BMC Network Interface Processor　指挥、控制、战场管理与通信网络接口处理器

CNRL　Communications and Navigation Research Laboratory　通信与导航研究实验室

CNS　Central Nervous System　中枢神经系统

CNS　Computer Network System　计算机网络系统

CNSR　Comet-Nucleus Sample Return　彗核取样返回任务

CNT　Carbon Nanotubes　碳纳米管

CNTB　Communications Network Test Bed　通信网络试验（测试）台

CNWDI　Critical Nuclear Weapons Design Information　关键核武器设计信息

CO　Cargo Operations　货物（有效载荷）操作

CO　Checkout　检测；检验；测试；校验；调整

CO　Circular Orbit　圆形轨道

CO　Comet Orbit　彗星轨道

CO　Cutoff　关机；断开；中止

CO-OP　Control of Operations & Paths　运行与轨道控制

COA　Center Operations Area　中心操作区

COAMPS　Combined Ocean Atmosphere Mesoscale Prediction System　海洋大气中等规模联合预报系统

COAMPS　Coupled Ocean Atmosphere Mesoscale Prediction System　大气海洋耦合中尺度预报系统

COAS　Crew Optical Alignment Sight　【美国航天飞机】航天员光学瞄准观察器

COAS　Crewman Optical Alignment Sight　【美国"阿波罗"飞船】航天员光学瞄准观察器

COAST　Computer Operation, Audit, and Security Technology　计算机操作、审查与安全技术

COAT　Corrected Outside Air Temperature　修正的外界大气温度

COATS　CCSM (Command & Control System Module) Offhull Assembly and Test Site　指挥与控制系统模块脱机装配与测试场

COBA　Computer-Oriented Bearing Response Analysis　面向计算机的方位响应分析

COBE　Cosmic Background Explorer　宇宙背景探测器

COBOL　Common Business-Oriented Language　面向商业的通用语言

COBRA　Co-Optimized Booster for Reusable Applications　可重用应用程序的联合优化助推器

COBRA　Computerized Boolean Reliability Analysis　计算机化布尔逻辑可靠性分析

COBS　Central On Board Software　中央机载软件

COC　Centrale d'Ordres Case (VEB Command Unit)　仪器舱指令装置

COC　Certification of Compliance　符合项认证

COCOMO　Constructive Cost Model　建造成本模型

COD Cockpit Only Deorbit 仅离轨时使用的座舱
CODAC Coordination of Operating Data by Automatic Computer 自动计算机控制的运算数据协调
CODEM Comet Dust Environment Monitor 彗星尘埃环境监测器
CoDR Conceptual Design Review 概念设计评审
COEA Cost and Operational Effectiveness Analysis 费用与操作成效分析
CODIS Controlled Orbital Decay and Impact System 受控轨道降落制动和着地系统
COECL Collector-Output Emitter Coupled Logic 集电极输出发射极耦合逻辑（电路）
COF Columbus Orbital Facility 【欧洲】"哥伦布"自由飞行实验舱轨道设施
COF Construction of Facilities 设施设备的建造
COF Cost of Facilities 设施设备的成本
COFEC Cause Of Failure, Effect and Correction 失败原因、影响及纠正；失事原因、影响及改正
COFI Checkout and Fault Isolation (on Board) 检查与故障隔离（箭/船上）
CoFR Certification of Flight Readiness 飞行准备状态验证
CoG Centre of Gravity 重心
COGENT Compiler and Generalized Translator 编译程序与广义翻译程序；自动编码器与广义译码器
COGS Combat Oriented General Support 面向作战的总体保障
COGS Continuous Orbital Guidance Sensor 连续轨道制导传感器
COGS Continuous Orbital Guidance System 连续轨道制导系统
COHOE Computer-Originated Holographic Optical Element 用计算机制成的全息光学元件
COI Coast Orbital Insertion 近岸入轨
COI Combat Operations Intelligence 战斗作战情报
COI Contingency Orbit Insertion 紧急入轨
COIC Combat Operations and Intelligence Center 战斗作战与情报中心
COIC Critical Operational Issues and Criteria 关键操作事项与标准
COIL Chemical Oxygen-Iodine Laser 化学氧碘激光器
COIM Checkout Interpreter (Software) Module 检测译码（软件）模块
COINS Computer Operated Instrument System 计算机控制仪表系统
COINS Cooperation Intelligence Network System 合作智能网络系统
COL Checkout Language 检测语言
COL Columbus Module "哥伦布"实验舱
COL-CC Columbus Control Centre 【欧洲】"哥伦布"控制中心
COLA Collision Avoidance 避撞
COLA Collision on Launch Assessment 发射撞击评估
COLD Center for Optical Logic Devices 光学逻辑设备中心
COLD-SAT Cryogenic on Orbit Liquid Depot Storage, Acquisition, Transfer 在轨低温液体的存储、采集与转换
COLD-SAT Cryogenic on Orbit Liquid Depot-Storage And Transfer 【美国国家航空航天局】低温液体在轨储存—贮存与转运

COLT Computerized On-Line Testing 计算机化联机测试

COLTS Contrast Optical Laser Tracking Subsystem 光学对比激光跟踪分系统

COM-M Common Mode 通用模式

COM³ Common Communications Components 通用通信部件

COMA Cometary Matter Analyzer 彗星物质分析器

COMAAS Committee on Meteorological Aspects of Aerospace System 航空航天系统气象学委员会

COMAS Combined Orbital Maneuvering and Abort System 轨道机动与取消综合系统

COMAT Compatibility of Materials 材料的兼容性

COMB Combustion 燃烧

COMBIC Combined Obscuration Model for Battlefield-Induced Contamination 战场引发污染的组合遮蔽模型

COMCM Communications Countermeasure(s) and Deception 通信干扰与欺骗

COMCS Command Control Software 指挥控制软件

COMDAC Command, Display and Control 指挥、显示与控制

COMDAC INS Command, Display and Control Integrated Navigation System 指挥、显示与控制集成导航系统

COMED Combined Map and Electronic Display 组合式地图与电子显示仪；地图与电子综合显示仪

COMEDS Continental Meteorological Data System 【美国】大陆气象数据系统

COMET Combined Optical Measurements Experiment Team 光学测量联合设备组

COMET Commercial Experiment Transporter 【美国】商用（微重力）实验运输器

COMET Compuscan Optical Message Entry Terminal 超声诊断系统光信息输入端口

COMET Conceptual Operations Manpower Estimating Tool 概念性操作人力预计工具

COMM Commercial Mission 商业性任务

COMMCMD Communications Countermeasure and Deception 通信对抗与欺骗；通信干扰与欺骗（诱惑）

COMO Combat-Oriented Maintenance Organization 面向作战维修组织

COMOL Connection Model-Description Language 连接模型描述语言

COMOPTEVFOR Commander, Operational Test and Evaluation Force 【美国海军】指挥官、操作试验与鉴定部队

COMP Computation of Rendezvous Targeting 交会靶标的计算

COMPACT Computer-Oriented Modular Planning and Control Technique 计算机专用模块化设计和控制技术

COMPACT(S) Computer-Programmed Automatic Checkout and Test System 计算机编程的自动检验和测试系统

COMPARE Computerized Performance and Analysis Response Evaluation 计算机化的性能与分析响应鉴定器

COMPARE Computer-Oriented Method of Program Analysis, Review and Evaluation 面向计算机的程序分析、评审与鉴定方法

COMPARE Console for Optical Measurement and Precise Analysis of Radiation From Electronics 电子设备辐射的光测和精确分析控制台

COMPASS Common Operational Mission Planning and Support System 【美国陆军】通用作战任务规划与保障系统

COMPASS Computerized Model for Predicting and Analyzing Support Structures 预测和分析支持结构的计算机制作模型

COMPES Contingency Operation/Mobility Planning and Execution System 应急作战/机动计划与执行系统

COMPLEX Committee on Planetary and Lunar Exploration 行星际与月球探索委员会

COMPOOL Common Data Pool 通用数据库

COMPTEL Compton Telescope 【美国】康普顿成像望远镜

COMR&DSAT Communication Research and Development Satellite 通信研究和发展卫星

COMSAT Communication Satellite 通信卫星

COMSEC Communications Security 通信安全

COMSS Coastal Ocean Monitoring Satellite System 沿岸海域监控卫星系统

COMSTAC Commercial Space Transportation Advisory Committee 商业空间运输咨询委员会

CON Connection-Oriented Network 面向连接网络

CON Controls of Navigation 导航控制

CONFIDAL Conjugate Filter Data Link 共轭滤波器数据线路

CONICET Consejo de Investigaciones Científicas y Técnicas 【法国】全国科学技术研究委员会

CONIE Comision Nacional de Investigaciones del Espacio 【西班牙】国家空间研究委员会

CONOPS Concept of Operations 操作概念

CONS Console 控制台；操纵台；仪表台

CONTAD Concealed Target Detection 隐蔽目标探测

CONTARAME Correlation Tracking and Ranging and Angle-Measuring Equipment 相关跟踪与测距角测量系统

COOS Chemical Orbit-to-Orbit Satellite 使用化学推进剂的轨道间卫星

COOS Chemical Orbit-to-Orbit Shuttle 使用化学推进剂的轨道间航天飞机

COP Coefficient of Performance 性能系数

COP Combined Operations Plan 联合操作计划

COP Command Operation Procedure 命令操作步骤；指令操作程序

COP Contingency Operations Plan 应急操作计划；应急行动计划

COP Continuous Optimization Program 连续优化程序

COP Control Optimization 控制优化

COP Co-Orbiting Platform 共轨平台

COPE Controlled Operating Pressure Engine 可控压比发动机

COPUOS Committee On the Peaceful Uses of Outer Space 和平利用外层

空间委员会
COPV Composite Overwrapped Pressure Vessel 复合材料包装压力容器
COQ Certificate of Qualification 资格认证
COR Concentric-Orbit Rendezvous 【美国国家航空航天局】同心轨道交会
CORAL Computer On-line Real Time Application Language 【英国国防部】联机实时应用计算机语言
CORDIC Coordinated Rotation Digital Computer 坐标旋转数字计算机
CORDS Coherent On Receive Doppler System 接收相干多普勒系统
CORE Common Operational Research Equipment 通用操作研究设备
CORE Cost-Oriented Resource Estimating Model 成本导向的资源估算模型
CORE SYSTEM Controlled Requirements Expression System 受控需求表述系统
CORSA Cosmic Radiation Satellite 【日本】宇宙辐射卫星（X射线天文卫星）
CORT Computation of Rendezvous Targeting 交会对接目标的计算
COS Carry-on Oxygen System 机载氧气系统
COS Celestial Observation Satellite 【欧洲】天文观测卫星
COS Console Operating System 控制台操作系统
COS Co-Orbiting Satellite 同轨卫星
COS Cosmic Origins Spectrograph 宇宙起源摄谱仪
COS Cosmic Ray Observatory Satellite 【欧洲空间局】宇宙射线观察卫星

COSA Communication Open System Architecture 通信开放式系统结构
COSE Common Open Software Environment 公用开放软件环境
COSI Closeout System Installation 整套关闭系统设备
COSIMA Crystallization of Organic Substances In Microgravity Application 微重力条件下的有机物结晶
COSIMI Corona Sounding Interplanetary Mission 日冕探测行星际任务
COSMIC Coherent Optical System of Modular Imaging Collectors 模块成像收集器的相关光学系统
COSMIC Computer Software Management and Information Center 【美国国家航空航天局】计算机软件管理与信息中心
COSPAR Committee On Space Research 【国际】空间研究委员会
COSS Centaur Operations at the Space Station 空间站上的"半人马座"上面级操作
COSTAR Corrective Optics Space Telescope Axial Replacement 【哈勃望远镜】太空望远镜轴向替换（象差）修正镜
COSTEP Comprehensive Suprathermal and Energetic Particle analyzer 全面高能粒子分析仪
CoSTER Consortium on Space Technology Estimating Research 空间技术预测研究联合会
COSTR Collaborative Solar-Terrestrial Research Programme 【美国和日本】日地空间合作研究计划
COT Cockpit Orientation Trainer 座舱定向训练装置
COTE Check-Out Terminal Equipment 终端测试设备

COTS Commercial Off The Shelf (hardware or software) 商业现用产品（硬件或软件）

COTS Commercial Off-the-Shelf 商用成品技术

COTS Commercial Orbital Transportation Services 商用轨道运输服务

COU Concept of Operation and Use 运行与利用概念

COV Cutoff Valve 关闭阀

COW Clean Operational Weight 净飞行重量

CP Center of Pressure 压力中心

CP Check Point 检测点

CP Coefficient of Performance 性能系数

CP Console Processor 控制台（操纵台）处理器

CP Control Panel 控制屏

CP Control Point 控制点

CP&AA Components, Parts, Accessories and Attachment 部件、零件、附件与附属装置

CP-DPO Deployable Payloads Projects Office 【美国肯尼迪航天中心】展开式有效载荷项目办公室

CP-FEO Cargo Facilities and GSE Projects Office 【美国肯尼迪航天中心】货物设备与地面支撑设备项目办公室

CP-PCO Cargo Projects-Program Control Office 货物计划－项目控制办公室

CP-SPO Spacelab Projects Office 【美国肯尼迪航天中心】"空间实验室"项目办公室

CP/OS Control Program/Operation System 控制程序／操作系统

CPA Closest Point of Approach 最接近点；最近进场点

CPA Contingency Planning Aid 应急规划辅助装置

CPA Critical Path Analysis 关键路径分析

CPAH Columbus Payload Accommodation Handbook 【欧洲】"哥伦布"实验舱有效载荷配置手册

CPAS CEV Parachute Assembly System 【美国乘员探索飞行器】降落伞组件系统

CPB Centaur Processing Building 【美国】"半人马座"上面级处理厂房

CPC Central Planning Center 中央规划中心

CPC Climate Prediction Center 【美国国家环境预报中心】气候预测中心

CPC Computer Program Component 计算机编程部件

CPCB Crew Procedures Control Board 乘员规程控制委员会

CPCEI Computer Program Contract End Item 计算机编程合同端项

CPCI Computer Program Change Instruction 计算机编程更改说明

CPCI Computer Program Configuration Item 计算机编程配置项

CPCL Computer Program Change Library 计算机编程更改信息库

CPCL Computer Program Control Library 计算机编程控制信息库

CPCR Computer Program Change Request 计算机编程更改请求

CPCR Crew Procedures Change Request 乘员程序更改请求

CPCS Cabin Pressure Control System 座舱压力控制系统

CPCS Coast Phase Control System 惯性飞行阶段控制系统；惯性跟踪

阶段控制系统

CPD Cryo Propellant Densification 低温推进剂致密化

CPDDS Computer Program Detail Design Specification 计算机编程详细设计说明

CPDO Kennedy's Center Planning and Development Office 【美国】肯尼迪航天中心的规划与开发办公室

CPDR Contractor's Preliminary Design Review 承包商初步设计评审

CPDS Command Processor Distributor Storage 指令处理分配存储器

CPDS Computer Program Design (or Development) Specification 计算机编程设计（或研发）说明

CPDS Crew Procedures Documentation System 乘员程序文件汇编系统

CPEI Computer Program End Item 计算机编程端项

CPES Crew Procedures Evaluation (Evaluator) Simulator 乘员程序评估模拟器

CPF Cargo Processing Facility 货物处理设备

CPF Central Processing Facility 中央处理设施（厂房）

CPF Cost Per Flight 每次飞行任务的成本

CPI Continuous Process Improvement 持续过程改善

CPL Capillary Pumped Loop 毛细抽吸两相液体回路

CPL Couple 耦合

CPLEE Charged Particle Lunar Environment Experiment 带电粒子月球环境实验

CPM Combustion Properties of Materials (for space applications) 【空间应用】材料燃烧特性

CPM Common Payload Module 通用有效载荷模块

CPM Computer Program Module 计算机编程模块

CPM Critical Path Method 关键路径方法

CPM Critical Phase Matching 临界相位匹配

CPM Core Processing Module 核心处理模块

CPMP Crew Procedures Management Plan 乘员程序管理计划

CPP Columbus Polar Platform 【欧洲】"哥伦布"极轨平台

CPP Columbus Preparatory Programme 【欧洲】"哥伦布"空间站筹备计划

CPR Critical Problem Report 关键问题通报

CPRV Cabin Pressure Relief Valve 增压舱降压阀门

CPS Cabinet Pressurization System 座舱增压系统

CPS Critical Path System 临界路径系统

CPSE Common Payload Support Equipment 通用有效载荷支撑设备

CPSS Cold Plate Support Structure 冷板支撑构件

CPSS Critical Phase System Software 关键阶段系统软件

CPT Cargo Processing Technician 货物（有效载荷）操作技术员

CPT Central Propellant Tank 中心推进剂贮箱

CPT Cockpit Procedure Trainer 座舱程序训练机

CPT Critical Path Technique 关键（临界）路径技术；统筹技术

CPU　Central Processing Unit　中央处理器
CPW　Critical Performance Weight　临界性能重量
CQA　Contract Quality Assurance　合同质量保证
CQCM　Cryogenic Quartz Crystal Microbalance　低温石英晶体微量秤
CQDR　Critical Qualification Design Review　关键鉴定设计评审
CR　Configuration Review　配置评审
CR　Control Room　控制室
CR　Critical Ratio　临界比
CR&T　Command, Ranging, and Telemetry　指挥、测距与遥测
CRA　Centro Richerche Aerospaziali　【意大利】航空航天研究中心
CRADA　Cooperative Research and Development Agreement　协同研究与发展协议
CRAF　Comet Rendezvous and Asteroid Flyby　【美国】彗星交会与小行星飞越探测器
CRAI　Capability, Requirements, Analysis, and Integration　能力、需求、分析与集成
CRAM　Cross-Tie Random Access Memory　交叉磁膜随机存取存储器
CRAMD　Cosmic Ray Antimatter Detector　宇宙射线反物质探测器
CRAS　Cost Reduction Alternative Study　降低成本方案性研究
CRaTER　Cosmic Ray Telescope for the Effects of Radiation　辐射效应宇宙射线望远镜
CRC　Cost Reduction Curve　成本降低曲线
CRCU　Centaur Remote Control Unit　【美国】"半人马座"上面级遥控装置
CRD　Cosmic Ray Detector　宇宙射线探测器
CRDM　Control Rod Drive Mechanism　控制杆驱动机械装置
CRE　Corrected Reference Equivalent　修正参考当量
CRE　Cosmic Ray Enhancement　宇宙线增强
CRES　Corrosion Resistant Steel　抗腐蚀性钢材
CRESDA　Center for Resources Satellite Data and Applications　资源卫星数据与应用中心
CRESS　Center for Research in Experimental Space Science　【加拿大约克大学】空间科学实验研究中心
CREST　Center for Research in Earth and Space Technology　地球与航天技术研究中心
CREST　Comprehensive Radar Effects Simulator Trainer　综合雷达效应模拟训练器
CREST　Crew Escape and Rescue Technics　【机组】乘员逃逸与救援技术
CREST　Crew Escape Technology　乘员逃逸技术
CREST　Crew Escape Technology　机组人员逃生技术（研制弹射座椅用）
CRF　Compressor Research Facility　压缩机研究设施
CRF　Cosmic Ray Flux　宇宙射线通量
CRFS　Common Radio Frequency Subsystem　共用射频分系统
CRG　Cosmic Ray Gas　宇宙线气
CRH　Critical Relative Humidity　临界相对湿度
CRI　Classification, Recognition and Identification　分级、识别与确认

CRIE Cosmic Ray Isotope Experiment 宇宙射线同位素实验

CRIFO Civilian Research, Interplanetary Flying Objects 星际间飞行体民间研究

CRIMS Comet Retarding Iron Mass Spectrometer 彗星减速离子质谱仪

CRIS Calibration Recall and Information System 【美国国家航空航天局】校准与信息系统

CRISM Compact Reconnaissance Imaging Spectrometers for Mars 火星磁盘侦察成像光谱仪

CRISTA Cryogenic Infrared Spectrometers and Telescopes for Atmosphere 【美国国家航空航天局】大气低温红外分光计与望远镜

CRL Common Rail Launcher 通用导轨式发射装置

CRL Configurable Rail Launchers 可配置导轨发射装置

CRM Chemical Release Module 化学释放模块

CRM Cockpit Resource Management 座舱资源管理

CRM Common Rocket Module 通用火箭模块

CRM Continuous Risk Management 连续性风险管理

CRM Crew Resource Management 乘员资源管理

CRMD Computer Resources Management Data 计算机资源管理数据

CRN Cosmic Ray Nuclei Experiment 宇宙射线晶核实验

CRO Chemical Release Observation 化学物质释放观测（卫星）

CRO Cosmic Ray Observatory 宇宙射线观测台

CRP Configuration Requirements Processing 配置要求处理

CRP Cryogenic Refrigerator Program 低温制冷器计划

CRPL Cosmic Ray Physics Laboratory 宇宙射线物理实验室

CRPM Completely Restrained Positioning Mechanisms 完全约束定位机构

CRPMD Combined Radar and Projected Map Display 雷达与投影地图综合显示

CRR Cosmodrome Readiness Review 发射场准备状态评审

CRR Critical Requirements Review 关键要求评审

CRRES Combined Radiation and Release Effects Satellite 【美国国家航空航天局】辐射与释放效应联合试验卫星

CRRF Cosmic Ray Radiation Facility 宇宙射线辐射设施

CRS Cargo Resupply Service 货物（有效载荷）再补给服务

CRS CO_2 Reduction Subsystem 降低二氧化碳分系统

CRS Command Remoting System 远程指挥系统

CRS Commercial Resupply Services 商业再补给服务

CRS Cosmic Ray Shower 宇宙射线簇射

CRS Cosmic Ray Source 宇宙射线源

CRS Cosmic Ray Subsystem 宇宙射线（探测）分系统

CRSI Ceramic Reusable Surface Insulation 陶瓷防热可重复使用表面

CRSS Consolidated Range Simulation System 统一靶场模拟系统

CRSS　Critical Resolved Shear Stress　临界分切应力

CRTC　Combat Readiness Training Center　战斗准备培训中心

CRTV　Composite Reentry Test Vehicle　再入（大气层）试验用组合飞行器

CRUP　Construction and Resource Utilization of the Probe　建造和资源利用探测器

CRUP　Cosmic Ray Upset Program　宇宙射线扰动观测计划

CRUX　Cosmic Ray Upset experiment　【美国航天飞机】宇宙射线扰动实验

CRV　Crew Rescue Vehicle　乘员（机组）搜救飞行器

CRV　Crew Return Vehicle　乘员返回飞行器

CRYO　Cryogenic　低温

Cryocoolers　Cryogenic Coolers　低温冷却器（装置）

CRYOSTAT　Cryogenic Propellant Storage and Transfer　低温推进剂存储与输送

CRYT　Cryogen Tank　低温贮箱

CS　Checkout Station　检测站

CS　Cockpit Simulation　座舱模拟

CS　Common Set　通用装置

CS　Communications Satellite　通信卫星

CS　Communications Support　通信保障

CS　Communications System　通信系统

CS　Control Segment　控制段

CS　Core Segment　芯级段

CS　Crew Station　乘员站

CS　STS Cargo Operations　【美国肯尼迪航天中心】航天飞机货物（有效载荷）操作

CS/CSS　Combat Support and Combat Service Support　作战保障与作战服务保障

CSA　Canadian Space Agency　加拿大航天局

CSAA　Committee on Space Astronomy and Astrophysics　天体物理与空间天文委员会

CSAR　Combat Search and Rescue　作战搜寻与救援

CSAR　Communication Satellite Advanced Research　通信卫星探索性研究

CSAR　Configuration Status Account Report　构型状态评估报告

CSAT　Combined Systems Acceptance Test　系统联合验收测试

CSAT　Crew System Associate Technology　乘员系统附带技术

CSB　Center for Space Biotechnology　太空生物技术中心

CSB　Common Support Building　通用保障厂房

CSC　Cargo System Controller　货物（有效载荷）系统控制装置

CSC　Command System Controller　指令系统控制器

CSC　Common Simulation Component　通用仿真组件

CSC　Computer Software Component　计算机软件部件

CSC　Contingency Support Center　【美国卡纳维拉尔空军基地】应急保障中心

CSCP　Cabin System Control Panel　座舱系统控制操作纵台（屏）

CSCS　Contingency Shuttle Crew Support　【美国】航天飞机乘员应急保障

CSCSAT Commercial Synchronous Communication Satellite 商用同步通信卫星

CSCSC Cost/Schedule Control System Criteria 费用/进度控制系统标准

CSD Control System Development 控制系统研发

CSD Crew Systems Division 乘员系统部

CSD Critical System Demonstration 关键系统演示验证

CSDRN Central Satellite Data Relay Network 【苏联】中央卫星数据中继网

CSDS Cargo Smoke Detector System 货物（有效载荷）烟探系统

CSE Catalytic Surface Effects Experiment 催化表面效应实验

CSE Common Support Equipment 共用支撑设备

CSE Communication Satellite for Experimental Purposes 【日本】实验通信卫星

CSE Computer Support Equipment 计算机保障设备

CSE Configuration Switching Equipment 配置转换设备

CSEDS Combat System Engineering Development Site (supporting the AEGIS Weapon System) 【美国】作战系统工程研发地（用于"宙斯盾"武器系统的保障）

CSF Central Supply Facility 中央保障设施

CSF Consolidated Support Facility 统一保障设施

CSF Cost Sensitivity Factor 成本敏感因子

CSF Critical Safety Functions 临界安全功能

CSF Cryo System Freezers 低温系统制冷机

CSF Customer Support Facility 用户保障设施

CSG Centre Spatial Guyanais (Guiana Space Center) 【法国】圭亚那航天中心

CSI Concentric Sequence Initiation 同心编序动作开始（航天术语，指登月舱与指挥、勤务舱会合前的准备飞行机动）

CSI Control Servo Input 控制伺服输入

CSI Controls Structure Interactions 【美国国家航空航天局】控制结构相互作用

CSI Crew Software Interface 乘员软件接口

CSINA Conseil Supérieur de l'Infrastructure et de la Navigation Aérienne 【法国】地面设施与导航最高委员会

CSIR Computer Systems (Hardware/Software) Integration Review 计算机系统（硬件与软件）综合评审

CSIS Center for Strategic and International Studies 战略与国际研究中心

CSIU Core Segment Interface Unit 芯级段接口装置

CSL Command Signal Limiter 指令信号限制器

CSL Computer Systems Laboratory 计算机系统实验室

CSL Crew Systems Laboratory 乘员系统实验室

CSLA Commercial Space Launch Act, Public Law 98-575 商业航天发射法案

CSLAA Commercial Space Launch

Amendments Act 商业航天发射修正案

CSLBM Conventionally-Armed Sea-Launch Ballistic Missile 海上发射常规弹头弹道导弹

CSM Apollo Command/Service Module 【美国】"阿波罗"飞船的指令舱/服务舱

CSM Common Support Module 通用支撑模块

CSM Communications Support Model 通信保障模型

CSM Computer Status Matrix 计算机状态矩阵

CSMA/CD Carrier Sense Multiple Access/Collision Detection 【美国国家航空航天局】用于多路存取/探测冲突的载波检测

CSNI Communications Shared Network Interface 通信共享网络接口

CSOC Consolidated Satellite Operations Center 综合卫星作战中心

CSOC Consolidated Space Operations Center 【美国】联合空间操纵中心

CSOP Crew Systems Operating Procedures 乘员系统操作程序

CSOSS Combat System Operational System Sequencing 作战系统操作系统时序

CSP Common Sensor Payload 通用感应器有效载荷

CSP Communications Support Processor 通信保障处理器（装置）

CSP Computer Support Program 计算机保障项目

CSP Cryogenic Solid Propellants 低温固体推进剂

CSPAAD Coarse Sun Pointing Attitude Anomaly Detection 低精度太阳定向姿态异常情况检测

CSPU Core Segment Processing Unit 芯级段处理装置

CSR Check Signal Return 检测信号返回

CSR Crew Station Review 乘员站评审

CSRC Contingency Sample Return Container 应急样品回收容器

CSRM Context-Based Software Risk Model 基于文本的软件风险模型

CSRR Combat Systems Readiness Review 作战系统准备状态评审

CSS Centaur Sun Shield 【美国】"半人马座"上面级的遮阳板

CSS Cislunar Swing Station 地球与月球轨道之间的回转站

CSS Coarse Sun Sensors 低精度太阳敏感器

CSS Cockpit Systems Simulator 座舱系统模拟器

CSS Completely Separating System 完全分离系统

CSS Computer Sighting System 计算机瞄准系统

CSS Computer System Simulation 计算机系统模拟

CSS Computer System Simulator 计算机系统模拟器

CSS Control Standard Shroud 【美国】"半人马座"上面级标准整流罩

CSS Control Stick Steering 控制杆操纵

CSS Conversational Software System 会话式软件系统

CSS Core Segment Simulator 芯级段模拟器

CSS Crew Safety System 乘员安全系统

CSS Critical Shearing Stress 临界剪应力

CSS-5 Medium-Range, Road-Mobile, Solid-Propellant Ballistic Missile 【中国】中程、道路移动式固体推进剂弹道导弹
CSSA Coarse Sun Sensor Assembly 低精度太阳敏感器设备
CSSCS Combat Service Support Control System 【美国】作战服务保障控制系统
CSSE Control System Simulation Equipment 控制系统模拟设备
CSSQT Combat System System Qualification Test 战斗系统系统鉴定试验
CSSS Cross Spin Stabilization System(s) 交叉自旋稳定系统
CSSTSS Combat Service Support Training Simulation System 【美国】作战服务保障培训模拟系统
CSO Center Support Operations 【美国肯尼迪航天中心】中心保障操作
CST Combined System Test 综合系统测试
CST Crew Space Transportation 乘员航天运输
CST Crew Station Trainer 乘员站培训装置
CST Crew Systems Trainer (One-G Trainer) 乘员系统训练机
CST Critical Solution Temperature 临界溶解温度
CST Cross-Software Test 交叉软件测试
CST Cycling Strength Test 循环强度试验
CSTA Crew Software Training Aid 乘员软件训练辅助设备
CSTAL Combat Surveillance and Target Acquisition Laboratory 战斗监察与目标捕获实验室
CSTAR Center for Space Transportation and Applied Research 太空运输与应用研究中心
CSTC Computer Security Technology Center 计算机安全技术中心
CSTC Consolidated Satellite Test Center 统一卫星试验中心
CSTC Consolidated Space Test Center 【美国】统一空间试验中心
CSTC Consolidated Space Test Center 统一航天试验（测试）中心
CSTI Civil Space Technology Initiative 【美国国家航空航天局】民用空间技术倡议
CSTR Committee on Solar-Terrestrial Research 【美国】日地研究委员会
CSTS Crew Space Transportation System 乘员空间运输系统
CSTS Cross-Software Test System 交叉软件测试系统
CSTS Cryogenic Storage and Transfer System 低温储存与转运系统
CSU Communications System Utilization 通信系统运用
CSU Computer Software Unit 计算机软件装置
CSV Cable Support Vehicle 线缆保障车
CSV Crew Support Vehicle 乘员保障运载器
CT Command Transmitter 指令发射机
CT Communications Terminal 通信终端（装置）
CT Crawler Transporter 履带运输车
CTA Centaur Truss Adapter 【美国】"半人马"上面级座桁架式有效载荷对接件
CTA Comando-Geral de Technologia Aerpspacial 巴西宇航技术中心

CTA Controlled Thrust Assembly 受控推力组件

CTACS Contingency Theater Air Control System 应急战区空中控制系统

CTACSAPS Contingency Theater Air Control System Automation Plan and System 战区紧急空中管制系统自动化计划与系统

CTAPS Contingency Tactical Air Combat System Automated Planning System 应急战术空中控制系统自动规划系统

CTAPS Contingency Theater Automated Planning System 应急战区自动规划系统

CTB Cargo Transfer Bag 货物（有效载荷）转移袋

CTB Communications Test Bed 通信试验（测试）台

CTB Comprehensive Test Ban 全面禁试

CTBE Cargo Transfer Bag Equivalent 货物（有效载荷）转移袋当量

CTBM Conventionally-Armed TBM 常规弹头战术弹道导弹

CTC Cargo Transportation Container 货物（有效载荷）运输容器

CTC Command Transmitter Controller 指令发射机控制器

CTC Crew Training Centre 乘员训练中心；航天员训练中心

CTC Crew Training Complex 【欧洲】航天员训练场，航天员训练设施

CTC Cryo Test Cavity 低温试验腔

CTCS Consolidated Telemetry Checkout System 联合遥测检查系统

CTD Communications Test Driver 通信试验（测试）驱动器

CTD Critical Technology Demonstration 关键技术演示验证

CTDA Cockpit Thrust Drive Actuator 座舱推力驱动促动器

CTDC Control Track Direction Computer 控制跟踪方向计算机

CTE Center for Test and Evaluation 试验（测试）与鉴定中心

CTE Central Timing Equipment 中央计时设备

CTE Coefficient of Thermal Expansion 热膨胀系数

CTE Common Test Environment 通用试验（测试）环境

CTEIP Central Test and Evaluation Investment Program 中央测试与评估投资项目

CTF Control Test Flight 控制测试飞行

CTF Controlled Test Flight 受控试验飞行

CTF Crew Training Facility 航天员训练设施

CTI Concept Technology Insertion 概念技术引入

CTI Critical Transportation Item 关键运输项

CTIS Crawler Transporter Intercom System 履带运输车内部通信系统

CTN Communications and Tracking Network 通信与跟踪网

CTOF Charge Time-Of-Flight 带电飞行时间

CTOL Conventional Takeoff and Landing 传统起飞与着陆

CTP Command and Telemetry Processor 指令和遥测处理器

CTP Core Technology Programme 核心技术项目

CTP Critical Technical Parameter 关键技术参数

CTPD Crew Training and Procedures Division 【美国约翰逊航天中心】乘员训练与过程管理部

CTPE Central Tactical Processing Element 中央战术处理部件（要素）

CTR Certification Test Requirement 合格试验要求

CTR Certified Test Record 鉴定试验记录

CTR Certified Test Report 鉴定试验报告

CTR Checkout Terminal Room 测试终端室

CTR Critical Temperature Range 临界温度范围

CTR Critical Temperature Resistor 临界温度电阻器

CTRD Central Transmitter/Receiver Device 中心发射/接收装置

CTRS Component Test Requirements Specifications 元器件测试要求说明

CTS Canadian Technology Satellite 加拿大技术卫星

CTS Canadian Test Satellite 加拿大试验卫星

CTS Captive Trajectory System 系留轨道系统

CTS Common Termination System 通用终端系统

CTS Communications and Tracking System 通信与跟踪系统

CTS Communications Technology Satellite 通信技术卫星

CTS Complex Test of System 综合系统测试

CTS Computerized Training System 计算机培训系统

CTS Coordinate Transformation System 坐标转换系统

CTS Crew Transportation System 乘员运输系统

CTS Critical Time Slice 临界时间限幅

CTS/TCTS Combat Training System/Tactical Combat Training System 战斗培训系统/战术战斗培训系统

CTTB Checkout Techniques Test Bed 检测技术测试台

CTTE Customer's Terminal Testing Equipment 用户终端检测设备

CTTO Concurrent Test, TCommunications Test Driverraining and Operations 并行试验（测试）、远程通信传动与操作

CTV Control Test Vehicle 控制测试飞行器

CTV Crew Transfer Vehicle 乘员转移飞行器

CTV Curved Trend Vector 曲向矢量

CU Charge Utile 有效载荷支架

CU Control Unit 控制装置

CUDIXS Common User Digital Information Exchange System 通用型用户数字信息交换系统

CUPID Computer for Uprange Point of Impact Determination 上靶场弹着点确定计算机

CURV Controlled Unmanned Research Vehicle 受控无人驾驶研究飞行器

CUS Cryogenic Upper Stage 低温上面级

CUST Cryogenic Upper Stage Technology 低温上面级技术

CUTE Common Use Terminal Equipment 通用终端设备

CV Carrier Vehicle 运载飞行器

CV Coefficient of Variation 变化系数

CV/BM Carrier Vehicle/ Battle Management 航母作战管理

CV/DFDR Cockpit Voice and Digital Flight Data Recorder 座舱声音与数字飞行数据记录装置

CVAP Centaur Vent and Pressurization 【美国】"半人马座"上面级排气与增压

CVAPS Centaur Vent and Pressurization System 【美国】"半人马座"上面级排气与增压系统

CVAS Configuration Verification Accounting System 配置验证统计系统

CVC Cryogenic Vacuum Calorimeter 低温真空量热计

CVCS Chemical and Volume Control System 化学与容量控制系统

CVDV Coupled Vibration-Dissociation-Vibration 耦合振动—分离—振动

CVE Complete Vehicle Erector 全飞行器起竖车

CVF Controlled Visual Flight 受控目视飞行

CVHG Carrier, Aircraft (V/STOL), Guided Missile 航母、飞机、制导导弹

CVI Chemical Vapor Infiltration 化学气相渗透

CVISC Combat Visual Information Support Center 作战可视化信息保障中心

CVM Control Valve Module 控制阀模块

CVMAS Continuously Variable Mechanical Advantage Shifter 连续可变的机械转换机构

CVR Change Verification Record 更改确认记录

CVR Configuration Verification Review 配置确认评审

CVRT Criticality, Vulnerability, Recoverability and Threats 【攻击目标的】紧要性、可接近性、可恢复性与威胁

CVSRF Crew-Vehicle Simulation Research Facility 【美国国家航空航天局】乘员飞行器模拟研究设施

CVT Concept Verification Testing 概念验证测试

CW Continuous Wave 连续波

CWA Controlled Work Area 受控工作区

CWAR Continuous Wave Acquisition Cycle 连续波捕获周期

CWAT Continuous Wave Angle Track 连续波角跟踪

CWBS Contract Work Breakdown Structure 承包工程分类结构

CWC Chemical Weapons Convention 《化学武器公约》

CWD Continuous Wave Detector 连续波检波器

CWD Cyclotron Wave Device 回旋波装置

CWDD Continuous Wave Deuterium Demonstrator 连续波氘演示验证器（装置）

CWE Caution and Warning Electronics 预防与告警电子设备

CWEA Caution and Warning Electronic Assembly 预防与告警电子设备组件

CWEU Caution and Warning Electronics Unit 预防与告警电子设备装置

CWFSP Caution and Warning/Fire Suppression Panel 预防与告警/灭火操纵台

CWID Coalition Warrior Interoperability Demonstration 联军勇士相互适应能力演示（项目）

CWLM Caution and Warning Limit Module 预防与告警限制模块

CWS Caution and Warning Status 预防与告警状态

CWS Caution and Warning System 预防与告警系统

CWS Collision Warning System 防撞报警系统；（导弹）截击空中目标警报系统

CWSU Caution and Warning Status Unit 预防与告警状态装置

CWSU Condensate Water Servicing Unit 冷凝水维修单位（美国国家航空航天局）

CXE Commercial Cross-Bay Carrier 【美国国家航空航天局】商业跨舱运载器

CxP Constellation Program 【美国】"星座"探月项目

CxP GO Constellation Program Ground Operations 【美国】"星座"探月项目地面操作

CxTF Constellation Training Facility 【美国】"星座"探月项目培训设施

CYSA Cape York Space Agency 【澳大利亚】约克角航天局

CZ Cursor Zero 光标归零

D

D Spec　Process Specification　工艺处理说明

D Star　Measure of Infrared Sensor Sensitivity　红外传感器灵敏度估算

D&B　Docking and Berthing　对接和停泊

D&C　Display and Control　显示与控制

D&C　Displays and Controls　显示和控制（装置）

D&CS　Display and Control Subsystem　显示与控制分系统

D&D　Decontamination and Demolition　净化与毁坏

D&D　Design and Development　设计和研制

D&D　Development and Design　发展与设计

D&E　Demonstration & Evaluation　演示与评估

D&I　Disassembly and Inspection　拆卸与检查

D&P　Drain and Purge　排放与吹扫

D&PD　Definition and Preliminary Design　技术条件确定和初步设计

D&PS　Design and Performance Specification　设计与性能说明

D&T　Detection and Tracking　探测与跟踪

D&V　Demonstration and Validation　演示与验证

D-IFOG　Depolarized-Interferometric Fiber Optic Gyro　去极干扰计的光纤陀螺仪

D/A　Digital to Analog　数字至模拟转换

D/C　Displays/Controls　显示／控制

D/D　Departure Delay　起飞延迟

D/L　Data Link　数据链

D/L　Deorbit/Landing　离轨／着陆

D/L　Downlink　下行链路

D/V　Demonstration and Validation Life Cycle Phase　演示与验证寿命周期阶段

D^2　Degrade and Destroy　削弱与摧毁

D^3　Degrade, Disrupt, Deny　削弱、扰乱与阻遏

D^4　Degrade, Disrupt, Deny, Destroy　削弱、扰乱、阻遏与摧毁

DA　Data Acquisition　数据采集

DA　Delayed Action　延时爆炸的（炸弹）

DA　Deployment Assembly　展开式装配

DA　Design Altitude　设计高度

DA　Detonation Altitude　起爆高度

DA　Device Attachment　装置附件（连接，接头）

DA　Digital-to-Analog　数字至模拟转换

DA　Distillation Assembly　蒸馏组件

DA　Distribution Assembly　分布式装配

DA　Double Amplitude　全幅；双幅

DA&D　Data Acquisition and Distribution　数据采集与分配

DA&P　Data Analysis and Processing　数据分析与处理

DAA　Demonstration of Advanced Avionics System　先进电子设备与控制

系统演示验证

DAALS Distributed Array Acoustic Locating System 分布式阵列声音定位系统

DAASAT Direct Ascent Anti-Satellite 直接升入轨道的反卫星武器

DAASM Doppler Angle of Arrival Spectral Measurements 多普勒着陆角光谱测量

DAASY Data Acquisition and Analysis System 数据采集与分析系统

DAAT Digital Angle of Attack Transmitter 数字式迎角发送器

DAB Data Acquisition Bus 数据采集总线

DAB Dynamically Allocated Bandwidth 动态分配带宽

DABNS Discrete Address Beacon and Navigation System 离散地址信标与导航系统

DABS Discrete Address Beacon System 离散地址信标系统

DABS Dynamic Air Blast Simulator 动态空气爆炸模拟器

DAC Damage Assessment Center 损毁评估中心

DAC Data Acquisition and Control 数据获取与控制

DAC Data Analysis and Control 数据分析与控制

DAC Design Analysis Cycle 设计分析周期

DAC Digital Analog Control 数字模拟控制

DAC Digital to Analog Converter 数字至模拟转换器

DAC Digital to Analytical Conversion 数字至分析转换

DAC Direct Access Control 直接存取控制

DAC Discriminating Attack Capability 判别攻击能力

DAC Display Analysis Console 显示分析控制台

DACAPS Data Capture System 数据采集系统

DACAPS Data Collection and Processing System 数据收集与处理系统

DACB Data Acquisition and Control Buffer 数据采集与控制缓冲器

DACB Data Adapter Control Block 数据适配器控制块

DACBU Data Acquisition and Control Buffer Unit 数据采集与控制缓冲器设备

DACC Direct Access Communications Channel 直接存取通信系统

DACE Data Acquisition and Control Executive 数据采集与控制执行（系统）

DACE Data and Command Equipment 数据与指挥设备

DACE Decision Support System Applied Center of Excellence 决策支持系统应用中心

DACL Dynamic Analysis and Control Laboratory 动力分析与控制实验室

DACM Data Adapter Control Mode 数据适配器控制模式

DACO Data Consistency Orbit 数据相容性轨道

DACON Data Controller 数据控制器

DACOS Data Communication Operation System 数据通信操作系统

DACS Data Acquisition and Command System 数据采集与指令系统

DACS Data Acquisition and Computer

System　数据采集和计算机系统
DACS　Data Acquisition and Control Subsystem　数据采集与控制子系统
DACS　Data Acquisition and Control System　数据采集和控制系统
DACS　Data Acquisition and Correction System　数据截获与校正系统
DACS　Data and Analysis Center for Software　软件数据与分析中心
DACS　Data Communication Operating System　数据通信操作系统
DACS　Digital Acquisition and Control System　数字化采集与控制系统
DACS　Directorate of Aerospace Combat Systems　【加拿大】航空航天作战系统管理局
DACS　Discrete Address Communication System　离散地址通信系统
DACS　Divert and Attitude Control System　转向与姿态控制系统
DACU　Data Acquisition and Control Unit　数据采集与控制装置
DAD　Damage Assessment Department　损伤评估部门
DAD　Discrimination Augmentation Devices　分类增长装置
DAD　Distributed Architecture Decoy　分散型布局诱饵
DAD　Dual Air Density　双重（层）空气密度（对高层大气层和低外逸大气层的全球空气密度计量）
DADC　Digital Air Data Computer　数字式大气数据计算机
DADC^2　Division Air Defense Command and Control　空中防御指挥与控制局
DADD　Data Archived and Distribution Device　数据存档与分配设备
DADE　Data Acquisition and Decommutation Equipment　数据采集与还原设备
DADE　Dual Air Density Explorer　双联大气密度探测器
DADEC　Design and Demonstration Electronic Computer　设计与演示验证电子计算机
DADS　Data Archive and Distribution System　数据档案与分配系统
DADS　Day-of-Launch Ascent Design System　发射日上升段设计系统
DADS　Digital Automated Data System　数字式自动化数据系统
DADS　Distributed Ada Development System　分布式 Ada 语言研发系统
DADS　Distributed Analysis for Decision Support System　决策支持系统的分布式分析
DADS　Dual Air Density Satellite　双联大气密度探测卫星；双重空气密度卫星
DADS　Dynamic Analysis and Design System　动态分析与设计系统
DADU　Data Accumulation and Distribution Units　数据积累与分配装置
DAE　Dual Acquisition Equipment　双联采集设备
DAE　Dynamic Augmentation Experiment　气动增强实验
DAEMON　Data Adaptive Evaluator and Monitor　数据自适应鉴定器和监控器
DAEZ　Downrange Abort Exclusion Zone　下靶场中止禁区
DAF　Data Acquisition Facility　数据采集设施
DAF　Data Analysis Facility　数据分析设施
DAFDTA　Dipole Antenna with Feed Point Displaced Transverse to its Axis

馈电点在轴的垂直方向有移动的偶极天线

DAFECS Digital Authority Full Engine Control System 数字管理局满功率控制系统

DAFICS Digital Automatic Flight Inlet Control System 数字式自动飞行入口控制系统

DAGC Delayed Automatic Gain Control 延迟自动增益控制

DAGC Digital Automated Gain Control 数字式自动增益控制

DAGGR Depressed Altitude Guided Gun Round 压低高度的制导炮弹

DAGMAR Drift and Ground Speed Measuring Airborne Radar 机载偏航角和地速测定雷达

DAGR Defense Advanced Global Positioning System (GPS) Receiver 先进防御全球定位系统接收机

DAI Damage Assessment Indicator 毁伤评估显示器（装置）

DAIRS Distributed Aperture Infrared Sensor 分布式孔径红外传感器（系统）

DAIS Data Acquisition and Interpretation System 数据采集和判读系统

DAIS Data Adapter Interface Software 数据转接接口软件

DAIS Data Avionics Information System 电子设备与控制系统数据信息系统

DAISS Digital Airborne Intercommunication Switching System 数字化机载内部通信切换系统

DAISY Decision Aided Information System 辅助决策信息系统

DAKOS Design Analysis Kit for Optimization Software 优化软件设计分析包

DAL Data Aided Loop 数据辅助回路

DALO Disconnect at Lift-off 在（导弹）起飞时分离

DAMA Demand Assigned Multiple Access 按需分配多址（通信）系统

DAMASK Direct Attack Munition Affordable Seeker 可承受的直接攻击弹药导引头

DAME Data Acquisition and Monitoring Equipment for Computer 计算机数据采集与监控设备

DAME Determination of Aircraft Missile Environment 飞机导弹环境确定

DAME Digital Automatic Measuring Equipment 数字式自动测量设备

DAME Drilling Automation for Mars Exploration 火星探索自动化钻孔项目

DAMP Detection Antimissile Program 反导弹探测计划

DAMP Preliminary Mission Analysis Document 初步任务分析文件

DAMS Defense Against Missile System 导弹防御系统

DAMS Defensive Antimissile System 防御性反导弹系统

DAMS Dynamic Airspace Management System 动态空域管理系统

DANA Delay and Network Analysis 延迟和网络分析

DANAC Data Analysis and Classification 数据分析与分类

DANLS Discrete Address Beacon Navigation and Landing System 离散地址信标导航和着陆系统

DANNASAT Direct Ascent Nonnuclear Antisatellite 直接上升非核反卫

星武器
DANS Doppler Airborne Navigation System 多普勒空中导航系统
DANTAN Detection Avoidance Navigation/Threat Avoidance Navigation 探测回避导航/威胁回避导航
DAP Digital Autopilot 数字化自动驾驶
DAPE Development Armament Probable Error 武器概率误差改进
DAPR Digital Automatic Pattern Recognition 数字式自动模式识别
DAR Data Aided Review 数据辅助评审
DAR Defense Acquisition Radar 防御目标探测雷达
DAR Designed Area for Recovery 指定的回收区
DAR Deviation Approval Request 偏差鉴定要求
DAR Digital Autopilot Requirements 数字化自动驾驶要求
DAR Drawing Analysis Record 图纸分析记录
DARAC Damped Aerodynamic Righting Attitude Control 有阻尼空气动力姿态校正控制
DARACS Damped Aerodynamic Righting Attitude Control System 有阻尼空气动力姿态校正控制系统
DARC Data Acquisition and Reports Control 数据测距与报告控制
DARDC Device for Automatic Remote Data Collection 自动遥测数据采集装置
DARE Doppler and Range Evaluation 多普勒导航与测距评价
DARE Doppler Automatic Reduction Equipment 多普勒自动处理装置
DARE DOVAP (Doppler Velocity and Position) Automatic Reduction Equipment 多普勒速度和位置自动简化装置
DARE Dynamic Abort Risk Evaluator Model 动态应急风险鉴别器模型
DARHT Dual Axis Radiographic Hydrodynamic Test 双轴射线流体力学试验
DARIN Display Attack and Ranging Inertial Navigation 显示攻击与测距惯性导航
DARP Defense Aeronautical Reconnaissance Program 【美国】国防部机载侦察计划
DARPA Defense Advanced Research Projects Administration 【美国】国防高级研究计划署
DARS Data Acquisition and Reduction System 数据采集与简化系统
DART Demonstration for Autonomous Rendezvous Technology 自主交会技术验证
DART Demonstration of Advanced Radar Technology 先进雷达技术演示
DART Demonstration of Autonomous Rendezvous Technology 自主交会技术演示验证
DART Directional Automatic Realignment of Trajectory 【弹射椅】轨迹方向自动重新对准
DARTS Digital Automated Radar Tracking System 数字自动化雷达跟踪系统
DARTS Digital Azimuth Range Tracking System 数字式方位和距离跟踪系统
DARTS Dynamic Avionics Real-Time Scheduling 动态（航空、导弹、航天）电子设备与控制系统实时调

度
DAS Data Acquisition System 数据采集系统
DAS Data Analysis Station 数据分析站
DAS Debris Assessment Software 碎片评估软件
DAS Defensive Avionics System 防御型航空电子系统
DAS Digital Address System 数字化访问系统
DAS Digital Avionics System 数字化电子设备与控制系统
DAS DME (Distance-Measuring Equipment)-Based Azimuth System (Landing System) 测距设备上安装的方位系统（着陆系统）
DAS Door Actuation System 舱门驱动系统
DAS/COTAR Data Acquisition System Correlation Tracking and Ranging 相关跟踪和测距数据采集系统
DAS^3 Decentralized (or Division) Automated Service Support System 分散自动化服务保障系统
DASA Dual Aerospace Servo Amplifier 航空航天双重伺服放大器
DASALS Distributed Aperture Semiactive Laser Seeker 分布式孔径半主动激光导引头
DASC Deep Air Support Center 深空支援（保障）中心
DASD Direct Access Storage Device 直接访问存储装置
DASO Demonstration and Shakedown Operation 演示与实地试验
DASP Discrete Analog Signal Processing 离散模拟信号处理
DASS Defense Aids Support System 防御手段保障系统
DASS Defensive Aids Sub-System 防御性辅助子系统
DASS Demand Assignment Signalling and Switching Unit 按需分配信令与交换单元
DASS Digital-Analog Servo System 数字—模拟伺服系统
DASS Doppler Acoustic Sounding System 多普勒声学探测系统
DASS Dynamic Assembly Scheduling System 动态装配进度系统
DAST Design, Architecture, Software and Testing 设计、构造、软件和试验
DASTAR Data Storage and Retrieval System 数据存储与检索系统
DAT Design Acceptance Test 设计验收试验
DAT Docking Alignment Target 【美国国家航空航天局】飞船对接校准目标
DAT Docking Alignment Test 【美国国家航空航天局】飞船对接试验
DAT Dynamic Address Translation 动态地址转换
DATAC Defense and Tactical Armament Control 防御与战术武器控制
DATACORTS Data Correlation and Transfer System 数据相关与传输系统
DATAR Digital Automated Tracking and Resolving System 数字自动化跟踪和解析系统
DATAR Digital Automatic Tracking and Ranging 数字式自动跟踪与测距
DATDC Data Analyses and Technique Development Center 数据分析与技术发展中心
DATE Dash Automatic Test Equip-

ment 控制板自动测试设备
DATE Decision Aided Test Environment 判定辅助测试环境
DATE Dynamics, Acoustics, and Thermal Environment 动力、声学及热环境
DATEE Dynamics, Acoustics, and Thermal Environment Experiment 动力、声学及热环境实验
DATIS Data Acquisition Telecommunication and Tracking Station 数据采集通信和跟踪站
DATIS Direct Access Technical Information System 直接存取技术信息系统
DATOM Data Aids for Training, Operations and Maintenance 训练、操作和维修的数据辅助设备
DATOR Digital Data, Auxiliary Storage, Track Display, Outputs and Radar Display 数字数据、辅助存储器、跟踪显示、输出与雷达显示
DATOS Detection and Tracking of Satellites 卫星探测与跟踪
DATS Despun Antenna Test Satellite 【美国空军】消旋天线试验卫星
DATS Dynamic Accuracy Test System 动态准确测试系统
DATSA Depot Automatic Test System for Avionics 【修理工厂】航空电子设备自动测试系统
DATSS Development and Test Support System 开发及测试保障系统
DATTS Data Acquisition, Telecommand and Tracking Station 【德国】数据采集、遥控和跟踪站
DATTS Data Acquisition, Telecommunication and Tracking Station 数据采集通信和跟踪站
DATU Direct Access Test Unit 直接存取测试设备
DAU Data Acquisition Unit 数据采集装置
DAU Digital Adapter Unit 数字适配器装置
DAUAS Direction and Range Acquisition System 方向与距离探测系统
DAV Data Available 可获取（利用）的数据
DAV Descent/Ascent Vehicle 下降/上升飞行器
DAVI Dynamic Antiresonant Vibration Isolator 动态反谐振缓冲器
DAVID Development of Advanced Very Long Wavelength Infrared Detector 【美国空军】先进甚长波红外探测器的研发
DAVID Digital Automatic Video Intrusion Detection 数字自动视频干扰检测
DAVO Dynamic Analog of Vocal Track 声道动态模拟（设备）
DAWCLM Data Automation Workload Control and Library Maintenance 数据自动化工作量控制与程序库维护
DAWS Defense Automated Warning System 自动化防御告警系统
DB Data Base 数据库
DB Data Bus 数据总线
DB Design Baseline 设计基线
DB Dry Bulb 【气象】干球
dBA Decibels Acoustic 分贝声学
DBASE Vehicle Base Diameter 飞行器底部直径
DBC Data Bus Control 数据总线控制
DBC Data Bus Coupler 数据总线耦合器
DBC Digital Battlefield Communications 数字化战场通信

DBCI DB with respect to a Circular Polarized Antenna 环形极化全向性天线分贝

DBE Data Bus Element 数据总线部件

DBE Droplet Burning Experiment 液滴燃烧实验

DBF Digital Beam Forming 数字波束形成

DBFRSC Dual Preburner Fuel Rich Staged Combustion 双预燃室富燃料级燃烧

DBG Data Bus Group 数据总线组

DBGMP Data Bus Generation and Maintenance Package 数据总线生成与维护程序包

DBI Differential Bearing Indicator 差动方位（角）指示器

DBIA Data Bus Interface Adapter 数据总线接口适配器

DBIA Data Bus Isolation Amplifier 数据总线隔离放大器

DBIU Data Bus Interface Unit-Launch 用于发射的数据总线接口装置

DBL Dynamic Balance Laboratory 动态平衡实验室

DBM Data Base Management 数据库管理

DBME Database Management Environment 数据库管理环境

DBMS Database Management System 数据库管理系统

DBN Data Bus Network 数据总线网

DBP Design Baseline Program 设计基线项目

DBP Design Burst Pressure 设计的爆裂压力

DBP Double Base Propellant 双基推进剂

DBS Data Base System 数据库系统

DBS Direct Broadcasting by Satellite 卫星直播

DBS Direct Broadcasting Satellite 直播（广播）卫星

DBS Distributed Broadcast Satellite 分布式广播卫星

DBS TVRO Direct Broadcast Satellite Television Receive Only 直播卫星电视单收站

DBSM Database System Management 数据库系统管理

DBSS Direct Broadcast Satellite Service 直播卫星广播业务

DBTF Duct Burning Turbofan 外涵道燃烧加力涡扇发动机

DBTM Design Bureau of Transport Machinery 【俄罗斯】运输机械设计局

DBUP Defense Buildup Plan 【日本】国防发展计划

DBW Differential Ballistic Wind 差动弹道风

DBWC Differential Ballistic Wind Computer 差动弹道风自动修正计算装置

DC Data Coordinator 数据协调装置

DC Development Center 研发中心

DC Display Coupler 显示耦合器

DC Drag Coefficient 阻力系数；牵引系数

DC Drag Control 阻力控制

DC-X Delta Clipper Experiment 【美国】"德尔它—快船"实验

DCA Deck Carrier Assembly 【美国空间站】上承货架装置

DCA Design Change Authorization 设计更改授权

DCA Digital Command Assembly 数字指令装置

DCA Drift Correction Angle 偏流

（航）修正角；偏移修正角

DCA EUR Defense Communication Agency Europe 欧洲国防通信局

DCAR Design Corrective Action Report 设计修正措施报告

DCAS Data Collection and Analysis System 数据收集与分析系统

DCCS Distributed Command and Control System 分布式指挥控制系统

DCCSR Dual Chamber Controllable Solid Rocket 双燃烧室可控固体（燃料）火箭

DCCU Digital Command and Control Unit 数字化指挥与控制设备

DCCU Digital Communications and Control Unit 数字化通信与控制设备

DCCU Display Computer Control Unit 计算机显示控制设备

DCDS Distributed Computing Design System 分布式计算设计系统

DCE Data Communications Equipment 数据通信设备

DCE Distributed Computer Environment 分布式计算机环境

DCE Drop Combustion Experiment 沉降燃烧实验

DCEC Defense Communications Electronics Command 国防通信电子设备司令部

DCEC Defense Communications Engineering Center 国防通信工程中心

DCF Dynamic Coupling Factor 动态耦合因数

DCGS Distributed Common Ground System 散布式通用地面系统

DCI Dual Channel Interchange 双信道交换

DCIB Data Communication Input Buffer 数据通信输入缓冲器

DCIM/DSCIM Display System Computer Input Multiplexer 显示系统计算机输入多路复用器

DCIS Distribution Construction Information System 分布结构信息系统

DCIU Digital Control and Interface Unit 数字控制接口部件

DCKNG Docking 对接；连接；会合

DCLS Data Collection and Location System 数据收集与定位系统

DCM Data Consolidation Manager 数据整合管理工具

DCM Display and Control Module 显示与控制模块

DCM Docking Cargo Module (Russian Segment) 对接货物舱（俄罗斯段）

DCMAO Defense Control Management Area Operations 国防控制管理区操作

DCMM Dedicated Control and Monitor Module 专用控制和监控模块

DCMS Data Communication Management System 数据通信与管理系统

DCO Delivery Configured Orbiter 【美国】运货轨道器

DCO Detailed Checkout 详细检测

DCOC Digital Combat Operations Center 数字化作战指挥中心

DCOM Distributed Component Object Model 分布式组件对象模型

DCOP Detailed Checkout Procedures 详细检测规程

DCOP Displays, Controls, and Operations Procedures 显示、控制与操作程序

DCOV Define, Characterize, Optimise, and Verify (6-sigma) 定义、描绘特征、优化和验证

DCP　Data Collection Platform　数据汇总平台

DCP　Display Control Panel　显示控制屏；显示控制操纵台

DCPDB　Distributed Characteristics and Performance Database　分布式特征与性能数据库

DCPG　Digital Clock Pulse Generator　数字时钟脉冲发生器

DCR　Design Certification Review　设计认证评审

DCR　Design Change Request　设计更改请求

DCR　Design Criteria Review　设计标准评审

DCS　Data Communication System　数据通信系统

DCS　Data Control System　数据控制系统

DCS　Defense Communication Satellite Program　【美国】国防通信卫星计划

DCS　Design Communication System　设计通信系统

DCS　Design Criteria Specification　设计标准说明

DCS　Detail Checkout Specification(s)　详细的检测说明

DCS　Digital Command System (Subsystem)　数字指令系统（分系统）

DCS　Display and Control System　显示与控制系统

DCS　Dual Checkout Station　双联检测站

DCS　Dynamic Coordinate System　动态坐标系统

DCSI　Data and Control Signal Interface　数据与控制信号接口

DCSP　Digital Control Signal Processor　数字控制信号处理器

DCSS　Defense Communication Satellite System　【美国】国防通信卫星系统

DCSS　"Delta" Cryogenic Rocket Second-Stage　"德尔它"火箭低温第二级

DCSU　Digital Computer Switching Unit　数字计算机切换装置

DCT　DSS (Deep Space Station) Communications Terminal Subsystem　太空站通信终端子系统（地面通信设备）

DCTN　Defense Commercial Telecommunications Network　国防商业电信网络

DCU　Digital Computer Unit　数字计算机装置

DCU　Display and Control Unit　显示与控制装置

DCVM　Design Centered Virtual Manufacturing　以设计为核心的虚拟制造

DCWCS　Directional Control and Warning Communications System　方向控制和报警通信系统

DCWS　Debris Collision Warning Sensor　【美国】空间碎片防撞报警敏感器

DD　Data Display　数据显示

DD　Deflection Difference　偏转差

DD　Mechanical and Facilities Engineering　【美国肯尼迪航天中心】机械与设施工程

DD&CS　Dedicated Display and Control Subsystem　专用显示与控制分系统

DDA　Digital Differential Analyzer　数据差分分析装置

DDAS　Digital Data Acquisition System　数字化数据采集系统

DDC Data Distribution Center 数据分配中心

DDC Detector Dewar Cooler 探测装置的杜瓦冷却器

DDCU DC to DC Convertor Unit 【美国空间站】直流－直流交换器装置

DDD Detection, Discrimination and Designation 探测、识别与标定

DDD Digital Data Display 数字式数据显示

DDD Digital Display Detection 数字显示探测

DDD Digital Display Driver 数字显示激励器

DDD Display Decoder Drive 显示器解码驱动

DDD Duplexed Display Distributor 双工显示分配器

DDDDD Demand Drive Direct Digital Dissemination 按需决定的直接数字式分发系统

DDDL Digital Data Downlink 数字数据下行线路

DDDR Decontamination, Decommissioning, Demolition and Restoration 净化、弃用、毁坏及恢复

DDDRSS Department of Defense Data Relay Satellite System 【美国】国防部数据中继卫星系统

DDDU Digital Decoder Driver Unit 数字解码器驱动装置

DDE Dynamic Data Evaluation 动态数据评价

DDE Dynamic Data Exchange 动态数据交换

DDEP Defense Development Exchange Program 国防发展交换计划

DDESB Department of Defense Explosives Safety Board 国防爆炸物安全委员会

DDG Guided Missile Destroyer 导弹驱逐舰

DDI Dedicated Display Indicator 专用显示指示器

DDI Discrete Data Input 离散式数据输入

DDI Discrete Digital Input 离散式数字输入

DDIC Digital Display Indicator Control 数字显示指示器控制

DDIS Digital Development and Integration System 数字式开发与综合系统

DDIS Document Data Indexing Set 文件数据索引编制装置

DDL Data Downlink 数据下行链路

DDM Data Display Module 数据显示模块

DDM Data Display Monitoring 数据显示监控

DDM Discrete Data Management 离散式数据管理

DDM Drop Dynamics Module 【欧洲空间实验室】液滴动力学实验舱

DDMS DOD Manager for Space Shuttle Support 【美国】为航天飞机项目保障的国防部经理

DDO Discrete Digital Output 离散式数字输出

DDOCE Digital Data Output Conversion Element 数字数据输出转换元件

DDOCE Digital Data Output Conversion Equipment 数字式数据输出转换设备

DDP Design Development Plan 设计研发计划

DDP Digital Data Processing (Proces-

sor) 数字化数据处理（处理器）
DDPC Digital Data Processing Center (Complex) 数字化数据处理中心（装置）
DDPF Dedicated Display Processing Function 专用显示处理功能
DDPS Digital Data Processing System 数字化数据处理系统
DDR Decoy Discrimination Radar 假目标识别雷达
DDR Design Development Record 设计研发记录
DDR Detail Design Review 详细设计评审
DDR Determination of Direction and Range 定向与测距装置
DDR&E Director, Defense Research and Engineering 【美国国防部】负责试验与鉴定的局长
DDS Data Display System 数据显示系统
DDS Deployable Defense System 可展开的防御系统
DDS Doppler Detection Station 多普勒探测站
DDS Doppler Detection System 多普勒探测系统
DDST Downsized Deployable Satellite Terminal 小型展开式卫星终端
DDSYS Digital Dynamic System Simulator 数字式动态系统模拟器
DDT Distributed Decision-Aid Terminal 分布式决策辅助终端
DDT Distributed Decision-Aid Tool 分布式决策辅助工具
DDT Downlink Data Transfer 下行链路数据转换
DDT Dynamic Debugging Technique 动态调试技术
DDT&E Design, Development, Test and Evaluation 设计、研制、试验和鉴定
DDTF Dynamic Docking Test Facility 动态停泊测试设备
DDTS Dynamic Docking Test System 动态停泊测试系统
DDU Data Display Unit 数据显示装置
DDU Data Distribution Unit 数据分配装置
DDU Display Driver Unit 显示驱动装置
DDV Discretionary Descent Vehicle 自由降落飞行器
DDVR Displayed Data Video Recorder 显示数据视频记录仪
DE Doppler Extractor 多普勒频率提取装置
DE Dynamics Explorers 【美国空间物理探测】动力学探险者卫星
DE Emergency Deceleration Flight Phase 应急减速飞行阶段
De Equivalent Diameter 当量直径
dea Aerospace Drift Error 航空航天偏移误差
DEA Data Envelopment Analysis 数据包线解析
DEA Data Exchange Agreement 数据交换协议
DEA Deployed Electronics Assembly 展开式电子设备
DEA Directed Energy Attack 定向能攻击
DEB Deadly Energy Blast 致命能量冲击波
DEBRA Debris, Radiance Model 碎片辐射模型
DEC Detached Experiment Carrier 离机实验台（美国航天飞机释放的小型遥控自由飞行实验平台）

DEC Dual Engine Centaur 【美国】双发动机型"半人马座"

DEC Dynamic Energy Conversion 动态能量转换

DECA Digital Electronic Countermeasure Analyser 数字式电子对抗分析仪

DECA Display/Electronic Control Assembly 显示器电子控制装置

DECAN Distance Measuring Equipment Command and Navigation 指挥与领航用测距设备

DECAT Dynamic Environment Communications Analysis Testbed 气动环境通信分析试验床

DECL Direct Energy Conversion Laboratory 【美国约翰逊航天中心】直接能量转换实验室

DECM Defense Electronic Countermeasures 防御电子对抗

DECO Direct Energy Conversion Operation 直接能量转换操作

DECTRA Decca Tracking and Ranging 台卡跟踪和测距导航系统

DECU Data Exchange Control Unit 数据交换控制装置

DECU Digital Engine Control Unit 数字式发动机控制单元；数字式发动机控制装置

DEDA Data Entry and Display Assembly 数据录入与显示组件

DEE Dexterous End Effector 【美国国家航天航空局】灵敏的末端执行器

DEEC Digital Electronic Engine Control 发动机数字电子控制；发动机数控调节

DEECS Digital Electronic Engine Control System 发动机数字电子操纵（系统）

DEEP Dangerous Environment Electrical Protection System 危险环境电气防护系统

DEER Directional Explosive Echo Ranging 定向爆炸回波测距（装置）

DEER Dual-Expansion Energy Recovery 对偶展开能量回收

DEES Dynamic Electronic Environment Simulator 动态电子环境模拟器

DEFCS Digital Electronic Flight Control System 数字式电子飞行控制系统

DEFSMAC Defense Special Missile and Astronautics Center 国防特种导弹与航天中心

DEFT Display Evaluation Flight Test 显示鉴定飞行试验

DEI Design Engineering Identification 设计工程鉴别

DEIMOS Development and Investigation of Military Orbital System 军用航天轨道系统研制（计划）

DEIS Defense Enterprise Integration Services 国防企业综合服务

DEIS Design Engineering Inspection Simulation 设计工程检测模拟

DEIS Design Evaluation Inspection Simulator 设计工程检测模拟装置

DEL Deorbit, Entry and Landing 离轨、进入及着陆

DELTA Differential Electronically Locking Test Accessory 差动电子锁定试验辅助设备

DELTA Distributed Electronic Test and Analysis 分布式电子试验与分析

Delta-DOR Delta Differential One-Way Ranging δ差分单向测距

DELTA-T Difference in Temperature 温度差分
DEM Digital Elevation Model 数字高程模型
DEM Discrete Element Modeling 离散单元模型
DEM Dynamic Effect Model 动力效应模型
DEM/EVAL Demonstration/Evaluation 演示验证／鉴定
DEM/VAL Demonstration and Validation 演示与确认；演示与验证
Demo Demonstration 演示验证
DEMON Diminishing Error Method for Optimization of Network 网络优化误差递减法
DEMS Digital Error Monitoring Subsystem 数字差错监控分系统
DEMS Dynamic Effectiveness Model Study 动效应模型研究
DEMS Dynamic Environment Measurement System 【美国航天飞机】动力环境测量系统
DEPCON Departure Control 起飞控制；离场管制；飞离控制；出发控制
DEPRESS Depressurize 减压
DEPTH Design Analysis for Personal Training and Human Factors 人员训练与人类生理因素设计分析
DES Data Enhancement System 数据增强系统
DES Data Exchange System 数据交换系统
DES Design Environmental Simulator 【美国空军】设计环境模拟器
DES Discrete Event Simulation 离散事件模拟
DESC Defense Electronics Supply Center 国防电子保障中心
DESC Descent 下降；下行；着陆
DESDynI Deformation, Ecosystem Structure and Dynamics of Ice 冰的变形、生态系统结构及动力学
DESPOT Design Performance Optimization 设计性能优化
DETAS Directed Energy Training Assessment Study 定向能训练评估研究
DETEC Defense Technology Evaluation Code 国防技术鉴定规范（标准）
DETL Digital Emulation Technology Laboratory 数字化仿真技术实验室
DEU Display Electronic(s) Unit 显示电子设备装置
DEUCE Digital Electronics Universal Calculation Engine 通用电子数字计算机发动机
DEV ENV Development Environment 研发环境
DEVELOP Digital Earth Virtual Environment and Learning Outreach Project 数字化地球虚拟环境和学习扩展计划
DEW Directed Energy Weapon 定向能武器
DEW Distant Early Warning 远程预警（系统）
DEW Line Defense Early Warning Line 防御早期告警线
DEW LINE Distance Early Warning Line 远程预警线
DEW/D Directed Energy Weapon/Discrimination 定向能武器／分类
DEWG-O Directed Energy Weapon Ground, Orbital 部署在轨道上的定向能武器
DEWIZ Distant Early Warning Identi-

fication Zone 远程预警识别区
DEWL Directed Energy Weapon, Laser 激光定向能武器（热或脉冲）
DEWP Directed Energy Weapon, Particle Beam (neutral or charged) 定向能武器（中性的或带电的粒子束）
DEWPOINT Directed Energy Weapon Power Integration 定向能武器能力集成
DF Development Flight 研发性飞行
DF Disassembly Facility 拆卸设施
DF-KBS Data Fusion Knowledge Based System 基于知识库的数据融合系统
DF-SPE-A Shuttle Project Engineering Office 航天飞机项目工程办公室
DFA Design For Assembly 面向装配/组装的设计
DFA Dynamic Force Analysis 动态受力分析
DFAD Digital Feature Analysis Data 数字特征分析数据
DFAR Defense Federal Acquisition Regulation 【美国】联邦国防采办条例
DFARS Defense Federal Acquisition Regulation Supplement 【美国】《联邦国防采办条例》补充条例
DFAS Defense Financing and Accounting Service 【美国】国防财政与会计局
DFC Digital Flight Controller 数字式飞行控制器
DFCC Digital Flight Control Computer 数字式飞行控制计算机
DFCS Digital Flight Control Software 数字化飞行控制软件
DFCS Digital Flight Control System 数字式飞行控制系统
DFDAF Digital Flight Data Acquisition Function 数字式飞行数据采集功能
DFDAMU Digital Flight Data Acquisition Management Unit 数字式飞行数据采集管理装置
DFDAU Digital Flight Data Acquisition Unit 数字式飞行数据采集组件
DFDP Distribution-Free Doppler Processor 自由配置的多普斯勒处理机
DFDR Digital Flight Data Recorder 数字式飞行数据记录仪
DFDU Digital Flight Data Unit 数字式飞行数据装置
DFE Direction Finding Equipment 测向设备
DFE Directional Frictional Effect 定向摩擦效应；方向性摩擦效应
DFG Display Format Generator 显示格式生成器
DFGC Digital Flight Guidance Computer 数字式飞行制导计算机
DFGC Digital Flight Guidance Control 数字式飞行制导控制
DFGS Digital Flight Guidance System 数字式飞行制导系统
DFI Development Flight Instrument 试制飞行仪器
DFI Development Flight Instrumentation 【美国航天飞机】研制性飞行仪器
DFIDU Dual Function Interactive Display Unit 双功能交互式显示装置
DFIU Digital Flight Instrument Unit 数字式飞行仪表装置
DFM Direct Flight Mode 直接飞行方式
DFMA Design for Manufacturing & Assembly 生产与装配设计

DFMR Design for Minimum Risk 最小风险设计
DFO Design for Operations/Operability 可操作性设计
DFOLS Depth of Flash Optical Landing System 闪光深度光学着陆系统
DFP Deviant Flight Plan 偏离飞行计划
DFPC Decoupled Flight Path Control 去耦飞行航迹控制
DFR Digital Flight Recorder 数字式飞行记录仪
DFRC Dryden Flight Research Center (Facility) 【美国国家航空航天局】德莱顿飞行研究中心
DFRN Differential 微分；差分；差动；差异；差别
DFRN Differential Velocity 差速
DFRN PRESS Differential Pressure 差压
DFRR Detailed Functional Requirements Review 性能要求详细方案的评审
DFS Demonstration Flight Satellite 演示飞行卫星；演示验证飞行卫星
DFS Design for Support/Supportability 可保障性设计
DFS Deutsche Fernmelde Satelliten 德国通信卫星
DFS Directional Finding System 测向系统
DFS Dynamic Flight Simulator 动力学飞行模拟器
DFT Demonstration Flight Test 演示验证飞行试验
DFT Design Feasibility Test 设计可行性试验
DFTI Distance From Touchdown Indicator 接地后滑跑距离指示器

DFVLR Federal German Aerospace Research Establishment 【德国】航空航天研究机构
DFWD Discrete Flight Warning Display 离散式飞行告警显示器
DG Displacement Gyro 偏移陀螺仪；角位移陀螺仪
DG Display Generator 显示生成器
DGC Digital Geo-Ballistic Computer 地球弹道数字计算机；数字式地球弹道计算机
DGC Dynamic Geospace Coupling 地球空间轨道动态耦合
DGDT Downsized Ground Data Terminal 小型地面数据终端
DGIF Deployable Ground Intercept Facility 可配置的地面拦截设备
DGIPS Defense Guidance Illustrative Planning Scenario 防御制导说明计划方案
DGM Directional Gyro Mode 定向陀螺模式
DGPS Differential Global Positioning System 差分式全球定位系统
DGRR Deutsche Gesellschaft für Raketentechnik und Raumfahrt 【德国】火箭技术和航天协会
DGSA Dynamic Ground Station Analysis 气动地面站分析
DGSE Deployable Ground Support Equipment 展开式地面保障设备
DGT Distributed Ground Test 分布式地面测试
DGVS Doppler Ground Velocity System 多普勒地速系统（仪）
DH Desired Heading 应飞航向、所需航向
DHA Design Hazards Analysis 设计危险性分析
DHDD Digital High Definition Dis-

play 高分辨率数字显示器
DHE Data Handling Electronics 数据处理电子设备
DHI Directional Horizon Indicator 水平方向指示器
DHMR Dry Heat Microbial Reduction 干热微生物减少
DHS Discrete Horizon Sensor 离散式水平传感器；离散式红外地平仪
DHUD Diffraction-Optics HUD 衍射光学平视显示器
DI Data Integrator 数据集成装置
DI Development Integration 研发一体化
DI Discrete Input 离散式输入
DI Display Interface 显示接口
DIAC Defense Intelligence Analysis Center 【美国】国防情报分析中心
DIAL Differential Absorption Lidar 差分吸收激光雷达
DIANE Digital Integrated Attack Navigation Equipment 数字式综合攻击导航设备
DIANE Distance Indicating Automatic Navigation Equipment 测距自动导航设备
DIBA Digital Integrated Ballistic Analyzer 数字式综合弹道分析器
DIBA Digital Internal Ballistic Analyzer 数字式内弹道分析器
DIBF Dummy Inertial Bomb Fuze 惯性炸弹教练引信
DIBTS Digital In-Band Trunk Signaling 数字带内中继线信令
DIC Deviation Indicating Controller 偏差指示控制器；偏向指示控制器
DICASS Directional Command Activated Sonobuoy System 主动式指令定向声呐浮标系统
DICBM Defensive Intercontinental Ballistic Missile 防御性洲际弹道导弹
DICBM Depressed Intercontinental Ballistic Missile 低弹道洲际导弹
DICBM Detection of Intercontinental Ballistic Missile (System) 洲际弹道导弹探测（系统）
DICBMS Detection ICBM System 洲际弹道导弹探测系统
DICCE Dynamic Inputs to Control Center Equipment 控制中心设备动态输入
DICE Data Integration and Collection Environment 数据综合与采集环境
DICE Digital Integrated Combat Evaluator 数字式综合作战鉴定器（装置）
DICE Digital Interface Control Equipment 数字接口控制设备
DICON Air Force Digital Communications through Orbiting Needles 【美国空军】轨道运行旋针式数字通信
DICON Digital Communications through Orbiting Needles 利用轨道运行制针状偶极反射带的数字通信
DICORAP Directional Controlled Rocket Assisted Projectile 定向控制火箭弹
DICS Digital Image Correction System 数字成像校准系统
DID Dust Impact Detection 【空间探测器】尘埃碰撞检测
DID Dynamic Interaction Diagnostic 动态交互作用诊断
DIDACS Digital Data Acquisition and Control System 数字数据采集和控制系统
DIDAS Dynamic Instrumentation Data Automobile System 动力仪表数据

自动系统

DIDAS Dynamic Instrumentation Data Autotest System 动力仪表数据自动测试系统

DIDDS Dynamic Integrated Data Display System 动态综合数据显示系统

DIDOCS Device Independent Display Operator Console Support 独立显示操作员控制台支持设备

DIDS Defense Integrated Data System 防御集成数据系统

DIDSY Dust Impact Detection System 尘埃碰撞检测系统

DIDU Defense Item Data Utilization 国防项目数据利用

DIEL Doppler Inertial Erection Loops 多普勒惯性修正回路

DIF Domsat Interface Facility 【美国】国内通信卫星接口处

DIFAR Directional Finding and Ranging 测向与测距

DIFAR Directional Frequency Analysis and Recording System 定向频率分析与记录系统

DIFAR Directional Low-Frequency Analysis and Ranging System 定向低频分析仪与测距系统

DIFAR Directional-Frequency Analysis and Ranging 定向频率分析与测距

DIFM Digital Instantaneous Frequency Measurement 数字瞬时频率测量

DIFMR Digital Instantaneous Frequency Measurement Receiver 数字式瞬时测频接收机

DIG Digital Image Generation 数字图像生成

DIGACC Digital Guidance and Control Computer 制导与控制用数字计算机

DIGNU Deeply Integrated Guidance and Navigation Unit 强集成化的制导与导航装置

DIGS Delta Inertial Guidance System 【美国】"德尔它"运载火箭惯性制导系统

DIGS Digital Image Generation Simulator 数字式图像生成模拟器

DIGS Digital Inertial Guidance System 数字惯性制导系统

DII Defense Information Infrastructure 【美国】国防情报基础设施

DIICOE Defense Information Infrastructure Common Operating Environment 【美国】国防部信息基础设施通用操纵环境计划

DIL Digitization Integration Laboratory 数字化集成实验室

DIL Doppler Inertial LORAN 多普勒惯性罗兰组合导航（系统）

DILS Doppler Instrument Landing System 多普勒仪表着陆系统

DIMACE Digital Monitor and Control Equipment 数字式监控和控制设备

DIME Dropping In a Microgravity Environment 微重力环境下的空投试验

DIMES Descent Image Motion Estimation System 连续图像运动（像移）预测系统

DIMS Distributed Intelligence Microcomputer System 分布式智能微计算机系统

DIMS Dynamic Inertial Measurement Systems 动态惯性测量系统

DIMUS³ Digital Multibeam Steering System Sonar 数字式多波束监控

系统声呐

DINAS Digital Inertial Nav/Attack System 数字式惯性导航/攻击系统

DINE Digital Inertial Navigation Equipment 数字式惯性导航设备

DINET Deep Impact Networking Experiment 深度撞击联网实验

DINS Digital Inertial Navigation System 数字式惯性导航系统

DINS Dormant Inertial Navigation System 休眠型惯性导航系统

DINS Dual Inertial Navigation System 双重惯性导航系统

DINU Dual Inertial Navigation Unit 复式惯性导航仪

DIOT&E Dedicated Initial Operational Test and Evaluation 专属初始作战试验与评估

DIP Depression Impact Point 下沉弹着点

DIP Designated Inspection Point 设计检测点

DIP Display Input Processor 显示输入处理器

DIP Display Interface Processing (Processor) 显示接口处理（处理器）

DIP Display Interface Processor (Software Functional Element) 显示接口处理器（软件功能要件）

DIP Displayed Impact Point 显示的弹着点

DIP Dual-In-Line Package 双列式数据包

DIPS Dynamic Isotope Power System 动态同位素发电系统

DIRBE Diffuse Infrared Background Explorer 漫射红外背景探测器

DIRCM Directional Infrared Countermeasures 定向红外对抗系统（技术）

DIRM Data Item Responsibility Matrix 数据项响应矩阵

DIRS Distributed Infrared Sensor 分布式红外传感器

DIS Distributed Interactive Simulation 分布式交互模拟

DISA Defense Information Systems Agency 【美国】国防信息系统局

DISCO Dual Spectral Irradiance and Solar Constant Orbiter 双光谱辐射和太阳常量轨道器（欧洲太阳天文卫星）

DISCOS Disturbance Compensation System 【卫星】扰动补偿系统

DISCOS Dynamic Interaction Simulation of Control & Structures 控制与结构的动态干扰模拟

DISDB Destruct Ignition Sequence Distribution Box 自毁引爆程序配电箱

DISE Deployable Intelligence Support Element 展开式智能支撑原件

DISN Defense Information System Network 国防信息系统网络

DISSP Defense-Wide Information Systems Security Program 【美国】国防级信息系统安全项目

DISTRAM Digital Space Trajectory Measurement System 数字空间轨道测量系统

DIT Dynamic Integrated Test 动力学综合试验

DITDS Defense Intelligence Threat Data System 【美国】国防情报威胁数据系统

DITFAC Development, Integration, and Test Facility 研制、集成、测试设施

DITS Data Information Transfer Sys-

tem 数据信息转换系统
DIU Data Interface Unit 数据接口装置
DK Docking 对接；连接；会合
DLA Drive Lock Assembly 驱动锁组件
DLAP Data Link Application Processor 数据链应用处理器
DLAT Destructive Lot Acceptance Testing 破坏性批量验收测试
DLB Data Link Buffer 数据链缓冲器
DLC Data Link Control 数据链控制
DLC Direct Lift Control 正向提升控制
DLCI Data Link Control Identifier 数据链控制识别器
DLL Data Link Layer 数据链层
DLL Design Load Limit 设计极限载荷
DLLF Design Limit Load Factor 设计极限载荷因素
DLM Data Link Management Unit 数据链管理装置
DLM Dead Load Moment 固定载荷力矩
DLM Design Level Maintenance 设计阶段的维护
DLP Data Link Processor 数据链处理器
DLPS Delta Launch Processing System 【美国】"德尔它"火箭发射操作系统
DLR Deutschen Zentrums für Luftund Raumfahrt (German Aerospace Center) 德国航空航天中心
DLRE Diviner Lunar Radiometer "占卜者"月球辐射计
DLS DME Landing System 测距仪着陆系统
DLS DME-Based Landing System 以测距设备为主的着陆系统
DLS Doppler Lidar System 多普勒测距系统
DLS Dynamic Load Simulator 动态载荷模拟器
DLS Dynamics Limb Sounder 【卫星】动力学临边探测器
DLSA Digital Linear Slide Switch Assembly 数字式线性滑动开关装置
DLSC Defense Logistics Service Center 防御后勤保障服务中心
DLSM Data Link Summary Message 数据链综合信息
DLTR Data Link Terminal Repeater 数据链终端转发器
DLTR Data Link Transmission Repeater 数据链传输转发器
DLW Design Landing Weight 设计着陆重量
DM Descent Module 降落舱
DM Development Motor 研发性发动机
DM Disconnected Mode 断开方式
DM Docking Mechanism 对接机构
DM Docking Module 对接舱
DM Doppler Missile 多普勒制导导弹
DM&O Data Management and Operations 数据管理和操作
DMA Data Management Analysis 数据处理分析
DMA Direct Memory Access 直接存储访问
DMA Drive Motor Assembly 驱动电机组件
DMAP Digital Missile Autopilot 数字式导弹自动驾驶仪
DMC Data Management Computer 数据管理计算机

DMC Data Management Controller 数据管理控制器（员）
DMC Degraded Mission Capability 飞行能力下降
DMC Direct Maintenance Cost 直接维修费用
DMC Direct Manufacturing Costs 直接制造成本
DMCF Deservicing, Maintenance, and Checkout Facility 停止检修、维护与检测设施
DMCO Delta Mission Checkout 【美国】"德尔它"火箭任务检查
DMCU Docking Mechanism Control Unit 对接机构控制装置
DMD Data Module Drive 数据模块驱动
DMD Dextrous Manipulator Demonstration 灵活操作装置演示验证
DMD Differential Mode Delay 差模时延
DMD Digital Micromirror Device 数字微镜装置
DMD Digital Missile Device 数字式导弹装置
DMDT Diagnostic Model Development Tool 诊断模型开发工具
DME Design Margin Evaluation 设计安全系数鉴定
DME Distance Measuring Equipment 距离测量设备
DME Distributed Management Environment 分布式管理环境
DME Dynamic Mission Equivalent 动态任务当量
DMEA Damage Modes and Effects Analysis 破坏模式与效应分析
DMEA Defect Mode and Effect Analysis 故障模式和影响分析
DMETS Distributed, Multi-Echelon Training System 分布式多层次训练系统
DMF Data Modeling Facility 数据建模设施
DMF DSIF (Deep Space Instrumentation Facility) Maintenance Facility 深空探测设备维护装置
DMGS Digital Missile Guidance Set 数字式导弹制导装置
DMI Diagnostic Monitor Interface 判断故障监视器接口
DMI Dual-Mode Interceptor 双模式拦截机（导弹）
DMIA Dual Multiplexer Interface Adapter 双多路传输接口适配器
DMICS Design Methods for Integrated Control System 综合控制系统设计方法
DMINS Dual Miniature Inertial Navigation System 微型双惯性导航系统
DMINS Dual Miniaturized Navigation System 双微型导航系统
DMIR Demand Mode Integral Rocket Ramjet 指令型整体式火箭冲压发动机
DMIS Data Management Information System 数据管理信息系统
DMIS DATICO (Digital Automatic Tape Intelligence Checkout) Missile Interface Simulator "达蒂科"（数字自动磁带情报检验）导弹接口模拟器
DMON Discrete Monitoring 离散式监控
DMOS Diffusive Mixing of Organic Solutions 【美国航天飞机】有机溶液扩散混合实验
DMP Deployable Maintenance Platform 展开式维修平台；伸缩式维

修平台

DMPI Desired Mean Point of Impact 预期平均弹着点

DMPI Desired Monition Point of Impact 预期弹着点

DMR Dual Mode Ramjet 复式冲压喷气发动机

DMR Dynamic Modular Radio 动态模块化无线电

DMRD Defense Management Review Decision 【美国】国防管理审查决策

DMRS Data Base Management and Retrieval System 数据库管理与检索系统

DMS Data Management System 数据管理系统

DMS Debris Monitoring Sensor 碎片监视感应器

DMS Defense Message System 【美国】国防报文系统

DMS Defense Meteorological Satellite 国防气象卫星

DMS Defensive Management System 防御管理系统

DMS Delta Modulation System 【美国国家航空航天局】增量调制系统

DMS Deviation, Mean Standard 平均标准偏差

DMS Differential Mobility Spectrometry 微分迁移光谱测定法

DMS Dissimilar Mission Simulator 不同任务模拟器（装置）

DMS Docking Mechanism System (Subsystem) 对接机构系统（子系统）

DMS Docking Module Subsystem 对接舱分系统

DMS Dynamic Missile Simulator 动态导弹模拟器

DMS Dynamic Motion Simulator 动态运动模拟器

DMS-R Data Management System for the Russian Service Module 俄罗斯服务舱的数据管理系统

DMSD DOD Mission Support Division 【美国约翰逊航天中心】国防部任务保障部门

DMSO Defense Modeling and Simulation Office 【美国国防部】国防建模和仿真办公室

DMSP Defense Meteorological Satellite Program 【美国】国防气象卫星计划

DMSP Defense Meteorological Support Program 国防气象支援计划

DMSS Data Management System Simulator 数据管理系统模拟器

DMSS Data Multiplex Subsystem 多路数据传输分系统

DMSS Distributed Mission Support Systems 分布式任务保障系统

DMSS Dual Mode Surveillance System 双型监视系统

DMTOGW Design Mission Takeoff Gross Weight 设计任务起飞总重

DMU Data Management Unit 数据管理装置

DMU Dual-Purpose Maneuvering Unit 【航天员】两用机动飞行装置

DMVR Digital Mass Memory Video Recorder 数字式大容量存储器视频记录仪

DMVS Dynamic Manned Vehicle Simulation 载人飞行器动力模拟

DMVS Dynamic Manned Vehicle Simulator 有人驾驶飞行器动态模拟器

DNA Deoxyribonucleic Acid 脱氧核糖核酸

DNAPL Dense, Non-Aqueous Phase Liquid 重质非水相液
DNCA Damage, Needs, and Capacities Assessment 损伤、需要与能力评估
DNCS Distributed Network Control System 分布式网络控制系统
DNL Dynamic Noise Limiter 动态噪声限制器
DNMS Distributed Network Management System 分布式网络管理系统
DNO Descending Node Orbit 降交点轨道
DNP Dynamic Nuclear Polarization 气动核极化
DNS Decentralized Data Processing Network System 分散型数据处理网络系统
DNS Direct Numerical Simulation 直接数值模拟
DNS Doppler Navigation System 多普勒导航系统
DNSIX DODIIS Network Security for Information Exchange 【美国】国防部情报信息系统信息交换的网络安全
DNSS Defense Navigation Satellite System 【美国】国防导航卫星系统
DNSS Doppler Navigation Satellite System 多普勒导航卫星系统
DNUC Digital Nonuniformity Corrector 数字式非均匀性校正
DNW Directional Network Waveform 定向网络波形
DO Discrete Optimization 离散最优化；分离最优化
DO Discrete Output 离散式输出
DOAMS Distant Objective Attitude Measurement System 远距离目标姿态测量系统
DOB De-Orbit Burn 脱（离）轨燃烧
DOC Delta Operations Center 【美国】"德尔它"火箭操作中心
DOC Designed Operational Capability 设计操作（作战）能力
DOC Display Operator Console 显示操作控制台
DOC Drift Orbit Correction 漂移轨道修正
DOCS Data Operations Control System 数据操作控制系统
DOD Department of Defense 【美国】国防部
DOD-STP Department of Defense-Standard Satellite 国防标准卫星部
DoDIIS Department of Defense Intelligence Information System 【美国】国防情报信息系统部
DODISS DOD Index of Specifications and Standard 【美国】国防部规范和标准索引
DODS Definitive Orbit Determination System 最终轨道测定系统
DOF Degree of Freedom 自由度
DOF Direction Of Flight 飞行方向
DOI Descent Orbit Insertion 进入下降轨道
DOI Docking Orbit Insertion 进入对接轨道
DOL Day of Launch 发射日
DOL De-Orbit/Landing 脱（离）轨/着陆
DOLARS Doppler Location and Ranging System 多普勒定位及测距系统
DOMSAT Domestic Satellite 家用卫星；民用卫星；本国卫星
DON Doppler Optical Navigation 多

普勒光学导航
DONS Doppler Optical Navigation System 多普勒光学导航系统
DOORS Dynamic Object-Oriented Requirements System 面向对象的动态需求系统
DOP Duct Overpressure 管道超压
DORACE Design Organization, Record, Analyze, Charge, Estimate 设计组织、记录、分析、费用和评价
DORAN Doppler Range and Navigation 多普勒测距与导航
DORIS Détermination d'Orbite et Radiopositionnement Intégrés par Satellite 卫星轨道测量及无线电定位综合系统
DORL Development Orbital Research Laboratory 发展用轨道研究实验室
DOSS Day Only SMS (Synchronous Meteorological Satellite) System 白天单一同步气象卫星系统
DOT Deep Space Optical Terminal 深空光学终端
DOT Designating Optical Tracker 【美国】目标指示光学跟踪器；激光照射仪
DOT&E Director, Operational Test and Evaluation 【美国】负责操作测试与鉴定的局长
DOT/CIAP DOT Climatic Impact Assessment Program 【美国】运输部气候影响评估项目
DOTMLPF Doctrine, Organization, Training, Material, Leadership, Personnel and Facilities 【美国国防部】条令、编制、训练、作战物资、领导者的培养、人员与设施
DOUSER Doppler Unbeamed Search Radar 多普勒非定向搜索雷达

DP Data Package 数据包
DP Data Processing 数据处理
DP Delivery of Payload 有效载荷运输
DP Deployable Payload 展开式有效载荷
DP Design Proof 设计验证
DP Development Phase 研发阶段；研制阶段
DP Development Prototype 研制样机
DP Differential Pressure 差分压力
DP Double Pole 双极
DP&S Data Processing and Software 数据处理与软件
DP&SS Data Processing and Software Subsystem 数据处理与软件分系统
DPA Data Processing Assembly 数据处理组件
DPA Destructive Physical Analysis 毁伤物理分析
DPA Dual Payload Adapter 双有效载荷适配器
DPA&E Director, Program Analysis and Evaluation 【美国】负责项目分析与鉴定的局长
DPAF Dual Payload Attach Fitting 双有效载荷附加配件
DPAS Defense Priorities and Allocation System 防御优先配置系统；国防优先配置系统
DPAT Dynamic Program Analysis Tool 动态程序分析工具
DPC Digital Pitch Control 数字化俯仰控制
DPC Dual Payload Carrier 双有效载荷支撑架
DPCC Data Processing Control Center 数据处理控制中心

DPCL Dedicated Payload Communications Link 【美国】专用有效载荷通信数据链
DPF Defense System Communications Satellite (DSCS) Processing Facility 国防通信卫星处理厂房
DPF Differential Pressure Feedback 差压反馈
DPF Dynamic Pressure Feedback 气动压力反馈
DPG Defense Planning Guidance 【美国】国防计划制定指导
DPI Designed Point of Impact 计算弹着点
DPI Desired Point of Impact 预期弹着点
DPL Development Prototype Launcher 开发模型发射装置
DPLCS Digital Propellant Level Control System 数字火箭推进剂水平控制系统
DPLR Doppler 多普勒效应；多普勒雷达
DPM Drop Physics Module 液滴物理模块
DPM Dynamics and Performance Missile 导弹动力学和性能（分析）
DPNL Distribution Panel 分布式操作台；分配操作台
DPP Dexterous Pointing Payload 灵活定位有效载荷
DPP Distributed and Parallel Processing 分布与并行处理
DPR Definition Phase Review 定认阶段评审
DPS Data Processing and Software 数据处理与软件
DPS Data Processing Software System 数据处理软件系统
DPS Data Processing System (Subsystem) 数据处理系统（分系统）
DPS Deorbit Propulsion Stage 离轨推进级
DPS Descent Propulsion System 【行星登陆舱或登月舱】下降推进系统
DPS Drag Parachute System 减速伞系统
DPS Drogue Parachute System 减速伞系统
DPSSL Diode-Pumped Solid State Laser 二极管泵浦（激励）固体激光器
DPT Descent Performance Test 下降性能试验
DPT Design Proof Test 设计验证试验
DPT Durability Proof Test 耐久性试验
DR Design Review 设计评审
DR&A Data Requirements and Analysis 数据要求与分析
DRA Design Reference Architecture 设计基准体系
DRADS Degradation of Radar Defense System 雷达防御系统性能降低
DRAF Data Reduction and Analysis Facility 数据简化与分析设备
DRAI Dead Reckoning Analogue Indicator 推测定位模拟指示器
DRAM Detection Radar Automatic Monitoring 探测雷达自动监视
DRAM Dynamic Random Access Memory 动态随机存取存储器
DRAM Dynamic Response of Articulated Machinery 铰接机械装置的动力响应
DRB Design Requirements Baseline 设计要求基线
DRC Damage Risk Criterion 损坏危险准则

DRC Data Reduction Center 数据简化中心

DRC Discrete Rate Command 离散率指令

DRD Data Requirement Description 数据要求说明

DRDO Defense Research and Development Organization 【印度】国防研究和开发机构

DRDP Detection Radar Data Processing 探测雷达数据处理

DRDT Digital Radar Data Transmission 数字式雷达数据传送

DRED Data Router and Error Detection 数据发送（指令）与误差探测

DRED Directed Rocket Engine Development 定向火箭发动机研制

DREN Defense Research and Engineering Network 国防研究与工程网

DRET Direct Reentry Telecommunications 直接再入（大气层）远距离通信

DRET Direct Re-Entry Telemetry 直接再入大气层遥测

DRETS Direct Reentry Telemetry System 直接再入遥测系统

DREWS Direct Readout Equatorial Weather Satellite 直接读出数据赤道气象卫星

DRF Depressing Range Finder 垂直基线测距仪；低气压区测位仪

DRF Digital, Radio Frequency 数字式射频

DRF Directional Radio Frequency 定向射频

DRFM Digital Radio Frequency Memory 数字式无线电频存储器；数字式射频存储器

DRG Digital Ranging Generator 数字式测距发动机

DRGS Direct Readout Ground Station 直接读取地面站

DRGS Direct Readout Ground System 直接读取地面系统

DRI Data Rate Indicator 数据率指示器

DRI Dynamic Response Index 动态响应指数

DRIFT Diversity Reliability Instantaneous Forecasting Technique 分集可靠性瞬时预测技术

DRIFT Dynamic Reliability Instantaneous Forecasting Technique 动态可靠性瞬时预报技术

DRIMS Delta Redundant Inertial Measurement System 三角冗余惯性测量系统

DRIPS Dynamic Real Time Information Processing Systems 动态实时信息处理系统

DRIRU Dry Rotor Inertial Reference Unit 干转子惯性基准装置

DRLMS Digital Radar Landmass Simulation 数字雷达陆地模拟

DRM Data Records Management 数据记录管理

DRM Design Reference Mission 设计基准任务；设计参考任务

DRM Ducted Rocket Motor 火箭冲压发动机；涵道火箭发动机

DROS Direct Read-out Satellite 直接读出卫星

DRPP Data Relay Satellite Preparatory Program 【欧洲空间局】数据中继卫星筹备计划

DRR Deployment Readiness Review 研发准备评审

DRR Design Readiness Review 设

计成熟度评审
DRR　Design Requirements Review　设计要求评审
DRS　Data Relay Satellite　【欧洲】数据中继卫星
DRS　Data Relay Station　数据中继站
DRS　Data Relay System　数据中继系统
DRS　Dead-Reckoning Subsystem　推测航行子系统
DRS　Delayed Repeater Satellite　延迟式中继卫星
DRS　Digital Range Safety　数字化靶场安全
DRS　Downrange Ship　下靶区遥测船
DRS/CS　Digital Range Safety/Command System　数字式靶场安全指挥系统
DRSS　Data Relay Satellite System　数据中继卫星系统
DRT　Data Relay Terminal　数据中继站终端
DRT　Data Relay Transponder　数据中继应答机
DRTS　Data Relay and Tracking Satellite　数据中继与跟踪卫星
DRTS　Data Relay Test Satellite　数据中继试验卫星
DRTS　Data Relay Tracking Satellite　数据中继跟踪卫星
DRV　Development Re-Entry Vehicle　重返大气层的试验飞行器
DRVID　Difference Range Versus Integrated Doppler　不同距离与综合的多普勒雷达
DRVS　Doppler Radar Velocity System　多普勒雷达测速系统
DRVS　Doppler Radar (Radial) Velocity Sensor　多普勒雷达（径向）速度传感器
DRVS　Doppler Radial Velocity System　多普勒径向速度测量系统
DRWP　Doppler Radar Wind Profiler　多普勒雷达风廓线仪
DS　Directional Solidification　方向固化
DS&R　Data Storage and Retrieval　数据存储与检索
DS-1　Category of Telecommunications Circuit Capability　远程通信回路性能分类
DS-3 LAN　Category of Telecommunications Circuit for A Local Area Network　局域网远程通信回路分类
DSA　Deep-Space Antenna　深空天线
DSA　Defense System Analysis　防御系统分析
DSAL　Decelerated Steep Approach and Landing　减速急剧到达与降落
DSAS　Digital Solar Aspect Sensor　数字式太阳方位遥感器
DSAT　Defensive Satellite　防御卫星
DSC　Direct Synchronized Control　直接同步控制
DSC　Dynamic Stability Control　动态稳定控制
DSCA　Differential Strain Curve Analysis　微分应变曲线分析
DSCB　Data Set Control Block　数据集控制部件（块）
DSCC　Deep Space Communications Complex　深空通信全套设备；太空通信中心
DSCF　Diffusion and Soret Coefficients Facility (for Columbus)　【欧洲】"哥伦布"舱的扩散和索雷特系统设备
DSCS　Defense Satellite Communications System　国防卫星通信系统
DSCSOC　Defense Satellite Communi-

cations System Operations Center 国防卫星通信系统操作中心

DSDCS Dynamic Sensor Display and Control Simulator 传感器动态显示与控制模拟器

DSDU Data Storage Distribution Unit 数据存储分配装置

DSE Data Storage Equipment 数据存储设备

DSE Data System Experiment 数据系统实验

DSE Distributed Sensing Experiment 分布式传感实验

DSEA Data Storage Electronics Assembly 数据存储电子设备组件

DSFC Direct Side-Force Control 直接侧力控制

DSG Digital Signal Generator 数字信号发生器

DSH Deep Space Habitat 深空适居性

DSH Deep Space Habitation 深空居住舱

DSI Defense Simulation Internet 国防模拟因特网

DSI Digital Satellite Interface 数字卫星接口

DSI Dynamic Simulation & Integration 动态模拟综合

DSIF Deep Space Instrumentation Facility 深空测量设备

DSIPS Digital Satellite Image Processing System 数字卫星图像处理系统

DSIS Defense Simulation Internet System 【美国】国防模拟互联网系统

DSIS Defense Special Intelligence System 【美国】国防特别情报系统

DSIS Digital Software Integration System 数字软件综合系统

DSL Deep Scattering Layer 深散射层

DSLARDTS Direction Station Low Accuracy Radar Data Transmission System 定向站低精确度雷达数据传输系统

DSM Decision Support Matrix 决策支持矩阵

DSM Deep Space Maneuver 深空探测机动

DSM Defense Suppression Missile 防御用压制导弹

DSM Docking Storage Module 对接存储舱

DSM Dutch Soyuz Mission 荷兰"联盟号"火箭任务

DSM Dynamic Scattering Mode 动态散射模式

DSMAC Digital Scene Matching Area Correlation 数字式景物匹配区域相关（技术）

DSMAC Digital Scene Matching Area Correlator 数字式景物匹配区域相关器

DSMS Deep Space Mission Systems 深空任务系统

DSN Deep Space Network 深空网

DSN Defense Switched Network 防御切换网

DSO Defensive System Operator 防御系统操作员

DSOTS Demonstration Site Operational Test Series 导弹发射阵地操作检验演习系统

DSP Data Storage & Data Processing 数据存储和数据处理

DSP Defense Support Program 国防保障项目

DSPN Deep Space Planetary Network

深空行星际跟踪网
DSPRTM Defense Support Program Real-Time Model 【美国】防御支持项目实时模型
DSR Dynamic Space Reproducer 动态空间再现装置
DSRCE Down Scoped Radio Control Equipment 缩小范围的无线电控制设备
DSRV Deep Submergence Rescue Vehicle 深潜救援艇
DSS Decision Support System 决策支持系统
DSS Deep Space Station 深空探测站
DSS Deep Space Surveillance 深空监视
DSS Dual Spacecraft System 双航天器系统
DSS Dynamic Support System 动态支持系统
DSSCTS Deep Space Station Communications Terminal Subsystem 太空站通信终端子系统
DST Decision Support Tool 决策支持工具
DST Dynamic Stability Test 气动稳定性测试
DSTAR Defense Strategic and Tactical Array Reproducibility 【美国】国防战略与战术阵列可再现性
DSTF Delta Spin Test Facility 【美国】"德尔它"火箭旋转测试设施
DSTL Defense Science and Technology Laboratory 国防科学技术实验室
DSTM Differential Space-Time Modulation 差分时空调制
DSTP Data System Technology Program 数据系统技术项目

DSTR Dynamic System Test Rigs 动态系统试验台
DSU Digital Service Unit 数字服务设备
DSUCR Doppler-Shifted Ultrasonic Cyclotron Resonance 多普勒漂移的超声回旋加速共振
DSUP Defense System Upgrade Program 防御系统改进计划
DSV Deep Space Vehicle 深空飞行器
DSV Deep Submergence Vehicle 深潜艇
DSV Douglas Space Vehicle 【美国】"道格拉斯"航天器
DSWS David's Sling Weapon System 【美国】"大卫"空投武器系统
DT Damage Tolerance 损伤容限
DT Dedicated Target 专用靶标
DT Depressed Trajectory 降低的轨迹
DT Digital Test Measurement System 数字化试验测量系统
DT&E Development, Test and Evaluation 发展、试验和评价
DT/OA Development Test/Operational Assessment 研发试验（测试）与操作评估
DT/OT Development Test/Operational Test 开发测试/操作测试
DTA Debris Transport Analysis 碎片移动分析
DTA Development Test Article 研发测试组件
DTA Drop Test Article 投放试验样机
DTA Dynamic Test Article 动力试验样机
DTAL Dual Thrust Axis Lander 双推力轴着陆器

DTB　Design Test Bench　设计试验台

DTC　Dead Time Correction　失效时间修正

DTCS　Digital Test Command System　数字测试指令系统

DTCW　Data Transfer Command Word　数据转换指令字段

DTD　Damage Tolerant Design　损伤容限设计

DTD　Digital Transfer Device　数字传输装置

DTE　Data Terminal Equipment　数据终端设备

DTE　Data Transmission Equipment　数据传输设备

DTE　Drop Tower Experiment　落塔（微重力）实验

DTED　Digital Terrain Elevation Data　数字式地形正视图数据

DTF　Demonstration Test Flight　验证性试飞；演练性试飞

DTF　Development Test Facility　研制试验设施

DTF　Development Test Flight　研制性试验飞行

DTG　Dynamically Tuned Gyro　动力调谐陀螺

DTH　Delayed Type Hypersensitivity　延迟型超过敏性；延迟型超感光性；延迟型超灵敏性

DTI　Design Technical Instruction　设计技术指令

DTI　Development Test Instrumentation　研发测试测量

DTIC　Defense Technical Information Center　国防技术信息中心

DTIS　Defense Technical Information Services　国防技术信息服务

DTLCC　Design to Life-Cycle Cost　按寿命同期成本设计；从设计到实用周期成本；定寿命周期费用设计

DTLS　Descriptive Top-Level Specification　描述性高级规程（说明）

DTM　Data Transfer Module　数据传输模块

DTM　Demonstration Test Milestone　演示验证试验新阶段

DTM　Digital Transient Model　数字式瞬态模型

DTMF　Data Tone Multiple Frequency　数据单多频

DTMO　Development, Test, and Mission Operations　研发、测试与任务操作

DTMS　Development, Test, and Mission Support　研发、测试与任务保障

DTMS　Digital Test Measurement System　数字式测试测量系统

DTMS　Digital Test Monitoring System　数字式测试监控系统

DTN　Data Transmission Network　数据传输网

DTN　Delay-Tolerant Network　容/延迟联网

DTN　Disruption (or Delay) Tolerant Networking　容中断/延迟联网

DTO　Development Test Objective　研发测试目标

DTOC　Division Tactical Operations Center　师战术作战中心

DTP　Distributed Targeting Processor　分布式靶标处理器

DTPR　Detailed Test Procedures　详细测试规程

DTR　Damage Tolerance Rating　损伤容限等级

DTR　Data Telemetry Register　数据遥测记录仪

DTRM　Dual Thrust Rocket Motor

双推力火箭发动机
DTS Data Transfer System 数据转换系统
DTS Data Transmission System 数据传输系统
DTS Defensive Technologies Study 防御技术研究
DTS Development Test Satellite 研制试验卫星
DTS Device Technology and Safety 设备技术和安全
DTS Diagnostic Test Facility 诊断测试设施
DTS Dynamic Test Station 动力试验台
DTS Dynamic Test System 动力测试系统
DTSE&E Director, Test Systems Engineering and Evaluation 【美国】负责试验（测试）系统工程与鉴定的局长
DTTE Development, Test, and Evaluation 研制、试验与评价
DTTS DAIS Integrated Test System 数字式航空电子设备信息系统的综合测试系统
DTTSCAP Department of Defense Information Technology Security Certification and Accreditation Process 【美国】国防部信息技术安全认证和认可过程
DTUL Deflection Temperature Under Load 载重情况下的偏差温度
DTV Design Test Vehicle 设计试验飞行器
DTV Development Test Vehicle 研发试验运载器
DTX Discrete Transmission 非连续发射
DU Display Unit 显示装置

DUA Design Upgrade Assessment 设计升级评估
DUA Digital Uplink Assembly 数字式上行链路组件
DUC Digital Uplink Command 数字式上行链路指令
DUNS Data Universal Numbering System 通用型数据编码系统
DUST Dual-Use Science and Technology 两用性科技
DV Designated Verification 指定性鉴定；指定性校验
DVAL Demonstration Validation 验证确认
DVD Delta Velocity Display 速度增量显示
DVF Demonstration and Validation Facility 演示验证与确认设施
DVOR Doppler VHF Omnidirectional Range 多普勒极低频全向范围
DVR Debris Verification Review 碎片演示验证评审
DVS Design Verification Specification 设计验证说明
DVS Digital Voice System 数字声音系统
DVT Design Verification Testing 设计验证测试
DVT Development Verification Test 开发验证测试
DVT Dynamic Vehicle Test 气动运载器测试
DWS Disaster Warning Satellite 灾难报警卫星
DWS Dispenser Weapon System 子母弹箱武器系统
DWSS Disaster Warning Satellite System 灾难报警卫星系统
DXS Diffuse X-Ray Spectrometer 弥散X射线分光仪

DYNAMO Dynamic Allocation Model 动态分配模型；动态定位模型

DYNASAR Dynamic Systems Analyzer 动力系统分析仪

DYNASOR Dynamic Nonlinear Analysis of Shells of Rotation 旋转壳体的动态非线性分析

DYRAD Dynamic Resolver Angle Digitizer 动态解算器角度转换器

DZT Digital Zero Trigger 数字零位触发器

E

E　Elevation Angle　仰角；高角
E　Engine　发动机；引擎
E Spec　Material Specification　材料说明（规格）
E&D　Engineering and Development　工程技术与研制；技术工艺与研发；技术装备与研制
E&D　Evaluation and Development　评估与研制
E&E　Electronics and Equipment　电子仪器与设备
E&I　Electrical and Instrumentation　电子测量
E&SP　Equipment and Spare Parts　设备与备（用零）件
E&ST　Employment and Suitability Test　使用性与适用性试验
E&ST　Employment and Suitability Test Program　【美国空军】使用性和适用性试验计划
E&V　Evaluation and Validation　评估与鉴定
E-LADAR　Enhanced Laser Detection and Ranging　增强型激光探测和测距装置
E-STARS　Electronic Suspense Tracking and Routing System　电子悬浮跟踪与航线飞行系统
E-UAV　Endurance-Unmanned Aerial Vehicle　长航时无人驾驶飞机
E/C　Encoder/Coupler　编码器/耦合器
E/D　End-of-Descent　下降段结束
E/L　Entry/Landing　进入/着陆
E/M　Engineering Model　技术工艺模型
E/O　Engineering/Operations　技术工艺/操作
E/O　Engine-Out　发动机停车；发动机熄火；发动机失效
E/R　Emergency Recovery　紧急回收
E/S　Earth Station　地面站
E/T; ET　Escape Tower　应急脱离塔；应急逃生塔
E/U　Erection Unit　架设组；竖起装置
E/W　Energy to Weight Ratio　能重比
E^2　Enhanced Effectiveness　效用增强
E^2E　End-to-End　端至端；端对端
E^2I　Endo-Exoatmospheric Interceptor　内外大气层拦截机（导弹）
E^2I　Exoatmospheric/ Endoatmospheric Interceptor　大气层外/大气层内拦截器
E^3　Electrical, Electronic, and Electromechanical　电气、电子和机电的
E^3　Electromagnetic Environmental Effect　电磁环境效应
E^3　Electronic Environmental Effect　电子环境效应
E^3A　Electronic Environmental Effect Analysis　电子环境效应分析
EA　Electronic Assembly　电子组件
EA　Electronic Attack　电子攻击
EA　Environmental Analysis　环境分析
EA　Environmental Assessment　环境评估
EA　Equalizingline Amplifier　均压线放大器

EA Extended Accumulator 扩充式累加器

EA External Access 外部存取；外部访问

EAA Earth Attitude Angle 地球姿态角

EAB Equipped Avionics Bay 仪器舱

EAC Early Analysis Capability 早期分析能力

EAC Effective Attenuation Coefficient 有效衰减系数

EAC Emergency Action Console 应急行动控制台

EAC Energy Absorbing Capacity 能量吸收容量；能量吸收能力

EAC Estimated Approach Control 估计进场[进近]控制

EAC European Astronauts Centre 【欧洲空间局】欧洲航天员中心（设在德国）

EAC Evaluation and Analysis Center 评估与分析中心

EAC Expected Approach Clearance 预期进场[进近]许可

EAC Experimental Apparatus Container 实验设备箱

EAC-TM European Astronaut Centre-Training Model 欧洲宇航员中心—训练模型

EACC Error Adaptive Control Computer 误差自适应控制计算机

EACS Electronic Automatic Chart System 自动电子地图显示系统

EAD Electrically Alterable Device 可改变电气装置

EAD Engine Alert Display 发动机报警显示

EAD Engineering Analysis and Design 工程分析与设计

EAD/D Engineering, Analysis, Design and Development 工程、分析、设计与研制

EADAS Engineering and Administrative Data Acquisition System 工程与管理数据采集系统

EADB Elevator-Angle Deviation Bar 升降舵角偏移杆

EADI Electronic Attitude Director (or display) Indicator 电子姿态指引仪；电子指引地平仪；电子垂直位置指示器

EADS Engineering Analysis and Data System 工程分析与数据系统

EADSC European Aeronautic Defense and Space Company 欧洲航空防务与航天公司

EADSIM Extended Air Defense Simulation 扩展型空中防御模拟

EADTB Extended Air Defense Testbed 延伸式防空试验台

EADTB Extended Air Defense Test Bed Program 扩展型空中防御试验（测试）台项目

EAF Effective Attenuation Factor 有效衰减系数

EAFB Edwards Air Force Base 【美国】爱德华兹空军基地

EAFR Enhanced Airborne Flight Recorder 增强型机载飞行记录装置

EAG Engine Accessory Generator 发动机辅助装置发电机

EAGE Electrical Aerospace Ground Equipment 航空航天地面电气设备

EAGLE Elevation Angle Guidance Landing Equipment 仰角导引着陆装置

EAGLE Extended Airborne Global Launch Evaluator 扩展型机载全球发射鉴定器（装置）

EAIC Engine Air Intake Controls 发动机进气道控制机构

EAID ESRO (European Space Research Organization) Advanced Imaging Detector 欧洲航天研究组织先进成像检测器

EAL Emergency Action Level 应急行动等级

EAL Evaluation Assurance Level 评估保证等级

EALS Ejector Augmented Lift System 弹射增升系统

EAM Electric Accounting Machine 电动计算机

EAM Emergency Action Message 应急行动讯息

EAP Engine Alert Processor 发动机报警处理器

EAP Equivalent Air Pressure 当量气压

EAP Etage d'Accélération à Poudre (Solid Rocket Booster) 【法属圭亚那航天中心】固体火箭助推器

EAP Expert Assessment Panel 专家估价小组

EAP External Auxiliary Power 外部辅助动力

EAPL Engineering Assembly Parts List 工程装配部件表

EAPS Engine Air Particle Separator 发动机进气颗粒分离器

EAPU External Auxiliary Power Unit 外部辅助动力装置

EAPV Earth Atmosphere Penetrating Vehicle 地球大气贯穿飞行器

EAR Engineering Analysis Report （工程）技术分析报告

EARL European Advanced Rocket Launcher 欧洲先进火箭运载器

EARSL European Association of Remote Sensing Laboratories 欧洲遥感实验室协会

EARTHNET European Earth Resources Satellite Data Network 欧洲地球资源卫星数据网

EAS Engine Analyzer System 发动机分析器系统

EAS^3 Engine Automatic Stop and Start System 发动机自动停车与起动系统

EAS Equivalent Air Speed 当量空速；等效空速

EAS Error Analysis Study 误差分析研究

EAS Escape, Equivalent Airspeed 等效空速逃离

EASEA Experimental Assembly of Structures in Extravehicular Activity 【美国航天飞机】舱外活动结构实验装置

EASEP Early Apollo Scientific Experiments Package 【美国】早期"阿波罗"飞船科学实验数据包

EASEP Early Apollo Scientific Experiments Payload 【美国】早期"阿波罗"飞船科学实验有效载荷

EASTT Experimental Army Satellite Tactical Terminals 【美国】陆军战术实验卫星终端

EASY Exception Analysis System 异常分析系统

EAT Environmental Acceptance Test 环境验收测试

EATCS External Active Thermal Control System 外部主动式热控制系统

EATHS Enhanced Airborne Target Handover System 增强型机载目标指挥交接系统

EATS Extended Area Tracking System

扩大的区域跟踪系统
EAV　Experimental Aerospace Vehicle　实验型航空航天运载器
EB　Electronic Beam　电子束
EB　Emergency Box　应急箱
EB　Equipment Bay　设备舱
EB　Explosive Bolts　爆炸螺栓
EB　Extraterrestrial Bases　星际站；航天站
EB-PVD　Electron Beam-Physical Vapor Deposition　电子束物理气相淀积（一种对陶瓷的喷涂方法）
Eb/No　Bit Energy to Noise Density Ratio　比特能量与噪声密度比
EBA　Engine Bleed Air　发动机放出（或引出）的空气
EBA/H　Engine Bleed Air and Hydrazine　发动机放出（或引出）的空气和肼
EBC　Emulated Buffer Computer　仿真缓冲计算机
EBCCD　Electron Bombarded Charge Couple Device　电子轰击的电荷耦合器件；电子轰击式电荷耦合器件
EBCDIC　Extended Binary Code Decimal Interchange Code　扩展型"二一十"进制交换码
EBE　Extraterrestrial Biological Entity　地球外生物实体
EBIS　Electron Beam Ion Source　电子束离子源
EBMA　Engine Booster Maintenance Area　发动机助推器维修场
EBPA　Electron Beam Parametric Amplifier　电子束参量放大器
EBR　Experimental Breeder Reactor　实验性增值反应堆
EBS　Emergency Breathing Subsystem　应急通气分系统
EBS　Engine Breather Separator　发动机通气装置隔板
EBS　European Broadcast Satellite　欧洲广播卫星
EBSV　Engine Bleed Shut Off Valve　发动机放气截流阀门
EBSW　Emulator Basic Software　仿真程序基础软件
EBU　Engine Build-Up　发动机配套装配
EBV　Engine Bay Ventilation　发动机舱通风
EBW　Effective Bandwidth　有效带宽
EBW　Electron Beam Welding　电子束焊接
EBW　Explosive Bridge Wire　爆炸电桥电路
EBZ　Effective Beaten Zone　有效弹着区
EC　Elasticity Coefficient　弹性系数
EC　Electrical Conductivity　导电性；电导率
EC　Electronic Combat　电子战
EC　Embedded Computer　嵌入式计算机
EC　Engine Control　发动机控制
EC　Engine Controller　发动机控制器
EC　Engine Cutoff　发动机关机
EC　Error Control　误差控制
EC　Events Controller　事件控制装置
EC　Events Coupler　事件耦合装置
Ec　Expectation of Casualty　伤亡预测
EC　Experiment Computer　实验型计算机
EC/EDI　Electronic Commerce/Electronic Data Interchange　电子商务/电子数据交换
EC/LS　Environmental Control and Life Support System　环境控制与生

命保障系统
ECA Electronics Control Assembly 电子设备控制组件
ECA Engineering Change Analysis 技术工艺更改分析
ECA Environmentally Controlled Area 环境受控区
ECA Epoxy Curing Agent 环氧树脂固化剂
ECA Equipment Condition Analysis 设备状态分析
ECA Expanded Chaff Adapter 扩展型箔片（条）适配器（装置）
ECAC Electromagnetic-Compatibility Analysis Center 电磁兼容性分析中心
ECAL East Coast Abort Landing 【美国】东海岸中止着陆
ECAM Electronic Caution Alert Module 电子警告与报警模块
ECAP Electronic Control Assembly-Pitch 俯仰电子控制组件
ECAR Electronic Control Assembly-Roll 横滚电子控制组件
ECAS Experiment Computer Applications Software 实验计算机应用软件
ECAY Electronic Control Assembly-Yaw 偏航电子控制组件
ECB Electronically Controlled Birefringence Mode 电子控制双折射模式
ECB Event Control Block 事件控制块
ECB Events Control Buffer 事件控制缓冲器
ECBRS Enhanced Concept Based Requirements System 强化概念基础需求系统
ECC Engineering Change Control 工程更改控制
ECC Engineering Critical Component 工程关键性部件
ECC Error Correcting Code 错误更正码
ECCCS Emergency Command Control Communications System 应急指挥、控制、通信系统
ECCM Electronic Counter-Countermeasures 电子反对抗措施；电子反干扰
ECCM Electronic Counter-Counter Measures 电子反对抗；反电子战
ECCS Emergency Core Cooling System 应急芯冷却系统
ECDR External Critical Damping Resistance 外部临界阻尼电阻
ECE Electron Cyclotron Emission 电子回旋加速器发射
ECE Environmental Control Equipment 环境控制设备
ECE Experiment Checkout Equipment 实验检测设备
ECE External Combustion Engine 外燃发动机
ECEP Experiment Checkout Equipment Processor 实验检测设备处理器
ECET Electronic Control Assembly-Engine Thrust 发动机推力电子控制部件
ECH Engine Compartment Heater 发动机舱加热器
ECHO Environmental Costs of Hazardous Operations 危险操作的环境成本
ECI Earth Centered Inertia 地心惯性
ECIC Earth-Centered Inertial Coordinates 地心惯性坐标
ECIC Equipment Calibration and In-

terference Check 设备校准和接口检查
ECIO Experiment Computer Input/Output 实验计算机输入/输出
ECL Entry Closed Loop 返回闭合回路
ECLS Environmental Control and Life Support 环境控制与生命保障
ECLS ERINT Command and Launch System 增程截击导弹指挥与发射系统
ECLSS Environment Control and Life Support System 环境控制与生命保障系统
ECLSS Environmental Control and Life Support System (Subsystem) 环境控制与生保系统（分系统）
ECM Electronic Control Module 电子控制模块
ECM Electronic Counter Measures 电子对抗；电子战
ECM Engine Condition Monitoring 发动机状态监控
ECM Engine Control Module 发动机控制模块
ECM Equipment Conditioning Monitoring 设备调节监控
ECM Experiment Control and Monitor 实验控制与监测
ECMS Electronic Component Management System 电子元器件管理系统
ECMS Engine Configuration Management System 发动机配置管理系统
ECN Engineering Change Notice 工程更改通知
ECO Engine Combustion 发动机燃烧
ECO Engine Cutoff 发动机停车
ECOM ESA Cost Model 欧洲空局成本模型
ECOMS Early Capability Orbital Manned Station 初期载人轨道站
ECOS Engine Checkout System 发动机检测系统
ECOS Experiment Computer Operating System 实验计算机操作系统
ECP Emergency Command Precedence 应急指挥优先（级）
ECP Evaporative Cooling Processor 蒸发冷却处理装置
ECPY Electronic Control Assembly-Pitch and Yaw 俯仰和偏航电子控制组件
ECR Electron Cyclotron Resonance 电子回旋共振（推进技术）
ECR External Channels Ratio 外通道比；外波道比；外电路比
ECRC Electronics Components Reliability Center 电子设备部件可靠性中心
ECRL Enhanced Common Rail Launcher 增强型通用轨道发射器
ECS Element Capability Specification 部队能力说明
ECS Engagement Control Station 攻击控制台
ECS Engine Control System 发动机控制系统
ECS Environmental Conditioning System 环境调节系统
ECS Environmental Control System (Subsystem) 环境控制系统（分系统）
ECS European Communications Satellite 【欧洲空间局】欧洲通信卫星
ECS Experiment Communication Satellite 【日本】实验通信卫星
ECS Experimental Communication Satellite 实验通信卫星

ECS² Executive Control and Subordinate System 执行控制与辅助系统
ECSC Environment Control System Computer 环境控制系统计算机
ECSC European Communication Satellite Committee 欧洲通信卫星委员会
ECSEL Electronic Combat Simulation and Evaluation Laboratory 电子战仿真与评估实验室
ECSO European Communication Satellite Organization 欧洲通信卫星组织
ECSS European Communication Satellite System 欧洲通信卫星系统
ECSS European Cooperation for Space Standardization 欧洲航天标准化合作（组织）
ECSS Extendable Computer System Simulation 可扩展计算机系统模拟
ECSS Extendable Computer System Simulator 可扩展计算机系统模拟器
ECT Engine Cutoff Timer 发动机停车计时器
ECT Error Control Translator 误差控制转发器
ECT Evaporative Cooling Techniques 蒸发冷却技术
ECTS Electronic Combat Training System 电子战训练系统
ECU Electronic Control Unit 电子控制装置
ECU Electronic Coupling Unit 电子耦合装置
ECU Environmental Control Unit 环境控制装置
ECUSAT Ecumenical Satellite Commission 世界卫星委员会
ECUT Energy Conversion and Utilization Technologies 能量转化与利用技术
ECWS Environment Control Workstation 环境控制工作站
ED End of Descent 下降结束
ED Engineering Division 工程部门
ED Evolutionary Development 渐进式发展
ED Explosive Device 爆炸装置
EDA Electronic Display Assembly 电子显示组件
EDA Elevation Drive Assembly 仰角传动装置
EDA European Defense Agency 欧洲防务局
EDA External Data Aiding 外部数据辅助设备
EDAC Earth Data Analysis Center 地球数据分析中心
EDAC Electronic Dive Angle Control 电子俯冲角控制
EDAC Error Detection and Correction 误差探测与校正
EDARS Environmental Data Acquisition and Recording System 环境数据采集与记录系统
EDASS Environmental Data Acquisition Subsystem 环境数据采集分系统
EDAU Engine Data Acquisition Unit 发动机数据采集装置
EDC Engine Diagnostic Computer 发动机诊断计算机
EDC Engineering Design Change 技术工艺设计更改
EDC European Defense Community 欧洲防务共同体
EDCC Environmental Detection Control Center 环境检测控制中心

EDCPF　Environmental Data Collection and Processing Facility　环境数据收集与处理设施

EDCS　Electronic Data Control System　电子数据控制系统

EDCS　Experimentation Site Exportable Data Collection System　实验站点可输出数据采集系统

EDDIC　Experimental Design Development Integration Center　实验设计发展综合中心

EDDS　Early Docking Demonstration System　早期对接演示验证系统

EDG　Emergency Diesel Generator　应急柴油发电机

EDGE　Enhanced Differential GPS for Guidance Enhancement　加强导航用的加强型差分全球定位系统

EDGES　Electronic Data/Guidelines for Element Survivability　电子数据/部队生存能力指南

EDHE　Experiment Dedicated Heat Exchanger　实验专用热交换器

EDI　Electronic Data Interchange　电子数据交换

EDI　Engine Data Interface　发动机数据接口

EDIF　Engine Data Interface Function　发动机数据接口功能

EDIPS　EROS Digital Image Processing System　【美国】地球资源观测系统数字成像处理系统

EDIU　Engine Data Interface Unit　发动机数据接口装置

EDL　Electrical Discharge Laser　放电激光（器）

EDL　Entry, Descent and Landing　进入、降落和着陆

EDLC　Electric Double Layer Capacitor　电双层电容器

EDLF　ESTEC Data Link Format　欧洲空间研究及空间技术中心的数据链接格式

EDLN　Engineering Development Logic Network　技术工艺研发逻辑网

EDM　Electro-Discharge Machining　放电机械加工（处理，操作）

EDM　Engineering Design Model　工艺设计模型

EDM　Engineering Development Model　工艺研发模型

EDMS　Electronic Data Management System　电子数据管理系统

EDNA　Enhanced Diagnostic Navigational Aid　增强型诊断导航设备

EDO　Extended Duration Orbiter　【美国航天飞机】延长运营时间轨道器

EDP　Electronic Data Processing　电子数据处理

EDP　Embedded Data Processor　嵌入式数据处理机

EDP　Engine Driven Pump　发动机驱动泵

EDP　Engineering Development Process　工艺研发过程

EDPE　Electronic Data Processing Equipment　电子数据处理设备

EDPM　Electronic Data Processing Machine　电子数据处理装置

EDR　Embedded Data Recorder　【美国"爱国者"导弹】嵌入式数据记录器

EDR　Engineering Design Review　技术工艺设计评审（审查）

EDR　Evolutionary Decision Review　渐进式决策评审（审查）

EDRL　Effective Damage Risk Level　有效损伤风险等级

EDRS　European Data Relay Satellite　欧洲数据中继卫星

EDRTS Experimental Data Relay and Tracking Satellite 【日本】实验型数据中继与跟踪卫星

EDS Earth Departure Stage 地球出发级

EDS Earth Departure Stage / Emergency Detection System 【美国】"阿波罗"飞船地球出发级/应急探测系统

EDS Electrical Distribution System 电子分布式系统

EDS Electronic Data Systems Corporation 电子数据系统公司

EDS Embedded Diagnostic System 嵌入式诊断系统

EDS Emergency Detection System 应急故障探测系统

EDS Explosive Detection System 爆炸物探测系统

EDS Explosive Device System 爆炸装置系统

EDSS Equipment Deployment & Storage System 装备调度与储备系统

EDT Electrodynamic Tether 电动力绳系

EDU Electronic Display Unit 电子显示装置

EDU Engineering Development Unit 工程研发装置

EDV EELV-Derived Vehicle 【美国】改进型一次性运载火箭的衍生运载器

EDV Electronic Depressurizing Valve 电子泄压阀

EDW Earth Departure Window 脱离地球窗口

EDWA Engagement Determination and Weapons Assignment 交战决定与武器配发

EDX Exoatmospheric Discrimination Experiment 外大气层识别实验

EE Earth Entry 进入地球大气层

EE Earth Explorer 地球探测器

EE Electrodynamic Explorer 电动力学探测卫星

EE Electronics Equipment 电子设备

EE Engineering Evaluation 工程鉴定

EE/CA Engineering Evaluation and Cost Analysis 工程鉴定与成本分析

EEAS Enhanced En-route Automation System 增强型飞行自动化系统

EEATCS Early External Active Thermal Control System 早期外部主动式热控制系统

EEB Equipped External Bay 外舱

EEC Earth Entry Capsule 进入地球大气层的舱

EEC Electronic Engine Control 电子发动机控制

EEC Electronic Equipment Compartment 电子设备舱

EEC Eurocontrol Experimental Center 欧洲管制的实验中心

EECOM Electrical, Environmental, and Communications 电力、环境与通信

EECOMS Electrical, Environmental, Consumables and Mechanical Systems 电力、环境、能耗与机械系统

EED Electro (Electrical) Explosive Device 电子（电气）爆炸装置

EED Emergency Escape Device 应急逃逸装置

EEE Electromagnetic Environment Experiment 电磁环境实验

EEE Electromagnetic Environmental Effect 电磁环境效应

EEE　Energy-Efficient Engine　节能发动机

EEEU　End Effector Electronics Unit　端部执行器电子装置

EEFI　Essential Elements of Friendly Information　己方信息基本要素

EEH　EMU Electrical Harness　【美国】航天服的电子线束

EEI　Essential Elements of Information　信息基本要素

EEIC　Element of Expense Investment Code　消耗因素/发明编码

EEIS　End-to-End Information System　端一端信息系统

EELP　Explosives, Energetic Liquids, and Pyrotechnics　爆炸、含能油液体与火工品

EELS　Early Entry, Lethality, and Survivability　早期进入、杀伤力与生存能力

EELS　Electron Energy Loss Spectroscopy　电子能量损失谱

EELV　Evolved Expendable Launch Vehicle　渐进型一次性运载火箭

EEM　Early Entry Module　早期返回地球舱

EEM　Earth Entry Module　进入地球大气舱；返回地球舱

EEM　Earth Environmental Monitoring　地球环境监测

EEM　Effective Elastic Modulus　有效弹性模量

EEP　Earth Equatorial Plane　地球赤道（平）面

EEP　Electronics Equipment Package　电子设备整套装置

EER　Explosive Echo Ranging　爆炸回波测距

EERC　Explosive Echo Ranging Change　爆炸回波测距变化

EERJ　External Expansion Ramjet　外膨胀冲压式喷气发动机

EEROC　Expedited Essential Required Operational Capability　快速主要需求作战能力

EES　Effectiveness Evaluation System　效能评价系统；有效性鉴定系统

EES　Ejection Escape Suit　弹射逃逸服

EES　Electro-Explosive Subsystem　电起爆分系统

EES　Emergency Egress System　紧急逃逸系统

EES　Emergency Ejection Suits　应急弹射服

EES　Escape Ejection Seat　弹射座椅

EES　External Environment Simulator　外部环境模拟器

EESS　Emergency Escape Sequencing System　应急逃逸定序系统

EESS　Environmental Effect on Space System　环境对空间系统的影响

EET　End-to-End Test　端到端测试

EET　Entry Elapsed Time　返回耗用时间

EET　Equator Earth Terminal　赤道地面站

EETB　Electronic Electrical Termination Building　电气电子终端厂房

EEU　Electronic Equipment Unit　电子设备装置

EEU　Extravehicular Excursion Unit　舱外偏移装置

EEV　Earth Entry Vehicles　进入地球（大气层）飞行器

EEV　Emergency Escape Vehicle　紧急逃逸飞行器

EEVD　Electronic Equipment Ventilation Device　电子设备通风装置

EEVT　Electrophoresis Equipment Ver-

ification Test 电泳现象设备确认试验
EF Experimental Flight 实验性飞行
EF Exposed Facility 暴露设施
EFA Experiment Flight Applications 实验性飞行应用
EFATO Engine Failure After Takeoff 起飞后发动机停车
EFB Electronic Flight Bag 电子飞行包
EFC Elevator-feel Computer 升降舵操纵感力计算机
EFC Emergency Fuel Control 应急燃料控制
EFC; EFCS Electrical Flight Control System 电气飞行控制系统
EFCS Earth Fixed Coordinate System 地球固定坐标系统
EFCS Electronic Flight Control System 电子飞行控制系统
EFCS Emergency Flight Control System 应急飞行控制系统
EFCU Electronic Flight Control Unit 电子飞行操作装置
EFD Early Failure Detection 早期探伤；早期故障探测
EFD Electronic Flight Display 电子飞行显示
EFD Experimental Fluid Dynamics 实验性流体动力学
EFDARS Expandable Flight Data Acquisition and Recording System 可扩展的飞行数据采集和记录系统
EFDAS Electronic Flight Data System 电子飞行数据系统
EFDC Early Failure Detection Centre 早期故障探测中心
EFDP European Flight Data Processing 欧洲飞行数据处理
EFDT Early Failure Detection Test 早期故障检测试验
EFE Extended Flight Envelope 延伸飞行包层
EFEX Endo-Aeromechanics Flight Experiment 内气动热飞行实验
EFEX Endo-Aerothermal Mechanics Flight Experiment 内气动热机械飞行实验
EFFTAS Effective True Airspeed 有效实际空速
EFGF Electrical Flight Grapple Fixture 电力飞行锚定装置
EFIC Electronic Flight Instrument Controller 电子飞行仪表控制器
EFID Electronic Flight Instrument Display 电子飞行仪表显示器
EFIP Electronic Flight Instrument Processor 电子飞行仪表处理器
EFIS Electronic Flight Instructions 电子飞行指令
EFIS Electronic Flight Instrument System 电子飞行仪表系统
EFL Engineering Field Laboratory 技术工程领域实验室
EFOGM Enhanced Fiber Optic Guided Missile 增强型光纤制导导弹
EFP Emergency Flight Phase 应急飞行阶段
EFP Explosively Formed Penetrator 爆炸成形穿甲弹
EFP Explosively-Formed Projectile 爆炸成形弹
EFRC Edwards Flight Research Center 【美国】爱德华兹飞行研究中心
EFRF Earth Fixed Reference Frame 地球固定参考系；地球固定坐标系统
EFSSS Engine Failure Sensing and Shutdown System 发动机故障指示和停车系统

EFT Emergency Flight Termination 飞行紧急中断
EFT Engine Fuel Tank 发动机燃料箱
EFTS Enhanced Flight Termination System 增强型飞行终端系统
EFV Expeditionary Fighting Vehicle 远征作战车辆
EFVS Enhanced Flight Vision System 增强型飞行可视系统
EG Earth Gravity 地心引力
EG Entry Guidance 返回制导
EGA Earth Gravity Assist 地球重力加速器
EGA Enhanced Graphic Adapter 扩展型图表适配器
EGA Evolved Gas Analysis 逸出气体分析
EGADS Electronic Ground Automatic Destruct Sequencer 电子地面自动炸毁程序装置
EGADS Electronic Ground Automatic Destruct System 电子地面自动炸毁系统
EGAL Elevation Guidance for Approach and Landing 进场[进近]和着陆仰角制导
EGC Experiments Ground Computer 地面实验计算机
EGCS Environmental Generation Control System 环境生成控制系统
EGECON Electronic Geographic Coordinate Navigation System 电气地理坐标导航系统
EGF Electrical Grapple Fixture 电动锚定装置
EGI Embedded GPS/INS 嵌入式全球定位/惯性导航系统
EGME Elevation Gimbal Mounted Electronics 万向支架俯仰电子设备
EGMTR Eglin Gulf Missile Test Range 【美国】埃格林湾导弹试验场
EGNOS European Geostationary Navigation Overlay Service/System 欧洲地球同步卫星覆盖面导航服务/系统
EGO Eccentric Geophysical Observatory 偏心轨道地球物理观测台
EGO Equator Geophysical Observatory 赤道地球物理观测台
EGO Experimental Geophysical Orbiting 实验型地球物理轨道
EGO Experimental Geophysical Orbiting Vehicle 实验型地球物理轨道车
EGPWS Enhanced Ground Proximity Warning System 增强型近地告警系统
EGR Embedded GPS Receiver 嵌入式全球定位系统接收器
EGR Engine Ground Run 发动机地面试车
EGR Enhanced GPS Receiver 增强型 GPS 接收器
EGRESS Emergency Global Rescue, Escape & Survival Systems 应急全球救援、脱险及获救系统
EGRESS Evaluation of Glide Re-Entry Structural System 下滑再入大气层结构系统估计
EGRET Energetic Gamma Ray Experiment Telescope 【美国】高能 γ 射线实验望远镜
EGRS Electronic Geodesic Ranging Satellite 电子测地测距卫星
EGRS Experimental Geodesic Research Satellite 实验测地研究卫星
EGS Experimental Geodesic Satellite 【日本】实验测地卫星;"紫阳花"

卫星
EGSE Electrical Ground Support Equipment 电气地面保障设备
EGSE Experiment Ground Support Equipment 地面实验保障设备
EGT Elapsed Ground Time 地面耗用时间
EGT Estimated Ground Time 地面预用时间
EGT Exhaust Gas Temperature 排气温度
EH&S Environmental Health and Safety 环境健康与安全
EHA Electrohydrostatic Actuator 电静液作动器
EHAS Electrohydrostatic Actuator System 电静液作动系统
EHC Electrical Heating Control 电气热控制
EHF Extremely High Frequency 极高频
EHFLDR Extremely High Frequency Low Data Rate 极高频低速率数据传输
EHL Effective Half Life 有效半衰期
EHL Environmental Health Laboratory 【美国空军】环境安全实验室
EHM Extra High Modulus 特高模量
EHOT External Hydrogen/Oxygen Tank 外储氢/氧箱
EHS Environmental Health Services 环境健康服务
EHS Environmental Health System 环境健康系统
EHSI Electronic Horizontal Situation Indicator 电子水平状态指示器
EHSV Electrohydraulic Servo Valve 电液压伺服阀
EHT Electrothermal Hydrazine Thruster 电热肼推力器
EHTV European Hypersonic Transport Vehicle 【德国】欧洲高超声速运输飞行器
EHV Electro-Hydraulic Valve 电动液压阀
EHVL Enhanced Hyper-Velocity Launcher 加强型超高速发射装置
EI Early Intercept 早期拦截
EI Earth Interface 地球（大气）界面
EI Electromagnetic Interference 电磁干扰
EI Electronic Interface 电子接口
EI Entry Interface 返回接口
EI Environmental Impact 环境影响
EIA Environmental Impact Assessment 环境影响评估
EIAP Environmental Impact Analysis Process 环境影响分析过程
EIC Experimental Intercom 实验内部通信；实验内部对讲
EICAS Engine Indication and Crew Alerting System 发动机（参数）指示与机组告警系统
EICO Element Integration and Checkout 部件装配与检测
EID Electrically Initiated Device 电气触动装置
EIDP End Item Data Package 端项数据包
EIDS Extremely Insensitive Detonating Substance 极不敏感的爆炸物
EIFA Element Interface Functional Analysis 要素（元器件、零部件）接口功能分析
EIOV Equivalent Input Offset Voltage 等效输入补偿电压
EIP Exoatmospheric Interceptor Propulsion 外大气层拦截机（导弹）

推进
EIR Element Integration Review 要素（元器件、零部件）综合评审
EIR Environmental Impact Report 环境影响报告
EIRP Earth Incident Radiated Power 地球事件辐射功率
EIRP Effective Isotropic Radiated Power 有效全向性辐射功率
EIS Ejection Initiation Subsystem 弹射启动分系统
EIS Electrical Integration System 电气集成系统；电气综合系统；电气一体化系统
EIS Electronic Instrument System 电子仪表系统
EIS Engine Indication System 发动机指示（读数，示数，示值）系统
EIS Environmental Impact Statement 环境影响说明
EIS Experiment Initiator System 实验起始器/点火器系统
EIS Explosive Initiation System 爆炸起动系统
EIS Extended Instruction Set 扩展型指令装置
EISA Extended Industry Standard Architecture 扩展型工业标准体系架构
EISE Extendable Integration Support Environment 可扩展综合保障环境
EISF Engine Initial Spares Factor 发动机原始零备件因数
EIT Entry Interface Time 返回接口时间
EIT Exoatmospheric Interceptor Technology 外大气层拦截器技术
EIU Electronic Interface Unit 电子接口装置
EIU Engine Interface Unit 发动机接口装置
EIVT Electrical and Instrumentation Verification Test 电气和仪表设备确认试验
EIVT Electrical Interface Verification Test 电气接口确认试验
EIVT Electronic Installation Verification Test 电子装置确认试验
EJN Ejection 弹射；喷射；推出；抛出
EJOTF Earth-Jupiter Orbiter Transfer Flight 地球－木星轨道器转移飞行
EJSM Europa-Jupiter System Mission 木卫二－木星系统任务
EK Engine Kerosene 发动机煤油
EKV Exoatmospheric Kill Vehicle 外大气层杀伤拦截器
EL Ejector Lift 弹射装置产生的升力（垂直推力）
EL Elastic Limit 弹性极限
EL Electroluminescent 电（致）光的；场致发光的；电荧光的
EL Elevation Angle 仰角；高程角；高度角；垂直角；目标角
EL Extraterrestrial Life 地球外生命
EL MECH Electromechanical 电机的；电动机械的；电力机械的
EL/AZ Elevation/Azimuth 俯仰－方位
EL/VT Ejector Lift, Vectored Thrust 引射升力，矢量推力
ELA Ensemble de Lancement Arian/Ariane Launch Site 【法国】"阿里安"运载火箭发射场
ELACS Extended Life Attitude Control System 扩展型生命姿态控制系统
ELB Electronic Logbook 电子飞行日志

ELC Emergency Launch Capability 应急发射能力

ELC Express Logistics Carrier 快速后勤运输装置

ELCA Earth Landing Control Area 返回地球着陆控制区

ELCG Energetic Liquid Compatibility Group 含能液体兼容组

ELCQ Earth Landing Control Qualified 经授权的返回地球着陆控制

ELDISC Electrical Disconnect 电气断开

ELDO European Launcher Development Organization 欧洲运载器开发组织

ELDT Early Launch Detection and Tracking 早期发射探测与跟踪

ELDV Electrically Operated Depressurization Valve 电气操作减压阀

ELF Experimental Launcher Facility 实验性发射设施

ELF Extremely Low Frequency (1 Hz to 3 kHz) 超低频 (1 Hz 至 3 kHz)

ELGO European Launcher for Geostationary Orbit 欧洲地球静止轨道运载器

ELGRA European Low Gravity Research Association 欧洲低重力研究协会

ELIAS Earth Limb Infrared Atomic Structure 地球临边红外原子结构

ELINFOSEC Electronic Information Security 电子信息安全

ELINT Electromagnetic Intelligence 电磁智能化

ELINT Electronic Intelligence 电子情报，电子侦察

ELINTS Electronic Intelligence Satellite 电子侦察卫星

ELIPS European Life and Physical Sciences (Program) 欧洲生命与物理科学（计划）

ELIS Emitter Location and Identification System 发射机定位与识别系统

ELM Experiment Logistics Module 实验后勤舱（日本空间站舱 JEM 的组成部分）

ELM Extended Lunar Module 【美国】扩展登月舱

ELMC Electrical Load Management Centers 电子载荷管理中心

ELMS Earth Limb Measurement Satellite 地球外缘测量卫星

ELMS Earth Limb Measurement System 地球外缘测量系统

ELMS Elastic Loop Mobility System 弹性回路机动系统

ELMS Electrical Load Management System 电气荷载管理系统

ELMS Engineering Lunar Model Surface 工程月球模型表面

ELMSEL Electrical Load Management System Electronics Unit 电气荷载管理系统电子装置

ELMSIM Engine Life Management Simulation Model 发动机寿命管理仿真模型

ELOISE European Large Orbiting Instrumentation for Solar Experiments 欧洲太阳实验大型轨道仪器

ELOR Extended Lunar Orbital Rendezvous 延长型月球轨道会合；延长型月球轨道会合点

ELORAN Enhanced Long Range Aid to Navigation 增强型长距离导航辅助

ELRAC Electronic Reconnaissance Accessory 电子侦察辅助设备

ELRAD Earth Limb Radiance Experi-

ment 地球临边辐射实验
ELRAS Electronic Reconnaissance Accessory System 电子侦察辅助设备系统
ELS Earth Landing System (Subsystem) 地球着陆系统（分系统）
ELS Earth Limb Sensor 地球临边传感器
ELS Eastern Launch Site 【美国】东部发射场
ELS Elastic Limit Under Shear 抗剪弹性极限
ELS Elevon Load System 升降副翼荷载系统
ELS Emergency Landing Site 应急着陆场
ELS Emergency Landing Strip 应急着陆跑道
ELS Emitter Location System 发射体定位系统
ELS Energy Loss Spectroscopy 能量损失光谱学
ELS Ensemble de Lancement Soyuz (Soyuz Launch Site) 【法属圭亚那航天中心】"联盟号"火箭发射场
ELS Equivalent Level of Safety 等效安全
ELSA Emergency Life Support Apparatus 应急生命保障装置
ELSC Earth Landing Sequence Controller 地球着陆序列控制器
ELSI Enhanced Longwave Spectrometer Imager 增强型长波光谱成像仪
ELSIE Emergency Life-Saving Instant Exit 应急救生瞬时逃逸口
ELSS Emergency Life Support System 应急生命保障系统
ELSS Emplaced Lunar Scientific Station 月球表面科学站
ELSS Extravehicular Life Support System 舱外生命保障系统
ELST Extremely Large Space Telescopes 极大型空间望远镜
ELT Emergency Locator Transmitter 应急定位发射机
ELV Earth Launch Vehicle 地球发射航天器
ELV ELV S.p.A. (European Launch Vehicle) 欧洲运载火箭
ELV Expendable Launch Vehicle 一次性运载火箭
ELV Experimental Launch Vehicle 实验型运载火箭
ELVIS Enhanced Launch Video Imaging System 增强型发射视频成像系统
ELVIS Expert Learning for Vehicle Instruction by Simulation 运载器指令模拟的专家学习
EM Electromagnetic 电磁的
EM Engineering Model 工程模型；工艺模型
EM Exception Monitor 异常状况监视器
EMA Electro Mechanical Actuation 电动机械致动（传动，促动）
EMA Electromagnetic Analysis 电磁分析
EMA Electromechanical Actuator 机电致动器
EMA Extended Mission Apollo 【美国】扩展任务用"阿波罗"飞船
EMAD Engine Maintenance Assembly and Disassembly 发动机维修、装配与拆卸
EMADB Engine Maintenance Assembly and Disassembly Building 发动机维修装卸厂房（间）
EMADS Emergency Malfunction and Display System 应急（备用）故障

与显示系统
EMAMDS　Electronic Master Monitor and Display Systems　主电子监视器与显示系统
EMATT　Enhanced Multi-Mission Advanced Tactical Terminal　增强型多任务先进战术终端
EMC　Elastic Memory Composite　弹性记忆复合材料
EMC　Electromagnetic Compatibility　电磁兼容
EMCC　Emergency Mission Control Center　应急任务控制中心
EMCD　Electromechanical Control Diagram　电子机械控制图
EMCI　Engineering Model Configuration Inspection　工程模型配置检测
EMCON　Emissions Control　发射控制
EMCP　Electromagnetic Containerless Processing　无模电磁处理
EMCPS　Electromagnetic Containerless Processing Science　无模电磁处理科学
EMCS　Enhanced Mission Communications System　增强型任务通信系统
EMCS　European Modular Cultivation System　欧洲舱生物培育系统
EMD　Engineering and Manufacturing Development　工程与制造开发
EMD　Entry Monitor Display　进入（返回）监控显示
EMDP　Engine Model Derivative Program　【美国】发动机改型计划
EMDR　Executive Mission Data Review　任务执行数据评审
EME　Electromagnetic Energy　电磁能
EME　Electromagnetic Environment　电磁环境
EME　Environmental Measurements Experiment　环境测量实验
EMEC　Electromagnetic Effects Capability　电磁影响力
EMEC　Electromagnetic Effects Compatibility　电磁效应兼容性
EMERAUDE　French Space Rocket　法国太空火箭
EMES　Electrical, Mechanical, and Environmental Systems　电气、机械与环境系统
EMES　Electromagnetic Environment Simulator　电磁环境模拟装置
EMETF　Electromagnetic Environmental Test Facility　电磁环境测试设施
EMF　Electrical Maintenance Facility　电气维护设施
EMF　Electromagnetic Force　电磁力
EMF　Electromotive Force　电动势
EMF　Explosive Metal Forming　金属爆炸成形
EMFF　Electromagnetic Form Factor　电磁波形因数
EMI　Electromagnetic Interference　电磁干扰
EMI/EMC　Electromagnetic Interference/ Electromagnetic Compatibility　电磁干扰/电磁兼容
EMIC　Electromagnetic Interference Control　电磁干扰控制
EMID　Electromagnetic Intrusion Detector　电磁干扰探测器
EMIP　Exoatmospheric Midcourse Interceptor Program　外大气层中程拦截机（导弹）项目
EMIR　European Microgravity Research　欧洲微重力研究
EMIS　Electromagnetic Isotope Sepa-

ration 电磁同位素分离
EMISF　EMI Safety Factor　电磁干扰安全系数
EMISM　Electromagnetic Interference Safety Margin　电磁干扰安全裕度
EMISS　Electro-Molecular Instrument Space Simulator　电分子仪表航天模拟器
EML　Electromagnetic Laboratory　电磁实验室
EML　Electromagnetic Launch　电磁发射
EML　Electromagnetic Levitator　电磁悬浮（漂浮）装置
EML　Equatorial Magnetosphere Laboratory　【美国】赤道磁层实验室
EMLDG　Emergency Landing　紧急着陆
EMMCC　Erection Mechanism Motor Control Center　【导弹】竖直（垂直）机械发动机控制中心
EMN　Engineering Management Network　工程管理网络
EMON　Exception Monitoring　异常状况监视（控）
EMOS　Earth's Mean Orbital Speed　地球平均轨道速度
EMOS　European Meteorological Satellite　欧洲气象卫星
EMP　Electro Magnetic Pulse (Generated By A Nuclear Explosion)　电磁脉冲（由核爆炸生成）
EMP　Environmental Measurement Payload　环境测量有效载荷
EMP　Equipment Mounting Plate　设备装配平台；设备安装板
EMP/EMI　Electromagnetic Pulse / Electromagnetic Interference　电磁脉冲/电磁干扰
EMPGS　Electrical/Mechanical Power Generation Subsystem　电气/机械发电分系统
EMPIRE　Early Manned Planetary-Interplanetary Round Trip Expedition　【美国】早期行星与星际载人往返探险
EMR　Eastern Missile Range　东部导弹靶场
EMR　Electromagnetic Radiation　电磁辐射
EMR　Engine Mixture Ratio　发动机燃料组分比
EMR　European Midcourse Radar　欧洲中程雷达
EMRLD　Excimer Moderate Power Raman-Shifted Laser Device　准分子中等功率喇曼频激光器
EMS　Electromagnetic Susceptibility　电磁敏感度（性）
EMS　Engineering Modeling System　工程建模系统
EMS　Entry Monitoring System (Subsystem)　进入（大气层）监测系统（分系统）
EMS　Environmental Management System　环境管理系统
EMS　Environmental Monitoring and Safety　环境监测与安全
EMS　Environmental Monitoring Satellite　环境监测卫星
EMS　Europe Mobile Satellite　欧洲移动卫星
EMS　European Missile System　欧洲导弹系统
EMS　Experimental Measurement Satellite　实验型测量卫星
EMS　Experimental Monitoring Satellite　实验型监控卫星；实验型监视卫星
EMS^3　Enterprise Mission Support Ser-

vices Solutions 【美国】"企业号"任务服务保障解决方案

EMSER Environmental Monitoring, Safety, and Emergency Response 环境监测、安全与应急反应

EMSF External Maintenance and Servicing Facility 外部维护与检修设施

EMSI European Manned Space Infrastructure 欧洲载人空间基础设施

EMSL Environmental Monitoring Systems Laboratory 环境监测系统实验室

EMSL European Material Science Laboratory 欧洲材料科学实验室

EMSP Enhanced Modular Signal Processor 增强型模块信号处理器

EMSS Electromagnetic Servoactuator System 电磁伺服拖动装置系统；电磁伺服执行机构系统；电磁随动操作机构系统

EMSS Emergency Mission Support System 紧急任务支援系统

EMSS Experimental Manned Space Station 实验型载人空间站

EMSS Experimental Mobile Satellite System 【日本】实验型移动通信卫星系统

EMT Electromechanical Test 电子机械测试（试验）

EMTBF Estimated Mean Time Between Failures 所估计的平均故障间隔时间；预计的平均故障间隔时间

EMTF Estimated Mean Time to Failure 所估计的平均故障时间；预计的平均故障时间

EMTTF Equivalent Mean Time To Failure 等效平均故障时间；等效（换算）的平均故障时间

EMU Engineering Model Unit 工程模型装置

EMU EVA(Extra Vehicular Activity) Mobility Unit 舱外活动移动装置（如：航天服）

EMU EVM (Extra-Vehicular-Maneuvering) Mobility Unit 舱外机动（操纵）/移动装置

EMU Expansion Module Unit 扩展模块组件

EMU Extended Memory Unit 扩展存储组件

EMUDS Extravehicular Maneuvering Unit Decontamination System 舱外机动飞行装置净化系统

EMUX Electrical Multiplexing 电多路传输

EMV Electromagnetic Vulnerability 电磁易损性

EMV Extended Mobility Vehicle 扩充机动性飞行器

ENA Engineering Architecture and Analysis 工程体系结构与分析

ENA Exhaust Nozzle Area 排气喷口面积

ENAP Energetic Neutral Atom Precipitation Experiment （美国航天飞机上的）高能中性原子淀析实验

ENCC Emergency Network Control Center 应急网络控制中心

ENCF Experimental Navigation Control Facility 实验导航控制设施

END EFF End Effector 端因子

ENDOSIM Endoatmospheric Simulation 大气层内模拟

ENEC Extendable Nozzle Exit Cone 可延伸喷管出口锥

ENH Earth Near Horizon 近地平线

ENMAP Environmental Mapping and Analysis Program 环境测绘与分析

规划（卫星）

ENNK Endoatmospheric Non-Nuclear Kill 大气层内非核杀伤

ENOC Engineering Network Operations Center 工程网络操作中心

ENORS Engine Not Operationally Ready-Supply 发动机不处于工作准备状态

ENP Exhaust Nozzle Position 排气喷口位置

ENR Equivalent Noise Ratio 等效噪声比

ENR Equivalent Noise Resistance 等效噪声电阻

ENSIP Engine Structural Integrity Program 【美国空军】发动机结构完整性计划

ENTC Engine Negative Torque Control 发动机负扭矩控制

ENviroTRADE Environmental Technologies for Remedial Actions Data Exchange 环境补救措施数据交换技术

ENVISAT Environment Observation Satellite 环境观测卫星

EO Earth Observation 对地球观测

EO Earth Orbit 地球轨道

EO Electro-Optics 电—光的

EO Engine Out 发动机停车；发动机熄火；发动机失效

EO Escape Orbit 逃逸轨道（摆脱天体引力的轨道）

EO DAS Electro-Optical Distributed Aperture System 电—光孔径分布系统

EO/IR Electro-Optical/Infrared 电—光/红外

EOA Early Operation Assessment 早期使用评估

EOARD European Office of Aerospace Research and Development 【美国空军】（驻）欧洲航空航天研究与开发处

EOC Early Operational Capability 初期作战能力

EOC Earth Orbital Capsule 环地球轨道舱

EOC Emergency Operations Center (U.S. Army) 【美国陆军】应急操作中心

EOC Engine-Out Capability 发动机停车（熄火，失效）性能

EOC Enhanced Operational Capability 增强作战能力

EOCT Element Operations Center Test Bed 部队作战中心试验（测试）台

EOD End of Descent 下降结束

EOD Estimated on Dock 泊靠预计

EOD Explosive Ordnance Disposal 爆炸物处理；爆炸性军械（易爆物资）处理

EODAP Earth and Ocean Dynamic Applications Program 【美国国家航空航天局】地球和海洋动力应用计划

EODB End of Data Block 数据块结束

EOE Earth Orbit Ejection 地球轨道入轨

EOED Earth Orbit Escape Device 地球轨道脱险装置

EOEP Earth Observation Envelope Programme 地球观测包络计划

EOF Earth Orbital Flight 地球轨道飞行

EOF Emergency Operations Facility 应急操作设施

EOF End Of Flight 飞行结束

EOF Experimenter's Operation Facili-

ty 实验者操作设施 （成分）
EOGO Eccentric Orbiting Geophysical Observatory 偏心轨道地球物理观测台
EOHT External Oxygen and Hydrogen Tank 外储氧氢箱
EOI Earth-Orbit Insertion 进入地球轨道
EOIATS Electro-Optical Identification and Tracking System 光电识别与跟踪系统
EOIEC Effects Of Initial Entry Condition 进入大气层的最初条件影响
EOIM Evaluation of Oxygen Interaction with Materials 氧与材料互作用的评估
EOL Earth Orbit Launch 地球轨道发射
EOL Engine-Off Landing 发动机关机着陆
EOLC Earth Orbital Launch Configuration 地球轨道发射配置
EOLIA European Pre-Operational Data Link Applications 欧洲预操作数据链应用
EOM Earth Observation Mission 地球观测任务
EOM Earth Orbital Mission 地球轨道飞行任务
EOM End of Mission 任务结束
EOM Engineering Operation Manual 工程操作手册
EOMA Emergency Oxygen Mask Assembly 应急氧气面罩组件
EOMD Earth Observation Marketing Direction 地球观测市场导向
EOMS Earth Orbital Military Satellite 地球轨道军用卫星
EOP Earth and Ocean Physics 地球与海洋物理；地球与海洋物理性质
EOP Earth Observation Program 地球观测项目
EOP Emergency Operating Procedures 应急操作规程
EOP Emergency Oxygen Pack 应急氧气包
EOP Engine Operating Point 发动机工作点
EOP Experiment of Opportunity Payload 搭载有效载荷实验
EOP Experiments of Opportunity 机会实验
EOPAP Earth and Ocean Physics Application Program 地球与海洋物理应用计划
EOPF End Of Powered Flight 动力飞行结束
EOPP Earth Observation Preparatory Program 地球观测筹备计划
EOR Earth-Orbit Rendezvous 地球轨道交会
EOR Extend/Off/Retract 放下（伸出）/关断/收起（缩入）
EORBS Earth Orbiting Recoverable Biological Satellite 地球轨道可回收生物卫星
EORSAT Electronic Intelligence Ocean Reconnaissance Satellite 电子情报海洋侦察卫星
EORSAT ELINT Ocean Reconnaissance Satellite 【俄罗斯】电子情报型海洋侦察卫星
EOS Earth Observation Satellite 地球观测卫星
EOS Earth Observing System 地球观测系统
EOS Earth Orbiting System 【美国国家航空航天局】地球轨道系统
EOS Earth to Orbit Shuttle 地球与轨

道间航天飞机

EOS Emergency Operation System 应急操作系统

EOS Emergency Oxygen Supply 应急供氧

EOS Emergency Oxygen System 应急供氧系统

EOS Extended Operating System 扩展操作系统

EOSD Earth Observation Sensor Development Laboratory 对地观测遥感器研制实验室

EOSDIS Earth Observation System Data and Information System 地球观测系统数据与信息系统

EOSDIS Earth Observing Station Data & Information System 地球观测站数据与信息系统

EOSPF Experimental Orbitography and Synchronisation Processing Facility 实验轨道测定技术与同步处理设备

EOSH Environmental Operational Safety and Health 环境操作安全与健康

EOSS Earth Orbiting Space Station 地球轨道空间站

EOSS Experimental Orbitography and Synchronisation Stations 实验轨道测定技术与同步站

EOT End of Transmission 传输结束；传送结束；发射结束

EOT&E Early Operational Test and Evaluation 早期操作的试验与鉴定

EOTAD Electrooptical Targeting and Designation (System) 光电瞄准与识别（系统）

EOTADS Electro-Optical Target Acquisition and Designation System 光电目标捕获与识别系统

EOTD Electro-Optical Tracking Device 光电跟踪设备

EOTDS Electrooptical Target Detection System 光电目标探测系统

EOTS Electrooptical Targeting System 光电瞄准系统

EOTS Electrooptical Tracking System 光电跟踪系统

EOTV Electric Orbit Transfer Vehicle 电推进轨道转移飞行器

EOV Electrically Operated Valve 电子操控阀

EOVAF Electro-Optical Vulnerability Assessment Facility 光电易损性评估设施

EOVL Engine-Out Vertical Landing 发动机停车垂直降落

EP Earth Probe 地球探测器

EP Electrical Power 电力

EP Electronic Protection 电子防护

EP Entry Point 进入（返回）点

EP Environmental Protective Plan 环境保护计划

EP Exposed Platform 暴露平台

EPAD Electrically Powered Actuation Device 电子制动装置

EPAK Enhanced Paveway Avionics Kit 增强型"宝石路"航空电子系统套件

EPB Equipped Propulsion Bay 推进舱

EPC Earth Prelaunch Calibration 地面发射前校准

EPC Error Protection Code 错误保护码

EPC Etage à Propergols Cryotechniques; Etage Principal Cryotechnique 【法语】低温主级

EPCDC Electrical Power Condition-

ing, Distribution, and Control 电力调节、分配与控制
EPCS　Engine Propulsion Control System　发动机推进控制系统
EPCU　Ensemble de Préparation des Charges Utiles / Payload Preparation Complex　【法语】有效载荷准备设施
EPD　Electrical Power Distribution　电力分配
EPD　Engineering Product and Development　工程产品与研制
EPDB　Electrical Power Data Base　电力数据库
EPDB　Electrical Power Distribution Box　电力分配箱
EPDB　Experiment Power Distribution Box　实验电源分配箱
EPDC　Electrical Power Distribution and Control　电力分配与控制
EPDCS　Electrical Power Distribution and Control System (Subsystem)　电力分配与控制系统（分系统）
EPDS　Electrical Power Distribution System　电力分配系统
EPDS　Experiment Power and Data System　实验电源与数据系统
EPE　Energetic Particle Explorer　高能粒子探险者卫星
EPEAT　Electronic Product Environmental Assessment Tool　电子产品环境评估工具
EPERA　Extractor Parachute Emergency Release Assembly　降落伞应急释放分离装置
EPES　Ejector-Powered Engine Simulator　喷射驱动的发动机模拟器
EPEV　Electro-Actuator Piloting Equipment VEGA　【法国】"织女星"运载火箭导航设备
EPF　Extended Payload Fairing　加长型整流罩
EPF　External Payload Facility　外部有效载荷设施
EPFCS　Electrical Primary Flight Control System　初级飞行电控系统
EPG　Electrical Power Generator　电力发生器；电力发电机
EPGS　Electrical Power Generation and Energy Storage　【美国空间站】发电与电能储存分系统
EPHV　Electronique Pilotage Hydraulique VEGA　【法国】"织女星"运载火箭液压电子导航设备
EPIC　Earth-Pointing Instrument Carrier　指向地球的仪器运载卫星
EPIC　Extended Performance and Increased Capabilities　延伸的性能与提高的能力
EPICC　Enhanced Processor, Interface Controller and Communications　增强型处理器、接口控制器和通信系统
EPICS　Experimental Physics and Instrument Control System　实验物理与仪表控制系统
EPIL　Experimental Preflight Inspection Letter　实验性飞行前检查书
EPIP　Evolutionary Phase Implementation Plan　渐进式阶段实施计划
EPIRB　Emergency Position Indicating Radio Beacon　紧急无线电示位标
EPITS　Essential Program Information Technology and Systems　基础项目的信息技术与系统
EPL　Electrical Power Level　电力级
EPL　Emergency Power Level　应急电源级
EPLD　Electronically Programmable Logic Device　电子程控逻辑装置

EPLRS Enhanced Position Location and Reporting System 增强型定位与报告系统

EPLRS Enhanced Position Location Reporting System 增强型位置定位通报系统

EPLRSGRU Enhanced Position Location & Reporting System Gridreference Unit 增强型定位报告系统坐标单元

EPM Enabling Propulsion Materials 能动推进材料

EPM Equipped Pressurized Module 增压舱

EPM European Physiology Module(s) (multi-user facility) 欧洲生理舱（多用户设施）

EPM-1 European Physiology Module(s) (launch configuration) 欧洲生理舱（发射结构型）

EPMARV Earth-Penetrating Maneuvering Reentry Vehicle 【美国空军】进入稠密大气层机动弹头；再入大气层机动飞行器

EPMS Engineering Performance Management System 工程效能管理系统；工艺性能（特性）管理系统

EPO Earth Parking Orbit 地球停泊轨道

EPOC External Payload Operations Center 外部有效载荷运行中心

EPOP European Polar-Orbiting Platform 欧洲极轨道平台

EPPVS Emergency Propulsive Propellant Venting System 应急推进剂排气系统

EPR Engine Pressure Ratio 发动机压力比

EPRL Engine Pressure Ratio Limit 发动机压力比极限

EPROM Electrically Programmable Read-Only Memory 电子编程只读存储（器）

EPS Electric Power System 电力系统

EPS Electrical Power System (Subsystem) 电力系统（分系统）

EPS Emergency Pressurization System 应急加压系统

EPS Emergency Propulsion System 应急推进系统

EPS Energetic Particle Satellite 高能粒子探测卫星

EPS Engineering Performance Specification 工程性能说明；技术工艺性能（特性）说明

EPS Escape Propulsion System 逃逸推进系统

EPS E'tage a Propergrds Sinkable 【法国】"阿里安"5上面级推进舱

EPS EUMETSAT Polar System 欧洲气象卫星极系统

EPS Evaluation Planning System 鉴定规划系统

EPS Experiment Prototype Silo 实验原型发射井

EPS Experimental Power Supply 实验供电

EPS Storable Propellant Stage 可贮存推进剂级

EPSOC Earth Physics Satellite Observation Campaign 地球物理卫星学观测运动

EPSP Experiment Power Switching Panel 实验电源切换控制屏（操纵台）

EPT Elevating Platform for Transportation 运输升降平台

EPT Elevating Platform Transporter 升降平台运输车

EPTME　Encapsulated Payload Transportation Mechanical Equipment　封装的有效载荷运输机械设备
EPU　Emergency Power Unit　应急动力装置
EPU　Experiment Preparation Unit　实验准备装置
EPUPS　Emergency Power Unit Pressures　应急动力装置压力
EPVS　Emergency Propellant Venting System　推进剂应急排气系统
EQ　Equivalent　当量；等效
EQC　Environmental Quality Control　环境质量控制
EQE　Event Queue Element　事件列元素（组件）
EQEA　Environmental Quality Economic Analysis　环境质量经济性分析
EQUIV　Equivalent　等效量，等效值，等效势；当量，等量
ER　Earth Research　地球研究
ER　Eastern Range　【美国】东部试验靶场
ER　Extended Range　扩展范围；扩展型靶场
ER-MLRS　Extended-Range Multiple-launched Rocket System　增程多管火箭发射系统
ERA　Electrical Replaceable Assembly　可替换的电子组件（装置）
ERA　European Robotic Arm　欧洲机器人臂
ERA　Exobiology and Radiation Assembly　外空生物学与辐射装置
ERAAM　Extended Range Air-to-Air Missile　增程空空导弹
ERAD　Explosive Release Atmospheric Dispersion　爆炸物释放大气弥散
ERADCO　Army Electronics Research and Development Command　【美国】陆军电子研究与发展司令部
ERAM　Extended Range Active Missile　增程型主动导弹
ERAPDS　Enhanced Recognized Air Picture Discrimination System　改进型已识别的空中目标分辨系统
ERASER　Elevated Radiation Seeking Rocket　加大仰角式辐射自导火箭
ERASER　Enhanced Recognition and Sensing Laser Radar　增强型识别与遥感激光雷达
ERBE　Earth Radiation Budget Experiment　地球辐射收支实验
ERBI　Earth Radiation Budget Instrument　地球辐射收支仪
ERBM　Extended Range Ballistic Missile　增程弹道导弹
ERBOS　Earth Radiation Budget Observation Satellite　地球辐射收支观测卫星
ERBS　Earth Radiation Balance Satellite　地球辐射平衡卫星
ERBS　Earth Radiation Budget Satellite　【美国】地球辐射收支测量卫星
ERBS　Earth Radiation Budget Sensing System　地球辐射收支遥感系统
ERC　Earth Rate Compensation　地球速率补偿
ERC　Electronics Research Center　【美国国家航空航天局】电子设备研究中心
ERC　Event Recorder　事件记录器（仪）
ERCS　Emergency Rocket Communications System　【美国空军】应急火箭通信系统
ERD　Emergency Return Device　【航天器】应急返回装置
ERD　Evoked Response Detector　受

激响应探测器
ERDS Explosive Residue Detention System 爆炸物残渣滞留系统
EREP Earth Resource Experiment Package 地球资源实验组件
EREP Earth Resource Experiment Program 【美国】地球资源实验计划
ERETS Edwards Rocket Engine Test Station 【美国】爱德华兹火箭发动机试验站
ERF Exponential Reliability Function 指数可靠性函数
ERFD Earth Resources Flight Data 地球资源飞行数据信息
ERI Extravehicular Reference Information 舱外活动参照信息
ERINT Extended Range Intercept Technology 增程拦截技术
ERINT Extended Range Interceptor 远程截击导弹
ERIP Experiment Requirement and Implementation Plan 实验要求与实施计划
ERIPS Earth Resources Interactive Processing System 【美国】地球资源交互式处理系统
ERIS Earth-Reflecting Ionospheric Sounder 地球反射电离层探测器
ERIS Exoatmospheric Reentry Vehicle Interceptor System 【美国】外大气层再入弹头拦截器系统
ERJ External Combustion Ramjet 外燃冲压式喷气发动机
ERL Environmental Research Laboratory 环境研究实验室
ERL (NASA) Environmental Resources Laboratory 【美国国家航空航天局】环境资源实验室
ERM Earth Reentry Module 返回地球舱；地球大气层再入舱
ERM Earth Return Module 重返地球舱
ERM Evaporated Rate Monitor 蒸发率监测器
ERM Experimental Rocket Motor 实验火箭发动机
ERNEE Energetic and Relativistic Nuclei and Electron Experiment 能量上的相对论核子与电子实验
EROPS Extended Range Operations 延距操作；延程操作
EROS Earth Resources Observation Satellite 【美国】地球资源观测卫星
EROS Earth Resources Observation System 地球资源观测系统
ERP Effected Radiative Power 实施的辐射功率
ERP Effective Radiation Power 有效辐射功率
ERPG Emergency Response Planning Guidelines 应急响应规划指南
ERPM Engineering Requirements and Procedures Manual 工艺要求与规程手册
ERPS Experimental Rocket Propulsion Society 实验性火箭推进器学会
ERR Element Requirements Review 元器（部）件需求评审
ERR CNTR Error Counter 错误计数器
ERRC Expendability, Recoverability, Repair Capability 消耗性、可复原性、可修复性
ERRC Expendability, Recoverability, Repairability Code 一次使用性、可回收性、可修复性准则
ERRC Expendability, Recoverability, Repairability Cost 消耗性、可复原

性、可修复性成本
ERRDF Earth Resources Research Data Facility 地球资源研究数据设施
ERS Earth Recovery Subsystem 地面回收子系统
ERS Emergency Response System 应急响应系统
ERS Engineering Release System 技术工艺发布系统
ERS Entry and Recovery Simulation 进入大气层和回收模拟
ERS Environmental Research Satellite 环境研究卫星
ERS ESA Remote Sensing Satellite 欧洲空间局遥感卫星
ERS European Remote Sensing Satellite 欧洲遥感卫星
ERS-1 Earth Resources Sensing Satellite-1 【日本】地球资源卫星1号
ERSATS Earth Resources Survey Satellites 地球资源勘测卫星
ERSI Elastomeric Reusable Surface Insulation 可再用弹性表面绝缘（材料）
ERSIR Earth Resources Shuttle Imaging Radar 地球资源航天飞机成像雷达
ERSL Enhanced and Redesigned Scripting Language 增强与再设计脚本语言
ERSOS Earth Resources Survey Operational Satellite 工作型地球资源勘测卫星
ERSR Equipment Reliability Status Report 设备可靠性状态报告
ERSS Earth Resources Survey Satellite 地球资源勘测卫星
ERSS ESA Remote Sensing Satellite 欧洲空间局遥感卫星

ERT Equivalent Random Theory 等效随机理论
ERTS Earth Resources Technology Satellites 【美国】地球资源技术卫星
ERTS Earth Resources Test Satellite 地球资源试验卫星
ERTS Embedded Realtime System 嵌入式实时系统
ERTS Emergency Remote Tracking Station 应急远程跟踪站
ERU Earth Rate Unit 地球转速单位；地球角速度单位
ERU Ejector Release Unit 弹射器；释放装置
ERU Engine Replay Unit 发动机继电器组件
ERU Equipment Replaceable Unit 设备可替换单元
ERV Earth Return Vehicles 地球返回飞行器
ERV Two Stage Expendable Rocket Vehicle 不可回收两级火箭飞行器
ES Earth Satellite 地球卫星
ES Earth Station 地面站
ES Electrical System 电气系统
ES Escape System 逃逸系统
ES Experiment Segment 实验舱段
ES Exposed Site 暴露场区
ESA Electronically Scanned Array 电子扫描阵列
ESA Engineering Supply Area 工程供应区
ESA Engineering Support Assembly 技术工艺保障装置（组件）
ESA European Space Agency 欧洲空间局
ESA Explosive Safe Area 爆炸安全区
ESA-IRS European Space Agency In-

formation Retrieval Service 欧洲空间局信息检索服务处

ESACCERT ESA Computer and Communications Emergency Response Team 欧洲空间局计算机与通信应急响应小组

ESACOM ESA Communication Corporate Wide Area Network 欧洲空间局通信合作宽域网

ESAD Electronic Safe and Arm Device 电子保险与解保装置

ESAHQ European Space Agency Head Quarters 欧洲空间局总部

ESAI Electronic Standby Attitude Indicator 姿态电子辅助指示器

ESAI Expanded Situational Awareness Insertion 增强态势感知插入

ESAISA Explosive Safety Internal Self-Assessment 爆炸物安全性内部自我评估

ESANET European Space Agency Information Network 欧洲空间局信息网

ESAR Extended Subsequent Application Review 扩大的后续应用评审

ESARS Earth Surveillance and Rendezvous Simulator 地球监视与交会模拟器

ESAS Electronic Situation Awareness System 电子态势感知系统

ESAS Enhanced Situational Awareness System 增强型态势感知系统

ESAS Expert System for Avionics Simulation 航空电子设备模拟专家系统

ESAS Exploration Systems Architecture Study 探索系统体系架构研究

ESBR Electronic Stacked Beam Radar 电子叠层波束雷达

ESC Electronic Security Command 电子安全指令

ESC Electronic System Center 电子系统中心

ESC Engine Start Command 发动机启动指令

ESC Engineering Support Center 工程保障中心

ESC Enhanced Satellite Capability 增强式卫星能力

ESC Error Status Code 【美国国家航空航天局】错误状态代码

ESC Etage Supérieur Cryotechnique 【法语】低温上面级

ESCA Electron Spectroscopy for Chemical Analysis 化学分析电子光谱仪（计）

ESCEC Experimental Satellite Communication Earth Center 实验卫星通信地面中心

ESCES Experimental Satellite Communications Earth Station 【印度】实验卫星通信地球站

ESCS Electronic Spacecraft Simulator 航天器电子模拟器

ESCS Emergency SATCOM System 应急卫星通信系统

ESCU Extended Service and Cooling Umbilical 扩展伺服与冷却脐带

ESD Electrostatic Discharge 静电放电

ESD Emergency Shutdown 紧急关机

ESD Equivalent Sphere Diameter 等效球体直径

ESD Event Sequence Diagram 事件序列图表

ESDAC European Space Data Analysis Centre 欧洲空间数据分析中心

ESDAC European Space Data Center 欧洲航天数据中心（设在德国）

ESDI External Spatial Data Infrastructure 外部空间数据基础设施

ESDIS Earth Science Data and Information System 地球科学数据与信息系统

ESDOC European Science Data and Operation Center 欧洲科学数据与运算中心

ESDP Evolutionary System for Data Processing 数据处理改进系统

ESDRN Eastern Satellite Data Relay Network 【俄罗斯】东部卫星数据中继网

ESDU Event Storage and Distribution Unit 事件存储与分配装置

ESE Earth Satellite Environment 地球卫星环境

ESE Electrical Support Equipment 电气保障设备

ESE Electronic Support Equipment 电子保障设备

ESE Extravehicular Support Equipment 舱外保障设备

ESEX Electric Propulsion Space Experiment 电子推进空间实验

ESF Environmental Storage Facility 环境存储厂房

ESF Explosive Safe Facility 爆炸安全设施

ESFAS Engineered Safety Features Actuation System 安全专用设备致动（驱动，触发）系统

ESH Environmental, Safety and Health 环境、安全与健康

ESI Electrical System Integration 电子系统集成（综合）；电子系统一体化

ESI Electro-Spray Ionization 电喷离子化

ESI Enterprise Software Initiative 企业软件倡议

ESI External System Interface 外部系统接口

ESID Engine and System Indication Display 发动机与系统指示显示

ESIHS Expert System Information Handling System 专家系统信息处理系统

ESIM Environment and Space Induction Mechanism 环境与太空感应检测装置

ESIS Electronic Standby Instrument System 电子备用仪表系统

ESIS Engine and System Indication System 发动机与系统指示系统

ESIT Electrical System Integrated Test 电气系统综合测试

ESL Earth Science Laboratory 地球科学实验室

ESL Estimated Service Life 估计使用年限

ESL Experimental Space Laboratory 实验性空间实验室

ESL External Sensors Lab 外部传感器实验室

ESLAB European Space Laboratory 欧洲空间实验室

ESLO European Satellite Launching Organization 欧洲卫星发射组织

ESLO European Space Launcher Organization 欧洲航天运载火箭组织

ESM Effectiveness Simulation Model 效率模拟模型；效率仿真模型

ESM Electronic Warfare Support Measures 电子战保障措施

ESM Energy Storage Module 能量存储模块

ESM Equipment Support Module 设备保障舱

ESM ESA Soyuz Mission 【欧洲空

间局】"联盟号"火箭发射任务
ESM Experimental Support Module 实验保障舱
ESMA Earth Science Modeling and Assimilation 地球科学建模及同化
ESMC Eastern Space and Missile Center 【美国空军】东部航天与导弹中心
ESMD Exploration Systems Mission Directions 探索系统任务指南
ESMD Exploration Systems Mission Directorate 【美国国家航空航天局】探测系统任务委员会
ESMOS Earth Science Mission Operations and Systems 地球科学任务操作与系统
ESMR Electrically Scanned Microwave Radiometer 电子扫描微波辐射计（仪）
ESMRO Experiments for Satellite and Materials Recovery from Orbit 从轨道回收卫星与材料的实验
ESNet Energy Sciences Network 能源科学网络
ESO European Southern Observatory 欧洲南部观测台；欧洲南部观测所
ESO Event Sequence Override 事件序列补偿
ESOC Environmental Satellite Operations Center 环境卫星作业中心
ESOC European Satellite Operations Center 【欧洲空间局】欧洲卫星操作中心
ESOC European Space Operations Centre 欧洲航天操作中心
ESP Encapsulating Security Payload 封装安全有效载荷
ESP Engine Service Platforms 发动机操作平台；发动机检修平台
ESP Experiment Sensing Platform 实验遥感平台
ESP External Stowage Platform 外部存储平台
ESP Extravehicular Support Pack 舱外活动保障组件
ESPA EELV (Enhanced Expendable Launch Vehicle) Secondary Payload Adapter 改进型一次性运载火箭第二有效载荷适配器
ESPOCC European Space Project Operations Control Center 欧洲空间项目运行管理中心
ESPRIT European Strategic Program of Research in Information Technology 欧洲信息技术研究战略项目
ESPS Experiment Segment Pallet Simulator 实验段平台模拟器
ESQD Explosive Safety Quantity Distance 爆炸安全量距
ESR Equivalent Service Rounds 当量发射弹数
ESR&T Exploration Systems Research and Technology 探索系统研究与技术
ESRANGE European Sounding Rocket Range 【欧洲空间局】欧洲探空火箭靶场
ESRANGE European Space Launching Range 欧洲航天发射试验场（位于瑞典的基律纳）
ESRANGE European Space Range 【瑞典】欧洲航天发射场
ESRANGE European Space Research Range 欧洲航天研究试验场
ESRIN European Space Research Institute 【欧洲空间局】欧洲空间研究所（设于意大利）
ESRLAB European Space Research Laboratory 【欧洲空间局】欧洲太空研究实验室

ESRO European Space Research Organization 欧洲空间研究组织
ESRP European Supersonic Research Program 欧洲超声速研究计划
ESS Electronic Switching System 电子切换系统
ESS Emplaced Scientific Station 安置在月球上的科学站
ESS Entry Survival System 进入（大气层）救生系统
ESS Environmental Support System 环境保障系统
ESS Environmental Survey Satellite 环境勘察卫星
ESS Equipment Support Section 设备保障部分
ESS European Space Station 欧洲空间站
ESS Expendable Second Stage 一次性使用的第二级（火箭）
ESS Experiment Subsystem Simulator 实验分系统模拟器
ESS Experiment Support System 【航天器】实验保障系统
ESS Experimental Synchronous Satellite 实验同步卫星
ESSA Environmental Survey Satellite 【美国】艾萨卫星；环境勘测卫星
ESSA Expand Space Sensor Architecture 扩展空间传感器体系结构
ESSDE European Space Software Development Environment 欧洲航天软件开发环境
ESSM Evolved (Enhanced) Sea Sparrow Missile 【美国】改进（增强）型"海麻雀"导弹
ESSME Expendable Space Shuttle Main Engine 一次性航天飞机主发动机
ESSP Earth System Science Pathfinder 地球系统科学探索者项目
ESSS European Space Suit System 欧洲航天服系统
ESSTE Embedded Software Simulation Testing Environment 嵌入式软件的仿真测试环境
EST Employment and Suitability Test 使用性与适用性试验
EST Engine System Test 发动机系统试验
ESTA Escape System Test Article 逃逸系统试验组件
ESTAR (NASA) Electronically Scanned Thinned Array Radiometer 【美国国家航空航天局】电子扫描薄阵列辐射计
ESTARS Electronic Suspense Tracking and Routing System 电子悬浮跟踪与航线飞行系统
ESTB EGNOS System Test Bed 欧洲静地星导航重叠服务系统测试台
ESTC European Space Technology Centre (ESA) 【欧洲空间局】欧洲空间技术中心
ESTC European Space Tribology Centre 欧洲空间磨损学中心
ESTE Electronic System Test Equipment 电子系统测试设备
ESTEC European Space Research and Technology Centre 【欧洲空间局】欧洲空间研究与技术中心（设于荷兰）
ESTEC European Space Technology Centre 欧洲航天研究与技术中心
ESTL Electronic Systems Test Laboratory 【美国国家航空航天局】电子系统测试实验室
ESTL European Space Tribology Laboratory 欧洲空间磨损学实验室
ESTP Electronic Satellite Tracking

Program 电子卫星跟踪计划
ESTRAC(K) European Space Satellite Tracking and Telemetry Network 欧洲卫星跟踪与遥测网
ESTRACK European Satellite Tracking Telemetry and Telecommunications Network 欧洲卫星跟踪、遥测与电信网
ESTSC Energy Science and Technology Software Center 能源科学与技术软件中心
ESV Earth Satellite Vehicle 地球卫星运载器
ESV Emergency Shutoff Valve 紧急关闭阀
ESVS Escape Suit Ventilation System 逃逸服排气系统
ESVS Escape System Ventilation System 逃逸系统通风系统
ESWS Earth Satellite Weapon System 地球卫星武器系统
ET Effective Temperature 有效温度
ET Elapsed Time 耗用时间
ET Elevated Temperature 温度升高
ET Embedded Test 嵌入式试验
ET Endurance Test 耐久性试验
ET Escape Tower 应急脱离塔；应急逃生塔
ET Event Timer 事件计时器
ET External Tank 【美国航天飞机】外贮箱
ET&C Extended Tracking and Control 扩展型跟踪与控制
ET-SEP External Tank Separation 【美国航天飞机】外储箱分离
ETA Ejector Thrust Augmentor 引射器加力装置
ETA Engine Temperature Alarm 发动机过热报警器
ETA Environmental Test Article 环境测试组件
ETA Estimated Time of Arrival 预计到达时间
ETA Estimated Time to Acquisition 预计采集时间
ETA Explosive Transfer Assembly 爆炸传输装置
ETA External Tank Attachment 外储箱附件
ETA Event Tree Analysis 事件树分析
ETAADS Engine Technical and Administrative Data System 发动机技术与管理数据系统
ETAS Elevated Target Acquisition System 大仰角目标捕获系统
ETB Equipment Transfer Bag 设备转移袋
ETB Extraterrestrial Base 星际站；星际基地
ETBA Energy Trace Barrier Analysis 能量痕量与屏障分析
ETC/LSS Environmental Thermal Control/Life Support System 环境热控制与生命保障系统
ETCA ET Intertank Carrier Plate Assembly 【美国航天飞机】外贮箱内顶（承）板组件（装配）
ETCS Exchange Terminal Circuit for Satellite Circuit 卫星电路交换机终端电路
ETCS External Thermal Control System 外部热控制系统
ETD Electrical Terminal Distributor 电子终端分配器
ETD Estimated Turnover Date 预计回转日期；预计周转日期；预计移交日期
ETD External Tank Door 【美国航天飞机】外贮箱门

ETDD Enabling Technology Development and Demonstration 使能技术发展与演示验证

ETDD Exploration Technology Development Demonstrations 探索技术发展演示验证

ETDL Electronic Technology and Devices Laboratory 电子技术与装置实验室

ETDP Exploration Technology Development Program 探索技术发展计划

ETDRS Experimental Track and Data Relay Satellite 【日本】实验型跟踪与数据中继卫星

ETE Element Test and Evaluation 元器（部）件测试与鉴定

ETE End-to-End 端至端

ETE Environmental Test and Evaluation 环境试验与评定

ETE Explosives Test Equipment 爆炸物测试设备

ETEC Expendable Turbine Engine Concept 一次性涡轮发动机设计方案

ETED External Thermal Environment Data 外部热环境数据

ETEDB External Thermal Environment Data Base 外部热环境数据库

ETERTS End-to-End Real Time Simulator 端对端实时模拟

ETESD End-to-End Sensor Demonstration 端对端传感器演示验证

ETF Engine Test Facility 发动机试车台

ETG Extra-Vehicular Thermal Garment 舱外隔热服

ETI Elapsed Time Indicator 耗用时间指示器

ETI Estimated Time of Interception 预计拦截时间

ETI-PD Early Transient Incapacitation and Performance Decrement 初期瞬时失能与减效

ETL Estimated Time of Launch 预计发射时间

ETLOW External Tank Lift-Off Weight 外储箱升空重量

ETM Elapsed Time Meter 耗用时间测定仪

ETM Electronic Trajectory Measurement 电子弹道测量

ETM Engineering Test Model 工程试验（测试）模型

ETM Engineering Test Motor 工程测试发动机

ETNO European Telecommunications Network Operators 欧洲通信网络操作员

ETO Earth to Orbit 地球至轨道

ETO Earth-to-Orbit (Launcher) 地球至轨道；地球至轨道（发射器）

ETO Estimated Takeoff 估计起飞时间；估计发射时间；预计起飞时间

ETP Experimental Test Procedure 实验性试验程序

ETP External Thermal Protection 外部防热系统

ETPAR External Tank Project Assessment Review 外储箱项目评估审查

EPB Equipped Propulsion Bay 推进舱

ETPY ECA Thrust Vector, Pitch and Yaw 推力矢量、俯仰与偏航的电子控制组件

ETR Eastern Test Range 【美国】东部试验靶场

ETR Engine Thrust Request 发动机推力要求

ETR Environmental Test Round 环境试验（测试）周期

ETRFB Engine Thrust Request Feedback 发动机推力要求反馈

ETRM External Tank Rocket Motor 外挂燃料箱式火箭发动机

ETRO Estimated Time of Return to Operation 估计返回战斗时间；估计重回运转时间；估计恢复运转时间

ETS Earth Technology Satellite 地球技术卫星

ETS Electrical Test Set 电子测试设备（装置）

ETS Elevated TOW System 增加仰角的"陶式"弹道系统

ETS Energy Transfer System 能量转移系统；能量传递系统

ETS Engineering Test Satellite【日本】工程试验卫星

ETS Experimental Test Site 实验导弹试验场

ETS External Tank System (Subsystem)【美国航天飞机】外储箱系统（分系统）

ETSS External Tank Separation Subsystem【美国航天飞机】外贮箱分离分系统

ETSTA External Tank Start Test Article【美国航天飞机】外贮箱启动试验件

ETSTA External Tank Static Test Article【美国航天飞机】外储箱静力试验件

ETU Engineering Test Unit 工程试验装置；技术工艺测试组件

ETV Engine Test Vehicle 发动机试验飞行器

ETVA External Tank Vent Arm 外储箱通气臂

ETVAS External Tank Vent Arm System 外储箱通气臂系统

ETW European Transonic Windtunnel 欧洲跨声速风洞

EU Electronic Unit 电子装置

EU Engineering Units 工艺组件

EU Experimental Unit 实验装置

EUAFS Enhanced Upper Air Forecast System 增强型上部大气预报系统

EUDDL Equivalent Uniformly Distributed Dead Load 等效均匀分布静载荷

EUE Experiment-Unique Equipment 实验特有设备

EULMS Engine-Usage Life Monitoring System 发动机使用寿命监控系统

EUMETSAT European Organization for the Exploitation of Meteorological Satellite 欧洲利用气象卫星组织

EUMILSAT European Military Communications Satellite 欧洲军用通信卫星

EUMILSATCOM European Military Satellite Communications 欧洲军用卫星通信

EUMS Engine-Usage Monitoring System 发动机使用监控系统

EURECA European Research Coordination Agency 欧洲研究协调机构（其研究计划即称为尤里卡计划）

EURECA European Retrievable Carrier 尤里卡平台；欧洲可回收平台

EUROCOMSAT European Consortium Communications Satellite 欧洲联合通信卫星

EUROSAT European Application Satellite System 欧洲应用卫星系统

EUROSAT European Regional Satellite 欧洲区域卫星

EUSC European Union Satellite Centre 欧洲联合卫星中心

EUSO Extreme Universe Space Observatory 极端宇宙空间天文台

EUTE Early User Test and Experimentation 早期用户测试与实验方法

EuTEF European Technology Exposure Facility 欧洲技术暴露设施

EUV Extreme Ultraviolet 极远紫外

EUVE Extreme Ultraviolet Explorer 【美国国家航空航天局】极远紫外探测器

EUVES Extreme Ultraviolet Explorer Satellite 【美国国家航空航天局】极远紫外探测器卫星

EV Expendable Vehicles 一次性使用运载器

EV Extravehicular 【航天】舱外的

EVA Elevation Versus Amplitude 仰角与振幅比较关系

EVA Extra Vehicular Activity 舱外活动；出舱活动

EVA WS EVA Workstation 舱外活动工作站；出舱活动工作站

EVA/IVA Extra-/Intra-Vehicular Activities 舱外和舱内活动

EVADE Escape Vehicle Analysis and Definition for EMSI 欧洲载人空间基础设施的乘员逃生飞行器分析与初步设计

EVAL Earth Viewing Applications Laboratory 地球观测应用实验室

EVAP Evaporator 蒸发器；汽化器；蒸发干燥器

EVAPSP Extravehicular Activity Physiology, Systems and Performance 出舱活动生理、系统和工作能力（性能）

EVAR Experimental Vehicle for Avionics Research 用于航空电子学研究的实验飞行器

EVAS Extravehicular Activity System 舱外活动系统

EVATA Extravehicular Activity Translational Aid 舱外活动平移辅助设备

EVC Error Vector Computer 误差矢量计算机

EVC Extravehicular Communications 舱外通信

EVCF Eastern Vehicle Checkout Facility 【美国】东部靶场飞行器检测厂房

EVCON Events Control (Subsystem) 事件控制（分系统）

EVCS Extravehicular Communications System 舱外通信系统

EVCTD Extravehicular Crew Transfer Device 航天器乘员舱外转移装置

EVCU Extravehicular Communications Umbilical 舱外通信脐带

EVD External Visual Display 外部可见显示；外部视觉显示

EVE Extraterrestrial Vulnerability Experiment 地球外易损性试验

EVEA Extra Vehicular Engineering Activities 舱外工程技术活动

EVLSS Extravehicular Life Support System 舱外生命保障系统

EVM Earth Viewing Module 对地观测舱

EVM Engine Vibration Monitoring 发动机振动监测

EVMS Earned Value Management System 收益值管理系统

EVMU Extra Vehicular Mobility Unit 舱外机动装置；舱外灵便式航天服

EVO Extravehicular Operation 舱外作业

EVPA Experimental Version Performance Assessment 实验型性能评估

EVPA/TEVS Experimental Version Performance Assessment Test Environment System 实验型性能评估与试验环境系统

EVR Extra Vehicular Robotics 舱外机器人技术；出舱机器人

EVR Extravehicular Robotics 出舱机器人

EVS Enhanced Verdin System 【美国】增强型"山雀"导弹系统

EVS Equipment Visibility System 设备可视化系统；设备能见度系统

EVS Extravehicular Suit 舱外活动服

EVSC Extravehicular Suit Communications 舱外活动服中的通话设备

EVSS Extravehicular Space Suit 舱外活动航天服

EVSTC Extravehicular Suit Telecommunications 舱外活动航天服的无线电通信

EVSTC Extravehicular Suit Telemetry and Communications 舱外活动航天服的遥测与通信

EVSTCS Extravehicular Suit Telemetry and Communications System 舱外活动服中的遥测和通信系统

EVSU Extravehicular Space Unit 舱外活动航天装置

EVT Engineering Verification Test 工程验证试验；工艺验证测试

EVT Extravehicular Transfer 舱外转运

EVVA Extravehicular Visor Assembly 舱外活动面盔装置

EW Early Warning 预警

EW Electronic Warfare 电子战

EW&C Early Warning and Control 预警与控制

EW/AA Early Warning and Attack Assessment 预警与攻击评估

EW/GCI Early Warning and Ground Controlled Intercept 预警与地面控制截击

EWAC Early Warning And Control 预警及控制

EWACS Early Warning and Control System 预警与控制系统

EWACS Emergency, Warning and Caution System 应急、告警与警报系统

EWAISF Electronic Warfare Avionics Integrated Support Facility 电子战航空电子设备综合保障设施

EWAMS Early Warning and Monitoring System 预警与监视系统

EWBMA Electronic Warfare Battle Management Aids 电子战战斗管理辅助设备

EWCC Enhanced Weapon Control Computer 增强型武器控制计算机

EWDT Early Warning Data Transmission 预警数据传输

EWE Emergency Window Escape 紧急窗口逃逸

EWFAES Electronic Warfare Flagging Analysis Expert System 电子战作用减弱分析专家系统

EWGETS Electronic Warfare Ground Environment Threat Simulator 电子战地面环境威胁模拟器

EWIS Electrical Wiring Interconnection System 电气线路互联系统

EWIS External Wireless Instrumentation System 外部无线测量系统

EWPE Electronic Warfare Pre-Processing Element 电子战预处要素

EWR Early Warning Radar 预警雷达

EWR　Eastern/Western Range　东部/西部靶场

EWT　Effects of Weapons on Targets　武器对目标的杀伤破坏作用

EWT　Emergency Water Tank　应急水箱

EWTES　Electronic Warfare Tactical Environmental Simulation　电子战作战环境模拟

EWTES　Electronic Warfare Threat Environmental Simulator　电子战威胁环境模拟器

EWVA　Electronic Warfare Vulnerability Assessment　电子战易损性评估

EXC　Experiment Computer　试验计算机

EXCEDE　Electron Accelerator Experiment　电子加速器试验

EXCITE　Estrack X.25 Communications to Internet Technology Evolution　欧洲空间跟踪遥测网X.25通信至因特网技术发展

EXIT　External Interface and Test　外部接口与试验（测试）

EXO　European X-Ray Observatory　欧洲X射线观测卫星

EXO　Experiment Operator (in Spacelab)　试验操作员（天空实验室）

EXOS　Exospheric Satellite　【日本极光研究卫星】外大气圈卫星

EXOS　Experiment X-Ray Observation Satellite　实验型X射线观测卫星

EXOSAT　European X-Ray Observatory Satellite　【欧洲空间局】欧洲X射线观测卫星

EXPEAV　Experimental Aeronautical Vehicle　试验型航空航天运载器

EXPRESS　Experiment Reentry Space System　【美国】"快车"卫星；实验型再入式航天系统

EXPRESS　Express Freight Box　快速集运箱

EXT　Experiment Terminal (Operator Console on Spacelab)　试验终端（天空实验室的操作员控制台）

EXTRADOVAP　Extended-Range Doppler Velocity and Position　测速与测位增程多普勒（雷达）

EXUV　Extreme Ultraviolet and X-Ray Survey Satellite　远紫外线与X射线勘察卫星

Ez　Vertical Electric Field　垂直电场

F

F	Failure	失效；故障；事故
F&E	Facilities and Equipment	设施与设备
F&E	Facility and Environment	设施与环境
F&R	Function and Reliability	功能与可靠性
F&S	Fire and Smoke Detection	火灾烟雾探测
F-T diagram	Flight Time Diagram	飞行时间图
F/A	Failure Analysis	故障分析
F/A	Final Assembly	总装；最后装配
F/A	Fuel-Air	燃料-空气
F/A	Fuel-Air (Ratio)	燃料-空气比
F/AP	Fragmentary/Armor-Piercing	破片杀伤/穿甲
F/C	Fit Check	配合检查
F/C	Flight Control	飞行控制；飞行管制
F/C	Flight Controller	飞行控制员；飞行控制器（装置）
F/C	Fuel Cell	燃料箱
F/C	Functional Checkout	功能检测
F/C VLV	Fill and Check Valve	加注止回阀
F/CD	Failure and Consumption Data	故障与损耗数据
F/CH	Flight Check	飞行检查
F/D	Fill/Drain	加注/排放
F/D	Flight Dynamics	飞行动力
F/D VLV	Fill and Drain Valve	加注排放阀
F/I	Flame/Incendiary	火焰/燃烧
F/O	Fiber Optic	光纤
F/O	Fuel/Oxidizer Ratio	燃料/氧化剂比
F/O	Fuel-to-Oxidizer	燃料氧化剂
F/R	Flared Rudder	喇叭形方向舵；喇叭形尾羽
F^2T^2	Find, Fix, Track and Target	探测、定位、跟踪和瞄准
F^3INS	USAF Standard Form, Fit and Function Inertial Navigation System	美国空军标准惯性制导系统（指形状、安装及功能均符合标准的惯导系统）
FA	Failure Analysis	失效分析；故障分析
FA	Fallback Area	安全区；（指发射场周围的安全区）隐蔽区
FA	Final Approach	最后进场；最终进场
FA	Final Assembly	总装；最后装配
FA	Flight Aft	飞行尾部
FA	Frequency Adjustment	调频
FA	Fully Automatic	全自动的
FA	Functional Analysis	功能分析；泛函分析
FA	Functional Area	作用面积
FA&C	Field Assembly and Checkout	现场装配与检查（检验）
FA&CO	Fabrication, Assembly and Checkout	制造、组装与检查
FA/COSI	Final Assembly and Checkout System Installation	总装与检验系统安装
FA/RD	Functional Analysis/Require-	

ments Definition　功能分析与要求定义

FAA　False Alarm Avoidance　避免误报警；避免假报警

FAA　Federal Aviation Agency　【美国】联邦航空总署

FAABMS　Forward Area Antiballistic Missile System　前沿地区反弹道导弹系统

FAAD　Forward Area Air Defense　【美国陆军】前沿区域空中防御

FAADC^2　Forward Area Air Defense Command and Control　前方地域防空指挥与控制（系统）

FAADC^{2}I　Forward Area Air Defense Command, Control and Intelligence　前方地域防空指挥、控制与情报（系统）

FAADC^{3}I　Forward Area Air Defense Command, Control, Communications and Intelligence　前方地域防空指挥、控制、通信与情报（系统）

FAADS　Forward Area Air Defense System　【美国陆军】前沿区域空中防御系统

FAATC　FAA Technical Center　【美国】联邦航空总署技术中心

FAB　Final Assembly Building　最后组装厂房；总装厂房

FAB　Fly Along Probe　沿探测器飞行

FAB　Functional Arm Brace　功能臂固定器

FAB　Functional Auxiliary Block　功能辅助舱

FAC　Field Assembly and Checkout　现场装配与检查

FAC　First Alert and Cueing　首次报警与提示

FAC　Forward Air Controller　前沿空中控制人员

FACD　Functional Analysis Concept Development　功能分析概念发展

FACEL　Feature Analysis, Comparison and Evaluation Library　特性分析、比较及评估程序库

FACET　Fluid Amplifier Control Engine Test　液压放大器控制发动机试验

FACGSE　Spaceport Facility and GSE Acquisition Cost Estimator　航天港设施与地面保障设备征用成本估计值（量）

FACI　First Article Configuration Inspection　产品配置首次检验

FACO　Fabrication and Acceptance Checkout　制造与验收检测

FACO　Factory Assembly and Checkout　工厂验收与检查

FACO　Final Assembly and Checkout　总装与检验

FACP　Fault and Caution Processing　故障与告警处理

FACP　Forward Area Control Post　前沿区域控制站

FACS　Fine Attitude Control System　精确姿态控制系统

FACS　Flight Augmentation Control System　飞行增稳操纵系统

FACT　Facility of Automation Control and Test　自动化控制与测试装置

FACT　Flexible Automatic Circuit Tester　灵活自动回路测试装置

FACT　Flight Acceptance Composite Test　飞行验收综合试验

FACTAR　Follow up Action on Accident Reports　根据事故报告采取的善后措施

FADD　Fatigue and Damage Data　疲劳与损伤数据

FADEC　Full-Authority Digital Electronic Controls 全权数字电子控制

FADS Framework for Atlas Detector Simulation 【美国】"土卫"十五探测器模拟框架

FAE Final Approach Equipment 最后进场（或进近）设备

FAE Fuel-Air Explosive 燃料一空气爆炸

FAF First Aerodynamic Flight 首次空气动力学飞行

FAF/L Final Approach Fix/Left 最后进场（或进近）定点 / 左转弯

FAF/R Final Approach Fix/Right 最后进场（或进近）定点 / 右转弯

FAFT Fore/Aft Fuselage Tankage 机体前后储箱

FAGC Fast Automatic Gain Control 快速自动增益控制

FAGS Federation for Astronomical and Geophysical Services 天文与地球物理联合会

FAHA Fault Hazard Analysis 失效危险分析；故障危险分析

FAID Fuel Assembly Identification 燃料组件识别

FAID Fuel Assembly Integrity Devices 燃料组件完整装置

FAIR Fabrication, Assembly and Inspection Record 制造、装配与检查记录

FAIR Fly Along Infrared 沿途红外线测量飞行试验

FAIT Fabrication, Assembly, Inspection/Integration, and Test 制造、装配、检测 / 组装与测试

FAITE Final Acceptance Inspection Test Equipment 最终验收检验试验设备

FAJ Final Assembly Jig 总装装配架

FAL First Approach and Landing (Test) 首次进场与着陆（试验）

FALCON Fission-Activated Light Concept 裂变激发光概念

FAM Facilities Analysis Model 设施分析模型

FAM Fine Acquisition Mode 精准捕获模式

FAM Functional Area Management 功能区管理

FAMAS Flutter and Matrix Analysis System 颤振与矩阵分析系统

FAME Fast Adaptive Maneuvering Experiment 快速适应机动实验

FAME Final Approach Monitoring Equipment 最后进场监控设备

FAME Forecasts, Appraisals and Management Evaluations 预测、鉴定与管理评估

FAMIS Financial Accounting Management Information System 财会管理信息系统

FAMOS Fast Multitasking Operating System 快速多任务处理系统

FAMOS Flight Acceleration Monitor Only System 飞行加速度检测器唯一系统

FAMS Fine Attitude Measurement System 最佳飞行姿态测量系统

FAMSNUB Frequencies and Mode Shapes of Non-Uniform Beams 非均匀波束型频率与波型

FAMTEC Flexible Automated Manufacturing Technology Evaluation Center 柔性自动化制造技术评价中心

FAMU Fuel Additive Mixture Unit 燃料添加剂混合装置

FANS Future Air Navigation System 未来大气导航系统

FAP Failure Analysis Program 故障

分析程序

FAP Fairing Acoustic Protection 整流罩声学防护
FAP Final Approach 最后进场
FAP Final Approach Plane 最后进场平面
FAP Flight Acceleration Profile 飞行加速剖面（曲线）
FAR Failure Analysis Report 故障分析报告
FAR False Alarm Rate 误报警率
FAR False Alarm Ratio 误报警比
FAR Federal Aviation Regulation 【美国】联邦航空条例
FAR Final Acceptance Review 最后验收评审
FAR Finned Air Rocket 装有尾翼的火箭；翼式火箭
FAR Flight Acceptance Review 飞行验收评审
FAR Fuelling Authorization Review 批准加注评审
FARADA Failure Rate Data Exchange Program 故障率数据交换程序
FARC Fast Accurate Refraction Correction 快速精确折射校正
FARC Federal Archives and Records Center 【美国】联邦档案与记录中心
FAROA Final Approach Runway Occupancy Awareness 最后进场跑道使用感知
FAS Fin Actuator System 尾翼作动器系统
FAS Final Assemble Schedule 总装进度计划
FAS Fuels Automated System 燃料自动化系统
FASA Final Approach Spacing Assignment 最后进场间隔分配
FASCOS Flight Acceleration Safety Cutoff System 飞行加速安全关闭系统；飞行加速安全中止系统
FASE Fast Auroral Snapshot Explorer 【美国】极光速摄探测器
FASS Frequency Agile Signal Simulator 频率捷变信号模拟器
FASSET Function Advanced Satellite System for Evaluation and Test 高性能卫星评估与检测系统
FASSP Flexible Adaptive Spatial Signal Processor 柔性自适应空间信号处理器
FASSTER Federal Aviation Surveillance System for Test and Evaluation Ranges 用于测试和鉴定靶场的联邦航空监视系统
FAST Facility for Adsorption and Surface Tension 吸收与表面拉力设施
FAST Fast Acquisition Search and Track 快速截获搜索与跟踪
FAST Fast Auroral Snapshot 快速拍摄极光号
FAST Field Data Applications, Systems and Techniques 数据组数据应用、系统和技术
FAST Flexible Access Secure Transfer 柔性接入与安全传输
FAST Flight Analogy Software Test 飞行模拟软件测试
FASTALS Force Analysis Simulation of Theater Administrative and Logistics Support 战区行政与后勤保障的部队模拟分析
FASTI Fast Access to Systems Technical Information 快速查明（导弹）系统故障技术信息
FAT Factory Acceptance Test 工厂验收试验
FAT Final Acceptance Test 最终验

收试验

FAT Flight Attitude Table 飞行姿态表

FATDL Frequency and Time Division Data Link 频率与时分数据链

FATE Fuze Arming Test and Evaluation 引信解除保险试验和鉴定

FATE Fuze Arming Test Experiment 引信解除保险试验

FAUST Far Ultraviolet Space Telescope 远紫外空间望远镜

FAWG Flight Assignment Working Group 飞行任务工作组

FB Feedback 反馈；回流；回输

FB Final Braking 最后制动

FBAS Fixed Base Aft Station 定基尾站

FBC Fluidized-Bed Combustion 流化床燃烧

FBC Forward Bay Cover 前端舱盖

FBCS Fixed Base Crew Station (SMS) 定基乘员站

FBIA Force-Balance Integrating Accelerometer 力平衡综合加速测量仪

FBL Friction Braked Landing 滑板制动着陆；摩擦制动着陆

FBM Fleet Ballistic Missile 舰队弹道导弹系

FBMS Fleet Ballistic Missile System 舰队弹道导弹系统

FBP Forward Based Probe 【美国】前沿配置的探测器

FBPT Force-Balance Pressure Transducer 力平衡式压力传感器

FBR Forward-Based Radar 【美国】前沿配置的雷达

FBR-T Forward Based Radar-Transportable 前沿配置的移动式雷达

FBS File Backup System 文档备用系统

FBS Flare Buildup Study 火炬建立研究；耀斑形成研究

FBS Forward Based Sensor 【美国】前沿配置的传感器

FBS Functional Breakdown Structure 功能分解结构

FBSA Filter Band Suppressor Assembly 滤波器频带抑制器组件

FBV Fuel Bleed Valve 燃料排泄阀

FBXR Forward-Based X-Band Radar 【美国】前沿配置的X波段雷达

FC Carrier Frequency 载波频率

FC Fit Check 适配性检查

FC Flight Capsule 飞行舱

FC Flight Computer 飞行计算机

FC Flight Control 飞行控制

FC Flight Controller 飞行控制器

FC Flight Critical 飞行临界（值）

FC Fuel Cell 燃料电池；燃料室；燃油舱

FC Functional Capability 功能能力

FCA Facility Condition Assessment 设施条件评估

FCA Flight Caution Area 飞行警告区

FCA Flight Configuration Audit 飞行配置检查（审查）

FCA Flight Critical Aft 飞行临界尾部（后部）

FCA Flow Control Assembly 流量控制设备

FCA Fluids Control Assembly 液体控制装置

FCA Frequency Control Analysis 频率控制分析

FCA Functional Compatibility Analysis 功能兼容性分析

FCA Functional Configuration Audit 功能配置检查（软件）

FCAC Frequency Control & Analysis Center 频率控制与分析中心
FCAF Flight Crew Accommodations Facility 飞行乘员食宿设施
FCAP Flight Control Applications Program 飞行控制应用项目
FCC Flat Conductor Cable 扁平导体带状电缆
FCC Flight Control Computer 飞行控制计算机
FCC Flight Crew Compartment 飞行乘员舱；驾驶舱
FCCS Fleet Command Centers 舰队指挥中心
FCCS Flight Control Computer System 飞行控制计算机系统
FCD Frequency Compression Demodulator 频率压缩解调器
FCD Function Control Document 功能控制文件
FCDAU Fibre Channel Data Acquisition Unit 光纤信道数据采集设备
FCDB Flight Control Data Bus 飞行控制数据总线
FCDL Flight Control Development Laboratory 飞行控制研制实验室
FCDR Failure Cause Data Report 故障原因数据报告
FCE Flexible Critical Experiment 柔性临界实验
FCE Flight Control Equipment 飞行控制设备
FCE Flight Crew Equipment 飞行乘员设备
FCEF Flight Crew Equipment Facility 飞行乘员设备设施
FCEI Facility Contract End Item 设备合同最终条款
FCES Flight Control Electronics System 飞行控制电子设备系统
FCEV Future Crew Exploration Vehicle 未来乘员探索飞行器
FCF Filling Configuration File 加注配置文件
FCF First Captive Flight 首次系留飞行
FCF Flight Critical Forward 飞行临界前部
FCF Fourier Coefficient Filter 傅里叶系数滤波器
FCF Functional Check Flight 功能（检查）试飞
FCFM Flight Combustion Facility Monitor 飞行燃烧设备监测器
FCH Flight Controllers Handbook 飞行控制员（装置）手册
FCHL Flight Control Hydraulics Laboratory 飞行控制液压系统实验室
FCHMT Flight Control and Hydraulics Maintenance Trainer 飞控与液压系统维修训练器
FCI Flight Control Indicator 飞行控制指示器
FCI Functional Configuration Identification 功能配置识别（鉴定，确定，检验）
FCIM Flight Control Interface Module 飞行控制接口模块
FCIP Flight Cargo Implementation Plan 飞行货物实施计划
FCL Flight Control Laboratory 飞行控制实验室
FCLA Ferrite-Core Loop Antenna 铁氧磁芯式环形天线
FCM Fine-Orbit Control Mode 精准轨道控制模型
FCM Future Concepts for Maintenance 未来维护方案
FCMU Foot Controlled Maneuvering Unit 脚控操纵装置

FCN Fully Connected Network　全联网络

FCO Final Checkout　最后检测

FCO Functional Checkout　功能检测

FCOH Flight Controllers Operational Handbook　飞行操作员（装置）操作手册

FCOS Flight Computer Operating System (Orbiter)　【美国航天飞机轨道器】飞行计算机操作系统

FCOS Flight Control Operating System　飞行控制操作系统

FCOS Flight Control Operational Software　飞行控制操作软件

FCP Failure Correction Panel　故障修正控制屏；故障修正操纵台

FCP Flight Control Panel　飞行控制操纵台

FCP Flight Control Processor　飞行控制处理器

FCP Flight Correction Proposal　飞行修正建议

FCP Fuel Cell Power　燃料电池动力装置

FCPS Fuel Cell Power System (Subsystem)　燃料电池动力系统（分系统）

FCR Final Configuration Review　最后配置评审

FCR Flight Configuration Review　飞行配置评审

FCR Flight Control Room　飞行控制室

FCS Flight Control System (Subsystem)　飞行控制系统（分系统）

FCS Flight Crew System　飞行乘员系统

FCS Future Combat Systems　未来战斗系统

FCS Future Concepts for Simulators　未来模拟装置概念

FCSG Flight Control Sensor Group　飞行控制传感器组

FCSM Flight Combustion Stability Monitor　飞行燃烧稳定性监控器

FCSMA Flexible Communication Satellite for Military Applications　军用机动通信卫星

FCSS Fuel Cell Servicing System　燃料舱伺服系统

FCT Fatigue Cracking Test　疲劳断裂试验

FCT Flight Control Team　飞行控制小组

FCT Flight Crew Trainer　飞行乘员教练员；飞行乘员训练设备

FCT Fuel Cell Test　燃料舱测试（试验）

FCTB Flight Crew Training Building　飞行乘员训练楼（室）

FCTF Fuel Cell Test Facility　燃料舱测试设施

FCTS Flight Crew Training Simulator　飞行乘员训练模拟器

FCU Fluid Checkout Unit　液体检测装置

FD Fault Detection　故障探测

FD Flight Day　飞行日

FD SBY Remote Flight Director Standby　远距飞行指挥仪备份

FD/FI Fault Detection/Fault Isolation　故障诊断、故障隔离

FD2 Flight Demonstration 2　飞行演示验证2

FDA Fault Detection & Annunciation　故障探测与通告

FDA Flight Deck Assembly　驾驶舱装置

FDAF Flight Data Acquisition Function　飞行数据采集功能

FDAI Flight Director Attitude Indicator 飞行指挥仪姿态指示器
FDAS Flight Data Acquisition System 飞行数据采集系统
FDAU Flight Data Acquisition Unit 飞行数据采集装置
FDCL Flight Control Development Laboratory 飞行控制研制实验室
FDD Flight Definition Document 飞行定义文件
FDDB Function Designator Data Base 飞行指示装置数据库
FDDI Fiber Distribution Data Interface 光纤分布数据接口
FDE Fault Detection and Exclusion 故障探测与排除
FDE Fluid Dynamics Experiment 流体力学实验
FDEP Flight Data Entry Panel 飞行数据进入面板
FDF Flight Data Files 飞行数据文件
FDF Flight Dynamics Facility 飞行力学设施
FDH Flight Deck Handset 飞行舱手持小型装置
FDI Failure Detector Indicator 故障探测装置指示器
FDI Fault Detection and Identification 故障探测与识别（确定）
FDI Fault Detection and Isolation 故障探测与隔离
FDIIR Fault Detection, Isolation, Identification, and Reconfiguration 故障检测、隔离、识别与重新配置
FDIMU Flight Data Interface Management Unit 飞行数据接口管理装置
FDIO Flight Data Input/Output 飞行数据输入/输出
FDIR Failure(Fault) Detection, Isolation and Recovery 失效（故障）探测、隔离与修复
FDIR Fault Detection, Identification/Isolation and Recovery/Recognition 故障探测、识别/隔离与修复/辨别
FDIR Fault Detection, Isolation, Identification and Recompensation 故障探测、隔离、识别与补偿
FDIS Flight Displays and Interface System 飞行显示器与接口系统
FDL Flight Director Loop 飞行指示仪回路
FDL Flight Dynamics Laboratory 飞行力学实验室
FDLN Feedline 馈线；供给线
FDLS Failure Detection Location System 故障探测定位系统
FDM Flight Data Monitoring 飞行数据监控
FDM Frequency Data Multiplexer 频率数据多路调制器（通道）
FDM Frequency Division Multiplex 频率分隔多路传输
FDMA Frequency Division Multiple Access 频率划分多路访问；频分多址
FDMA Full Diameter Motorized Door Assembly 全直径电动门组件
FDOR Flight Design Operations Review 飞行设计操作评审
FDP Flight Data Processor 飞行数据处理器
FDP Flight Demonstration Program 飞行演示验证项目
FDPS Flight Data Processor System 飞行数据处理器系统
FDR Final Design Review 最终设计评审
FDR Final/Formal Design Review 最终/正式设计评审
FDR Flight Data Recorder 飞行数据

记录仪
FDR Functional Design Review 功能设计评审
FDRD Flight Definition and Requirements Document 飞行定义与要求文件
FDRFA Flight Data Recorder and Fault Analyzer 飞行数据记录仪与故障分析仪
FDRI Flight Director Rate Indicator 飞行指挥仪速率指示器
FDRS Flight Data Recorder System 飞行数据记录仪系统
FDS Fire Detection and Suppression 火灾探测与抑制
FDS Flight Demonstration System 飞行演示验证系统
FDS Flight Design and Scheduling 飞行设计与计划编制
FDS Flight Design System 飞行设计系统
FDS Flight Dynamics Simulator 飞行动力模拟器（装置）
FDS Flight Dynamics Software/System 飞行动力软件（系统）
FDS Fluid Distribution System 流体分布系统
FDS Functional Design Specifications 功能设计说明
FDSC Flight Dynamics Simulation Complex 飞行动力学仿真综合装置
FDSC Flight Dynamics Situation Complex 飞行动力学状态综合装置
FDSS Fine Digital Sun Sensor 精密数字式太阳传感器
FDSS Flight Dynamics Support System 飞行动力保障系统
FDSSR Flight Dynamics Staff Support Room 【美国国家航空航天局】飞行动力人员保障室
FDSV Flight Demonstration Space Vehicle 飞行演示验证航天器
FDT&E Force Development Test and Evaluation 【美国陆军】部队发展试验与鉴定
FDTS Fault Detection Test Set 故障检测测试装置
FDU Fluid Distribution Unit 流体分布装置
FE Far Encounter 远交会阶段（美国"旅行者"航天器与天王星交会过程的一个阶段）
FEA Failure Effect Analysis 故障影响分析
FEA Finite Element Analysis 有限元分析；有限元分析法
FEA Fluids Experiment Apparatus 流体实验设备
FEA Front End Analysis 前端分析
FEAT Final Engineering Acceptance Test 最后工程验收测试（试验）
FEB Forward Equipment Bay 前端设备舱
FEC Forward Error Correction 前端错误修正
FEC Forward Events Controller 前端事件控制器（装置）
FECA Front-End Cost Analysis 前端成本分析
FED Flight Events Demonstration 飞行事件演示验证
FEDCAC Federal Computer Acquisition Center 【美国】联邦计算机采集中心
FEDP Facility and Equipment Design Plan 设施设备设计计划
FEDS Flexible Engine Diagnostic System 柔性发动机诊断系统

FEDSIM　Federal Emergency Management Agency　【美】联邦应急管理局

FEE　Failure Effect Evaluation　故障影响评估

FEET　Flight Emergency Egress Test　飞行应急出舱试验

FEID　Flight Equipment Interface Device　飞行设备接口装置

FEID　Functional Engineering Interface Device　功能工艺接口装置

FEL　First Element Launch　【美国空间站】第一批组件发射

FEL　Free Electron Laser　自由电子激光器

FELTS　Federal Evaluation of Lightning Tracking System　【美国】联邦机构对雷电跟踪系统的评估

FEM　Finite Element Method　有限元方法

FEM　Finite Element Model　有限元模型

FEMA　Federal Emergency Management Agency　【美国】联邦应急管理局

FEMDB　Finite Element Model Database　有限元模型数据库

FEMP　Federal Energy Management Program　【美国】联邦能源管理项目

FEP　Flight Evaluation Plan　飞行评估计划

FEP　Front End Processor　前端处理器（装置）

FERD　Facility and Equipment Requirements Document　设施设备要求文件

FERNS　Far East Radio Navigation Service　远东无线电导航服务

FES　Facility Engineering Surveillance Plan　设施工程监视计划

FES　Flash Evaporator System　快速蒸发器系统

FES　Flight Element Set　飞行部件装置

FES　Fluid Experiment System　流体实验系统

FES/VCGS　Fluid Experiment System/Vapor Crystal Growth System　流体实验系统/蒸汽晶体生长系统

FESS　Flight Experiment Shielding Satellite　辐射屏蔽飞行实验卫星

FESS　Flight Experiment Subsatellite　【美国】飞行实验子卫星

FESTIP　Future European Space Transportation Investigations Program　欧洲未来航天运载研究计划

FET　Facility Evaluation Test　设备鉴定试验

FET　Flight Elapsed Time　飞行耗用时间

FET　Flight Environment Test　飞行环境试验

FET　Field Extraction Thrusters　场提取推进器

FETF　Flight Engine Test Facility　发动机飞行试验设备

FETS　Field Evaluation and Test System　外场评价与试验系统

FEU　Flight Equivalent Unit　飞行等效部件

FEU　Flight Evaluation Unit　飞行鉴定装置

FEWG　Flight Evaluation Working Group　飞行评估工作组

FEWS　Follow-on Early Warning System　【美国】后继早期预警系统

FF　First Flight　首次飞行

FF　Flight Forward　飞行前端

FF　Flip Flop　触发器；振荡器

FF Forward Fairing 前端整流罩
FF Free Flight 自由飞行
FF Free Flyer 【航天器】自由飞行器
FF Fuel Flow 燃料流动；燃料流量
FFAR Fikdung-Fin Aerial Rocket 菲克顿芬航空火箭
FFAR Folding-Fin Aerial Rocket 折叠尾翼空射火箭
FFAR Forward Firing Aerial Rocket 前射航空火箭
FFAR Free Flight Aerial Rockets 自由飞行（无控）航空火箭弹
FFAR Free-Flight Aircraft Rocket 无控航空火箭弹；自由飞行航空火箭弹
FFBD Functional Flow Block Diagram 功能流程框图
FFC Final Flight Certification 最后飞行确认（鉴定）
FFCS Free-Fall Control System 自由下落控制系统
FFD Free Flight Data 自由飞行数据
FFDP Final Flight Data Package 最后飞行数据包
FFE Full Flight Envelope 全飞行包线
FFF Fast Forward Flight 快速前向飞行
FFFT Forced Flow Flame-Spreading Test 强制流火焰扩张试验
FFL Fuel Fill Line 燃料加注管
FFM Free-Flying (Experiment) Module 自由飞行（实验）舱
FFM Fuel Fill to Missile 导弹加注燃料
FFPA Final Flight Plan Approval 最后飞行计划审批
FFR Flight Feasibility Review 飞行可行性评估；飞行可行性评审
FFR Free Flight Rocket 自由飞行火箭；非制导火箭
FFRDC Federally Funded Research and Development Center 【美国】联邦政府资助的研发中心
FFS Full Flight Simulator 全飞行模拟器（装置）
FFTO Free Flying Tele-operator (program) 自由飞行遥控操作器；自由飞行遥控操作程序
FFTO Free-Flying Teleoperator 自由飞行遥控操作器（操纵装置）；自由飞行遥控机器人（机械手）
FFTV Free Flight Test Vehicle 自由飞行试验飞行器
FGC Flight Guidance and Control 飞行制导与控制
FGC Flight Guidance Computer 飞行制导计算机
FGCS Flight Guidance and Control System 飞行制导与控制系统
FGE Fine Guidance Electronics 精确制导电子设备
FGGE First GARP (Global Atmospheric Research Program) Global Experiment 全球大气研究计划的首次全球实验
FGS Fine Guidance Sensors 精确制导传感器
FGS Fine Guidance System 精确制导系统
FGS Flight Guidance System 飞行制导系统
FGSE Fine Guidance System Electronics 精确制导系统电子设备
FH Flight Hardware 飞行硬件
FHA Facility Hazard Analysis 设施危险性分析
FHA Flight Hazard Area 飞行危险地区

FHA Functional Hazard Assessment 功能危险性评估
FHB Flight Hardware Building 飞行硬件厂房
FHB Fuel Handling Building 燃料处理厂房
FHC Flight Half Coupling 飞行半耦合
FHF First Horizontal Flight 首次水平飞行
FHP Fuel High Pressure 燃料高压
FHS Forward Heat Shield 前端热防护
FHTE Flight Hardware Test Equipment 飞行硬件测试设备
FI Fault Identification 故障识别
FI Fault Isolation 故障隔离
FI Flight Instrumentation 飞行仪表工具；飞行测量装置
FI&A Fault Isolation and Analysis 故障隔离与分析
FIA Functional Interoperability Architecture 功能相互适应能力体系结构
FIA Future Imaging Architecture 【美国】未来成像结构（美国未来成像卫星计划）
FIAD Flame Ionization Analyzer and Detector 火焰电离分析器与探测器
FIAR Failure Investigation Action Report 失效（故障）调查措施报告
FIARE Flight Investigation of Apollo Reentry Environment 【美国】"阿波罗"飞船再入环境飞行研究
FIC Fault Isolation Checkout (System) 故障隔离测试（系统）
FICO Flight Information and Control of Operations 飞行信息与飞行控制

FICS Fault Isolation Checkout System 故障隔离检查系统
FID Failure Identification 失效（故障）识别
FID Flight Implementation Directive 飞行执行指令
FIELD First Integrated Experiment for Lunar Development 首次月球开发综合实验
FIES Factor of Initial Engine Spares 发动机初始备件系数
FILE Feature Identification and Location Experiment 【美国国家航空航天局】特征识别与定位实验
FIPS Federal Information Processing Standards 【美国】联邦信息处理标准
FIPS Flight Inspection Positioning System 飞行检查定位系统
FIR Fuel Indicating Reading 燃料显示读数
FIRE Flight Investigation Reentry Environment 再入大气层环境飞行研究
FIREX Fire Extinguisher System 灭火系统
FIRMR Federal Information Resources Management Regulation 【美国】联邦信息资源管理条例
FIRST Fabrication of Inflatable Reentry Structures for Test 测试用充气再入大气层结构的制备
FIRST Far Infrared and Submillimeter Space Telescope 【欧洲空间局】远红外亚毫米波空间望远镜
FIRST Far Infrared Space Telescope 【欧洲空间局】远红外空间望远镜
FIRST Flight-Oriented Integrated Reliability and Safety Tool 飞行导向综合可靠性与安全工具

FIS Flight Information Services 飞行信息服务

FIS Flight Instrument System 飞行仪表系统

FISS Foreign Intelligence and Security Services 外国情报与安全服务

FISSP Federal Information System Support Program 【美国】联邦信息系统保障项目

FIT Fault Isolation Test 故障隔离测试（试验）

FITVC Fluid Injection Thrust Vector Control 射流推力向量控制

FIV Fault Isolation Valve 故障隔离阀

FIWC Fleet Information Warfare Center 舰队信息战中心

FKF Flight Kits Facility 飞行配套设施

FL Flight Level 飞行高度

FL MECH Fluid Mechanical 流体机构

FLAG Fixed Link Aerospace to Ground 航空航天器至地面的固定通信线路

FLAGE Flexible Lightweight Agile Guided Experiment 轻型机动灵活制导实验

FLAME Fast, Lightweight, Agile Missile 【英国】轻型快速敏捷导弹（SRAAM 导弹的改进型）

FLAP Flight Application Software 飞行应用软件

FLAPS Flexible Acquisition Processing System 灵活捕获处理系统

FLARE Fault Locating and Reporting Equipment 故障定位与报告设备

FLASER Forward Looking Infrared Laser Radar 前视红外激光雷达

FLASH Fault Location and Simulation Hybrid 故障定位与仿真混合电路

FLAT Flight-Plan-Aide Tracking 飞行计划辅助跟踪

FLC Forward Load Control 前端荷载控制

FLCA Forward Load Control Assembly 前端荷载控制组件（装置）

FLCH Flight Level Change 飞行高度改变

FLEEP Flying Lunar Excursion Experiment Platform 月球旅行实验平台

FLEXBEAM Flexible Beam Experiment 【美国空军】柔性梁架实验

FLIDRAS Flight Data Replay and Analysis System 飞行数据重放与分析系统

FLIP Flight Launched Infrared Probe 飞行中发射的红外探测器

FLIP Floated Lightweight Inertial Platform 轻型液浮惯性平台

FLIPS Floating Instrument Platform 悬浮式仪表平台；浮动仪表平台

FLIR Forward Looking Infrared Radar 前视红外雷达

FLIR Forward Looking Infrared (System) 前视红外探测（系统）

FLN Fuel Line 燃料线路

FLPS Flight Load Preparation System 飞行荷载准备系统

FLS Future Launch System 未来（航天）发射系统

FLSC Flexible Linear-Shaped Charge 柔性聚能炸药索（导弹间分离用）；挠性线型空心装药

FLT/HDWE Flight Hardware 飞行硬件

FLTCK Flight Check 飞行检测

FLTSATCOM Fleet Satellite Communications 卫星群间通信

FLtWinds Flight and Weather Infor-

mation and Decision Support 飞行与气象信息决策支持
FM　Failure Mechanics 失效力学
FM　Failure Mode 故障模式
FM　Flight Mission 飞行任务
FM　Flight Module 飞行舱
FM　Frequency Modulation (Modulated) 频率调制
FM　Frequency Multiplexing 频率多路调制
FMA　Failure Mode Analysis 故障模式分析
FMA　Flight Mode Annunciator 飞行模式信号指示
FMAHTS　Flight Manifest and Hardware Tracking System 飞行舱单和硬件跟踪系统
FMANTS　Flight Manifest 飞行乘员舱单
Fmax　Maximum Thrust 最大推力
FMC　Flexible Manufacturing Cell 灵活机动舱
FMC　Flight Management Computer 飞行管理计算机
FMC　Flight Management Control 飞行管理控制
FMC　Fully Mission Capable 全任务能力
FMCC　French Mission Control Centre 法国任务控制中心
FMCDU　Flight Management Control and Display Unit 飞行管理控制与显示装置
FMCF　First Manned Captive Flight 首次载人系留飞行
FMCF　Flight Management Computer Function 飞行管理计算机功能
FMCS　Flight Management Computer System 飞行管理计算机系统
FMEA　Failure Modes and Effects Analysis 失效模式与效应分析
FMEC　Forward Master Events Controller 前端事件主控制器（装置）
FMECA　Failure Modes, Effects and Criticality Analysis 故障模式、影响与危害性分析
FMES/FCI　Full Mission Engineering Simulator/Flight Controls Integration 全任务工程模拟器/飞行控制综合
FMET　Failure Mode and Effects Testing 失效模型与影响试验
FMF　Flight Management Function 飞行管理功能
FMGC　Flight Management Guidance Computer 飞行管理制导计算机
FMGEC　Flight Management Guidance Envelope Computer 飞行管理制导包线计算机
FMHEA　Failure Mode and Hazardous Effect Analysis 故障模式与危险影响分析
FMOF　First Manned Orbital Flight 首次载人在轨飞行；首次载人轨道飞行
FMOFEV　First Manned Orbital Flight with EVA 首次载人轨道飞行带舱外活动
FMOFPL　First Manned Orbital Flight with Payload 首次载人轨道飞行带有效载荷
FMP　Flight Mode Panel 飞行模式操纵台
FMRAAM　Future Medium Range Air-to-Air Missile 未来中程空空导弹
FMS　Fixed and Mobile Standard Segments 【美国】固定与移动标准段
FMS　Flight Management System 飞行管理系统
FMS　Fluid Management System 流体管理系统

FMS　Full Mission Simulator　全任务模拟器（装置）

FMSP　Frequency Modulation Signal Processor　频率调制信号处理器（装置）

FMSR　French Meteorological Sounding Rocket　法国气象探空火箭

FMT　Flight Management Team　飞行管理组

FMTR　Florida Missile Test Range　【美国】佛罗里达导弹试验靶场

FMTV　Family of Medium Tactical Vehicles　中等战术车辆族

FMU　Fuel Metering Unit　燃料计量装置

FNA　Function Need Analysis　功能需求分析

FO　Forward Observer　前端观测员；前端观测装置

FO　Functional Objectives　功能目标

FOB　Flight Operations Building　飞行操作厂房

FO Link　Fiber Optic Link　光纤链路

FOBS　Fractional Orbit Bombardment System　部分轨道轰炸系统（俄罗斯亚轨道武器）

FOC　Final Operational Capability　最终操作能力

FOC　Flight Operations Center　飞行操作中心

FOC　Full Operational Capability　全操作能力

FOC　Future Operational Capability　未来作战能力

FOCAS　Fiber Optic Communications for Aerospace System　【美国空军】航空航天光纤通信系统

FOD　Foreign Objects and Debris　外来物体与碎片

FODA　Flight Operations Data Assurance　飞行操作数据保证

FOF　First Operational Flight　首次操作飞行

FOF　Flight Operations Facilities　飞行操作设施

FOFA　Follow-On Force Attack　【美国空军】后续部队攻击

FOG　Fiber-Optic Gyroscope　光纤陀螺仪

FOIH　Flight Operations Integration Handbook　飞行操作综合手册

FOL　Forward Operating Location　前部操作位置

FOLAN　Fiber Optic Local Area Network　光纤局域网

FOMMS　Flight Operations Maintenance Management System　飞行操作维护管理系统

FOMR　Flight Operations Management Room　飞行操作管理室

FON　Fiber Optic Network　光纤网络

FOP　Flight Operations Panel　飞行操作控制屏

FOP　Flight Operations Plan　飞行操作计划

FOP　Follow on Production　改进型生产

FOPG　Flight Operations Planning Group　飞行操作规划组

FOPP　Follow on Parts Production　改进型零部件生产

FOPS　Flight Operations Planning Schedule　飞行操作规划进度

FOQA　Flight Operational Quality Assurance　飞行操作质量保证

FOR　Flight Operations Review　飞行操作评审

FORTRAN　Formula Translation Language　公式翻译程序

FOS　Flight Operations Segment　飞

行操作段
FOS Flight Operations Support 飞行操作保障
FOS Fuel Oxygen Scrap 燃料氧混合
FOSP Flight Operations Support Personnel 飞行操作保障人员
FOSPLAN Formal Space Planning Language 正式航天规划语言
FOSS Fiber-Optic Sensor System 光纤传感器系统
FOST Flight Operations Support Team 飞行操作保障组
FOT Final On Trajectory 轨迹终点；弹道终点
FOT Follow-On Technologies 后续技术
FOT&E Follow-on Operational Test and Evaluation 后续操作试验与评估
FOT&E Follow-On Test and Evaluation 改进型测试与评估；后续试验与鉴定
FOTC Force Over-the-Horizon Track Coordinator 部队超视距跟踪目标指示协调官
FOV Field of View 视场
FOV First Orbital Vehicle 首个在轨运载器
FOXSI Focusing Optics X-Ray Solar Imager 调焦光学系统 X 射线太阳成像仪
FP Fine Pointing 精确定点
FP Fixed Point 固定点
FP Flashless Propellant 无焰推进剂
FP Flight Path 飞行轨迹；飞行路线
FP Flight Plan 飞行计划
FP Fuel Pressure 燃料压力
FP Function Path 功能路径

FP/DF Fluid Physics/Dynamics Facility 【美国空间站】流体物理学/动力学设施
FPA Failure Probability Analysis 故障概率分析
FPA Flight Path Accelerometer 飞行轨迹（航迹）加速度计
FPA Flight Path Analysis 飞行轨迹（航迹）分析
FPA Flight Path Angle 飞行轨迹（航迹）倾角
FPA Focal Plane Array 焦平面阵列
FPA Focal Plane Assembly 焦平面组件
FPAA Flight Path Analysis Area 航迹分析区
FPAC Flight Path Analysis and Command 飞行航迹分析与指挥
FPB Fuel Preburner (Space Shuttle Main Engine) 燃料预燃室（航天飞机主发动机）
FPBL Flight Parameter Boundary Limitation 飞行参数边界限制
FPBOV Fuel Preburner and Oxidizer Valve 【美国航天飞机】燃料预燃室与氧化剂阀
FPC Flight Path Control [Command] 飞行轨迹控制（指令）
FPC Fluids Pressure Control 液压控制
FPC Forward Power Controller 前端动力控制器（装置）
FPCA Forward Power Control Assembly 前端动力控制组件
FPCC Flight Propulsion Control Coupling 飞行推进控制耦合
FPDS Feasibility Pre-Definition Study 可行性先期论证
FPE Functional Program Element 功能编程要素

FPGA Field Programmable Gate Array 现场可编程门阵列

FPGS Flight Parameter Ground Station 飞行参数地面设备（站）

FPHB Flight Procedures Handbook 飞行规程手册

FPL Flight Propulsion Laboratory 飞行推进装置实验室

FPL Full Power Level 全推力水平；（火箭的）全功率

FPL Full Power Load 全动力荷载

FPLC Full Power Level Certification 全动力级确认（鉴定）

FPM Flight Path Marker 飞行轨迹标志

FPM Fluid Physics Module 【空间实验室】流体物理舱

FPM Folding Platform Mechanism 折叠式平台机械装置

FPOT Flight Power Operating Times 飞行动力运行次数

FPOV Fuel Preburner Oxidizer Valve 【美国航天飞机】燃料预燃室与氧化剂阀

FPOVA Fuel Preburner Oxidizer Valve Actuator 【美国航天飞机】燃料预燃室与氧化剂阀致动器

FPPP Flight Preparation and Process Planning 飞行准备与操作规划

FPR Flight Performance Reserve 飞行性能准备

FPR Full Propellant Requirement 推进剂需求总量

FPS Floating Point System 漂浮点系统

FPS Focal Plane Structure 焦平面结构

FPS Forward Power Supply 前端供电

FPS Full Pressure Suit 全压飞行服；高空密闭服

FPTOC Force Projection Tactical Operations Center 【美国】兵力投送战术作战中心

FPU Floating Point Unit 漂浮点装置

FPV Flow Proportioning Valve 流量比例阀

FPV Flight Path Vector 飞行路径矢量

FQ Flight Qualification 飞行条件；飞行技能；飞行评定

FQI Flight Qualification Instrumentation 飞行条件测量装置

FQIS Fuel Quality Indicating System 燃料质量指示系统

FQPU Fuel Quality Processor Unit 燃料质量处理器装置

FQR Flight Qualification Recorder 飞行条件记录器（仪）

FQR Flight Qualification Review 飞行条件评审

FQR Formal Qualification Review 正式资质评审

FQT Formal Qualification Test 正式评定测试（试验）

FR Firing Room 点火间

FR Flight Rules 飞行规则

FR Functional Requirements 功能要求

FRACAS Failure Reporting, Analysis and Corrective Action System 故障报告、分析与纠正措施系统

FRACAS Failure Reporting and Corrective Action System 故障通报与修正措施系统

FRACAS Filter Response Analysis for Continuously Accelerating Spacecraft 连续加速航天器的滤波器响应分析

FRACS Forward Reaction Altitude Control System 前向反作用高度控制系统

FRAG-HE Fragmentation-High Explosive 高爆杀伤弹

FRAGROC Fragmentation Warhead Rocket 破片杀伤弹头火箭

FRAM Flight-Releasable Attachment Mechanism 飞行可释放连接机构

FRB Failure Review Board 故障审查委员会

FRC Flight Control Room 飞行控制间

FRC Flight Research Center 【美国】飞行研究中心（设在加利福尼亚州）

FRCI Fibre Reinforced Composite Insulation 纤维增强复合材料绝热层

FRCI Fibrous Refractory Composite Insulation 耐高温纤维复合材料绝热层

FRCI Flexible Reusable Carbon Insulation 柔性可重复使用碳绝热层

FRCMC Fiber Reinforced Ceramic Matrix Composite 纤维增强陶瓷基复合材料

FRCS Forward Reaction Control System (Subsystem) 【美国航天飞机】正向反应控制系统；前向反作用控制系统（分系统）

FRD Flight Requirements Document 飞行要求文件

FRDT Facility Requirements Definition Team 设施要求定义组

FRE Flight Related Element 飞行相关零部件；飞行相关要素

FREQ CONV Frequency Converter 频率转换器（装置）

FREQ DIV Frequency Divider 频率分配器

FRF Flight Readiness Firing 【火箭】飞行前点火试验；【发动机】飞行前检查起动

FRFT Flight Readiness Firing Test 导弹飞行准备的发射试验

FRGF Flight-Releasable Grapple Fixtures 飞行释放抓钩装置

FRI Facility Risk Indicator 设施风险指示器（装置）

FRJCS Flight Weight Reaction Jet Control System 飞行重量反作用喷气控制系统

FRJD Forward Reaction Jet Driver 前向反作用喷气驱动器

FRL Flame Retardant Latex 火焰抑制橡胶（胶浆）

FRL Fuselage Reference Line 机身（弹体、壳体）基准线

FRM Fluid Resupply Module 液体再补给模块

FROD Functionally Related Observable Differences 与功能相关的明显差别

FROG Free Rocket Over Ground 地面上空的非制导火箭

FRP Fuselage Reference Plane 机身（弹体、壳体）基准平面

FRPA Fixed Reception Pattern Antenna 固定接收式天线

FRR Flight Readiness Review 待飞状态评审；飞行准备状态评审

FRRID Flight Readiness Review Item Description 待飞状态（飞行准备状态）评审项说明

FRRID Flight Readiness Review Item Disposition 待飞状态（飞行准备状态）评审项配置（处理）

FRS Ferret Reconnaissance Satellite 搜索侦察卫星

FRSSACS Free Reaction Sphere Satellite Attitude Control System 自由反应球形卫星姿态控制系统

FRT	Failure Rate Test	故障率试验
FRT	Fatigue Resistance Test	抗疲劳试验
FRT	Flight Readiness Test	飞行准备状态测试（试验）
FRT	Flight Readiness Training	飞行准备状态培训
FRT	Frequency Response Test	频率响应测试（试验）
FS	Factor of Safety	安全因子
FS	Feasibility Study	可行性研究
FS	Fire Suppression	火力压制；灭火
FS	First Stage	一子级
FS	Flight Simulator	飞行模拟器（装置）
FS	Flight System	飞行系统
FS	Full Scale	足尺（比例）；全尺寸；满标
FS&E	Facilities, Siting & Environment	设施、建设地点与环境
FS^3	Future Strategic Strategy Study	未来战略性战略研究
FSA	Federal Space Agency	【俄罗斯】联邦航天局
FSA	Florida Space Agency	【美国】佛罗里达航天局
FSA	Fuel Storage Area	燃料存储区
FSAA	Fairing Storage and Assembly Area	整流罩存储与装配区
FSAA	Flight Simulator for Advanced Aircraft	【美国国家航空航天局】高级飞机飞行模拟器
FSAR	Final Safety Analysis Report	最终安全分析报告
FSAS	Flight Service Automation System	自动飞行检修系统
FSAT	Ford Satellite	【美国通信卫星】福特卫星
FSATCOM	Fleet Satellite Communications	舰队卫星通信
FSB	First Stage Booster	一子级助推器
FSB	Five-Segment Booster	五段式助推器
FSC	Fixed Satellite Communications (Terminal)	固定卫星通信（终端）
FSCL	Fire Support Coordination Line	火力支援协调线
FSCP	Fire Sensor Control Panel	火灾感应器控制屏（板）
FSCW	Fast Space Charge Wave	快速太空电荷波
FSD	Full Scale Development	全尺寸研发
FSDA	Fail Safe Design Analysis	故障安全设计分析
FSDP	Facilities Safety Data Package	设施安全数据包
FSDP	Final Safety Data Package	最终安全数据包
FSDPS	Flight Service Data Processing System	飞行勤务数据处理系统
FSDVF	Flight Software Development and Verification Facility	飞行软件研发与验证设施
FSDWS	Fixed Site Detection and Warning System	固定位置探测与报警系统
FSE	Flight Support Equipment (Environment)	飞行保障设备（环境）
FSEMC	Flight Simulator Engineering and Maintenance	飞行模拟器（装置）工程与维护
FSF	Fairing Support Facility	整流罩保障厂房
FSF	First Static Fairing	第一代静态整流罩
FSF	Flight Safety Foundation	飞行安全基础

FSI Final Systems Installation 系统总装

FSI Flame Spread Index 火焰扩散指数（系数，指标）

FSI Fluid-Structure Interaction 流体结构互作用

FSIM Functional Simulation 功能仿真

FSIM Functional Simulator 功能仿真器

FSIWG Flight System Interface Working Group 飞行系统接口工作组

FSL Flight Simulation Laboratory 飞行模拟实验室

FSL Flight Systems Laboratory 飞行系统实验室

FSLP First Spacelab Payload 首个天空实验室有效载荷

FSLT First Sea Level Test 首个海平面测试（试验）

FSM Flight Schedule Monitoring 飞行进度监控（视）

FSM Flight Synchronizer Module 飞行同步器舱

FSM Fluid System Module 流体系统模块

FSM Fuel Supply Module 燃料供给模块

FSMC First-Stage Motor Container 一子级发动机外壳

FSMDB Flight Support Maintenance Data Base 飞行保障维护数据库

FSMF Flight Software Maintenance Facility 飞行软件维护设施

FSMMS Flight Support Maintenance Management System 飞行保障维护管理系统

FSMP Facility Safety Management Plan 设施安全管理计划

FSMS Flight Structural Monitoring System 飞行结构监控系统

FSO Functional Supplementary Objectives 功能补充目标

FSOH Flight Support Operations Handbook 飞行保障操作手册

FSPOC Flight Safety Project Officer Console 飞行安全项目官操控台

FSR Fin Stabilized Rocket 尾翼稳定火箭

FSR Final System Release 最终系统发布

FSR Flight Safety Region 安全飞行区

FSR Flight Safety Reporting 飞行安全报告

FSR Flight Specific Requirements 具体飞行要求

FSRM First Stage Rocket Motor 一级火箭发动机

FSRR Flight Software Readiness Review 飞行软件准备状态评审

FSRR Flight System Readiness Review 飞行系统准备状态评审

FSRS Flight System Recording System 飞行系统记录系统

FSS Fire Suppression System 灭火系统

FSS Fixed Service Structure 固定勤务塔

FSS Flight Safety System 飞行安全系统

FSS Flight Service Station 飞行检修站；飞行勤务站

FSS Flight Support Station 飞行保障站

FSS Flight Support Structure 飞行支撑结构

FSS Flight Support System 飞行保障系统；飞行支撑系统

FSS Flight Systems Simulator 飞行

系统模拟器（装置）
FSS Freedom Space Station 【美国】"自由号"空间站
FSSA Fine Sun Sensor Assembly 精确太阳感应器组件
FSSD First-Stage Separation Device 【多级火箭】第一级分离装置
FSSP Forward Scattering Spectrometer Probe 前向散射粒谱仪探头
FSSR Flight Systems Software Requirements 飞行系统软件要求
FSSR Functional Subsystem Software Requirements 功能分系统软件要求
FSSS Future Security Strategy Study 未来安全战略研究
FSSSAT Flight Support System Servicing Aid Tool 飞行支援系统辅助服务工具
FSST Forward Space Support in Theater 战区前沿空间支援
FST Flight Simulation Test 飞行模拟试验
FST Flight System Testbed 飞行系统试验（测试）台
FSTD Flight Simulation Test Data 飞行模拟试验数据
FSTE Factory Special Test Equipment 工厂专用测试（试验）设备
FSTS Future Space Transportation System 未来航天运输系统
FSTV Full Scale Test Vehicle 全尺寸试验飞行器
FSW Flight Software 飞行软件
FSW Friction Stir Welding 摩擦搅拌焊接
FT Flight Termination 飞行终止
FT Flight Test 飞行测试（试验）
FT Flight Time 飞行时间
FT Functional Test 功能测试（试验）

FT&C Formal Training and Certification 正式培训与认证
FT/ADIRS Fault Tolerant/Air Data Inertial Reference System 故障容限/大气数据惯性参照系统
FTA Fatigue Test Article 疲劳试验试件
FTA Fault Tree Analysis 故障树分析
FTA Flight Test Article 飞行试验件；飞行测试项目
FTA/RA Facility Tempest Assessment/Risk Analysis 瞬间电磁脉冲辐射保障设备的评估/威胁分析
FTC Fault Tolerant Computing 故障容限计算
FTC Florida Test Center 【美国】佛罗里达州测试（试验）中心
FTC Fuel Transfer Canal 燃料转换管道
FTD Flagship Technology Demonstration 旗舰技术演示验证
FTD Flagship Technology Demonstrator 主要技术示范者
FTE Factory Test Equipment 工厂测试（试验）设备
FTE Flight Technical Error 飞行技术错误
FTE Full-Time Equivalent 全时等量（等效值，当量）
FTF Flexibility Target Family 灵活目标族
FTF Functional Test Flight 功能试验飞行
FTI Fixed Target Indicator 固定目标指示器（装置）
FTIS Flight Test Instrumentation System 飞行测试（试验）仪器仪表系统
FTMD Flight Torque Measurement

Demonstration 飞行扭矩测量演示验证

FTMO Flight Test Mission Objectives 飞行试验任务目标

FTMS Fluid Transfer Management System 流体传输管理系统

FTO Flight Test Objective 飞行测试（试验）目标

FTO Functional Test Objectives 功能测试（试验）目标

FTOH Flight Team Operations Handbook 飞行组操作手册

FTOH Flight Test Operations Handbook 飞行测试（试验）操作手册

FTOS Flight Termination Ordnance System 飞行终止军械系统

FTP Flight Test Procedure 飞行测试（试验）规程

FTP Functional Test Program 功能测试（试验）项目

FTP Functional Test Progress 功能测试（试验）进程

FTPP Fault Tolerant Power Panel 故障容限配电屏

FTR Flight Test Requirement 飞行测试（试验）要求

FTR Flight Test Round 飞行试验（测试）周期

FTR Functional Test Requirement 功能测试（试验）要求

FTRD Flight Test Requirements Document 飞行测试（试验）要求文件

FTRD Functional Test Requirements Document 功能测试（试验）要求文件

FTRFME Flight Test Rocket Facilities Mechanical Engineering 飞行测试火箭设备机械工程

FTS Federal Telecommunications System 【美国】联邦通信系统

FTS Flight Telerobot Servicer 飞行遥控机器人服务器（美空间站服务操作系统）

FTS Flight Telerobotic Servicer 飞行遥控机器人服务程序

FTS Flight Termination System 飞行终止系统

FTS Flight Test Station 飞行测试（试验）站

FTS Flight Test System 飞行测试（试验）系统

FTS Functional Test Specifications 功能测试（试验）说明

FTS-DTF Flight Telerobot Servicer-Demonstration Test Flight 【美国国家航空航天局】飞行遥控机器人服务器—验证（性）试飞

FTSIL Flight Test System Integration Lab 飞行试验系统综合实验室

FTU Flight Termination Unit 飞行终止装置

FTV Flight Test Vehicle 飞行测试（试验）运载器

FTV Functional Technology Validation 功能性技术验证

FU Flight Unit 飞行装置

FU Fuel 燃料；燃油；燃烧剂；可燃物

FUO Fuel Mass at Lift Off 【火箭】离地时的燃料质量

FUT Fixed Umbilical Tower 固定脐带塔

Fvac Vacuum Thrust 真空推力

FVF First Vertical Flight 首次垂直飞行

FVMS Fluid Volume Measurement System 液体容积测量系统

FVP Flight Verification Payload 飞行验证有效载荷

FVS Flight Software Verification Sys-

tem 飞行软件校验系统
FVS Flight Vehicle Structure 飞行器结构
FVT Functional Validation Test 功能鉴定试验
FVV Facility Verification Vehicle 设施验证运载器
FWAS Failure Warning and Analysis System 故障告警与分析系统
FWAS Flight Warning and Analysis System 飞行告警与分析系统
FWC Flight Warning Computer 飞行告警计算机
FWD HT SHLD Forward Heat Shield 前端热防护

FWHM Full Width at Half Maximum 半峰全宽（半峰值全宽度，一半峰值全带宽）
FWS Flight Warning System 飞行告警系统
FWV Fixed Wing Vehicle 固定翼飞行器
FXBR Forward-Based X-Band Radar 前沿配置的 X 波段雷达
FYS Fixed Yaw Stabilizer 固定偏航稳定器
FZCG Float Zone Crystal Growth 漂浮区晶体生长
FZF Fussy Zero Function 模糊零函数

G

G　Gravity　万有引力；地心吸力；重力

G&C　Guidance and Command　制导与指令

G&C　Guidance and Control　制导与控制

G&CC　Guidance Control Coupler　制导与控制耦合器

G&CEP　Guidance and Control Equipment Performance　制导与控制设备性能

G&CFAP　Guidance & Control Flight Analysis Program　制导与控制飞行分析计划

G&N　Guidance and Navigation　制导与导航

G&NS　Guidance and Navigation Subsystem　制导与导航子系统

G&O　Goals and Objectives　目标与目的

G-II　Gulfstream II (Shuttle Training Aircraft)　【美国】湾流2（航天飞机训练飞机）

G-V　Gravity-Velocity　重力—速度

G/A　Ground-to-Air　地对空

G/AIT　Ground/Airborne Integrated Terminal　地面/机载综合终端

G/B　Ground Beacon　地面信标

G/CS　Guidance and Control System　制导与控制系统

G/E　Graphite Epoxy　石墨环氧化合物

G/G　Ground-to-Ground　地对地

G/S　Guided Steering　制导操纵

G/VLLD　Ground/Vehicle Laser Locator Designator　地面/飞行器激光定位（照射）器

GA　General Assembly　总装配

GA　Gimbal Angle　万向架转角

GA　Glide Angle　滑行角

GA　Ground Antenna　地面天线

GA　Gyro Assembly　陀螺仪组件（装配）

GA&CS　Ground Acquisition and Command Station　地面采集与指挥站

GAC　General Automatic Control　【飞行操纵系统】全自动操纵

GAC　Ground Analysis Centre　地面分析中心

GAC　Ground Attitude Control　地面姿态控制

GACC　Ground Attack Control Center　地面攻击控制中心

GACC　Guidance Alignment and Checkout Console　制导校准与测试控制台

GACIAC　Guidance and Control Information and Analysis Center　制导与控制信息与分析中心

GACM　Global Atmospheric Composition Mission　全球大气成分项目

GACU　Ground Air Conditioning Unit　地面空气调节装置

GACU　Ground Avionics Cooling Unit　地面电子设备与控制系统冷却装置

GADRAS　Gamma Detector Response and Analysis Software　伽马射线探测器响应与分析软件

GAELIC　Grumman Aerospace Language for Instructional Checkout　用

于结构测试的格鲁门航空航天语言
GAEO General Assembly Engineering Order 通用装配工程指令
GAIA Global Astrometry Interferometer for Astrophysics 用于天体物理学研究的全球天体测量干涉仪
GAINS Gimballess Analytic Inertial Navigation System 无框架解析惯性导航系统
GAINS Global Air-Borne Integrated Navigation System 全球机载综合导航系统
GAINS GPS, Air Data and Inertial Navigation System 全球定位系统、大气数据与惯性导航系统
GAIT Ground-Based Augmentation and Integrity Technique 陆基增强与综合型技术
GAN Geostationary Augmentation Network 地球同步增强型网络
GAN Gyro Automatic Navigation System 陀螺自动导航系统
GANDER Guidance and Navigation Development and Evaluation Routine 制导与导航发展及评定程序
GAO Government Accountability Office 【美国】联邦审计署
GAP GOAL Automatic Procedure 航空航天地面操作语言自动化规程
GAPC Ground Attitude and Positioning Control 地面姿态与位置控制
GARD-TRAK Gamma Absorption and Radiation Detection Tracking 伽马射线吸收与辐射探测跟踪
GARDAE Gathers, Alarms, Reports, Displays and Evaluates 收集、警报、报告、展示与评估
GARDIAN General Area Defense Integrated Antimissile Laser System 通用区防御综合反导弹激光系统
GARS Gyrocompassing Attitude Reference System 陀螺罗盘姿态基准（参考）系统
GAS-can Get-Away Special Canister 【美国航天飞机】有效载荷搭载容器
GASES Gravity Anchored Space Experiment Satellite 重力稳定空间实验卫星
GASP Grand Accelerated Space Platform 大型加速空间平台
GASP Gravity-Assisted Space Probe 【美国国家航空航天局】重力辅助式太空探测
GASPI Guidance Attitude Space Position Indicator 制导姿态空间位置显示器
GASS Geophysical Airborne Survey System 地球物理航空测量系统
GATE Get-Away Tether Experiment 【美国】搭载容器施放系绳卫星实验
GATE Graphic Analysis Tool Environment 图形分析工具环境
GATP Ground Acceptance/Article Test Procedure 地面验收/组件测试（试验）规程
GATS GPS-Aided Target System 全球定位系统辅助目标系统
GATS Guidance Acceptance Test Set 制导系统验收试验设备
GATV Gemini Agena Target Vehicle "双子星座—阿吉纳"目标飞行器
GAUSS Galileo and UMTS Synergetic System 伽利略与通用移动通信系统的协作系统
GAVRS Gyrocompassing Altitude and Velocity Reference System 陀螺罗盘高度与速度参考系
GBAS Ground-Based Augmentation

System　陆基增强型系统
GBDL　Ground-Based Data Link　地基数据链
GBDLS　Ground-Based Doppler Lidar System　陆基多普勒激光测距系统
GBEV　Ground-Based Experimental Version　地基实验型
GBFEL　Ground-Based Free Electron Laser　地基自由电子激光器
GBHE　Ground-Based Hypervelocity Gun Experiment　地基超高速炮实验
GBI　Ground Based Interceptor　陆基拦截导弹
GBI-P　Ground-Based Interceptor – Prototype　地基拦截机（导弹）—样机
GBI-X　Ground-Based Interceptor–Experiment　地基拦截机（导弹）—实验型
GBKV　Ground-Based Kinetic Kill Vehicle　地基动能杀伤飞行器
GBL　Ground-Based Laser　地基激光（器）
GBLD　Ground-Based Launcher Demonstration　地基发射装置（火箭）演示验证
GBLRS　Ground-Based Laser Repeater Station　地基激光器重发站
GBMD　Global Ballistic Missile Defense　全球弹道导弹防御
GBMI　Ground-Based Midcourse Interceptor　地基中程拦截机（导弹）
GBOS　Ground-Based Optical System　地基光学系统
GBPST　Ground-Based Passive Signal Tracking　地基被动信号跟踪
GBR-P　Ground Based Radar-Prototype　地基雷达—样机
GBR-T　Ground-Based Radar-Theater　地基雷达—战区
GBR-X　Ground-Based Radar –Experimental　地基雷达—实验型
GBRF　Ground-Based Radio Frequency　地基无线电频率
GBS　Ground Based Software　陆基软件
GBRI　Ground-Based Rocket Interceptor　地基火箭拦截机（导弹）
GBRT　Ground-Based Radar–Terminal　地基雷达—终端型
GBS　Ground-Based Sensor　地基传感器
GBSAA　Ground-Based Sense and Avoid　陆基感应与躲避
GBST　Ground-Based Software Tool　陆基软件工具
GBSTS　Ground Based Surveillance Tracking System　地面监视跟踪系统
GBT　Global Ballistic Transport　环球弹道运输
GBT　Ground-Based Test　陆基测试（试验）
GBT　Ground-Based Transceiver　陆基无线电收发机
GBU　Guided Bomb Unit　制导炸弹装置
GC　Ground Control　地面控制
GC　Guidance Control　制导控制
GC&A　Guidance, Control, and Avionics　制导、控制和电子设备与控制系统
GC-MS　Gas Chromatography-Mass Spectroscopy　气相色谱—质谱
GC³　Ground, Command, Control and Communication　地面、指挥、控制与通信
GCA　Ground-Controlled Approach　地面控制进场（或进近）（系统）

GCA Guidance and Control Assembly 制导控制组件
GCAS Ground Collision Avoidance System 地面避撞系统
GCAU Ground Control Approach Unit 地面控制进场（或进近）装置
GCC Ground Cluster Controller 地面簇控制器（装置）
GCC Ground Communications Coordinator 地面通信协调员；地面通信协调装置
GCC Ground Component Commander 地面部队指挥官
GCC Ground Control Center 地面控制中心
GCC Guidance & Control Computer 制导与控制计算机
GCC Guidance Control Computer 制导控制计算机
GCCC Ground Communications, Command and Control 地面通信、指挥与控制
GCCC Ground Control Computer Center 地面控制计算机中心
GCCD Guidance and Control Concept Demonstrator 制导与控制概念演示器
GCCO Ground Control Checkout 地面控制检测
GCCS Geostationary Communication and Control Segment 地球同步通信与控制部分
GCCS Global Command and Control System 【美国国防部】全球指挥控制系统
GCCS-J Global Command and Control System-Joint 联合全球指挥控制系统
GCD GAS Control Decoder 有效载荷搭载容器控制译码器

GCDC Ground Checkout Display and Control 地面检验显示与控制
GCDCS Ground Checkout Display and Control System 地面检验显示与控制系统
GCE Ground Checkout Equipment 地面检测设备
GCEL Ground Control Experiment Laboratory 地面控制试验实验室
GCF Ground Communications Facility 地面通信设施
GCHX Ground Cooling Heat Exchanger 地面冷却热交换器（装置）
GCI Ground Controlled Interception 地面控制拦截
GCIL Ground Control Interface Logic 地面控制接口逻辑
GCILC Ground Command Interface Logic Controller 地面指令接口逻辑控制器（装置）
GCILU Ground Control Interface Logic Unit 地面控制接口逻辑装置
GCL Ground Controlled Landing 地面控制着陆
GCL Ground Coolant Loop 地面冷却剂回路
GCN Global Command Network 全球指挥网络
GCN GMD Communications Network 陆基导弹防御通信网
GCN Ground Communications Network 地面通信网
GCN Ground Control Network 地面控制网
GCO Ground Checkout 地面检测
GCOS General Computer Operational System 通用型计算机操作系统
GCOS Ground Computer Operating System 地面计算机操作系统
GCP Geostationary Communications

Platform 静止轨道通信平台
GCP　Glare-Shield Control Panel　遮光控制屏（操纵台）
GCP　Ground Control Point　地面控制点
GCQA　Government Contract Quality Assurance　政府合同质量保证
GCR　Galactic Cosmic Ray　银河宇宙射线
GCR　Ground Controlled Radar　地面控制雷达
GCS　Global Control System　全球控制系统（不扩散导弹和导弹技术）
GCS　Ground Clutter Suppression　地面离合抑制
GCS　Ground Command System　地面指挥系统
GCS　Ground Communications System　地面通信系统
GCS　Ground Control Station　地面控制站
GCS　Guidance Cutoff Signal　制导关闭信号
GCSS　Global Communication Satellite System　全球通信卫星系统
GCSS　Ground-Controlled Space System　地面控制航天系统
GCTS　Gas Component Test Stand　气体元器件测试（试验）台
GCTS　Ground Communications Tracking System　地面通信跟踪系统
GCU　General Control Unit　通用控制装置
GCU　Generator Control Unit　发生器（发电机，振荡器）控制装置
GCU　Ground Cooling Unit　地面冷却装置
GCU　Gyro Coupling Unit　陀螺仪耦合装置
GCWS　Ground Collision Warning System　防撞地告警系统
GD　Ground Distance　地面距离
GDA　Gimbal Drive Actuator/Assembly　万向架驱动致动器／组件
GDAP　GEOS (Geodetic Earth-Orbiting Satellite) Data Adjustment Program　地球轨道大地测量卫星数据调整计划
GDBS　Generalized Data Base System　通用数据库系统
GDC　Gyro Display Coupler　陀螺仪显示耦合器
GDCP　GOES Data Collection Platform　地球静止环境业务卫星数据收集平台
GDDS　GOES Data Distribution System　地球静止环境业务卫星数据分发系统
GDE　Ground Demonstration Engine　地面演示验证发动机
GDES　Geostationary Operational Environmental Satellite　地球同步操作环境卫星
GDFA　Gas Jet Diffusion Flame Apparatus　气体喷射火焰扩散装置
GDIP　General Defense Intelligence Program　【美国】国防情报总计划
GDL　Gas Dynamic Laser　气动激光器
GDLP　Ground Data Link Processor　地面数据链处理器（装置）
GDMS　Ground Data Management System　地面数据管理系统
GDP　Ground Delay Program　地面延迟项目
GDS　GNC Dynamic Simulator　制导、导航与控制动力模拟器（装置）
GDS　Goldstone Deep Space　【美国国家航空航天局】"金石"深空项

目

GE General Electric 通用电子设备

GEAV Guidance Error Analysis Vehicle 制导误差分析（模型）飞行器

GECS Gemini Environmental Control System 【美国】"双子星"飞船环境控制系统

GECS Ground Environmental Control System 地面环境控制系统

GEDAC General Electric Detection and Automatic Correction 通用电气探测与自动修正

GEDI Ground-Based Electromagnetically-Launched Defensive Impactors 地基电磁发射防御性碰撞器

GEM General Electronics Module 通用电子设备与控制系统模块

GEM Graphite Epoxy Motor 石墨环氧树脂（固体火箭）发动机

GEM Guidance Enhanced Missiles 制导增强型导弹

GEM-FLO Generic Environment for Modeling Future Simulation Launch Vehicle 未来运载火箭模拟建模的通用环境

GEM-FLO Generic Model for Future Launch Operations 未来发射操作的通用模型

GEMACS General Electromagnetic Model for Analysis of Complex Systems 用于综合系统分析的通用电磁模型

GEMAN General Electric Miniature Aerospace Navigator 【美国】通用电气公司微型航空航天导航仪

GEMS Geostationary European Meteorological Satellite 【欧洲空间研究组织】欧洲地球同步气象卫星

GEMS Grouped Engine Monitoring System 组合发动机监控系统

GEMSIP Gemini Stability Improvement Program 【美国】"双子星"飞船稳定性改进计划

GENOPAUSE Geodetic Satellite in Polar Geosynchronous Orbit 极轨同步轨道测地卫星

GENSAA Generic Spacecraft Analysis Assistant 通用航天器分析助手

GenX Generation-X Vision 生成X射线景象

GEO Geostationary Earth Orbit 地球静止轨道

GEO Geosynchronous Earth Observatory 【美国】同步地球观测台

GEO Geosynchronous Earth Orbit 地球同步轨道

GEO-CAPE Geostationary Coastal and Air Pollution Events 地球同步轨道沿海和空气污染事件

GEODSS Ground-Based Electro-Optical Deep Space Surveillance 陆基光电深空监视

GEODSS Ground-Based Electro-Optical Deep Space Surveillance (System) 陆基光电深空监视（系统）

GEON Gyro-Erected Optical Navigation 陀螺仪稳定光学导航系统

GEONS GPS-Enhanced Onboard Navigation System GPS增强星载导航系统

GEOP Geostationary Orbit Phase 地球静止轨道段

GEOS Geodetic Earth Orbiting Satellite 地球轨道测地卫星

GEOS Geodetic Orbit Satellite 大地测量轨道卫星；测地轨道卫星

GEOS Geodetic Survey Satellite 【美国国家航空航天局】大地测量卫星

GEOS Geodynamics Experimental Ocean Satellite 【美国】地球动力

学实验海洋卫星

GEOS Geostationary Earth Orbiting Satellite 【欧洲空间局】对地静止轨道卫星；地球同步卫星

GEOS Geostationary Scientific Satellite 对地静止科学卫星

GEOS Geosynchronous Earth Observation System 地球同步观测系统

GEOS-2 Geodetic Earth-Orbiting Satellite 2 测地地球轨道卫星—2

GEOSAR Geosynchronous Earth Orbit Synthetic Aperture Radar 地球同步轨道合成孔径雷达

GEOSAT Geological Satellite 地质卫星

GEOSCAN Ground-Based Electronic Omnidirectional Satellite Communications Antenna 陆基电子全向卫星通信天线

GEOSEPS Geosynchronous Solar Electric Propulsion Stage 太阳同步电推进阶段

GEOSS Global Earth Observing System of System 全球对地观测综合系统

GEOSS Global Earth Observation System of Systems 全球观测系统集成

GEP Ground Entry Point （航天器）入地点；地面入口点

GERB Geostationary Earth Radiation Budget Experiment 地球同步辐射平衡试验

GERSIS General Electric Range Safety Instrumentation System 【美国】通用电气公司靶场安全测试设备系统

GERT Graphical Evaluation and Review Technique 图形评估与评审技术

GERTS General Remote Terminal System 通用远程终端系统

GES Goddard Experiment Support System 【美国】戈达德航天中心实验保障系统

GES Ground Engineering System 地面工程系统

GESOC General Electric Satellite Orbit Control 【美国】通用电气公司卫星轨道控制系统

GEST Gemini Slow-Scan Television 【美国】"双子星"飞船慢扫描电视

GET Ground Elapsed Time 地面耗用时间

GETI Ground Elapsed Time of Ignition 地面点火耗用时间

GET1L Ground Elapscd Time of Landing 地面着陆耗用时间

GETS Generalized Electronic Trouble Shooting 综合电子故障排除

GETS Ground Equipment Test Sets 地面设备测试装置

GEV Ground Effect Vehicle 地效运载器；地效飞行器

GEV Generalized Extreme Value 通用极值

GF Gauge Factor 压力计因子

GF&P Gases, Fluids, and Propellants 气体、液体与推进剂

GFAE Government Furnished Aerospace Equipment 政府提供的航空航天设备

GFC/C GMD Fire Control and Communications 陆基导弹防御火控与通信

GFCC Ground Facilities Control Centre 地面设施控制中心

GFD Government Furnished Data 政府提供的数据

GFE Government Furnished Equip-

ment 政府提供的设备
GFFC Geophysical Fluid Flow Cell Experiment 地球物理流体电池实验
GFM Government Furnished Material 政府提供的材料
GFP Government Furnished Property 政府提供的物资
GFRP Glass Fiber Reinforced Plastics 玻璃纤维强化塑料
GFRP Graphite Fiber Reinforced Plastics 石墨纤维强化塑料
GFS Global Forecast System model 全球预报系统模型
GFS Government Furnished Software 政府提供的软件
GFV Guided Flight Vehicle 制导飞行器
GG Gas Generator 气体发生器
GG Gravity Gradient 重力梯度
GGEM Gravity Gradiometer Explorer Mission 【美国国家航空航天局】重力梯度仪探测器飞行任务
GGI&S Global Geospatial Information and Services 全球地球空间信息与服务
GGL Gravity-Gradient Libration 重力梯度摆动；重力梯度振动
GGM Gravity Gradient Mission 重力梯度研究（任务）
GGP GPS Guidance Package 全球定位系统制导组件
GGS Global Geospace Science 全球地球空间科学任务（美国、西欧、日本合作）
GGS Global Geosynchronous Science 全球地球同步科学
GGS Gravity-Gradient Satellite 重力梯度卫星
GGS Gravity-Gradient Sensor 重力梯度传感器
GGS Ground Guidance System 地面制导系统
GGS GPS Ground Station 全球定位系统地面站
GGSE Gravity Gradient Stabilization Experiment 【美国空军、海军】重力梯度稳定性实验；重力梯度稳定性实验卫星
GGSTIDE Global Geosynchronous Science Thermal Ions Dynamics Experiment 全球地球同步科学热离子动力实验
GGTS Gravity Gradient Test Satellite 【美国】重力梯度试验卫星
GGVM Gas Generator Valve Module 气体发生器阀模块
GH Gemini Hatch 【美国国家航空航天局】"双子星座"飞船舱口
GH Ground Handling 地面处理
GH$_2$ Gaseous Hydrogen 气态氢
GHC Ground Half Coupling 地面半耦合
GHe Gaseous Helium 气态氦
GHF Gradient Heating Facility 梯度加热设施
GHG Greenhouse Gas 环保型气体
GHOST Global Horizontal Sounding Technique 全球水平回声探测技术
GHX Ground Heat Exchanger 地面热交换器（装置）
GI&S Geospatial Information and Services 地球空间信息与服务
GIA GPC Interface Adapter 通用计算机接口适配器
GIB Global Navigation Satellite System Integrity Broadcast 全球导航卫星系统统一预报
GIC Global Navigation Satellite System Integrity Channel 全球导航卫

星系统统一频道
GICO GEOS Index of Cloud Opacity 地球同步卫星云阻光指数
GICS Global Instrumentation Control System 全球观测控制系统
GICS Ground Instrumentation and Communication System 地面仪表与通信系统
GIDEP Government-Industry Data Exchange Program 政府－工业数据交换项目
GIE Ground Instrumentation Equipment 地面仪器仪表设备
GIFC Global Integrated Fire Control 全球综合火力控制
GIGS Gemini Inertial Guidance System 【美国】"双子星"飞船惯性制导系统
GII Global Information Infrastructure 全球信息基础设施
GIM Generalized Information Management 通用信息管理
GIM Ground Intervention Mode 地面干预模式
GIMS Ground Identification of Missions in Space 航天任务地面识别
GIMU Gimballess Inertial Measuring Unit 无框架惯性测量装置
GINS GPS Inertial Navigation System 全球定位系统惯性导航系统
GIP Ground Impact Point 地面碰撞点
GIRAS Geographic Information Retrieval and Analysis System 地理信息检索与分析系统
GIRD Ground Integration Requirements Document 地面集成要求文件
GIS Graphical Information System 图形信息系统
GISAT Ground Identification of Satellite 卫星地面识别
GIT Geospace Interorbital Transportation 地球太空轨道间的运输
GITG Ground Interface Technical Group 地面接口技术组
GITIS Government Integrated Technical Information System 【美国】政府综合技术信息系统
GITIS Ground Integrated Target Identification System 地面综合目标识别系统
GIWG Ground Interface Working Group 地面接口工作组
GLANCE Global Lightweight Air Navigation Computing Equipment 轻便式全球空中导航计算设备
GLASS Global Area Strike System 全球范围攻击系统（一种将高能激光、动能武器和跨大气层飞行器合并在一起的系统）
GLAST Gamma-Ray Large Area Space Telescope 伽马射线大视野空间望远镜
GLCM Ground-Launched Cruise Missile 地面发射巡航导弹
GLDS Gemini Launch Data System 【美国】"双子星"飞船发射数据系统
GLE Gemini Laser Experiment 【美国】"双子星"飞船激光通信实验
GLEP Group for Lunar Exploration and Planning 月球探险与计划小组
GLFC Graphite Lunar Fuel Cask 登月舱石墨燃料箱
GLINT Geo Light Imaging National Testbed 地球同步轨道光成像国家试验台
GLIPA Guide Line Identification Program for Antimissile Research 反导弹主探测方向识别程序；反导弹研

究的制导线识别程序

GLMS Ground Lightning Monitoring System 【美国国家航空航天局】地面雷电监测系统

GLNS GPS Landing and Navigation System 全球定位系统的着陆与导航系统

GLOBIXS Global Information Exchange System 全球信息交换系统

GLOMRS Global Low Orbiting Message Relay Satellite 【美国】全球低轨道信息中继卫星

GLONASS Global Navigation Satellite System 【俄罗斯】全球导航卫星系统

GLOPAC Gyroscopic Low Power Attitude Control 陀螺小功率姿态控制

GLORS Global Low Orbiting Relay Satellite 全球低轨道中继卫星（同 GLOMRS）

GLOV Global Observational Vehicle 全球观测飞行器

GLOW Goddard Lidar Observatory for Wind 戈达德激光雷达测风天文台

GLOW Gross Lift-Off Weight 起飞总重；起飞总重量

GLOW Ground Lift-Off Weight 地面起飞重量

GLP GOAL Language Processor 航空航天地面操作语言处理器（装置）

GLP Ground Launched Probe 地面发射的探测器

GLRS Geodynamics Laser Ranging System 地球动力学激光测距系统

GLS Glide Slop 滑行着陆

GLS Government Launch Services 政府发射服务

GLS Ground-Launched Sensor 地面发射的传感器

GLS Ground Launch Sequence (Sequencer) 地面发射顺序（序列装置）

GLS GNSS Landing System 全球导航卫星系统的着陆系统

GLS GPS Landing System 全球定位系统的着陆系统

GLSP Ground and Launch Systems Processing 地面与发射系统操作

GLSTE Ground Launch Support and Test Equipment 地面发射支撑（保障）与测试设备

GLTD Ground and Launch Technology Demonstration 地面与发射技术演示验证

GLU GPS Landing Unit 全球定位系统的着陆装置

GLV Gemini Launch Vehicle 【美国】"双子星"飞船运载火箭

GM Gaseous Mixture 气态混合物；混合气体

GMAB Guided Missile Assembly Building 导弹装配间

GMACC Ground Mobile Alternate Command Center 地面移动式备用指挥中心

GMAL General Electric Macro Assembly Language 通用电气宏装配语言

GMAOC Ground Mobile Alternate Operations Center 地面移动式备用操作中心

GMC Ground Movement Control 地面移动控制

GMCC Ground Mobile Command Center 地面移动式指挥中心

GMCF Guided Missile Control Facility 导弹控制设施

GMCL Ground Measurements Command List 地面测量指令清单

GMCP Ground Mobile Command Post 地面移动式指挥所
GMD Global Missile Defense 全球导弹防御
GMD Ground-Based Midcourse Defense 陆基中程防御
GMD Ground-Based Midcourse Defense System 陆基中段防御系统；陆基中程防御系统
GMD Ground-Based Missile Defense 陆基导弹防御
GMD Ground Meteorological Device 地面气象装置
GMDS Ground Mobile Data Service 地面移动数据服务
GMESS Global Monitoring for Environment and Security 全球环境与安全监测（计划）
GMF Ground Measurement Facility 地面测量设施
GMFC Guided Missile Fire Control 导弹发射控制
GMFCS Guided Missile Fire Control System 导弹火控系统；导弹发射控制系统
GMFP Guided Missile Firing Panel 导弹发射仪表板
GMI Ground Measurement Infrastructure 地面测量基础设施
GML General Measurement Loop 通用测量回路
GML Guided Missile Launcher 导弹发射架（车）
GMLA Guided Missile Launch Assembly 导弹发射装置
GMLOF Guided Missile Line of Flight 导弹飞行弹道
GMLR Guided Missile Launch Rocket 导弹运载火箭
GMLR Guided Missiles and Large Rockets 导弹与大型火箭
GMLRS Guided Multiple Launch Rocket System 多管制导火箭发射系统
GMLS Guided Missile Launching System 导弹发射系统
GMM Geometric Math Model 测地数学模型
GMMMU Ground Mounted Manned Maneuvering Unit 地面装配的载人机动装置
GMR General Modular Redundancy 通用模块化冗余（技术）
GMR Ground Mapping Radar 地形测绘雷达
GMR Ground Mobile Radio 地面移动式无线电
GMR Ground Movement Radar 地面活动雷达
GMS Gemini Mission Simulator 【美国】"双子星"飞船飞行模拟器
GMS Geostationary Meteorological Satellite 【日本】地球静止轨道气象卫星
GMS Guided Missile System 导弹系统
GMSQUAD Guided Missile Squadron 导弹中队
GMSS Geo-Stationary Meteorological Satellite System 地球同步气象卫星系统
GMSTS Guided Missile Service Test Station 导弹勤务试验站
GMSTS Guided Missile System Test Set 导弹系统测试装置
GMT Ground Moving Target 地面活动目标
GMT Guided Missiles Target 可控导弹靶
GMTD Guided Missile Training Device 导弹训练装置

GMTI Ground Moving Target Indication 地面移动目标指示
GMTI Ground Moving Target Indicator 地面动目标指示器
GMTS Guided Missile Test Set(s) 导弹测试装置（仪）
GMTT Ground Moving Target Track 地面动目标跟踪
GMTT Ground Moving Target Tracker 地面活动目标跟踪装置
GMTT&C Ground Mobile Tracking, Telemetry, and Control 地面移动跟踪、遥测与控制
GMTU GPS Metric Tracking Unit 全球定位系统测量跟踪装置
GMU GPS Monitoring Unit 全球定位系统监视装置
GN Gain 增益（系数）；放大（系数）；增量（大，进，加）
GN Gas Nitrogen 氮气
GN Gaseous Nitrogen 气态氮
GN Ground Navigation 地面导航
GN Ground Network 地面网络
GN&C Guidance, Navigation and Control 制导、导航与控制
GN&C Guidance, Navigation and Control Subsystem 制导、导航与控制分系统
GN$_2$ Gaseous Nitrogen 气态氮
GNA Global Network Architecture 全球网络体系架构
GNC Global Network Operations Center 全球网络操作中心
GNC Guidance and Navigation Computer 制导与导航计算机
GNC&P Guidance, Navigation, Control and Propulsion 制导、导航、控制与推进
GNCFTS GN&C Flight Test Station 制导、导航与控制飞行试验站
GNCIS GN&C Integration Simulator 制导、导航与控制综合模拟器
GNCS Guidance, Navigation and Control System 制导、导航与控制系统
GNCST Global Net-Centric Surveillance and Targeting 全球监视与目标分配中心网
GNCTS GN&C Test Station 制导、导航与控制试验站
GND Ground Stabilization 地面稳定状态
GND C/O Ground Checkout 地面检测
GNE Guidance and Navigation Electronics 制导与导航电子设备
GNP GPS Navigational Processor 全球定位系统导航处理机
GNR Global Navigation Receiver 全球导航接收器（装置）
GNS Ground Network Services 地面网络服务
GNSS Global Navigation Satellite System 【俄罗斯】全球导航卫星系统
GNSSU Global Navigation Satellite Sensor Unit 全球导航卫星传感器装置
GNT Ground Test 地面试验
GO Geostationary Orbit 地球静止轨道；对地静止轨道
GO Ground Operations 地面操作
GO$_2$ Gaseous Oxygen 气态氧
GOAL General Organization Analysis Language 通用结构分析语言
GOAL Ground Operations Aerospace Language 航空航天地面操作语言
GOB Ground Operation Bus 地面操作总线
GOC Ground Operations Center 地面操作中心

GOC Ground Operations Coordinator 地面操作协调员；地面操作协调装置

GOCA Ground Operations Control Area 地面操作控制区；地面作战控制区

GOCE Gravity Field and Steady-State Ocean Circulation Explorer 地球重力场与海洋环流探测卫星

GOCVG Gaseous Oxygen Control Valve Guard 气态氧控制阀防护罩

GODAE Global Ocean Data Assimilation Experiment 全球海洋数据同化实验

GODAS Global Ocean Data Assimilation System 全球海洋数据同化系统

GODSEP Guidance and Orbit Determination for Solar Electric Propulsion 太阳能电力推进的制导与轨道测定

GOE Ground Operational Equipment 地面操作设备

GOES Geostationary Operational Environmental Satellite 【美国】地球静止环境业务卫星

GOES Geostationary Orbiting Environmental Satellite 地球静止轨道环境卫星

GOES Geosynchronous Orbiting Environmental Satellite 地球同步在轨环境卫星

GOETS Ground Operations Estimating Techniques System 地面操作技术评估系统

GOH Geosynchronous Orbiting Habitat 地球同步轨道生长环境

GOI Ground Object Identification 地面目标识别

GOIB Group Orbit Injection Bay 组入轨舱

GOMMS Ground Operations and Material Management System 地面操作与材料管理系统

GOMS Geostationary Operational Meteorological Satellite 对地静止业务气象卫星

GOMS Geostationary Orbit Meteorological Satellite 【俄罗斯】地球静止轨道气象卫星

GOMS Ground Operations Management System 地面操作管理系统

GOODS Great Observatories Origins Deep Survey 大天文台宇宙起源深空观测

GOP Ground Operations Panel 地面操作控制台

GOPG Ground Operations Planning Group 地面操作规划组

GOR Ground Operations Review 地面操作评审

GORA Ground Operations Risk Analysis 地面操作风险分析

GORP Ground Operations Requirements Plan 地面操作要求计划

GORP Ground Operations Review Panel 地面操作评审委员会

GORR Ground Operations Readiness Review 地面操作准备评审

GORS Ground Observer RF System 地面观测装置射频系统

GOSC Ground Operational Support Center 地面（飞行）操作保障中心

GOSP Government Open System Protocol 【美国】政府开放式系统协议

GOSS Ground Operational Support System 地面操作保障系统；地面作战支援系统

GOST Ground Operations Simulation

Technique 地面操作模拟（仿真）技术
GOTCha Goals, Objectives, Technical Challenges and Approaches 目的、目标、技术挑战与解决方法
GOTS Government Off-the-Shelf 政府性成品；官方现货
GOTS Gravity-Oriented Test Satellite 研究重力的试验卫星；重力定向测试卫星
GOWG Ground Operations Working Group 地面操作工作组
GOX Gaseous Oxygen 气态氧
GP Galactic Probe 银河系探测器
GP Gaseous Propellant 气态推进剂
GP Gravity Probe 【美国国家航天局】重力探测器
GP&C Global Positioning and Communications 全球定位与通信
GP-B Gravity Probe B 【美国国家航空航天局】重力探测器B
GPA General Purpose Amplifier 通用放大器；通用增幅器
GPA Guidance Platform Assembly 制导平台组件
GPADIRS Global Positioning, Air Data, and Inertial Reference System 全球定位、大气数据与惯性参照系统
GPALS Global Protection Against Limited Strikes 对付有限打击的全球保护系统（现称为BMD—弹道导弹防御系统）
GPAS General Purpose Airborne Simulator 通用机载模拟器（装置）
GPB General Purpose Bomb 通用炸弹
GPBIM General Purpose Buffer Interface Module 通用缓冲接口模块
GPC General Purpose Computer 通用型计算机
GPCB GOAL Program Control Block 航空航天地面操作语言程序控制块
GPD Gimbal Position Display 万向架定位显示
GPES Goal Performance Evaluation System 目标性能评估系统
GPETE General Purpose Electronic Test Equipment 通用电子测试设备
GPF Gas Processing Facility 气体处理设施
GPGPU General Purpose Graphics Processing Unit 通用图形处理装置
GPHS General Purpose Heat Source 通用热源
GPI Gimbal Position Indicator 万向架定位指示器
GPI Ground Point of Impact 地面碰撞点；落地点
GPI Ground Position Indicator 地面定位指示器
GPIB General Purpose Interface Bus 通用接口总线
GPIRS Global Positioning Inertial Reference System 【美国】全球定位惯性参照（或基准）系统
GPL Gemini Programming Language 【美国】"双子星"飞船程序设计语言
GPL General Purpose Laboratory 通用型实验室
GPLS General Purpose Logic Simulator 通用逻辑模拟器
GPLS Glide Path Landing System 下滑道着陆系统
GPM Global Precipitation Measurement 全球沉降测量
GPME General Purpose Mission Equipment 通用任务设备
GPOT Ground Power Operating Times

地面动力操作次数
GPPF Gravitational Plant Physiology Facility 重力装置生理学设施
GPR Ground Penetrating Radar 穿地雷达；探地雷达
GPRSS General Purpose Remote Sensor System 通用遥感系统
GPRT General Purpose Radio Transmitter 通用无线电发射机
GPS General Processing Subsystem 综合处理分系统
GPS Global Positioning (Satellite) System 全球定位（卫星）系统
GPS Ground Power Supply 地面动力供应
GPS Ground Processing Simulation 地面工艺模拟；地面操作（处理）模拟（仿真）
GPSF General Purpose Simulation Facility 通用模拟设备
GPSPAC Global Positioning System Package 全球定位系统组件
GPSR Global Positioning System Receiver 全球定位系统接收机
GPSS General Purpose System Simulator 通用系统模拟器
GPSS Global Purpose Simulation Software 通用模拟软件
GPSS Global Purpose Simulation System 通用模拟系统
GPSS Ground Processing Scheduling System 地面处理调度系统
GPSSU Global Positioning System Sensor Unit 全球定位系统敏感部件
GPTE General Purpose Test Equipment 通用测试（试验）设备
GPU Graphical Processing Unit 图形处理装置
GPU Ground Power Unit 地面动力装置
GPU Guidance Processor Unit 制导处理器装置
GPUR GOAL Test Procedure Update Request 航空航天地面操作语言试验程序更新要求
GPV General Purpose Vehicle 通用飞行器
GPWS General Purpose Work Station 通用型工作站
GPWS Ground Proximity Warning System 地面邻近告警系统
GQR Ground Qualification Review 地面鉴定评审
GR Ground Relay 地面中继
GR Ground Rule 地面规则
GR&TE Ground Receiver & Transmitter Equipment 地面接收机和传输设备
GrabS Galactic Radiation and Background Satellite 银河辐射与背景卫星
GRACE Gravity Recovery and Climate Experiment 重力回溯（重建）与气候实验
GRACE Gravity Research And Climate Experiment 重力研究与气候试验
GRACS Gas-Reaction Attitude Control System 燃气反作用姿控系统
GRAIL Gravity Recovery and Interior Laboratory 重力回溯（重建）与内部结构实验室
GRAM Global Reference Atmosphere Model 全球参考大气模式
GRAM GPS Receiver Application Module 全球定位系统接收机应用模块
GRAMPA General Analytical Model for Process Analysis 过程分析的通用解析模型

GRAPES Graphical Route Analysis and Penetration Evaluation System 图解航迹分析与突防评估系统

GRAS Ground-Based Regional Augmentation System 陆基区域增强型系统

GRASP Generalized Reentry Application Simulation Program 【美国国家航空航天局】通用再入大气层应用模拟程序

GRAVSAT Gravity Survey Satellite 【美国】地球重力测量卫星

GRBDS Gyroscopes Rate Bomb Direction System 陀螺速率炸弹定向系统

GRC Glenn Research Center 【美国国家航空航天局】格伦研究中心

GRE Gamma-Ray Explorer 【美国】γ射线探险者卫星

GREBS Galactic Radiation Experiment Background Satellite 【美国】银河辐射实验背景卫星

GREM Geopotential Research Explorer Mission 【美国国家航空航天局】重力势研究探测器飞行任务

GRFL Groundwater Remediation Field Laboratory 水下保留现场实验室

GRID Graphic Retrieval and Information Display 图形恢复与信息显示

GRIPS Gamma-Ray Imager/Polarimeter for Solar 太阳伽马射线成像仪/偏振计

GRITS Gamma-Ray Imaging Telescope Study 伽马射线成像望远镜研究

GRITS Goddard Range Instrumentation Tracking System 【美国国家航空航天局】戈达德靶场仪表跟踪系统

GRM Global Range Missile 环球射程导弹

GRO Gamma-Ray Observatory 伽马射线观测台

GRS Galaxy Rocket System 银河系火箭系统

GRS General Range Support 靶场总体保障

GRTLS Glide Return to Landing Site 滑行返回至发射场

GS Gas Servicer 气体燃料加注车；气体服务车

GS Geosurvey Satellite 测地卫星

GS Glare Shield 遮光罩；闪光屏挡

GS Glide Slope 滑行倾斜角（斜度）

GS Ground Segment 地面段

GS Ground Speed 地面速度

GS Ground Station 地面站

GS Ground System 地面系统

GSA Gas Supply Assembly 气体供给组件（装置）

GSA Guiana Space Center 【法国】圭亚那航天中心

GSAT General Satellite 通用型卫星

GSC Geodetic Spacecraft 测地宇宙飞船

GSCD Ground Systems Control Document 地面系统控制文件

GSCU Ground Service Cooling Unit 地面伺服冷却装置

GSDC Ground Station Demonstration Lab 地面站演示验证实验室

GSDL Ground Software Development Laboratory 地面软件研发实验室

GSDO Ground Systems Development and Operations Program 地面系统研发与操作项目

GSE Geocentric Solar Ecliptic 地心太阳黄道系统

GSE Government Supplied Equipment 政府供给设备

GSE Ground Servicing Equipment 地面勤务设备

GSE Ground Support Electronics 地面保障电子设备

GSE Ground Support Equipment 地面保障设备

GSFC Goddard Space Flight Center 【美国国家航空航天局】戈达德航天飞行中心

GSFSR Ground Safety and Flight Safety Requirements 地面安全与飞行安全要求

GSI Glide Slope Indicator 滑行倾斜角（斜度）指示器

GSIF Ground Station Information Frame 地面站信息框架

GSII Government Services Information Infrastructure 【美国】政府服务信息基础设施

GSIP Ground Segment Implementation Plan 地面段执行计划

GSIU Ground Standard Interface Unit 地面标准接口装置

GSLV Geosynchronous Satellite Launch Vehicle 【印度】地球同步卫星运载火箭

GSM Ground Station Module 地面站模块

GSMS Ground Station Management System 地面站管理系统

GSO Geostationary Earth Orbit 地球静止轨道；对地静止轨道

GSO Geostationary Satellite Orbit 地球同步卫星轨道

GSO Ground Support Operations 地面保障操作

GSO Ground Systems Operations 地面系统操作

GSOC German Science Operations Center 德国科学操作中心

GSOC German Space Operations Center 德国航天操作中心

GSOV Ground Segment Operational Validation 地面段操作验证

GSP Ground Safety Plan 地面安全计划

GSP Ground Support Personnel 地面保障人员

GSP Ground Support Position 地面保障定位

GSR Grid Space Relay 格网坐标航天中继站

GSR Ground Station Radar 地面站雷达

GSRP Ground Safety Review Process 地面安全评审进程

GSRPS Group for the Study of Rocket Propulsion System 火箭推进系统研究组

GSRR Ground Segment Readiness Review 地面段准备状态评审

GSRR Ground Segment Requirements Review 地面段需求评审

GSRS General Support Rocket System 普通的辅助火箭系统；通用保障火箭系统

GSRS Ground Support Rocket System 地面保障火箭系统

GSS Geodetic Stationary Satellite 静止测地卫星

GSS Geospace Swing Station 摆动（的）地球空间站

GSS Ground Support Software 地面保障软件

GSS Ground Support System 地面保障系统

GSSA Ground Support Systems Activation 地面保障系统激活

GSSC Ground Support Simulation Computer 地面支援（或保障）模

拟计算机

GSSC Ground Support Systems Contractor 地面保障系统承包商

GSSI Ground Support System Integration 地面保障系统集成

GSSL Geospace Science Laboratory 地球空间科学实验室

GSSP Generally-Accepted Systems Security Principles 普遍接受的系统安全准则

GSSR Ground System Support Requirements 地面系统保障要求

GSSS Ground Support Software System 地面保障软件系统

GSSS Guide Star Selection System 导航星选取系统

GSST Ground Segment System Test 地面部分系统测试

GST Gemini System Trainer 【美国】"双子星"飞船系统训练设备

GST Global Space Transport 环球空间运输（工具）

GST Ground System Test 地面系统测试

GSTB Galileo System Test Bed 【欧洲】"伽利略"系统测试台

GSTDN Ground Satellite Tracking and Data Network 地面卫星跟踪与数据网

GSTDN Ground Spaceflight Tracking and Data Network 【美国国家航空航天局】地面航天飞行跟踪与数据网

GSTDN Ground Spacecraft Tracking and Data Network 地面航天器跟踪与数据网

GSTF Ground Systems Test Flow 地面系统测试流程

GSTS Ground-Based Surveillance and Tracking System 陆基监视与跟踪系统

GSTU Guidance System Test Unit 制导系统试验装置

GSU Gas Servicer Unit 气体燃料加注车装置

GSUIT Gravity Suit 抗过载飞行服

GSUS Ground Support Unit System 地面保障设备系统

GSV Guided Space Vehicle 制导航天器

GSVP Ground Support Verification Plan 地面保障检验计划

GT Ground Test 地面测试（试验）

GT Gyro Torque 陀螺仪扭矩

GT&A Ground Test and Acceptance 地面测试与验收

GTA Ground Test Accelerator 地面试验（测试）加速器

GTA Ground Test Access 地面测试（试验）数据访问

GTA Ground Test Article 地面测试（试验）部件

GTA Ground Torquing Assembly 地面扭矩组件

GTACS Ground Theater Air Control System 地面战区空中控制系统

GTC Ground Terminal Computer 地面终端计算机

GTCU Ground Thermal Conditioning Unit 地面热调节装置

GTDS Goddard Trajectory Determination System 【美国国家航空航天局】戈达德航天中心轨迹演示验证系统

GTE Ground Transport Equipment 地面运输设备

GTH Gross Thrust 总推力

GTI Ground Test Instrumentation 地面测试（试验）仪器仪表

GTM Ground Test Missile 导弹地面

试验

GTM Ground Test Motor 地面测试（试验）发动机

GTO Geostationary Transfer Orbit 地球静止转移轨道

GTO Geosynchronous Transfer Orbit 地球同步转移轨道

GTOS Ground Terminal Operation and Support 地面终端操作与保障

GTOW Gross Takeoff Weight 起飞总重；起飞总重量

GTS General Test Support 综合测试（试验）保障

GTS Geostationary Technology Satellite 地球静止轨道技术卫星

GTS GN&C Test Station 制导、导航与控制测试（试验）站

GTS GNS Test Station 制导与导航系统测试（试验）站

GTS Ground Telemetry Station 地面遥测站

GTSF Guidance Test and Simulation Facility 制导试验（测试）与模拟设施

GTU Ground Test Unit 地面测试（试验）装置

GTV Ground Test Vehicle 地面试验火箭

GTV Ground Test Verification 地面测试（试验）确认（鉴定）

GTV Ground Transport Vehicle 地面运输车辆

GTV Guidance Test Vehicle 制导试验飞行器

GUCA Ground Umbilical Carrier Assembly 【美国航天飞机】地面脐带承载组件

GUCP Ground Umbilical Carrier Panel 【美国航天飞机】地面脐带承载栅栏

GUCP Ground Umbilical Carrier Plate (interface between ET and hydrogen vent line) 【美国航天飞机】地面脐带承载板（外贮箱与氢排放管线之间的接口）

GUI Graphical User Interface 图形用户接口

GUID Guidance 制导；导航；引导

GUIDAR Guided Intrusion Detection and Ranging 制导干扰探测与测距

GUISE Guidance System Evaluation 制导系统评估（试验）

GUL GSE Utilization List 地面保障设备利用清单

GVA GOX Vent Arm 气态氧排放臂

GVT Ground Vibration Test 地面振动测试（试验）

GVTA Ground Vibration Test Article 地面振动测试（试验）组件

GVW Gross Vehicle Weight 飞行器总重量

GW Gross Weight 毛重；总重

GWA General Work Area 综合工作区

GWDS Graphic Weather Display System 气象图形显示系统

GWEF Guided Weapons Evaluation Facility 制导武器鉴定设施

GWEN Ground Wave Emergency Network 地波应急通信网

GWM Guam STDN Station 【美国国家航空航天局】关岛航天器跟踪数据网站

GWT Ground Winds Tower 地面风塔

GX Galaxy Express 【日本】"银河快车"运载火箭

GZ Ground Zero （核弹）爆心地面投影点

GZN Grid Azimuth 坐标方位角

H

H　Altitude　（飞行，海拔）高度；高空（程，线，位）；标高；垂直距离；地平纬度
H　Altitude Rate　高度比
H&I　Harassing and Interdicting　袭扰与遮断
H&T　Handling and Transportation　装卸与运输
H&V　Hazard and Vulnerability　危险性与易损性
H-CITE　Horizontal-Cargo Integration Test Equipment　水平货物综合测试设备
H-SAT　Heavy Satellite　重型卫星
H/A　Hazardous Area　危险区
H/C　Hydrogen-Carbon Radio　氢碳比
H/D　Height-to-Diameter Ratio　高度直径比
H/E　Heat Exchanger　热交换器（装置）
H/F　High Frequency　高频
H/L　Hardline　硬线
H/P　Horizontal Polarization　水平偏振
H/S　Heat Shield　热屏蔽；隔热罩
H/S IR　Hardware/Software Integration Review　硬件／软件综合评审
H/S/O　Hardware/Software/Operations　硬件／软件／操作
H/W　Hardware　硬件
HA　Apogee Altitude　远地点高度
HA　Hazard Analysis　危险分析
HA　Head Amplifier　前置放大器
HA　Heat-Shield Assembly　热屏障装配
HA　Hydraulic Actuator　液压传动装置
HAA　High Angle of Attack　大迎角
HAADS　High Altitude-Altitude Determination System　高空高度测定系统
HAAM　High Altitude Ablative Materials　高空烧蚀材料
HAARP　High Altitude Aurora Research Program　高空极光研究计划
HAARP　High-Frequency Active Aurora Research Program　高频活动极光研究计划
HAAT　High Ambient Air Temperature　大气环境高温试验
HAB　Habitat Module　居住模块
HABE　High Altitude Balloon Experiment　高空气球实验
HABT　Habitability Technology　可居住性技术
HAC　Heading Alignment Center　【美国航天飞机】航向校准中心
HAC　Heading Alignment Circle　【美国航天飞机】航向校准圆
HAC　Heading Alignment Cone　【美国航天飞机】航向校准锥面
HAC　Heading Alignment Cylinder　【美国航天飞机】航向校准柱面
HACBSS　Homestead and Community Broadcasting Satellite Service　住宅与社区广播卫星服务
HACCP　Hazard Analysis and Critical Control Point　危害分析和临界控制点

HACM High Altitude Cruise Missile 高空巡航导弹

HACP Hard Core Penetrator 硬心穿甲弹

HACS Hazard Assessment Computer System 事故估计计算机系统

HAD Heat Actuated Device 热致动装置

HAD High Altitude Diagnostic (Launcher) 高空故障诊断（发射装置）

HAD High-Altitude Density (Rocket) 【澳大利亚】高空大气层密度探测火箭

HAD Hybrid Detective Assembly 复合探测组件

HAD Hypersonic Aerothermal Dynamics 高超声速气动热力学

HADIOS Honeywell Analogous Digital Input-Output Subsystem 霍尼韦尔模拟数字输入—输出分系统

HADOPAD High Altitude Delayed Opening Parachute Actuation Device 高空延迟开伞作动设备

HADS High Accuracy Data System 高精度数据系统

HADS High Accuracy Digital Sensor 高精度数字传感器

HADS High Altitude Defense System 高空防御系统

HADS Hypersonic Air Data Sensor 高超声速飞行数据传感器

HADTS High Accuracy Data Transmission System 高精度数据传输系统

HAE Height Above Ellipsoid 椭圆面之上的高度

HAEMP High Altitude Electromagnetic Pulse 高空电磁脉冲

HAENS High Altitude Exoatmospheric Nuclear Survivability 高空外大气层核生存能力

HAES High Altitude Effects Simulation 高空效应模拟

HAF High Altitude Facility 高空（飞行）设备

HAF Hypersonic Aerothermodynamics Facility 高超声速气动热力设备

HAG Height Above Geoid 大地水准面之上的高度

HAINS High Accuracy Inertial Navigation System 高精度惯性领航系统

HAIR High Accuracy Instrumentation Radar 高精度测量雷达

HAISS High Altitude IR Sensor System 高空红外传感器系统

HAK Horizontal Access Kit 水平检修工具

HAL Hardware Adaptation Layer 硬件适配层

HAL High Order Articulated Language 高级组装语言

HAL High Order Assembly Language 高级汇编语言

HAL Houston Aerospace Language 【美国】休斯敦航空航天语言

HAL/S High Order Assembly Language for Shuttle Flight Computer 【美国】航天飞机飞行计算机高级汇编语言

HAL/S High Order Programming Language for Spacelab Usage "空间实验室"使用的高级程序设计语言

HALA Height Above Landing Area 着陆区域上空高度

HALE High Altitude Long-Endurance 高空长航时的

HALE UAV High Altitude Long-Endurance Unmanned Aerial Vehicle 高空长航时无人机

HALEX　Halogen Lamp Experiment　【美国航天飞机】卤灯实验

HALO　High Altitude Observatory　高空观测台；高空天文台

HALO　Hughes Automated Lunar Observer　"休斯"自动月球观测器

HALOE　Halogen Occultation Experiment　卤素掩星实验（装置）

HALSIM　Hardware Logic Simulator　硬件逻辑模拟器

HALT　Highly Accelerated Life Testing　高加速寿命测试

HALTATA　High and Low Temperature Accuracy Testing Apparatus　高低温精密度测试仪

HAM　Height Adjustment Maneuver　高度修正机动

HAMOS　High-Altitude Synoptic Meteorological Observation　高空天气气象观测

HAMOTS　High Accuracy Multiple Object Tracking System　高精度多目标跟踪系统

HAMP　High Altitude Measurement Probe　高空测量探测器

HAMS　Hardness Assurance, Maintenance and Surveillance　核防护能力确保、维护与监视；硬度确保、维护与监视

HAN　Hydroxyl Ammonium Nitrate　硝酸羟胺

HANE　High-Altitude Nuclear Effects Study　高空核效应研究

HANE　High Altitude Nuclear Explosion　高空核爆炸

HAO　High Altitude Observatory　高空观测台；高空天文台

HAOI　High Altitude Optical Imaging　高空光学成像

HAOIS　High Altitude Optical Imaging System　高空光学成像系统

HAOSS　High Altitude Orbital Space Station　高纬度轨道空间站

HAP　Hardware Allocation Panel　硬件配置操纵台

HAP　Hazardous Air Pollutant　危险气体污染物

HAP　High-Altitude Probe　高空探测器

HAPDAR　Hard Point Defense Array Radar　硬点防御阵列雷达

HAPI　High Altitude Plasma Instrument　高空等离子体测量仪

HAPS　Hydrazine Auxiliary Propulsion System　肼辅助推进系统

HAR　Hazardous Activities Restriction　危险活动限制

HARDS　High Accuracy Radar Data System　高精度雷达数据系统

HARE　High Altitude Ramjet Engine　高空冲压喷气发动机

HARE　High-Altitude Recombination-Energy Propulsion　高空复合能力装置（飞行器）

HARE　Hydrazine Auxiliary Rocket Engine　肼辅助火箭发动机

HARIS　High-Altitude Radiological Instrumentation System　高空辐射测量仪器系统

HARM　High Altitude Anti-Radiation Missile　高空反辐射导弹；哈姆导弹

HARM　High-Altitude Recovery Mission　高空回收任务

HARM　High-Speed Advanced Radiation Missile　先进高速辐射导弹

HARM　High-Speed Anti-Radiation Missile　高速反辐射导弹

HARP　High-Altitude Rocket Probe　高空火箭探测器

HARPVSS Homing Anti-Radiation Remotely Piloted Vehicle Sensing System 自寻的反辐射遥控飞行器探测系统
HARV High Altitude Research Vehicle 高空研究验证机
HAS Heading and Attitude Sensor 航向与姿态传感器
HAS Holddown Alignment Support 牵制校准支撑
HAS Hydraulic Actuation System 液压致动系统
HAS Hydrogen Actuation System 氢驱动系统
HASP Hardened Ada Signal Processor 加固的 Ada 语言信号处理器
HASP High Altitude Space Platform 高空空间平台；高空航天站；高空间站
HASP High Altitude Space Probe 高度空间探测器
HASR High Altitude Sounding Rocket 高空探测火箭
HASS Highly Accelerated Stress Screening 高加速应力屏蔽
HASSS High Accuracy Spacecraft Separation System 高精度航天器分离系统
HAST High Altitude Supersonic Target 高空超声速目标
HASTI High Altitude Strike Indicator 高空打击指示器
HASVR High Altitude Space Velocity Radar 高纬度空间测速雷达
HAT Australian High Altitude Testing Sounding Rocket 澳大利亚高空试验气象探测火箭
HAT Height Above Touchdown Zone 着地区之上的高度
HAT High Altitude Target 高空目标
HATI Hazard Analysis Tracking Index 危险分析跟踪指数
HATELM Highspeed Anti-TEL (Transport-Erect-Launch) Missile 高速反运输－起竖－发射车导弹
HATMD High Altitude Tactical Missile Defense 高空战术导弹防御
HATOL Horizontal Attitude Takeoff and Landing 水平姿态起飞与着陆
HATRS High Altitude Transmit/Receive Satellite 高空收发卫星
HATV High Altitude Test Vehicle 高空试验飞行器
HAV High Altitude Vehicle 高空飞行器
HAW Hawaii STDN Station 【美国国家航空航天局】航天器跟踪数据网夏威夷站
HAZOP Hazardous Operating Permit 危险操作许可
HAZOP Hazardous Operation 危险性操作
HAZOPS Hazard and Operability Study 危险与操作性研究
HB High Bay 高跨区；高跨间
HBM Heavy Ballistic Missile 重型弹道导弹
HBT Heflex Bioengineering Test 【美国航天飞机】"Heflex"生物工程实验
HC Critical Height 临界高度
HC Hazardous Classification 危险性分级
HC Hybrid Computer 混合型计算机
HCCF Host Command Control Facility 主命令控制装置
HCED Hand Controller Engage Driver 手控连接驱动装置
HCHE High Capacity HE 高能高爆炸弹

HCI Human-Computer Interaction 人—机交互；人机对话
HCI Human-Computer Interface 人机接口
HCL Horizontal Centerline 水平中心线
HCM Hard Copy Module 硬拷贝模块
HCMM Heat Capacity Mapping Mission 【美国国家航空航天局】热容量测绘卫星（1978年发射的一种以粗空间分辨率对地球进行精确辐射测量的传感器）
HCP HARM Control Panel 高速反辐射导弹控制板
HCR High Cross Range 高空侧向；高空横向
HCS Hydrological Communication Satellite 水文通信卫星
HCSL Hybrid Computation and Simulation Laboratory 混合计算与模拟（仿真）实验室
HCU Hydraulic Charger Unit 液压加载装置组件；液压装弹器装置
HCU Hydraulic Charging Unit 液压装填装置；液压加注（料）装置
HCV Hydrogen Check Valve 氢止回阀
HCV Hypersonic Cruise Vehicle 超声速巡航飞行器
HD High-Definition 高清
HD Holddown 牵制；固定；抑制
HD Horizontal Distance 水平距离
HD Horizontal Drain 水平排放
HD Hydrogen Drain 氢排放
HDA Hybrid Detector Assembly 混合探测器组件
HDBTD Hard and Deeply Buried Target Defeat (Program) 摧毁硬目标和深埋目标（计划）；坚固与深层地下目标摧毁（计划）
HDBTDC Hard and Deeply Buried Target Defeat Capability 坚固和深层地下目标摧毁能力
HDC Hybrid Device Controller 混合装置控制器
HDCS Hipparcos Dedicated Computer System 高精度视差收集卫星专用计算机系统
HDD Hazardous Debris Distance 危险物残骸距离
HDD Heads Down Display 俯视显示；俯视显示器
HDDR High-Density Digital Recorder 高密度数字记录仪（装置）
HDG Heading 航向
HDL Hybrid Data Link 混合数据链
HDLC High-Level Data Link Control 高级数据链控制
HDLMS Hybrid Data Link Management System 混合数据链管理系统
HDMI High-Definition Multimedia Interface 高清多媒体接口
HDOP Horizontal Dilution of Precision 水平位置精度冲淡系数
HDP High Detonation Pressure 高爆压力
HDP Holddown Post 牵制杆
HDPF Holographic Data Processing Facility 全息数据处理设施
HDR Hardware Delivery Review 硬件递交评审
HDR High Data Rate 高数据率
HDRA High Data Rate Assembly 高数据率组件
HDRM High Data Rate Multiplexer 高数据率多路调制器
HDRR High Date Rate Recorder 高数据率记录仪（装置）
HDRS High Data Rate System 高数

据率系统
HDRSS High Data Rate Storage System 高数据率存储系统
HDRTSAT High Data Rate Tactical Satellite Terminal 高数据率战术卫星终端
HDS High Density Satellite 高密度卫星
HDS Hydrodynamic Support System 流体动力保障系统
HDU Hydraulic Drive Unit 液压驱动装置
HDW Hardware 硬件
HE Heat Exchange (Exchanger) 热交换（器，装置）
He Helium 氦
HE High Eccentricity Orbit 高偏心率轨道
HE High Energy Astrophysics 高能天体物理学
HE High Explosive 高爆炸药
HEA Human-Error Analysis 人－错误分析
HEAO High Energy Astronomic Observatory Satellite 高能天文观察卫星
HEAO High-Energy Astronomy Observatory 【美国国家航空航天局】高能天文观测台
HEAP High Energy Aim Point 高能瞄准点
HEAT High-Enthalpy Ablation Test 高焓（热函）烧蚀试验
HEAT High-Explosive Anti-Tank 高爆反坦克炸药（成形炸药）
HEATE High Energy Astronomical Transient Explorer 高能天体物理顺变现象探测卫星
HEC High End Computing 高端计算
HECAD Human Engineering Computer Aided Design 人机工程计算机辅助设计
HECRE High-Energy Cosmic Ray Experiment 高能宇宙射线实验
HECV Helium Check Valve 氦止回阀
HEDI High Endoatmospheric Defense Interceptor 高外大气层防御拦截机（导弹）
HEDM High Energy Density Materials 高能密度材料
HEDM High-Energy Density Material 高能密度材料
HEDP High Explosive Dual-Purpose 两用高爆缓弹药
HEDR High Endoatmospheric Defense Radar 高外大气层防御雷达
HEDS High Endoatmospheric Defense System 高外大气层防御系统
HEDS Human Exploration and Development of Space 【美国国家航空航天局】人类探索与空间开发
HEDS Human Exploration and Development of Space 载人航天探索与发展
HEDS Human Exploration Destination Systems 载人探索目的地系统
HEE Hydrogen Environmental Embrittlement 氢环境性脆化（裂）
HE-FRAG-FS High-Explosive Fragmentation-Fin-Stabilized 高爆破片稳定翼（弹）；尾翼稳定高爆破片杀伤弹
HEFT Human Exploration Framework Team 载人探索框架小组
HEG Helium Gauge 氦压力计
HEGV Helium Gauge Valve 氦压力计阀
HEHIO Highly Excentric/Highly In-

clined Orbits 大倾角轨道

HEI High Endoatmospheric Interceptor 高外大气层拦截机（导弹）

HEIAP High Explosive Incendiary Armour Penetrating 高爆燃烧弹穿透装甲

HEIAS Human Engineering Information Analysis Service 人机工程学信息分析业务

HEIE High Energy Isotope Experiment 高能同位素实验

HEL High Energy Laser 高能激光

HELIOS Hetero-Power Earth Launch Interplanetary Orbital Spacecraft "太阳神号"太阳探测器；异动地球发射行星际轨道航天器

HELIOS Heteropowered Earth-Launched Interorbital Spacecraft 混合动力地球发射星际轨道航天器

HELKS High Energy Laser Kill System 高能激光杀伤系统

HELLO High Energy Laser Light Opportunity 高能激光器发光时机

HELOS High Eccentric Lunar Occultation Satellite 高偏心月球掩遮卫星；大偏心度月球卫星

HELP High Energy Lightweight Propellant 轻型高能推进剂（火箭燃料）

HELSTF High Energy Laser Systems Test Facility 高能激光器系统试验设施

HELWS High Energy Laser Weapon System 高能激光武器系统

HEM Hitchhike Experiment Module 搭载试验舱

HEMEF HE Missile Effective Unit 高爆导弹的有效部件

HEML High Energy Microwave Laboratory 高能微波实验室

HEMP High Altitude Electromagnetic Pulse 高空电子脉冲

HEMV Helium Manual Valve 氦人工阀

HEO High Earth Orbit 高地球轨道

HEO High-Energy (High-Eccentricity) Orbit 高能（高偏心率）轨道

HEO Highly Elliptical Orbit 大扁率椭圆轨道

HEO Highly-Inclined Elliptical Orbit 大椭圆轨道

HEOS Highly Eccentric Orbit Satellite 高偏心率轨道卫星；大椭圆轨道卫星

HEP Hall Effect Probe 霍尔效应探测器

HEP-T High-Explosive Plastic Tracer 曳光碎甲弹

HEPA High Efficiency Particle Accumulator 高效粒子累加器

HEPA High Efficiency Particle Air (Filter) 高效粒子空气（过滤器）

HERA Hermes Remote Manipulator Arm 【欧洲空间局】"赫尔姆斯号"航天飞机遥控机械臂

HERF High-Energy Radio Frequency 高能射频

HERFS High Energy Rate Forming System 高能快速成形系统；高能快速成形方法

HERMES Hermes Spaceplane 【欧洲空间局】"赫尔姆斯号"航天飞机；"使神号"航天飞机

HERO Hazards of Electromagnetic Radiation to Ordnance 电磁辐射对军械的危害

HERO High-Energy Replicated Optics 高能可复制光学系统

HERSCP Hazardous Exposure Reduction and Safety Criteria Plan 危险

暴露减少与安全标准计划
HESP High Efficiency Solar Panel 高效能太阳能电池板
HESS High Energy Squib Simulator 高能爆管模拟器（装置）
HESS Human Engineering System Simulation 人机工程系统模拟
HETGS High Energy Transmission Grating Spectrometer 高能传输光栅分光计
HETM Hybrid Engineering Test Model 混合工程试验模型
HETS High Environment Test System 高空环境试验系统
HETS High Equivalent to a Theoretical Stage 理论级等效高度
HETS Hyper Environmental Test System 高空环境试验系统
HEVART High Explosive Vulnerable Areas and Repair Time 高爆易损区与修理时间
HEX High Frequency Transceiver 高频无线电收发机
HEXE High Energy X-Ray Experiment 高能 X 射线实验
HEXTE High Energy X-Ray Timing Experiment 高能 X 射线定时（计时）实验
HF Hard Failure 硬故障（失效）
HF High Frequency 高频率
HF Horizontal Flight 水平飞行
HF Hot Firing 热点火
HF Hydrogen Fill 氢加注
HF Hyperfiltration 超（过）滤
HF Human Factors 人为因素；人员因素
HFA High Frequency Accelerometer 高频加速计
HFACS Human Factors Analysis and Classification System 人为因素分析与分类系统
HFC Hydraulic Hight Control 液压约定控制
HFCC Hermes Flight Control Center 【欧洲空间局】"赫尔姆斯号"航天飞机飞行控制中心
HFCNR High Frequency Combat Net Radio 高频作战无线电通信设备
HFCT Hydraulic Flight Control Test 液压飞行控制测试
HFCV Helium Flow Control Valve 氦流量控制阀
HFD Hazardous Fragment Distance 危险碎片距离
HFDL High-Frequency Data Link 高频数据链
HFDM High-Frequency Data Modem 高频数据调制解调器
HFDR High Frequency Data Radio 高频数据无线电传送
HFDS Hydrogen Fluid Distribution System 氢液分配系统
HFDS Human Factors Design Standards 人为因素设计标准
HFE High Frequency Executive 高频实施
HFE Human Factors Engineering 人体因素工程学
HFEA Human Factors Engineering Analysis 人为因素工程分析
HFFF Hypervelocity Free Flight Facility 超高速自由飞行设备
HFI Human Factors Integration 人为因素综合
HFNPDU High Frequency Network Protocol Data Unit 高频网络协议数据装置
HFPA Horizontal Flight Path Angle 水平航迹倾角
HFS High Frequency System 高频

系统

HFS Horizontal Flight Simulator 水平飞行模拟器

HFT Horizontal Flight Test 水平飞行试验

HFTF Horizontal Flight Test Facility 水平飞行试验设施

HFTS Horizontal Flight Test Simulator 水平飞行试验模拟器

HFV Horizontal Flight Vector 水平飞行矢量

HG High Gain 高增益

HGA High Gain Antenna 高增益天线

HGAS High Gain Antenna System 高增益天线系统

HGDS Hazardous Gas Detection System 危险气体探测系统

HGDS High Gradient Directional Solidification 高梯度方向固化

HGF Hot Gas Facility 热气体设施

HGLDS Hazardous Gas Leak Detection Subsystem 危险气体泄漏探测分系统

HGMS Hellfire Guided Missiles System 【美国】"地狱火"导弹系统

HGR Hangar 飞机（设备）库；飞机栅；大库栅

HGR Hypervelocity Guided Rocket 超高速制导火箭

HGR&SPTAC Hangar and Support Facility 机库与保障设施

HGS Hydrogen Gas Saver 氢气体回收器（装置）

HGSI Hot Gas Secondary Injection 热气二次喷射（推力向量控制）

HGSITVC Hot Gas Secondary Injection Thrust Vector Control 热燃气二次喷射推力矢量控制

HGV Hydrogen Gas Valve 氢气阀

HGVT Horizontal Ground Vibration Test 地面水平振动测试（试验）

HH Hook Height 吊钩高度

HHC Hammer Head Crane 锤式起重机

HHL Hand Held Laser 手控激光装置

HHLS&HS Human Health, Life Support and Habitation Systems 人类健康、生命保障与适居系统

HHMU Hand Held Maneuvering Unit 手控机动装置

HHP Human Health and Performance 乘员健康与绩效

HHRS Hardware History Retrieval System 硬件历史检索系统

HHSS Human Health Support System 乘员健康保障系统

HIAD High Altitude Defense 高空防御

HIAD Hypersonic Inflatable Aerodynamic Decelerator 高超可充气动力减速器

HIAS High Incidence Auto Stabilizer 大迎角自动稳定器

HIBEX High Acceleration Experiment (missile booster) 高加速导弹助推器实验

HIBEX High Impulse Booster Experiment 高冲量助推器实验

HIBEX High-Acceleration Boost Experiment 高加速度助推实验

HIBREL High Brightness Relay 高亮度中继

HICANS High Speed Collision Avoidance and Navigation System 高速防撞与导航系统

HICTB Human-in-Control Test Bed 人参与控制的试验（测试）台

HID Hardware Interface Device 硬

件接口装置

HID Host Interface Display 主接口显示

HIDACZ High Density Aerospace Control Zone 高密度航空航天控制区

HIF Horizontal Integration Facility 水平组装厂房

High Q High Dynamic Pressure 高气动压力

HIL Hardware-in-the-Loop 回路中的硬件

HIL Horizontal Integrity Limit 水平整合容限

HIL Human-in-the-Loop 回路中的人

HILAT High Latitude Ionospheric Research Satellite 【美国】高纬度电离层研究卫星

HILV Heavy-Lift Launch Vehicle 重型运载火箭

HIM Halley's Intercept Mission 哈雷彗星交会任务

HIM Hardware Interface Module 硬件接口模块

HIMAD High to Medium Altitude Air Defense 中高空空中防御

HIMAG High Mobility Agility (Vehicle) 高机动灵活性（飞行器）

HIMAG High Mobility and Agility 高机动性与灵敏性

HIMAT High Manoeuvrable Advanced Technology 高机动性先进技术

HIMES High Manoeuvrable Experimental Space (Vehicle) 【日本】高机动实验航天器

HIMEZ High Altitude Missile Engagement Zone 【防空作战】高空导弹攻击区

HIMORS Highly Mobile Rocket System 高机动火箭系统

HIMS Heavy Interdiction Missile System 重型遮断导弹系统

HIMSS High Resolution Microwave Spectrometer Sounder 高分辨率微波光谱探测器

HIP Hardware Interface Program 硬件接口程序

HIP Hot Isostatic Processing 热均衡处理

HIPAR High Power Acquisition Radar 大功率目标探测雷达

HIPAS High Performance Armament System 高性能武器系统

HIPC High Chamber Pressure 高燃烧室压力

HIPPAG High Pressure Pure Air Generator 高压纯净空气发生器

HIPPARCOS High Precision Parallax Collecting Satellite 高精度视差采集卫星；"伊巴谷"卫星（西欧天文卫星）

HIPPS Highly Integrated Pluto Payload System 高度集成"冥王星"有效载荷系统

HIPRA High Speed Digital Processor Architecture 高速数字化处理器结构

HIRAM High Resolution Infrared Auroral Measurements 高分辨率红外极光测量

HIRAM Hybrid In-Bore Rocket Assisted Motor 混合膛内火箭助推发动机

HIRF High-Intensity Radiated Field 高密度辐射场

HiRISE High Resolution Imaging Science Experiment 高分辨率成像科学实验

HIRS High Impulse Retrorocket Sys-

tem 高冲量反推火箭系统
HIS Horizontal Situation Indicator 水平状态指示器
HIS Human System Integration 人-系统一体化
HISEM High Speed Environmental Multi-Burst Model 高速环境多波群模型
HISRAN High-Precision Short-Range Navigation 高精度"肖兰"系统；高精度近程导航系统
HIT Hybrid/Inertial Technology 混合/惯性技术
HIT Hypergravity Isolation Training 超重力隔离训练
HIT Hypersonic Interference Technique 高超声速干扰技术
HIT Hypervelocity Impulse Tunnel 超高速脉冲风洞
HITEMP High Temperature Engine Materials Technology Program 高温发动机材料技术计划
HITL Hardware-in-the-Loop 回路中的硬件
HITL Human in the Loop 人在回路中
HITS High Rate Multiplexer Input/Output Test System 高速率多路调制器输入/输出测试系统
HIV Helium Isolation Valve 氦隔离阀
HIVOS High Vacuum Orbital Simulator 高真空轨道模拟器
HL Hydrogen Line 氢气管路
HLA Hazardous Launch Areas 危险发射区
HLA Hydraulic Launch Assist 液压发射助推器
HLAL High Level Assembler Language 高级装配语言
HLCS High Lift Control System 高推力控制系统
HLD Hardware Description Language 硬件说明语言
HLDS Hydrogen Leak Detection System 漏氢检测系统
HLHS Human Health, Life Support, and Habitation Systems 乘员健康、生命保障与居住系统
HLL Hard Lunar Landing 月球硬着陆
HLL High Level Language 高级语言
HLLV Heavy Lift Launch Vehicle 重型运载火箭
HLPS Hot Liquid Process Simulator 热液处理模拟器（装置）
HLPT Heavy Lift Propulsion Technology 重型推进技术
HLR Human Lunar Return 载人登月返回（地球）
HLS Heavy Lift Shuttle 【俄罗斯】重型航天飞机
HLTL High Level Test Language 高级测试语言
HLV Hybrid Launch Vehicle 混合动力运载火箭
HM Habitat Module 居住舱
HM&E Human Spaceflight Microgravity & Exploration 载人飞行微重力及探索
HMA Hypergol Maintenance Area 自燃燃料维护区
HMaG Heliospheric Magnetics 太阳风层磁学
HMC Hybrid Microcircuit 混合型微电路
HMC Hypergolic Maintenance and Checkout 自燃燃料维护与检查
HMC&M Hazardous Material Control

and Management 危险材料控制与管理

HMCC Hypergolic Maintenance and Checkout Cell 自燃燃料维护与检查室

HMCF Hypergolic Maintenance and Checkout Facility 自燃燃料维护与检查设施

HME Handheld Microgravity Experiment 【美国国家航空航天局】手持微重力实验（仪）

HMEA Hazard Mode and Effects Analysis 事故模式与后果分析；险情模式与后果分析

HMF Horizontal Mating Facility 水平对接设施

HMF Hypergolic Maintenance Facility 自燃燃料维护设施

HMGE High Modulus Graphite Epoxy 高模数石墨环氧树脂

HMGF High Modulus Glass Fiber 高模数玻璃纤维

HMI Hazardously Misleading Information 危险误导信息

HMI Human-Machine Interface 人－机接口

HMMWV High Mobility Multipurpose Wheeled Vehicle 高机动性中型轮式车辆

HmNT Hydrazine milli-Newton Thruster 肼微推力器

HMP Hydrazine Monopropellant (RCS Propellant) 联氨单基推进（卫星推进控制系统推进剂）

HMPC Hazardous Maintenance Procedure Code 危险维护程序代码

HMS Hail Monitor System 【美国肯尼迪航天中心】冰雹监控系统

HMS Health Monitoring System 健康监控系统

HMSC Houston Manned Space Center 【美国】休斯敦载人航天中心

HMSF Hazardous Material Storage Facility 危险物料贮存设备

HMU Hardware Mockup 硬件模型

HMU Height Monitoring Unit 高度监视装置

HMU Human Maneuvering Unit 载人机动设备

HMV Hydrogen Manual Valve 氢手动阀

HO Hydrogen-Oxygen 氢一氧

HOBS High Orbital Bombardment System 高轨道轰炸系统

HOE Homing Overlay Experiment 大气层外自动寻的实验

HOFS Hydrogen-Oxygen Fuel System 氢氧燃料系统

HOGS Homing Optical Guidance System 自动寻的光学制导系统

HOL High Order Language 高阶语言

HOLC High Order Language Computer 高阶语言计算机

HOM High Orbit Mission 高轨道飞行任务

HOMS Hellfire Optimized Missile System 【美国】"地狱火"优化导弹系统

HOP Hazardous Operating Procedure 危险操作规程

HORUS Hypersonic Orbital Rocket Upper Stage 高超声速轨道器火箭的上面级；【欧洲空间局】霍鲁斯轨道器

HOS High Order Software 高阶软件

HOSC Huntsville Operations Support Center 【美国】亨茨维尔操作保障中心

HOST Hardened Optical Sensor Test-

bed 加固光学传感器试验（测试）台
HOTOL Horizontal Takeoff and Landing 水平起飞与着陆的航天器；霍托尔空天飞机；水平起飞与着陆
HOTPC Higher Operating Temperature Propulsion Components 【美国】更高工作温度推进系统部件计划
HOUSE Human Occupation and Utilization of Space Environment 人类居住与利用空间环境系统
Hp Height of Perigee 近地点高度
HP High Pressure 高压
HP Horizontal Plane 水平面
HP Hydrogen Purge 氢气吹洗
HP Perigee Altitude 近地点高度
HP-MSOGS High Performance-Molecular Sieve Oxygen Generation System 高性能的分析筛氧发生系统
HPA High Power Amplifier 大功率放大器
HPA High-Power Antenna 大功率天线
HPAC Hazards Prediction and Assessment Capability 危险预测和评估能力
HPAF Hydraulic Performance Analysis Facility 液压性能分析设施
HPAG High Performance Air-to-Ground 高性能空对地（导弹）
HPAG High Performance Air-to-Ground Rocket 高性能空对地火箭
HPC High Performance Computing 高性能计算
HPC High Pressure Compressor 高压压缩器
HPC Hydropneumatic Constant 液压气动常数
HPCC High Performance Computing and Communications 高性能计算与通信
HPCC High-Pressure Combustion Chamber 高压燃烧室
HPCC&S High Performance Computing, Communication, and Simulation 高性能计算、通信和仿真
HPCG Handheld Protein Crystal Growth Middeck Experiment 【美国国家航空航天局】手持式蛋白质晶体生长中甲板实验
HPDE High Performance Demonstration Experiment 高性能演示验证实验
HPF Horizontal Processing Facility 水平处理设施
HPFD High Pressure Fuel Duct 高压燃料管道
HPFT High-Pressure Fuel Turbopump 【美国航天飞机主发动机】高压燃料涡轮泵
HPG High Pressure Gas 高压气体
HPGS High Pressure Gas System 高压气体系统
HPI High Performance Insulation 高性能绝缘
HPI High Performance Intercept 高性能拦截；高性能截击
HPICS Heat Pipe Instrument Control System 热管仪器控制系统
HPIR High Power Illuminator Radar 大功率照明雷达
HPL Horizontal Protection Limit 水平防护容限
HPM High Performance Motor 高性能发动机
HPM High Power Microwave 高功率微波
HPOP High Pressure Oxidizer Pump 高压氧化剂泵
HPOT High Pressure Oxidizer Tur-

bopump 【美国航天飞机主发动机】高压氧化剂涡轮泵
HPOX High Pressure Oxygen 高压氧
HPPF Horizontal Payloads Processing Facility 水平有效载荷处理设施
HPR Hydrogen Pressure Regulator 氢气压力调节器
HPRA Heat Pipe Radiator Assembly 热管辐射器组件
HPRL Human Performance Research Laboratory 【美国国家航空航天局】人类行为特性研究实验室
HPRV Hydrogen Pressure Relief Valve 氢气压力排泄阀
HPS Hybrid Propulsion System 混合推进系统
HPS Hydraulic Power System 液压动力系统
HPS Hydrazine Propellant System 肼推进剂系统
HPSC Hydraulic Package Storage Container 液压组件存贮容器
HPSI High Power System Integration 大功率系统集成
HPSI High Pressure Safety Injection 高压安全喷射
HPSLT High Power Semiconductor Laser Technology 大功率半导体激光技术
HPSRM High Performance Solid Rocket Motor 高性能固体火箭发动机
HPT Heading Pitching and Twisting 航向俯仰和扭转
HPT High Pressure Test 高压测试（试验）
HPTE High Performance Turbine Engine(s) 高效能涡轮发动机
HPTE High-Precision Tracking Experiment 【美国国家航空航天局】高精确度跟踪实验
HPTF Hazardous Processing Testing Facility 危险操作测试厂房
HPU Hydraulic Power Unit 液压动力装置
HPV Helium Precharge Valve 氦预加压阀
HPV High Pressure Valve 高压阀
HR Hazard Report 危险报告
HR Hydrogen Relief 氢排泄
HRA Human Reliability Analysis 人员可靠性分析
HRA Human Reliability Assessment 人员可靠性评估
HRAA High Rate Acquisition Assembly 高速率采集部件
HRAP High Resolution Accelerometer Package 高分辨率加速计装置
HRDA High Rate Data Assembly 高速率数据装置
HRDI High Resolution Doppler Image 高分辨率多普勒成像
HRDS High Resolution Display System 高分辨率显示系统
HRE Hydrazine Rocket Engine 肼火箭发动机
HRE Hypersonic Ramjet Engine 高超声速冲压式喷气发动机
HRE Hypersonic Research Engine 高超声速研究发动机
HRFE High Resolution Flight Element 高分辨率飞行元器件
HRI Human-Robotic Interaction 人—机器人接口
HRIA High Resolution Imager Assembly 高分辨率成像装置
HRIR High Resolution Infrared Radiometer 高分辨率红外辐射计
HRIS High Resolution Imaging Spec-

trometer 高分辨率成像光谱仪

HRL Horizontal Reference Line 水平基线

HRM Hermes Resource Module 【欧洲空间局】"赫尔姆斯号"航天飞机资源舱

HRSI High-Temperature Reusable Surface Insulation 高温重复使用表面绝热材料

HRSO High Resolution Solar Observatory 【美国】高分辨率太阳观测台

HRST Highly Reusable Space Transportation 重复可使用率高的空间运输

HRV Hypersonic Research Vehicle 高超声速研究用飞行器

HSB Hypergolic Support Building 自燃燃料保障大楼

HSBA Horizontal Static Balancing Adjustment 水平静平衡调整

HSC Human Systems Center 人体系统中心

HSCT Hughes Satellite Communications Terminal 【美国】休斯公司卫星通信终端

HSCU Hydraulic Supply and Checkout Unit 液压传送与检测装置

HSD Horizontal Situation Display 水平状态显示

HSDB High Speed Data Bus 高速数据总线

HSDE Hermes Software Development Environment 【欧洲空间局】"赫尔姆斯号"航天飞机的软件开发环境

HSDL High Speed Data Line 高速数据线

HSE Heat Shield Entry 使用防热层进入大气层

HSF Hazardous Storage Facility 危险品贮存设施

HSF High Speed Flight 高速飞行

HSF Human Space Flight 载人空间飞行

HSF Hypergol Servicing Facility 自燃燃料检修设施

HSF Hypergolic Storage Facility 可自燃物品贮存设施

HSI Horizontal Situation Indicator 水平状态指示器

HSI Human System Integration 人与系统的综合

HSIF Hardware/Software Integration Facility 硬件/软件综合设施

HSIT Hardware and Software Integration Test 硬件与软件集成测试（试验）

HSL Hardware Simulation Laboratory 硬件仿真实验室

HSO Habitation/Station Operations 居住/空间站操作

HSOM Habitation/Station Operations Module 居住/空间站操作舱

HSPR High Speed Propulsion Research 高速推进研究

HSQ Heat Shield Qualification 热屏蔽鉴定试验；隔热层鉴定试验

HSRB Hypersonic Suborbital Reusable Booster 高超声速亚轨道可重复使用助推器

HSRT Human Systems Research and Technology 载人系统研究与技术

HSS Hydraulic Subsystem Simulator 液压分系统模拟器（装置）

HSS Hydrological Sensing Satellite 水文遥感卫星

HSSINPM High Speed Serial Interface Network Processor Module 高速串行接口网络处理器模块

HST Hubble Space Telescope 【美国】

哈勃太空望远镜
HSTCXO High-Stability Temperatu-recompensated Crystal Oscillator 高稳定性稳定补偿式晶体振荡器
HSTS Horizontal Stabilizer Trim Setting 水平尾翼配平调正
HSTSM Hubble Space Telescope Salvage Mission "哈勃"太空望远镜抢修任务
HSTV High Survivability Test Vehicle 高生存性试验飞行器
HSTV-L High Survivability Test Vehicle-Light Weight 轻型高生存性试验飞行器
HSU Helium Service Unit 氦伺服装置
HT Heat Transfer 热传输
HT EXCH Heat Exchanger 热交换器（装置）
HTA Hermes Training Aircraft 【欧洲空间局】"赫尔姆斯号"航天飞机训练机
HTC Hybrid Technology Computer 混合技术计算机
HTDE High Technology Demonstrator Engine 高技术演示验证发动机
HTF Horizontal Test Facility 水平测试设施
HTI Hyper Temporal Infrared Sensor 超瞬时红外传感器
HTMIAC High Temperature Materials Information Analysis Center 高温材料信息分析中心
HTO Horizontal Takeoff 水平起飞
HTPB Hydroxy Terminated Polybutadiene (Propellant) 端羟基聚丁二烯（推进剂）
HTS HARM Targeting System 高速反辐射导弹目标攻击系统
HTSC Huntsville Space Center 【美国】亨茨维尔航天中心
HTSS Hardened-Sub-Miniature Telemetry and Sensor System 加固的超小型遥测与传感器系统
HTSSE High Temperature Superconductivity Space Experiment 高温超导太空实验
HTV H-Ⅱ Transfer Vehicle 【日本】H-2 转移飞行器
HTV High-Altitude Test Vehicle 高空试验飞行器
HTV Homing Text Vehicle 自导引的试验飞行器
HTV Hypersonic Test Vehicle 高超声速试验飞行器
HUD Head-Up Display 平视显示
Humint Human Intelligence 人工情报
HUS Hypergolic Umbilical System 自燃燃料脐带系统
HUT Hard Upper Torso 【航天服】上部玻璃纤维硬壳
HUT Hopkins Ultraviolet Telescope "霍普金斯"紫外望远镜
HV High Velocity 高速
HV High Voltage 高电压
HV Hydrogen Vent 氢排放
HVAC Heating, Ventilating and Air Conditioning 采暖、通风与空调调节
HVDS Hypergolic Vapor Detection System 自燃燃料蒸汽探测系统
HVES Hypergolic Vent Exhaust System 自燃燃料通气管排气系统
HVI Hyper Velocity Impact 超速撞击
HVPS High Voltage Power Supply 高压供电
HVPS High Volume Particle Sampler 高容量粒子样品（取样器）
HVSF Honeywell Verification Simulation Facility 【美国】"霍尼韦尔"

验证仿真设施
HW/SW Hardware/Software 硬件/软件
HWIL Hardware In-the-Loop 回路中硬件；半实物仿真
HWILT Hardware-in-the-Loop Test 回路硬件试验（测试）；半实物仿真试验
HWS Hazard Warning Systems 危险告警系统
HWSTF Hazardous Waste Storage and Transfer Facility 危险废物贮存和转移设施
HX Heat Exchanger 热交换器（装置）
HXIS Hard X-Ray Imaging Spectrometer 硬X射线成像分光计
HXLV Hyper-X Launch Vehicle 高超声速运载器
HXRV Hyper-X Research Vehicle 高超声速研究机
HYD Hydraulic Systems (Subsystem) 液压系统（分系统）
HYDIM HYD Interface Module 液压系统接口模块
HYFAC Hypersonic Research Facilities 高超声速研究设施
HYFES Hypersonic Flight Environment Simulator 高超声速飞行环境模拟器
HYFLG Hypersonic Flight 高超声速飞行（验证计划）
HYFLTE Hypersonic Flight Test Experiment 高超声速试飞实验
HYGL Hypergolic 自发火的，自行着火的；可燃的
Hyp Hypersonic 高超声速的
HYPACE Hybrid Programmable Attitude Control Electronics 混合编程姿态控制电子设备与控制系统
HYPERION Hypermedia Intelligence Organization and Navigation 超媒体智能结构与导航
HYPR Supersonic/Hypersonic Transport Propulsion System 超声速/高超声速运输推进系统
HYPTV Hypersonic Test Vehicle 高超声速试验飞行器
HYSET Hydrocarbons Scramjet Engine Technology 碳氢燃料超燃冲压发动机技术
HyspIRI Hyperspectral Infrared Imager 超光谱红外成像仪
HySTP Hypersonic Systems Technology Program 超声速系统技术项目
HYSTRU Hydraulic System Test and Repair Unit 液压系统测试和修理设施
HyTech Hypersonic Technology 超声速技术
HYTEX Hypersonic Technology Experimental Vehicle 【德国】高超声速技术实验飞行器
HYWARDS Hypersonic Weapon Research and Development Supporting System 高超声速武器研究与开发支持系统

I

I LOAD Initialization Load 初始化载荷
I&C Identification and Control 识别与控制
I&C Installation and Calibration 安装与校准
I&C Installation and Checkout 安装与检测
I&C Instrumentation and Communication 测量装置与通信
I&C Instrumentation and Control 测量装置与控制
I&C Instruments and Controls 仪表和操纵机构
I&C/O Installation and Checkout 安装与检测
I&E Integration and Evaluation 综合与评估
I&I Installation and Integration 安装与集成
I&M Improvements and Maintenance 改进与维修
I&O Intake and Output 进入与输出
I&OP In and Out Processing 输入和输出（信息）处理
I&PA Integration and Performance Analysis 集成与性能分析
I&R Interchangeability and Replaceability 互换性与可换性
I&RS Instrumentation and Range Safety 仪表装置和靶场安全
I&S Interchangeability and Substitutability 互换性与可修复性
I&T Identification and Traceability 识别与跟踪能力
I&T Integration and Test 集成（综合）和测试
I&T Interceptor Integration and Testing 拦截机（导弹）装配与测试
I&TRR Integration and Test Readiness Review 综合与试验准备状态评审
I&W Indications and Warning 指示与报警
I-CASE Integrated Computer-Aided Systems Engineering 综合计算机辅助系统工程
I-MOSC Integrated Mission Operations Support Center 综合任务操作保障中心
I/A Interface Adapter 接口适配器
I/E Input Electronics 输入电子设备
I/F Interface 接口
I/FU Interface Unit 接口装置
I/O Input/Output 输入/输出
I/OB Input/Output Bus 输入/输出总线
I/OC Input/Output Controller 输入/输出控制器（装置）
I/OMI Integration/Operations and Maintenance Instruction 集成（综合）/操作与维护说明
I/OT Input/Output Test 输入/输出测试（试验）
I/OU Input/Output Unit 输入/输出装置
I/R Interchangeability and Repairability 互换性与可修复性
I/S Ignition and Separation Assembly 点火和分离装置
I/S Impulse Per Second 脉冲/秒

I/S　Interstage　级间段
I/T　Intertank　罐（槽，箱，容器）间的
I/U　Instrumentation Unit　仪表装置，仪表舱
IA　Independent Assessment　独立评估
IA　Information Assurance　信息保证
IA　Input Axis　输入轴
IA　Installation and Assembly　安装和装配
IA　Interface Adapter　接口适配器
IA　Interoperability Architecture　互操作性结构
IA&C　Integration, Assembly and Checkout　集成、总装和检查
IA&T　Integration, Assembly, and Test　集成、总装和测试；集成、组装与试验
IAA　Indian Astronautical Association　印度航天协会
IAA　International Academy of Astronautics　国际宇航科学院
IAAA　Integrated Advanced Avionics for Aerospace　航空航天用先进集成航空电子设备
IAAC　Integrated Application of Active Controls　主动控制系统综合应用
IAAD　Incremental Attitude Anomaly Detector　增加的姿态异常探测器
IAAI　Indonesian Aeronautical and Astronautical Institute　印度尼西亚航空航天研究院
IAASM　International Academy of Aviation & Space Medicine　国际航空航天医学院
IABS　Integrated Apogee Boost Subsystem　【美国】整体远地点助推子系统（第三代国防卫星通信系统）
IAC　Integrated Analysis Capability　综合分析能力
IAC　Integrated Assembly and Checkout　整体装配和检查
IAC　Interface Adapter Unit　接口适配器单元
IACS　Inertial Attitude Control System　惯性姿态控制系统
IACS　Initially-Referenced Attitude Control System　惯性基准姿态控制系统
IACS　Integrated Avionics Control System　综合航空电子控制系统
IAD　Inflatable Aerodynamic Decelerator　可充气动力减速器
IADP　Integrated Analysis Data Package　综合分析数据包
IADS　Integrated Air Defense System　综合空中防御系统
IADT(S)　Integrated Automatic Detection and Tracking (System)　综合自动探测与跟踪（系统）
IAE　Infrared Astronomy Explorer　红外天文探测卫星
IAE　Integrated Application Environment　综合应用环境
IAEDS　Integrated Advanced Electronic Display System　先进综合电子显示系统
IAGC　Instantaneous Automatic Gain Control　即时自动增益控制
IAL　Instrument Approach and Landing　仪表（指引）进场和着陆
IALC　Instrument Approach and Landing Chart　仪表（指引）进场与着陆图
IALE　Integral, Absolute Linear Error　线性绝对误差积分
IAM　Initially-Aided Munition　惯性弹药
IAM　Instrument Approach Minima

仪表进场最低气象条件
IAM　Iterative Array Model　迭代阵列模型
IAMAP　International Association of Meteorological & Atmospheric Physics　国际气象与大气物理协会
IAMD　Integrated Air and Missile Defense　综合空中和导弹防御
IAP　Integrated Actuator Package　综合致动器组件
IAP　Integrated Avionics Package　综合航空电子系统包
IAPS　Integrated Actuator Package System　综合致动器组件系统
IAPS　Integrated Avionics Processing System　综合航空电子设备处理系统
IAPS　Ion Auxiliary Propulsion System　离子辅助推进系统
IAR　Instrument Acceptance Review　器械（仪器）验收评审
IAR　Integrated Assessment Report　综合评估报告
IARS　Independent Air Revitalization System　独立大气再生系统
IAS　Indicated Airspeed　计示大气速度
IAS　Inertial Active System　惯性有源系统
IAS　Integral Application Software　集成应用软件
IAS　Integrated Avionics System　综合电子设备与控制系统
IAS　International Application Satellite　国际应用卫星
IAS　Interplanetary Automated Shuttle　行星间自动穿梭飞行器
IASA　Institute of Space and Astronautical Science　【日本】宇宙科学研究所
IASF　Instrumentation in Aerospace Simulation Facilities　航空航天模拟设施的测量设备
IASR　Intermediate Altitude Sounding Rocket　中（高）空探空火箭
IASSC　Instrument Approach System Steering Computer　仪表进场系统操纵计算机
IAT　Integrated Acceptance Test　综合验收试验
IAT　Integrated Assembly Test　综合装配测试
IAT　Integrated Avionics Test　综合电子设备与控制测试（试验）
IAT-VC　Integrated Aerodynamic Thrust-Vector Control　推力矢量与气动力复合控制
IATACS　Improved Army Tactical Communications System　【美国】改进的陆军战术通信系统
IATACS　Integrated Acquisition Tracking and Air Point Control System　综合（目标）捕获跟踪与瞄准控制系统
IATCO　Integration, Assembly, Test & Checkout　集成、总装、测试与检查
IB　Inert Building　惰性气体厂房
IBA　Igniter Booster Assembly　点火器助推器组件
IBCS　Integrated Battle Command System　综合作战指挥系统
IBDA　Indirect Bomb Damage Assessment　间接炸弹破坏评估
IBDM　International Berthing Docking Mechanism　国际停泊对接机构
IBE　Inboard Booster Engine　内侧助推器发动机
IBECO　Inboard Booster Engine Cutoff　内侧助推发动机关机

IBEX Interstellar Boundary Explorer 星际间的探测

IBI Interim Ballistic Instrumentation 临时弹道仪表测量

IBID Integrated BMC^3 Infrastructure Demonstration 综合弹道导弹指挥、控制、通信基础设施演示验证

IBIT ICBM (Intercontinental Ballistic Missile) Blast Interference Test 洲际弹道导弹爆炸冲击波干扰试验

IBLS Integrity Beacon Landing System 整体信标着陆系统

IBM Inertial Biased Mode 惯性偏心状态

IBMP Integrated Ballistic Missile Picture 综合弹道导弹图像

IBPDSMS Improved Basic Point Defense Surface Missile System 改进型基点防御水面导弹系统

IBR Initial Baseline Review 初期基线评审

IBR Integrated Baseline Review 综合基线评审

IBS INTELSAT Business Service 国际通信卫星商用业务

IBS Ionospheric Beacon Satellite 电离层信标卫星

IBSS Infrared Background Signature Survey 【美国国家航空航天局】红外背景特征探测器

IC Intercom (Orbiter to Ground via Hardline) 内部通信（通过硬线将轨道器与地面连接）

IC-AS Independent Collision-Avoidance System 独立防撞系统

IC/ES Intercommunication/Emergency Station 内部通信与应急站

ICA Independent Cost Analysis 独立成本分析

ICA Interface Compatibility Analysis 接口兼容性分析

ICAD Integrated Control and Display 综合控制与显示

ICAD Intelligent Computer Aided Design 智能计算机辅助设计

ICADS Integrated Correlation and Display System 综合相关与显示系统

ICAE Integrated Computer-Aided Engineering 综合计算机辅助工程

ICAE International Commission on Atmospheric Electricity 国际大气电子委员会

ICAPS Interactions in Cosmic and Atmospheric Particle Systems 宇宙和大气微粒系统的相互作用

ICAR Interim Capability Assessment Report 临时能力评估报告

ICAs Industrial Capability Assessments 工业化能力评估

ICAS Integrated Condition Assessment System 综合条件评估系统

ICASE Integrated Computer Assisted Software Engineering 综合计算机辅助软件工程

ICBM Inter-Continental Ballistic Missile 洲际弹道导弹

ICBT Interactive Computer-Based Training 交互式计算机辅助训练

ICC Interface Control Chart 接口控制图表

ICCP Interface Coordination and Control Procedure 接口协调与控制规程

ICCS Inter-Site Control and Communication System 【导弹部队】发射阵地间控制与通信系统

ICCS ITW/AA Configuration Control System 综合战术预警与攻击效果评估配置控制系统

ICD Interface Control Document 接口控制文件

ICDCS Integrated Control Display and Communications Subsystem 综合控制显示与通信子系统
ICDR Incremental Critical Design Review 增量关键性设计评审
ICDU Inertial Coupling Data Unit 惯性耦合数据装置
ICDU Inertial Coupling Display Unit 惯性耦合显示装置
ICE Improved Combat Efficiency 提高作战效能
ICE Independent Cost Estimate 独立成本估算
ICE Index of Combat Effectiveness 战斗效能指标
ICE In-Flight Contamination Experiment 【欧洲空间局】飞行中污染物（控制）实验
ICE Input Control Element 输入控制组件
ICE Instrument Checkout Equipment 仪器检测设备
ICE Instrument/Communication Equipment 仪器/通信设备
ICE Inter-Connecting Element 连接舱（欧洲未来空间站多用途舱段）
ICE International Cometary Explorer 【美国】国际彗星探测器
ICESat Ice, Cloud, and Land Elevation Satellite 冰层、云及地面俯仰角测量卫星
ICF Inertial Confinement Fusion 惯性约束聚变
ICF Interface Control Function 界面控制功能
ICG In-Flight Coverall Garment 飞行中使用的衣裤相连工作服
ICIO Interim Cargo Integration Operations 临时货物综合操作
ICM Improved Capabilities Missile 能力增强型导弹
ICM Increased Capability Missile 性能提高的导弹
ICM Interim Control Module 【国际空间站】临时控制模块
ICM/MIRV Intercontinental Missile/Multiple Independently-Guided Re-entry Vehicle 洲际导弹/多弹头分导再入系统
ICME Interplanetary Coronal Mass Ejection 行星间的顶部集中排出
ICMS Indirect Cost Management System 间接成本管理系统
ICMS Integrated Configuration Management System 综合配置管理系统
ICMS Intercom Master Station 内部通信主站
ICNI Integrated Communication, Navigation and Identification 综合通信、导航与识别
ICNIA Integrated Communication, Navigation and Identification Avionics 综合通信、导航与识别航空电子系统
ICNIS Integrated Communication, Navigation and Identification Subsystem 综合通信、导航与识别分系统
ICNIS Integrated Communication, Navigation and Identification System 通信、导航与识别综合系统
ICO Ignition Cut-Off 点火断开
ICO Integrated Checkout 综合检测
ICOM Intercommunications 内部通信
ICOS Improved Crew Optical Sight 改进型乘员光学瞄准具
ICOT Isothermal Corrosion Oxidation Test 等温腐蚀氧化试验
ICP Ignition Control Programmer 点火控制程序装置

ICPS　Interim Cryogenic Propulsion Stage　过渡低温推进级
ICR　Interface Compatibility Record　接口兼容性记录
ICRH　Ion Cyclotron Resonant Heating　离子回旋波加热
ICS　Instrumentation Control System　测量装置控制系统
ICS　Integrated Checkout Station　综合检测站
ICS　Integrated Control System　综合控制系统
ICS　Intercommunication System　内部通信系统
ICS　Interface Control Specification　接口控制说明
ICS　Interpretive Computer Simulator　释义计算机模拟器（装置）
ICS　Instrument Command Sequences　仪器指令序列
ICSS　Interim Contractor Support System　临时承包商保障系统
ICST　Information, Communication, and Space Technology　信息、通信与空间技术
ICT　Influence Coefficient Tests　干扰系数测试（试验）
ICT　Interface Control Tooling　接口控制工具（仪器）
ICTC　Inertial Components Temperature Controller　惯性部件温度控制器（装置）
ICTP　Information Collection, Transfer, and Processing　信息收集、转让与处理
ICU　Interface Control Unit　接口控制装置
ICW　Interrupted Continuous Wave　间歇性连续波
ICWC　Integrated Comprehensive Weaponeering Capability　综合全面武器效应量化能力
ICWDS　Infrared Concealed Weapon Detection System　红外线隐藏武器探测系统
ICWS　Integrated Weapons Control System　综合武器控制系统
ID&CA　Inverter Distribution and Control Assembly　反相器分布和控制组件
ID/ATS　Integrated Diagnostic/Automatic Test System　综合诊断/自动测试系统
IDA　Integrated Data Architecture　综合数据体系结构
IDA　Interactive Debugging Aid　交互式调试辅助工具；交互式排除故障辅助设备
IDA　Interface Display Assembly　接口显示组件
IDAC　Integrated Digital-Analog Converter　集成数字－模拟转换器
IDAL　Integrated Defense Avionics Laboratory　综合防御性（航空、导弹、航天）电子设备与控制系统实验室
IDAP　Integrated Data Analysis Plans　综合数据分析计划
IDAS　Integrated Data Acquisition System　综合数据采集系统
IDAS　Interactive Defensive Avionics System　交互式航空电子设备防御系统（后改为"特种作战部队航空电子设备防御系统"）
IDASC　Improved Direct Air Support Center　改进的直接空中保障中心
IDASS　Intelligence Data Analysis for Satellite Systems　卫星系统智能数据分析
IDASS　Intelligence Data Analysis System for Spacecraft　航天飞机的情

报数据分析系统
IDB In-Suit Drink Bag 航天服内饮水袋
IDC Initial Demonstration of Capability 初始能力演示验证
IDC Integrated Displays and Controls 综合显示与控制
IDC Interface Document Control 接口文件控制
IDC² Intelligent Distributed Command and Control 智能分布式指令与控制
IDCA Inverter Distribution and Control Assembly 变换器分配与控制组件
IDCSP Initial Defense Communication Satellite Program 【美国】初级国防通信卫星计划
IDD Interface Definition Document(s) 接口定义文件
IDE Initial Design Evaluation 初步设计评估
IDEA Integrated Design and Evaluation of Advance Spacecraft 高级航天器的综合设计与估算
IDEA Integrated Dose Environmental Analysis （辐射）剂量环境综合分析
IDEA Intelligence Diagnosis Expertise Administration 智能诊断专家管理系统
IDEAS Interactive Design and Evaluation of Advanced Spacecraft 先进航天器交互式设计与评估
IDECM Integrated Defensive Electronics Countermeasures 综合防御性电子对抗
IDEF Intercept During Exoatmospheric Fall 外大气层下落段拦截
IDES Intrusion Detection Expert System 入侵探测专家系统
IDG Integrated Drive Generator 集成驱动生成器（装置）
IDGE Isothermal Dendritic Growth Experiment 等温树状生长实验
IDHS Intelligence Data Handling System 智能数据处理系统
IDI Instrumentation Data Items 测量装置数据项
IDNE Inertial Doppler Navigation Equipment 惯性多普勒导航设备
IDO Initial Defensive Operations 初始防御作战
IDP Integrated Data Processor 综合数据处理器（装置）
IDP Interface Digital Processor 接口数字处理器（装置）
IDR Initial Design Review 初步设计评审
IDR Intercept During Re-Entry 再入大气层段拦截
IDR Interim Design Review 临时性设计评审
IDR Intermediate Design Review 过渡性设计评审
IDRD Information Definition Requirements Document 信息定义要求文件
IDRD Internal Data Requirement Description 内部数据要求说明
IDS Interdiction Strike 遮断（性）攻击
IDS Interface Design Standards 接口设计标准
IDS Intrusion Detection System 闯入检测系统
IDSA Institute of Defence and Strategic Analysis 【美国】国防与战略分析所
IDSCS Initial Defense Satellite Com-

munication System　初级防御卫星通信系统

IDSS　Intelligent Decision Support System　智能决策支持系统

IDT　IFICS Data Terminal　飞行拦截机（导弹）通信系统数据终端

IDTS　Improved Doppler Tracking System　改进型多普勒跟踪系统

IDU　Interface Demonstration Unit　接口演示装置

IDU　Intelligent Data Understanding　智能数据解读

IE　Ionosphere Explorers　电离层探险者卫星

IEA　Integrated Electronic Assembly　集成电子设备组件

IEC　Inertial Electrostatic Confinement　惯性静电约束

IECM　Induced Environmental Contamination Monitor　【美国航天飞机】诱导环境污染监测器

IECMS　In-Flight Engine-Condition Monitoring System　飞行中发动机状态监控系统

IECO　Inboard Engine Cutoff　内侧发动机关机

IECS　Igloo Environment Control Subsystem　【航天轨道试验室】小型加压试验室环境控制分系统

IED　Impact Energy Density　冲击能量密度

IED　Improvised Explosive Device　简易爆炸装置

IED　Independent Exploratory Development　独立的探索性研究

IED　Ionospheric Electron Density　电离层电子密度

IEF　Inertial Electrodynamic Fusion　惯性电动聚变

IEF　Isoelectric Focusing　【美国航天飞机】等电聚焦（技术、方法）

IEF　Isoelectric Focusing Experiment　等电聚焦实验

IEH　International Extreme-UV/Far-UV Hitchhiker　国际远紫外搭载星（美国电子侦察卫星）

IEI　Integrated Engineering Infrastructure　综合工程基础设施

IEIS　Integrated Engine Instrument System　综合发动机仪表系统

IEM　Integrated Engine Mode　综合发动机模式

IEO　Intermediate Earth Orbit　中高度地球轨道

IESA　Integrated Earth System Analysis　综合地球系统分析

IESS　Integrated Electromagnetic System Simulator　综合电磁系统模拟器（装置）

IET　Initial Engine Test　发动机初期测试

IET　Integration Event Matrix　综合事件矩阵

IETD　Interactive Electronics Technical Data　交互式电子技术数据

IETE　Initial Engine Test Facility　初始发动机试验设备

IETF　Initial Engine Test Firing　发动机初期点火试验

IETM　Integrated Electronic Technical Manual　综合电子技术手册

IETV　Interoperability Experimentation, Testing, and Validation　互用性实验、测试和确认

IEW　Intelligence and Electronic Warfare　情报与电子战

IF　Integration Facility　综合性设施

IF　Intermediate Frequency　过渡性频率

IFA　In-Flight Analysis　飞行中的（数

据）分析

IFA Interface Functional Analysis 接口功能分析

IFA In-flight Anomaly 飞行中的异常

IFC Initial Flight Clearance 初始飞行许可

IFCAS Integrated Flight Control and Augmentation System 综合飞行控制与增稳系统

IFCS In-Flight Checkout System 飞行检查系统

IFCS Integrated Flight Control System 综合飞行控制系统

IFCS Intelligent Fight Control System 智能飞行控制系统

IFD Integrated Flight Demonstration 综合飞行验证

IFDF Ideal Frequency Domain Filter 理想频率域滤波器

IFDPS Integrated Flight Data Processing System 综合飞行数据处理系统

IFDS Inertial Flight Data System 惯性飞行数据系统

IFE Isoelectric Focusing Experiment 【美国国家航空航天局】等电位聚集实验

IFI Inflight Insertion 飞行中插入；飞行中入轨

IFICS In-Flight Interceptor Communications System 飞行拦截机（导弹）通信系统

IFIP Integrated Flight Instrument Panel 综合飞行仪表板

IFIS Integrated Flight Information System 综合飞行信息系统

IFM In-Flight Maintenance 飞行中的维护

IFMEA Intelligent FMEA 智能故障模式影响分析

IFMS Impact Force Measuring System 冲击力测量系统

IFNC Integrated Flight/Navigation Control 综合飞行及导航控制

IFO Infrastructure, Facilities, and Operations 基础结构、设施设备与操作

IFOG Interferometric Fiber Optic Gyroscope 干涉光纤陀螺仪

IFOT In-Flight Operations and Training 飞行中的操作与培训

IFP Initial Flight Path 起始飞行路线

IFPC Integrated Flight/Propulsion Control 综合飞行/推进控制

IFPL Inflight Power Loss 飞行中的功率损失

IFPM Inflight Performance Monitor 飞行中性能监控器

IFPM Inflight Performance Monitoring 飞行中性能监视

IFPS Inflight Performance Signal 飞行中性能信号

IFR Instrument Flight Rules 飞行仪表规则

IFRU In-Flight Replacement Unit 进入飞行的替换装置

IFSAR Interferometric Synthetic Aperture Radar 干涉式合成孔径雷达

IFT In-Flight Test 进入飞行的测试（试验）

IFT Integrated Flight Test 综合飞行试验（测试）

IFT Interface Tool 接口工具

IFTA Inflight Thrust Augmentation 飞行中增大推力

IFTC Integrated Flight Trajectory Control 综合飞行轨迹控制

IFTM In-Flight Test Maintenance 飞行中的测试维护

IFTS In-Flight Test System 飞行中的测试（试验）系统
IFTS Integrated Function Test System 综合功能测试系统
IFTU In-Flight Target Update 飞行中目标更新
IFU Interface Unit 接口装置
IG Inertial Guidance 惯性制导
IGA Inner Gimbal Angle 内万向角
IGACS Integrated Guidance and Control System 综合制导与控制系统
IGAS Integrated GPS Anti-Jam System 全球定位系统综合抗干扰系统
IGAX Inner Gimbal Axis 内万向轴
IGDA Interactive Graphics and Data Analysis 交互式图形与数据分析
IGDS Integrated Graphics Design System 集成图形设计系统
IGDS Iodine Generating and Dispensing System 碘生成与弥散系统
IGEMP Internally Generated Electromagnetic Pulse 内部生成的电磁脉冲
IGES Initial Graphics Exchange Software 原图形交换软件
IGES International Graphic Exchange Specification 国际图形交换规格
IGM Interactive Guidance Mode 交互式制导模式
IGMF Inertial Guidance Maintenance Facility 惯性制导系统维修设施
IGOR Intercept Ground Optical Recorder 地面拦截光学记录仪
IGOS Integrated Global Observing Strategy 综合全球观测战略
IGRAL International Gamma-Ray Astrophysical Laboratory 国际伽马射线天体物理实验室
IGS Inertial Guidance System 惯性制导系统
IGS Instrumentation Ground System 地面仪表测量系统
IGS Interconnection Ground Subnet 互联地面子网
IGSCC Intergranular Stress Corrosion Cracking 粒间应力腐蚀性裂纹
IGSE In-Space Ground Support Equipment 空间地面保障设备
IGSE Integrated Ground Support Equipment 综合地面保障设备
IGSM Interim Ground Station Module （联合监视目标攻击雷达系统）临时地面站模块
IGSS Inertial Guidance System Simulator 惯性制导系统模拟器
IGT Integrated Ground Test 综合地面试验
IGV Incremental Growth Vehicle 渐进式发展飞行器
IH/SR Integrated Hardware and Software Review 综合硬件与软件评审
IHLLV Interim Heavy Lift Launch Vehicle 过渡性大推力运载火箭
IHM Integrated Health Management 一体化健康管理
IHPRPT Integrated High Payoff Rocket Propulsion Technology 一体化高回报火箭推进技术；综合高性能火箭推进系统
IHPTET Improved High Performance Turbine Engine Technology 改进的高性能涡轮发动机技术
IHPTET Integrated High Performance Turbine Engine Technology 综合高性能涡轮发动机技术
IHTV Interim Hypersonic Test Vehicle 过渡性高超声速试验飞行器
IHW International Halley Watch 国际哈雷彗星观测
IIP Instantaneous Impact Point 瞬时

弹着点；连续弹着点
IIP Instantaneous Impact Prediction 瞬时弹着点预测；连续弹着点预测
IIP Instantaneous Impact Predictor 瞬时弹着（点）预测器
IIP Interim Impact Predictor 临时弹着（点）预测器
IIPACS Integrated Information Presentation and Control System 综合信息显示与控制系统
IIRS Infrared Imaging Spectrometer 红外成像光谱仪
IIS/G Internal Integral Starter/Generator 内装启动/发电综合装置
IJC³S Initial Joint Command, Control and Communications System 初始联合指挥、控制与通信系统
IJC³S Integrated Joint Command, Control and Communications System 综合联合指挥、控制与通信系统
IJS Interior Joint Space 地球轨道内侧的共同空间；太阳和地球轨道之间的共同空间
IJSOW Improved Joint Stand Off Weapon 改进型联合防区外武器
IKAROS Interplanetary Kite-Craft Accelerated by Radiation of the Sun 伊卡洛斯航天器（太阳辐射加速星际风筝航天器）
ILA Instrument Landing Approach 仪表引导着陆进场
ILAS Instrument Landing Approach System 仪表引导着陆进场系统
ILC Initial Launch Capability 初始发射能力
ILCC Integrated Launch Control and Checkout 综合发射控制和检测
ILCCS Integrated Launch Control and Checkout System 综合发射控制与检测系统

ILD Intraline Distance 线内距离
ILFLCFCS Intra Launch Facility and Launch Control Facility Cabling Subsystem 发射设施与发射控制设施间的布缆子系统
ILIDS International Low Impact Docking System 国际低撞击对接系统
ILL Impact Limit Line 撞击边界线
ILL International Lunar Laboratory 国际月球实验所
ILM Independent Landing Monitor 独立式着陆监控设备
ILMS Improved Launcher Mechanical System 改进型发射装置机械系统
ILMS Inner Layer Missile System 内层导弹系统
ILOAD Initialization Load 初始载荷
ILP Integrated Logistics Panel 综合后勤保障委员会
ILRV Integrated Launch and Recovery Vehicle 综合发射与回收航天器着陆系统
ILRV Integrated Launch and Reentry Vehicle 综合发射与再入飞行器
ILRVS Integrated Launch-and-Recovery Vehicle System 发射与回收两用飞行器系统
ILRVS Integrated Launch-and-Reentry Vehicle System 发射与再入两用飞行器系统
ILS Inertial Landing System 惯性着陆系统
ILS Instrument Landing System 仪表引导着陆系统
ILS Integrated Logistics Support 综合后勤保障
ILS Integrated Logistics System 综合后勤保障系统
ILS International Launch Service Company 国际发射服务公司

ILS　International Launch Services　国际发射服务

ILS/LAR　Integrated Logistics System and Logistics Assessment Review　综合后勤系统和后勤评定评审

ILS/VOR　Instrument Landing System/ VHF Omnidirectional Range　仪表着陆系统/甚高频全向（无线电）信标

ILSC　Integrated Launch Support Center　发射综合保障中心

ILSD　Integrated Logistics Support Division　【美国】一体化后勤保障部门

ILSP　Integrated Logistics Support Review　装备综合保障（综合后勤保障）评审

ILSREM　Instrument Landing System Radio Environmental Monitor　仪表着陆系统的无线电环境监控器

ILSSE　Integrated Life Science Shuttle Experiments　综合生命科学飞行器试验

ILSTAN　Instrument Landing System and Tactical Air Navigation　仪表着陆系统与战术空中导航

ILV　Integrated Launch Vehicle　组装完毕的运载火箭

IM　Injection Module　入轨舱

IM　Instrument Module　仪器舱

IM　Intercept Missile　截击导弹

IM　Interface Module　接口模块

IM　Interim Mission　过渡性任务

IM　Inverse Maneuver　反向机动

IM/IT　Information Management/Information Technology　信息管理/信息技术

IMA　Integrated Modular Avionics　综合模块化航空电子系统

IMAGE　Intruder Monitoring and Guidance Equipment　入侵飞行体监视与制导设备

IMAP　Intergrated Mission Analysis Planning　综合任务分析与规划

IMBLMS　Integrated Medical and Behavioral Laboratory Measurement System　综合医学和行为实验室测量系统

IMC　Image Motion Compensation　图像运动补偿

IMC　Instrument Meteorological Conditions　仪表飞行气象条件

IMCC　Integrated Mission Control Center　综合飞行任务控制中心；联合航天任务控制中心

IMCE　Image Motion Compensation Electronics　图像运动补偿电子设备

IMCO　Improved Combustion　改进型燃烧

IMCP　Integrated Monitor and Control Panel　综合监测与控制面板

IMCSRS　Integrated Monitor and Control Condition Status Reporting System　综合设施装备状况报告系统

IMCSS　Interim Military Communication Satellite System　临时军用通信卫星系统

IMDB　Integrated Maintenance Data Base　综合维护数据库

IMDSPO　Integrated Maintenance Data SPO　综合维护性数据系统计划官

IME　International Magnetospheric Explorer　国际磁层探测器

IMEWS　Integrated Missile Early Warning Satellite　【美国】综合导弹预警卫星

IMF　Interplanetary Magnetic Field　行星际磁场

IMFSS　Integrated Missile Flight Safety System　综合导弹飞行安全系统

IMIC Integrated Management Information Computer 综合管理信息计算机

IMID Inadvertent Missile Ignition Detection 导弹意外点火探测

IMINT Imagery Intelligence 图像情报

IMIS Integrated Maintenance Information System 综合维护信息系统

IML International Microgravity Laboratory 【美国国家航空航天局】国际微重力实验室

IMLEO Initial Mass in Low Earth Orbit 近地轨道初始质量

IMLS Interim Microwave Landing System 临时微波着陆系统

IMLSS Integrated Maneuvering and Life Support System 综合机动和生命保障系统

IMMAPDA Interacting Multiple Model Adaptive Probabilistic Data Association 交互式多模型自适应概率数据关联

IMMS Integral Maintenance Optimization 综合（整体）维修最优化

IMMU Integrated Man Maneuvering Unit 综合载人机动装置

IMOS Interactive Multi-Programming Operating System 交互式多道程序设计操作系统

IMP Impact Prediction 弹着点预测

IMP Intermetallic Phase 金属间相

IMP Interplanetary Monitoring Platforms 行星际监测平台

IMP Inter-Planetary Monitoring Probe （行）星际监测器

IMP-ATACMS Improved Army Tactical Missile System 改进的陆军战术导弹系统

IMPACT International Multi-User Plasma, Atmospheric, and Cosmic dust Twin facility 国际多用户等离子、大气及宇宙尘双设施

IMPCS Integrated Monitoring and Power Control Subsystem 【卫星通信系统】综合监测与动力控制子系统

IMPF International Microgravity Plasma Facility 国际微重力等离子设施

IMPS Inter-Planetary Monitoring Probe Satellite （行）星际监测卫星

IMRPV Interim Multi-Mission Remotelypiloted Vehicle 临时多任务遥控飞行器

IMS Information Management System 信息管理系统

IMS Integrated Management Subsystem 综合管理子系统

IMS Interim Meteorological Satellite 临时气象卫星

IMS Inter-Planetary Mission Support 星际（飞行）任务支援

IMS Inter-Planetary Monitor Satellite 星际监视卫星

IMSAT Imagery Satellite 成像卫星

IMSC Independent Modal Space Control 独立空间控制

IMSCCC Initial Military Satellite Command and Control Center 初始军用卫星指挥控制中心

IMSEP Improved Modular Scientific Experiments Package 改进型模块化科学实验软件包

IMSFP Integrated Manned Space Flight Program 综合载人航天飞行计划

IMSR Interplanetary Mission Support Requirements 行星际航行任务保障需求

IMSS Integrated Manned System Sim-

ulator 载人系统综合模拟器
IMT Integrated Modeling and Test 综合建模与试验
IMTS Improved Moving Target Simulator 改进型移动目标模拟器
IMU Inertial Measurement Unit 惯性测量装置
IMUGSE Inertial Measurement Unit Ground Support Equipment 惯性测量装置地面保障设备
IN LINAC Induction Linear Accelerator 感应线性加速器（装置）
INAS International Near-Earth Asteroid Search 国际近地小行星搜索
INCA Ion-Neutral Coupling in the Atmosphere 大气中离子—中性粒子耦合
INCS Integrated Network Control System 综合网络控制系统
INCU Inertial Navigation Control Unit 惯性导航控制部件
INDCU Inertial Navigation Display and Control Unit 惯性导航显示和控制装置
INETS Integrated Effects Tests for Survivability 可生存性综合效用测试
INETS Integrated Nuclear Environment Testbed Simulator 综合核环境试验（测试）台模拟器（装置）
INEWS Integrated Electronic Warfare System 【美国海军】综合电子战系统
INFI Integrated Navigation and Flight Inspection 综合导航与飞行检查
INFIS Inertial Navigation Flight Inspection System 惯性导航飞行检验系统
INFIS Integrated Navigation Flight Inspection System 综合导航飞行检查系统
INFOES In-Flight Operational Evaluation of a Space System 航天系统飞行工作状态评估
INFORSAT Information Transfer Satellite 信息传输卫星
InGaAs Indium Gallium Arsenide 砷化铟镓
INMARSAT International Maritime Satellite 国际海事卫星
INMS Ion and Neutral Mass Spectrometer 离子和中性粒子质谱仪
INR Inertial Navigation Reliability 惯性导航可靠性
INRTL VEL Inertial Velocity 惯性速度
INS Improved Navigation Satellite 改进型导航卫星
INS Inertial Navigation Set 惯性导航仪
INS Inertial Navigation System 惯性导航系统
INS Information Network System 【日本】信息网络系统
INS Internal Navigation System 内部导航系统
INSAT Indian National Satellite 印度卫星（印度通信、气象多用途卫星）
INSCS Integrated Navigation Steering and Control System 综合导航操纵与控制系统
INSE Integrated Network Support Environment 综合网络保障环境
INSICOM Integrated Special Intelligence Communications 综合特别情报通信
INSPEX Indonesia Space Experiment 印度尼西亚空间实验
INSSCC Interim National Space Sur-

veillance Control Center 【美国】国家临时空间监测控制中心
INST SYS Instrumentation System 仪器仪表系统
INST/COMM Instrumentation and Communication 仪器仪表与通信
INSTAR Inertialess Scanning, Tracking and Ranging 非惯性扫描、跟踪与测距（雷达）
INSTL&C/O Installation and Checkout 安装与检测
INSURE Integrated Survivability Experiments 综合可生存性实验
INTAC Intercept Tracking and Control 截击跟踪与控制
INTC/O Integrated Checkout 综合检测
INTELSAT Intelligence Satellite 情报卫星
INTELSAT International Telecommunication Satellite 国际通信卫星
INTV Interim Hypersonics Test Article 过渡性超声速测试（试验）部件
IOA Input/Output Adapter 输入/输出适配器
IOA Input/Output Assembly 输入/输出组件
IOB Input/Output Box 输入/输出箱
IOB Input/Output Buffer 输入/输出缓冲器
IOBPS Input/Output Box and Peripheral Simulator 输入/输出箱与边界模拟器（装置）
IOC Indirect Operating Costs 间接操作成本
IOC Initial Operational Capability 初始操作能力
IOC In Orbit Checkout 在轨测试
IOC In-Orbit Checkout and Calibration 在轨检测与校准
IOC Input/Output Controller 输入/输出控制器（装置）
IOC Intelligence Operations Center 情报操作中心
IOC INTELSAT Operations Center 国际通信卫星操作中心
IOC Interorbit Communication Experiment 【欧洲】轨道间通信实验
IOCM Interim Operational Contamination Monitor 使用间歇期污染监测（测量航天飞机舱内分子及微粒污染的实验）
IOCU Input/Output Control Unit 输入/输出控制装置
IODE Issue Of Data Ephemeris 卫星轨道参数的期号
IODS Intruder and Obstacle Detection System 入侵者和障碍物探测系统
IODT Inter-Operability Developmental Testing 互用性发展测试
IOF Initial Operational Flight 初始操作飞行
IOGE Integrated Operational Ground Equipment 综合操作地面设备
IOHA Integrated Operational Hazards Analysis 综合使用危险分析
IOI In Orbit Infrastructure 在轨基础设施
IOI Initial Operations Inspection 初始使用检查
IOL Inter-Orbit Optical Links 【欧洲】轨道间光学数据链
IOM Inert Operational Missile 惰性装药练习导弹
IOM Input/Output Module 输入/输出模块
IONCOMPSAT Ionospheric Component Satellite 电离层成分探测卫星
IONDS Integrated Operational Nuclear Detection System 综合操作核探测

系统
IOOSF Integrated Orbital Operation Simulation Facility 综合轨道操作模拟设施
IOP In-Orbit Plane 在轨平面
IOP Input/Output Port 输入/输出端口
IOP Input/Output Processor 输入/输出处理器（装置）
IOP Integrated Operation Plan 综合操作计划
IOP Integrate-On-Pad 发射台总装
IOS Integrated Operation Scenario 综合操作方案
IOSC Integrated Operations Support Center 综合操作保障中心
IOSS Integrated Orbital Service System 轨道集成服务系统
IOSV Inter-Orbital Space Vehicle 轨道间空间飞行器
IOT Initial Orbit Time 初始轨道飞行时间
IOT&E Initial Operational Test and Evaluation 初期操作试验（测试）与鉴定
IOU Input/Output Unit 输入/输出装置
IOV In-Orbit Validation 在轨验证
IP Impact Point 弹着点；命中点；碰撞点
IP Impact Prediction 撞击预测
IP Impact Probability 碰撞概率；弹着概率
IP Inertial Platform 惯性平台
IP Inertial Processing 惯性处理
IP Initial Point 起始点
IP Instrumentation Payload 仪器仪表有效载荷
IP Integration Phase 装配阶段
IP Internet Protocol 互联网协议

IPA Integrated Program Assessment 综合项目评估
IPAC Infrared Processing and Analysis Center 【美国国家航空航天局】红外处理与分析中心
IPAC Integrated Power Attitude Control System 综合动力姿态控制系统
IPACS Integrated Power and Attitude Control System 综合动力与姿态控制系统
IPACS Interactive Pattern Analysis and Classification System 交互式图形分析和分类系统
IPAD Integrated Program for Aerospace Vehicle Design 航空航天运载器设计一体化项目
IPAD Integrated Programmes for Aerospace Vehicle Design 航空航天器设计综合计划
IPAS Integrated Pressure Air System 综合压缩空气系统
IPB Intelligence Preparation of the Battlefield or Battlespace 战场情报准备
IPBM Interplanetary Ballistic Missile 行星际弹道导弹
IPC Integrated Payload Carrier 有效载荷综合载体
IPC Intelligence Propulsion Control 智能推进控制
IPC Interplanetary Communications 行星际通信
IPCCS Information Processing Command and Control System 信息处理指令和控制系统
IPCE Interface Power and Control Equipment 接口动力与控制设备
IPCS Integrated Propulsion Control System 综合推进控制系统

IPD Impact Prediction Data 弹着预测数据

IPD Initial Performance Data 初始性能数据

IPD Integrated Product (Process) Development 产品综合开发；集成式产品研制

IPDCMM Integrated Product Development Capability Maturity Model 综合产品研制能力成熟模型

IPDP Integrated Product and Development Process 综合产品与研制过程

IPDSMS Improved Point Defense Surface Missile System 改进的点防御地面导弹系统

IPE Improved Performance Engine 性能改进型发动机

IPEC Integrated Power & Environmental Control System 综合动力与环境控制系统

IPF Instrument Processing Facility 仪器处理设施

IPF Integrated Payload Facility 有效载荷对接厂房

IPF Integrated Processing Facility 综合处理设施

IPL Integrated Payload 集成有效载荷

IPL Interplanetary Physics Laboratory 美国行星际物理实验室

IPM Integration and Processing Facility 组装与处理厂房

IPM Interplanetary Medium 行星际介质

IPMP Investigation into Polymer Membranes Processing 【美国航天飞机】聚合物薄膜加工研究

IPMS Impact Predictor Monitor Set 弹着点预测仪监视装置

IPOMS International Polar Orbiting Meteorological Satellite 国际极轨气象卫星

IPOTP Integrated Payload Operations Training Plan 综合有效载荷操作培训计划

IPPD Integrated Process and Product Development 产品与过程的综合开发

IPRD Integrated Payload Requirements Document 综合有效载荷要求文件

IPRR Initial Production Readiness Review 初期生产成熟度评审

IPS Induced Pluripotent Stem Cells 诱发型多能干细胞

IPS Inertial Positioning System 惯性定位系统

IPS Instrument Pointing System (Subsystem) 仪表定位系统（分系统）

IPS Instrumentation Power Subsystem 仪器仪表动力分系统

IPS Integral Propulsion Subsystem 集成推进分系统

IPS Integrated Payload Requirements Review 综合有效载荷要求评审

IPS Integrated Power System 综合动力系统

IPSC Liquid Propulsion System Center 液体推进系统中心

IPSEAM Interactive Product Simulation Environment for Assembly and Maintenance 交互式产品装配与维修模拟环境

IPSRU Inertial Pseudo-Star Reference Unit 惯性伪星参考装置

IPSS Information Processing System Simulator 信息处理系统模拟器

IPT Integrated Product Team 一体化产品小组

IQSYE International Quiet Sun Year

Explorer 国际宁静太阳年探测器
IR Inclination of the Ascending Return 上升返回的倾角
IR International Rendezvous 国际性交会
IR SPECT Infrared Spectrometer 红外光谱仪
IR&D Independent Research and Development 独立性研发
IR-IE Infrared Imaging Equipment 【美国国家航空航天局】红外成像设备
IRAD Independent Research and Development 独立研究与发展
IRADS Infrared Attack & Designation System 红外攻击与目标指定系统
IRAMMP Infrared Analysis, Measurement and Modeling Program 红外分析、测量与建模程序
IRAS Infrared Astronomical Satellite 红外天文卫星
IRAT Integrated Reliability Analysis Tool 集成的可靠性分析工具
IRBM Intermediate Range Ballistic Missile 中远程弹道导弹
IRBS Infrared Background Sensor 红外背景传感器
IRBS Intermediate-Range Booster System 中远程助推器系统
IRCFE Infrared Communications Flight Experiment 【美国航天飞机】红外通信飞行实验
IRCM Infrared Countermeasures 红外对抗
IRCMAS Intelligent RCM Analysis System 智能化可靠性为中心的维修分析系统
IRCMS Integrated RCM System 综合的以可靠性为中心的维修系统
IRCS Inter-Site Radio Communication System 发射场间无线电通信系统
IRDS Infrared Detection System 红外探测系统
IRDS Integrated Reliability Data System 综合可靠性数据系统
IRE Integral Rocket Engine 一体化火箭引擎
IRE Ion Rocket Engine 离子火箭发动机
IREMBASS Improved Remotely Monitored Battlefield Sensor System 改进型远程监视战场传感器系统
IRFPA Infrared Focal Plane Array 红外焦平（面）阵列
IRIA Infrared Information Analysis Center 红外信息分析中心
IRIG Inter-Range Instrumentation Group 靶场间仪表小组
IRIS Incident Reporting Information System 事故通报信息系统
IRIS Infrared Instrumentation System 红外测量系统
IRIS International Radiation Investigation Satellite 国际辐射研究卫星
IRISTM Interactive Radar Information System 交互式雷达信息系统
IRL Integration Readiness Level 集成成熟度
IRLA Item Repair Level Analysis 物件修理级别分析
IRLS Interrogation, Recording and Location System 【美国"雨云"气象卫星】查询、记录与定位系统
IRM Information Resources Management 信息资源管理
IRM Ion Release Module 离子释放舱
IRMA Infrared Modeling and Analysis 红外建模与分析
IRMA Integrated Risk Management

Assessment 综合风险管理评估
IRMLD Infrared Missile Launch Detection 红外导弹发射探测
IROS Increased Reliability of Operational System 已增进的操作系统可靠性；已增进的作战系统可靠性
IRR Integral Rocket Ramjet 整体火箭冲压喷气发动机
IRR Integrated Rocket Ramjet 一体化火箭冲压式喷气发动机
IRR Integration Readiness Review 综合成熟度评审
IRRAS Integrated Reliability and Risk Analysis System 可靠性与风险综合分析系统
IRRS Information Resources Requirements Study 信息资源要求研究
IRS Indian Remote Sensing Satellite 印度遥感卫星
IRS Indian Remote Sensing System 印度遥感系统
IRS Integrated Radiator System 集成辐射计系统
IRST Infrared Search and Track 红外搜索与跟踪
IRST Infrared Sensor and Tracker 红外传感器与跟踪器
IRSS Infrared Sensor System 红外传感器系统
IRSS Instrumentation Range Safety System 仪表装置与靶场安全系统
IRSS Integrated Range Safety System 综合靶场安全系统
IRT Infrared Telescope 红外望远镜
IRT Integrated Rendezvous Radar Target 会合点综合雷达目标（用于测试航天飞机在轨道上的会合技术及能力的目标）
IRT Integrated Rendezvous Target 综合交会目标
IRTCM Integrated Real-Time Contamination Monitor 综合实时污染监控
IRTP Integrated Reliability Test Program 可靠性综合试验计划
IRU Inertial Reference Unit 惯性参照装置
IRV Isotope Reentry Vehicle 同位素再入飞行器
IRVE Inflatable Reentry Vehicle Experiment 充气再入飞行器试验
IS Installation Support 安装保障
IS Interconnecting Station 互联站
IS&T Innovative Science and Technology 创新科学与技术
IS&T Integrated Science & Technology 综合科学与技术
ISA Industry Standards Architecture 工业标准体系结构
ISA Inertial Sensor Assembly 惯性传感器组件
ISA Information System Architecture 信息系统体系结构
ISA Instruction System Architecture 指令系统体系结构
ISA Interactive State Analysis 交互式状态分析
ISA Interim Stowage Assembly 过渡性存储组件
ISA Interstage Adapter 级间段适配器
ISAC India Space Research Organization Satellite Center 印度空间研究组织卫星中心
ISAC International Telecommunications Satellite Solar Array Coupon 国际通信卫星太阳阵试样
ISACC Initial Satellite Control Center 初始卫星控制中心
ISADS Integrated Strapdown Air Data

System 综合捷联式大气数据系统
ISAFR In-Space Assembly, Fabrication and Repair 空间集成、装配和维修
ISAGE Imager/Sounder Analysis Groundsupport Equipment 成像器/探测器分析地面保障设备
ISAGEX International Satellite Geodesy Experiment 国际卫星测地实验
ISAMS Improved Stratospheric and Mesospheric Sounder 改进型平流层与中间层探测器
ISAR International Surveillance of Atmospheric Radioactivity 大气层放射性国际监测
ISAR Inverse Synthetic Aperture Radar 反向合成孔径雷达
ISARC Installation Shipping and Receiving Capability 设施装运和接收能力；设施的运送与接收能力
ISAS Institute of Space and Aeronautical Sciences 【日本】宇宙与航空科学研究院
ISC Information Systems Command 信息系统指令
ISC Intelligence Support Cell 情报保障室
ISC Interstellar Communications （恒）星际通信
ISC² Integrated Space Command and Control 综合空间指挥与控制
ISCCC Initial Satellite Communications Control Center 初期卫星通信控制中心
ISCD Integrated System Configuration Database 综合系统配置数据库
ISCO Innermost Stable Circular Orbit 最内侧稳定圆轨道
ISD Integrated Strategic Defense 综合战略防御

ISDB Integrated SATCOM (satellite communication) Database 综合卫星通信数据库
ISDE Inter-System Data Exchange 系统间数据交换
ISDN Integrated Services Digital Network 综合业务数字网
ISDN Space-Based Integrated Services Digital Network 天基综合业务数字网
ISDOS Information Systems Design and Optimization System 信息系统的设计与优化系统
ISDS Inadvertent Separation Destruct Subsystem 故障分离自毁分系统
ISDS Inadvertent Separation Destruct System 故障分离自毁系统
ISDU Inertial System Display Unit 惯性系统显示部件
ISE Integrated Space Experiment 综合空间试验
ISEC Information Systems Engineering Command 信息系统工程指令
ISECG International Space Exploration Coordination Group 国际空间探索合作组织
ISEE Integrated Software Engineering Environment 集成化软件工程环境
ISEE International Sun-Earth Explorers 国际日地探测器
ISEE NASA/ESA International Sunearth Explorer 【美国国家航空航天局/欧洲空间局】日一地探测器
ISEMS Improved Spectrum Efficiency Modeling and Simulation 改进式频谱功效建模与模拟
ISEPS International Sun-Earth Physics Satellite 国际日地物理探测卫星
ISERV International Space Station Environment Research and Visualization

System 国际空间站环境研究可视化系统
ISETS Integrated Support Equipment Tracking System 综合保障设备跟踪系统
ISF Industrial Space Facility 【美国国家航空航天局】工业航天设施
ISFDS-SBB Integrated Software Fault Diagnosis System Based on the SBB 基于软件黑匣子的集成软件故障诊断系统
ISFSI Independent Spent-Fuel Storage Installation 独立废燃料贮存设备
ISGEX International Satellite Geodesy Experiment 国际卫星测地实验
ISHM Integrated System Health Management 集成系统健康管理；综合系统健康管理
ISHM Integrated System Health Monitoring 一体化系统健康监测
ISHM Integrated (Intelligent) System Health Management (Monitoring) 综合（智能）系统健康管理（监测）
ISI Initial Systems Installation 初始系统安装
ISI Infrared Space Interferometer 红外空间干涉仪
ISI In-Service Inspection 在役检查
ISI Instrumentation Support Instruction 仪器仪表保障说明
ISIS Integrated Satellite Information Service 综合卫星信息服务
ISIS Integrated Sensor in Structure 结构体系中的一体化传感器
ISIS International Satellite for Ionospheric Studies 用于电离层研究的国际卫星
ISL Inertial Systems Laboratory 惯性系统实验室
ISL Inter-Satellite Link 卫星间（通信）链路
ISL Intersystem Link 系统间数据链
ISL/LAR Integrated Logistics System and Logistics Assessment Review 综合后勤保障系统与后勤保障评估评审
ISL^2 Inter-Satellite Laser Link 卫星间激光链路（使用二氧化碳激光器的高容量抗干扰卫星通信系统）
ISLSCP International Satellite Land Surface Climatology Project 国际陆地表面气候卫星计划
ISLSWG International Space Life Science Working Group 国际空间生命科学工作组
ISM Igniter Safety Mechanism 点火器安全机构
ISM Integrated Structures Model 综合体系结构模型
ISMLS Interim Standard Microwave Landing System 临时标准微波着陆系统
ISMMS Integrated Stores Monitor and Management Set 综合存储监控与管理装置
ISMS Improved Stratospheric and Mesospheric Sounder 改进型平流层与中间层探测器
ISN Information System Network 信息系统网络
ISO Imaging Spectrometric Observatory 成像光谱观测台
ISO Infrared Space Observatory 红外空间观测台
ISO Infrared Space Observatory 红外空间观测台（欧洲与美国合作的天文卫星）
ISO International Organization for Standardization 国际标准化组织
ISO International Organization of Quali-

ty Standards 国际质量标准组织
ISP　Inertially Stabilized Platform　惯性稳定平台
ISP　Internally Stored Program　内部存储程序
ISP　Internet Service Provider　国际互联网接入服务供应商
ISP　Specific Impulse　比冲
ISPAN　Integrated Strategic Planning and Analysis Network　综合战略计划与分析网
ISPE　Improved Sonar Processing Equipment　改进型声呐处理装置
ISPM　International Solar Polar Mission　国际太阳极轨探测卫星（欧洲与美国合作）
ISPP　In-Situ Producted Propellant　就地生产的推进剂（美国国家航空航天局的火星利用方案）
ISPR　International Standard Payload Rack　国际标准有效载荷架
ISPS　Instruction Set Processor Specification　指令系统处理机规范
ISPS　Interface Skirt and Payload Structure　接口裙部与有效载荷结构
ISPSTA　In-Space Propulsion Systems Technology Area　空间推进系统技术领域
ISPT　In-Space Propulsion Technology　在轨推进技术
ISR　Information Storage and Retrieval　信息存储与检索
ISR　Intelligence, Surveillance, and Reconnaissance　情报、监视与侦察
ISR　Interim Systems Review　中间（过渡期）系统评审
ISRI　Information Storage and Retrieval Interface　信息存储与检索接口
ISRMD　In-Service Reliability and Maintainability Demonstrations　使用中的可靠性与维修性验证
ISRO　Indian Space Research Organization　印度空间研究组织
ISRS　Inertial Space Reference System　惯性空间基准系统
ISRU　In Situ Resource Utilization　原位资源利用
ISS　Ignition Shielding System　点火屏蔽系统
ISS　Indirect Sighting System　间接瞄准系统
ISS　Inertial Sensor System　惯性传感器系统
ISS　Information System Security　信息系统安全
ISS　Infrared Surveillance System　红外监视系统
ISS　Installation Support Services　安装保障服务
ISS　Instrument Subsystem　仪表分系统
ISS　Integrated Satellite System　综合卫星系统
ISS　Integrated Scheduling System　综合进度系统
ISS　Integrated Sensor System　综合传感器系统
ISS　Integrated Source Sensor　综合资源传感器
ISS　Integrated Spacecraft System　综合航天器系统
ISS　Integrated Support Stand　集成支撑台
ISS　Integrated Support System　综合保障系统
ISS　Integrated System Schematic　综合系统方案
ISS　Intelligent Support System　智能支持系统
ISS　International Space Station　国际

空间站
ISS Ionospheric Sounding Satellite 【日本】电离层探测卫星
ISSA International Space Station Application 国际空间站应用程序
ISSAC International Space Station Agricultural Camera 国际空间站农业照相机
ISSE Information System Security Engineering 信息系统安全工程
ISSF Industry Satellite Services Facility 工业卫星服务设施
ISSMP International Space Station Medical Research Projects 国际空间站医学研究计划
ISSO International Small Satellite Organization 国际小型卫星组织
ISSO International Space Station Organization 国际空间站组织
ISSP Indian Scientific Satellite Project 印度科学卫星计划
ISSP International Space Station Program 国际空间站项目
ISSS Interactive Subscriber Service Subsystem 交互式用户服务子系统
ISST ICBM SHF Satellite Terminal 洲际弹道导弹超高频卫星终端
ISST ICBM Silo Superhardening Technology 洲际弹道导弹发射井超级加固技术
IST Initial System Test 系统初始测试（试验）
IST Innovative Science and Technology 创新科学与技术
IST Integrated Systems Test 综合系统测试
IST Interstellar Travel 星际航行
ISTA Intertank Structural Test Assembly 罐间结构试验组件
ISTA Integrated Stage Test Article 综合芯级测试（试验）部件
ISTAR Integrated System Test of an Airbreathing Rocket 吸气式火箭的综合系统测试
ISTAR Intelligence, Surveillance, Target Acquisition and Reconnaissance 情报、监视、目标截获与侦察
ISTB Integrated Subsystem Test Bed 综合子系统试验台
ISTC Integrated System Test Capability 综合系统试验（测试）能力
ISTD Integrated Space Technology Demonstration 综合航天技术演示验证
ISTEC Information Systems Test and Evaluation Center 信息系统试验与鉴定中心
ISTEF Innovative Science and Technology Experiment Facility 创新科学与技术实验设施
ISTF Installed System Test Facility 安装好的系统试验设施
ISTF Integrated Services and Test Facility 综合维护与测试设施
ISTF Integrated Space Technology Flights 综合航天技术飞行
ISTF Integrated System Test Flow 综合系统测试流程
ISTMC Instrumentation Section Test and Monitor Console 仪表舱测试与监控台
ISTOS Integrated Space Technologies Operational System 综合空间技术操作系统
ISTP International Solar-Terrestrial Physics Programme 国际日地物理计划（美国、欧洲、日本合作）
ISTRAC ISRO Telemetry Tracking and Command Network 印度空间研究组织测控网

ISU　Ignition Safety Unit　点火安全装置
ISU　Instrument Switching Unit　仪表切换装置
ISU　International Space University　国际空间大学
ISUS　Integrated Solar Upper Stage　综合太阳上面级
ISV　Interceptor Sensor Vehicle　拦截机（导弹）传感器车
IT　Identification Transponder　识别应答器（装置）
IT　Information Technology　信息技术
IT　Installation Test　安装测试（试验）
IT　Integrated Test　综合试验（测试）
IT　Interoperability Test　互操作性（互用性）试验
IT　Interplanetary Trajectories　行星际轨道
IT&E　Independent Test and Evaluation　独立试验与评价
IT&E　Integration, Test and Evaluation　集成、试验与评价
IT&V　Integration,Test and Validation/Verification　集成、测试和确认/鉴定
IT&VE　Integrated Test and Verification Environment　综合试验与鉴定环境
ITA　Integrated Test Area　综合测试（试验）区
ITA　Integrated Thruster Assembly　综合推进器装置
ITA　Interface Test Adapter　接口测试适配器
ITA　Intermediate Thrust Arc　中间推力段（弹道）弧
ITAC　Intelligence Threat Analysis Center　情报战区分析中心
ITALSS　Integrated Test and Logistic Support System　综合试验与后勤支援系统
ITAM　Integrated Test and Maintenance　综合试验与维修
ITAMS　Integrated Test and Maintenance System　综合试验和维修系统
ITANS　Inertial Terrain-Aided Navigation System　惯性地形辅助导航系统
ITAP　Integrated Technical Assessment Panel　综合技术评估委员会
ITAR　International Traffic in Arms Regulations　国际武器交易规则
ITB　Burn Time Impulse　燃烧时间冲量
ITB　Integrated Test Bed　综合试验（测试）台
ITC　Integrated Telemetry and Command　综合遥测与指挥
ITCE　International Cooperative Experiment　国际合作实验
ITCP　Integrated Test and Checkout Procedures　综合试验与检验过程
ITCS　Integrated Target Control System　综合目标控制系统
ITD　Integration Technology Demonstration　综合技术演示验证
ITDAP　Integrated Test Data Analysis Plan　综合试验（测试）数据分析计划
ITDAP　Integrated Test Design and Assessment Plan　综合试验（测试）设计与评估计划
ITDC　Interoperability Technology Demonstration Center　互用性技术验证中心
ITE　Inner Thermal Enclosure　内热控隔层
ITE　Instrumentation Test Equipment

仪器仪表测试（试验）设备
ITE Integration Test Equipment 综合测试（试验）设备
ITE Intersite Transportation Equipment 场内运输设备
ITEA Integrated Test, Evaluation and Acceptance Plan 试验、评价与接收综合计划
ITEA International Test and Evaluation Association 国际试验与评价协会
ITEC Involute Throat and Exit Cone 火箭渐伸线喷管喉道与尾喷管
ITED Integrated Trajectory Error Display 综合弹道误差显示；综合弹道误差显示器
ITER International Tokomak Experimental Reactor 国际托卡马克实验反应堆
ITERS Improved Tactical Events Reporting System 改进的战术事件通报系统
ITF Impulse Transfer Function 脉冲传递函数
ITF Integration and Test Facility 综合与测试（试验）设施
ITF Integration Test Facility 综合试验设施
ITF Interactive Terminal Facility 交互式终端设备
ITGS Integrated Track Guidance System 综合跟踪制导系统
ITI Inspection and Test Instruction 检查与测试说明
ITIS Integrated Technical Information System 综合技术情报系统
ITL Information Trouble Locator 信息故障定位器
ITL Integrate-Transfer-Launch 垂直组装、输送、发射

ITLC Integrate-Transfer-Launch Complex 组装－输送－发射成套设备
ITLF Integration, Transfer and Launch Facility 组装－输送－发射设施
ITMG Integrated Thermal Micrometeoroid Garment-Outside Layer of EMU 防热微流星体整体航天服－舱外灵便式航天服的外层
ITOC Ion Thruster On a Chip 芯片离子推力器
ITOF Interceptor Total Time of Flight 拦截机（导弹）总飞行时间
ITOS Improved TIROS Operational Satellite 【美国】爱托斯卫星；改进型泰罗斯业务卫星
ITOS Iterative Time Optimal System 迭代时间最优系统
ITOSS Intent Trajectory Operated Signal Source 意向弹道操作信号源
ITP Inflation Technology Program 暴涨技术计划
ITPAC In-Transport Payload Air Conditioning 有效载荷运输中空气调节
ITR Information Technology Resources 信息技术资源
ITR Initial Technical Review 初始技术评审
ITR Integration Testing Review 集成测试评审
ITRR Integrated Launch Vehicle Transfer Readiness Review 组装完毕的运载火箭转运准备评审
ITS Information Technology Service 信息技术服务
ITS Information Transfer Satellite 信息传输卫星
ITS Instrumentation Telemetry Station 仪器仪表遥测站
ITS Integrated Teleprocessing System 综合远程信息处理系统；综合遥控

处理系统
ITS　Integrated Trajectory System　综合弹道测量系统
ITTB　Integrated Technology Testbed　综合技术试验平台
ITV　Inert Test Vehicle　无动力试验飞行器
ITV　Instrument Target Vehicles　【美国】装有仪表的靶标导弹
ITV　Instrument Test Vehicle　仪表试验飞行器
ITV　Interface Verification Test　接口验证试验
ITVE　Integrated Test and Verification Environment　综合测试与验证环境
ITVETS　Improved TOW (Tubelaunched Optically Tracked Wireguided) Vehicle Evasive Target Simulator　【美国】改进型"陶"式导弹运载车规避目标模拟器
ITVF　Integration, Test and Verification Facility　组装、测试与检验设施
ITW　Integrated Tactical Warning　综合战术预警
ITW&AA　Integrated Tactical Warning and Attack Assessment　综合战术警报与攻击判断
ITW/AA　Initial Threat Warning/Attack Assessment　初始威胁告警/攻击评估
IU　Instrument Unit　仪表装置
IU　Interface Unit　接口装置
IUA　Inertial Unit Assembly　惯性装置组件
IUA　Interface Unit Adapter　接口单元适配器
IUCS　Instrumentation Unit Update Command System　测量装置升级指令系统
IUE　International Ultraviolet Explorer　国际紫外探测器（美国、欧洲空间局和英国共同研制）
IUS　Inertial Upper Stage　惯性上面级
IUS　Interim/Intermediate Upper Stage　临时/过渡性上面级
IV　Initial Velocity　初始速度
IV　Integrated Vehicle　一体化运载器
IV　Interceptor Vehicle　拦截机（导弹）车
IV　Intravehicular　【航天器】舱内的
IV&V　Independent Verification and Validation　独立验证与确认
IV&VF　Independent Verification and Validation Facility　独立鉴定（验证）与确认设施
IVA　Intra-Vehicular Activity　【航天员】舱内活动
IVALA　Integrated Visual Approach and Landing Aids　综合目视进场与着陆辅助系统
IVAM　Inter-Orbital Vehicle Assembly Mode　轨道间航天器装配方式
IVBC　Integrated Vehicle Baseline Configuration　一体化运载器基线配置
IVC　Intervehicular Communications　舱内通信
IVDP　Initial Vector Display Point　初始矢量显示点
IVDS　Internet Voice Distribution System　互联网音频分配系统
IVE　Interface Verification Equipment　接口验证设备
IVGVT　Integrated Vehicle Ground Vibration Test　一体化运载器地面振动测试（试验）
IVHM　Integrated Vehicle Health Managment　一体化运载器健康管理
IVHM　Integrated Vehicle Health Monitoring　综合飞行器状态（健康）

监控

IVHM Integrated (Intelligent) Vehicle Health Management (Monitoring) 集成(智能)运输器健康管理(监测)

IVHMS Integrated Vehicle Health Management System 综合飞行器状态(健康)管理系统

IVIS Inter-Vehicular Information System 车内信息系统；飞行器内信息系统

IVMS Integrated Vehicle Management System 综合飞行器管理系统

IVoDS Internet Voice Distribution System 互联网语言分配系统

IVSC Integrated Vehicle System Controller 集成飞行器系统控制器

IVSI Instantaneous Vertical Speed Indicator 瞬时垂直速度指示器

IVT Interface Verification Test 接口验证测试(试验)

IVT Intra-Vehicular Transfer 【航天器】舱内转移

IVTE Integration and Verification Test Environment 集成和确认测试环境

IVV Instantaneous Vertical Velocity 瞬时垂直速度；瞬间升降速度

IVVI Instantaneous Vertical Velocity Indicator 瞬时垂直速度指示器

IVVS Instantaneous Vertical Velocity Sensor 瞬时垂直速度传感器

IWBS Indirect Work Breakdown Structure 间接工作分类结构

IWBS Integral Weight and Balance System 整体称重与平衡系统

IWBS Integrated Weight and Balance System 整体重量与平衡系统；综合重量与平衡系统

IWCS Integrated Work Control System 集成工作控制系统

IWCD Integrated Wavefront Control Demonstration 波前控制综合演示验证

IWEDA Integrated Weather Effect Decision Aid 综合气象效应决策辅助(设备)

IWFS Integrated Waste Fluid System 【美国空间站】综合废液处理系统

IWS Indications and Warning System 显示与预警系统

IWS Integrated Water System 【美国空间站】水综合处理系统

IWSDB Integrated Weapon Systems Database 综合武器系统数据库

IWSM Integrated Weapon Support Management 武器保障综合管理；综合武器支援管理

IWSM Integrated Weapon Systems Management 综合武器系统管理

IWT Inland Waterway Transport 内河(航道)运输

IWW Intra-Coastal Waterway 沿海水道；沿海航道

IXO International X-Ray Observatory 国际X射线天文台

IXS Information Exchange System 信息交换系统

IXV Intermediate Experimental Vehicle 过渡性试验飞行器

Iy Pitch Inertia 俯仰惯性；俯仰惯性矩

Iz Yaw Inertia 偏航惯性；偏航惯性矩

IZLID Infrared Zoom Laser Illuminator Designator 红外移向目标激光照射指示器

J

J-CALS Joint-Computer-Aided Acquisition and Logistic System 联合计算机辅助采办与后勤保障系统

J-MASS Joint Modeling and Simulation System 联合建模与仿真系统

J-SSOD JEM Small Satellite Orbital Deployer 小卫星轨道释放装置

J/IST JSF/Integrated Subsystem Technology 联合打击战斗机/综合分系统技术

J/M Jettison Motor 弹射发动机

JABMD Japan BMD 目标弹道导弹防御

JACD Joint Architectural Control Document 联合体系架构控制文件

JACMAS Joint Approach Control Meteorological Advisory Service 联合进场管制气象咨询勤务

JAD Joint Application Development 联合应用软件开发

JAD Joint Assembly Demonstration 联合装配验证

JADE Joint Analysis Data Engine 联合分析数据工具

JADO Joint Air Defense Operation 联合防空作战行动

JAGUAR US Research Rocket 美国研究型火箭

JAIC Joint Air Intelligence Center 联合空中情报中心

JALCMC Joint Ammunition Life Cycle Management Command 联合弹药寿命周期管理指令

JAMES Joint Automated Message Editing System 联合自动化报文编辑系统

JAMS Jamming Analysis Measurement System 干扰分析测量系统

JAMTRAC Jammers Tracked by Azimuth Crossings 以方位交叉法跟踪的干扰台

JANUS Joint Analog Numerical Understanding System 联合模拟数字判断系统

JAO Joint Area of Operations 联合操作区

JAOC Joint Air Operations Center 联合空域操作中心

JARRS Joint Advanced Range Safety System 先进联合靶场安全系统

JARSS-MP JARSS Mission Planning element 先进联合靶场安全系统任务规划要素

JARSS-RT JARSS Real Time element 先进联合靶场安全系统实时要素

JASS Joint Anti-Satellite Study 联合反卫星研究

JASS Joint Autonomic Sustainment System 联合自主维持系统

JASSM Joint Air-to-Surface Standoff Missile 联合空对地阻击导弹

JAST Joint Advance Strike Technology 联合先进打击技术

JAT Joint Application Testing 联合应用测试

JATO Jammer Technique Optimization 干扰技术优化

JATO Jet Assisted Takeoff 喷气助推起飞

JATS Jamming Analysis and Transmission Selection 干扰分析与发射选择

JAWOP Joint Automated Weather Observation Program 联合自动气象观测系统

JAWS Jamming and Warning System 干扰与告警系统

JAXA Japanese Aerospace and Exploration Agency 日本宇宙航空研究开发机构

JB Jet Booster 喷气助推器

JBP Jettison Booster Package 弹射助推器装置；投掷加速器组件

JC²WC Joint Command and Control Warfare Center 联合指挥控制作战中心

JCCC Joint Command Control Center 联合指挥控制中心

JCDB Joint Common Database 联合通用数据库

JCEOI Joint Communications-Electronics Operation Instructions 联合通信电子操作指南

JCIDS Joint Capabilities Integration and Development System 联合能力集成和研发系统

JCM Joint Conflict Model 联合冲突模型

JCMP Joint Cruise Missile Project 联合巡航导弹项目

JCP Joint Power Conditions 联合动力条件

JCR Jet Control Rocket 喷射控制火箭

JCS Japanese Communication Satellite 日本通信卫星

JCTEA Joint Cost and Training Effectiveness Analysis 联合成本与训练效果分析

JCTN Joint Composite Tracking Network 联合合成跟踪网络

JCTV Joint Control Test Vehicle 联合控制测试车

JDAC Joint Data Analysis Center 联合数据分析中心

JDAM Joint Direct Attack Munition 联合直接打击弹药

JDCU Jamming Detection Control Unit 干扰探测控制单元

JDEP Joint Distributed Engineering Plant 联合分布式工程设施

JDIL JDIS Integration Laboratory 联合分布式信息系统综合实验室

JDIS Joint Distributed Information System 联合分布式信息系统

JDISS Joint Deployable Intelligence Support System 联合部署智能保障系统

JDMTA Jonathan Dickinson Missile Tracking Annex 【美国】乔纳森·迪金森导弹跟踪设施

JDN Joint Data Network 联合数据网

JDSS Joint Decision Support System 联合决策支持系统

JDSSC Joint Data Systems Support Center 联合数据系统支持中心

JDT&E Joint Development Test & Evaluation 联合研制试验与评价

JDTR Joint Deficiency Tracking and Reporting 联合缺陷跟踪与报告

JDX Joint Damping Experiment 联合阻尼实验

JEA Joint Effectiveness Analysis 联合效能分析

JECEWSI Joint Electronic Combat Electronic Warfare Simulation 联合电子作战电子战模拟

JELM Japanese Experiment Logistics Module 日本实验后勤舱

JEM Japanese Element Module 日本元器件舱

JEM Japanese Experiment Module 日本实验舱（与美国空间站对接的）

JEM-EF Japanese Explorer Module-Exposed Facility 日本探测舱－暴露设施

JEM-X Joint European Monitor of X-Rays 欧洲X射线联合监测器

JEMRMS Japanese Experiment Module Remote Manipulator System 日本实验舱遥控操纵装置系统

JEOS Japanese Earth Observation Satellite 日本地球观测卫星

JEP Joint Experiments Program 联合实验项目

JERS Japanese Earth Resources Satellite 日本地球资源卫星

JES Joint Environment Simulator 联合环境模拟器

JETEC Joint Expendable Turbine Engine Concepts 联合消耗性涡轮发动机设计

JETTA Joint Environment for Testing, Training, and Analysis 联合试验（测试）、训练与分析环境

JEWC Joint Electronic Warfare Center 联合电子战中心

JEWL Joint Early Warning Laboratory 联合预警实验室

JEZ Joint Engagement Zone 联合交战区

JFACT Joint Flight Acceptance Composite Test 联合飞行验收综合试验

JFC Johnson Space Flight Center 约翰逊航天飞行中心

JFC Joint Forces Command 联合部队司令部

JFCC SPACE Joint Functional Component Command for Space 太空联合职能司令部

JFCC-IMD Joint Functional Component Command-Integrated Missile Defense 负责综合导弹防御的联合部队下属职能部队司令部

JFLC Joint Force Land Component 联合部队地面部队

JFS Jet Fuel Starter 喷气（发动机）燃油起动机

JFSOC Joint Forces Special Operations Component 联合部队特种作战组成部队

JFTO Joint Flight Test Organization 联合飞行试验组织

JFTOT Jet Fuel Thermal Oxidation Tester 喷气燃油热氧化试验装置

JFV Jupiter Flyby Vehicle 飞越木星航天器

JHSV Joint High Speed Vessel 联合高速船艇

JI Jupiter Inlet 【美国国家航空航天局】进入木星轨道

JICM Joint Integrated Contingency Model 联合一体化应急作战模型

JICPAC Joint Intelligence Center, Pacific 太平洋联合情报中心

JILL Jet-Induced Lift Loss 喷流引起的升力损失

JIN Japanese Institute of Navigation 日本导航研究所

JINTACCS Joint Interoperability of Tactical Command and Control Systems 战术指挥控制系统的联合互操作性

JIOP Joint Interface Operational Procedures 联合接口操作规程

JIS Joint Integrated Simulation 联合综合仿真

JISP Joint Integrated Simulation Procedures 联合综合仿真程序

JISS Japanese Ionospheric Sounding Satellite 日本电离层探测卫星
JITC Joint Interoperability Test Center 联合互操作性试验中心
JLASS Joint Land, Aerospace and Sea Simulation 联合陆地、太空与海洋模拟
JLENS Joint Land Attack Cruise Missile Defense Elevated Netted Sensor System 联合对地攻击巡航导弹空中防御网传感器系统
JLOTS Joint Logistics Over-the-Shore 联合岸上后勤保障
JLRPG Joint Long Range Proving Grounds 联合远距离试验靶场
JLTV Joint Light Tactical Vehicle 联合轻型战术车；联合轻型战术飞行器
JM&S Joint Modeling and Simulation 联合建模与模拟
JMAAT Joint Mission Area Analysis Tool 联合任务区分析工具
JMASS Joint Modeling and Simulation System 联合建模与仿真系统
JMATMCLU Javelin Medium Antitank Missile Control Launch Unit 【英国】"标枪"中程反坦克导弹控制发射装置
JMCC Johnson Mission Control Center 【美国】约翰逊飞行任务控制中心
JMCCOC Joint MILSTAR Communications Control and Operations Concept 联合军事战略与战术中继系统通信控制与操作方案
JMCIS Joint Maritime Collaborative Information System 联合海战合作信息系统
JMCIS Joint Maritime Command (Combat) Information System 联合海战信息系统
JMCP Joint Mission Capability Package 联合任务能力包
JMDN Joint Missile Defense Network 联合导弹防御网
JMNA Joint Military Net Assessment 联合军事网络评估
JMRC Joint Mobile Relay Centre 联合机动中继站
JMSP Joint Multispectral Sensor Program 联合多频谱传感器项目
JMTSS Joint Multichannel Trunking and Switching System 联合多路中继与交换系统
JNESSY JNTF Electronic Security System 【美国】国家联合试验设施电子保密系统
JNIC Joint National Integration Center 【美国】联合国家综合中心
JNN Joint Network Node 联合网络节点
JNTF Joint National Test Facility 联合国家试验设施
JNTFOMC Joint National Test Facility Operations and Maintenance Contractor 【美国】联合国家试验设施操作与维护承包商
JNTFRDC Joint National Test Facility Research and Development Contractor 联合国家试验设施研究与发展承包商
JNTFUSLA Joint National Test Facility Unclassified Standalone and Laptop Access 【美国】联合国家试验设施非保密（无类别）单独与重复访问
JOC Joint Operations Center 联合操作中心
JOERAD Joint Spectrum Center Ordnance Electromagnetic Environmental

Effect Risk Assessment Database 联合光谱中心电磁环境效应风险评估数据库

JOI Jupiter Orbit Insertion 木星轨道入轨

JOIN Joint Optical Information Network 联合光学信息网络

JOIP Joint Operations Interface Procedure 联合操作接口规程

JOP Joint Operating Procedure 联合操作规程

JOPES Joint Operational Planning and Execution System 联合作战（操作）规划与执行系统

JOPS Joint Operations Planning System 联合作战（操作）规划系统

JOR Joint Operational Requirements 联合作战（操作）要求

JOSS JTF Operational Support System 联合特遣部队作战支援（保障）系统

JOT&E Joint Operational Test Evaluation 联合使用试验与评价

JOTS Joint Operational Tactical System 联合作战（操作）战术系统

JOVE Jupiter Orbiting Vehicle for Exploration 木星轨道探测飞行器

JP Jet Propellant 喷气式发动机推进剂（燃烧剂）

JP Jet Propulsion 喷气（发动机）推进的；装喷气发动机的

JPALS Joint Precision Approach Landing System 联合精确进场（进近）着陆系统

JPDO Joint Projects Development Office 联合项目研发办公室

JPDRD Joint Program Definition and Requirements Document 联合项目定义与要求文件

JPL Jet Propulsion Laboratory 【美国国家航空航天局】喷气推进实验室

JPOC JSC Payload Operations Center 【美国约翰逊航天中心】有效载荷处理中心

JPOPS Japan Polar Orbit Platform Satellite 日本极轨平台卫星

JPR Joint Program Review 联合项目评审

JPS Jet Propulsion System 喷气推进系统

JPS Joint Precision Strike 联合精确打击

JPSD Joint Precision Strike Demonstration 联合精确打击演示验证

JPTL Jet Pipe Temperature Limiter 尾喷管温度限制器

JPTO Jet Propelled Takeoff 喷气推进起飞

JRCC Joint Rescue Coordination Center 联合救援协调中心

JRCS Jet Reaction Control System 喷气反作用控制系统

JRD Joint National Integration Center Research and Development 【美国】联合国家综合中心研究与发展

JRDOD Joint Research and Development Objective Document 联合研究与开发目标文件

JRE Joint Range Extension 联合扩程

JRMET Joint Reliability and Maintainability Evaluation Team 可靠性与维修性联合鉴定组

JRRC Joint Rapid Response Center 联合快速反应中心

JRSOI Joint Reception, Staging, Onward Movement and Integration 联合接收、中间整备、作战区运动和整合

JRV Javelin Rocket Vehicle 【英国】

"标枪"型火箭飞行器

JS&MDWC Joint Space and Missile Defense Warfare Center 【美国】联合航天与导弹防御战中心

JSAC Japan's Space Activities Commission 日本空间活动委员会

JSAS Jammer System Analysis Simulator 干扰系统分析模拟器

JSAS Joint Strike Analysis System 联合打击分析系统

JSASS Japan Society for Aeronautical and Space Science 日本航空航天学会

JSB Joint Synthetic Battlespace 联合模拟战场空间

JSC Johnson Space Center 约翰逊航天中心

JSCP Joint Space Capabilities Plan 联合航天能力计划

JSECST Joint Service Electronic Countermeasures System Tester 联合军种电子对抗系统试验装置

JSEP Joint Service Electronics Program 军种联合电子设备项目

JSF Joint Strike Fighter 联合打击战斗机

JSIC Joint SPACECOM Intelligence Center 【美国】联合航天司令部情报中心

JSIMS Joint Simulation System 联合模拟系统

JSIPS Joint Service Imagery Processing System 军种联合图像处理系统；联合勤务图像处理系统

JSMTS Joint Services Mobile Tactical System 联合勤务移动战术系统

JSOC Joint Space Operations Center 联合航天操作中心

JSOC Joint Special Operations Command 【美国】联合特种作战司令部

JSOW Joint Standoff Weapon 联合防区外（远射）武器

JSpOC Joint Space Operations Center 联合空间操作中心

JSPS Joint Strategic Planning System 联合战略规划系统

JSS Joint Surveillance System 【美国】联合监视系统

JSSEE Joint Services Software Engineering Environment 【美国】三军共用软件工程环境

JSST Joint Space Support Team 联合航天支援（保障）小组

JSTARS Joint Surveillance and Target Attack Radar System 联合监视与目标攻击雷达系统

JSTARS Joint Surveillance and Target Attack Reconnaissance System 联合监视与目标打击侦察系统

JSTARS Joint Surveillance Tracking and Reporting System 联合监视跟踪与通报系统

JSUP Japanese Space Utilization Promotion Center 日本空间应用促进中心

JT&E Joint Test and Evaluation 联合试验与评估

JTA Joint Technical Architecture 联合技术体系结构（文件）

JTACMS Joint Tactical Cruise Missile System 联合战术巡航导弹系统

JTAG Joint Test Action Group 联合试验（测试）行动组

JTAGG Joint Turbine Advanced Gas Generator 联合涡轮机高级燃气发电机

JTAGS Joint Tactical Ground Station 联合战术地面站

JTAMD Joint Theater Air and Missile

Defense 联合战区防空与导弹防御
JTAMDO Joint Theater Air and Missile Defense Organization 联合战区防空与导弹防御组织
JTAOM JTIDS-Equipped Tactical Air Operations Module 用联合战术信息分发系统装备的战术空中作战模块
JTASC Joint Training Analysis and Simulation Center 联合训练分析与模拟中心
JTDE Joint Technology Demonstrator Engine 【美国空、海军】联合技术演示装置发动机（引擎）
JTE Joint Targeting Element 联合判定目标摧毁要件
JTG Joint Technology Group 联合技术组
JTIDS Joint Tactical Information Data System 联合战术信息数据系统
JTIDS Joint Tactical Information Distribution System 联合战术信息分布系统
JTMD Joint Theater Missile Defense 联合战区导弹防御
JTMS Joint Tactical Missile System 联合战术导弹系统
JTRS Joint Tactical Radio System 联合战术无线电系统
JTS Jet Thrust Stoichiometric 喷气推力化学当量
JTT Joint Tactical Terminal 联合战术终端
JTTC Joint Interoperability Test Center 联合互用性试验中心
JTTP Joint Tactics, Techniques, and Procedures 联合战术、技术与规程
JULIE Joint Utilization of Laser Integrated Experiments 激光综合实验的联合应用
JVIDS Joint Visually Integrated Display System 联合视觉综合显示系统
JVSEAS Joint Virtual Security Environment Assessment System 联合虚拟安全环境评估系统
JWARS Joint Warfare Analysis and Requirements System 联合作战分析和要求系统
JWARS Joint Warfare Simulation 联合作战模拟
JWARS Joint Warfighting System 联合战斗系统
JWC Joint Warfare Center 联合作战中心
JWCA Joint Warfare Capabilities Assessment 联合作战能力评估
JWCO Joint Warfare Capability Objective 联合作战能力目标
JWFC Joint Warfighting Center 联合战斗中心
JWICS Joint Worldwide Intelligence Communications System 联合全球情报通信系统
JWID Joint Warfighting Interoperability Demonstration 联合作战互通性验证
JWST James Webb Space Telescope 【美国】詹姆斯·威布太空望远镜
JWSTP Joint Warfighting Science and Technology Plan 联合战斗科学与技术计划
JZ Jump on Zero 遇零则转移

K

K-APP KSC Automated Payloads Plan/Requirement 【美国】肯尼迪航天中心自主有效载荷计划/要求

K-APPS KSC Automated Payloads Project Specification 【美国】肯尼迪航天中心自主有效载荷计划说明

K-DPM KSC DOD Payloads Plan/Requirement 【美国】肯尼迪航天中心/国防部自主有效载荷计划/要求

K-DPPS KSC DOD Payloads Project Specification 【美国】肯尼迪航天中心/国防部自主有效载荷计划说明

K-SLM KSC Spacelab Plan/Requirement 【美国肯尼迪航天中心】空间实验室计划/要求

K-SLPS KSC Spacelab Project Specification 【美国肯尼迪航天中心】空间实验室计划说明

K-SM KSC Shuttle Management Document 【美国肯尼迪航天中心】航天飞机管理文件

K-SPS KSC Shuttle Project Specification 【美国肯尼迪航天中心】航天飞机计划说明

K-SSS KSC Shuttle Project Station Set Specification 【美国肯尼迪航天中心】航天飞机计划站点设置说明

K-STSM KSC STS Plan/Requirement 【美国肯尼迪航天中心】空间运输系统计划/要求

K/S Kick Stage 反冲芯级

KAC Kagoshima Space Center 【日本】鹿儿岛航天中心

KADS Knowledge Acquisition Data System 知识获取数据系统

KAI Korean Aerospace Industries Ltd. 韩国航空航天工业公司

KARI Korean Aerospace Research Institute 韩国航空航天研究所

KASC Knowledge Availability Systems Center 信息提供系统中心

KASD Kinematics and Sensor Dynamics 动力学与传感器动态特性(学)

KASIMA Karlsruhe Simulation Model of the Middle Atmosphere 卡尔斯鲁厄中层大气层模拟模型

KAT Kill Assessment Technology 毁伤效应技术

KATE Knowledge-Based Automatic Test Equipment 基于知识的自动测试设备

KATS Kennedy Avionics Test Set 【美国】肯尼迪航天中心航空电子设备测试装置

KB Knowledge Base 知识库

KBAC Kennedy Booster Assembly Contractor 【美国】肯尼迪航天中心助推器组件承包商

KBS Knowledge Based System 人工智能系统

KBSA Knowledge-Based Software Assistant 基于知识的辅助软件/人工智能软件辅助装置

KCAS Calibrated Air Speed in Knots 以节表示的校准空速

KCCS Kennedy Complex Control System 【美国】肯尼迪航天中心发射场综合控制系统

KDEC Kinetic Energy Weapon Digital Emulation Center 动能武器数字仿真中心

KDMS Kennedy Data Management System(s) 【美国】肯尼迪航天中心数据管理系统

KDP Kennedy Documented Procedure 【美国】肯尼迪航天中心文字记录的程序

KDP Key Data Processor 关键数据处理器

KDS Kwajalein Discrimination System 【美国】夸贾林导弹靶场识别系统

KDT Knowledge Development Tools 知识开发工具

KE Kinetic Energy 动能

KE ASAT Kinetic Energy Anti-Satellite Weapon 动能反卫星武器

KEASAT Kinetic Energy ASAT 动能反卫星（武器）

KHILS Kinetic Kill Vehicle Hardware in-the-Loop Simulator 动能杀伤半实物模拟（仿真）器（装置）；动能杀伤回路内硬件模拟（仿真）器（装置）

KEI Kinetic Energy Intercept 动能拦截；动能截击

KEI Kinetic Energy Interceptor 动能截击导弹

KEII Khrunichev-Energia International Incorporated 赫鲁尼切夫－能源国际公司

KEK Kinetic Energy Kill 动能杀伤

KEM Kinetic Energy Missile 动能导弹

KEP Key Emitter Parameters 发射机主要参数

KEP Kinetic Energy Penetrator 动能穿透器

KEW Kinetic Energy Weapon 动能武器

KEWC Kinetic Energy Weapon, Chemical (Propulsion) 化学（推进）动能武器

KEWE Kinetic Energy Weapon, Electromagnetic (Propulsion) 电磁（推进）动能武器

KEWG Kinetic Energy Weapon, Ground 地面动能武器

KEWO Kinetic Energy Weapon, Orbital 轨道动能武器

KH Keyhole 【美国】"锁眼"卫星（侦察卫星代号）

KHIL KEW (Kinetic Energy Weapon) Hardware Integration Lab 动能武器硬件综合实验室

KHIL Kinetic Hardware-in-the-Loop (Test) 动能硬件在回路（试验）

KHILS KKV (Kinetic Kill Vehicle) Hardware-in-the-Loop Simulator 动能杀伤飞行器半实物仿真器；动能杀伤回路内硬件模拟器

KHIT Kinetic Hover Interceptor Test 动能悬浮拦截器试验

KhSC Khrunichev State Research and Production Space Center 【俄罗斯】赫鲁尼切夫国家科研生产航天中心

KICS Kennedy Integrated Communications System 【美国】肯尼迪航天中心综合通信系统

KIDD Kinetic Impact Debris Distribution 动能撞击碎片分布

KIDDS Kwajalein Instrumentation Data Distribution System 【美国】夸贾林仪表测量数据分配系统

KIFIS Kollsman Integrated Flight Instrument System 科尔斯曼综合飞行仪表系统

KITE Kinetic Isolation Tether Experiment 系绳动力隔离实验

KITE KKV (Kinetic Kill Vehicle) Integrated Technology Experiment 【弹道导弹防御组织】动能杀伤飞行器综合技术试验

KKV Kinetic Kill Vehicle 【美国 SDI 计划】动能杀伤飞行器

KKVHILS Kinetic Kill Vehicle Hardware in the Loop Simulation 回路模拟中的动能杀伤飞行器硬件

KKVWS Kinetic Kill Vehicle Weapon System 动能杀伤武器系统

KKW Kinetic Kill Weapon 动能杀伤武器

KLC Kodiak Launch Complex 【美国】科迪亚克航天发射场

KM Kick Motor 入轨发动机；脉冲式发动机（短时间工作的）

KMAN Kennedy Metropolitan Area Network 【美国】肯尼迪城市区域网络

KMCC Kwajalein Mission Control Center 【美国】夸贾林导弹靶场任务控制中心

KMI KSC Management Instruction 【美国】肯尼迪航天中心管理条例

KMOR Keep Missile on the Rail 使导弹保持在导轨上

KMR Kwajalein Missile Range 【美国】夸贾林导弹靶场

KMRSS Kwajalein Mobile Range Safety System 【美国】夸贾林移动靶场安全系统

KMU Guidance Unit for Guided Bombs 制导炸弹的制导装置

KNET Kennedy Institutional Network 【美国】肯尼迪航天中心指令网

KOCOA Key Terrain, Observation and Fields of Fire, Cover and Concealment, Obstacles, and Avenue of Approach 关键性地形、观察和火力地域、覆盖和隐蔽、障碍物和接近通路

KOD Kick-Off Drift 消除偏流修正角

KOI Kennedy Operating Instructions 【美国】肯尼迪（航天中心）操作指令

Kosmos USSR Polar Orbiting Satellites 苏联极轨卫星

KP Kill Probability 击毁概率

KPI Key Performance Indicator 主要性能指示器

KPP Key Performance Parameter(s) 关键性能参数

KPP Key Production Process 关键生产工艺

KPS Key Parameters 关键参数

KREMS Kiernan Reentry Measurement Site 【美国】基尔南再入测量站

KREMS Kiernan Reentry Measurement System 【美国】基尔南再入测量系统

KSC Kagoshima Space Center 【日本】鹿儿岛航天中心

KSC Kennedy Space Center 【美国】肯尼迪航天中心

KSCNF KSC News Facility 【美国】肯尼迪航天中心新闻中心

KSDI Key System Development & Integration 关键系统开发与综合

KSRC Khrunichev State Research and Production Space Center 【俄罗斯】赫鲁尼切夫国家科研生产航天中心

KTM Kineto Tracking Mount 【美国】"Kineto"跟踪装置

KTS Kineto Tracking System 【美国】"Kineto"跟踪系统

KTP Key Technical Parameters 关键技术参数

KUBIK Incubator with Centrifuge (Russian for "cube") 带有离心分离机的培养箱（俄罗斯的"立方体"）

KVETS KKV Validation and Evaluation Test System 【综合】动能杀伤飞行器验证和评价试验系统

KVRB General Cryogenic Upper Stage 通用低温上面级

KVRB LO_2/LH_2 Upper Stage; Oxygen-Hydrogen Upper Stage 液氧/液氢上面级

KY Kapustin Yar 【俄罗斯】卡普斯金亚尔发射场

KZ Killing Zone 毁伤范围；诱歼区

L

L　Landing Flight Phase　着陆飞行阶段
L　Launch　发射；弹射；起飞；开始
L　Launcher　发射装置；发射架，运载火箭弹射器
L&C　Laboratory and Checkout　实验室与检测
L&D　Landing and Deceleration　着陆与减速
L&D　Launch and Defend　发射与防御
L&D　Launch and Defense　发射与防御
L&D　Loss and Damage　损耗与毁坏
L&ES　Laser and Electronic System　激光和电子系统
L&I　Launch and Impact　发射与冲击；发射与命中；发射与弹着（点）
L&L　Launch and Landing　发射与着陆
L&M　Logistics and Maintenance　后勤保障和维修
L&MM　Logistics and Material Management　后勤物资管理
L&MR　Logistics and Materials Readiness　后勤与装备处于备战状态
L&PP　Lunar and Planetary Program　月球和行星计划
L&R　Landing and Recovery　着陆与救援；降落与回收
L&RS　Launch and Recovery System　着陆与回收（救援）系统
L&S　Launch and Service　发射与检修
L&S　Logistics and Support　后勤与保障
L&T　Laboratory and Test　实验室与测试（试验）
L&TH　Lethality and Target Hardening　杀伤力与目标加固
L&V　Lethality and Vulnerability　毁坏性与弱点
l.o.　Local Oscillator　本机振荡器
L.R.　Liquid Rocket　液体火箭
L/D　Length-to-Diameter (Ratio)　长度直径比
L/D　Lift-to-Drag Ratio　升阻比
L/DOS　Launch/Deploy Operation Segment　发射/展开作业段
L/F　Launch Facility　发射设施设备
L/HIRF　Lightning/High Intensity Radiated Field　闪电/高密度辐射场
L/M/L　Operations/Maintenance/Logistics　使用、维修、后勤
L/MC　Launch/Missile Control　发射与导弹控制
L/MCC　Launch/Mission Control Center　发射/任务控制中心
L/MCP　Launch/Missile Control Processor　发射与导弹控制处理器
L/O　Lift Off　起飞；发射
L/O　Light Off　【发动机】熄火
L/R　Launch / Recovery　发射与回收
L/S　Landing Site　着陆场
L/S　Load System　载荷系统；负载系统
L/S　Logistics Support Analysis　保障性分析；后勤保障分析
L/T　Load Test　载荷测试

L/T　　Ratio of Lift to Thrust　　升推比
LA　　Launch Abort　　发射失败；紧急中断发射；发射故障处理
LA　　Launch Analysis　　发射分析
LA　　Launch Area　　发射区（域）
LA　　Launch Azimuth　　发射方位（角）
LA　　Lethal Area　　杀伤范围；杀伤区域
LA　　Lightning Arrester　　避雷器；避雷针
LA　　Limited Availability　　有限的可用性
LA　　Linear Accelerator　　线性加速度
LA　　Load Analysis　　载荷分析
LA　　Logistics Assistance　　后勤支援
LA　　Lower Assembly　　下部组装
LAA　　Launch Area Antenna　　发射区天线
LAA　　Limited Access Area　　有限访问区域
LAA　　Live Assemble Area　　【导弹等】实弹装配区
LAAD　　Low Altitude Air Defense　　低空空中防御
LAAR　　Liquid Air Accumulator Rocket　　液体空气蓄压火箭
LAAT　　Laser-Augmented Airborne Tow (Sight)　　机载激光增强型"陶"式反坦克导弹（瞄准器）
LAB　　Laboratory Module　　实验舱
LABOCC　　Laboratory Module Operations Control Center　　实验舱操作控制中心
LABRV　　Large Advanced Ballistic Reentry Vehicle　　大型先进弹道再入飞行器；大型高级弹道式再入飞行器
LAC　　Launch Analysis Console　　发射分析操控台
LAC　　Launch Azimuth Corridor　　发射方位角走廊（通道）
LAC　　Launcher Assignment Console　　发射器分配控制台
LAC　　Lightning Arrester Connector　　避雷器接头
LAC　　Limiting Admissible Concentration　　允许极限浓度
LAC　　Line of Actual Control　　实际控制线
LAC　　Lunar Aeronautical Chart　　月球航行图
LAC　　Lunar Astronautical Chart　　月球航天图；月球航行图
LAC　　Lunar Atlas Chart　　月球全图
LACA　　Linear Actuating Cylinder Assembly　　线性致动圆筒设备
LACB　　Landing Aids Control Building　　【美国肯尼迪航天中心】着陆辅助设备控制大楼
LACB　　Look Angles of Celestial Bodies　　天体观察角
LACE　　Laser Aerospace Communication Experiment　　航空航天激光通信实验
LACE　　Laser Atmospheric Compensation Experiment　　激光大气补偿实验
LACE　　Launch Angle Condition Evaluator　　发射角度状态鉴定器
LACE　　Launch Automatic Checkout Equipment　　发射自动检测设备
LACE　　Liquid Air Cycle Engine　　液态空气循环发动机
LACE　　Low Power Atmospheric Compensation Experiment　　【美国】低功率大气补偿实验（卫星）
LACE　　Low Power Atmospheric Compensation Experiment　　低功率大气补偿实验（卫星）
LACE　　Lunar Atmosphere Composition Experiment　　【美国】月球大气

成分实验
LACE Lunar Atmosphere Composition Experiment 月球大气成分实验
LACES Low Altitude Conical Earth Sensor 低空圆锥地球传感器
LACIE Large Area Crop Inventory Experiment 【美国陆地卫星】大面积作物估产实验
LACM Land Attack Cruise Missile 对地攻击巡航导弹
LAD Landing Assist Device 着陆辅助装置
LAD Landing Distance Available 可用着陆距离
LAD Large Area Detector 大范围探测器（装置）
LAD Liquid Acquisition Device 液体采集装置
LAD Location Aid Device 定位辅助设备
LAD Lunar Atmosphere Detector 月球大气探测器
LADAR Laser Detection and Ranging 激光探测与测距
LADD Lens Antenna Deployment Demonstration 镜面天线研制验证
LADEE Lunar Atmosphere and Dust Environment Explorer 月球大气和尘埃环境探测器
LADFU Large Area Detector Flight Unit 大范围探测器（装置）飞行单元
LADGNSS Local Area Differential GNSS 局域差分全球导航卫星系统
LADIR Low-Cost Arrays for Detection of Infrared 低成本红外探测天线阵列
LADL Lightweight Air Defense Launcher 轻型空防导弹发射装置

LADO Launch, Anomaly Resolution, and Disposal Operations 发射、异常解决与处理操作
LADS Laser Area Defense System 激光区域防御系统
LADS Life Assessment Detector System 生命判定探测系统
LADS Linear Analysis and Design of Structures 结构线性分析和设计
LADS Low Altitude Demonstration System 低空演示验证系统
LAE Liquid Apogee Engine 液体远地点发动机
LAES Landing Aids Experiment Station 着陆辅助装置实验站
LAFTS Laser and FLIR Test Set 激光与前视红外探测（系统）试验设备
LAGEOS Laser Geodynamics Satellite 【美国国家航空航天局】激光地球动力学卫星 [一种完全被动（无源）的球状卫星，表面覆盖反射器，由陆基激光器照亮，对地壳运动进行精确测量]
LAGM Long-Range Air-to-Ground Missile 远程空对地导弹
LAGS Laser Activated Geodetic Satellite 激光驱动的大地测量卫星
LAGS Launch Abort Guide Simulation 发射中断制导模拟
LAGUMS Laser-Guided Missile System 激光制导导弹系统
LAHAWS Laser Homing and Warning System 激光寻的与告警系统
LAI Lot Acceptance Inspection 成批验收检验
LAIR Liquid Air 液体空气
LAIU Launch Abort Interface Unit 发射中断接口装置
LAL Launch and Leave 发射后离开

LAL Launch Area Lateral 发射区侧向（侧部）
LAL Lock After Launch 发射后锁定
LALMS Low Altitude Laser Measurement System 低空激光测量系统
LAM Launch Alert Message 发射准备通报
LAM Launch Area Monitor 发射区监控台；发射区监视器
LAM Liquid Apogee Motor 液态远地点发动机
LAM Logistics Attrition Model 后勤损耗模型
LAM Loitering Attack Missile 待机攻击导弹
LAMAR Large Amplitude Modular Array 大型调幅模块阵列
LAMARS Large Amplitude Multimode Aerospace Research Simulator 大型调幅多模式航空航天科研模拟器
LAMC Last Maneuver Calculation 【轨道识别】最后机动计算
LAMP Lehmann Alpha Mapper 莱曼阿尔法测绘仪
LAMP Lockheed Adaptive Modular Payload 【美国】洛克·希德任务适应模式化有效载荷
LAMP LODE Advanced Mirror Program 【美国 SDI】大型光学器件实验先进反射镜计划
LAMP Lunar Analysis and Mapping Program 月球（面）分析和测绘计划
LAMPS Light Airborne Multipurpose System 轻型机载多用途系统
LAMS Launch Acoustic Measuring System 发射声音测量系统
LAMSIM Launcher and Missile Simulator 发射器与导弹模拟器

LAMTS Launcher Adapter Missile Test Set 发射架适配器导弹试验装置
LANDSAT Land Observation Satellite 大地观测卫星
LANDSAT Land Remote-Sensing Satellite 大地遥感卫星
LANDSAT Land Resources and Mapping Satellite 陆地资源与测绘卫星
LANDSAT Land Satellite 陆地卫星；地球资源卫星
LANDZ Landing Zone 着陆区
LANE Local Area Network Emulation 局域网网络仿真
LANL Los Alamos National Laboratory 【美国】洛斯·阿拉莫斯国家实验室
LANS Land Navigation System 陆地导航系统
LANS Local Area Network System 区域网络系统
LANTIRN Low Altitude Navigation and Targeting Infrared for Night 夜间红外低空导航与目标判定
LANTIRNS Low Altitude Navigation and Targeting Infrared Night System 夜间红外低空导航与目标判定系统
LANTISN Low-Altitude Navigation and Targeting Infrared System for Night 夜间低空导航与红外目标锁定系统
LAOS Large Astronomical Observatory Satellite 大型天文观测卫星
LAP Laboratory for Atmosphere Probing 大气探测实验室
LAP Laboratory for Atmospheric Probing 大气探测实验室
LAP Launch Analysis Panel 发射分析控制台；发射分析组
LAP Launcher Avionics Package 运载火箭航空电子设备组件

LAP Lightning Advisory Panel 雷电咨询组

LAPADS Low Altitude Parachute Retrorocket Airdrop System 低空降落伞制动火箭空投系统

LAPC Large Area Proportional Counter Array 大面积正比计数管阵列

LAPE Laser Altimeter for Planetary Exploration 行星探测的激光高度计

LAPS Launch Analysis Production System 发射分析生产系统

LAPS Left Aft Propulsion System (Subsystem) 左后部推进系统（分系统）

LAPS Logistics Analysis and Provisioning System 后勤分析和供给系统

LAPSS Large Area Pulsed Solar Simulator 大面积脉冲太阳模拟器

LAR Launch Acceptability Region 可发射区；发射可承受地区

LAR Liquid Argon 液氩

LAR Logistics Assessment Review 后勤评估审核

LAR Long-Range Aircraft Rocket 远程机载火箭

LAR Low-Angle Reentry 低角度再进入；小角度再入（大气层）

LARA Launch Risk Analysis 发射风险分析

LARC Langley Research Center 【美国国家航空航天局】兰利研究中心

LARGOS Laser-Activated Reflecting Geodetic Optical Satellite 激光反射式光学测地卫星

LARP Launch and Recovery Platform 发射和回收平台

LARS Laboratory for Applications of Remote Sensing 【美国珀杜大学】遥感应用实验室

LARS Large Amplitude Resonance Simulator 大幅度共振模拟器

LARS Laser Aerial Rocket System 激光航空火箭系统

LARS Laser-Aided Rocket System 激光辅助火箭系统

LARS Launch and Recovery System 发射与回收系统

LARS Launch Area Recovery System 发射区回收系统

LARS Leak Alarm and Response System 泄漏报警与响应系统

LARS Low Atmosphere Research Satellite 低大气层研究卫星；低层大气研究卫星

LARS Lower Atmosphere Research Satellite 低层大气层研究卫星

LARTTRS Low-Altitude Radar Tracking and Target Recognition System 低空雷达跟踪与目标识别系统

LARV Low Angle Reentry Vehicle 小角度再入飞行器；低角度再入飞行器

LARV Low-Altitude Research Vehicle 低空研究飞行器

LAS LAGEOS (Laser Geodynamics Satellite) Apogee Stage 【美国】激光地球动力学卫星远地点级

LAS Land Sat Sensor 【美国】陆地卫星遥感器

LAS Landing Aid System 着陆辅助系统

LAS Landing Approach Simulator 着陆进场模拟器

LAS Landing Area Security 着陆区安全防护措施

LAS Large Astronomic(al) Satellite 大型天文卫星

LAS Launch Abort System 发射中

断系统；发射故障处理系统
LAS Launch Area Steep 发射区陡前沿
LAS Launch Auxiliary System 发射辅助系统
LAS Long-Living Autonomous Station 【苏联火卫一着陆探测器】长寿命自主站
LAS Low Altitude Satellite 低轨卫星
LAS Lunar Attitude System 月球姿态系统
LASA Large Aperture Seismic Array 大孔径地震探测阵—核爆观测装置
LASA LIDAR Atmospheric Sounder and Altimetry 激光雷达大气探测与测高
LASAM Laser Semi-Active Missile 激光半主动寻的导弹
LASAR Logic Automatic Stimulus and Response 逻辑自动激励与响应
LASCO Large Angle And Spectrometric Coronagraph 大角度分光日冕观测仪
LASE Large Aperture Speckle Experiment 大孔径斑点实验
LASE LIDAR Acquisition and Sizing Experiment 【美国海军航天司令部】激光雷达捕获与测量实验
LASERCOM Laser Communications 激光通信
LASI Landing Site Indicator 着陆场指示灯
LASM Laser Semi-Active Missile 激光半主动式导弹；激光半主动式自导引导弹
LASM Long-Range Anti-Ship Missile 远程反舰导弹
LASP Laboratory for Atmosphere and Space Physics 大气与空间物理实验室
LASP Launch Abort Subpanel 发射中止取消（装置）的辅助面板
LASP Low Altitude Surveillance Platform 【美国】低空侦察平台
LASP Low-Altitude Space Platform 低高度空间平台
LASPAC Landing-Gear Avionics System Package 起落架电子设备系统组件
LASRM Low-Altitude Supersonic Research Missile 低空超声速研究导弹
LASS Land Surveillance Satellite 陆地监视卫星
LASS Large Advanced Space System 大型先进空间系统
LASS Large Area Space Simulator 大型空间模拟器
LASS Laser Spark Spectroscopy 激光感应荧光光谱
LASS Lateral Acceleration Sensing System 横向加速传感系统
LASS Logistics Automated Support System 后勤自动化保障系统
LASS Lunar Applications of a Spent Stage 燃尽芯级在月球的应用；失效火箭芯级在月球的应用
LASSII Low-Altitude Satellite Studies of Ionospheric Irregularities 电离层不规则性低空卫星研究
LASSO Landing and Approach System, Spiral Oriented 盘旋定向式着陆与进场系统
LASSO Laser Synchronization from Synchronous Orbit 【欧洲空间局】激光与地球同步轨道同步
LASSO Logistics Automation Systems Support Office 后勤自动化系统保

障办公室
LASSO Lunar Applications of a Spent Stage in Orbit 轨道内失效级火箭的月球应用；在轨燃耗火箭芯级的月球应用
LASSV Land and Approach System for Space Vehicle 航天飞行器着陆与进场系统
LASV Low Altitude Supersonic Vehicle 低轨超声速飞行器
LAT Large Angle Torque 大角度扭矩
LAT Lot Acceptance Test 批量验收测试（试验）
LATAR Laser-Augmented Target Acquisition and Recognition 激光增强目标截获与识别
LATR Large Area Tracking Range (e.g. missile range) 大区域跟踪靶场（如导弹靶场）
LATS Launcher Automatic Test Set 发射架自动测试设备
LATS LDEF (Long Duration Exposure Facility) Assembly and Transportation System 【美国】长期暴露设施组装与运输系统
LATS Longwave Infrared Advanced Technology Seeker 长波红外先进技术寻的器
LATS Low Altitude Target Satellite 低空目标卫星
LAU Launcher Adapter Unit 发射装置适配器
LAU Launcher Armament Unit 发射武器装置
LAU Launcher Unit 发射架设备
LAV Launch Abort Vehicle 发射中止飞行器
LAV Light Armored Vehicle 轻型装甲车

LAWDS LORAN Aided Weapon Delivery System 罗兰（远程导弹系统）辅助武器投射系统
LAWMS Light All-Weather Missile System 轻型全天候导弹系统
LB Launch Boost 发射推进
LB Launch Bunker 发射掩体
LB Launch Bus 发射总线
LB Low Bay 低跨间
LB Lower Brace 下部支架
LB/TS Large Blast/Thermal Simulator 大爆炸/热模拟器
LBC Laboratoire Banc de Contrôle/Check Out Equipment Room 测试设备间
LBD Laser Beam Detector 激光束探测器（装置）
LBDT Low Bay Dolly Tug 低跨间的平台拖车
LBL Laminar Boundary Layer 层流边界层
LBM Liquid Boost Module 液体助推舱
LBM Load Buffer Memory 载荷缓冲存储
LBM Lunar Breaking Module 登月分离舱
LBR Liquid Rocket Booster 液体火箭助推器
LBRV Large Ballistic Recovery Vehicle 大型弹道回收飞行器
LBS Launch Blast Simulator 发射喷射流模拟装置
LBT Land Based Transponder 陆基发射器应答机
LBTS Land-Based Test Site 陆基试验场
LBTS Land-Based Test System 陆基试验系统
LBU Launcher Booster Unit 发射助

推器装置；发射器助推装置
LC　Landing Chart　着陆图
LC　Landing Control　着陆控制
LC　Launch Complex　发射场；发射综合设施（工位）
LC　Launch Control　发射控制
LC　Launch Cost　发射成本
LC　Launch Countdown　发射倒计时
LC　Launch Cycle　发射周期
LC　Loop Check　环路检查；回路检验；循环检查
LC-39　Launch Complex 39 (A or B)　【美国肯尼迪航天中心】39号发射工位
LCA　Launch Control Amplifier　发射控制放大器
LCA　Launch Control Analyst　发射控制分析员
LCA　Launch Control Area　【导弹】发射控制区
LCA　Life Cycle Analysis　寿命周期分析
LCA　Life Cycle Assessment　寿命周期评价（法）；寿命周期评估
LCA　Load Controller (Control) Assembly　载荷控制器（控制）组件
LCAR　Launch Complex Assessment Report　发射场评估报告
LCAS　Low-Cost Access to Space　低成本进入空间
LCB　Launch Control Box　发射控制箱
LCB　Launch Control Building　发射控制大楼
LCB　Light Case Bomb　薄壳炸弹
LCC　Landing Command and Control Ship　登陆指挥与控制船
LCC　Landing Control Center　着陆控制中心
LCC　Launch Command and Control　发射指挥与控制
LCC　Launch Commit Criteria　发射限制标准
LCC　Launch Control Car　发射控制车
LCC　Launch Control Center　发射控制中心
LCC　Launch Control Console　发射控制操控台
LCC　Life Cycle Cost　寿命周期成本（核算）；全寿命费用
LCC　Life Cycle Costs　寿命周期成本（费用）
LCCA　Lateral-Control Central Actuators　横向操纵中央致动器
LCCA　Life Cycle Costs Analysis　寿命周期费用分析；寿命周期成本分析
LCCBA　Life Cycle Cost Benefits Analysis　寿命周期成本效益分析
LCCC　Launch Control Center Computer　发射控制中心计算机
LCCD　Launch Commit Criteria Document　发射限制标准文件
LCCD　Linear Charge Cord Device　线型装药爆炸索装置
LCCDP　Low Cost Combustor Development Program　低成本燃烧室发动机研制计划
LCCE　Life Cycle Cost Estimate　寿命周期费用评估；全寿命费用估算；使用周期费用估算
LCCEV　Low Cost Cryogenic Expendable Vehicle　低成本低温一次性运载器
LCCFC　Launch Control Complex Facility Console　发射控制综合设施设备控制台
LCCM　Life Cycle Cost Management　寿命周期费用管理

LCCM Life Cycle Cost Model 寿命周期费用模型

LCCS Launch Checkout and Countdown System 发射检测和倒计时系统

LCCS Launch Control and Checkout System 发射控制与检测系统

LCCS Launcher Captain Control System 发射台指挥员控制系统

LCCV Low Cost Concept Validation 低成本方案论证

LCCVT Life Cycle Cost Verification Test 寿命周期费用验证试验

LCD Launch Control Design 发射控制设计

LCD Launch Countdown 发射倒计时

LCD Lunar Module Change Directive 登月舱改变指令

LCDE Laser Communication Demonstration Experiment 激光通信实验装置

LCDP Lateral Control Departure Parameter 横向控制漂移参数

LCDS Low Cost Development System 低成本开发系统

LCE Laboratory Component Evaluation 实验室元件评估

LCE Large Cryogenic Engine 大型低温发动机（欧洲空间局"阿里安"系列火箭）

LCE Launch Complex Equipment 发射场设施；发射场设备

LCEB Launch Control Equipment Building 发射控制设备楼；发射控制设备间

LCEF Launch-Centered Earth-Fixed 地面固定发射中心

LCELV Low Cost Expendable Launch Vehicle 低成本一次使用运载火箭

LCES Limited-Capability Earth Station 有限能力地面站

LCET Logistics Cost Estimating Tool 后勤费用估算工具

LCF Launch Control Facilities 发射控制设施设备

LCF Launch Control Facility 发射控制设备

LCFG Low Cost Flight Guidance 低成本飞行制导

LCFS Launch Control Facility Simulator 发射控制设备模拟器；发射控制设施模拟器

LCG Launch Control Group 发射控制组

LCG Liquid Cooling Garment 液体冷却服

LCGA Low Concentration Gas Analyzer 低浓度气体分析仪

LCGR Launch Control Group Replacement 发射控制组替换

LCHST Launching Site 发射场

LCI Launch Complex Instrumentation 发射场仪器设备

LCI Launch-Centered Inertial 发射中心惯性系统

LCI Launcher Control Indicator 发射器控制指示器；发射架控制指示器

LCLM Low-Cost Lightweight Missile 低成本轻型导弹

LCLS Low Cost Launch System 低成本发射系统

LCLU Landing Control and Logic Units 着陆控制与逻辑装置

LCLV Low Cost Launch Vehicle 低成本运载火箭

LCM Land Combat Missile 地面作战导弹

LCM Launch Confirmation Message

发射确认通报（靶场间作业发射前1天通报）
LCM Launch Control Monitor 发射控制监视器
LCM Life Cycle Maintenance 寿命周期维修
LCM Life Cycle Management 寿命周期管理；全寿命管理
LCM Life Cycle Model 寿命周期模型；寿命周期模式
LCM Lightweight Communications Module 轻型通信模块
LCMIMS Logistics Capability Assessment Models Information Management System 后勤能力评估模型信息管理系统
LCMM Life Cycle System Management Model 寿命周期系统管理模型
LCMR Light Weight Counter-Mortar Radar 轻型反迫击炮雷达
LCMS Launch Control and Monitoring System 发射控制与监视系统
LCMS Low Cost Missile System 低成本导弹系统
LCMS Low Cost Modular Spacecraft 低成本模件式航天器
LCN Launch Commit Criteria Change Notice 发射放行标准更改通知
LCNT Link Celestial Navigation Trainer 林克式天文导航训练器
LCO Launch Control Operation 发射控制操作
LCOC Launch Control Officer's Console 发射控制官操控台
LCOLNT Low Coolant 低温冷却剂
LCOM Logic Control Output Module 逻辑控制输出模块
LCOM Logistics Composite Model 后勤保障复合模型

LCOS Launch Checkout Stations 发射检查台
LCOSE Launch Complex Operational Support Equipment 发射场操作支撑设备
LCOT Laser Station Load Terminals 激光通信站载终端
LCP Launch Control Post 发射控制站；发射指挥所
LCQ Launch Crew Quarters 发射组人员住所
LCR Launch Control Room 发射控制室（间）
LCR Level Controller and Recorder 液面控制器与记录仪
LCR Line Control Register 线路控制寄存器
LCRM Launch Control Room 发射控制室
LCROSS Lunar Crater Observation and Sensing Satellite 月球陨石坑观测与传感卫星
LCRU Lunar Communication Relay Unit 月球通信中继装置
LCS Landing Control System 着陆控制系统
LCS Laser Communications System (Subsystem) 激光通信系统（分系统）
LCS Laser Crosslink System 【美国】（国防支援计划）激光交叉链接系统
LCS Launch Complex Set 发射场综合设施；发射场全套装置
LCS Launch Control Sequence 发射控制程序
LCS Launch Control Simulator 发射控制模拟器
LCS Launch Control Station 发射控制站

LCS Launch Control System 发射控制系统

LCS Lincoln Calibration Sphere 【美国】林肯校准球卫星；林肯校准用球体；林肯校准天体仪

LCS Longitudinal Control System 纵向控制系统

LCS Loral Communications System 【美国】洛拉尔公司通信系统

LCS Lunar Communication System 月球通信系统

LCSB Launch Control Support Building 发射控制保障大楼（厂房）

LCSCU Launch Coolant System Control Unit 发射冷却剂系统控制装置

LCSE Laser Communication Satellite Experiment 激光通信卫星实验

LCSE Life Cycle Software Engineering 全寿命软件工程

LCSEC Life Cycle Software Engineering Center 全寿命软件工程中心

LCSI Launch Critical Support Item 发射关键保障项目

LCSMM Life Cycle System Management Model 全寿命系统管理模型

LCSS Laser Communication Spacecraft (Satellite)System 激光通信航天器（卫星）系统

LCSS Launch Control and Sequencer System 发射控制与定序系统

LCSS Launch Control System Simulator 发射控制系统模拟器

LCSS Life Cycle Software Support 寿命周期软件支持

LCSSC Life Cycle Software Support Center 全寿命软件支持中心

LCT Laser Communications Terminal 激光通信终端

LCT Launch Control Trailer 发射控制拖车

LCT Launch Countdown 发射倒计时

LCT Lunar Cycle Test 登月循环测试

LCTA Landing Condition Trend Analysis 地面条件趋势分析

LCTB Launch Control Training Building 发射控制训练大楼

LCTI Large Component Test Installation 大型组件试验设施

LCTL Large Component Test Loop 大型部件测试回路

LCTP Launcher Control Test Panel 发射器控制操作台；发射器控制试验板

LCTSU Launch Control Transfer Switching Unit 发射控制转换开关装置

LCTT Launch Complex Telemetry Trailer 发射场遥测拖车

LCU Launch Correlation Unit 发射关联装置

LCU Launcher Control Unit 发射器控制装置；发射架控制装置

LCU Line Coupling Unit 线路耦合装置

LCU Load Control Unit 【全球定位系统】负载控制装置

LCVG Liquid Cooling and Ventilation Garment 流体冷却与通风服

LCWF Launch Complex Work Flow 发射场工作流程图；全套发射设备工作流程图

LCXT Large Cosmic X-Ray Telescope 大型宇宙X射线望远镜

LD Landing Distance 着陆距离

LD LEM (Lunar Excursion Module) Docking 登月舱对接

LDA Landing Distance Available 着

陆滑行可用距离；可用着陆滑跑距离

LDA Large Deployable Particle Detector Array 【美国】大型展开式粒子阵列探测器

LDA Launch Danger Area 发射危险区

LDACS Liquid Divert and Attitude Control System 液体转向与姿态控制系统

LDAM Local Damage Assessment Model 局部毁伤评估模型；局部损坏评估模型

LDAPS Long Duration Auxiliary Power System 长期使用辅助动力系统

LDAR Laser Detection and Ranging 激光探测与测距

LDAR Laser Doppler and Radar 激光多普勒仪与雷达

LDAR Lightning Detection and Ranging 闪电探测与测距

LDARS Laser Detection and Ranging System 激光探测与测距系统

LDAS Laser Detector and Analysis System 激光探测器与分析系统

LDAS Logic Design Automation System 逻辑设计自动化系统

LDASE Large Deployable Antenna Shuttle Experiment 【美国】航天飞机大型展开式天线实验

LDB Launch Data Bus 发射数据母线

LDB Lighting Distribution Box 照明配电箱

LDB Logistics Data Bank 逻辑数据库

LDBLC Low Drag Boundary Layer Control 低阻力边界层控制

LDC Launch Detection Center 发射检查中心

LDC Limited Defensive Capability 有限防御能力

LDCM Landsat Data Continuity Mission 地球资源探测卫星数据连续性任务

LDD Lunar Dust Detector 月球尘埃探测器

LDDS Low Density Data System 低密度数据系统

LDE Lifting Devices and Equipment 提升装置与设备

LDEC Lunar Docking Events Controller 月球对接活动控制器

LDEF Long Duration Exposure Facility 【美国国家航空航天局】长期暴露设施

LDFS Landing Direction Finding Station 着陆定向台

LDG GEAR Landing Gear Control 起落架操纵；起落架控制

LDGE Lunar Excursion Module Dummy Guidance Equipment 登月舱模拟制导设备

LDGPL Landing Place 着陆地点；降落点

LDGPS Local Differential GPS 本地差分全球定位系统

LDI Landing Direction Indicator 着陆方向指示器

LDI Life Detection Instrument 生命探测仪

LDI Local Data Interface 本地数据接口

LDI Long Dwell Imager (Satellite) 长驻留成像器（卫星）

LDIU Launch Data Interface Unit 发射数据接口装置

LDMS Lunar Distance Measuring System 月球距离测量系统

LDNA Long-Distance Navigation Aid

远程导航（辅助）设备
LDNS Laser Doppler Navigation System 激光多普勒导航系统
LDNS Lightweight Doppler Navigation System 轻型多普勒导航系统
LDOC Long Distance Operational Control 远距离操作控制
LDOS Long Duration Orbital Simulator 长期轨道模拟器
LDR Landing Distance Required 着陆距离要求
LDR Large Deployable Reflector 【美国】大型展开式反射望远镜
LDRS LEM Data Reduction System 登月舱数据压缩系统
LDRS Lighting Detection and Ranging System 照明探测与测距系统
LDS Landing, Deservicing, and Safing 着陆、停止工作并解保
LDS Landing/Deceleration Subsystem 着陆/减速分系统
LDS Laser Docking Sensor 激光对接感应装置
LDS Launch Data System 发射数据系统
LDS Launch Detection Satellite 发射探测卫星（独联体用于提供美国洲际弹道导弹发射预警）
LDS Layered Defense System 分层配置防御系统
LDS Lunar Drill System 月球钻探系统
LDSFD Laser Docking Sensor Flight Demonstration 【美国国家航空航天局】激光对接传感器飞行演示验证
LDTM Lander Dynamic Test Model 着陆舱动态测试模型
LE Launch Effectiveness 发射效率；发射有效性；发射效果

LE Launch Emplacement 发射掩体
LE Launch Escape 发射逃逸
LE Launch(ing) Equipment 发射设备；发射装置
LE Launcher Electronics 发射器电子设备
LE Limit of Error 误差极限；误差范围
LE&S Logistics Engineering and Support 后勤工程与保障
LEA Launch Enable Alarm 发射启动警报
LEA Launch, Entry, and Abort 发射、进入与中止
LEA Launch Escape Assembly 发射逃逸设备
LEA Launcher Electronics Assembly 发射器电子设备
LEA Linear Error Analysis 线性错误分析
LEA Logistics Engineering Analysis 后勤工程分析
LEA SAT Leased Communications Satellite 租赁的通信卫星
LEAE Low Energy Astrophysics Explorer 低能天体物理探测器
LEAG Lunar Exploration and Analysis Group 月球探测与分析组
LEAM Lunar Ejecta and Meteorites 月球喷出物与陨石
LEAP Liftoff Elevation and Azimuth Programmer 弹射高度与方位程序装置
LEAP Lightweight Exo-Atmospheric Projectile 【弹道导弹防御组织】轻型外大气层拦截弹
LEAP Lunar Escape Ambulance Pack 月球逃逸救护包
LEAPFROG Air Force Program for Testing Satellite Communications

【美国】空军卫星通信试验计划
LEAR Logistics Evaluation and Review 后勤鉴定与评审
LEART Logistics Evaluation and Review Techniques 后勤鉴定与评审技术
LEASAT Leased Satellite 【美国军用通信卫星】租赁卫星
LEB Lower Equipment Bay 【航天器】下部设备舱
LEC Launch Escape Control 发射逃逸控制
LEC Lunar Equipment Conveyor 月球设备传送器
LECA Launch Escape Control Area 发射逃逸控制区域
LECOS Lunar Environment Construction and Operations Simulator 月球环境构建与操作模拟器
LECS Launching Equipment Checkout Set 发射设备检测装置
LECS Liquid Effluent Control System 废液控制系统
LED Light-Emitting Diode 发光二极管
LED Low Energy Detector 低能探测器
LEDI Low Endoatmospheric Defense Interceptor 大气层内低空防御拦截机（导弹）
LEDS Link Eleven Display System 【美国】11号链路显示系统
LEDS Low Endoatmospheric Defense System 大气层内低空防御系统
LEED Leadership in Energy and Environmental Design 能源与环境设计的领先地位
LEF Leading Edge Flap 前缘襟翼
LEFAS Leading Edge Flap Actuation System 前缘襟翼作动系统

LEGACI Life Protection, Explore the navigation Algorithm and Consumables for Answering Machine 生命保障，探索导航算法和消耗品询问应答机
LEGM Low Energy Gamma Monitor 低能伽马射线监测器
LEGOS Laboratoire d'Etudes en Géophysique et Océanographie Spatiales 【法国】地球物理及海洋空间研究实验室
LEI Low Endoatmospheric Interceptor 大气层内低空拦截机（导弹）
LEIS Low Energy Ion Scattering 低能离子散射
LEL Low Energy Laser 低能激光
LEL Lower Explosive Limit 低爆极限；爆炸下限
LELTS Light Weight Electronic Locating and Tracking System 轻型定位与跟踪电子系统
LELU Launch Enable Logic Unit 发射启用逻辑装置
LELWS Low Energy Laser Weapon System 低能激光武器系统
LEM Laboratory Environment Model 实验室环境模型
LEM Launch Enclosure Maintenance 发射隔离设施的维护保养
LEM Launch Escape Monitor 发射逃逸（救生）监视器
LEM Launch Escape Motor 发射逃逸发动机
LEM Launcher Electronics Module 发射器电子设备模块
LEM Local Exponential Model 【大气密度】局部指数模型
LEM Lunar Excursion Module (or Lunar Module) 月球旅行舱（又称登月舱）

LEM Lunar Exploration Module 月球探测舱

LEMDE Lunar Excursion Module Descent Engine 登月舱下降发动机

LEMDRS Lunar Excursion Module Data Reduction System 登月舱数据压缩系统

LEMF Leading-Edge Maneuver Flap 前缘机动襟翼

LEMGC Lunar Excursion Module Guidance Computer 登月舱制导计算机

LEMT Lunar Excursion Module Track 登月舱轨道

LEMTV Lunar Excursion Module Television 登月舱电视

LEMTV Lunar Excursion Module Test Vehicle 登月舱试验车

LEND Lunar Exploration Neutron Detector 月球探索中子探测器

LENPD Apollo Low Energy Nuclear Particles Detector 【美国】"阿波罗"低能核粒子探测器

LENR Low-Energy Nuclear Reactions 低能核反应

LENS Lightweight Exhaust Nozzle Structure 轻型尾喷管结构

LEO Launch and Early Orbit 发射与早期轨道

LEO Low Earth Orbit 近地轨道

LEO Lunar Exploration Operations 月球探测作业；月球探测工作

LEOP Launch and Early Operations Phase 发射和早期操作阶段

LEOP Launch and Early Orbit Phase 【对地静止卫星】发射与早期轨道阶段

LEOPCS Launch and Early Orbit Phase Computer System 发射和早期轨道阶段计算机系统

LEOS Low Earth Orbit Satellite 近地轨道卫星

LEOX Low Earth Orbiting Experiment 近地轨道实验

LEP Lunar Exploration Plan 月球探测计划

LEPEPS Lunar Equipment Prime Electrical Power System 登月设备主电子系统

LEPS Launch Escape Propulsion System 发射逃逸推进系统

LEPS Launcher Environmental Protective System 【美国第20航空队】发射场环境保护系统

LER Launch Equipment Room 发射设备室；发射设备间

LER Launch Evaluation Report 发射评估报告

LERC Lewis Research Center 【美国国家航空航天局】刘易斯研究中心

LES Laser Experimental Satellite 激光实验卫星

LES Launch Enabling System 发射保障系统；防止发射失误系统

LES Launch Entry Suit 发射进入（航天）服

LES Launch Environment Simulator 发射环境模拟器

LES Launch Escape System 【美国"阿波罗"飞船】发射逃逸（救生）系统

LES Lincoln Experiment Satellite 林肯实验卫星（美国军用通信试验卫星）

LES Liquid Oxygen Expert System 液氧加注智能系统（美国航天飞机发射前加注液氧时的故障检测系统）

LES LOX Expert System 液氧专家系统

LES Lunar Escape System 月球逃逸系统

LESA Lunar Exploration Stay, Apollo 【美国】"阿波罗"飞船在月球延长停留

LESA Lunar Exploration System for Apollo 【美国】"阿波罗"飞船月球探测系统

LESC Launch Escape System Control 发射逃逸系统控制

LESCS Launch Escape Stabilization and Control System 发射逃逸稳定与控制系统

LESOC Lincoln Experimental Satellite Operation Center 林肯实验卫星操作中心

LESS Launch Escape System Simulator 发射逃逸系统模拟器

LESS Leading Edge Structure Subsystem 【美国航天飞机】前缘结构子系统

LESS Lunar Escape System Simulator 月球救生系统模拟器

LESS Lunar Escape-to-Orbit System Simulation 从月球逃逸到进入轨道的系统模拟

LEST Large Earth Survey Telescope 大型地球测量望远镜

LEST Launch Enable System Turret 保障发射系统塔

LET Large Earth Terminal 大型地面终端

LET Launch and Escape Time 发射与逃逸时间

LET Launch Effects Trainer 发射效应训练器

LET Launch Escape Tower 发射逃逸塔台

LET Lincoln Experimental Terminal 【美国】林肯实验室实验终端

LET Linear Energy Transfer 能量线性转移

LET Live Environmental Testing 实弹（发射）环境试验

LET Loading Evaluation Team 装弹评定小组

LETF Launch Equipment Exposure Facility 发射设备暴露设施

LETF Launch Equipment Test Facility 【美国肯尼迪航天中心】发射设备测试厂房

LETS Launch Equipment Test Set 发射设备测试装置

LETS Lunar Experiment Telemetry System 月球实验遥测系统

LETS LWIR Environment and Threat Simulation 长波红外环境与威胁模拟（仿真）

LEV Logistics Entry Vehicle 【进入大气层的】航天飞行器后勤保障舱

LEV Lunar Escape Vehicle 月球逃逸飞行器

LEVA Lunar Extravehicular Visor Assembly 登月舱外罩组件；月球舱外活动遮光设备

LEZ Lunar Equatorial Zone 月球赤道带

LF Launch Facility 发射装置；发射设备

LF Launch Forward 发射前置端

LF Load Factor 载荷因子

LF Low Frequency 低频率

LFA Landing Fuel Allowance 着陆燃油许可量

LFAF Low Frequency Accelerometer Flutter 低频加速装置颤振

LFAM Low Frequency Accelerometer Modes 低频加速装置模式

LFBB Liquid Fly-Back Booster 液体倒转助推器

LFC Large Format Camera 【航天飞机】大幅面摄影机

LFC Lunar Facsimile Capsule 月球传真通信舱

LFC Lunar Farside Chart 月球背面图

LFCC Lightning Flight Commit Criteria and Associated Definitions 【美国联邦航空总署】雷电飞行放行标准及相关定义

LFCC Lightning Flight Commit Criteria 飞行雷电限制标准

LFD Load Fault Detection Protection 载荷故障检测防护

LFDF Load Frequency Design Factor 载荷频率设计系数

LFEB Launch Facility Equipment Building 发射设施设备厂房

LFF Liquid Fueling Facility 液体加注设施

LFF Load Factor for Flight 飞行载荷因数

LFH Lunar Far Horizon 月球远视距界；月球远地平线

LFNA Launch Facility Not Authenticated 发射设施未经确认

LFNG Launch Facility No Go 发射设施不能使用

LFOP Landing and Ferry Operations Panel 着陆与渡运操作控制台

LFOS Launch and Flight Operation System 发射与飞行操作系统

LFPL Lewis Flight Propulsion Laboratory 【美国】刘易斯飞行推进实验室

LFR Launch and Flight Reliability 发射与飞行可靠性

LFRED Liquid-Fuelled Ramjet Engine Development 液体燃料冲压（式）喷气发动机的研制

LFRT Launch Facility Radio Test 发射设施无线电测试

LFS Launch Facility Simulator 发射设施模拟器

LFSMS Logistic Force-Structure Management System 后勤保障部队结构管理系统

LFSR Linear Feedback Shift Register 线性反馈移位寄存器

LFSS Launch Facility Security System 发射设备安全系统

LFT Launch Facility Trainer 发射设施教练员；发射设施训练器

LFT Live Fire Test 实弹发射试验；实弹射击试验

LFT&E Live-Fire Test and Evaluation 实弹测试与评估；实弹射击试验与评估

LFTE Live-Fire Test and Evaluation 实弹测试与评估；实弹射击试验与评估

LFU Lunar Flying Unit 【美国国家航空航天局】月球飞行装置

LFV Lunar Exploration Flying Vehicle 月球探测飞行器

LFV Lunar Flying Vehicle 月球飞行舱；月球飞行器

LG Landing Ground 着陆场；飞机场；登陆场

LGA Low Gain Antenna 低增益天线

LGAS Low-Gain Accelerometer System 低增益加速装置系统

LGAS Lunar Get-Away-Special 【美国航天飞机】月球探测器搭载容器

LGC Lunar Excursion Module Guidance Computer 登月舱制导计算机

LGC Lunar Geological Camera 月球地质摄像机

LGCIU Landing Gear Control and In-

terface Unit 起落架控制与接口组件
LGE Landing Gear Extending Speed 起落架允许速度
LGE Lunar Excursion Module Guidance Equipment 登月舱制导设备
LGE Lunar Geological Equipment 月球地质设备
LGE Lunar Geological Experiment 月球地质实验
LGEC Lunar Geological Exploration Camera 月球地质勘探摄影机
LGGE Laboratoire de Glaciologie et Géophysique de l'Environnement【法国】冰川及地球物理环境实验室
LGI Lunar Geology Investigation 月球地质勘察
LGIU Laser Gyro Interface Unit 激光陀螺仪接口装置
LGM Laser Guided Missile 激光制导导弹
LGM Logistics Module 后勤保障舱
LGM Loop Group Multiplexer 回路群多路转换器
LGO Low Gravity Orbit 低重力轨道
LGO Lunar Geoscience Observer【美国】月球地质科学探测器
LGS Landing Guidance System 着陆引导系统
LGS Lunar Geophysical Surface 月球地质物理表面
LGS Lunar Gravity Simulator 月球重力模拟器
LGSM Light Ground Station Module 轻型地面站模块
LGTAS Low Gain Telemetry Antenna System 低增益遥测天线系统
LGTR Laser Guided Training Round 激光制导训练周期
LH$_2$ Liquid Hydrogen 液氢
LHASA Logic and Heuristics Applied to Synthesis Analysis 用于综合分析的逻辑和直观推断法
LHe Liquid Helium 液氦
LHOX Low and High Pressure Oxygen 低压与高压氧设备
LHP Launcher Handling Procedure 发射器操作程序
LHP Loop Heat Pipe 回路热管
LHS Lightweight Hydraulic System 轻型液压系统
LHS Lunar Horizon Sensor 月球地平线传感器
LHSC Liquid Hydrogen System Complex 液氢系统综合设备
LHT Liquid Hydrogen Tank 液氢贮箱
LHV Liquid Hydrogen Vessel 液氢容器
LHX Load Heat Exchanger 载荷热交换器
LIB Launcher Integration Building 运载火箭总装厂房
LIBS Laser Induced Breakdown Spectrometer 激光诱导击穿分光计
LIBS Laser-Induced Breakdown Spectroscopy 激光诱导击穿光谱学
LIC Lunar Instrument Carrier 登月仪器运输车
LICAT Large IC Automatic Test System 大规模集成电路自动测试系统
LID Lunar Ionosphere Detector 月球电离层探测器
LIDAR Laser Identification and Ranging 光探测与测距；光学雷达；激光雷达
LIDAR Laser Imaging Detection and

Ranging 激光成像检测与测距
LIDAR Light Detection and Ranging 光学检测和测距
LIDEA Laser Identification Experiment Airborne Technology Demonstration 激光识别试验机载技术论证
LIDS Lead System Integrator (LSI) Integrated Distributed Simulation 引线系统积分仪集成分布模拟
LIDS Low Impact Docking System 低撞击对接系统
LIEF Launch Information Exchange Facility 发射信息交换设备
LIF LANDSAT Image Formater 陆地卫星图像分幅器
LIF Low Insertion Force 【人造卫星射入轨道】低射入力
LIFE Laser Infrared Flyout Experiment 激光红外试飞实验
LIFE Linear Integrated Flight Equipment 线性综合飞行设备
LIFESAT Life Sciences Satellite 【美国】生命科学卫星（用于μ重力环境）
LIFLEX Lifting Body Flight Experiment 【日本】升力体试验飞行器
LIL Lunar International Laboratory 国际月球实验室；国际月球研究所
LILT Low Intensity/Low Temperature 低强度/低温
LIMA Long-Range Laser Induced Mass Analysis 远距离激光诱导质量分析
LIME Laser Induced Microwave Emissions 激光引导微波发射
LIMP Lunar Interplanetary Monitoring Probe 月球行星际监视探测器
LIN Liquid Nitrogen 液氮
LINS Laser Inertial Navigation System 激光惯性导航系统
LIOC Limited Initial Operational Capability 有限的初始作战能力
LIP Launch In Process 正在进行发射
LIP Lunar Impact Probe 月球硬着陆探测器
LIPA List of Interchangeable Parts and Assemblies 可互换零件与组件清单
LIPS Laboratory Integration and Prioritization System 实验室集成与优先系统
LIPS Living Plume Shield 【美国】防火箭尾焰卫星
LIRTS Large Infrared Telescope on Spacelab 空间实验室大型红外望远镜
LIS Launch Instant Selector 发射瞬间选择器
LIS Liaison Inter Satellite 卫星间联络
LIS Lightning Imaging Sensor 雷电成像感应装置
LIS Logistics Information System 后勤保障信息系统
LISA Laser Interferometer Space Antenna 激光干涉仪空间天线
LIST Lidar Surface Topography 激光雷达表面地形学
LITAS Low Intensity Two-Color Approach Slope System 低亮度双色进场（进近）下滑道系统
LITE LIDAR In-Space Technology Experiment 【美国国家航空航天局】太空激光雷达技术实验
LITT LEO Little Launch Vehicle for Low-Earth Orbit 【英国】小型近地轨道运载火箭
LITTLE JOE Early Development

Space Rocket 【美国国家航空航天局】早期研制小约瑟式航天火箭

LITVC Liquid Injection Thrust Vector Control 液体喷射推力矢量控制；液体注入推力矢量控制

LIV Lunar and Interplanetary Vehicle 月球和星际间飞行器

LIVE Lunar Impact Vehicle 月球硬着陆飞行器

LIWA Land Information Warfare Activity 地面信息战活动

LJE Liquid Rocket Engine 液体火箭发动机

LJS Liquid Jet System 液体喷气式发动机系统

LKEI Lockheed-Khrunichev-Energia International 国际洛克希德－赫鲁尼切夫－能源公司

LL Launch and Landing 发射与着陆

LL Lunar Landing 月球着陆

LLB Large Liquid Booster 大型液体助推火箭

LLC Logic Link Control 数据链逻辑控制

LLCC Lightning Launch Commit Criteria 航天发射雷电放行标准

LLCCA Long Life Cycle Cost Avionics 长寿命周期费用航空电子设备

LLCCA Low Life Cycle Cost Avionics 低寿命周期费用航空电子设备

LLCD Lunar Laser Communications Demonstration 月球激光通信演示验证

LLCF Launch and Landing Computational Facilities 发射与着陆计算设备

LLFBB Large Liquid Fast-Burn Booster 大型液体速燃助推火箭

LLM Lunar Landing Mission 登月飞行任务；月球着陆任务

LLM Lunar Landing Module 登月舱

LLNL Lawrence Livermore National Laboratory 【美国】劳伦斯·利弗莫尔国家实验室

LLO Low Lunar Orbit 低环月轨道；低月球轨道

LLP Lunar Landing Program 登月计划

LLPS Lightning Location and Protection System 雷电定位与防护系统

LLR Lunar Laser Ranging 月球激光测距

LLRF Lunar Landing Research Facility 登月研究设施

LLRS Laser Lightning Rod System 激光避雷针系统

LLRV Lunar Landing Research Vehicle 【美国】登月研究飞行器

LLS Launch and Landing Site 发射和着陆场

LLS Lightning Locator System 闪电定位系统；雷电定位系统

LLS Lunar Landing Simulator 月球着陆模拟器

LLS Lunar Logistics System 月球后勤系统

LLSAR Logical Local Storage Address Register 逻辑局部存储器的地址寄存器

LLSS Long Lived Space System 使用寿命长的空间系统

LLSV Lunar Logistics Support Vehicle 月球后勤供应飞行器

LLTOW Landing Limiting (Limited) Takeoff Weight 受着陆限制的起飞重量

LLTV Lunar Landing Training Vehicle 登月训练飞行器；月球着陆训练飞行器

LLV Light Launch Vehicle 小型运载火箭
LLV Lockheed Launch Vehicle 洛克希德公司的运载火箭
LLV Lunar Landing Vehicle 月球着陆飞行器
LLV Lunar Launch Vehicle 月球探测（使用的）运载火箭
LLV Lunar Logistics Vehicle 月球后勤保障飞行器
LM Landing Module 登陆舱；着陆舱
LM Launch Mount 发射架
LM Launcher Mechanism 发射器机械装置
LM Long Module 长舱
LM Long Module (NASA) 【美国国家航空航天局】空间实验室乘员舱长舱
LM Lunar Module 登月舱
LM/GES Lockheed Martin/Government Electronic Systems 【美国】洛克希德·马丁公司/政府电子设备系统
LMA Lunar Meteoroid Analyzer 月球流星体分析器
LMA Lunar Module Adapter 登月舱对接器
LMAE Lunar Module Ascent Engine 登月舱上升发动机
LMANS Lockheed Martin Aeronautic and Naval Systems 【美国】洛克希德·马丁公司/航空航天与海军系统
LMC Launch Monitor Console 发射监视器操控台
LMC Launch Monitor Control 发射监视器控制
LMCLS Lockheed Martin Commercial Launch Services 洛克希德·马丁公司商业发射服务
LMCLS Lockheed Martin Commercial Launch System 洛克希德·马丁公司商业发射系统
LMCMS Launch and Mission Control Monitoring System 发射与任务控制监测系统
LMCN Launch Maintenance Coordination Network 发射维修协调网
LMCN Launch Missile Control Network 导弹发射控制网
LMD Lunar Meteoroid Detector 月球流星体探测器
LMD Lunar Meteoroid Detector-Analyzer 月球流星体探测器－分析器
LMDE Lunar Module Descent Engine 登月舱下降发动机
LMDP Logistics Management and Data Processing 后勤管理和数据处理
LME Launch Monitor Equipment 发射监控设备
LME Lunar Module Engine 登月舱发动机
LMES Laboratory for Meteorology and Earth Sciences 【美国国家航空航天局】气象学和地球科学实验研究所
LMF Liquid Methane Fuel 液态甲烷燃料
LMFBR Liquid-Metal Fast-Breeder Reactor 液态金属快中子增殖反应堆
LMI Logistics Management Information 后勤管理信息
LMIS Logistics Management Information System 后勤管理信息系统
LMLF Limit Manoeuvre Load Factor 限制机动飞行载荷因数
LMM Light Microscopy Module 光镜模块

LMMA　Lockheed Martin Missiles and Space　【美国】洛克希德·马丁公司/导弹与航天

LMMS　Land Mobile Missile System　陆地机动导弹系统

LMP　Liquid Monopropellant　液体单组元推进剂

LMP　Lunar Module Mission Programmer　登月舱任务程序设计器

LMP　Lunar Module Pilot　登月舱驾驶员

LMPS　Lunar Module Procedure Simulator　登月舱程序模拟器

LMR　Launch Mission Rules　发射任务规则

LMR　Launch Monitor Room　发射监控室

LMR　Lunar Module Rendezvous　登月舱交会

LMRA　Lunar Module Replaceable Assemblies　登月舱可更换装置

LMRD　Launcher Missile Round Distributor　发射器导弹弹数配置

LMRR　Lunar Module Rendezvous Simulator　登月舱交会模拟器

LMS　Load Measurement System　载荷测量系统

LMS　Lunar Excursion Module Mission Simulator　登月舱任务模拟器

LMS　Lunar Mass Spectrometer　月球质谱仪

LMS　Lunar Module Simulator　登月舱模拟器

LMSP　Logistics Modeling and Simulation Panel　后勤建模与仿真组

LMSS　Lunar Mapping and Survey System　月球测绘系统

LMSSC　Lockheed Martin Space Systems Company　【美国】洛克希德·马丁航天系统公司

LMST　Lightweight Multi-Band Satellite Communications Terminal　轻型多波段卫星通信终端

LMT　Launch Motor Test　发射发动机试验

LN　Liquid Nitrogen　液氮

LNAV　Lateral Navigation　侧向导航

LNE　Land Navigation Equipment　陆上导航设备；地面导航设备

LNG　Liquefied Natural Gas　液化的天然气

LNM　Launch Notification Message　发射通告信息

LNP　Lunar Neutron Probe　月球中子探测器

LNPS　Low Noise Power System　低噪声动力系统

LNS　Land Navigation System　地面导航系统

LNT　Launch Network Test　发射网络测试

LNTS　Liquid Nitrogen Transfer System　液氮转移系统

LNV　Lift Nozzle Vector　升力喷管矢量

LNVT　Launch Network Verification Test　发射网络验证测试

LO　Landing Operations　着陆作业；登陆作战

LO　Launch Operations　发射操作

LO　Launch Operator　发射操作员

LO　Lunar Observer　【美国】月球观测器

LO　Lunar Orbiter　月球轨道器（美国"阿波罗"飞船登月前的无人月球探测器）

LO　Lunar Orbits　月球轨道；环月轨道

LO&SC　Launch Operations and Support Contract　发射操作与保障合同

LO/LOX Liquid Oxygen 液氧
LO/TO Lockout/Tag out 锁定/标识
LOA Landing Operations Area 着陆操作区
LOA Launch on Assessment 评估发射
LOA Launch Operations Area 发射操作区
LOACS Launch Operations Access Control System 发射操作入口控制系统
LOAD Low Altitude Defense 低空防御
LOAF Launch on Anticipated Failure 在预期故障基础上的发射
LOAL Lock-on after Launch 发射后跟踪；发射后锁定
LOAS Lift Off Acquisition System 起飞探测系统；升空探测系统
LOB Launch Operation Branch 发射操作组
LOB Launch Operations Building 发射操作厂房
LOBL Lock-on Before Launch 发射前锁定
LOC Launch Operation Console 发射操控台
LOC Launch Operation Control 发射操作控制
LOC Launch Operations Center 发射操作中心
LOC Launch Operations Complex 发射操作综合设施
LOC Limited Operational Capability 有限操作能力
LoC Line of Control 控制线
LOC Localize Transmitter 无线电着陆信标发信机
LOC Logistics Operations Center 后勤保障操作中心
LOCAAS Low Cost Autonomous Attack System 【美国空军】低成本自主攻击系统
LOCC Launch Operations Control Center 发射操作控制中心
LOCE Large Optical Communications Experiment 大型光学通信实验
LOCINS Low Cost Integrated Navigation System 低成本综合导航系统
LOCS Logic and Control Simulator 逻辑与控制模拟器
LOD Launch on Demand 按需发射
LOD Launch Operations Directive 发射操作指令
LODCS Lunar Orbiter Data Conversion System 月球轨道飞行器数据转换系统
LODE Large Optics Demonstration Experiment 大型光学器件演示验证实验
LODE Laser Optics Demonstration Experiment 激光光学演示验证实验
LODUHFS Low Data-Rate Ultra High Frequency Satellite 低数据率超高频卫星
LOE Low-Earth Orbit Environments 近地轨道环境
LOERO Large Orbiting Earth Resources Observatory 大型地球资源轨道观测台
LOFAR Low Frequency Acquisition and Ranging 低频采集与测距
LOGACS Low Gravity Accelerometer Calibration System 低重力加速度计校准系统
LOGFAC Logistics Feasibility Analysis Capability 【美国空军】后勤保障可行性分析能力
LOGIC Laser Optical Guidance Inte-

gration Concept　激光光学制导综合方案

LOGMR　Logistics Maintenance and Repair　后勤维护与修理

LOGS　Logistics Supportability　后勤保障性

LOGSACS　Logistics Structure and Composition System　后勤结构与组成系统

LOGSAFE　Logistics Sustainability Analysis and Feasibility Estimator　后勤持续能力分析与可行性评估

LOGSIM　Logistics Simulation Model　后勤仿真模型

LOI　Lunar Orbit Insertion　进入奔月轨道；射入月球轨道

LOLA　Lunar Orbit and Landing Approach　月球轨道与着陆方式

LOLA　Lunar Reconnaissance Orbiter Laser Altimeter　月球轨道勘测器激光测高仪

LOLS　Location of Launching Site　发射场定位；发射场位置

LOM　Lunar Orbit Map　月球轨道图

LOM　Lunar Orbit Mission　月球轨道飞行任务

LOMAR　Logistics Maintenance and Repair Satellite　后勤维修卫星

LOMEZ　Low Altitude Missile Engagement Zone　低空导弹交战区

LON　Launch on Need　按需发射

LONGFOG　Long Range Fiber Optic Guided (Missile)　远距离光纤制导（导弹）

LOP　Launch Operator's Panel　发射操作人员操控版

LOP　Lunar Orbit Plane　月球轨道机

LOP　Lunar Orbiting Photography　月球轨道摄影

LOPAC　Load Optimization and Passenger Acceptance Control　载荷最佳数值与乘员验收控制

LOPC　Lunar Orbit Photo Craft　月球轨道摄影飞行器

LOPP　Lunar Orbit Photographic Project　月球轨道摄影计划

LOPS　Lunar Orbital Photographic Spacecraft　月球轨道摄影航天器

LOR　Launch on Remote　远程发射

LOR　Lunar Orbit Rendezvous　月球轨道交会

LORAC　Long-Range Accuracy Navigation System　远程精确（雷达）导航系统

LORAD　Long-Range Active Detection　远程主动探测

LORAD　Long-Range Detection (System)　远程探测（系统）

LORADAC　Long-Rang Active Detection and Communications System　远程有源探测与通信系统

LORADS　Long Range Radar and Display System　远距雷达与显示系统

LORAN　Long Range Aids to Navigation (System)　远程辅助导航（系统）

LORAN　Long Range Electronic Aids to Navigation　远距离电子辅助导航

LORAN　Long Range Navigation　远距离导航

LORBI　Locked-on Radar Bearing Indicator　跟踪雷达方位指示器

LORL　Large Orbital Research Laboratory　大型轨道研究实验室

LOROPS　Long-Range Oblique Optical System　远距倾斜光学系统

LORS　Lunar Excursion Module Optical Rendezvous System　登月舱光学交会系统

LORS Lunar Orbiting Reconnaissance System 月球轨道侦察系统

LORV Low Orbital Reentry Vehicle 低轨道再入飞行器

LOS Land Observation Satellite 【日本】陆地观测卫星

LOS Launch Operation Station 发射操作站；发射操作台

LOS Launch Operation System 发射操作系统

LOS Launch Optional Selector 发射选择器

LOS Lift-Off Simulator 起飞模拟器

LOS Line of Sight 可视线

LOS Local Operating System 局部操作系统；本地操作系统

LOS Loss of Signal 【美国国家航天局】（航天器、卫星等）信号丢失；信号消失

LOS Loss of Spacecraft 航天器失控丢失

LOS Loss of Synchronization 同步消失

LOS Lunar Orbital Station 月球轨道站

LOS Lunar Orbiter Spacecraft 月球轨道航天器

LOS Lunar Orbiting Satellite 月球轨道卫星

LOSM Launch Operating Simulation Model 发射操作模拟模型

LOSS Launch Operations Support Services 发射操作保障服务

LOSS Lunar Orbit Space Station 月球轨道空间站

LOSS Lunar Orbit(al) Survey System 月球轨道勘察系统

LOT Large Orbiting Telescope 【卫星】大型轨道望远镜

LOT Life-of-Type 型号寿命

LOT Lift-Off Time 起飞时间

LOT Lock-on and Tracking 锁定与跟踪

LOTAS Large Optical Tracker Aerospace 【航空航天】大型光学跟踪仪

LOTEX Life-of-Type Extension 延长型号装备寿命

LOTREX Longitudinal Land-Surface Traverse Experiment 陆地表面纵向移动实验

LOTS Launch Operations Television System 发射操作电视系统

LOTS Launch Optical Trajectory System 发射光学弹道系统

LOTS Lunar Excursion Module Optical Tracking System 登月舱光学跟踪系统

LOTV Launch Operations and Test Vehicle 发射作业与试验飞行器

LOV Loss of Visibility 【美国国家航空航天局】失去能见度

LOVER Lunar Orbiting Vehicle for Emergency Rescue 月球轨道紧急救援飞行器

LOVISIM Low-Visibility Landing Simulation 能见度差的条件下着陆模拟

LOW Launch on Warning 警告发射

LOWKATRER Low Weight Kinetic Energy Active Tracker 轻型动能武器主动跟踪器

LOWTRAN Atmospheric and Interstellar Background Signature Model 大气和星际背景信号特征模型

LOX Liquid Oxygen 液氧

LOX/LH Liquid Oxygen/Liquid Hydrogen 液氧和液氢

LOXT Large Orbital X-Ray Telescope 轨道X射线大型望远镜

LP Landing Performance 着陆性能
LP Landing Point 着陆点
LP Launch Package 发射包
LP Launch Pad 发射台
LP Launch Platform 发射平台
LP Launch Point 发射点
LP Liquefied Propane 液化丙烷
LP Liquid Propellant 液体推进剂
LP Load Package 载荷舱
LPA Launcher-Plant Assembly 发射器动力装置
LPAR Large Phased Array Radar 大型相控阵雷达
LPARM Liquid-Propellant Applied Research Motor 液体推进剂应用研究发动机
LPASA Linear Pulse Height Analyzer Spectrum Analysis 线性脉冲振幅分析仪的光谱分析
LPB Liquid Propellant Booster 液体推进剂助推器
LPCA Lunar Pyrotechnic Control Assembly 登月火工品信号控制装置
LPCC Low Pressure Combustion Chamber 低压燃烧室
LPCP Launching Preparation Control Panel 发射准备控制板
LPCVD Low Pressure Chemical Vapor Deposition 低压化学气相淀积
LPD Landing Parachute Demonstration 着陆降落伞演示验证
LPD Landing Point Designator 着陆点指示器
LPD Launch Point Determination 发射点测定
LPD Low Probability of Detection 低发现概率；低探测概率
LPE Launch Point Estimate 发射点估计
LPE Launch Preparation Equipment 发射准备设备
LPEC Launch Preparation Equipment Compartment 发射准备设备舱
LPES Launch Preparation Equipment Set 全套发射准备设备
LPF Large Payload Fairing 大型有效载荷整流罩
LPFTP/AT Low-Pressure Fuel Turbo-Pump/Alternate Turbopump 【美国国家航空航天局】低压燃料涡轮泵/备用涡轮泵
LPGE Lunar Excursion Module Partial Guidance Equipment 登月舱部分制导装置
LPGG Liquid Propellant Gas Generator 液体推进剂气体发生器
LPI Launch(ing) Position Indicator 发射位置指示器
LPI Low Probability of Intercept 低拦截率
LPI Lunar and Planetary Institute 【美国】月球与行星研究所
LPI/D Low Probability of Intercept/Detection 低拦截/探测概率
LPIA Liquid Propellant Information Agency 【美国】液体推进剂信息所
LPL Lunar and Planetary Laboratory 月球与行星实验室
LPL Lunar Projects Laboratory 月球计划实验室
LPLV Large Payload Lift Vehicle 大型有效载荷升力飞行器
LPLWS Launch Pad Lightning Warning System 发射台雷电报警系统
LPM Landing Path Monitor 着陆航迹监视器
LPM Liquid Propulsion Module 【美国】（火箭）液体推进剂舱
LPM Lower Payload Module 下部有

LPM Lunar Payload Module 月球有效载荷舱

LPM Lunar Portable Magnetometer 便携式月球磁强计

LPO Lunar Parking Orbit 月球停泊轨道

LPO Lunar Polar Orbiter 月球极地轨道器

LPOX Low Pressure Oxygen 低压氧

LPP Launch Point Prediction 发射点预测

LPP Launching Preparation Panel 发射准备控制板;发射准备小组

LPP Lean Premixed Prevaporized 贫油预混预蒸发

LPR Liquid Propellant Rocket 液体燃料火箭;液体推进剂火箭

LPRC Launch Pitch Rate Control 发射俯仰速率控制

LPRE Liquid Propellant Rocket Engine 液体推进剂火箭发动机

LPRE Lunar Polar Resource Extractor 月球两级资源提取器

LPS Launch Phase Simulator 发射阶段模拟器

LPS Launch Processing System 发射处理系统;(航天飞机)发射(数据)处理系统

LPS Lightning Protection System 雷电防护系统

LPS Liquid Propulsion System 液体推进系统

LPS/CDS LPS/Central Data Subsystem 【美国肯尼迪航天中心】发射操作分系统/中心数据分系统

LPS-2 Second-Generation Launch Processing System 第二代发射处理系统—航天飞机与空间站人工智能地面处理系统

LPSAC Life and Physical Science Advisory Committee 生命与物理科学咨询委员会

LPSE Launch Platform Support Equipment 发射平台支撑设备

LPSI-A Low Power System Integration – Active 低功率系统集成—主动(有源)

LPSTA Launch Propulsion Systems Technology Area 发射推进系统技术领域

LPT Laser Propulsion Test 激光推进试验

LPT Lightning Protection Tower 避雷塔

LPTV Large Payload Test Vehicle 大型有效载荷试验飞行器

LQAP Laboratory Quality Assurance Program 实验室质量保证计划

LR Landing Radar 着陆雷达

LR Launch Reliability 发射可靠性

LR Liquid Rocket 液体燃料火箭

LR Load Ratio 载重比;载荷比

LR Long Radius 大活动半径

LR Lunar Rendezvous 月球(轨道)交会

LRA Line-Replaceable Assembly 线路可更换组件;外场可更换组件

LRALS Long Range Approach and Landing System 远距离进场(进近)与着陆系统

LRALT Long Range Air Launched Target 远距离空中发射目标

LRB Liquid Rocket Booster 液体火箭助推器

LRBM Long Range Ballistic Missile 远程弹道导弹

LRCS Launch and Recovery Control Station 发射与回收控制站

LRD Launch Readiness Demonstration

发射准备验证
LRD Lightning and Radio Detector 闪电与无线电探测器
LRD Lightning and Radio Emissions Detector 闪电与射电发射探测器
LRDPF Low Rate Data Processing Facility 低速数据处理设备
LRE Launch and Recovery Element 发射与回收设备
LRE Launch and Recovery Equipment 发射与回收设备
LRE Liquid Rocket Engine 液体火箭发动机
LRE Lunar Retrograde Engine 登月反向发动机
LRECM Liquid Rocket Engine Cost Model 液体火箭发动机成本模型
LRES Linear Rocket Engine System 线性火箭发动机系统
LRF Launch Rate Factor 发射率因素
LRF Liquid Rocket Fuel 液体火箭燃料
LRFS Long Range Forecast System 长期预报系统
LRGM Launch Region Gravity Models 发射区重力模型
LRI Lunar Rover Initiative 月球漫游车方案
LRIA Level Removable Instrument Assembly 水平可拆除仪器组装
LRL Lightweight Rocket Launcher 轻型火箭发射器
LRL Lunar Receiving Laboratory 月球采集样品实验室（密闭无菌室，登月返回人员及带回标本均在此室检疫）
LRM Liquid Rocket Engine 液体火箭发动机
LRM Lunar Receiving Laboratory 月球样品采集研究室
LRM Lunar Reconnaissance Mission 月球侦察任务
LRM Lunar Reconnaissance Module 月球探测舱
LRM Lunar Rendezvous Mission 月球（轨道）交会任务
LRML Long Range Missile Launcher 远程导弹发射器
LRMTR Laser Ranger and Marked Target Receiver 激光测距仪和标定目标接收机
LRMTS Laser Range Measurement Target System 激光测距目标系统
LRNIEP Long-Range Navigation Integrated Engineering Program 远程导航综合工程计划
LRNIS Long Range Navigation Inertial System 远程惯性导航系统
LRO Large Radio Observatory 大型无线电观测台
LRO Lunar Reconnaissance Orbiter 月球勘测轨道飞行器
LROC Lunar Reconnaissance Orbiter Camera 月球勘测轨道飞行器照相机
LRP Launch Reference Point 发射参考点；发射基准点
LRPDS Long-Range Position Determining System 远程测位系统
LRPG Long-Range Proving Ground 远程试验靶场
LRPL Liquid Rocket Propulsion Laboratory 【美国陆军】液体火箭推进实验室
LRPS Lunar Radioisotope Power System 月球放射性同位素电源系统
LRR Launch Readiness Report 发射准备状态报告
LRR Launch Readiness Review/Revue

d'aptitude au lancement 发射准备状态评审；发射准备状态检查
LRR Launch Requirements Review 发射需求评审
LRR Long Range Rocket 远程火箭
LRS Launch and Recovery Shelter 发射与回收掩体
LRS Launch and Recovery Site 发射与回收场
LRS&T Long Range Surveillance and Tracking 远距离监视与跟踪
LRSBDS Long Range Standoff Biological Detection System 远程防区外生物探测系统
LRSI Low Temperature Reusable Surface Insulation 低温可再次使用的表面隔热瓦
LRSS Limb Radiance Stratospheric Sounder 临边辐射平流层探测器
LRT Launch, Recover, Transport 发射、回收、转运
LRTIR Long Range Tracking and Instrumentation Radar 远距离跟踪和测量雷达
LRU Line Replaceable Units 线路可更换装置
LRU Lowest Replaceable Unit 最低可更换装置
LRV Launch and Recovery Vehicle 发射与回收飞行器
LRV Launch Readiness Verification 发射准备状态验证
LRV Lunar Rover (Roving)Vehicle 【美国】月行车；月球（漫游）车
LS Landing Site 着陆场
LS Landing Strip 着陆跑道
LS Launch Section 发射部门
LS Launch Sequence 发射程序
LS Launch Service 发射勤务
LS Launch Silo 发射井
LS Launcher Site 发射架位置
LSA Large Space Antenna 大型空间天线
LSA Latent Semantic Analysis 潜在的语义分析
LSA Launch Services Agreement 发射服务协议
LSA Logistics Support Analysis 后勤保障分析
LSADB LSA Data Base 后勤保障分析数据库
LSAM Lunar Surface Access Module 月球表面着陆模块；月球表面入口舱
LSAP Launch Sequence Applications Program 发射序列应用项目
LSAR Logistics Support Analysis Records 后勤保障分析记录
LSAR Logistics Support Analysis Report 后勤保障分析报告
LSAT Large Satellite 大型卫星
LSB Launch Service Building 发射勤务厂房
LSB Leased Spacecraft Bus 租借航天器总线
LSB Lunar Surface Base 月面基地
LSC Landing Schedule 着陆时间表
LSC Large Solar Concentrator 大型太阳能聚集器
LSC Launch Sequence Control 发射程序控制
LSC Launch Support Center 发射保障中心
LSCA Logistics Support Cost Analysis 后勤保障费用分析
LSCAD Lightweight Standoff Chemical Agent Detector 轻型远距离毒剂遥测仪
LSCD Laser Standoff Chemical Detector 激光毒剂遥测仪

LSCE　Launch Sequence and Control Equipment　发射程序与控制设备
LSCI　Launch Critical Support Item　发射关键保障项目
LSCRD　Lunar Surface Cosmic Ray Detector　月面宇宙射线探测仪
LSD　Landing Site Determination　着陆场测定
LSD　Launch Site Determination　发射场测定
LSD　Launch System Data　发射系统数据
LSD　Low Speed Data　低速数据
LSD　Lunar Surface Drill　月面钻探；月面钻机
LSDB　Launch Support Data Base　【火箭】发射保障数据库
LSDS　Large-Scale Dynamical System　大型动力系统
LSDU　Launching Station Diagnostic Unit　发射站诊断装置
LSE　Launch Sequence Equipment　发射程序装置
LSE　Launch Support Equipment　发射保障设备
LSE　Life Science Experiment　生命科学实验
LSE　Life Support Equipment　生命保障设备
LSE　Lifetime Support Engineering　寿命周期保障工程
LSE　Lunar Support Equipment　登月支撑设备
LSE　Lunar Surface Experiment　月面实验
LSECS　Life Support and Environment(al) Control System　生命保障与环境控制系统
LSEP　Lunar Surface Experiment Package　月面实验装置
LSEV　Lunar Surface Exploration Vehicle　月球表面勘探车
LSF　Laboratory Simulation Facility　实验室模拟设施
LSF　Launch Support Facility　发射保障设施
LSFE　Life Sciences Flight Experiment　生命科学飞行实验
LSFR　Launch Site Flow Review　发射场流程评审
LSG　Landing Site Guidance Radar　着陆场制导雷达
LSG　Life Science Glove Box　生命科学手套式操作箱
LSG　Lunar Surface Gravimeter　月球表面重力仪；月面比重计；月面重差计
LSGR　Launch Site Guidance Radar　发射场制导雷达
LSI　Large Scale Integration　大型集成
LSI　Launch Success Indicator　发射成功指示器
LSI　Lead System Integrator　引线系统积分仪
LSI　Lunar Science Institute　月球科学研究所
LSI　Lunar Surface Instrument　月面测量仪表
LSID　Launch Sequence and Interlock Document　发射顺序与内部连接文件
LSIN　Logistics System Information Network　后勤系统信息网
LSITVC　Liquid Secondary Injection Thrust Vector Control　液体二次喷射推力矢量控制
LSL　Life Sciences Laboratory　生命科学实验室
LSLE　Life Sciences Laboratory Experi-

ment 生命科学实验室试验
LSM Large Solid Motor 【火箭】大型固体燃料发动机
LSM Launch Site Maintenance 发射场维护保养
LSM Life Science Module 生命科学舱
LSM Logistics Simulation Model 后勤模拟模型
LSM Lunar Surface Magnetometer 月面磁强计
LSMI Logistics Support Management Information 后勤保障管理信息
LSMIP Launch Status Missile Indicator Panel 导弹发射状态指示台
LSMU Laser Communications Space Measurement Unit 激光通信太空测量装置
LSNLIS Lunar Science Natural Language Information System 月球科学自然语言信息系统
LSO Large Solar Observatory 大型太阳观测台
LSOP Lunar Surface Operations Planning 月面操作计划
LSP Light Space Plane 轻型空天飞机
LSP LM (Lunar Module) Specification 登月舱技术要求
LSP Lunar Seismic Profiling 月球震型断面
LSP Lunar Spectral Photometrics 月球光谱技术
LSP Lunar Surface Probe 月面探测器
LSP Lunar Survey Probe 月球勘测器
LSPBP Large Solid Propellant Booster Program 大型固体推进剂助推器计划

LSPDF Life Science Payloads Development Facility 生命科学有效载荷开发设施
LSPET Lunar Sample Preliminary Examination Team 月球样品初步检查组
LSPSS Low Shock Payload Separation System 低冲击有效载荷分离系统
LSR Laboratory of Space Research 空间研究实验室
LSR Launch Site Recovery 发射场恢复；发射场回收
LSR Launch Support Room 发射保障间
LSR Lunar Surface Rendezvous 月面交会
LSRBM Long-and Short-Range Ballistic Missiles 长、短距弹道导弹
LSRD Large Scale Reusable Demonstrator 大尺寸可重复使用演示器；大型可重复使用演示器
LSRL Lunar Sample Receiving Laboratory 【美国】月球标本接收实验室
LSRM Life Science Research Module 生命科学研究舱
LSRR Launch Site Readiness Review 发射场准备状态评审
LSRS Loral Space and Range Systems 【美国】洛拉尔公司航天与靶场系统
LSRV Lunar Surface Roving Vehicle 月面漫游车
LSS Landing, Separation Simulator 着陆、分离模拟器
LSS Large Space Simulator 大型空间模拟器
LSS Large Space Structure 【美国国家航空航天局】大型航天结构
LSS Launch Sequence Simulator 发

射程序模拟器
LSS Launch Service Structure 发射勤务塔
LSS Launch Site Support 发射场保障
LSS Launch Status Summarizer 发射状态总计器
LSS Launch Support Section 发射保障组
LSS Launch Support Services 发射保障服务
LSS Launch Support Shelter 发射保障掩体
LSS Launch Support System 发射保障系统
LSS Life Support System (Subsystem) 生命保障系统（分系统）
LSS Lightning Sensor System 雷电传感系统
LSS Load Sensing System 载荷感应系统
LSS Logistic Support System 后勤保障系统
LSS Lunar Surveying System 月球勘测系统
LSSA Launch Site Safety Assessments 发射场安全评估
LSSC Lincoln Space Surveillance Complex 【美国】林肯实验室空间监视综合设施
LSSCV Large Scale Structure Control Verification 大尺寸结构件控制验证
LSSD Lunar Surface Sampling Device 月面取样设备
LSSE Life Support System Evaluator 生命保障系统鉴定器
LSSF Life Sciences Support Facility 生命科学保障设施
LSSI Logistics Support Systems and Integration 后勤保障系统和集成

LSSIP Launch Site Systems Integration Plan 发射场系统综合计划
LSSL Life Sciences Space Laboratory 生命科学空间实验室
LSSM Lunar Surface Scientific Module 月面科研舱
LSSO Launch Site System Operations 发射场系统操作
LSSP Launch Site Support Plan 发射场保障计划
LSSP Lunar Surveying System Program 月球勘测系统计划
LSSR Landing Site Support Review 着陆场保障评审
LSSR Life Science and Space Research 生命科学与空间研究
LSSRC Life Sciences Shuttle Research Centrifuge 航天飞机生命科学研究使用的离心机
LSSS Lightweight SHF SATCOM System 轻型超高频卫星通信系统
LSST Large Space System Technology 大型空间系统技术
LSST Launch Site Support Team 发射场保障组
LST Large Space Telescope 大型空间望远镜
LST Large Stellar Telescope 大型恒星望远镜
LST Large Structure Technology Group 【美国】大型结构技术组
LST Laser Spot Tracker 激光光斑跟踪器（装置）
LST Life Support Technologies 生命保障技术
LST Liquid Storage Tank 液体（燃料）贮箱
LST Lunar Surface Telescope 月面望远镜
LST Lunar Surface Transponder 月

球表面应答机

LSTD Lunar Satellite Tracking Data 月球卫星跟踪数据

LSTE Large Structures Technology Experiment 大型结构技术实验

LSTE Launch Site Transportation Equipment 发射场运输设备

LSTP Linear Surveillance and Tracking Processor 线性监视跟踪处理器

LSTR Launch System Test Rack 发射系统试验架

LSTS Launcher Station Test Site 发射车测试场（区）

LSTS Lunar Surface Thermal Simulator 月面温度模拟器

LSU Life Support Umbilical 生命保障（供氧）脐带线缆

LSU Life Support Unit 生命保障装置

LSV Lunar Surface Vehicle 月球表面车

LSWG Life Science Working Group 【欧洲空间局】生命科学工作组

LT Lander Tankers 【美国火星着陆器】着陆器储箱

LT Launch Table 发射台体

LT Launch Tube 发射管；管式发射导向器

LT Launcher Test 发射装置试验

LT Light Terminal (Satellite Ground) 【卫星地面】轻型终端

LT Lunar Trajectories 月球轨道

LTA Launcher Tube Azimuth Datum Line 发射器导管方位基准线

LTA LEM (Lunar Excursion Module) Test Article 登月舱试验项目

LTA Lower Torso Assembly 【航天服】躯干下部装置

LTA Lunar Test Article 月球试验样品

LTAS Launch Trajectory Acquisition System 发射轨迹采集系统

LTC Lunar Topographic Camera 月球地形摄像机

LTCS Location Tracking Control System 定位跟踪控制系统

LTD Laser Target Designator 激光目标指示器

LTD Launch Test Directive 发射试验指令

LTDR Laser Target Designator Rangefinder 激光目标指示器与测距仪

LTDR Laser Target Designator Receiver 激光目标指示器接收机

LTDS Launch Tracking Data System 发射跟踪数据系统

LTDS Launch Trajectory Data System 发射弹道数据系统

LTE Long Term Evolution Towards European Manned Spaceflight 【欧洲空间局】欧洲载人航天事业的长远发展研究

LTE Low Thrust Engine 低推力发动机

LTGCC Lightning Cloud-to-Cloud 云间闪电

LTGCG Lightning Cloud-to-Ground 云对地闪电

LTGCW Lightning Cloud-to-Water 云—水面闪电

LTGIC Lightning in Cloud 云中闪电

LTIV Lunar Trajectory Injection Vehicle 射入月球轨道飞行器

LTLA Launcher Tube Longitudinal Axis 发射器导管纵向轴线

LTM LAN Traffic Monitor 局域网传输监视器

LTM Laser Transmitter Module 激

光发射机舱

LTM Low Temperature Moulding 低温固化成形

LTMCC Large-Throat Main Combustion Chamber 大喉道主燃烧室

LTMPF Low Temperature Microgravity Physics Facility 低温微重力物理学设备

LTMR Laser Target Marker and Receiver 激光目标指示器与接收机

LTMS Lunar Terrain Measuring System 月球地形测量系统

LTO Landing-Take Off 着陆－起飞

LTOT Latest Time Over Target 最迟飞越目标上空时间

LTP Laser Technology Program 激光技术项目

LTP LEM (Lunar Excursion Module) Test Procedure 登月舱试验程序

LTR Lander Trajectory Reconstruction 着陆器轨道重显

LTRF Low-Temperature Research Facility 低温研究设施

LTRS Launch Test Range System 发射试验靶场系统

LTS Lateral Test Simulator 横向试验模拟器

LTS Launch Telemetry Station 发射遥测站

LTS Launch Tracking System 发射跟踪系统

LTS Link Terminal Simulator 数据链终端模拟器

LTS Lunar Touchdown System 登月着陆系统

LTT Lunar Test Table 月球试验台

LTTAT Long Tank Thrust Augmented Thor 【美国】长贮箱推力增大的"雷神"火箭

LTTS Logistics Information Technology Strategy 后勤信息技术策略

LTU Lateral Thrust Unit 横向推力装置

LTV Land Transport Vehicle 地面运输车

LTV Launch Test Vehicle 发射试验车；发射试验飞行器

LTV Life Test Vehicle 生命科学试验飞行器

LTV Lunar excursion module Test Vehicle 登月舱试验车；登月舱试验飞行器

LTVC Launcher Tube Vertical Centerline 发射器导管垂直中线

LTVS Lunar Transport Vehicle System 月球运输车系统

LUA Launch Under Attack 进攻性发射

LUCOM Lunar Communication System 月球通信系统

LUCOM Lunar Communications Relay Unit 月球通信中继装置

LULS Lunar Logistic System 登月后勤（保障）系统

LUM Living Utility Module 【航天飞行器】通用起居舱

LUMAS Lunar Mapping System 月球测绘系统

LUNANAUT Lunar Astronaut 登月宇航员

LURE Lunar Ranging Experiment 【美国】月球测距实验

LUS Liquid Upper Stage 【火箭】液体上面级

LUSEX Lunar Surface Explorer 月球表面探测器

LUSI Lunar Surface Inspection 月球表面考察；月面观测

LUT Launcher Umbilical Tower 【火箭】发射脐带塔

LUT Limited User Testing 有限用户测试
LV Landing Vehicle 着陆飞行器；着陆车
LV Launch Vehicle 运载火箭
LV&P Launch Vehicles and Propulsion 运载火箭及推进装置
LV/SC Launch Vehicle/Spacecraft 运载火箭／航天器
LVA Launch Vehicle Adapter 运载火箭适配器
LVA Launch Vehicle Architecture 运载火箭结构
LVA Launch Vehicle Availability 运载火箭可用性
LVAR Launch Vehicle Assessment Report 运载火箭评估报告
LVAR Launch Vehicle Assessment Review 运载火箭评估审查
LVAS Launch Vehicle Alarm System 运载火箭警报系统
LVC Launch Vehicle Cost 运载器成本
LVCM Launch Vehicle Cost Model 运载火箭成本模型
LVCS Launch Vehicle Checkout System 运载火箭检测系统
LVCS Launch Vehicle Coordinate System 运载火箭协调系统
LVDA Launch Vehicle Data Adapter 运载火箭数据转接器
LVDC Launch Vehicle Data Center 运载火箭数据中心
LVDC Launch Vehicle Digital Computer 运载火箭数字计算机
LVDC/DA Launch Vehicle Digital Computer/Data Adapter 运载火箭数字计算机／数据转接器
LVE Launch Vehicle Engine 运载火箭发动机
LVFC Launch Vehicle Flight Control 运载火箭飞行控制
LVFCS Launch Vehicle Flight Control System 运载火箭飞行控制系统
LVG Low Viscosity Gyro 低滞性陀螺仪
LVGC Launch Vehicle Guidance Computer 运载火箭制导计算机
LVHM Launch Vehicle Health Management 运载火箭状况管理
LVID Low Velocity Impact Damage 低速率冲击损伤
LVIS Low Voltage Ignition System 低电压点火系统
LVLS Low Visibility Landing System 低能见度着陆系统
LVNRA Study of Launch Site Processing and Facilities for Future Launch Vehicles 未来运载火箭发射场操作与设施的研究
LVO Launch Vehicle Operations 运载火箭操作
LVOR Low Powered Very High Frequency Omnidirectional Range 低功率甚高频全向信标
LVOS Launch Vehicle Operations Simulator 运载火箭操作模拟器
LVPD Launch Vehicle Pressure Display 运载火箭压力显示
LVPP Launch Vehicle and Propulsion Program 运载火箭与推进计划
LVPS Low Voltage Power Supply 低压供电
LVRS Launch Vehicle Recovery System 运载火箭回收系统
LVRT Launch Vehicle Readiness Test 运载火箭准备状态测试
LVS Launch Vehicle Simulator 运载火箭模拟器
LVS Launch Vehicle System 运载火

箭系统
LVSSTS Launch Vehicle Safety System Test Set 运载火箭安全系统试验装置
LVTC Launch Vehicle Test Conductor 运载火箭试验导体；运载火箭试验导线
LW Landing Weight 着陆重量
LW Launch Window 发射窗；最佳发射时间
LWABTJ Light Weight Afterburning Turbojet 轻型加力燃烧室涡轮喷气发动机
LWC Liquid Water Content 液体水含量
LWD Launch Window Display 发射窗显示（器）；最佳发射时间显示（器）
LWHC Landing Gear Warning Horn Control 起落架告警器控制
LWIR Long Wavelength Infrared 长波红外
LWIR FPA Long Wavelength Infrared Focal Plane Array 长波红外焦平（面）矩阵
LWO Launch Window Open 发射窗口打开

LWS Lightning Warning System 闪电预警系统；雷电告警系统
LWT Launch Weather Team 发射气象小组
LWT Launch Window Time 发射窗口时间
LWT Light Weight Tank 轻型储箱
LWTAS Lightweight Target Acquisition System 小型目标捕获系统
LWTS Liquid Waste Treatment System 废液处理系统
LWW Launch Window Width 发射窗口宽度；最佳发射窗口
LZC Landing Zone Construction 着陆区准备
LZCC Landing Zone Control Center 着陆区控制中心
LZD Launch Zone Display 发射区显示（器）
LZE Luminous Zone Emissivity 发光区辐射率
LZF Launch Zone Flag 发射区标志旗
LZSA Landing Zone Support Area 着陆区（前沿）保障场；着陆地带支援区

M

M&A Maintenance and Assembly 维修与装配

M&A Manufacturing and Assembly 制造与装配

M&C Maintenance and Checkout 维修和检查

M&C Monitor and Control 监测和控制

M&D Maintenance and Diagnostics 维修与诊断

M&E Maintenance and Equipment 维修与设备

M&F Materials and Facilities 器材和设施

M&I Modification and Installation 改装与安装

M&LC Mission and Launch Control 任务与发射控制

M&M Materials and Maintenance 器材和维修

M&O Maintenance and Operations 维修与操作

M&P Materials and Processing 材料与处理

M&R Maintainability and Reliability 维修性与可靠性

M&R Maintenance and Refurbishment 维修和整修

M&R Maintenance and Repair 维护与修理

M&RF Maintenance and Refurbishing Facility 维修和整修设施

M&RO Maintenance and Refurbishment Operations 维修和整修操作

M&S Maintenance and Supply 维护和供给

M&S Materials and Structure 材料与结构件

M&S Modeling and Simulation 模拟与仿真

M&SA Modeling, Simulation and Analysis 模拟、仿真和分析

M&SS Mapping and Survey System 测绘与测量系统

M&T Maintenance and Test 维修与试验

M&T Manufacturing and Test 制造和试验

M&T Monitor and Test 监视与试验

M&TE Measurement and Test Equipment 测量与试验设备

M-FASP Midcourse Fly Along Sensor Package 沿传感器设备的中距飞行

M-M-L-S Model-Modes-Loads-Stresses 模型－模态－载荷－应力

M-SPAREA Multi-Spares Priority and Resource Evaluation Availability 多种备件优先次序与资源评价的有效度

M-T-M Model-Test-Model 模型－测试－模型

M.S.I Maintenance Service Item 维修服务项目

M/A Maintenance Analysis 维护分析

M/DF Mating/Demating Facilities 对接/拆卸设施

M/LWIR Medium/Long Wavelength Infrared 中长波红外

M/M Master Model 主模型

M/OD Meteoroid/Orbital Debris 流星体/轨道残片

M/P Main Parachute 主降落伞

M/R Maximum Range 最大航程；最大射程

M/R Missiles/Rockets 导弹与火箭

M/R Mixture Ratio (Fuel to Oxidizer) 混合比（燃料—氧化剂）

M/S Mainstage 主芯级

M/SCI Mission/Safety Critical Item 任务/安全关键项目

M/TRMS Mobile/Tracked Remote Manipulator System 导轨移动式遥控机械臂系统

M/U Mockup 实体模型；样品机；同实物等大的研究用模型

M0 Maneuver Control System Workstation Class Computer Unit 机动控制系统站级计算机设备

M³V Mobile Medical Monitoring Vehicle 移动式医学监测车

MA Maintenance Ability 维修能力

MA Master Alarm 主报警装置

MA Mechanical Assembly 机械装置/机械组装

MA Memory Address 存储器地址

MA Mercury-Atlas 【美国】"水星"号飞船/"宇宙神"运载火箭

MA Mission Abort 执行任务中断

MA Mission Analysis 任务分析

MA Mission Assignment 任务分配

MA Mission Assurance 任务保障

MA Multiple Access 多个访问

MA&P Maintenance Analysis and Planning 维修（维护）分析与计划

MA&PP Mission Analysis and Performance Program 执行任务分析与执行计划

MA&T Manufacturing Assembly and Test 机械组装与测试

MA&T Missile Assembly and Test 导弹组装与试验

MA/ML Missile Alert/Missile Launch 导弹警告/导弹发射

MAA Maximum Authorized Altitude 最大授权高度

MAA Mechanical Arm Assembly 机械臂装置

MAA Minimum Approach Altitude 最低进场（进近）高度

MAA Minimum Approach Angle 最小进场角度

MAA Missile Assembly Area 导弹装配区；导弹组装区

MAA Mission-Area Analysis 任务区域分析

MAADP Mission Area Analysis Deficiency Program 任务范围分析缺陷弥补方案

MAADP Mission Area Analysis Development Plan 任务区域分析发展计划

MAATE Multiple Application Automatic Test Equipment 多用途自动测试设备

MAB Master Acquisition Bus 主控采集总线

MAB Minimum Altitude Bombing 最小高度投弹

MAB Missile Assembly Building 导弹装配厂房

MABES Magnet Bearing Flywheel Experimental System 磁轴承飞轮试验系统卫星（日本军事通信技术试验卫星）

MABLE Miniature Automatics Base Line Equipment 小型自动控制基线设备

MAC Main Display Console 主显示控制台
MAC Maintenance Allocation Chart 维修项目分配图
MAC Maintenance Analysis Center 【美国联邦航空总署】维修分析中心
MAC Maximum Allowable Concentration 最大允许浓度
MAC Maximum Concentration of Organics 最大有机物浓度
MAC Measurement and Analysis Center 测量分析中心
MAC Measurement and Control 测量与控制
MAC Mechanical Analogue Computer 机械模拟计算机
MAC Missile Analysis Center 导弹分析中心
MAC Monitor and Control 监视与控制
MAC Monitor, Alarm and Control 监视、告警与控制
MAC Monitoring and Checking 监控与检查
MAC Multiaccess Computer 多路访问计算机
MAC/SM Maintenance Allocation Chart/System Maintenance 维修配置表/系统维修
MACA Materials and Components Aging 器材和部件老化
MACA Missile Assembly and Checkout Area 导弹装配与检测区
MACB Missile Assembly Control Building 导弹装配控制大楼；导弹装配控制厂房
MACC Multiple Applications Control Center 【美国】多种应用控制中心
MACCK Multi-Application Command and Control Kit 多用途指挥与控制装置
MACDS Monitor and Control Display System 监视与控制显示系统
MACE Maintenance Analysis Checkout Equipment 维修分析检测设备
MACE Mechanical Antenna/Array Control Electronics 机械天线/阵列控制电子设备
MACE Missile and Control Equipment 导弹与控制设备
MACE Multiple Application Core Engine 多用途核心发动机
MACET Modular Algorithm Concept Evaluation Tool 模块化运算概念评估工具
MACF Missile Assembly and Checkout Facility 导弹组装与检查设施
MACH Modular Avionics Component Hardware 模块化电子设备硬件
MACI Military Adaptation of Commercial Items 商用产品的军用改型
MACI Monitor, Access, and Control Interface 监视、访问与控制界面
MACL Minimum Acceptable Compliance Level 最低可接受合格标准
MACMIS Maintenance and Construction Management Information System 维护和工程管理信息系统
MACMT Mean Active Corrective Maintenance Time 平均实际修复性维修时间
MACO Major Assembly Checkout 主要装配检测
MACOM Maintenance Assembly and Checkout Model 维修装置与检测模型
MACRP Maintenance Monitor and Control Representation 维修监视器与控制显示
MACS Medium Altitude Communica-

tion Satellite　中高度通信卫星
MACS　Merchant Airship Cargo Satellite　商用飞艇运载卫星
MACS　Missile Air Conditioning System　导弹空气调节系统
MACS　Modular Attitude Control System (Subsystem)　模块式姿态控制系统（分系统）
MACS　Momentum and Attitude Control System　动量和姿态控制系统
MACS　Monitor and Control Station　监测与控制站
MACS　Monitor and Control System　监测与控制系统
MACS　Multiple Applications Control System　多用途控制系统
MACS　Multiproject Automated Control System　多元自控系统
MACS　Multipurpose Acquisition and Control System　多用途目标捕捉与控制系统
MACS　Multipurpose Application and Control System　通用采集和控制系统
MACSAT　Multiple Access Communications Satellite　多址通信卫星（美国小型军用通信卫星）
MACSS　Medium Altitude Communication Satellite System　中高度通信卫星系统
MAD　Magnetic Anomaly Detector　磁场异常探测器（装置）
MAD　Maintenance Analysis Data　维修分析数据
MAD　Maintenance, Assembly and Disassembly　维修、装配与拆卸
MAD　Mathematical Analysis of Downtime　故障时间数学分析；停工期数学分析
MAD　Missile Assembly Data　导弹装配数据
MAD　Mission Area Deficiency　执行任务地域不足
MAD　Motor Assembly and Disassembly　发动机装配与拆卸
MAD　Mutual Assured Destruction　相互确保摧毁（战略）
MADAM　Multi-Purpose Automatic Data Analysis Machine　多用途自动数据分析机
MADAPS　Management Data Processing System　管理数据处理系统
MADAR　Maintenance Analysis, Detection and Recorder　维修分析、检测与记录器
MADAR　Maintenance Analysis, Detection and Recording　维修分析、检测与记录
MADAR　Malfunction Analysis Detection and Recording　故障分析检测与记录
MADARS　Malfunction Analysis Detection and Recording System　故障分析检测和记录系统
MADC　Multiplexer Analog to Digital Converter　多路模拟－数字转换器
MADCAP　Mobilization and Deployment Capability Assurance Project　机动与部署能力保证计划
MADCAP　Mosaic Array Data Compression and Analysis Program　镶嵌阵列数据压缩与分析项目
MADCAP　Mosaic Array Data Compression and Processing　镶嵌阵列数据压缩与处理
MADDAM　Micromodule and Digital Differential Analyzer Machine　微型组件与数字微分分析机
MADP　Material Acquisition Decision Process　装备（器材）采办决策过程

MADRE Manufacturing Data Retrieval System 制造数据检索系统

MADREC Malfunction Detection and Recording System 故障检测与记录系统

MADRS Maintenance Analysis, Detection and Reporting System 维修分析、检测与报告系统

MADS Air Force Missile Attitude Determination System Using Laser 【美国空军】导弹姿态激光测定系统

MADS Mars Atmosphere Density Sensor 火星大气密度传感器

MADS Missile Attitude Determination System 导弹姿态测定系统

MADS Modeling and Analysis Data Set 建模与分析数据集

MADS Modified Air Defense System 改进的空中防御系统

MADS Modular Auxiliary Data Systems 模块辅助数据系统

MAE Mean Absolute Error 平均绝对误差

MAE Mean Area of Effectiveness 平均有效面积

MAE Missile Assembly Equipment 导弹装配设备

MAEB Mean Area of Effectiveness for Blast 平均面积爆炸效应

MAECHAM/CHEM Middle Atmosphere European Centre Hamburg Model with Chemistry 【欧洲】汉堡中层大气层化学模型中心

MAEP Minimum Autoland Entry Point 最小自动落地进入点

MAESTRO Mission Analysis Evaluation and Space Trajectory Operations 执行任务分析评估与太空轨道操作

MAESTRO Multiple Autonomous Experimental Spacecraft for Telecommunications, Recording and Observations 通信、记录与观测用多功能自主式实验卫星

MAF Michoud Assembly Facility 米休德组装厂房

MAF Missile Alert Facility 导弹警戒设施

MAF Missile Assembly Facility 导弹装配设施

MAFET Microwave and Analog Front-End Technology Program 微波与模拟前端技术项目

MAGE Magellan Shuttle-Launched Venus Radar Mapper Interplanetary Mission 【美国】航天飞机发射的"麦哲伦"金星雷达制图仪星际飞行任务

MAGE Mechanical (or Mobile) Aerospace Ground Equipment 机械（或移动式）航空航天地面设备

MAGGE Medium Altitude Gravity Gradient Experiment 中高度重力梯度实验

MAGIC Magnetospheric Intercosmos 磁差-国际宇宙卫星

MAGIC Matrix Analysis via Generative and Interpretive Computation 通过生成与解释计算的矩阵分析

MAGIC Multimission Advanced Ground Intelligent Control 多任务先进地面情报控制

MAGION Magnetospheric and Ionospheric Satellite 【捷克】磁层电离层卫星

MAGNOLIA CNES/NASA Project to Measure the Earth's Magnetic Field 【法国航天研究中心/美国国家航空航天局联合制定的】测量地球磁场的计划

MAGR Miniaturized Airborne GPS

Receiver　小型机载全球定位系统接收机
MAGS　Mars Global Surveyor　【美国】"火星全球勘测者"探测器
MAGSAT　Magnetic Field Satellite　【美国】磁场卫星；地磁卫星
MAGSAT　Magnetic Particle Mapping Satellite　磁粒测绘卫星
MAGSAT　Magnetic Satellite　地磁卫星
MAGSAT　Magnetometer Satellite　磁力计卫星
MAGSI　Minimum Altitude at Glide Slope Intersection Inbound　返航下滑线交点的最低高度
MAGSS　Maintenance and Ground Support System　维修与地面保障系统
MAHLI　Mars Hand Lens Imager　火星手持透镜成像仪
MAHRSI　Middle Atmospheric High Resolution Spectrograph Investigation　中层大气高分辨率摄谱仪研究
MAIDS　Management Automated Information Display System　管理信息自动显示系统
MAIDS　Multi-Purpose Automatic Inspection and Diagnostic System　多用途自动检查和诊断系统
MAILS　Multiple Antenna Instrument Landing System　复合天线仪表着陆系统
MAIR　Machine Analysis and Instruction Reentry　机器分析与指令再进入
MAIR　Manufacturing and Inspection Record　生产（制造）与检查记录
MAIS　Major Automated Information System　重大自动化信息系统
MAIS　Mobile Automated Instrumentation Suite　移动式自动化测量成套设备
MAIS　Mobile Automated Instrumentation System　机动的自动测试系统
MAIT　Manufacture, Assembly, Integration and Test　制造、组装、总装和测试
MAJAC　Monitor, Anti-Jam and Control　监控、抗干扰与控制
MAKE　Missile Agility/Kinematic Enhancement　导弹机动/运动性能增强
MAL　Mobile Airlock　移动式气闸；移动式密封（压差隔离，气压过渡）舱
MALLAR　Manned Lunar Landing and Return　载人登月和返回
MALN　Minimum Air Low Noise　最小空中低噪声
MALS　Max Abort Launch System　大型发射后异常中断系统
MALS　Medium Intensity Approach Light System　中等亮度进场（进近）照明系统
MALSR　MALS with Runway Alignment Indicator Lights　有对准跑道指示灯的中等亮度进场照明系统
MALV　Miniature Air-Launched Vehicle　小型空射飞行器
MAM　Maintenance Assistance Modules　维修辅助模块（舱）
MAM　Maintenance Avionics Module　维修航空电子设备舱
MAM　Missile Alarm Monitor　导弹报警监视器
MAMB　Missile Assembly and Maintenance Building　导弹装配和维修间；导弹装配和维修厂房
MAMDT　Mean Active Maintenance Downtime　平均有效维修停用时间

MAMS Microgravity Acceleration Measurement System 微重力加速测量系统
MAMS Military Airspace Management Systems 军事航空航天管理系统
MAMS Missile Altitude Measurement System 导弹高度测量系统
MAMS Missile Assembly and Maintenance Shop 导弹装配与保养车间
MAMS Missile Assistance Maintenance Structure 导弹辅助保养厂房
MAMS Multispectral Atmospheric Mapping Sensor 多光谱大气测绘感应器（装置）
MAN Maintenance Alert Network 维修告警网络
MAN Microwave Aerospace Navigation 微波航空航天导航
MAN OPER Manual Operation 手动操作
MANIAC Mathematic Analyzer, Numerical Integrator and Computer 数字分析器、数值积分仪与计算机
MANS Microcosm Autonomous Navigation System 微型自动导航系统
MANS Missile and NUDET Surveillance 导弹与核爆炸探测系统
MANS Mission Analysis for Missile and NUDET Surveillance 导弹和核爆炸监视任务分析
MANSAT Manned Satellite 载人卫星
ManTech Manufacturing Technology 生产（制造）技术
MANTRAC Manual Angle Tracking Capability 手控角跟踪能力
MANUPACS Manufacturing Planning and Control System 制造计划与控制系统
MAO Manned Apollo Operations 【美国】"阿波罗"飞船载人飞行
MAO Mars Aeronomy Observer 【美国】火星大气物理观测器
MAOC Modular Air Operations Center 模块化空中作战（操作）中心
MAOPR Minimum Acceptable Operational Performance Requirements 最低作战性能要求
MAOT Maximum Allowable Operation Time 最大允许运行时间；最大允许操作时间
MAOT Missile Auxiliary Output Tester 导弹辅助输出检测器
MAP Maintenance Analysis Program 维修分析程序
MAP Malfunction Analysis Procedure 故障分析过程
MAP Manageability, Availability and Performance 可管理性、可用性和（整体）性能
MAP Mars Atmosphere Probe 火星大气探测器
MAP Microgravity Applications Program 微重力应用项目
MAP Microwave Anisotropy Probe 微波各向异性探测器
MAP Middle Atmosphere Programme 中层大气（探测）计划
MAP Missed Approach Point 错失的进场点
MAP Missile and Package Acceptance Test 导弹与组件验收实验
MAP Missile and Package Tester 导弹与组件测试器
MAP Missile Application Propulsion 导弹应用推进
MAP Missile Assembly Procedure 导弹装配程序
MAP Missile Assignment Program 导弹分配计划

MAP Mission Application Program 任务应用项目

MAP Mode and Program 模型与程序

MAP Modular Architecture Processor 模块体系结构处理器（装置）

MAP Modular Avionics Packaging 模块式航空电子组件

MAPG Maximum Allowable Power Gain 最大有效功率增益

MAPL Manufacturing Assembly Parts List 制造装配部件表

MAPP Missile Accident Prevention Program 导弹事故预防计划

MAPS Maintainability Analysis and Prediction System 维修性分析和预计系统

MAPS Major Assembly Performance System 主装配作业系统；主要组件性能系统

MAPS Management Analysis and Planning System 管理分析与规划系统

MAPS Measurement of Air Pollution Satellite 测量大气污染卫星

MAPS Measurement of Air Pollution Sensor 【航天飞机】传感器测量空气污染

MAPS Measurement of Atmospheric Pollution from Satellites 卫星测量大气污染

MAPS Missile Application Propulsion Study 导弹应用推进研究

MAPS Mission Air Purge System 执行任务使用的气体吹除系统

MAPS Modular Azimuth Position System 模块式方位角与定位系统

MAPS Monopropellant Accessory Power Supply 【火箭】单元推进剂辅助电源

MAPS Multi-Satellite Attitude Program System 多卫星姿态程序系统

MAPS/ALPS Multi Aim Point System/Alternative Launch Point System 多瞄准点系统／交替发射点系统

MAPSE Minimal Ada Programming Support Environment 最低限度 Ada 程序设计支持环境

MAPTIS Materials and Processes Technical Information Services 材料与工艺方法技术信息服务

MAR Maintenance and Refurbishment 维修与整修

MAR Malfunction Array Radar 故障阵列天线雷达

MAR Management Assessment Review 管理评估评审

MAR Marginal Age Relief 临界安全使用期限

MAR Missile Active Range 导弹自主飞行距离

MAR Missile Analysis Report 导弹分析报告

MAR Modular Acceptance Review 模块验收评审

MAR Multi-Function of Arrayed Radar 多功能阵列天线雷达；多功能相控阵雷达

MARC-DN Measurement of Atmospheric Radiance Camera-Day-Night 大气辐射率昼夜摄像机测量（航天飞机进行的实验，旨在测试电视摄像机在各种光照条件下对天空、地球外缘及地面目标的摄像能力）

MARCAS(S) Manoeuvring Reentry Control and Ablation Studies (System) 机动再入（大气层）控制与烧蚀研究（系统）

MARCEP Maintainability and Reliability Cost-Effectiveness Program 维修性与可靠性成本效益计划

MARECS Maritime European Communication Satellite 【欧洲空间局】欧洲海事通信卫星

MARENTS Modified Advanced Research Environmental Test Satellite 改进型高级研究环境试验卫星

MARES Muscle Atrophy Research and Training System 肌肉萎缩研究和训练系统

MARGI Methodology for Analyzing Reliability and Maintainability Goals and Investments 可靠性与维修性目标及投资的分析方法

MARGO Manned Apollo Recovery Ground Optics 【美国】"阿波罗"飞船载人回收地面光学（设备）

MARINER Unmanned Interplanetary Spacecraft 【美国国家航空航天局】无人驾驶星际飞船

MARISAT Marine Communications Satellite 【美国】海事通信卫星

MAROTS Marine Orbital Technical Satellite 【欧洲】海洋轨道技术卫星

MAROTS Maritime Orbital Test Satellite 海事轨道试验卫星

MARRS Multiple Application Reusable Rocket System 多应用可重复使用火箭系统

MARS Manned Aerodynamic Reusable Spaceship 可重复使用气动载人飞船

MARS Manned Astronautical Research Station 载人空间航行研究站

MARS Meteorological Automatic Reporting System 自动气象报告系统

MARS Microgravity Advanced Research and Support 微重力先进研究和支持

MARS Microgravity Advanced Research and Support Center 微重力先进研究和支持中心

MARS Microwave Atmospheric Remote Sensor 微波大气遥感器

MARS Modular Airborne Recorder System 模块式机载记录仪系统

MARS Modular Airborne Recording System 模块化星（箭）载记录系统

MARS Multiwarfare Assessment and Research System 多次交战评估与研究系统

MARSAT Maritime Satellite System 海事卫星系统

MARSSS Meteorological and Range Safety Support System 气象和靶场安全保障系统

MARSYAS Marshall System for Aerospace Systems Simulation 【美国】马歇尔航天中心航空航天模拟系统

MART Mean Active Repair Time 平均有效维修时间

MART Missile Automatic Radiation Tester 导弹辐射自动测试仪

MARTE Modeling and Analysis of Real-Time and Embedded (Systems) 实时及植入（系统）的建模和分析（系统）

MARTI Missile Alternative Range Target Instrument 导弹替代远程目标装置

MARV Maneuverable Anti-Radiation Vehicle 机动反辐射飞行器；机动反辐射运载工具

MARV Maneuverable Reentry Vehicle 机动再入飞行器；机动再入弹头；（洲际弹道导弹）分弹头

MARV Maneuvering Reentry Vehicle 机动再入飞行器（火箭、弹头）

MARVIS Mid-Apogee Reentry Vehicle Intercept System 中等弹道高度再入飞行器拦截系统

MAS Mars Approach Sensor 火星接近传感器

MAS Mercury-Atlas Spacecraft 【美国】"水星—宇宙神"号飞船

MAS Minor-Autonomous Satellite 【法国】小型自主卫星

MAS Missile Alignment System 导弹校准系统

MAS Missile Assembly Site 导弹装配场

MAS Missile Auxiliary System 导弹辅助系统

MAS Mobility Assessment Stage 机动性评估等级

MAS Modal Analysis System 模态分析系统

MAS Monitor and Alarm System 监控与报警系统

MASA Modular Avionics System Architecture 模块式航空电子系统结构

MASAMBA Mobile Advanced Support and Maintenance Base 机动的高级保障与维修基地

MASAMD Machine Aids to Surface-to-Air Missile Development 地对空导弹计算机辅助设备的研制

MASCON Mass Concentrations of Dense Material on Lunar Surface 月球表面密致材料的质量浓度

MASCOT Manned Shuttle Comprehensive Optimizations and Targeting 载人航天飞机总体最优与目标确定

MASCOT Modular Approach to Software Construction Operation and Test 模块化软件结构操作测试法

MASD Mobile Air and Space Defense 航空与航天机动防御

MASFM Maintenance and Supply Facility Management 维修与供应设施管理

MASG Missile Auxiliary Signal Generator 导弹辅助信号发生器

MASINT Measurement and Signature Intelligence 测量与特征情报；测量与信号情报

MASIS Mercury Abort Sensing Instrumentation System 【美国】"水星"飞船故障检测仪器系统

MASMP Mission Analysis and Support Management Plan 任务分析与支持管理计划

MASS Manned Activity Scheduling System 【美国国家航空航天局】有人操作的活动调度（程序）系统

MASS Mesoscale Atmospheric Simulation System 中尺度大气模拟系统

MASSDAR Modular Analysis, Speed-up, Sampling and Data Reduction 模块分析、加速、取样和数据简化处理

MASSOP Multi-Automatic System for Simulation and Operational Planning 模拟与操作设计用多路自动化系统

MAST Manned Astronomical Space Telescope 有人控制的天文太空望远镜

MAST Manual Acquisition Satellite Track 手动式卫星跟踪

MAST Measurement and Simulation Technology 测量与模拟技术

MAST Missile Automatic Supply Technique 导弹自动供应技术

MASTIF Multiple-Axis Spin Test Inertia Facility 多轴旋转试验惯性设备

MASTIFF Modular Automated System to Identify Friend from Foe 模

块化自动敌我识别系统

MAT Missile Acceptance Test 导弹验收试验

MAT Missile Adaptor Tester 导弹（发射）架适配器测试

MAT Mission Allowable Temperature 任务可允许的温度

MAT Modification Approval Test 改型鉴定试验

MAT Multiple Actuator Test 多致动器（装置）测试（试验）

MATB Missile Auxiliary Test Bench 导弹辅助试验台

MATC Missile Auxiliary Test Console 导弹辅助试验操作台

MATCH Model of Atmospheric Transport and Chemistry 大气与化学传送模型

MATCO Materials Analysis,Tracking, and Control 材料分析、跟踪和控制

MATCON Microwave Aerospace Terminal Control 航空航天微波终端控制

MATE Materials for Advanced Turbine Engines 先进涡轮发动机材料

MATE MDM Application Test Environment 维修诊断手册应用软件测试环境

MATE Modular AUTODIN Terminal Equipment 模块式自动数字网络终端设备

MATE Modular Automatic Test Equipment 模块式自动测试设备

MATILDE Microwave Analysis Threat Identification and Launch Decision Equipment 微波分析威胁识别与发射判定设备

MATP Missile Auxiliary Test Position 导弹辅助测试阵地

MATPS Machine Aided Technical Processing System 计算机（机器）辅助技术处理系统

MATS Maintenance Analysis Test Set 维修分析测试装置

MATS Missile Auxiliary Test Set 导弹辅助试验装置

MATS Mission Analysis and Telemetry Simulation 任务分析与遥测模拟

MATS Mission Analysis and Trajectory Simulation 任务分析与轨道模拟

MATS Mobile Automatic Test Set 移动式测试装置

MATT Multi-Mission Advanced Tactical Terminal 多任务先进战术终端

MATTS Multiple Airborne Target Trajectory System 多目标轨道测量系统

MATV Multi-Axis Thrust Vectoring 多轴推力矢量

MAU Modular Avionics Units 模块化航空电子装置

MAV Mars Aerocapture Vehicle 火星大气制动飞行器

MAV Mars Ascent Vehicle 火星上升飞行器

MAV Military Aerospace Vehicle 军用航空航天飞行器

MAVES Manned Mars and Venus Exploration Studies 载人火星与金星探测研究

MAVIS Microprocessor-Based Audio-Visual Information System 微处理机声频目视信息系统

MAW Maintainability Analysis Workspace 维修性（工作空间）分析

MAW Missile Approach Warning (System) 导弹接近目标告警（系统）

MAW Missile Attack Warning 导弹攻击告警

MAWA Missile Attack Warning and Assessment 导弹攻击告警与估计
MAWD Mars Atmospheric Water Detector 火星大气水分探测器
MAWP Maximum Allowable Working Pressure 最大允许工作压力
MAWR Missile Approach and Warning Receiver 导弹临近与警报接收器
MAWS Missile Approach Warning System 导弹临近告警系统
MAWS Modular Automated Weather System 模块式自动气象观测系统
MAX Mars Astrobiological Explorer 火星天体生物探索器
MAX CLB Maximum Engine Thrust for Two Engines Climb 双发动机爬升的最大发动机推力
MAX-C Mars Astrobiological Explorer-Cacher 火星天体生物探索器
MaxCMT Max Corrective Maintenance Time 最大修复性维修时间
MaxCMTOMF Max Corrective Maintenance Time for Operational Mission Failure 作战任务故障的最大修复性维修时间
MAXDP Maximum Dynamic Pressure 最大动态压力
MAXI Modular Architecture Exchange of Information 模块结构信息交换
MAXI Monitor of All-Sky X-Ray Image (Japanese external ISS facility) 全天X射线成像监视器 (日本外部国际空间站设施)
MAXIE Magnetospheric Atmospheric X-Ray Imaging Experiment 磁层性大气层X射线成像实验
MaxIIP Maximum Instantaneous Impact Point 最大瞬间撞击点
MAXORD Maximum Ordinate 弹道最高点
MAZH Missile Azimuth Heading 导弹方位航向
MAZO Missile Azimuth Orientation 导弹方位定向
MB Main Bus 主总线
MB Management Baseline 管理基线
MB Manned Base 载人基地
MB Mesa Basic Model Musa 基本执行时间模型
MB Missile Base 导弹基地
MB Missile Body 导弹弹体
MB Multiplexer Buffer 多路缓冲器
MBAR Main Belt Asteroid Rendezvous 【美国】主带小行星交会任务
MBC Master Bus Controller 主总线控制器
MBC Missile to Booster Connector 导弹—助推器连接器
MBCM MUX Bus Communication Module 多路传输总线通信模块
MBCS Motion-Base Crew Station 移动式乘员站
MBD Missile Base Depot 导弹基地补给站
MBE Missile-Borne Equipment 导弹弹载设备；导弹弹上设备
MBF Missile Beacon Filter 导弹信标滤波器
MBGE Missile Borne Guidance Equipment 导弹弹载制导设备
MBGS Missile Borne Guidance Set 导弹上的制导装置
MBGTS Missile Borne Guidance Test Set 弹载制导试验装置
MBI Maintenance Bus Interface 维修总线接口
MBI Multibus Interface 多总线接口
MBIT Maintenance Build in Test 维修自检查

MBL Missile Base Line 导弹基线

MBMCS Model-Based Mission Control System 以模型基础的任务控制系统

MBO Maintenance Based Optimization 基于维修的优化

MBO Management by Objective 目标型管理

MBO Motor Burnout 【导弹】发动机燃料耗尽熄火

MBOL Motor Burn Out Locking 【导弹】发动机熄火锁定

MBRL Multiple Ballistic Rocket Launcher 多联装弹道式火箭发射器

MBRLS Multi-Barrel Rocket Launching System 多管火箭发射系统

MBRV Manoeuvrable Ballistic Reentry Vehicle 机动弹道再入飞行器；弹道导弹分弹头

MBRV Matching Ballistic Reentry Vehicle 匹配的弹道再入飞行器（火箭、弹头）

MBS Missile Bunker Storage 导弹贮藏库

MBSE Model-Based Systems Engineering 基于模拟的系统工程

MBSI Missile Battery Status Indicator 导弹电池状态指示器

MC Mid-Course Correction 弹道中段修正

MC Midcourse Correction Maneuver 中期轨道纠正机动

MC Missile Control 导弹控制

MC Mission Capability 任务能力

MC Mission Completion/Continuation 任务完成/持续性

MC Modular Computer 模块化计算机

MC&C Measurement, Command, and Control 测量、指挥和控制

MC&G Mapping, Charting and Geodesy 测绘、制图与大地测量

MC&W Master Caution and Warning 主预警与告警（系统）

MCA Master Control Assembly 主控制组件

MCA Maximum Crossing Altitude 最大横切高度

MCA Monitoring and Control Assembly 监测和控制组件

MCA Motor Circuit Analysis 发动机电路分析

MCA Motor Control Assembly 发动机控制组件

MCA Multichannel Analyzer 多通道分析器（装置）

MCB Missile Control Box 导弹控制盒

MCBF Mean Cycles Between Failures 平均故障间隔循环（次）数（可靠性水平的指标）

MCC Main Combustion Chamber 主燃烧室

MCC Maintenance Control Center 维护控制中心

MCC Mercury Control Center 【美国】"水星"飞船控制中心

MCC Midcourse Correction 弹道中段修正

MCC Missile Control Center 导弹控制中心

MCC Missile Control Console 导弹控制台

MCC Mission Control Center 【美国约翰逊航天中心】任务控制中心

MCC Mission Control Complex 任务控制设施

MCC Mission Control Console 任务控制台

MCC Motor Control Center 【美国约

翰逊航天中心】发动机控制中心

MCC-DOD Mission Control Center-DOD 【美国国防部】任务控制中心

MCC-H Mission Control Center in Houston 【美国国家航空航天局】休斯敦任务控制中心

MCC-K Mission Control Center-Kennedy 【美国国家航空航天局】肯尼迪航天中心的任务控制中心

MCC-M Mission Control Center-Moscow 【俄罗斯】莫斯科任务管理中心

MCC-NASA Mission Control Center-NASA 【美国国家航空航天局】任务控制中心

MCCC Mission Control and Computing Center 任务控制与计算中心

MCCC Mobile Consolidated Command Center 移动式统一指挥中心

MCCIS Maritime Command and Control Information System 海上指挥与控制信息系统

MCCR Mission Critical Computer Resources 任务关键计算机资源

MCCS Missile Critical Circuit Simulator 导弹标准电路模拟器

MCCS Mission Control Center Simulation (System) 任务管理中心仿真（系统）

MCCS Mobile Command and Control System 移动式指挥与控制系统

MCCS Multifunction Command and Control System 多功能指挥与控制系统

MCCV Multi-Control Configured Vehicle 多路操纵系统飞行器

MCD Magnetic Chip Detector 磁片探测器

MCD Maintainability Computer Demonstration 维修性计算机演示

MCD Minimum Cost Design 最低成本设计

MCD Mission and Communication Display 任务与通信显示器

MCDF MIDIAS (Missile Defense Alarm System) Control and Display Facilities 导弹防御警报系统控制与显示设备

MCDP Maintenance Control and Display Panel 维修控制与显示面板

MCDP Master Control and Display 主控显示台

MCDS Maintenance Control and Display System 维修控制与显示系统

MCDS Management Communications and Data System 管理通信与数据系统

MCDS Multifunction CRT Display System 多功能CRT显示系统

MCDU Multifunction CRT Display Unit 多功能CRT显示装置

MCE Mechanism Control Electronics 机械控制电子设备

MCE Mission Control Element 任务控制要素（组件）

MCE Modular Control Equipment 模块式控制设备

MCELV Modified Current Expendable Launch Vehicle 改进型一次性运载火箭

MCF Main Control Facility 主控设施

MCF Maintenance and Checkout Facility 维修与检查设施

MCF Major Component Failure 主要部件故障（失效）

MCF Master Control Facility 【印度卫星的地面控制设施】主控站

MCF Modular Combustion Facility

模块式燃烧装置
MCG Mobile Command Guidance 移动式指令制导
MCG&I Mapping, Charting, Geopositioning and Imagery 测绘、制图、测地与成像
MCH Mission Critical Hardware 任务关键硬件
MCHFR Minimum Critical Heat Flux Ratio 最小临界热通量比
MCI Maintenance Console Interface 维护控制台接口
MCIB Maintenance Console Interface Bus 维护控制台接口总线
MCIC Maintenance Console Interface Circuit 维护控制台接口电路
MCIC Maintenance Console Interface Controller 维护控制台接口控制器
MCIDAS Man Computer Interactive Data Access System 人机交互式数据访问系统
MCIDAS Man-Computer Interactive Data Analysis System 人机交互数据分析系统
MCIF Man-Computer Interactive Function 人机交互式功能
MCIL Maintenance Console Interface Load 维护控制台接口负荷
MCIS Maintenance Console Information System 维修控制信息系统
MCIS Maintenance Console Interface Software 维修控制台接口软件
MCIU Manipulator Controller Interface Unit 遥控机械臂控制器接口装置
MCIU Master Control and Interface Unit 主控制与接口装置
MCIU Mission Control and Interface Unit 任务控制与接口设备
MCL Mission-Configured Load 任务设定的弹药携带量
MCM Maintenance Control Module 维修控制模块
MCM Manipulation and Control Mechanization 操作与控制机械化
MCM Manned Circumlunar Mission 载人环月飞行任务
MCM Mars Cruise Module 火星巡航舱
MCM Master Control Module 主控制舱；主控制模块
MCM Mean Common Mode 平均指令模式
MCM Missile Control Module 导弹控制舱
MCM Mission Control Module 任务控制舱
MCM Multichip Module 多芯片模块
MCO Mapping and Communication Orbiter 测绘和通信轨道器（美国火星探测计划轨道器之一）
MCO Missile Checkout 导弹检查
MCO Mission Control Operations 任务控制操作
MCOAM Material Control Order Additional Material 补充材料控制指令
MCOP Mission Control Operations Panel 任务控制操作操纵台
MCOT Missile Checkout Trailer 导弹检测车；导弹测试拖车
MCOTEA Marine Corps Operational Test and Evaluation Activity 【美国】海军陆战队作战测试与鉴定部门
MCP Master Computer Program 主计算机程序
MCP Measurements Control Procedure 测量控制规程
MCP Message Control Program 信息控制程序

MCP　Microchannel Plate　微通道板
MCP　Missile Command Post　导弹指挥所
MCP　Missile Control Panel　导弹控制板
MCP　Mission Control Programmer　任务控制程序员（装置）
MCP　Monitoring and Control Panel　监测与控制操纵台
MCP　Multibeam Communication Package　多波束通信装置
MCPF　Modular Containerless Processing Facility　【美国空间站】模块式无容器加工装置
MCR　Mission Capable Rate　能执行任务率
MCR　Mission Completion Rate　任务成功率
MCR　Mission Concept Review　任务设计概念评审
MCR　Mission Control Room　任务控制间
MCS　Maintenance and Checkout Station　维修与检测站
MCS　Management Control System　管理控制系统
MCS　Maneuver Capability Study　机动能力研究
MCS　Maneuver Control System　机动控制系统
MCS　Measurements Calibration System　测量校准系统
MCS　Missile Calibration Station　导弹校准站
MCS　Missile Checkout Set　导弹检测设备
MCS　Missile Checkout Station　导弹检测站
MCS　Missile Commit Sequence　导弹参战程序
MCS　Missile Compensating System　导弹补偿系统
MCS　Missile Control Station　导弹控制站
MCS　Missile Control System　导弹控制系统
MCS　Mission Control System　任务控制系统
MCS　Multipurpose Communications and Signaling　多用途通信与信号设备
MCSE　Missile Control System and Equipment　导弹控制系统与设备
MCSP　Maintenance Control and Statistics Process　维护控制和统计处理
MCSP　Mission Completion Success Probability　执行任务成功概率
MCSS　MATE Control and Support Software　模块化自动测试设备控制和支持软件
MCSS　Midcourse Surveillance System　中程监视系统
MCSS　Military Communication Satellite System　军用通信卫星系统
MCSS　Missile Checkout System Selector　导弹检测系统选择器
MCT　Maximum Climb Thrust　最大爬升推力
MCT　Missile Compensating Tank　导弹补偿燃料箱
MCTE　Mission, Course of Action, Task, and Element Control Directives　任务、行动方案、作业与部队控制指令
MCTP　Missile Control Test Panel　导弹控制试验操控板
MCTR　Missile Control Technology Regime　导弹控制技术制度
MCTV　Man-Carrying Test Vehicle

载人试验飞行器
MCU　Master Control Unit　主控制装置
MCU　Mission Control Unit　任务控制装置
MCV　Mission Capable Vehicle　能执行任务的飞行器（火箭、弹头）
MCVP　Materials Control and Verification Program　材料控制与验证项目
MD　Malfunction Detection　故障探测
MD　Manual Disconnect　人工断开
MD　Missile Defense　导弹防御
MD　Missile Detect　导弹探测
MDA　Main Distribution Assembly　主分配组件
MDA　Maintainability Design Approach　维修性设计方法
MDA　Material Diffusion System　材料扩散装置
MDA　Motorized Door Assembly　配备电机的舱门组件
MDA　Multiple Docking Adapter (Skylab)　【美国"天空实验室"】多用途对接舱
MDAP　Major Defense Acquisition Program　【美国】大型国防采办计划；重要防务采办计划；主要国防采购计划
MDAR　Malfunction Detection Analysis and Recording　故障检测分析与记录
MDAS　Malfunction Detection and Analysis System　故障检测和分析系统
MDAS　Meteorological Data Acquisition System　气象数据采集系统
MDAS　Miniature Data Acquisition System　小型数据采集系统
MDAS　Mission Data Acquisition System　任务数据获取系统
MDAU　Maintenance Data Acquisition Unit　维修数据获取单元（装置）
MDB　Modified Double Base Propellant　改良的双基推进剂
MDB　Multiplex Data Bus　多路传输数据总线
MDBIC　Missile Defense Battle Integration Center　导弹防御作战综合中心
MDC　Main Display Console　主显示控制台
MDC　Maintainability Design Criteria　维修性设计准则
MDC　Maintenance Data Center　维修数据中心
MDC　Maintenance Data Collection　维修数据收集
MDC　Maintenance Diagnostic Computer　维修诊断计算机
MDC　Maintenance Direct Cost　直接维修费用
MDC　Missile Development Center　【美国空军】导弹研制中心
MDCA　Main Distribution Control Assembly　主分配控制组件
MDCA-FLEX　Multi-Purpose Droplet Combustion Apparatus—Flame Extinguishment Experiment　多用途液滴燃烧装置－阻燃灭火实验
MDCHECK　Mission Design Check　飞行任务设计检查程序
MDCI　Multi-Discipline Counterintelligence　多种训练反情报
MDCS　Maintainability Data Collection System　维修性数据收集系统
MDCS　Maintenance Data Collection System　维修数据收集系统
MDCS　Malfunction Display and Control System　故障显示和控制系统
MDCS　Master Digital Command Sys-

tem 主数字指令系统
MDCS Material Data Collection System 材料数据采集系统
MDD Mate/Demate Device 对接/拆卸装置
MDD Mate-Demate Device 轨道器装卸设备
MDDC Missile Defense Data Center 导弹防御数据中心
MDE Mission Dependent Equipment 与任务相关的设备
MDE Modular Design of Electronics 电子设备的模块化设计
MDE Modular Display Electronics 模块显示电子设备
MDF Main Distribution Frame 主要分配框架
MDF Manipulator Development Facility 机械臂研制设施
MDF Minimum Duration Flight 最小持续飞行
MDGW Maximum Design Gross Weight 最大设计总重
MDGW Mission Design Gross Weight 飞行总重量
MDHS Meteorological Data Handling System 气象数据处理系统
MDI Missile Distance Indicator 导弹距离指示器
MDIOC Missile Defense Integrated Operations Center 导弹防御综合作战（操作）中心
MDM Manipulator Deployment Mechanism 机械臂展开机构
MDMT Minimum Design Metal Temperature 金属最小设计温度
MDP Management Development Program 管理研发项目
MDPRF Mimic Display Performance Record File 模拟显示性能记录文件
MDPS Mechanized Data Processing System 机械数据处理系统
MDPS Metric Data Processing System 米制数据处理系统
MDPS Mission Data Planning System 任务数据计划系统
MDR Maintenance Demand Rate 维修需求率
MDR Major Design Review 主要设计评审
MDR Mission Data Reduction 任务数据简化
MDR Mission Data Review 任务数据评审
MDR Mission Definition Review 任务定义评审
MDR Mission Dress Rehearsal 任务合练
MDRD Mission Data Requirements Document 任务数据要求文件
MDRS Manufacturing Data Retrieval System 生产（制造）数据查询系统
MDRS Mission Data Retrieval System 任务数据查询系统
MDS Malfunction Detection System 故障探测系统
MDS Management Data System 管理数据系统
MDS Meteoroid Detection Satellite 流星体探测卫星
MDS Minimum Discernible Signal 最小识别信号
MDS Minimum Discernible System 最小识别系统
MDS Mission Demonstration Satellite 任务演示卫星
MDS Mission Development Simulator 任务开发模拟器

MDS Motion Detection System 动作探测系统

MDSC Missile Defense Scientific and Technical Information Center 导弹防御科学技术信息中心

MDSD Mate/Demate Stiffleg Derrick 对接/拆卸刚性支架转臂起重机

MDSE Missile Defense System Exerciser 导弹防御系统训练器

MDSEC Missile Defense Space Experimentation Center 导弹防御与航天实验中心

MDSF Manipulator Development and Simulation Facility 机械臂研制与模拟设施

MDSF Manipulator Dynamics Simulator Facility 【加拿大】机械臂动力学模拟设施

MDSS Management Decision Support System 管理决策支持系统

MDSS Manual Diagnosis Support System 人工诊断支持系统

MDSTC Missile Defense and Space Technology Center 导弹防御与航天技术中心

MDSV Manned Deep Space Vehicle 载人深空飞行器

MDT Maintenance Demand Time 维修需求时间

MDUC Meteorological Data Utilization Center 【印度】气象数据利用中心

MDW Mars Departure Window 飞离火星窗口

MDW Mass Destruction Weapon 大规模杀伤性武器

ME Main Engine 主发动机

ME Management Engineering 管理工程

ME Manoeuvre Enhancement Mode 机动性能增强模态

ME Meteoroid Environment 陨星群环境

ME Micrometeoroid Explorer 【美国】微流星体"探险者"卫星

ME Mission Effectiveness 任务效能

ME-GAS Main Engine Gimbaled Actuator System 主发动机摆动致动器系统

ME/VA Mission Essential/Vulnerable Area 任务要素（基础）/易损区

MEA Main Electronics Assembly 主电子设备组件

MEA Maintenance Engineering Analysis 维修工程分析

MEA Materials Experiment Assembly 【美国航天飞机】材料实验装置

MEA Membrane Electrode Assembly 膜电极组件

MEA Mission Effectiveness Analysis 任务效能分析

MEADS Maintenance Engineering Analysis Data System 维修工程分析数据系统

MEADS Medium Extended Air-Defence System 中空扩大空防系统

MEAP Maintainability Effectiveness Analysis Program 维修性效益分析程序

MEAR Maintenance Engineering Analysis Request 维修工程分析请求

MEB Main Electronics Box 主电子设备箱

MEB Missile Equipment Building 导弹设备厂房

MEBO Main Engine Burnout 主发动机燃烧中止（熄火）

MEC Main Engine Controller 主发动机控制器（装置）

MEC Manual Emergency Control 应急手控

MEC Materials Experiment Carrier 【美国航天飞机】材料实验台

MEC Mission Events Controller 任务事件控制器（装置）

MECA Main Engine Controller Assembly 主发动机控制器组件

MECA Microscopy, Electrochemistry, and Conductivity Analyzer 显微镜、电化学与传导性分析仪

MECA Missile Electronics and Computer Assembly 导弹电子设备与计算机组件

MECCA Master Electrical Common Connector Assembly 主通用电子接头设备

MECF Main Engine Computational Facilities 主发动机计算设备

MECO Main Engine Cutoff 主发动机停车；主机停车

MECOM Missile Equipment Command 导弹装备司令部

MECS Maritime European Communications Satellite 欧洲海事通信卫星

MECU Main Engine Electronic Control Unit 主发动机电子控制装置

MED Manual Entry Device 人工返回装置

MED Momentum Exchange Devices 动量交换装置

MEDIA Missile Error Data Integration Analysis 导弹误差综合分析

MEDIA Modular Electronic Digital Instrumentation Assemblies 模块式电子数字仪器组件

MEDICS Medical Information Computer System 医学信息计算机系统

MEDILAB Medical Laboratory 医学实验舱（与空间站对接的舱段之一）

MEDOC Motion of the Earth By Doppler Observation Campaign 多普勒观测地球运动计划

MEE Mission Essential Equipment 任务基本设备

MEERS Maximum Effective Echo Ranging Speed 最大有效回波测距速度

MEF Mission Effectiveness Factor 任务效益因子

MEGALOS Multimission European Geostationary Ariane Launched Orbital Station 【欧洲空间局】"阿里安"火箭发射的欧洲多任务对地静止轨道站

MEI Main Engine Ignition 主发动机点火

MEI Mission Effectiveness Inspection 任务效能检查

MEIG Main Engine Ignition 主发动机点火

MEISR Minimum Essential Improvement in System Reliability 系统可靠性最低限度的改进

MEIT Multi-Element Integration Testing 多组件综合测试

MEIU Main Engine Interface Unit 主发动机接口装置

MEL Mobile Erector Launcher 移动式起竖发射装置

MEL Modular Electromagnetic Levitator 模块化电磁悬浮器

MELEO Material Exposure in Low Earth Orbit 近地轨道的材料暴露

MELH Missile Elevation Heading 导弹仰角航向；导弹射角方向

MELV Medium (Class) Expendable Launch Vehicle 一次性中级运载火箭

MEM Mars Excursion Module 【美国】

火星登陆舱
MEM Materials Experimentation Module 【美国航天飞机】材料实验舱
MEM Meteoroid Engineering Model 陨星群工艺模型
MEM Meteoroid Exposure Module 陨星群暴露舱
MEM Middeck Electronics Module 中甲板电子设备模块
MEM Mission Effectiveness Model 任务效能模型
MEM Mission Equipment Modernization 任务设备现代化改造
MEM Module Exchange Mechanism 模块交换机械装置
MEML Molecular Engineering and Material Laboratory 分子工程和材料实验室
MEMO Maximizing the Efficiency of Machine Operations 最大限度提高机器运行效率
MEMO Model for Evaluating Missile Observation 导弹观测鉴定模型
MEMOC Mars Express Mission Operations Centre 【美国】"火星快车"任务操作中心
MEMPW Nuclear Electromagnetic Pulse Weapon 核电磁脉冲武器
MEMS Micro Electro Mechanical Systems 微机电系统
MEMS Missile Equipment Maintenance Sets 导弹设备保养装置
MEMU Manned Extravehicular Manipulating Unit 载人舱舱外操控装置
MEO Manned Earth Observation 载人地球观测台
MEO Manned Earth Orbit 载人地球轨道
MEO Manned Extra-Vehicle Operation 载人航天器舱外操作
MEO Mass in Earth Orbit 地球轨道上的质量
MEO Medium Earth Orbit 中地球轨道
MEOL Manned Earth Orbiting Laboratory 载人地球轨道实验室
MEOM Manned Earth Orbit Mission 载人地球轨道飞行任务
MEOP Maximum Expected Operating Pressure 【固体火箭发动机】预期的最大工作压力
MEOS Multidisciplinary Earth Observation Satellite 【欧洲空间局】多学科地球观测卫星
MEOTBF Mean Engine Operating Time Between Failures 发动机平均故障间隔时间
MEP Main Engine Propellant 主发动机推进剂
MEP Main Entry Point 主返回点
MEP Management Engineering Program 管理工程项目
MEP Mars Entry Probe 进入火星探测火箭
MEP Mars Exploration Program 火星探索计划
MEP Mean Effective Pressure 平均有效压力
MEP Meteosat Exploitation Programme 气象卫星开发计划
MEP Minimum Entry Point 最小返回点
MEPED Medium Energy Proton and Electron Detector 中能质子与电子探测器
MEPF Multiple Experiment Processing Furnace 多实验性处理炉
MEPF-CGF Multiple Experiment Processing Facility-Crystal Growth

Furnace 多实验性处理设施—晶体生长炉

MEPF-MAS Multiple Experiment Processing Facility-Metal Alloy Solidification 多实验性处理设施—金属合金固化

MEPS Manned Electric-Propulsion Ship 电推进载人飞船

MEPS Medium Energy Particle Spectrometer 中能粒子质谱仪

MEPSS Maintenance Electronic Performance Support System 维修电子性能保障系统

MEPU Monofuel Emergency Power Unit 单元燃料应急动力装置

MER Machinery Effectiveness Review 机械效能评审

MER Maintenance Effectiveness Review 维修有效性评审

MER Manned Earth Reconnaissance 载人地球侦察（卫星）

MER Mars Exploration Rover 火星探索漫游车

MER Mission Evaluation Room 【美国约翰逊航天中心】飞行任务评估机房

MERA Multi-Flow Expert Resource Assessment 多流式专家源评估

MERGV MARTIAN Mars Exploratory Rocket Glide Vehicle 火星探测火箭滑翔飞行器

MERIS Medium Resolution Imaging Spectrometer 中分辨率成像光谱仪

MERL Materials Engineering Research Laboratory 材料工程实验室

MERM Material Evaluation of Rocket Motor 火箭发动机材料鉴定

MERSAT Meteorology and Earth Observation Satellite 气象与地球观测卫星

MES Main Engine Start 主发动机启动

MES Manned Escape System 载人逃逸（救生）系统

MES Manufacturing Execution System 生产（制造）实施系统

MES Mated Elements Simulator 对接部件模拟器（装置）

MES Mated Events Simulator 对接事件模拟器（装置）

MES Missile Electrical Simulator 导弹电气模拟器

MES Mission Events Sequence 任务事件序列

MES Mobile Earth Station 移动式地球站

MES MOL (Manned Orbiting Laboratory) Environmental Shelter 载人轨道实验室环境屏蔽

MESA Manned Environment System Assessment 载人环境系统鉴定

MESA Mobile Exploration System for Apollo 【美国】"阿波罗"飞船移动探测系统

MESA Modular Equipment Storage Assembly 【美国"阿波罗"登月舱】模块式设备存放装置

MESA Modular Equipment Stowage Area 【美国国家航空航天局】模块式设备存放区

MESA Modular Equipment Stowage Assembly 模块化设备存储组装

MESA MSFC Engineering Support Area 【美国】马歇尔航天飞行中心工程保障区

MESAR Multifunction Electronically Scanned Adaptive Radar 多功能电子扫描自适应雷达

MESC Master Events Sequencer Controller 主事件序列器控制装置
MESF Mobile Earth Station Facility 移动式地球站设施
MESSAGE Modular Electronic Solid State Aerospace Ground Equipment 组合式航空航天固体电子地面设备
MESSENGER Mercury Surface, Space Environment, Geochemistry and Ranging 【美国】"信使号"水星探测器（水星表面，太空环境，地球化学和广泛探索）
MESSOC Model to Estimate Space Station Operations Costs 空间站运行成本估算模型
MEST Missile Electrical System Test 导弹电气系统测试；导弹电气系统检测
MEST Mobile Earth Station Facility 移动式地球站设施
MESTS Missile Electrical System Test Set 导弹电气系统检测装置
MESUR Mars Environmental Surveyor 火星环境探测者（美国的火星探测器）
MET Master Events Timer 主事件计时器（装置）
MET Mission Elapsed Time 任务耗用时间
MET Mission Events Timer 任务事件计时器（装置）
MET Mobile Equipment Transporter 机动设备运输器（美国航天员在月球表面使用）
MET Momentum Exchange Tether 动量交换绳系
META Megachannel Extraterrestrial Assay 强通道外星探测
METADS Meteorological Acquisition and Display System 气象采集与显示系统
METATS Meteorological Applications Technology Satellite 【俄罗斯】气象应用技术卫星
METEOR Multiple Experiment Transporter to Earth Orbit and Return 进入地球轨道并返回的多项实验运输飞船
METEOSAT European Meteorological Satellite 欧洲气象卫星
METOP Meteorology Operational Satellite 气象业务卫星
METOXI Military Effectiveness in a Toxin Environment 有害环境中的军事效能；染毒环境下的战斗能力
METROC Meteorological Rocket 气象火箭
METS Mobile Engine Test Stand 活动式发动机试车台
METS Modular Engine Test System 模块式发动机试验系统
METVC Main Engine Thrust Vector 主发动机推力矢量
MEV Manned Entry Vehicle 可进入大气层的载人飞行器
MEZ Missile Engagement Zone 导弹交战区
MF Adsorption and Microfiltration 过滤吸附
MFA Manned Flight Awareness 载人飞行感知
MFAR Modular Multifunction Phased Array Radar 模块化多功能相控阵列雷达
MFAT Maintenance-Free Await Order Time 无维修待命时间
MFC Microgravity Facilities for Columbus 【欧洲空间局】"哥伦布"舱的微重力设备
MFC Mission Flight Control 任务飞

行控制

MFC　Multiple Flight Computer　多飞行任务计算机

MFC　Multiple Flight Controller　多用途飞行控制器

MFCA　Missile Flight Caution Area　导弹飞行警戒区

MFCC　Mission Flight Control Center　任务飞行控制中心

MFCS　Manual Flight Control System　人工飞行控制系统

MFCS　Missile Fire Control System　导弹发射控制系统

MFCS　Missile Flight Control System　导弹飞行控制系统

MFD　Main Flame Deflector　主导流器

MFD　Malfunction Detection　故障探测

MFD　Multifunction Display　多功能显示

MFDS/FWS　Multifunction Display System/Flight Warning System　多功能显示系统、飞行告警系统

MFE　Magnetic Field Explorer　磁场探测器（美国地磁场探测卫星）

MFES　Main Fixed Earth Station　固定式主地球站

MFHBCF　Mean Flight Hours between Critical Failures　平均关键故障间隔飞行小时

MFHBEM　Mean Flight Hours between Essential Maintenance　平均基本的维修间隔飞行小时

MFHBF　Mean Flight Hours between Failures　平均故障间隔飞行小时

MFHBHMA　Mean Flight Hours between Hardware Maintenance Action　平均硬件维修活动间隔飞行小时

MFHBHMF　Mean Flight Hours between Hardware Mission Failure　平均硬件任务故障间隔飞行小时

MFHBMA　Mean Flight Hours between Maintenance Action　平均维修活动间隔飞行小时

MFHBMCF　Mean Flight Hours between Mission Critical Failure　平均任务严重（关键）故障间隔飞行小时

MFHBMCF　Mean Flight Hours between Mission-Critical Failures　平均关键任务故障间隔飞行小时

MFHBME　Mean Flight Hours between Maintenance Events　平均维修事件间隔飞行小时

MFHBOMA　Mean Flight Hours between Operation Maintenance Action　平均作战（使用）维修活动间隔飞行小时

MFHBUMA　Mean Flight Hours between Unscheduled Maintenance Action　平均非预定维修活动间隔飞行小时

MFHOMF　Mean Flight Hours between Operation Mission Failure　平均作战（使用）任务故障间隔飞行小时

MFI　Module Fail Interrupt　模块故障中断

MFM　Missile Field Monitor　导弹发射场监视器

MFMR　Multifrequency Microwave Radiometer　多频率微波辐射仪

MFMS　Fine Attitude Measurement System　精确姿态测量系统

MFOD　Manned Flight Operation Division　载人飞行活动处

MFOD　Manned Flight Operations Directive　载人飞行操作指令

MFOP　Missile Fuel Operating Program 导弹燃料使用程序

MFR Manipulator Foot Restrain 【美国航天飞机】机械臂足固定器

MFR Maximum Flight Rate 最大飞行率

MFR Multifunctional Receiver 多功能接收装置

MFR Multifunctional Review 多用途评审

MFS Manned Flying Simulator 载人飞行模拟器

MFS Manned Flying System 载人飞行系统

MFS Module Fail Status 模块故障状态

MFSC Missile Flight Safety Center 导弹飞行安全中心

MFSG Missile Firing Safety Group 导弹发射安全组

MFSIM Multifunction Simulation 多功能模拟

MFSS Missile Flight Safety System 导弹飞行安全系统

MFT Mean Fault Time 平均故障时间

MFT Mean Flight Time 平均飞行时间

MFT Missile Fire Test 导弹点火试验

MFTGS Missile Flight Termination Ground System 导弹飞行终端地面系统

MFV Main Fuel Valve 主燃料阀

MFV Mars Flyby Vehicle 火星飞越航天器

MFVA Main Fuel Valve Actuator 主燃料阀致动器（装置）

MG Missile Guidance 导弹制导

MGAM Microgravity Accelerometer Measurement System 微重力加速仪测量系统

MGC Manual Gain Control 人工增益控制

MGC Missile Guidance Computer 导弹制导计算机

MGCC Missile Guidance and Control Computer 导弹制导与控制计算机

MGCO Mars Geoscience and Climatology Orbiter 【美国】火星地质学与气候学轨道器（后改名为"火星观测器"）

MGCS METEOSAT Ground Computer System 【欧洲】气象卫星地面计算机系统

MGCS Missile Guidance and Control System 导弹制导和控制系统

MGDF Modified Granular Diffusion Flame (propellant) 改进的粒状扩散火焰（推进剂）

MGE Maintenance Ground Equipment 地面维修设备

MGE Missile Ground Equipment 导弹地面设备

MGEP Mobile Ground Entry Point 移动式地面进入点

MGLI Midcourse Ground Launched Interceptor 地面发射中程拦截机（导弹）

MGM Mobile-Launched Guided Missile 【地面】机动发射导弹

MGO Main Geophysical Observatory 【俄罗斯】地球物理观测总台

MGO Mars Geoscience Orbiter 火星地质学轨道器

MGPCU Missile Ground Power Control Unit 导弹地面电源控制装置

MGPS Mobile Ground Processing System 移动式地面处理系统

MGROSS Stage Gross Liftoff Mass 子级起飞时的总质量

MGROSS-VEH Vehicle Gross Liftoff Mass 运载器/飞行器起飞时的总质量
MGS Missile Guidance Set 导弹制导装置
MGS Missile Guidance System 导弹制导系统
MGSC Missile Guidance Set Control 导弹制导装置控制
MGSE Maintenance Ground Support Equipment 地面维修辅助设备
MGSE Mechanical Ground Support Equipment 地面机械保障设备
MGSE Missile Ground Support Equipment 导弹地面支撑设备
MGSE Mobile Ground Support Equipment 移动式地面保障设备
MGSTS Missile Guidance System Test Set 导弹制导系统试验装置
MGT Motor Gas Temperature 【固体火箭】发动机燃气温度
MGTOW Maximum Gross Takeoff Weight 最大起飞总重量
MGTP Main Gear Touchdown Point 主起落架着陆接地点
MGTS Mobile Ground Telemetry Station 移动式地面遥测站
MHC Manipulator Hand Controller 机械臂手部控制器
MHD Magneto-Hydrodynamics 磁流体力学
MHE Material Handling Equipment 材料处理设备
MHE Missile Handling Equipment 导弹装卸设备
MHF Medium High Frequency 中高频率
MHR Monopropellant Hydrazine Rocket 单元推进剂肼火箭
MHS Missile Hazard Space 导弹危险区间
MHSA Mars Horizon Sensor Assembly 火星水平传感器组件
MHSC Manipulator Handset Controller 机械臂的手持小型装置控制器
MHT Missile Handling Trailer 导弹装卸拖车
MHTV Manned Hypersonic Test Vehicle 载人高超声速试验飞行器
MHV Manned Hypersonic Vehicle 载人高超声速飞行器
MHV Miniature Homing Vehicle 小型自动寻标飞行器（美国的一种反卫星武器）
MI Management Information 管理信息
MI Military Intelligence 军事情报
MIA Multiplex Interface Adapter 多路复用接口适配器
MIB Motor Inspection Building 发动机检测厂房
MIC Management Information Center 管理信息中心
MIC Mandatory Access Control, Identification and Authentication, Discretionary Access Control 强制访问控制、鉴别与认证、自主访问控制
MIC Measuring Instruments Compartment 测量仪器舱
MICCS Minuteman Integrated Command and Control System 【美国】"民兵"导弹综合指挥与控制系统
MICFAC Mobile Integrated Command Facility Ashore Center 岸上移动综合指挥设施
MICIS Material Information Control and Information System 材料信息控制与信息系统
MICNLS Modular Integrated Communications and Navigation Link System

模块式综合通信与导航数据链路系统

MICROLAB Micro-Laboratory 【美国】微型实验室（微型太空轨道实验台）

MICS Management Information and Control System 管理信息与控制系统

MICS MVS (Missile Velocity Servo System) Integrated Control System 导弹速度伺服系统综合控制系统

MIDAS Maintenance Inspection Data Analysis System 维修检查数据分析系统

MIDAS Malfunction Identification Data Acquisition System 故障识别数据采集系统

MIDAS Management Information and Development Aids System 管理信息与开发辅助系统

MIDAS Management Information Database Automation System 管理信息数据库自动化系统

MIDAS Missile Defense Alarm System 导弹防御报警系统

MIDAS Missile Defense Alert Satellite 导弹防御警戒卫星；"米达斯"警戒卫星

MIDAS Missile Detection Alarm System 导弹探测报警系统

MIDAS Modular Integrated Design Automation System 模块综合设计自动化系统

MIDAS Multifunctional Integrated Defense Avionics System 多功能综合防御航空电子系统

MIDDS Meteorological Integrated Distribution and Display System 气象综合分配与显示系统

MIDDS Meteorological Interactive Data Display System 交互式气象数据显示系统

MIDES Missile Detection System 导弹探测系统

MIDEX Medium Class Explorer 中级探测器

MIDIS Multifunctional Integrated Defense Information System 多功能综合防御信息系统

MIDOT Multiple Interference Determination of Trajectory 复式干涉仪轨道测定法

MIDRAM Micro-Dynamic Angle and Rate Monitoring (System) 微型动态角度与速率监控（系统）

MIDS Management Information and Decision Support 管理信息与决策支持

MIDS Man-Machine Integration Design and Analysis System 人机综合设计和分析系统

MIDS Missile Ignition and Destruct Simulator 导弹点火与自毁模拟器

MIDS Multi-Functional Information Distribution System 多功能信息配发系统

MIDT Mean Detonating Time 平均引爆时间

MIDU Missile-Ignition Delay Unit 导弹点火延迟装置

MIF Module Integration Facility 舱段总装设施

MIFI Missile Flight Indicator 导弹飞行指示器

MIFSA Missile In-Flight Safety Approval 导弹飞行安全措施核准

MIJI Meaconing, Intrusion, Jamming, and Interference 【美国】模拟干扰、无意干扰、人为干扰与电磁干扰（保护频带免受干扰的电子战计划）

MIK Assembly and Integration Building 【俄罗斯】组装与总装测试厂房

MIK Spacecraft Integration Facility for ROCKOT in Plesetsk 【俄罗斯】普列谢茨克航天发射场"轰鸣号"运载火箭总装设施

MIL Man-in-the-Loop 回路中的人

MIL Spaceflight Tracking and Data Network Station 【美国国家航空航天局】戈达德航天飞行中心的空间飞行跟踪与数据网络站

MIL-HDBK Military Handbook 军事手册

MIL-STD Military Standard 【美国】军用标准

MILA Merritt Island Launch Area 【美国国家航空航天局】梅里特岛发射区

MILCOMSAT Military Communication Satellite 军用通信卫星

MILCON Military Construction 军事施工

MILOGS Marine Integrated Logistics System 【美国】海军陆战队综合后勤系统

MILS Missile Impact Location System 导弹弹着点定位系统

MILS Missile Location System 导弹定位系统

MILSAT Military Satellite 军用卫星

MILSATCOM Military Satellite Communication 军用卫星通信（系统）

MILSTAR Military Strategic Tactical and Relay System 军事战略战术中继系统（美国军用通信卫星）

MILSTRIP Military Standard Requisitioning and Issue Procedures 军事申请与发放标准规程

MILU Missile Interface and Logic Unit 导弹接口与逻辑装置

MIM Mobile Maintenance Interface Module 移动维修接口模块

MIMIC Microwave/Millimeter Wave Monolithic Integrated Circuit 微波/毫米波单片集成电路

MIMOSA Mission Modes and Space Analysis 任务模式与空间分析

MIMOSA Mission Modes and System Analysis 【登月】任务方式与系统分析

MIMS Medical Information Management System 医学信息与管理系统

MINDAC Miniature Inertial Navigation Digital Automatic Computer 小型惯性导航数字自动计算机

Mini-DAMA Miniature Demand Assigned Multiple Access 小规模按需分配多路存取（访问）

Mini-MADS Minimodular Auxiliary Data System 小型模块辅助数据系统

Mini-RF Mini-RF Technology Demonstration 微型射频技术展示器

Mini-SAR Mini-Synthetic Aperture Radar 小型合成孔径雷达

MinIIP Minimum Instantaneous Impact Point 最小瞬间撞击点

MINITRACK (NASA) Minimum Weight Tracking System 【美国国家航空航天局】最轻型跟踪系统

MINTATS Mobile Integrated Telemetry and Tracking System 移动式综合遥测与跟踪系统

MINW Master Interface Network 主接口网络

MINWR Minimum Weapon Radius 武器最小作用半径

MIOQR Mission In Orbit Qualification Review 轨道任务鉴定评审

MIP　Military Intelligence Program　军事情报项目

MIP　Minimum Impulse Pulse　最小冲击脉冲

MIP　Missile Impact Prediction　导弹弹着（点）预测

MIP　Missile Impact Predictor　导弹弹着（点）预测器

MIP　Moon Impact Probe　撞月探测器

MIPAS　Michelson Interferometer Passive Atmospheric Sounder　【美国】"迈克逊"无源大气探测干涉仪

MIPE　Magnetic Induction Plasma Engine　磁感应等离子体发动机

MIPIR　Missile Precision Instrumentation Radar　导弹精确测量雷达

MIPS　Missile Impact Prediction System　导弹弹着（点）预测系统

MIPS　Missile Impact Predictor Set　导弹弹着点预测器装置

MIPS　Multiband Imaging Photometer for SIRTF　【美国空军】用于美国空间红外望远镜设施上的多谱段成像光度计

MIR　Microwave Imaging Radiometer　微波成像辐射计

MIRACL　Mid-Infrared Advanced Chemical Laser　中红外先进化学激光器（装置）

MIRADCOM　Missile R&D Command　【美国陆军】导弹研究与发展司令部

MIRADS　Marshall Information Retrieval and Display System　【美国国家航空航天局】马歇尔航天中心信息查询与显示系统

MIRAS　Mir Infrared Atmospheric Spectrometer　【俄罗斯】"和平号"空间站红外大气光谱仪

MIRCE　Management of Industrial Reliability, Cost and Effectiveness　工业可靠性、费用和效益管理

MIRFS　Multifunction Integrated Radio Frequency System　多功能综合射频系统

MIRFS/MFA　Multifunction Integrated Radio Frequency System/Multifunction Array　多功能综合射频系统/多功能阵列

MIRS　Management Information and Reporting System　管理信息与通报系统

MIRTS　Modularized Infrared Transmitting Set　模块式红外发射装置

MIRV　Multiple Independently Reentry Vehicle　多弹头分导式再入飞行器

MIRV　Multiple Independently Targeted Reentry Vehicles　多弹头分导再入（大气层）飞行器；分导式多弹头

MIS　Management Information System (Subsystem)　管理信息系统（分系统）

MIS　Man-In-Space　太空人计划

MIS　MDSEC Interchange System　【美国】导弹防御与航天实验中心内部交换系统

MIS　Meteorological Impact Statement　气象影响通报

MISDAS　Mechanical Impact System Design for Advanced Spacecraft　先进航天器的硬着陆系统设计

MISER　Microwave Space Electronic Relay　空间微波电子中继器

MISHAP　Missile High-Speed Assembly Program　导弹高速组装计划

MISIM　Missile Simulator　导弹模拟器

MISS　Man-In-Space Simulator　载人

航天（飞行）模拟器
MISS Missile Intercept Simulation System 导弹截击模拟系统
MISS Model Integrated Suspension System 模型综合悬挂系统
MISSE Materials International Space Station Experiment 国际空间站材料实验
MISSI Multilevel Information Systems Security Initiative 多级信息系统安全倡议
MIST Modular Interoperable Surface Terminal 模块式互操作地面终端
MISTE Miniature Integrated Satellite Terminal Equipment 微型综合卫星终端设备
MISTRAM Missile Trajectory Measurement System 导弹弹道测量系统
MIT Miniature Interceptor Technology 微型拦截机（导弹）技术
MITE Missile Integration Terminal Equipment 导弹综合终端设备
MITRV Multiple Independently Targeted Reentry Vehicles 多弹头分导式再入飞行器
MITS Marshall Integrated Telecommunications System 【美国国家航空航天局】马歇尔航天中心综合远距离通信系统
MITS Micro Instrumentation and Telemetry Systems 微型仪表与遥测系统
MITS Missile Ignition Test Simulator 导弹点火试验模拟器
MIU Missile Interface Unit 导弹接口装置
MIU Multiplex Interface Unit 多路接口装置
MIUS Modular Integrated Utility Systems 模块化综合使用系统
MJO Mission Jupiter Orbit 环木星轨道飞行
MJS Manipulator Jettison System (Subsystem) 机械臂抛掉系统（分系统）
MJS Mariner Jupiter/Saturn 【美国】"水手"号木星—土星探测器
MKIDS Microwave Kinetic Inductance Detectors 微波动力学电感探测器
ML Missile Launcher 导弹发射装置
ML Mission Life （飞行）任务期限；（卫星）使用寿命
ML Mobile Launcher 机动发射装置（架）
ML PED Mobile Launcher Pedestal 移动发射架基座
MLAF Missile Loading Alignment Fixture 导弹负载校准装置
MLAF Missile Lot Acceptance Firing 导弹批量验收发射
MLAS Max Launch Abort System 最大的发射中止系统
MLAW Missile Launch and Approach Warning 导弹发射与接近告警
MLC Maneuver Load Control (System) 机动载荷控制（系统）
MLC Minimum Launch Capability 最小发射能力
MLC Mixed Logistic Carrier 混合型货舱
MLC Mobile Launcher Computer 移动发射装置计算机
MLCB Missile Launch Control Blockhouse 导弹发射控制掩体
MLCC Mobile Launch Control Center 机动发射控制中心
MLCP Mission Launch Control Processor 任务发射控制处理器（装

置）

MLCS Maneuver Load Control System 机动载荷控制系统

MLD Missile Launch Detector 导弹发射探测器

MLDT Mean Logistic Delay Time 平均后勤保障延迟时间

MLDT Mean Logistics Down Time 平均后勤保障停机时间

MLDT Missile Downlink Transmitter 导弹下链路发射机（装置）

MLE Manned Lunar Exploration 载人月球探测

MLE Mesoscale Lightning Experiment 【美国国家航空航天局】中尺度闪电实验

MLE Missile Launch Envelop 导弹发射包线

MLE Mobile Launcher Equipment 机动发射器设备；移动发射器装置

MLEP Manned Lunar Exploration Program 载人探月计划

MLEV Manned Lifting Entry Vehicle 载人升力进入大气层飞行器

MLF Mobile Launcher Facility 机动发射器设施

MLF MOL Launch Facility 载人轨道实验室发射设施

MLG Main Landing Gear 主着陆起落架

MLGS Microwave Landing Guidance System 微波着陆制导系统

MLI Multilayer Insulation 多层绝缘

MLL Manned Lunar Landing 载人登月

MLL Manned Lunar Launching 载人登月飞船发射

MLLP Manned Lunar Landing Program 载人登月计划

MLLR Manned Lunar Landing and Return 载人登月并返回地球

MLLV Medium-Lift Launch Vehicle 中型运载火箭；中等推力运载火箭

MLM Multi-Function Module 多功能舱

MLMS Multipurpose Lightweight Missile System 多用途轻型导弹系统

MLN Main Lift Nozzle 主升力喷管

MLO Manned Lunar Orbiter 载人月球轨道飞行器

MLOADS Maneuver Loads 机动载荷；操纵载荷

MLP Missile Launching Platform 导弹发射平台

MLP Mobile Launcher Platform 【美国航天飞机】活动发射平台

MLR Monodisperse Latex Reactor 【美国航天飞机】单弥散乳胶反应器

MLRB Medium-Life Reentry Body 中等寿命再入体（弹体）

MLRS Multiple Launch Rocket System 多管火箭发射系统

MLRV Manned Lunar Roving Vehicle 载人月行车

MLS Main Liquid Separator 主液分离器

MLS Microwave Landing System 微波着陆系统

MLS Model Lunar Service 月球模型设施

MLS Moon Landing Site 月球着陆场

MLS/ILS Microwave Landing System/Instrument Landing System 微波着陆系统/仪表着陆系统

MLSC Minuteman Launch Support Center 【美国】"民兵"导弹发射保障中心

MLSC Missile Launching System

Control　导弹发射系统控制
MLSP　Multiple Link Satellite Program　多路（通信）卫星计划
MLSS　Missile Launching Surveillance System　导弹发射监视系统
MLT　Manned Lunar Test　载人登月试验
MLTU　Missile Loop Test Unit　导弹回路测试装置
MLUT　Mobile Launch Platform Umbilical Tower　活动发射平台脐带塔
MLV　Manned Lunar Vehicle　载人月球飞行器
MLV　Medium Launch Vehicle　中型运载火箭
MLV　Medium Lift Variant (of EELV)　中型运载火箭变型（改进型一次性运输火箭）
MLV　Missile Launch Vehicle　导弹发射车
MLV　Mobile Launch Vehicle　【导弹】活动发射车；活动发射装置
MLV　Multi-Purpose Launch Vehicle　多用途发射车
MLVP　Manned Lunar Vehicle Program　载人月球飞行器计划
MLVPS　Manual Low Voltage Power Supply　手控低压电源
MLW　Maximum Landing Weight　最大着陆重量
MLW　Maximum Permitted Landing Weight　最大允许着陆重量
MLWIR　Medium-Long Wavelength Infrared　中长波红外
MM　Main Mode　主模式
MM Project　Marine-Mars Project　【美国】"水手号"火星探测器计划
MM/OD　Micrometeoroid Object Damage　微流星体碎片损伤
MMA　Mass Memory Assembly　大规模存储器组件
MMA　Missile Main Assembly Area　导弹主要装配区
MMA　Missile Maintenance Area　导弹维修区
MMACS　Maintenance Management and Control System　维修管理与控制系统
MMAS　Material Management Accounting (Accountability) System　材料管理统计系统
MmaxCT　Maximum Corrective Maintenance Time　最大修正维修时间
MMC　Missile Measurement Center　导弹测量中心
MMC　Modular Mission Computer　模块化任务计算机
MMCC　Mission Management and Control Centre　【欧洲空间局】任务管理和控制中心
MMCCS　Milstar Mobile Constellation Control Station　【美国】"军事星"机动星座控制站
MMD　Missile Miss Distance　导弹误差距离
MMDB　Master Measurement Data Base　主测量数据库
MME　Manned Mars Expedition　载人火星探险
MME　Missile Maintenance Equipment　导弹保养设备
MMES　MSFC Mated Element Systems　【美国国家航空航天局】马歇尔航天中心对接部件系统
MMFV　Manned Mars Flyby Vehicle　载人火星飞越飞行器
MMIC　Microwave Monolithic Integrated Circuit　微波单一综合电路
MMIC　Multi-Mission Integration Cell　多任务综合舱（室）

MMICS Maintenance Management Information and Control System 【美国空军】维修管理信息与控制系统

MMIPS Multiple Mode Integrated Propulsion System 多模式综合推进系统

MMIS Manpower Management Information System 人力管理信息系统

MML Man-Tended Multipurpose Laboratory 有人照料的多用途实验室

MML Multi-Missile Launcher 复式导弹发射架

MMM Manned Mars Mission 载人火星任务

MMM Mars Mission Module 火星探测任务舱

MMM Multi-Mode Missile 多模式导弹

MMMS Militarized Multimission Modular Spacecraft 军事化多任务模块化航天器

MMO Mercury Magnetospheric Orbiter 水星磁层轨道器

MMOD Micro Meteoroid/Orbital Debris 微流星体/轨道碎片

MMODS Modular Mechanical Ordnance Destruct System 模块化机械火工品自毁系统

MMOS Multimode Optical Sensor 多模式光学感应器（装置）

MMPA Materials, Mechanical Parts and Processes Assurance 材料、机械零部件和工艺保证

MMPF Microgravity and Materials Processing Facility 微重力材料加工装置

MMPSE Multiuse Mission Payload Support Equipment 多用途任务有效载荷支撑设备

MMRC Mars Mission Research Center 火星任务研究中心

MMS Magnetospheric Multiscale 磁层多尺度

MMS Meteorological Measuring System 气象测量系统

MMS Missile Management Station 导弹管理站

MMS Mission Modular Spacecraft 任务模块式航天器

MMS Modular Multiband Scanner 模块式多波段扫描装置

MMS Modular Multimission Satellite 【美国】模块式多用途卫星

MMS Multimission Modular Spacecraft 多用途模块化太空船；多执行任务（或用途）组合式航天器

MMS Multimission Spacecraft 【美国】多用途航天器

MMS-CP Missile Management Station-Control Panel 导弹管理站一控制板

MMSE Multiuse Mission Support Equipment 多用途任务保障设备

MMSEV Multi Mission Space Exploration Vehicle 多任务航天探索飞行器

MMSL Microgravity Materials Science Laboratory 微重力材料科学实验室

MMSS Manned Maneuverable Space System 载人机动航天系统

MMSS Multi-Module Space Station 多舱空间站

MMSSG Moon-Mars Science Linkage Steering Group 月球一火星科学联动督导小组

MMTD Miniaturized Munition Technology Demonstration 小型化军用品技术验证

MMTS Manned Military Test Station 载人军事试验空间站

MMU Manned Maneuvering Unit 载人机动装置（美国航天员舱外机动飞行装置，形状像特大座椅，是一种便携"背负"式推力装置，可使宇航员及其装备在舱外活动期间在太空移动）

MMU Model Management Utility 模型管理设施

MMU Modular Maneuvering Unit 模块式机动飞行装置

MN-ED Materiel Need-Engineering Development 装备需求－工程研制

MNORP Missile Not Operationally Ready for Lack of Parts 因缺少零件未处于战备状态的导弹

MNTNGFL Maintaining Flight Level 保持飞行高度

MNTV Mercury Network Test Vehicle 【美国】"水星号"飞船跟踪网试验飞行器

MNTVC Move Nozzle Thrust Vector Control 摆动喷管推力矢量控制（器）

MO Manned Orbiter 载人轨道器

MO Manual Orientation 人工定位

MO Mars Observer 【美国】火星观测器

MO Mission Operations 任务操作

MOA Military Operating Area 军事操作区

MOAB Missile Optimized Anti-Ballistic 优化的反弹道导弹

MOAO Manned Orbiting Astronomical Observatory 载人轨道天文观测台

MOB Main Operations Base 主作战（操作）区

MOBCOMSAT Mobile Communication Satellite 移动通信卫星

MOBIDIC Mobile Digital Computer 移动式数字计算机

MOC Mars Observer Camera 火星观测器摄影机

MOC Mars Orbiter Camera 火星轨道器摄像机

MOC Milstar Operations Center 军用卫星操作中心

MOC Minimum Operational Characteristics 最低工作性能；最低限度作战特性

MOC Missile Defense Agency Operations Center 导弹防御局操作中心

MOC Mission Observation Center 任务观察中心

MOC Mission Operation Center 任务操作中心

MOC Mission Operations Computer 任务操作计算机

MOC Missions Operations Center 任务操作中心

MOC Mobile Operations Center 移动式作战（操作）中心

MOCC Maintenance and Operation Coordination Center 【美国西靶场】维修和作业协调中心

MOCC METEOSAT Operations Control Centre 欧洲气象卫星运行控制中心

MOCC Mission Operations Control Center 任务操作控制中心

MOCC Multi-Satellite Operation Control Center 多卫星操作控制中心

MOCC Western Range Maintenance and Operations Coordination Center 西部试验场维修和作业协调中心

MOCF Mission Operations Computational Facilities 任务操作计算设施

MOCR Mission Operations Control

Room 【美国约翰逊航天中心】任务操作控制室

MOCS Multichannel Ocean Color Sensor 多通道海洋水色感应器（装置）

MOCVD Metal Organic Chemical Vapor Deposition 金属有机化学汽相沉积

MOD Maintenance, Operation and Diagnosis 维修、使用和诊断

MOD Manned Orbital Development (Station) 载人轨道开发研究（站）；载人轨道研究（站）

MOD Mission Operations Director 飞行操作导向装置；飞行操作指挥仪

MOD/SIM Modeling/Simulation 建模与仿真

MODACS Modular Data Acquisition and Control Subsystem 模块式数据采集与控制子系统

MODAP Modified Apollo 【美国】改进型"阿波罗"飞船

MODAP Modified Apollo Logistics Spacecraft 改进型"阿波罗"后勤飞船

MODAPS Maintenance and Operational Data Presentation Study 维修与操作数据显示研究

MODAPS Modal Data Acquisition and Processing System 模态数据获取和处理系统

MODART Methods of Defeating Advanced Radar Threats 击败先进雷达威胁的手段

MODAS Maintenance and Operational Data Access System 维修与使用数据存取系统

MODAS Modular Data Acquisition System 模块式数据采集系统

MODB Master Object Data Base 主目标数据库

MODIL Manufacturing Operations, Development, and Integration Laboratory 生产（制造）操作、研制与组装实验室【美国战略防御计划组织术语】

MODILS Modular Instrument Landing System 模块式仪表着陆系统

MODIS Moderate Resolution Imaging Spectrometer 【美国国家航空航天局】中等分辨率成像光谱仪

MODLAN Mission Operations Division Local Area Network 【美国国家航空航天局】飞行任务行动管理处局域网

MODS Manned Orbital Development Station 【美国国家航空航天局】载人轨道研究工作站

MODS Military Orbital Development Station 军用轨道研究站

MODS Military Orbital Development System under Study for the Air Force 空军在研的军用轨道研究系统

MODS Missile Offense/Defense System 导弹进攻与防御系统

MODS Mission Operations and Data System 任务操作与数据系统

MOE Mars Orbit Ejection 射入火星轨道

MOERO Medium Orbiting Earth Resources Observatory 中空轨道地球资源观测卫星

MOF Manned Orbital Flight 载人轨道飞行

MOF Mobile Operations Facility 移动式操作设施

MOHOLE Project to Explore the Inner Layers of the Earth's Surface 地球内层探测计划

MOI Mars-Orbit-Insertion 进入火星轨道

MOL Manned Orbital Laboratory 载人轨道实验室

MOL/ACTS Manned Orbiting Laboratory/Altitude Control and Transmission System 载人轨道实验室/高度控制与传输系统

MOLA Mars Orbiter Laser Altimeter 火星轨道激光高度计

MOLAB Mobile Lunar Laboratory 活动式月球实验室

MOLC Multiple Operational Launch Complex 多发射阵地综合设施

MOLEM Mobile Lunar Excursion Module 活动式登月舱

MOM Maintenance and Operation Module 维修和使用模块

MOM Manned Orbiting Mission 载人轨道飞行任务

MOM Mars Orbital Module 火星轨道舱

MOM Mass Optical Memory 大容量光存储器

MOMS Modular Optoelectronic Multispectral Scanner 模块式光电多谱段扫描仪

MOMV Mini Orbital Maneuvering Vehicle 小型轨道机动飞行器

MONO Monopropellant 【航天器】单元推进剂

MONSEE Monitoring of the Sun-Earth Environment 日地环境监测

MOO Manned Orbital Observatory 载人轨道观测台

MOOSE Manned Orbital-Operations Safety Equipment 载人轨道操作安全设备

MOOSE Man-Out-of-Space-Easiest 最简便空间救生（设备）

MOOSE Man-Out-of-Space-Escape (System) 航天应急逃逸（系统）

MOOSSE Manned Orbital Oceanographic Survey System Experiment 载人轨道（运行）海洋测量系统实验

MOP Manned Orbital Platform 载人轨道平台

MOP METEOSAT Operational Programme Satellite 欧洲气象卫星业务计划卫星

MOP Mission Operations Plan 任务操作计划

MOPA Master Oscillator Power Amplifier 主控振荡器功率放大器

MOPF Missile On-Loading Prism Fixture 导弹装填棱柱固定装置

MOPR Minimum Operational Performance Requirements 最低使用性能要求

MOPR Mission Operations Planning Review 任务操作规划评审

MOPS Minimum Operational Performance Standard 作战性能最低标准

MOPS Mission Operations Planning System 任务操作规划系统

MOPTAR Multiple Object Phase Tracking and Ranging 多目标相位跟踪与测距

MOPTARS Multiple Object Phase Tracking and Ranging System 多目标相位跟踪与测距系统

MOR Manufacturing Operation Record 生产（制造）操作记录

MOR Mars Orbital Rendezvous 火星轨道交会

MOR Missile Operationally Ready 导弹处于准备发射状态

MOR Mission Operations Report 任

务操作报告

MORDS Manned Orbital Research and Development System 载人轨道研究与开发系统

MORE Maintenance on Reliable Engine 可靠性发动机的维修

MORF Manned Orbital Research Facility 载人轨道研究设施

MORL Manned Orbital Research Laboratory 载人轨道研究实验室

MORL Medium Orbital Research Laboratory 中型轨道（空间）研究实验室

MORO Moon Orbiting Observatory 环月轨道观测台

MORT Management Oversight and Risk Tree 管理监督与风险树

MORT Missile Operational Readiness Test 导弹战备状态试验

MOS Manned Orbital Station 载人轨道空间站

MOS Marine Observation Satellite 海洋观测卫星

MOS Missile Operations Station 导弹发射操作工作站

MOS Mission Operations System 任务操作系统

MOS Multi-Satellite Operating System 多卫星操作系统

MOS/PIM Multiple Orbit Satellite/Program Improvement Module 多轨道卫星/项目改进模块

MOSA Mission Operations Support Area 任务操作保障区

MOSART Modular Open System Architecture 模块化开放系统结构

MOSB Multi Operations Support Building 多作业保障厂房

MOSES Molecules in Outer Space and Earth Stratosphere 【瑞典】外层空间与平流层分子探测卫星

MOSHED Multiplanar Organic Scintillator High Energy Detector 多面体闪烁器高能探测器

MOSL Manned Orbital Space Laboratory 载人轨道空间实验室

MOSP Multi-Optical Stabilized Payload 【美国】多光学传感器装置稳定有效载荷

MOSPO Mobile Satellite Photometric Observatory 移动卫星测光观测台

MOSS Manned Orbital Space Station 载人轨道空间站

MOSS Manned Orbital Space System 载人轨道空间系统

MOSS Military Operational Satellite System 军事业务卫星系统

MOSS Military Orbital Space System 军用轨道航天系统

MOST Manned Orbital Solar Telescope 载人轨道太阳观测台望远镜

MOST Multiple Target Tracking Optical Sensor Array Technology 多目标跟踪光学传感器阵列技术

MOSTT Mosaic Optical Sensor Technology Testbed 光学镶嵌传感器技术测试台

MOT Manned Orbital Telescope 载人轨道望远镜

MOT&E Multiservice Operational Test and Evaluation 多军种使用试验与评价

MOTIF Maui Optical Tracking and Identification Facility 【美国"航天监视网"】毛伊岛光学跟踪和识别设施

MOTNE Meteorological Operational Telecommunications Network in Europe 欧洲气象业务通信网

MOTR Multiple Object Tracking Radar

多目标跟踪雷达
MOTS Military Off the Shelf 军队提供的现成品
MOTS Minitrack Optical Tracking System 人造卫星跟踪系统光学跟踪站
MOTS Mobile Optical Tracking System 机动光学跟踪系统
MOTS Modified (or Modifiable) Off the Shelf (Hardware or Software) 改进的（或可更改的）现用产品（硬件或软件）
MOTV Manned Orbital Transfer Vehicle 【美国】载人轨道转移飞行器
MOUSE Minimum Orbital Unmanned Satellite of the Earth 最小轨道无人地球卫星
MOV Main Oxidizer Valve 主氧化剂阀
MOV Manned Orbiting Vehicle 载人轨道飞行器
MOVA Main Oxidizer Valve Actuator 主氧化剂阀致动器（装置）
MOVE Mission Operation Voice Enhancement Project 任务操作语言增强项目
MOW Mission Operations Wing 作战使命翼
MOWS Manned Orbital Weapon System 载人轨道武器系统
MP Maintenance Period 维修周期
MP Maintenance Platform 维修平台
MP Maintenance Prevention 预防检修
MP Measuring Point 测量点
MP Medium Pressure 中度压力
MP&A Mission Planning and Analysis Area 任务计划与分析区域
MP&C Maintenance Planning and Control 维修规划和控制
MPA Main Propulsion Assistant 主推动辅助装置
MPA Maintenance Planning Analysis 维修规划分析
MPA Maneuvering Propulsion Assembly 机动推进装置
MPA Mass Processing Analysis 质量操作分析
MPA Mission Payload Analysis 任务有效载荷分析
MPA Mission Phase Analysis 任务阶段分析
MPA Mission Profile Analysis 任务剖面分析
MPA Multiple Payload Adapter 多个有效载荷连接器
MPAC Multipurpose Application Console 多用途应用控制台
MPAR Modified Precision Approach Radar 改进型精确进场雷达
MPC Maximum Permissible Concentration 最大容许浓度
MPC Meteorological Prediction Center 气象预测中心
MPC Multiple Payload Carrier 【美国】多个有效载荷运输器
MPCP Missile Power Control Panel 导弹电源控制板
MPCS Maintenance/Power Control System 维修/电源控制系统
MPCV Multi-Purpose Crew Vehicle 多用途乘员飞行器
MPCV/LAS Multi-Purpose Crew Vehicle/Launch Abort System 多用途载人飞船/发射中止系统
MPD Magnetic Plasma Dynamics 磁等离子体动力学
MPD Magnetosphere Particle Detector 磁层粒子探测器
MPDE Multipurpose Cockpit Display

Electronics 多功能座舱显示电子设备
MPDR Mission-Peculiar Design Review 特殊任务设计评审
MPDT Magneto-Plasma Dynamic Thruster 磁等离子动力推进器
MPE Maximum Predicted Environment 预测的最有利环境
MPE Mission Peculiar Equipment 任务专用设备
MPE Mission to Planet Earth 地球使命计划（美国全球环境监视计划）
MPE Module Powering Electronics 模块式电力电子设备
MPEP Multi-Purpose Experimental Platform 多用途实验平台
MPES Maximum Performance Ejection Seat 最佳性能弹射座椅
MPESS Mission Peculiar Experiment Support Structure 【美国航天飞机货舱】任务专用设备支撑架
MPF Mars Pathfinder 火星探路者
MPF Materials Process Facility 【美国航天飞机】材料加工装置
MPF Medium Payload Fairing 中型有效载荷整流罩
MPG Main Propulsion Gear 主推进器传动装置
MPG Multipoint Grounding 多点接地
MPGHM Mobile Payload Ground Handling Mechanism 移动式有效载荷地面装卸机构
MPH Missile Potential Hazard 导弹潜在危险
MPHE Material and Personnel Handling Equipment 材料与人员操作设备
MPI Mean Point of Impact 平均弹着点；平均落点

MPI Mission Payload Integration 任务有效载荷一体化
MPL Mars Probe Lander 火星探测着陆车
MPL Materials Processing Laboratory 材料加工实验室
MPL Multi-Platform Launcher 多平台发射装置
MPL Multiple Pulse Laser 多脉冲激光
MPL Multiprogramming Level 多程序设计计级
MPLG Materials Processing in Low Gravity 低重力下的材料处理
MPLM Mini Pressurized Logistics Module 【意大利】微型增压后勤舱（系"阿尔法"国际空间站的后期）
MPLM Multi Purpose Logistic Module 多用途后勤模块（舱）
MPLM Multi-Purpose Pressurized Logistics Module 多用途增压后勤舱
MPLS Multi-Protocol Label Switching 多协议标签交换
MPM Manipulator Positioning Mechanism 机械臂定位机构
MPM Missile Performance Monitoring System 导弹性能监控系统（惯性制导分系统）
MPM Multipurpose Missile 【美国空军】多用途导弹
MPMS Missile Performance Measurement System 弹道性能测量系统
MPMS Missile Performance Monitoring System 导弹性能监控系统（惯性制导分系统）
MPMSE Multiuse Payload and Mission Support Equipment 多用途有效载荷和任务保障设备
MPMT Mean Preventive Maintenance Time 平均预防性维修时间

MPOS Multiprogramming Operating System 多级程序设计运算系统
MPP Materiel, Processes and Parts 材料、工艺与零部件
MPP Meteoroid Penetration Probe 流星体穿入探测器
MPP Modernization Planning Process 现代化规划过程
MPP Multiple Payload Program 多种有效载荷计划
MPPF Multi-Payload Processing Facility 多有效载荷操作厂房
MPPG Mars Project Planning Groups 火星项目计划组
MPPSE Multipurpose Payload Support Equipment 多用途有效载荷保障设备
MPR Maintainability Problem Report 维修性故障报告
MPR Management Program Review 管理项目评审
MPR Mission Planning Room 任务规划室
MPRS Mission Planning Rehearsal System 任务规划演练系统
MPS Main Propulsion System (Stage) 主推进系统（级）
MPS Material Processing System 材料处理系统
MPS Materials Processing in Space 太空材料加工
MPS Materials Processing System 材料加工系统
MPS Mercury Procedures Simulator 【美国】"水星号"飞船程序模拟器
MPS Mission Payload Subsystem 任务有效载荷子系统
MPS Mixed-Propellant System 混合推进剂系统
MPS Modular Power System 模块式动力系统
MPS Multiple Protective Structure 【洲际弹道导弹】多层防护结构
MPSBLS Man-Portable Scanning Beam Landing System 便携式扫描（束）着陆系统
MPSOC Multipurpose Satellite Operations Center 多用途卫星作战中心
MPSR Mission Profile Storage and Retrieval 任务分布图存储与查询
MPSR Multipurpose Support Room 多用途保障间
MPSS Main Parachute Support Structure 主降落伞支撑结构
MPSS Mission Planning and Scheduling System 任务规划与调度系统
MPSS Multiple Payload Support Structure 多有效载荷支承结构
MPT Main Propulsion Test 主推进测试（试验）
MPT Main Propulsion Turbine 主推进器涡轮机
MPT Mercury Procedures Trainer 【美国】"水星号"飞船程序训练机
MPT Missile Preflight Tester 导弹飞行前检测器
MPT Missile Procedures Trainer 导弹（测试）程序训练器
MPTA Main Propulsion Test Article 【美国航天飞机】主推进系统试验件
MPTF Main Propulsion Test Facility 主推进试验设施
MPTP Main Propulsion Test Program 主推进测试项目
MPTS Missile/Precision Orbit Determination Tracking System 导弹/轨道精确测定跟踪系统
MPTS Multi Purpose Tracking System 多用途跟踪系统

MPTS Multipurpose Tool Set 多用途工具集

MPVA Main Propellant Valve Actuator 主推进剂阀门执行机构

MQF Mobile Quarantine Facility 流动式隔离设施（登月归来的航天员先到此室隔离检疫）

MQT Model Qualification Test 模型合格性试验

MR Manned Reentry 载人再入（大气层）

MR Material Review 材料评审

MR Mercury-Redstone 【美国】"水星—红石"运载火箭

MRA Main Ring Assembly 主环组件

MRA Mechanical Readiness Assessment 机械装备状态评估

MRA Milestone Review and Approval 阶段决策评审与批准

MRA Minimum Release Altitude 最低投弹高度

MRAC Model Reference Adaptive Control 模型参考自适应控制

MRBF Mean Rounds between Failures 故障间平均发弹数

MRBM Medium Range Ballistic Missile 中程弹道导弹

MRBS Mean Rounds between Stoppages 两次故障间平均发（弹）数

MRC Material Readiness Capability 器材战备完好能力

MRC Micro-Gravity Research Centre 微重力研究中心

MRC Moon's Radar Coordinates 月球雷达坐标

MRC Multirole Recoverable Capsule 多用途返回舱（英国微重力实验与航天员救生舱）

MRCP Mobile Radar Control Post 移动式雷达控制站

MRCS Missile Range Calibration Satellite 导弹射程校准卫星

MRCTS Missile Round Cable Test Set 成套导弹全备弹电缆测试设备；导弹弹头电缆试验箱

MRD Mission Requirements Document 飞行任务要求文件

MRD/FT Missile Restraint Device/Field Tester 导弹固定装置/野外试验台

MRDA Mission Requirements and Definition Analysis 任务要求与定义分析

MRDC Military Research and Development Center 军事研究开发中心

MRE Motor Requirement Evaluation 发动机要求鉴定

MRF Meteorological Research Flight 气象研究飞行

MRF Module Repair Facility 舱室组件维修设备

MRH Module de Resources Hermes 【欧洲空间局】"赫尔姆斯"资源舱

MRI Measurement Requirements and Interface 测量要求与接口

MRICC Missile and Rocket Inventory Control Center 导弹与火箭物资控制中心

MRIR Medium Resolution Infrared Radiometer 中分辨率红外辐射仪

MRIS Missile Range Instrumentation Ship 导弹靶场测量船

MRIS Mobile Range Instrumentation System 移动式靶场测量系统

MRL Missile Reference Line 导弹发射参考（基准）线

MRL Missions Rail Launcher 任务导轨发射装置

MRL Modular Rocket Launcher 模块式火箭发射器

MRL Multiple Rocket Launcher 多管火箭发射器

MRLC Moon Rocket Launching Center 月球火箭发射中心

MRMDF Mobile Remote Manipulator Development Facility 移动式远程机械臂研制设施

MRMMS Millimeter Wave Reflectivity Measurement System 毫米波反射侦测（测量）系统

MRMS Mobile Remote Manipulator System 移动式远程机械臂系统

MRMU Mobile Remote Manipulating Unit 移动式遥控机械臂装置

MRO Maintenance, Repair and Overhaul 维修、修理与大检修

MRO Mars Reconnaissance Orbiter 火星探测轨道器

MROC Mobile Regional Operations Center 移动式区域作战（操作）中心

MROC Multiple Required Operational Capabilities 多种需求作战（操作）能力

MRP Manned Rotating Platform 载人旋转平台

MRP Missile Round Pallet 圆形导弹托架

MRR Mechanical Reliability Report 机械可靠性报告

MRR Mechanized Repair & Recovery Vehicle 机械修理与回收车

MRR Mission Readiness Review 任务准备状态评审

MRRV Manoeuvrable Reentry Research Vehicle 【美国】机动再入研究飞行器

MRS Management Review System 管理评审系统

MRS Manned and Retrievable System 载人与回收系统

MRS Manned Reconnaissance Satellite 载人侦察卫星

MRS Manned Reusable Spacecraft 可重复使用载人航天器

MRS Minimum Residual Shutdown 最小残余量关机

MRS Missile Reentry System 导弹再入系统

MRS Multiple Reusable Spacecraft 可多次使用的航天器

MRSE Microwave Remote Sensing Experiment 微波遥感实验

MRSM Maintenance and Reliability Simulation Model 维修与可靠性仿真模型

MRSR Mars Recover Sample Return 【美国】火星车取样返回

MRSS Mobile Range Safety System 移动式靶场安全系统

MRSSS Manned Revolving Space Systems Simulator 载人旋转式航天系统模拟器

MRSV Maneuverable and Recoverable Space Vehicle 机动与可回收的航天飞行器

MRT Maneuverable Reentry Technology 机动再入（大气层）技术研究

MRT Medium Range Target 中距离目标

MRTC Multiple Real-Time Commands 多个实时指令

MRTF Mean Residual Time to Failure 平均失效前剩余时间

MRTFB Major Range and Test Facility Base 主要靶场与试验设施基地

MRV Maneuverable Reentry Vehicle 机动再入飞行器（火箭、弹头）

MRV Maneuvering Reentry Vehicle 机动再入飞行器（火箭、弹头）

MRV Mars Roving Vehicle 火星漫游车

MRV Missile Recovery Vessel 导弹回收船

MRV Missile Reentry Vehicle 火箭再入（大气层）飞行器

MRV Multiple Reentry Vehicles 多弹头再入飞行器

MRVIS Mid-Apogee Reentry Vehicle Intercept System 中等弹道高度再入飞行器拦截系统

MRVL(L)P Maneuvering Reentry Vehicle for Low-Level Penetration 低空突防的机动再入飞行器（火箭、弹头）

MRWS Manned Remote Work Station 【美国航天飞机】载人遥控工作站

MS Man Station 载人空间站

MS Manned Satellite 载人卫星

MS Mass Spectrometry 质谱仪

MS Mating Sequence and Control 对接序列与控制

MS Mercury-Scout 【美国】"水星—侦察兵"航天器

MS&PA Mission Success and Product Assurance 任务成功率和产品保证

MS/MS Material Science/Manufacturing in Space 材料科学与空间制造

MSA Major Satellite Anomaly 【美国国防气象卫星计划】卫星严重异常

MSA Material Service Area 材料服务区

MSA Minimum Safe Altitude 最低安全高度

MSA Missile Security Area 导弹安全区

MSA Missile Storage Area 导弹储存区

MSA Mission Support Area 任务保障区域

MSA Mission Systems Avionics 任务系统航空电子设备

MSAA Mars Surface Anomaly Analysis 火星表面异常分析

MSAD Materials Summary Acceptance Document 材料汇总验收文件

MSAD Multi-Satellite Attitude Determination 多卫星姿态确定

MSAP Multi-Satellite Attitude Prediction 多卫星姿态预测

MSBLMS Multi Station Boundary Layer Model System 多站边界层模型系统

MSBLS Microwave Scanning Beam Landing System 【美国航天飞机着陆场】微波扫描波束着陆系统

MSC Master Sequence Controller 主序列控制器（装置）

MSC Military Space Command 军事航天司令部

MSC Mission Support Center 任务保障中心

MSC Mobile Satellite Communication 机动卫星通信

MSC Mobile Servicing Center 移动式检修服务中心

MSC Motor Speed Control 发动机转速控制

MSCB Missile Site Control Building 导弹发射场控制厂房

MSCC Manned Spaceflight Control Center 载人航天飞行控制中心

MSCC Manned Systems Control Center 载人系统控制中心

MSCC Missile Site Control Center 导弹发射场控制中心

MSCC Monitoring System Control Console 监控系统控制台；监控系统操作台

MSCE Mission Operations and Satel-

lite Control Element　任务操作与卫星控制元件
MSCOP　Missile System Checkout Program　导弹系统检测程序
MSCP　Master Shielding Computer Program　主屏蔽计算机程序
MSCP　Missile System Checkout Program　导弹系统检测程序
MSCP　Missile System Checkout Programmer　导弹系统检测程序装置
MSCS　Manned Spaceflight Control Squadron　载人航天飞行控制中队
MSD　Minimum Safe Distance　最小安全距离
MSD　Modular Security Device　模块化安全装置
MSDP　Missile Site Data Processor　导弹发射场数据处理机
MSDPS　Missile Site Data Processing System　导弹发射场数据处理系统
MSDRCS　Meteorological Satellite Data Receiving Ground Station　气象卫星数据接收地面站
MSDS　Missile Static Development Site　导弹静态试验场
MSDS　Missile System Development Stand　导弹系统研制试验台
MSDS　Multispectral Scanner and Data System　多频谱扫描器与数据系统
MSE　Maintenance Support Equipment　维修支援设备；保养支援设备
MSE　Mean Squared Error　平均方差
MSE　Measuring and Stimuli Equipment　测量与刺激设备
MSE　Mechanical Support Equipment　机械保障设备
MSE　Medical Support Equipment　医疗保障设备
MSE　Mobile Subscriber Equipment　移动用户设备
MSE　Multiple Sample Exchanger　多次采（抽）样器
MSE　Multiple Simultaneous Engagements　多枚同时交战
MSEC　Master Separation Events Controller　主分离事件控制器
MSECU　Missile Setting Equipment Control Unit　导弹就位设备控制装置
MSEF　Missile System Evaluation Flight　导弹系统鉴定飞行
MSEG&C　Multi-Spectral Environmental Generator and Chamber　多光谱环境生成器和实验舱
MSEL　Master Scenario Events List　主场景事件清单
MSEPS　Modular Space Electrical Power Station　积木式空间发电站；积木式空间电力站
MSER　Material Science Experiment Rocket　材料科学实验火箭
MSER　Multiple Stores Ejection (Ejector) Rack　弹射式多外挂物挂架；多外挂弹射架
MSF　Manned Space Flight　载人航天飞行
MSFC　Manned Space Flight Center　载人航天飞行中心
MSFC　Marshall Space Flight Center　【美国国家航空航天局】马歇尔航天中心
MSFCC　Manned Space Flight Control Center　载人航天飞行控制中心
MSFDPS　Manned Space Flight Data Processing System　载人航天飞行数据处理系统
MSFH　Manned Space Flight Headquarters　载人航天飞行司令部
MSFL　Manned Space Flight Laboratory　载人航天飞行实验室

MSFLS Microwave Scanning Beam Land Station 微波扫描波束着陆站
MSFLS Microwave Scanning Beam Landing System 微波扫描波束着陆系统
MSFLV Manned Space Flight and Launch Vehicle 载人航天飞行与运载火箭
MSFN Manned Space Flight Network 载人航天飞行网络
MSFNOC Manned Space Flight Network Operation Center 载人航天飞行网络操作中心
MSFO Manned Space Flight Operations 载人航天飞行运营
MSFP Manned Space Flight Program 载人航天计划
MSFS Manned Space Flight Simulator 载人航天模拟器
MSFS Manned Space Flight System 载人航天飞行系统
MSFSG Manned Space Flight Support Group 【美国空军】载人航天保障大队
MSG METEOSAT Second Generation 第二代气象卫星（欧洲气象卫星的后继型号）
MSG Microgravity Science Glove Box 微重力科学手套式操作箱
MSGP Microwave Sounder for Geostationary Platform 【美国国家航空航天局】对地静止平台的微波探测器
MSHI Medium-Scale Hybrid Integration 中等规模混合集成（电路）
MSI Manned Satellite Interception 载人卫星拦截
MSI Man-System Integration 人与系统的结合；人机一体化
MSI Medium Scale Integration 中等规模集成电路
MSI Missile Status Indicator 导弹状态指示器
MSI Missile Subsystem Integration 导弹子系统总合
MSI Moon's Sphere of Influence 月球的引力范围
MSIA Multispectral Image Analyzer 多谱成像分析器
MSIAC Modeling and Simulation Information Analysis Center 建模与仿真信息分析中心
MSIC Missile and Space Intelligence Center 导弹与航天情报中心
MSID Measurement Stimulation Identification 测量刺激识别
MSIL Mission Systems Integration Laboratory 任务系统综合实验室
MSIS Man Systems Integration Standard 载人系统集成标准
MSIS Manned Satellite Inspection System 载人检查卫星系统
MSITP Modeling, Simulation, Information Technology and Processing 建模、仿真、信息技术及处理
MSK Mechanical Steering Kit 机械操纵设备
MSL Manned Space Laboratory 载人航天实验室
MSL Mapping Sciences Laboratory 【美国国家航空航天局】地图绘制科学实验室
MSL Mars Science Laboratory 火星科学实验室
MSL Materials Science Laboratory 【美国航天飞机】材料科学实验室
MSL Mechanical System(s) Laboratory 机械系统实验室
MSL Meteorological Satellite Laboratory 【美国】气象卫星实验室

MSLD Mass Spectrometer Leak Detector 质谱分光漏泄检测器
MSLEML Materials Science Laboratory Electromagnetic Levitator 材料科学实验室电磁悬浮器
MSLS Maneuverable Satellite Landing System 机动卫星着陆系统；可机动卫星着陆系统
MSLS Multi-Service Launch Systems 多种服务发射系统
MSM Manned Support Module 载人保障舱
MSM Mars Surface Module 火星表面（观测）舱
MSM Mission Simulation Model 任务模拟模型
MSMCP Maintenance Support Mission Completion Probability 维修保障任务完成概率
MSME Medical, Space, and Mission Electronics 医疗、空间与任务电子设备
MSMM Materials, Structures, Mechanical Systems and Manufacturing 材料、结构、机械系统和制造
MSMS Maximum Safety Mach System 最大安全马赫数系统
MSMV Monostable Multi-Vibrator 单稳态多谐振荡器
MSN Manned Space Network 载人航天网
MSN Message Switching Network 信息转换网络
MSN Multiple Subscriber Number 多（重）用户号码
MSO Manned Spacecraft Operation 载人航天器操作
MSO Military Satellite Communications Systems Organization 军用卫星通信系统组织
MSO Model for Spare(s) Optimization 备件优化模型
MSOB Manned Spacecraft Operations Building 载人航天器操作厂房
MSOCC Multi-Satellite Operation Control Center 【美国国家航空航天局】多卫星运行控制中心
MSOE Multi-Band Spectral Observation Equipment 多频带谱线观测设备
MSOGS Molecular-Sieve Oxygen Generation System 分子筛制氧系统
MSOL Manned Scientific Orbital Laboratory 载人轨道科学实验室
MSOM Manned Solar Observatory Mission 载人太阳观测任务
MSOP Measurement System Operation Procedure 测量系统操作程序
MSP Mars Surface Probe 火星表面探测器
MSP Material Space Platform 空间材料平台
MSP Maximum Structural Payload 最大结构有效载荷
MSP Military Space Plane 军事航天飞机
MSPITT Military Space Plane Integrated Technology Testbed 军用航天飞机综合技术试验台
MSPR Multi-Purpose Small Payload Rack 多用途小型有效载荷机柜
MSPSP Missile System Pre-launch Safety Package 导弹系统射前安全数据包
MSR Mars Sample Return 火星取样返回；火星标本采集
MSR Meteorological Sounding Rocket 气象探测火箭
MSR Minimum Security Requirements 最低安全要求

MSR Missile Simulation Round 导弹模拟弹

MSR Missile Site Radar 导弹发射场雷达

MSR Mobile Sea Range 移动式海上靶场

MSR Multi-Stage Rocket 多级火箭

MSRE Moon Signal Rejection Equipment 月球信号抑制设备

MSRF Microwave Space Research Facility 微波航天研究设施

MSRM Mars Sample Return Mission 【苏联】火星取样返回任务

MSRP Massive Selective Retaliatory Power 大规模选择性报复打击力量

MSRSIM Missile Site Radar Simulation 导弹发射场雷达模拟

MSRT Multistage Shuttle Run Test 多级航天飞机运行测试

MSS Magnetic Shield Simulator 磁屏蔽模拟器

MSS Magnetic Storm Satellite 磁暴探测卫星

MSS Maintenance Status System 维修状态系统

MSS Manned Space Station 载人空间站

MSS Manned Space System 载人航天系统

MSS Maritime Satellite Service 海事卫星服务

MSS Material Science in Space 空间材料科学；太空材料科学

MSS Mechanical Support Systems 机械保障系统

MSS (NASA) Message Switching System 【美国国家航空航天局】信息交换系统

MSS Meteorological Sounding System 气象探测系统

MSS Midcourse Surveillance System 导弹中段监视系统

MSS Missile Sensor Study 导弹传感器研究

MSS Missile Sight Subsystem 导弹瞄准具子系统

MSS Missile Stabilization System 导弹稳定系统

MSS Missile Station Select 导弹阵地选择

MSS Missile Support Subsystem 导弹辅助子系统

MSS Missile Suspension System 导弹悬挂系统

MSS Mobile Satellite System 移动卫星系统；机动卫星系统

MSS Mobile Service Structure 活动勤务结构；活动勤务塔

MSS Mobile Servicing System 移动式维修系统（美国空间站机械臂操作系统）

MSS Modeling and Simulation Support 建模与模拟支持（保障）

MSS Modular Space Station 模块式空间站

MSS Moored Sonobuoy System 系留声呐浮标系统

MSS Multi-Satellite System 多卫星系统

MSS Multispectral Scanner System 多光谱扫描器系统

MSS Munitions Survivability Software 弹药可靠性软件

MSSB Missile Servicing and Storage Building 导弹勤务与贮存厂房；导弹维护与储存大楼

MSSCC Military Space Surveillance Control Center 军事航天监视控制中心

MSSCS Manned Space Station Com-

munication System 载人空间站通信系统
MSSE Missile System Support Equipment 导弹系统保障设备
MSSF Man Machine System Simulation Facility 人机系统模拟设施
MSSL Mullard Space Science Laboratory 【英国】马拉德太空科学实验室
MSSM Mars Spinning Support Module 火星飞船旋转辅助舱
MSSR Mars Soil Sample Return 火星土壤取样返回
MSSR Mars Soil (Surface) Sample Return 火星土壤（表面）取样回返
MSSR Mars Surface Sample Return 火星表面取样返回
MSSS Man/Seat Separation System 人椅分离系统
MSSS Manned Space Station Simulator 载人空间站模拟器
MSSS Manned Static Space Simulator 静态载人航天模拟器
MSSS Maui Space Surveillance Site 【美国】毛伊岛航天监视站
MSSS Maui Space Surveillance System 毛伊岛航天监视系统
MSSS Mobile Space Support System 移动航天保障系统
MSSS Multi Satellite Support System 多卫星支持系统
MST Main Satellite Thruster 主卫星推力器
MST Mercury System Test 【美国】"水星"系统测试
MST Missile Service Tower 导弹勤务塔
MST Missile Surveillance Technology 导弹监视技术
MST Missile System Test 导弹系统测试
MST Mission Sequence Test 任务序列测试（试验）
MST Mobile Service Tower 活动勤务塔
MST Module Service Tool 模块勤务工具
MSTAR Missile Defense Science, Technology & Research 导弹防御科学、技术与研究
MSTAR MLRS Smart Tactical Rocket 多管火箭发射系统战术制导火箭
MSTAR Moving and Stationary Target Acquisition and Recognition 移动与静止目标截获与识别
MSTC Manned Spacecraft Test Center 【美国】载人航天器试验中心
MSTER Meteoroid and Space Debris Terrestrial Environment Reference 流星体和空间碎片陆地环境参考（基准）
MSTI Miniature Seeker (Sensor) Technology Integration 小型搜索器（遥感器）技术综合卫星
MSTI Miniature Sensor Technology Integration Satellite 微型传感器技术综合卫星
MSTM Missile Service Test Tool 导弹技术勤务试验模型
MSTP Manned Space Transportation Program(me) 载人航天运输计划
MSTR Mobile Sea Test Range 流动海上试验靶场
MSTRS Miniature Satellite Threat Reporting System 微型卫星威胁通报系统
MSTRS Miniaturized Satellite Threat Reporting System 小型卫星威胁报告系统
MSTS Meteorology System Test Satellite 气象系统试验卫星

MSTS Micro Satellite Target System 微型卫星目标系统
MSTS Mid-Course Surveillance and Tracking System 中程监视与跟踪系统
MSTS Missile Static Test Site 导弹静态试验场
MSTS Multi Source Tactical System 多源战术系统
MSTS Multi-Source Targeting System 多源目标确定系统
MSTV Manned Supersonic Test Vehicle 载人超声速试验飞行器；有人驾驶超声速试验飞行器
MSU Module Support Unit 模块式保障设备
MSV Manned Space Vehicle 载人航天飞行器
MSV Mars Surface Vehicle 火星表面飞行器
MSV Modular Support Vehicle 组合式保障车辆
MSX Midcourse Space Experiment 【美国】空间中段监视试验卫星
MT Maximum Torque 最大扭矩
MT Mean Time 平均时间
MT Missile Test 导弹试验
MT Missile Tracker 导弹跟踪系统
MT Mission Trajectory 任务轨迹；任务弹道
MT Mobile Transporter 【美国空间站】移动运输器
MT&CE Missile Test and Checkout Equipment 导弹试验与检测设备
MT&L Manufacturing,Test,and Logistics 制造、试验与后勤保障
MTA Magnetic Torquer Assembly 磁扭矩装置组件
MTA Maintenance Task Analysis 维修任务分析

MTA Major Test Article 主要测试（试验）部件
MTA Mass Thermal Analysis 质量热分析
MTA Missile Transfer Area 导弹转运区
MTACCS Marine Tactical Air Command and Control System 【美国】海军战术空中指挥与控制系统
MTAD Multitrace Analysis Display 多轨迹分析显示
MTASS Multi-Mission Three-Axis Stabilized Satellite 多任务三轴稳定控制卫星
MTBA Mean Time between Alarm [Warning] 平均告警间隔时间
MTBAA Mean Time between Avionics Anomaly 航空电子系统平均故障间隔时间
MTBCF Mean-Time-before-Critical-Failure 平均关键故障前的时间
MTBD Mean Time between Demand 平均需求间隔时间
MTBF Mean Time between Failures 故障间的平均时间
MTBFS MTBF Software 故障间平均时间软件
MTBM Mean Time between Maintenance 平均维修间隔时间
MTBM Mean-Time-between Unscheduled Maintenance 平均无计划性维修间隔时间
MTBMA Mean Time between Maintenance Action 平均维修措施间隔时间
MTBO Mean Time between Overhauls 平均大修间隔时间
MTBOMF Mean Time between Operational Mission Failures 平均操作执行任务失败时间

MTBR Mean Time between Replacement　平均替换间隔时间
MTBR Mean-Time-between-Removal　平均拆除间隔时间
MTBUM Mean Time between Unscheduled Maintenance　平均非计划内维修间隔时间
MTC Mars Transport Craft　【美国拟议中的飞向火星的载人飞船】火星运输飞船
MTC Master Thrust Control　主推力控制
MTC Master Thrust Controller　主推力控制器（装置）
MTC Missile Test Center　导弹试验中心
MTC Missile Transfer Car　导弹运输车
MTC Mission Training Center　任务训练中心
MTCA Monitor and Test Control Area　监测与测试控制区
MTCR Missile Technology Control Regime　导弹技术控制方法
MTCT Manipulator/Teleoperator Control Technology　机械臂与遥控操作器控制技术
MTD Maintenance Task Distribution　维护任务分配
MTD Missile Technology Demonstration　导弹技术演示验证
MTDE Modern Technology Demonstrate Engine　现代技术演示验证发动机
MTDS Minimum Technical Data Set　最小技术数据设备（装置）
MTDS Missile Trajectory Data System　导弹弹道数据系统
MTE Maintenance Test Equipment　维修测试设备
MTE Maximum Tracking Error　最大跟踪误差
MTE Mesosphere-Thermosphere Explorer　【美国】中间层—热层探险者（卫星）
MTE Missile Targeting Equipment　导弹目标（瞄准）设备
MTE Missile Test Equipment　导弹试验设备
MTE Multi-System Test Equipment　多系统测试设备
MTEC Maintenance Test Equipment Catalog　维修测试设备目录
MTEE Electrical Maintenance Test Equipment　电气维修测试设备
MTEEC Electronic Maintenance Test Equipment　电子维修测试设备
MTEF Fluid Maintenance Test Equipment　液体维修测试设备
MTEM Maintenance Test Equipment Module　维修测试设备模块（舱）
MTEM Mechanical Maintenance Test Equipment　机械维修测试设备
MTEO Optical Maintenance Test Equipment　光学维修测试设备
MTEWS(A) Mobile Tactical Early Warning System for Air Defense　移动式战术防空预警系统
MTEX Mission Template Expert　飞行任务模型专家
MTF Maintenance Test Flight　维修测试飞行
MTF Mean Time to Failures　平均失效（故障）前时间
MTF Medium Time to Failures　失效（故障）前中位时间
MTF Missile Training Facility　导弹训练设施
MTFD Minimum Tracking Flux Density　最小跟踪通量密度

MTFF Man-Tended Free Flyer 【欧洲空间局】有人照料的自由飞行平台

MTFO Modular Training Field Option 模块化培训场方案

MTGCF Mobile Transportation Ground Command Facility 机动运输地面指挥设施

MTGS Mid-Course and Terminal Guidance System 【导弹】中段与末段制导系统

MTI Missile Training Installation 导弹训练装置

MTI Moving Target Indicator 移动目标指示器

MTIE Micro-Thrust Ion Engine 微推力离子发动机

MTIL Maxim Tolerable Insecurity Level 最大可容忍的不安全度

MTIS Missile Tracking Instrumentation System 导弹跟踪测量系统

MTL Main Transfer Line 主要输送线路

MTL Materials Technology Laboratory 材料技术实验室

MTLB Soviet Universal Combat Vehicle 苏联太空作战飞行器

MTM Maneuvering Tactical Missile 机动战术导弹

MTM Maximum Takeoff Mass 最大起飞质量

MTM Mechanical Test Model 机械试验模型

MTM Mission Test Module 任务测试模块（舱）

MTM Module Test and Maintenance 模块测试与维修

MTMC Military Traffic Management Control 军事交通管理控制

MTMIU Module Test and Maintenance Bus Interface Unit 模块测试与维修总线接口单元

MTOGW Maximum Takeoff Gross Weight 最大起飞总重量

MTOW Maximum Takeoff Weight 最大起飞重量

MTP Manufacturing Technical Procedure 生产（制造）技术规程

MTP Master Test Plan 主测试（试验）计划

MTP Meteosat Transition Programme 气象卫星转移计划

MTP Missile Tube Pressurization 导弹发射管增压

MTP Mission Test Plan 任务测试计划

MTPE Mission To Planet Earth 行星—地球任务

MTPU Missile Tank Pressurization Unit 导弹贮箱增压装置

MTR Maximum Tracking Range 最大跟踪距离

MTR Minimum Turning Radius 最小转弯半径

MTR Missile Tracking Radar 导弹跟踪雷达

MTRE Missile Test and Readiness Equipment 导弹测试与准备状态试验设备

MTRE Missile Test and Readiness Evaluation 导弹测试与准备状态鉴定

MTRF Modular Test and Repair Facility 模块化测试与修理设备

MTRI Missile Test Range Instrumentation 导弹试验靶场仪器设备；导弹试验靶场测量仪表

MTRS Mobile Telemetry Receiving System 移动式遥测接收系统

MTS Master Timing System 主计时系统

MTS　Meteoroid Technological Satellite　【美国】流星体技术卫星（太空物理卫星）

MTS　Microwave Temperature Sounder　微波温度探测器

MTS　Missile Test Set　导弹试验装置

MTS　Missile Test Stand　导弹试验台

MTS　Missile Test Station　导弹试验站

MTS　Missile Tracking Sensor　导弹跟踪传感器

MTS　Missile Tracking System　导弹跟踪系统

MTS　Mobile Test Set　移动试验装置

MTS　Mobile Tracking Station　移动式跟踪站

MTS　Mobile Training Set　移动训练装置

MTS　Module Test System　模块测试系统

MTS　Module Testing System　组件试验系统

MTS　Movement Tracking System　运动跟踪系统

MTS-B　Multispectral Targeting System, Version B　B版多频谱目标判定系统

MTSAT　Multifunctional Transport Satellite　【日本】多功能传送卫星

MTSC　Micro Satellite Target System　微型卫星目标系统

MTSF　Mean Time to System Failure　系统故障前平均时间

MTSL　Microelectronics Technology Support Laboratory　【欧洲空间局】微电子技术支持实验室

MTSS　Maintenance Test and Simulation System　维修前检测和仿真系统

MTSS　Manned Test Space Station　载人试验航天站

MTSS　Manned Test Space System　载人试验航天系统

MTSS　Military Test Space Station　军用试验空间站

MTT　Maximum Touch Temperature　最大触摸温度

MTTA　Mean Time to Accomplish　平均完成时间

MTTE　Mean Time to Exchange　平均交换时间

MTTF　Mean Time to Failure　平均故障（失效）时间

MTTFF　Mean Time to First Failure　首次故障（失效）平均时间

MTTR　Mean Time to Repair　平均修理时间

MTTRS　Mean Time to Restore System　存储系统平均时间

MTTS　Mobile Target Tracking System　【美国】机动式目标跟踪系统

MTTV　Maneuvering Tactical Target Vehicle　机动战术目标飞行器（火箭、弹头）

MTV　Maneuvering Target Vehicle　机动目标飞行器（火箭、弹头）

MTV　Mars Transfer Vehicle　火星运输车

MTV　Missile Test Vehicle　导弹试用车；导弹试验飞行器

MTV　More Electric Test Vehicle　多电子设备测试车

MTV　Multipurpose Transatmospheric Vehicle　多用途穿越大气层航天器

MTV　Munition Test Vehicle　弹药测试车

MTVC　Main Thrust Vector Control　主推力矢量控制

MTWS　MAGTF (Marine Air-Ground Task Force) Tactical Warfare Simulation　【美国】海军陆战队空地特遣

队战术作战模拟（仿真）

MTVC Manual Thrust Vector Control 人工推力矢量控制；推力矢量手控

MTVC Motor Thrust Vector Control 【导弹】发动机推力矢量控制

MU-SERIES Japanese Space Launch Vehicle 日本 MU 航天运载火箭系统

MUA Maximum Usable Altitude 最大可用高度

MUF Maximum Usable Frequency 最大可用频率

MUF Model Uncertainty Factor 模型不确定因子

MULTICS Multiplexed Information and Computing System 信息多路转接与计算系统

MULTOS Multi-Application Operating System 多应用程序操作系统

MULTOTS Multiple Unit Linktest and Operational Training System 多单元链路测试与操作训练系统

MULTS Mobile Universal Link Translator System 机动通用链路转发系统

MUMS Multiple Use Marc System 多用途不可溶物质系统

MUOS Mobile User Objective System 移动用户目标系统

MUPID Multipurpose Universally Programmable Intelligent Decoder 多功能通用可编程智能解码器

MURP Manned Upper Stage Reusable Payload 可重复使用载人上面级有效载荷

MUS Mission Unique Software 任务专用软件

MUSAP MultiSatellite Augmentation Program 多卫星增强项目

MUSAT Canadian Government Satellite 加拿大政府卫星

MUSAT Multipurpose UHF Satellite 多用途超高频卫星

MUSE Mobile Utility Support Equipment 机动的多用途保障设备

MUSE Monitor of Ultraviolet Solar Energy 紫外太阳能监测器

MUSR Monitor of Ultraviolet Solar Radiation 太阳紫外辐射监视仪

MUST Multimission UHF SATCOM Terminal 多任务超高频卫星通信终端

MUSTAD MultiUnit Space Transport and Recovery Device 多组件航天运输与回收装置

MUSTRAC Multiplesimultaneous-Target Steerable Telemetry Tracking System 可同时操纵的多个目标遥测跟踪系统

MUT Modular Universal Terminal 模块化通用终端

MUTE Mobile Universal Test Equipment 移动通用试验设备

MUX-AEC Multiplex Automatic Error Correction 多通道自动误差修正

MV ManufacturingVerification 生产（制造）验证

MVA Machinery Vibration Analysis 机械装备振动分析

MVA Main Valve Actuator 主阀门致动器（装置）

MVA Measurement of Variable Activity 可变活性测量；可变辐射性（强度）测量

MVDF Modular VHF Direction Finder 模件式甚高频测向仪

MVF Manned Vertical Flight 载人垂直飞行

MVFV Manned Venus Flyby Vehicle 飞越金星的载人飞行器

MVGVT Mated Vertical Ground Vibration Tests 【美国航天飞机】整

机垂直地面振动试验

MVM Mariner-Venus-Mercury 【美国】"水手号"金星—水星探测器

MVOD Minimum Variance Orbit Determination 最小偏差轨道测定

MVS Minimum Vector Speed 最小矢量速度

MVS Missile Velocity Servo 导弹速度伺服机构

MVS Mission Video System 飞行任务视频信号系统

MVS Modularized Vehicle Simulation 模块化飞行器模拟；组件式飞行器模拟

MVS Multiple Virtual Storage 复式虚拟存储

MVSM Mixed Vector Space Model 混合的矢量空间模型

MVT Malfunction Verification Test 故障验证试验

MVT Mission Verification Test 任务验证测试（试验）

MVT Multi-Programming with a Variable Number of Tasks 可变任务数多道程序设计系统

MVULE Minimum Variance Unbiased Linear Estimate 极小方差无偏线性估计量

MVW Missile Viewing Window 导弹观察窗

MW-RAILS Microwave Remote Area Instrument Landing System 微波远方仪表降落系统

MWAE Minim-Weight-Absolute-Error 最小加权绝对误差

MWBS Manufacturing Work Breakdown Structure 制造执行任务分解结构（一种统筹方法）

MWC Missile Warning Center 导弹预警中心

MWCS Missile Weapons Control System 导弹武器控制系统

MWDS Missile Warning Display Subsystem 导弹警报显示子系统

MWE Maximum Weight Empty 最大空重

MWHS Modified Warhead Section 改进型弹头部分

MWIR Midwavelength Infrared 中波红外

MWNT Multiwall Carbon Nanotubes 多壁碳纳米管

MWP Maneuvering Working Platform 【美国空间站】机动飞行工作平台

MWP Maximum Working Pressure 最大工作压力

MWR Mean Width Ratio 平均宽度比

MWS Maintenance Workstation 内场用的维修工作站

MWS Maximum Wale Spacing 最大横撑间隙

MWS Maximum Working Space 最大工作空间

MWS Modular Workstation 模块化工作站

MWV Maximum Working Voltage 最大工作电压

MWVS Mission Weapon Visionics System 任务武器视觉系统

MX Peacekeeper Ballistic Missile 【美国】"卫士"弹道导弹

MYTA Maintainability Task Analysis 维修性任务分析

MYVAL Maintainability Evaluation 维修性鉴定

MZFW Maximum Zero Fuel Weight 最大无燃油重量

MZFW Maximum Zero-Fuel Weight 无燃料时最大重量

N

N&G Navigation and Guidance 导航和制导

N/C Nose Cone 鼻锥，头锥；(导弹、火箭) 圆锥形弹头

N/UWSS NORAD/USSPACECOM Warfighting Support System 北美航空航天防御司令部/美国航天司令部作战支援系统

N/W Network 网络

N₂ Nitrogen 氮（N）；氮气

N₂H₄ Hydrazine 肼

N₂O Nitrous Oxide 一氧化二氮；氧化亚氮

N₂O₄ Nitrogen Tetroxide 四氧化二氮

NA Numerical Analysis 数值分析

NA Numerical Aperture 数值孔径

NAA National Aerospace Association 【美国】国家航空航天协会

NAAD North America Aerospace Defense 北美航空航天防御

NAAL North American Aerodynamic Laboratory (Wind Tunnel) 北美空气动力学实验室（风洞）

NAAPS Nozzle Actuator Auxiliary Power Supply 喷管作动器辅助电源

NAARS NASA Aviation Anomaly Reporting System 美国国家航空航天局异常通报系统

NAC NASA Advisory Council 美国国家航空航天局咨询委员会

NACA National Advisory Committee for Aeronautics 【美国】国家航空咨询委员会

NACE Neutral Atmospheric Composition Experiment 中间大气层结构实验

NACES Naval Aircrew Common Escape System 【美国】海军空勤人员通用逃生系统

NACS Nonlinear Automatic Control System 非线性自动控制系统

NACSI National Communications Security Instruction 【美国】国家通信安全指南

NADAC Navigation Data Assimilation Computer 导航数据类比计算机

NADGE NATO Air Defense Ground Environment 北约空防地面设施网

NADIR Network Anomaly Detection Intrusion Reported 网络异常探测与侵入通报（侵入发现系统）

NADUC Nimbus/ATS Data Utilization Center 【美国】"雨云"卫星/应用技术卫星数据利用中心

NAEDS Non-Aqueous Equipment Decontamination System 无水装备洗消系统

NAFCOM NASA and Air Force Cost Model 美国国家航空航天局与空军成本模型

NAFIS Navigational Air Flight Inspection System 导航飞行检查系统

NAGC Noise Automatic Gain Control 噪声自动增益控制

NAI National Aerospace Initiative 【美国】国家航空航天倡议

NAIC National Astronomy and Ionosphere Center 【美国】国家天文与电离层中心

NAIF Navigation Ancillary Information Facility 导航辅助信息设备
NAIP Nanoscale Aluminum-Ice Propellant 纳米级铝-冰火箭推进剂
NAL National Aerospace Laboratory of Japan 日本国家航空航天实验室
NAMES NAVDAC (Navigation Data Assimilation Computer) Assembly, Monitor, Executive System 导航数据类比计算机汇编、监控和执行系统
NAMFI NATO Missile Firing Installation 北约导弹发射场
NAMIS NASA Aircraft Maintenance Information System 美国国家航空航天局飞机维修信息系统
NAMT National Advanced Manufacturing Test-Bed 国家先进制造试验平台
NAOC National Airborne Operations Center 【美国】国家机载作战（操作）中心
NAP Navigation Analysis Program 导航分析程序
NAR Numerical Analysis Research 数值分析研究
NARCOM North Atlantic Relay Communication Satellite 北大西洋中继通信卫星
NAREL National Air and Radiation Environmental Laboratory 【美国】国家大气与辐射环境实验室
NARF Nuclear Aerospace Research Facility 【美国空军】航天核能研究设施
NARRDS National Advanced Reactor Reliability Data System 【美国】国家先进反应堆可靠性数据系统
NAS NASA Advanced Supercomputer 美国国家航空航天局的先进超大规模计算机
NAS National Aerospace Standards 【美国】国家航空航天标准
NAS National Airspace System 【美国】国家领空体系
NAS Navigation and Avoidance System 导航与防撞系统
NAS Numerical Aerodynamic Simulator 数字空气动力学模拟器（美国国家航空航天局的一种大型计算机）
NAS Numerical Aerospace Simulation 航空航天数字仿真
NASA National Aeronautic and Space Administration 美国国家航空航天局
NASA National Aerospace and Space Act 【美国】《国家航空和航天法》
NASA DFRC NASA Dryden Flight Research Center 美国国家航空航天局德莱顿飞行研究中心
NASA ERC NASA Electronic Research Center 美国国家航空航天局电子研究中心
NASA KSC NASA Kennedy Space Center 美国国家航空航天局肯尼迪航天中心
NASA MSC NASA Manned Spacecraft Center 美国国家航空航天局载人航天器中心
NASA RECON NASA Remote Console 美国国家航空航天局遥控台
NASA STIF NASA Scientific and Technical Information Facility 美国国家航空航天局科技情报（或信息）所
NASA TM NASA Tech Memo 《美国国家航空航天局技术备忘录》
NASA TR NASA Test Reactor 美国国家航空航天局试验反应堆

NASA/STIMS NASA Scientific and Technical Information Modular System 美国国家航空航天局科技情报模块系统

NASAAT NASA Apollo Trajectory 美国国家航空航天局的"阿波罗"飞船轨道

NASACOM NASA Communications Network 美国国家航空航天局通信网

NASACOP NASA Communications Operating Procedures 美国国家航空航天局通信操作规程

NASAP Network Analysis for Systems Applications Program 系统应用程序网络分析

NASAPR NASA Procurement Regulation 美国国家航空航天局采购条例

NASCAD NASA Computer Aided Design 美国国家航空航天局计算机辅助设计

NASCOM NASA Communications (Network) 美国国家航空航天局通信（网络）

NASCOM NASA Worldwide Communications Network 美国国家航空航天局全球通信网络

NASCOP NASA Communications Operating Procedures 美国国家航空航天局通信操作程序

NASDA National Space Development Agency 【日本】国家空间发展机构

NASDAC National Aviation Safety Data Analysis Center 国家航空安全数据分析中心

NASIP Navy Airframe Structural Integrity Program 【美国】海军飞机结构完整性计划

NASIRC NASA Automated Systems Internet Response Capability 美国国家航空航天局自动系统互联网响应能力

NASIS NASA Aerospace Safety Information System 美国国家航空航天局航空与航天安全情报系统

NASM National Air and Space Museum 美国国家航空航天局博物馆

NASM National Air and Space (Warfare) Model 【美国】国家航空航天（作战）模型

NASM National Air and Space (Warfare) Model 国家航空与航天（作战）模型

NASO National Astronomical Space Observatory 【美国】国家空间天文观测台

NASREM NASA Standard Reference Model 美国国家航空航天局标准参考模型

NASTAD National Aerospace Laboratory, STOL Aircraft Design Consortium 【美国】国家航空航天实验室短距离起降设计财团

NASTRAN NASA Structural Analysis 美国国家航空航天局结构分析程序

NAT NASA Apollo Trajectory 美国国家航空航天局"阿波罗"飞船轨道

NATC Naval Air Test Center 【美国】海军空中试验中心

NATOPS Naval Air Training and Operating Procedures Standardization 【美国】海军空中训练与操作规程标准化

NATOSAT NATO Satellite 北约卫星

NATS NMCS AUTODIN Terminal Subsystem 【美国】国家军事指挥系统自动数字信息网络终端子系统

NAVACO Navigation Action Cut Out

Switchboard 导航操作断流开关板
NAVAID Navigation Aid 导航辅助设备
NAVAIRSYSCOM Naval Air Systems Command 【美国】海军空中系统指挥部（指令）
NAVAR Navigation and Ranging 导航与测距
NAVAR Navigation Radar 导航雷达
NAVARHO Navigation and Radio Homing 导航与无线电寻的系统
NAVCOM Navigation and Communication Equipment 导航通信设备
NAVDAC Navigation Data Assimilation Center 导航数据同化中心
NAVDAC Navigation Data Assimilation Computer 导航数据同化计算机
NAVDSOC Navy Defense System Operations Center 【美国】海军防御系统作战（操作）中心
NAVEX Navigation Experiment (Spacelab D-1 Experiment) 导航试验（空间实验室 D-1 试验）
NAVFAC Navy Facilities Engineering Command 【美国】海军设施工程司令部
NAVGS Navigation Glide Slope 航行下滑坡度
NAVHARS Navigation, Heading and Altitude Reference System 导航、航向与高度参考系统
NAVIDS/SIS Navigation Aids/Selective Identification System 导航设备/选择识别系统
NAVMACS Navy Modular Automated Communications System 【美国】海军模块化自动通信系统
NAVPOOL Navigation Parameter Common Pool 导航参数共用

NAVS Navigation System 导航系统
NAVSAT Navigation Satellite 导航卫星
NAVSAT Navigation-Communication Satellites 导航通信卫星
NAVSAT United States Navy Navigation Satellite System 美国海军导航卫星系统
NAVSATS Navigational Satellite System 导航卫星系统
NAVSPACE Naval Space Command 【美国】海军航天司令部
NAVSPOC Naval Space Operations Center 【美国】海军航天作战（操作）中心
NAVSTAR Navigation System Using Timing and Ranging 导航星（美国 GPS 全球定位系统卫星）
NAVSTAR Navigational System Tracking and Range 导航系统跟踪及范围
NAVSUP Naval Supply Systems Command 【美国】海军保障系统司令部
NavTEL Navigational Test and Evaluation Laboratory 导航试验（测试）与鉴定实验室
NAVWASS Navigation and Weapon-Aiming Subsystem 导航与武器瞄准子系统
NAWAS National Warning System 【美国】国家预警系统
NB Neutral Buoyancy 中性浮力
NB Nimbus 【美国】"雨云"（气象）卫星
NBA Node Buffer Allocation 节点缓冲器分配
NBC Nuclear, Biological, Chemical 核、生物、化学
NBCC Nuclear, Biological and Chem-

ical Contamination 核、生物、化学污染

NBD Network Based Defense 基于网络的国防

NBDN Nuclear Burst Detector Network 核爆炸探测网

NBDS Nuclear-Burst Detection System 核爆炸探测系统

NBF Neutral Buoyancy Facility 中性浮力设施

NBILST Narrow Beam Interleaved Search and Track 窄波束交叠搜索与跟踪

NBL Neutral Buoyancy Laboratory 中性浮力实验室

NBM Nuclear Ballistic Missile 核弹道导弹

NBS Neutral Buoyancy Simulator 中性浮力模拟器

NBSR NASA Biological Specimen Repository 美国国家航空航天局生物样本仓库

NBT Neutral Buoyancy Trainer 中性浮力培训装置（员）

NBT New Boost Technique 新型助推技术

NC Normal Corrective Maneuver 正常修正

NC Nozzle Control 喷口控制器

NC Numerical Control 数字（值）控制

NCADE Net Centric Airborne Defense Element 网络中心机载防御单元（武器系统）

NCAR National Center for Atmospheric Research 【美国】国家大气研究中心

NCAR Non-Conformance Analysis Report 不一致性分析报告

NCC Navigation Computer Control 导航计算机控制

NCC Network Control Center 【美国国家航空航天局】网络控制中心

NCCIS NASA (National Aeronautical and Space Administration) Command, Control and Information System 美国国家航空航天局指挥控制与信息系统

NCCOSC Naval Command, Control, and Ocean Surveillance Center 【美国】海军指挥、控制与海洋监视中心

NCCR New Construction/Conversion Requirements (System) 新建和改装技术要求系统

NCCS NASA Center for Computational Sciences 美国国家航空航天局计算科学中心

NCCS Navy Command and Control System 【美国】海军指挥、控制系统

NCCT Network Centric Collaborative Targeting 网络中心协同目标定位系统

NCDCS Narrow Band Coherent Data Collection System 窄带相关数据收集系统

NCDS NASA Climate Data System 美国国家航空航天局气候数据系统

NCEP National Centers for Environmental Prediction 【美国国家海洋与大气管理局】国家环境预测中心

NCES Network Centered Enterprise System 以网络为中心的企业系统

NCES Network Centric Enterprise Services 网络中心企业服务

NCIU Network Common Interface Unit 网络通用接口装置

NCIU NEXRAD Communications Interface Unit 改进型气象雷达通信

接口单元

NCOS National Commission on Space 【美国】国家空间委员会

NCOS Network Computer Operating System 网络计算机操作系统

NCP (NASA) Network Consolidated Plan 【美国国家航空航天局】网络综合计划

NCP Network Control Processor 网络控制处理器

NCS National Communications System 【美国】国家通信系统

NCS Net Control Station 网络控制站

NCS Node Control Software 节点控制软件；网点控制软件

NCSC National Computer Security Center 【美国】国家计算机安全中心

NCSRS National Cyberspace Security Response System 国家网络空间安全反应系统

NCTR Non-Cooperative Target Recognition 【轨道】非预期会合目标识别；独立目标识别

NDBC National Data Buoy Center 【美国】国家数据浮标中心

NDC Navigational Digital Computer 导航数字计算机

NDE Nondestructive Evaluation 无损评估

NDET Nondestructive Electric Test 无损电气试验

NDEW Nuclear Directed-Energy Weapon 核定向能武器

NDEWG Nuclear Directed Energy Weapon-Ground-Based 陆基核定向能武器

NDHF Nimbus Data Handling Facility 【美国】"雨云"气象卫星数据处理中心

NDHS Nimbus Data Handling System 【美国】"雨云"气象卫星数据处理系统

NDI Non-Destructive Inspection 无损检验（探伤）

NDI Non-Developmental Item 非研制产品

NDI/E Non-Destructive Inspection/Evaluation 非毁伤性检测与鉴定

NDPF NASA Data Processing Facility 美国国家航空航天局数据处理中心

NDR Navigation Doppler Radar 导航多普勒雷达

NDS National Defense System 国家防御系统

NDS Navigational Development Satellite 【美国】导航发展卫星

NDS Non-Developmental Software 非研制型软件

NDS Nuclear Detection Satellite 核爆炸探测卫星

NDS Nuclear Detection System 核爆炸探测系统

NDSC Network for the Detection of Stratospheric Change 同温层变化探测网

NDSS NASA Document Storage System 美国国家航空航天局文件存储系统

NDT Non-Destructive Test 非毁伤性试验（测试）

NDTC Non-Destructive Testing Center 无损检测中心；无损探伤中心

NDTF Nondestructive Test Facility 无损探伤测试设施

NDTI Non-Destructive Testing and Inspection 无损检测与检查；无损探伤与检查

NDTL Nondestructive Test Laboratory

无损探伤测试实验室
NDV　NASP Derived Vehicle　【美国】国际空天飞行衍生运载器
NEA　Near Earth Asteroid　近地小行星
NEA　Noise Equivalent Angle　噪声当量角
NEACP　National Emergency Alternate Command Post　【美国】国家应急备用指挥所
NEAP　Near Earth Asteroid Prospector　近地小行星"探险者"航天器
NEAR　Near Earth Asteroid Rendezvous　【美国】近地小行星交会探测任务
NEASIM　Network Analysis Simulator　网络分析模拟器
NEAT　Near Earth Asteroid Tracker　近地球小行星追踪系统
NEB　Nuclear Exo-Atmospheric Burst　大气层外核爆炸
NEC　Network Enabled Capability　网络使能能力
NECAP　Navigation Equipment Capability Analysis Program　导航设备能力分析程序
NECC　National Enabled Command Capability　国家使能控制能力
NECCP　Net Enabled Command Capability Program　网络使能控制能力项目
NED　NASA Extragalactic Database　美国国家航空航天局银河系外数据库
NEDS　Nonviolent Explosive Destruct(ive) System　非强烈爆炸破坏系统
NEDT　Noise Equivalent Differential Temperature　不同温度噪声系数
NEEDS　NASA End-To-End Data System　美国国家航空航天局端至端数据系统
NEEDS　Neighborhood Environmental Evaluation and Decision System　邻域环境评估与判定系统
NEEMO　Extreme Environment Mission Operations　极端环境任务行动
NEF　NASA Extended FORTRAN　美国国家航空航天局扩展型公式翻译程序语言
NEF　Noise Exposure Forecast　噪声暴露预测
NEFD　Noise Equivalent Flux Density　噪声等效通量密度
NEI　Noise Equivalent Intensity　噪声等效强度
NEMP　Nuclear Electromagnetic Pulse　核电磁脉冲
NEMS　NASA Equipment Management System　美国国家航空航天局设备管理系统
NEMS　Near-Earth Magnetosphere Satellite　近地磁层探测卫星
NEMS　Near-Earth Magnetospheric Satellite　近地磁层探测卫星
NEMS　Nimbus E Microwave Spectrometer　【美国】"雨云 E"卫星微波光谱仪
NEMSS　National Environmental Monitoring Satellite System　国家环境监测卫星
NEN　NASA Near Earth Network　美国国家航空航天局近地网络
NEO　Near Earth Object　近地目标
NEO　Near-Earth Orbit　近地轨道
NEOF　National Emergency Operations Facilities　国家应急行动设施
NEP　Near Earth Phase　近地球阶段
NEP　Noise Equivalent Power　噪声等效功率
NEP　Nominal Entry Point　【美国航天

飞机】标称进入点
NEP Nuclear Electric Power 核电源
NEP Nuclear Electric Propulsion 核电推进
NEP Nuclear Environment Protection 核环境防护
NEPA National Environmental Policy Act 国家环境（保护）政策法
NEPA National Environmental Protection Act 国家环境保护法
NEPAL Network Processor Application Language 网络处理器应用描述语言
NEPN Near Earth Phase Network 近地球段网络
NEPSTP Nuclear Electric Propulsion Space Flight Test Program 核电推进航天飞行试验（测试）项目
NEQ Net Explosive Quantity 爆炸品净重
NEQA NASA Engineering and Quality Audit 美国国家航空航天局工程与质量审核
NERAC New England Research Applications Center 【美国国家航空航天局】新英格兰研究应用中心
NERAM Network Reliability Assessment Model 网络可靠性评估模型
NERDAS NASA Earth Resources Data Annotation System 美国国家航空航天局地球资源数据注释系统
NERO Near-Earth Rescue and Operations 近地（轨道）救援和操作
NERV Nuclear Energy Research Vehicle 核发动机研究（用）飞行器
NERV Nuclear Engine Recovery Vehicle 核发动机回收飞行器
NERVA Nuclear Engine for Rocket Vehicle Application 火箭飞行器使用的核发动机（星际飞行器）

NESC NASA Engineering and Safety Center 美国国家航空航天局工程与安全中心
NESC National Electrical Safety Code 【美国】国家电气安全规范
NESEAD Naval Electronic Systems Engineering Activity Detachment 【美国】海军电子系统工程处分遣队
NESHAP National Emission Standards for Hazardous Air Pollutants 【美国】国家危险性空气污染物质辐射标准
NESR Natural Environments Support Room 自然环境保障间
NESSUS Numerical Evaluation of Stochastic Structures Under Stress 应力作用下随机结构数值计算（法）
NESTAR Nuclear Energy Waste Space Transportation and Removal 核能废弃物太空运输和迁移
NETCOM Network Communications 网络通信
NETCOS NASA ESRO Traffic Control Satellite 美国国家航空航天局欧洲空间研究处交通管制卫星
NETDS Near Earth Tracking and Data System 近地球跟踪与数据系统
NETPARS Network Performance Analysis Reporting System 网络性能分析报告系统
NETWAR Network Effects-Based Targeting with Adversarial Reasoning 按敌方推理以网络效应为基础确定目标
NEUS Nuclear-Electric Unmanned Spacecraft 核电不载人航天器
NEW Net Explosive Weight 净爆炸重量
NEW MOONS NASA Evaluation with Models of Optimized Nuclear Space-

craft 美国国家航空航天局最佳核（动力）航天器模型

NEWQD Net Explosive Weight for Quantity Distance 量子点的爆炸物净重

NEWRADS Nuclear Explosion Warning and Radiological Data System 核爆炸警报与辐射数据系统

NEXRAD Next Generation Radar-Doppler 下一代雷达—多普勒

NEXUS NASA Engineering Extendible United Software System 美国国家航空航天局工程可扩展的软件系统

NFAC National Full Scale Aerodynamic Complex 【美国国家航空航天局】国家全尺寸空气动力综合实验设施

NFAC National Full-Scale Aerodynamic Complex 国家大型空气动力综合设施

NFC Nose Fairing Container 头部整流罩外壳

NFCS Navigation and Flight Control System 领航及飞行控制系统

NFCS Nuclear Forces Communications Satellite 核动力通信卫星

NFESC Naval Facilities Engineering Service Center 【美国】海军设施工程服务中心

NFIRE Near Field Infrared Experiment 【美国】近地红外实验卫星

NFPM Nuclear Flight Propulsion Module 核动力飞行推进舱

NFQT Nuclear Field Qualification Test 核场合格测试

NFR Near Field Range 近场距离

NFS Near Field Scattering 近场散射

NFS Nozzle Flow Sensor 喷口流量传感器；喷管流量传感器

NFSP Nonflight Switch Panel 非飞行切换控制台

NG&ST National Geophysical and Solar Terrestrial Data Center 【美国】国家地球物理与日地资料中心

NGDB National Geodetic Data Bank 【美国】国家大地测量数据库

NGL Next Generation Launcher 下一代运载器

NGLT Next Generation Launch Technology 下一代发射技术

NGM Nitrogen Generation Module 制氮模块

NGN Next-Generation Network 下一代网络

NGP NASA Geodetic Program 美国国家航空航天局的大地测量计划

NGPS Nationwide-Differential Global Positioning System 全国性差分全球定位系统

NGPS NAVSTAR Global Positioning System 导航星全球定位系统

NGR NAVSTAR Geodetic Receiver 导航星测地接收机

NGSDC National Geographic and Solar-Terrestrial Data Center 【美国】国家地球物理与日地数据中心

NGSLV New Generation Space Launch Vehicle 新一代航天运载器

NGSO Non-Geosynchronous Orbits 非地球同步轨道

NGSP National Geodetic-Satellite Program 【美国国家航空航天局】国家测地卫星计划

NGST Next Generation Space Telescope 新一代太空望远镜

NGT NASA Ground Terminal 美国国家航空航天局地面终端

NH&S Nuclear Hardening and Survivability 核加固和生存能力；核（爆炸）强化防护设施及生存能力

NHELTR National High Energy Laser Test Range 国家高能激光试验靶

场

NHMT Nuclear-Hardened Mosaic Technology 核加固镶嵌技术

NHQ NASA Headquarters 美国国家航空航天局总部

NIA (Boeing) National Institute of Aerospace 【美国波音公司】国家航空航天研究所

NIBS Non Invasive Back Scattering 非损伤性后散射

NIC Network Interface Controller 网络接口控制器

NID Naval Intelligence Database 【美国】海军情报数据库

NIDS Network Interface Data System 网络接口数据系统

NIF National Ignition Facility 【美国】国家点火设施

NIF Network Interface Function 网络接口功能

NIGE Netlander Ionosphere and Geodesy Experiment 欧洲火星探测项目电离层和大地测量实验

NIGS Non-Inertial Guidance Set 无惯性制导装置

NIH National Institute of Health 国家健康研究所

NII National Information Infrastructure 【美国】国家信息基础设施

NIIKS Scientific Institute of Cosmic System 【日本】宇宙系统科学研究所

NILES NATO Improved Link Eleven System 北约改进型1号链路系统

NIM Nuclear Instrumentation Module 核测量模块（舱）

NIMIT Nimbus Integrated and Test 【美国】"雨云"卫星总装与试验

NIMS NASA Interface Monitoring System 美国国家航空航天局接口监测系统

NIMS Near Infrared Mapping Spectrometer 近红外测绘光谱仪

NIP Network Input Processor 网络输入处理器（装置）

NIP Network Interface Processor 网络接口处理器（装置）

NIP Normal Impact Point 正常弹着点；原定弹着点

NIPS NTCS Intelligence Processing Service 【美国】海军战术指挥系统情报处理机构

NIR Near Infrared 近红外

NIR Nose-Fuse Impact Rocket 弹头引信火箭

NIRTS New Integrated Range Timing System 新综合靶场计时系统

NISC Naval Intelligence Support Center 【美国】海军情报保障中心

NISN NASA Integrated Service Network 美国国家航空航天局综合业务网

NIST National Institute of Standards and Technology 【美国】国家标准与技术研究所

NITES Naval Integrated Tactical Environmental Subsystem 【美国】海军综合战术环境分系统

NITROX Nitrogen Oxygen 氮-氧

NJAD Nozzle Joint Assembly Demonstration 发动机喷嘴联合装配演示验证

NJES Nozzle Joint Environment Simulator 发动机喷嘴组合环境模拟器（装置）

NKEW Nuclear Kinetic Energy Weapon 核动能武器

NLAP Non-Linear Analysis Program 非线性分析程序

NLDM Non-Linear Delay Model 非

线性延迟模型
NLDN National Lightning Detection Network 【美】国家雷电探测网
NLF Navigation Light Flasher 导航灯闪光装置
NLFFL Non-Linear Feed-Forward Logic 非线性前馈控制逻辑
NLOS Nonlinear Optical System 非线性光学系统
NLOS-LS Non-Line of Sight Launch System 非瞄准发射系统
NLR National Lucht-en Ruimtevaart laboratorium 【荷兰】国家航空航天实验室
NLS National Launch System 【美】国家发射系统
NLVP NASA National Launch Vehicle Planning Program 美国国家航空航天局国家运载火箭计划
NMC National Meteorological Center 【美】国家气象中心
NMC Naval Missile Center 【美】海军导弹中心
NMCC National Military Command Center 【美】国家军事指挥中心
NMCS National Military Command System 国家军事指挥系统
NMD National Missile Defense 国家导弹防御系统
NMD GBR National Missile Defense Ground-Based Radar 国家导弹防御系统地基雷达
NMD/TRP National Missile Defense Technology Readiness Program 【美】国家导弹防御技术成熟度项目
NMDPS Network Management Data Process System 网络管理数据处理系统
NMFT New Material Flight Tests 新材料飞行试验
NMI NASA Management Instruction 美国国家航空航天局管理说明
NMILA NASA Merritt Island Launch Area 美国国家航空航天局梅里特岛发射区
NML Neighborhood Matching Logic 领域匹配逻辑线路
NMM NMD Maturity Matrix 【美国】国家导弹防御系统成熟矩阵
NMMT Network Management Module Testing 网络管理模块测试
NMP Near-Term Modernization Programme 近期现代化（改造）项目
NMPG New Mexico Proving Ground 【美】新墨西哥试验场
NMR National Missile Range 国家导弹靶场
NMRF Mesosphere-Stratosphere-Troposphere Radar Facility 中间层－同温层－对流层雷达设备
NMS National Missile Strategy 【美】国家导弹战略
NMS Navigation Management System 导航管理系统
NMS Network Management System 网络管理系统
NMS Neutral Mass Spectrometer 中性质谱仪
NMSA New Mexico Space Authority 【美国】新墨西哥州航天管理局
NMSS National Meteorological Satellite System 全国气象卫星系统
NMSS National Multipurpose Space Station 国家多用途航天站
NMST New Material System Test 新材料系统试验
NMTC National Missile Test Center 国家导弹试验中心
NNDC National Nuclear Data Center 国家核数据中心

NNRDC National Nuclear Rocket Development Center 国家核火箭发展中心

NNRDF National Nuclear Rocket Development Facility 国家核火箭研制设施

NOAA National Oceanographic and Atmosphere Administration 【美国】国家海洋和大气局

NOC Network Operation Control 网络运行控制

NOC Network Operations Centre 网络操作中心

NOCC Navigation Operational Checkout Computer 导航操作检测计算机

NOCC Network Operations Control Center 网络操作控制中心

NODC National Oceanographic Data Center 【美国】国家海洋数据中心

NODIS NASA Online Directives Information System 美国国家航空航天局在线定义信息系统

NODS Night of Observation and Detection System 夜间观察与探测系统

NOESS National Operational Environmental Satellite System 【美国】国家业务环境卫星系统

NOIC Naval Operational Intelligence Center 【美国】海军作战（操作）情报中心

NOMAD Nozzle Motion Attenuation Device 喷管运动衰减装置

NOMR Nitrous Oxide Monopropellant Rocket 一氧化氮单元推进剂火箭

NOMSS National Operational Meteorological Satellite System 【美国】国家气象卫星系统

NOMTS Naval Ordnance Missile Test Center 【美国】海军军械导弹试验（测试）中心

NONAP Non-Linear Adaptive Processor 非线性自适应处理器

NONLISA Nonlinear Network Simulation and Analysis Program 非线性网络模拟和分析程序

NOOS Nuclear Orbit-to-Orbit Shuttle 【美国国家航空航天局】核动力轨道间航天飞机

NORAD North American Aerospace Defense Command 【美国】北美航空航天防御司令部

NOS Network Operating System 网络操作系统

NOS Nimbus Operational System 【美国】"雨云"气象卫星操作系统

NOS Norwegian Domestic Satellite Communications Network 挪威国内卫星通信网

NOSC Naval Ocean Systems Center 【美国】海军海洋系统中心

NOSC-D Network Operations and Security Center, Deployable 可配置的网络作战与安全中心

NOSIC Naval Ocean Surveillance Information Center 【美国】海军海洋监视信息中心

NOSL Nighttime/Daylight Optical Survey of Lightning 【美国航天飞机】闪电昼夜光学研究

NOSP Network Operations Support Plan 网络操作保障计划

NOSS National Ocean Surveillance Satellite (White Cloud) 海军海洋监视卫星（又名"白云"卫星）

NOSS National Oceanic Satellite System 【美国】国家海洋卫星系统

NOSS National Oceanic Survey Satellite 【美国】国家海洋勘察卫星

NOSS National Orbiting Space Station 【美国】国家轨道空间站
NOSS National Orbiting Space Station 国家轨道空间站
NOSS Network Operating Support Systems 网络使用保障系统
NOSS Nimbus Operational Satellite System 【美国】"雨云"实用气象卫星系统
NOTHS Network Operations Trouble Handling System 网络操作故障处理系统
NOTIS Network Operations Trouble Information System 网络操作故障信息系统
NOVA Networked Open Versatile Architecture 联网开放式通用体系结构
NPA Non-Precision Approach 非精确进场
NPA Non-Propulsion Attachment 无推力控制的附件;非推进附属装置
NPC Near Point of Convergence 近会合点;近会聚点;集合近点
NPC Nitrogen Purge Control 氮气净化控制
NPDS Nuclear Particle Detection Subsystem 核粒子探测子系统
NPDU Network Protocol Data Unit 网络协议数据单元
NPE Non-Propulsive Payload Element 非推进有效载荷组件
NPF Navstar Processing Facility 导航星操作厂房
NPG Nevada Proving Ground 【美国】内华达试验场
NPI Nozzle Position Indicator 喷口位置指示器
NPOES National Polar-Orbiting Environmental Satellite 【美国】国家极轨道环境卫星
NPOESS National Polar-Orbiting Operational Environmental Satellite System 【美国】国家两极轨道环境卫星运行系统
NPOESS National Polar-Orbiting Operational Environmental Satellite System 国家极轨道操作环境卫星系统
NPOL NASA Polar-Metric Radar 美国国家航空航天局测定偏振雷达
NPP NPOESS Preparatory Project 国家极轨道地球卫星系统预计划
NPR Noise Power Ratio 噪声功率比
NPR Nozzle Pressure Ratio 喷管压力比
NPRDS Nuclear Plant Reliability Data System 核电站可靠性数据系统
NPRV Negative Pressure Relief Valve 负压减压阀
NPRV Nitrogen Pressure Relief Valve 氮气减压阀
NPSS Numerical Propulsion Simulation System 推进系统数值仿真
NPSVI Non-Public Space Vehicle Information 非公众运载火箭信息
NPTR National Parachute Test Range 国家降落伞试验靶场
NPU Nitrogen Purge Unit 氮气净化装置
NPV Nitrogen Pressure Valve 氮气压力阀
NPW Net Propellant Weight 推进剂净重
NQC NASA Quality Control 美国国家航空航天局质量控制
NRDS Nuclear Rocket Detection System 核火箭探测系统
NRE Non Recoverable Engineering 不可恢复工程
NREDC National Rocket Engine De-

velopment Complex 【美国】国家火箭发动机研发设施

NRL Naval Research Laboratory 【美国】海军研究实验室

NRLA Network Repair Level Analysis 网络式修理级别分析；修理级别分析网络（模型）

NRM Non-Recurring Maintenance 一次性维护

NRM Nuclear Radiation Monitor 核辐射监测

NRRM Nuclear Risk Reduction Measures 核风险降低措施

NRS NASA Range Safety 美国国家航空航天局靶场安全

NRS Nuclear Rocket Shuttle 核火箭航天飞机

NRT Near Real-Time 近实时

NRT DAS Non-Real-Time Data Automation System 非实时数据自动系统

NRTF National Radar Cross Section Test Facility 【美国空军】国家雷达截面积试验设施

NRTS National Reactor Testing Station 【美国】国家反应堆测试站

NRX NERVA Reactor Experiment 【美国】火箭飞行器核发动机反应器实验

NS Near Space 近空间

NSAWC Navy Strike Aircraft Weapons Center 【美国】海军作战飞机武器中心

NSBF National Scientific Balloon Facility 【美国】国家科学探测气球中心

NSC NASA Safety Center 美国国家航空航天局安全中心

NSCAT NASA Scatterometer System 美国国家航空航天局散射测量（卫星）系统

NSCORT NASA Specialized Centers of Research and Training 美国国家航空航天局特种研究与训练中心

NSCI NASACOM System Control Interface 美国国家航空航天局通信网系统控制接口

NSDC National Space Development Center (Japan) 【日本】国家航天发展中心

NSDD National Security Decision Directive 【美国】国家安全决策条令

NSDS-E Navy Satellite Display System-enhanced 改进型海军卫星显示系统

NSE Navigation System Error 导航系统误差

NSESCC NASA Space and Earth Science Computing Center (GSFC) 美国国家航空航天局太空与地球科学计算中心

NSF National Science Foundation 【美国】国家科学基金会

NSI NASA Standard Initiative/Initiator 美国国家航空航天局标准起爆器

NSIE Network Security Information Exchange 网络安全信息交换

NSIS NASA Software Information System 美国国家航空航天局软件信息系统

NSLD NASA Shuttle Logistics Depot 美国国家航空航天局航天飞机后勤保障站

NSMV Near Space Maneuvering Vehicle 近太空机动飞行器

NSN NASA Science Network 美国国家航空航天局科学网络

NSO NASA Support Operation 美国国家航空航天局的保障性运营

NSOC National Signals Intelligence

Operations Center 【美国】国家信号情报操作中心
NSOC Navy Satellite Operations Center 【美国】海军卫星操作中心
NSP NASA Support Plan 美国国家航空航天局保障计划
NSP Navigational Satellite Program 导航卫星计划
NSP Network Signal Processor 网络信息处理器（装置）
NSPIRES NASA Solicitation and Proposal Integrated Review and Evaluation 美国国家航空航天局征求与建议综合评审与鉴定
NSR Noise-to-Signal Ratio 噪信比
NSRL NASA Space Radiation Laboratory 美国国家航空航天局空间辐射实验室
NSRS NASA Safety Reporting System 美国国家航空航天局安全通报系统
NSS NASA Safety Standard 美国国家航空航天局安全标准
NSS National Security Strategy 【美国】国家安全战略
NSS Navigation Satellite System 导航卫星系统
NSS Nitrogen Supply Subsystem 氮供给分系统
NSSC NASA Standard Spacecraft Computer 美国国家航空航天局标准航天器计算机
NSSC National Space Surveillance Center 【美国】国家空间监视中心
NSSCC National Space Surveillance Control Center 【美国】国家空间监视控制中心
NSSDC National Space Science Data Center 【美国】国家航天科学数据中心
NSSPS New Space Signals Processing Stations 新航天信号处理站
NSSS National Security Space Strategy 国家安全空间战略
NSSS National Space Surveillance System 国家空间监视系统
NSSSC National Software Simulation Support Center 国家软件仿真保障中心
NST Noise, Spikes and Transients 噪声、峰值与瞬值
NSTAP National Strategic Target and Attack Policy 【美国】国家战略目标与进攻策略
NSTART Nuclear Space Technology Applications and Research Teams 航天核技术应用与研究组
NSTC NASA Safety Training Center 美国国家航空航天局安全培训中心
NSTC National Sustainment Technology Center 国家持续（发展）技术中心
NSTL National Software Testing Lab 国家软件测试实验室
NSTL National Software Testing Labs 【美国】国家软件测试实验室
NSTL National Space Technology Laboratory 【美国】国家航天技术实验室
NSTL National Space Transportation Laboratory 【美国】国家空间运输实验室
NSTP National Software Testing Procedure 国家软件测试程序
NSTR National Software Testing Report 国家软件测试报告
NSTS NASA Shuttle Transportation System 美国国家航空航天局航天飞机运输系统
NSTS National Space Transportation System 国家空间运输系统

NSTS National Space Transportation System Program Office 【美国】国家空间运输项目办公室

NSV Non-Spinning Vehicle 不自旋飞行器

NTACS Navy Tactical Air Control System 【美国】海军战术空中控制系统

NTB National Testbed 国家试验台（美国模拟"星球大战"计划的计算机和视频操作系统）

NTB/WAN National Test Bed/Wide Area Network 【美国】国家试验（测试）台/广域网络

NTBI National Test Bed Integration 【美国】国家试验（测试）台集成

NTBN National Testbed Network 国家试验台网络

NTC National Test Center 【美国】国家试验（测试）中心

NTC National Training Center 【美国】国家训练（培训）中心

NTCC Nimbus Technical Control Center 【美国】"雨云"气象卫星技术控制中心

NTCS Naval Tactical Command System 【美国】海军战术指挥系统

NTDS Naval Tactical Data System 【美国】海军战术数据系统

NTE Node Test Environment 节点试验（测试）环境

NTF Joint National Test Facility 【美国】联合国家试验（测试）设施

NTF National Test Facility 国家试验设施（美国"星球大战"计划模拟试验场）

NTF National Transonic Facility 【美国】国家跨声速设施

NTIAC Non-Destructive Testing Information Analysis Center 非破坏性测试信息分析中心

NTIC National Technical Information Center 【美国】国家技术信息中心

NTIC Navy Tactical Intelligence Center 【美国】海军战术情报中心

NTIS National Technical Information Service 【美国】国家技术信息服务

NTIS Non-Destructive Testing Information System 非破坏性测试信息系统

NTP Nuclear Thermal Propulsion 热核推进

NTR Nuclear Thermal Rocket 热核火箭

NTS Navigation Technology Satellite 【美国】导航技术卫星

NTS Nozzle Thrust Stand 喷管推力试验台

NTTC National Technology Transfer Center 【美国国家航空航天局】国家技术转移中心

NTTF National Tracking and Test Facility 【美国国家航空航天局】国家跟踪与测试中心

NTTF Network Test and Training Facility 【美国国家航空航天局】网络测试与训练中心

NTTF Network Testing and Training Facility 网络测试和培训设施

NTTR Nellis Test Training Range 【美国】内利斯试验训练靶场

NTWD(S) Navy Theater-Wide Defense (System) 【美国】海军全战区防御（系统）

NUDETS Nuclear Detonation Detection and Reporting System 核爆炸探测与报告系统

NUICCS NORAD and USSPACECOM Integrated Command and Control

System 北美航空航天防御司令部/美国航天司令部综合指挥与控制系统

NULACE Nuclear Liquid-Air Cycle Engine 液态空气循环式核发动机

NUROC Nuclear Rocket Project 核火箭计划

NUSAT Northern Utah Satellite 北犹他卫星（美国雷达校准卫星）

NUSAT Nuclear Saturn 【美国】核动力"土星"运载火箭

NuSTAR Nuclear Spectroscopic Telescope Array 原子光谱望远镜阵列

NV&EOL Night Vision and Electroptics Laboratory 夜视与电子光学实验室

NVIS Near Vertical Incidence System 近垂直入射系统

NVP National Verification Plan 国家验证计划

NVW Net Vehicle Weight 飞行器净重

NWAI Nuclear Weapons Acceptance Inspection 核武器验收检查

NWDC/S Navigation and Weapons Delivery Computer/System 导航与武器投放计算机/系统

NWDS Navigation and Weapons Delivery System 导航与武器投放系统

NWE Nuclear Weapon Effects 核武器效能

NWE Nuclear Weapons Employment 核武器使用

NWEF Naval Weapons Evaluation Facility 【美国】海军武器鉴定设施

NWEF Nuclear Weapon Evaluation Facility 核武器鉴定设施

NWFZ Nuclear Weapons Free Zone 无核武器区

NWP Numerical Weather Prediction 数值天气预报

NWSC Naval Weapons Support Center 【美国】海军武器保障中心

NWTC National Wind Tunnel Complex 国家风洞综合设施

NWTI Nuclear Weapon Technical Inspections 核武器技术检查

NYIAS New York Institution of the Aerospace Sciences 【美国】纽约航空航天科学研究所

NZ Nike-Zeus 【美国】"奈基—宙斯"地空导弹

O

O&C　Operation and Checkout Building　操作与检测厂房
O&C　Operations and Control　操作与控制
O&C　Operations and Control Building　操作与控制厂房
O&CO　Operations and Checkout　操作与检测；运转与检测
O&FS　Operations and Flight Support　操作与飞行保障
O&M　Operation and Maintenance　运行与维修
O&MF　Operation and Maintenance of Facilities　设施运行与维修
O&P　Operations and Procedures　操作与程序
O&P　Operations and Robotics　操作与机器人技术
O&R　Overhaul and Repair　大检修与修理
O&S　Operations and Support　操作与支援（保障）
O&S　Operations and Sustainment　操作与持续性
O&SC　Operations and Support Cost　操作与保障费用
O&SHA　Operating and Support Hazard Analysis　操作与保障危险分析
O-BIT　Operating Built-In Test　运行自检测
O-NAV　Onboard Navigation　机载导航；箭载导航
O/A　Omnirange Antenna　全向式无线电信标天线
O/C　Overcharge　过载；增压
O/E　Output Electronics　输出电子设备
O/ET　Orbiter/External Tank　【美国航天飞机】轨道（飞行）器/外贮箱
O/F　Orbital Flight　轨道飞行
O/F　Oxidizer-to-Fuel (Ratio)　氧化剂燃料（比）
O/L　Overload　过载
O/L-RC　Overload-Reverse Current　过载—反向电流
O/O　On Orbit　在轨
O/R　Oxygen Relief　氧气释放
O/S　Operation Signal　操作信号
O/S　Operations/Support Phase　运行与保障阶段
O/V-U/V　Over Voltage/Under Voltage　超电压/低电压
O^2　Object Oriented　面对目标；面向对象
OA　Obstacle Avoidance　障碍规避
OA　Open Architecture　开放式体系结构
OA　Operational Assessment　作战评估
OA　Operational Availability　操作（运行）可利用性
OA　Options Assessment　方案评估
OA　Orbital Assembly　轨道装配
OA　Orbital Availability　在轨可用性
OAA　Open Application Architecture　开放式应用体系结构
OAA　Orbiter Access Arm　【美国航天飞机】轨道（飞行）器入舱臂
OAA　Orbiter Alternate Airfield　【美国航天飞机】轨道（飞行）器备用机

场

OAB Ordnance Assembly Building 火工品组装厂房

OACS Orbit and Attitude Control System 轨道和姿态控制系统

OAD Orbital Aerodynamic Drag 轨道气动阻力

OADS Omnidirectional Air Data System 全向大气数据系统

OAE Orbiting Astronomical Explorer 轨道天文探测卫星

OAEE Overall Application Effectiveness and Evaluating of Number Control Equipment 数控设备综合应用效率与测评

OAF Orbital Antenna Frame 轨道天线架

OAFD Orbiter Air Flight Deck 【美国航天飞机】轨道（飞行）器飞行甲板

OAH Overhead Air Hoist 高架气动起重机

OAI Omnidirectional Airspeed Indicator 全向空速指示器

OAI Open Applications Interface 开放的应用软件界面

OAIDE Operational Assistance and Instructive Data Equipment 操作辅助与指令数据设备

OALS Orbiter Automatic Landing System 【美国航天飞机】轨道（飞行）器自动着陆系统

OAM&P Operations, Administration, Maintenance and Provisioning 运行、管理、维护和供应

OAME Orbital Attitude and Maneuvering Electronics 轨道姿态控制与机动飞行电子设备

OAMP Optical Analogue Matrix Processing 光学模拟矩阵处理

OAMS On-Orbit Alignment Measurement System 在轨（道上）对准测量系统

OAMS Orbit Altitude Maneuvering System 轨道高度机动系统

OAMS Orbit Attitude and Maneuvering System 轨道姿态控制与机动飞行系统

OAN Orbital Area Network 轨道区域网络

OANS Orbiter Ancillary 轨道飞行器辅助设备

OAO Orbital Assembly Operation 轨道组装作业

OAO Orbiting Astronomical Observatory (Satellite) 【美国】轨道天文观测台（卫星一）

OAP Offset Aiming Point 辅助瞄准点

OAPS Orbit Adjust Propulsion System 轨道调整推进系统

OAR Operational Availability and Reliability 操作（运行）可利用性与可靠性

OAR Optical Automatic Ranging 光学自动测距

OARF Outdoor Aerodynamic Research Facility 【美国国家航空航天局】室外气动力研究装置

OARS Ocean Area Reconnaissance Satellite 【反潜战专用】海域侦察卫星

OARS Ocean Area Reconnaissance Satellite 海洋侦察卫星

OAS Operational Announcing System 操作通告系统

OAS Optical Acquisition System 光学数据采集系统

OAS Optical Alignment Sights 光学准直瞄准具

OAS Orbit Adjust Subsystem 轨道调整分系统

OAS Orbiter Aeroflight Simulator 【美国航天飞机】轨道（飞行）器飞行模拟器

OAS Orbiter Atmospheric Simulator 【美国航天飞机】轨道（飞行）器大气层模拟器

OAS Orbiter Avionics System 【美国航天飞机】轨道（飞行）器电子设备与控制系统

OAS Output Amplitude Stability 输出振幅稳定性

OASCB Orbit Avionics Software Control Board 轨道电子设备软件控制台

OASCB Orbiter Avionics Software Control Board 【美国航天飞机】轨道（飞行）器航空电子软件控制台

OASF Orbiting Astronomical Support Facility 轨道天文观测保障设备

OASIS Oceanic and Atmospheric Scientific Information System 【美国国家海洋与大气管理局】海洋与大气科学信息系统

OASIS Oil Analysis Standard Interservice System 油液分析跨军种标准系统

OASIS Open Architecture System for Integrated Systems 集成系统的开放型结构体系

OASIS Open Avionics System Integration Study 开放式航空电子系统综合研究

OASIS Operational Analysis and Simulation of Integrated Systems 集成系统的作战分析和模拟

OASIS Operational Analysis Strategic Interaction Simulator 【洲际弹道导弹】作战分析研究用战略交互作用模拟器

OASIS Operational Application of Special Intelligence Systems 特种情报系统作战应用

OASIS Optimized Action Sequence Interpreter System 最佳行动程序翻译系统

OASIS Optimized Air-to-Surface Infrared Seeker 最佳空对地（或空对舰）红外自导导弹

OASIS Orbiter Autonomous Supporting Instrumentation System 【美国航天飞机】轨道（飞行）器自主保障仪表系统

OASIS Organically Assured and Survivable Information System 建制上确保的和生存力强的信息系统

OASP Optical Adaptive Signal Processor 光学自适应信号处理机

OASV Orbital Assembly Support Vehicle 轨道装配保障飞行器

OASYS Obstacle Avoidance System 障碍规避系统

OAT Onto Atmospheric Temperature 上层大气温度

OAT Operating Ambient Temperature 工作周围温度；工作环境温度

OAT Operational Acceptance Test 操作验收测试

OAT Optional Auxiliary Terminal 可选辅助终端

OAT Outer Atmospheric Temperature 外层大气温度

OATA Optical Acquisition and Tracking Aid 光学目标捕捉与跟踪辅助装置

OATS Optical Attitude Transfer System 光学姿态变换系统

OATS Optimum Aerial Targeting System 最佳空中目标选定系统

OATS Orbit and Attitude Tracking Subsystem 轨道与姿态跟踪子系统

OB Observatory 观测台阶段（美国"旅行者"号太空探测器与天王星交会过程的第一阶段）

OB Orbital Booster 轨道助推器

OBA Optical Bench Assembly 光学试验台设备

OBA Oxygen Breathing Apparatus 供氧设备

OBC On Board Checkout 箭上检测

OBC On Board Computer 箭载计算机

OBC On Board Controller 箭载控制器

OBC Onboard Computer 星（箭）载计算机

OBCO On Board Cargo Operations (System) 箭载货物操作（系统）

OBCO Onboard Checkout (Instrumentation) 星（箭）载检查（仪器）

OBCOE On Board Checkout Equipment 箭上检测设备

OBCP On-Board Computer Program 星上（航天器上）计算机程序

OBCS Onboard Checkout Subsystem 星（箭）载检查子系统

OBCS Onboard Computer System 星（箭）载计算机系统

OBD On Board Database 箭上数据库

OBD On Board Diagnostics 箭载自动诊断系统

OBD (SYSTEM) Omnibearing Distance System 全向距离导航系统

OBDC On-Board Digital Computer 箭上数字计算机

OBDH On Board Data Handling 箭上数据处理；星上数据处理

OBDMS Onboard Data Monitoring System 箭载数据监控系统

OBDP Onboard Data Processor 箭载（星载、机载）数据处理器

OBE Offboard Expendables 非箭载消耗品

OBE Outboard Booster Engine 机外助推发动机

OBECO Outboard Engine Cutoff 机外发动机停车

OBFCO On Board in-Flight Checkout 箭上飞行中检查

OBFCS On-Board Fire Control System 机（车、星、箭）载点火控制系统

OBGS Orbital Bombardment Guidance System 轨道轰炸武器制导系统

OBIGGS On-Board Inert Gas Generating System 箭载（星载、机载）惰性气体生成系统

OBM On-Board Monitoring 机载监控；箭载监视

OBOGS On-Board Oxygen Generation System 机（车、船、箭、星）载制氧系统

OBP Onboard Processing 机（车、船、箭、星）载处理

OBP Onboard Processor 机载处理机

OBR Output Buffer Register 输出缓冲寄存器

OBRP On Board Repair Part 机（车、星、箭）上的修理备件

OBS On Board Simulation 箭载模拟器

OBS On-Board Software 机载（箭载）软件

OBS On-Board System 机（车、船、箭）载系统

OBS Operational Bioinstrumentation System 【美国航天飞机】运转的生物仪表系统
OBS Operational Biomedical System 实用（或作战）生物医学系统
OBS Operational Flight Program Build System 作战飞行程序构建系统
OBS Optical Boresight System 光学枪械准星校正系统
OBS Orbital Bombardment System 轨道轰炸系统
OBSS Orbiter Boom Sensor System 轨道飞行器臂传感器系统
OBT On-Board Training 机载培训
OBV Objective Boost Vehicle 目标导飞行器
OBV Oxidizer Bleed Valve 氧化剂放气阀
OC Obstacle Clearance 离障碍物高度
OC Orbital Check 在轨检查
OC Open Cycle 开循环
OCA Obstacle Clearance Allowance 离障碍物高度余量
OCA Obstacle Clearance Altitude 越障高度
OCA Orbit Communication Adapter 轨道通信适配器
OCA Orbital Communication Adapter 在轨通信适配器
OCA Overall Configuration Analysis 全面技术状态分析
OCAC Operational Control and Analysis Center 作战控制与分析中心
OCAMS Onboard Checkout and Monitoring System 箭载检测与监控系统
OCC Onboard Complex Control 机载综合设施控制
OCC Operation Communication Center 操作通信中心
OCC Operation(al) Control Center 操作控制中心；作战控制中心
OCCS Onboard Complex Control System 机（车、船、箭、星）载综合控制系统
OCD Operational Capability Development 操作能力研发
OCDF Operations Control and Display Facility 操作控制与显示装置
OCDMS On Board Checkout and Data Management System 箭载检测与数据管理系统
OCDR Orbiter Critical Design Review 【美国航天飞机】轨道（飞行）器关键设计审查
OCDU Optics Coupling Data Unit 光学耦合数据装置
OCE Ocean Color Experiment 【美国航天飞机】海洋水色实验
OCE Optical Control Electronics 光学控制电子设备
OCF Onboard Computational Facility 机（车、船、箭）载计算设备
OCF Operation Check Flight 操作检查飞行
OCF Operational Control Facility 操作控制设备
OCF Orbiter Computational Facilities 【美国航天飞机】轨道（飞行）器计算设备
OCGF Organic Crystal Growth Laboratory Facility 有机晶体生长实验室设备
OCH Orbiter Common Hardware 【美国航天飞机】轨道（飞行）器通用硬件
OCI Ocean Color Imager 海洋水色成像仪
OCI DCI Offensive and Defensive

Counter-Information Operations　进攻性和防御性反信息作战

OCL　Operation Control Language　操作控制语言

OCM　Optimal Control Model　最佳控制模型

OCM　Optimality Criterion Method　优化准则法

OCM　Orbit Control Mode　轨道控制模式

OCM　Overland Cruise Missile　横跨大陆的巡航导弹

OCMD　Overland Cruise Missile Defense　横跨大陆的巡航导弹防御

OCMS　Onboard Checkout and Monitoring System　箭载检测与监视系统

OCNIS　Optimal Communications, Navigation and Identification System　最佳通信、导航与识别系统

OCO　Orbiting Carbon Observatory　轨道碳观测

OCO　Overseas Contingency Operations　海面应急操作（作战）

OCOE　Overall Checkout Equipment　整套检查设备；总测试设备

OCP　Open Cherry Picker　【美国航天飞机】敞开式升降台

OCP　Open Control Platform　开放式控制平台

OCP　Orbital Control Program　轨道控制程序

OCP　Output Control Pulse　输出控制脉冲

OCR　Operations Capability Reference　作战能力参考；操作能力参考

OCR　Operations Capability Review　操作能力评审

OCRM　Orbiter Crash and Rescue Manuals　【美国航天飞机】轨道（飞行）器失事与救援手册

OCS　Ocean Color Scanner　海洋水色扫描仪

OCS　Onboard Checkout System　星上检查系统；箭上检测系统

OCS　Onboard Computer System　星（机、箭）载计算机系统

OCS　Operational Control Segment　操作控制段

OCS　Operational Control System　操作控制系统

OCS　Operations and Control Software　操作与控制软件

OCS　Optical Control System　光学控制系统

OCS　Orbit Control System　轨道控制系统

OCSF　Operational Control Segment of the Future　未来作战控制区

OCSTL　Onboard Checkout System Test Language　箭上检查系统测试语言

OCSV　Oxidizer Control Solenoid Value　氧化剂控制电磁阀

OCTS　Ocean Color and Temperature Scanner　海洋水色与温度扫描仪

OCU　Operators Console Unit　操作员控制台装置

OCU　Operators Control Unit　操作员控制装置

OCZ　Operational Control Zone　操作控制区

OD　Orbital Debris　轨道碎片

OD　Operational Demonstration　作战演示；操作演示

OD　Operations Division　操作部门

OD　Orbit Determination　轨道测定；轨道确定

OD　Oxygen Drain　氧气排放

ODA　Optical Data Analysis　光学数

据分析

ODA Optical Discrimination Algorithms/Architecture 光学识别算法/体系结构

ODA Orbital Debris Assessment 轨道碎片评估

ODADS Omnidirectional Air Data System 全向大气数据系统

ODALS Omnidirectional Approach Lighting System 全向进场照明系统

ODAP Operation Data Analysis Program 运转数据分析程序

ODAPS OGE Data Acquisition and Patching System 运行地面装备数据采集及修补系统

ODAPS Overocean Display and Positioning System 越洋显示及定位系统

ODAR Orbital Debris Assessment Report 轨道碎片评估报告

ODARS On-Line Diagnostic and Reporting System 在线诊断与报告系统

ODAS Ocean Data Acquisition System 【美国国家航空航天局】海洋数据收集系统

ODAS Onboard Data Acquisition System 机（车、船、箭、星）载数据采集系统

ODBC Open Database Connectivity 开放数据库连接

ODBMS Onboard Data Base Management System 星上（航天器上）数据管理系统

ODC Object-Oriented Database Connectivity 面向对象的数据库连接

ODEA Orbit Data Editor Assembly 轨道数据编辑程序汇编

ODERACS Orbital Debris Radar Calibration Spheres 轨道碎片雷达校准球

ODES Operational Deployment Experiments Simulator 作战使用实验模拟器

ODES Optical Discrimination Evaluation Study 光学辨别力鉴定研究

ODESSA Oceanographic Data and Environmental Satellite System Application 海洋图表数据与环境卫星系统应用

ODF Onboard Data File 星（箭）载数据文件

ODF Orbit Determination Facility 轨道测定设施

ODG Orbit Data Generator 轨道数据发生器

ODI Onboard Digital Computer Control 箭载数字计算机控制

ODIN Onboard, Data, Interfaces and Networks 箭载、数据、接口和网络

ODIN Orbital Design Integration (System) 轨道设计综合（系统）

ODLI Open Data Link Interface 开放式数据链路接口

ODM Object Definition Language 对象定义语言

ODM Operational Development Model 操作（运行）研发模型

ODM Orbital Determination Module 轨道测定舱

ODM Orbiter Deflection Maneuver 【美国航天飞机】轨道（飞行）器偏移机动飞行

ODMS Oxygen Deficiency Monitoring System 缺氧监控系统

ODP Operational Display Procedure 操作显示规程

ODP Ozone Depletion Potential 臭

氧层消耗潜能值

ODPO NASA Orbital Debris Program Office 美国国家航空航天局轨道碎片项目办公室

ODR Omnidirectional Range 全向无线电信标（导航台）；全向测距

ODR Output Data Redundancy 输出数据冗余

ODRN Orbital Data Relay Network 轨道数据中继网

ODRS Orbiting Data Relay System 轨道数据中继系统

ODRSS Orbital Data Relay Satellite System 轨道数据中继卫星系统

ODS Operational Data Store 操作数据存储

ODS Optical Detection System 【美国】光学探测系统

ODS Orbiter Docking System 【美国航天飞机】轨道（飞行）器对接系统

ODSI Orbital Deep Space Imager 轨道深空成像仪

ODSRS Orbiting Deep Space Relay Station 轨道深空中继站

ODT Operational Demonstration Test 操作演示试验

ODT Orbit Delay Time 轨道延迟时间

ODTM Orbiter Dynamic Test Model 【美国航天飞机】轨道（飞行）器动力学试验模型

ODTS Offset Doppler Tracking System 频移多普勒跟踪系统

ODTS Operational Development Test Site 操作研制试验场

ODTS Optical Data Transmission System 光学数据传输系统

ODTS Orbit Determination and Time Synchronization 轨道测定与时间同步

ODU Output Display Unit 输出显示装置

ODVAR Orbit Determination and Vehicle Attitude Reference 轨道测定与飞行器姿态基准

OEAS Orbital Emergency Arresting System 轨道紧急拦阻减速系统

OEC Operational Evaluation Command 操作鉴定指令；作战鉴定指挥部

OECO Outboard Engine Cutoff 星（箭）外发动机停车

OED Operational Evaluation Demonstration 使用评价（飞行）演示验证

OEFD Orbiter, Electrical Field Detector 【美国航天飞机】轨道（飞行）器电场探测器

OEM Original Equipment Manufacturer 原设备生产方

OEP Operations and Engineering Panel 操作与工程委员会

OER Operational Effectiveness Rate 使用效能变化率

OERC Optimum Earth-Reentry Corridor 最佳重返大气层走廊

OES Orbital Escape System 轨道逃逸系统；轨道救生系统

OES Orbiter Emergency Site 【美国航天飞机】轨道（飞行）器应急场区

OESC Operational Employment and Support Concept 作战使用与保障方案

OESS Orbiter/ET Separation Subsystem 【美国航天飞机】轨道（飞行）器/外贮箱分离子系统

OET Orbital Environments Test 在轨环境测试（试验）

OET&E Operational Employment Testing and Evaluation 作战使用试验与评价

OEW Operating Empty Weight 有效（使用）空重

OEW Overall Empty Weight 空载总重

OEX Observation Experiment Target 观测实验目标卫星

OEX Orbital Experiment Target 轨道实验目标卫星

OEX Orbiter Experiment 【美国国家航空航天局】轨道（飞行）器实验

OEX Target Observation Experiment Target 观测试验目标卫星（美国反卫星试验目测目标卫星）

OEX Target Orbiter Experiment Target 【美国航天飞机】轨道（飞行）器实验目标

OF Orbital Flight 轨道飞行

OF Oxidizer-to-Fuel Ratio 氧化剂与燃料比率

OF Oxygen Fill 氧加注

OFDS Orbiter Flight Dynamics Simulator 【美国航天飞机】轨道（飞行）器动态模拟器

OFDS Oxygen Fluid Distribution System 氧液体分配系统

OFI Opacified Fibrous Insulation 乳白纤维绝缘

OFI Operational Flight Instrumentation 操作飞行仪表

OFI Orbital Flight Instrumentation 轨道飞行仪表

OFK Official Flight Kit 正式飞行装备

OFK Optical Flight Kit 光学飞行工具包

OFO Orbiting Frog Otolith "蛙耳石"卫星（美国的生物卫星）

OFP Operational Flight Profile 作战（操作）飞行剖面图

OFP Operational Flight Program 操作飞行程序

OFP Orbiter Flight Program 【美国航天飞机】轨道（飞行）器飞行程序

OFR Operational Failure Report 操作故障报告

OFS Ocean Ferret Satellite 海洋侦察卫星

OFS Operation Flight Software 操作飞行软件

OFS Operational Flight Simulation 操作（作战）飞行模拟

OFS Orbital (Orbiter) Flight System 轨道（飞行器）飞行系统

OFS Orbiter Functional Simulator 轨道飞行器功能模拟器

OFT Orbital Flight Test 轨道飞行试验

OFT Orbiter Flight Test 【美国航天飞机】轨道（飞行）器飞行试验

OFTDS Orbital Flight Test Data System 轨道飞行试验数据系统

OFTM On-Orbit Flight Technique Meeting 在轨飞行技术会议

OFTR Orbital Flight Test Requirement 轨道飞行试验要求

OG Oxygen Gauge 氧压力计

OG Zero Gravity 零重力；失重

OG^2 Communications Satellite Constellation 通信卫星星座

OGA Outer Gimbal Angle 外部万向角

OGA Outer Gimbal Axis 外部万向轴

OGE Operating Ground Equipment 操作地面设备

OGMT Orbiter Greenwich Mean Time

【美国航天飞机】轨道（飞行）器的格林尼治时间

OGO Orbiting Geophysical Observatory 【美国】轨道地球物理观测卫星

OGS Optical Grating Spectrometer 栅形光谱

OGS Oxygen Generation System 制氧系统

OGSA Open Grid Service Architecture 开放的栅格（坐标网）服务体系结构

OGSE Operational Ground Support Equipment 地面操作辅助设备

OGV Oxygen Gauge Valve 氧压力阀

OHA Operational Hazard Analysis 操作危险性分析；运行危险性分析

OHA Orbital Height Adjustment Maneuver 轨道高度调整机动

OHDETS Over Horizon Detection System 超视距探测系统

OHGVT Orbiter Horizontal Ground Vibration Test 【美国航天飞机】轨道（飞行）器水平地面振动试验

OI Operational Instrumentation 操作仪器仪表

OI Operations Interface 操作接口

OI Orbit Insertion 入轨

OI Orbiter Insertion 【美国航天飞机】轨道（飞行）器入轨

OI Orbiter Instrumentation 【美国航天飞机】轨道（飞行）器测试设备

OI&S Operational Integration and Support 作战一体化和保障

OIA Operations Impact Assessment 操作影响评估

OIA Orbiter Interface Adapter 轨道（飞行）器接口适配器

OIB Orbiter Interface Box 【美国航天飞机】轨道（飞行）器接口箱

OIC Orbiter Integrated Checkout 【美国航天飞机】轨道（飞行）器综合检测

OICETS Optical Interorbit Communications Engineering Test Satellite 【日本】轨道间光通信工程试验卫星

OII Operations Integration Instruction 操作集成（一体化）指南

OIL Orbital International Laboratory 国际轨道实验室

OIM Open Information Model 开放式信息模型

OIM Optical Image Processor 光学图像处理器

OIM Orbit Insertion Maneuver 入轨机动飞行

OIM Orbit Insertion Module 入轨舱

OIM Oxygen Interaction Mechanism 【美国国家航空航天局】氧相互作用机制

OIR Operations Integration Review 操作综合评审

OIS Onboard Information System 箭载信息系统

OIS Operational Intercommunication System 作战内部通信系统

OIS Orbit(al) Insertion Stage 入轨子级

OIS Orbiter Instrumentation Systems 【美国航天飞机】轨道（飞行）器测试设备系统

OIS Orbiting Insertion System 入轨系统

OIT Object Identification Test 目标识别测试

OIT Orbiter Integrated Test 【美国】轨道（飞行）器综合测试

OIU Orbiter Interface Unit 【美国】轨道（飞行）器接口装置

OIV Oxidizer Isolation Valve 氧化剂隔离阀

OIVS Orbiter Interface Verification Set 【美国】轨道（飞行）器接口验证装置

OIW Offensive Information Warfare 进攻性信息战

OJT On-the-Job Training 在职培训

OL Open Loop 开放式回路

OL Orbit Launch 轨道发射

OLAFS Orbiting and Launch Approach Flight Simulator 轨道飞行与发射接近飞行模拟器

OLAN Onboard LAN 箭载局域网

OLAP Online Analytical Processing 联机（或在线）分析处理（方法、技术）

OLBM Orbit-Launch Ballistic Missile 轨道发射的弹道导弹

OLC² Operational Level Command and Control 操作（作战）级指挥与控制

OLCC Optimum Life Cycle Costing 最佳寿命周期成本

OLCOP On-Line Control and Operational Program 在线控制与操作程序

OLD Open Loop Damping 开环阻尼

OLD On-Line Debugging 联机调试

OLD On-Line Diagnostics 在线诊断；联机诊断

OLDB Online Data Bank 联机数据库

OLDI On-Line Data Interchange 在线数据交换

OLDI On-Line Display System 联机显示系统

OLDP On-Line Data Processing (or Processor) 在线数据处理（或处理机）

OLE Object Linking and Embedding 对象链接与嵌入

OLE Orbital Life Extension 轨道寿命延期

OLEO Orbiting Large Engineering Observatory 轨道大型工程观测台

OLF Orbital Launch Facility 轨道发射设施

OLF Orbiter Landing Facility 【美国航天飞机】轨道（飞行）器着陆设施

OLF Orbiter Lifting Frame 【美国航天飞机】轨道（飞行）器吊运架

OLIF Orbiter Landing Instrumentation Facilities 【美国航天飞机】轨道（飞行）器着陆仪表设施

OLO Orbital Launch Operation 轨道火箭发射活动；轨道发射操作

OLOGS Open-Loop Oxygen Generating System 开环式制氧系统

OLOS Operational Land Observation Satellite 地面观测运行卫星

OLOW Orbiter Lift-Off Weight 【美国航天飞机】轨道（飞行）器起飞重量

OLP Off-Line Processing 离线（脱机）处理

OLPARS On-Line Pattern Analysis and Recognition System 联机模式分析与识别系统

OLRTO On-Line Real Time Operation 联机实时操作

OLRTS On-Line Real Time System 联机实时系统

OLS Operational Launch Station 发射操作台

OLS Operational Linescan System 线性扫描操作系统

OLS Optical Landing System 光学

着陆系统

OLS Orbiting Lunar Station 环月轨道站

OLS Order Location System 指令存储系统

OLS Osaki Launch Site 【日本】大崎发射场

OLSA Orbiter/LPS Signal Adapter 【美国航天飞机】轨道（飞行）器/发射（数据）处理系统信号转换器

OLSCA Orientation Linkage for Solar Cell Array 太阳能电池阵列的定位联动装置

OLSP Orbiter Logistics Support Plan 【美国航天飞机】轨道飞行器后勤保障计划

OLST Orbital Launch Star Tracker 轨道发射星体跟踪器

OLV Orbital Launch Vehicle 轨道运载火箭

OM Operations Module 作业舱

OM Orbit Modification 轨道修正

OMA Object Management Architecture 对象管理体系结构

OMA Operations Maintenance Area 操作维护区

OMA Operations Management Application 操作管理应用

OMA Optical Multichannel Analyzer 光学多通道分析仪

OMA Orbiter Maintenance Area 【美国航天飞机】轨道（飞行）器保养区；轨道飞行器维修场

OMA Organizational Maintenance Activity 初级维修机构；外场维修机构

OMBI Observation-Measurement-Balancing and Installation 观测、测量、平衡与安装

OMBUU Orbiter Midbody Umbilical Unit 【美国航天飞机】轨道（飞行）器中段脐带装置

OMC Onboard Maintenance Computer 箭载维修计算机

OMC Optical Monitor Camera 光学监视器照相机

OMC Orbiter Maintenance and Checkout 【美国航天飞机】轨道（飞行）器维修与检测

OMCF Operations and Maintenance Control File 操作与维护控制文件

OMCF Orbiter Maintenance and Checkout Facility 【美国航天飞机】轨道（飞行）器维修与检测设施

OMD Orbiter Mating Device 【美国航天飞机】轨道（飞行）器对接装置

OMDP Orbiter Maintenance Depot Processing 【美国航天飞机】轨道（飞行）器维修基地处理

OMDP Orbiter Maintenance Down Period 【美国航天飞机】轨道（飞行）器维修停飞期

OMDR Operations and Maintenance Data Record 操作与维修数据记录

OME Operational Mission Environment 作战任务环境

OME Orbital Main Engine 轨道主发动机

OME Orbital Maneuvering Engine 轨道机动发动机

OME Orbiter Main Engine 【美国航天飞机】轨道（飞行）器主发动机

OMEC Optimized Microminiature Electronic Circuit 最佳超小型电子电路

OMEGA Optimized Method for Estimating Guidance Accuracy (VLF Navigation System) 【甚低频导航系统】推测制导准确度优选法

OMET Orbiter Mission Elapsed Time 【美国航天飞机】轨道（飞行）器任务运行时间

OMEWG Orbiter Maintenance Engineering Working Group 【美国航天飞机】轨道飞行器维护工程工作组

OMF Object Management Framework 目标（或对象）管理框架

OMF Operational Mission Failure 作战（操作）任务故障

OMF Orbiter Mid-Fuselage 【美国航天飞机】轨道（飞行）器机身中部

OMFCA Operation and Maintenance of Facilities Cost Account 设备使用与维修成本计算

OMI On-Orbit Maintainable Item 在轨维修产品（项目）

OMI Operations and Maintenance Instruction 操作与维护指南

OMIS Operations and Maintenance Information System 操作与维修信息系统

OML On-Orbit Maneuver Lifetime 在轨机动（飞行）寿命期

OML Orbiter Mold Line 【美国航天飞机】轨道（飞行）器模线

OML Orbiting Military Laboratory 军用轨道实验室

OMLRS Operations, Maintenance and Logistics Resources Simulation 操作、维修与后勤资源模拟

OMM Onboard Mission Management 箭载任务管理

OMM Optical Mass Memory 光学高密度存储系统

OMM Orbital Maintenance Mission 轨道维修任务

OMM Orbiter Major Maintenance 【美国航天飞机】轨道（飞行）器重大维修

OMM Orbiter Major Modification 【美国航天飞机】轨道（飞行）器重要改装（改进）

OMMH Orbiter Maintenance Man-Hours 【美国航天飞机】轨道（飞行）器维修工时

OMOB Offensive Missile Order of Battle 进攻性导弹战斗序列

OMPR Operational Maintainability Problem Reporting 操作维护性故障通报

OMPRA One Man Propulsion Research Apparatus 【美国国家航空航天局】单人推进研究装置

OMPS Orbit Maneuvering Propulsion System 轨道机动推进系统

OMPT Observed Mass Point Trajectory 被观察质点轨迹

OMR Operations and Maintenance Requirements 操作与维修要求

OMR Orbiter Management Review 【美国航天飞机】轨道（飞行）器管理评审

OMRC Operational Maintenance Requirements Catalog 【美国航天飞机】轨道（飞行）器维护要求目录

OMRF Orbiter Maintenance and Refurbishment Facility 【美国航天飞机】轨道（飞行）器维护与整修设施

OMRP Operations and Maintenance Requirements Plan 操作与维护要求规划

OMRS Operations and Maintenance Requirements Specification 操作与维护要求说明

OMRSD Operational Maintainability Reporting Systems Document 操作维护性通报系统文档

OMRV Operational Maneuvering Re-

entry Vehicle 作战机动再入飞行器
OMRV Orbiting Maneuverable Reentry Vehicle 轨道机动再入飞行器
OMRVS Operational Maneuvering Reentry Vehicle Study 作战机动再入飞行器研究
OMS Onboard Maintenance System 箭载维修系统
OMS Operational Meteorological Satellite 【美国国家航空航天局】业务气象卫星
OMS Operational Monitoring System 操作监控系统
OMS Operations Management System 操作管理系统
OMS Orbital Maneuvering System 【美国航天飞机】轨道机动系统
OMS Orbital Multifunction Satellite 轨道多功能卫星
OMTSS Ordnance Multiple-Purpose Tactical Satellite System 军用多用途战术卫星系统
OMU Optical Measuring Unit 光学测量装置
OMU Orbital Maneuvering Unit 轨道交会装置
OMUX Satellite Transponder Output Multiplexer Filter 卫星转发输出（多路）复用器滤波器
OMV Orbital Maneuvering Vehicle 【美国】轨道机动飞行器
OMV Oxygen Manual Valve 氧气人工阀
OMVCC Orbital Maneuvering Vehicle Control Center 【美国】轨道机动飞行器控制中心
OMWS Orbiter Maintenance Workstation 【美国航天飞机】轨道（飞行）器维修工作站
ONA Optical Navigation Attachment 光学导航辅助装置
ONAP Orbit Navigation Analysis Program 轨道导航分析程序
ONBT Orbiter Neutral Buoyancy Trainer 轨道飞行器中性浮力训练机
ONEP Origins of Near Earth Plasma 近地等离子体起源
ONLAS Optical Night Landing Approach System 光学夜间着陆系统
OOA Object-Oriented Analysis 面向对象的分析
OOA Optimum Orbital Altitude 轨道最佳高度
OOA Optimum Orbital Attitude 最佳轨道姿态
OOAD Object-Oriented Analysis and Design 面向对象的分析与设计
OOAM&S On-Orbit Assembly, Maintenance and Services 轨道上装配、维修与保养
OOCF Out of Control Flight 失控飞行
OOD Object-Oriented Design 面向对象设计
OOD Orbiter On-Dock 【美国航天飞机】对接的轨道（飞行）器
OODA Observation, Orientation, Determinant and Action 观察、定位、判断与行动
OODB Object-Oriented Database 面向对象数据库
OODBMS Object Oriented Data Base Management System 面向对象的数据库管理系统
OODM Object-Oriented Data Model 模型对象数据模型
OOM Orbital Operation Module 轨道工作舱
OOMM Organizational Operations

and Maintenance Manual 基层级使用与维修手册
OOMS On-Orbit Maintenance/Servicing 在轨维修/保养
OOS On Orbit Servicing 在轨勤务
OOS On Orbit Support 在轨保障
OOS On-Orbit Station 在轨站
OOS Orbit-to-Orbit Shuttle 【美国国家航空航天局】轨道间往返飞行器；航天飞机
OOS Orbit-to-Orbit Stage 轨道—轨道级；轨道间过渡阶段
OOSDP On-Orbit Station Distribution Panel 在轨站分配控制台
OOTF On-Orbit Test Facility 在轨测试设施
OOTP On-Orbit Test Plan 在轨测试计划
OOTP On-Orbit Test Report 在轨测试报告
OP Orbit Parameter 轨道参数
OP Orbital Period 轨道周期
OP Oxygen Purge 氧气吹除
OPA Operations Planning Analysis 操作规划分析
OPA Optical Parametric Amplification 光学参量放大
OPAL Orbiting Picosatellite Automated Launcher 轨道微型卫星自动发射装置
OPB Oxidizer Preburner 氧化剂预燃室
OPBOV Oxidizer Preburner Oxidizer Valve 氧化剂预燃室氧化剂阀门
OPCC Offut Processing and Correction Center 【美国】奥弗特处理与相关中心
OPCG Organic Polymer Crystal Growth 有机与聚合物晶体生长
OPCGE Organic and Polymer Crystal Growth Experiment 有机与聚合物晶体生长实验
OPCGF Organic and Polymer Crystal Growth Facility 有机与聚合物晶体生长设备
OPCON Operational Control 作战控制；操作控制
OPCON Optimizing Control 最优控制
OPD Orbiting Propellant Depot 【美国国家航空航天局】轨道推进剂库
OPDAR Optical Detection and Ranging 光学探测与测距
OPE Outer Planets Explorer 外行星探测器
OPEN Origin of Plasma in the Earth's Neighborhood 【美国】近地等离子体起源计划（航天飞机进行的实验，对穿越地球空间环境的能量流进行测估）
OPEP Orbit Plane Experiment Package 轨道（飞行）器实验（仪表）舱
OPEVAL Operational Evaluation 操作性评估
OPF Orbiter Processing Facility 【美国国家航空航天局】轨道（飞行）器处理设施
OPFAC Operational Facility 操作设施
OPFC Orbiter Preflight Checklist 【美国航天飞机】轨道（飞行）器预飞行清单
OPFM Outer Planet Flagship Mission 外部行星首要任务
OPGUID Optimum Guidance (Technique) 最佳制导（技术）
OPH Operational Propellant Handling 实用推进剂处理
OPHTS Operational Propellant Han-

dling Test Site 实用推进剂处理试验场

OPI Orbit Position Indicator 轨道位置指示器

OPI Orbiter Payload Interrogator 【美国航天飞机】轨道（飞行）器有效载荷讯问器

OPIDF Operational Planning Identification File 操作规划识别文件

OPINE Operations in the Nuclear Environment 核环境操作

OPINTEL Operational Intelligence Processor 操作（作战）情报处理器

OPIS Orbiter Prime Item Specification 【美国航天飞机】轨道（飞行）器初始（主要）项说明

OPLF Orbiter Processing and Landing Facility 【美国航天飞机】轨道（飞行）器处理与着陆设施

OPM Orbital Parameter Message 轨道参数信息

OPM Outer Planet Mission 外行星飞行任务

OPO Optical Parametric Oscillation 光学参量振荡

OPO Orbiting Planetary Observatory 轨道行星观测卫星

OPOS Optical Properties of Orbiting Satellite 轨道卫星光学特性

OPOV Oxidizer Pre-Burner Oxidizer Value 氧化剂预燃室氧化剂阀门

OPOVA Oxidizer Preburner Oxidizer Valve Actuator 氧化剂预燃室氧化剂阀门致动器（装置）

OPPOSITE Optimization of a Production Process by an Ordered Simulation and Iteration Technique 应用秩序模拟与迭代技术的生产过程最佳化

OPPS Oxygen Partial Pressure Sensor 氧分压传感器

OPR Operations Phase Review 操作阶段评审

OPR Operations Planning Review 操作规划评审

OPR Orbiter/Payload Recorder 【美国航天飞机】轨道（飞行）器/有效载荷记录器

OPR Overall Pressure Ratio 总压比

OPR Oxygen Pressure Regulator 氧气压力调节器

OPRV Oxygen Pressure Relief Valve 氧气释压阀；氧气减压阀

OPS Operations Sequence 操作顺序

OPS Orbiting Primate Spacecraft 灵长类动物轨道实验航天器

OPS Oxygen Purge System 氧气吹除系统

OPS Spatial Operations 空间操作

OPSA Optimal Pneumatic System Analysis 最佳气动系统分析

OPSB Orbiter Processing Support Building 【美国航天飞机】轨道（飞行）器处理保障大楼

OPSCAP Operations Capabilities 操作（作战）能力

OPSEC Operations Security 操作安全；作战安全措施

OPSF Orbital Propellant Storage Facility 轨道推进剂储存设施

OPSMOD Operations Module 操作模块

OPSNET Operations Network 地面操作网

OPSS Orbital Propellant Storage Subsystem 轨道推进剂储存子系统

OPSV Oxidizer Purge Solenoid Valve 氧化剂吹除电磁阀

OPT Operational Pressure Transducer

操作压力转换器（装置）
OPTE Operational Test and Evaluation 操作试验与鉴定
OPTEC Operational Test and Evaluation Command 操作性试验（测试）与鉴定指令；操作试验（测试）与鉴定司令部
OPTRAK Optical Tracking and Ranging Kit 光学跟踪与测距装置
OPTS On-Line Program Testing System 联机程序测试系统
OQF Orbiting Quarantine Facility 轨道检疫设施
OR Operational Readiness 操作（作战）成熟度（准备）
OR Operational Reliability 使用可靠性；操作可靠性
OR Oxygen Relief 氧气释放
OR&PE Object Recognition and Pose Estimation 目标识别与姿势估测
OR/SA Operations Research/Systems Analysis 操作研究与系统分析
ORACL Overtone Research Advanced Chemical Laser 先进化学激光器泛音研究
ORACL HYLTE Overtone Research Advanced Chemical Laser Hypersonic Low Temperature 先进化学激光器低温超声频泛音研究
ORACLE Optimized Reliability and Component Life Estimator 最佳可靠性与部件寿命估计值
ORAD Orbiter Radar 【美国国家航空航天局】轨道（飞行）器雷达
ORADS Optical Ranging and Detecting System 光学测距与探测系统
ORAN Orbiting Analysis 轨道分析
ORAO Orbiting Radio Astronomical Observatory 轨道射电天文观测台
ORASA Operational Research and System Analysis 作战研究与系统分析
ORASIS Optical Real-Time Adaptive Spectral Identification System 光学实时自适应光谱识别系统
ORATS Operational Readiness Assessment and Training System 备战状态评估与训练系统
ORB Orbiter 【美国航天飞机】轨道（飞行）器
ORBCOMM Orbital Communications 轨道卫星通信（系统）
ORBCOMM Orbital Communications Satellite 美国轨道通信卫星
ORBCOMM-X Orbital Communications-X 轨道通信X卫星
ORBIS Orbiting Radio Beacon Ionospheric Satellite 【美国】轨道无线电信标电离层卫星
ORBIS-CAL Orbiting Radio Beacon Ionospheric Satellite-Calibration 【美国】轨道无线电信标电离层校正卫星
ORBIT On-Line Real Time Branch Information Transmission 在线实时转移信息传送
ORBIT Orbit Ballistic Impact and Trajectory 轨道的弹道导弹落点与轨迹
ORBS Orbital Rendezvous Base System 轨道交会基地系统
ORC Operational Readiness Condition 操作成熟度（准备）条件
ORCA Air Force Advanced Ballistic Missile Floating Capsule Concept Study 空军高级弹道导弹浮舱概念研究
ORCA Operational Requirements Continuity Assessment 操作（作战）需求连续性评估

ORCA Ordnance Remote Control Assembly 火工品遥控装置
ORD Operational Readiness Date 操作准备就绪日期；装备可供使用的日期
ORD Operational Ready Data 操作准备就绪数据
ORD Orbital Refueling Demonstration 【美国航天飞机】轨道燃料加注演示验证
ORDC Orbiter Data Reduction Center 【美国航天飞机】轨道（飞行）器数据处理中心
ORDEAL Orbital Rate Display Earth and Lunar 地球和月球的轨道速度显示
ORDEAP Orbital Rate Drive Electronics for Apollo and LM 【美国】"阿波罗"飞船和登月舱轨道速度驱动电子设备
ORE On-Orbit Repair Experiment 在轨修理实验
OREM Objective Reference Equivalent Measurement 目标基准等效测量；目标参考等效测量
OREO Orbiting Radio Emission Observatory 轨道无线电辐射观测卫星
OREX Orbiting Reentry Experiment 【日本】轨道再入飞行实验（飞行器）
ORFEUS Orbiting and Retrievable Far and Extreme Ultraviolet Spectrometer 【美国国家航空航天局】在轨运行的可回收远紫外和极远紫外光谱仪
ORI Operational Readiness Inspection 工作准备状态检查
ORI Orbital Replacement Instruments 轨道更换仪器
ORION Air Force Manned Nuclear Propulsion Space Vehicle Study 【美国】空军核推进载人航天飞行器研究
ORL Orbital Research Laboratory 轨道研究实验室
ORLA Optimum Repair Level Analysis 优化性修理分析
ORMU Orbital Remote Maneuvering Unit 轨道遥控机动装置
ORP Orbital Radiation Program 轨道辐射计划
ORP Orbital Rendezvous Procedure 轨道交会程序
ORPICS Orbital Rendezvous Positioning, Indexing and Coupling System 轨道交会、定位标定与连接系统
ORR Operational Readiness Review 操作成熟度评审
ORR Operations Requirements Review 操作要求评审
ORR Orbital Rendezvous Radar 轨道交会雷达
ORS Occurrence Reporting System 偶发事件报告制度
ORS Octahedral Research Satellite 八面体研究卫星（美国的军事科学研究卫星）
ORS Operational Response 作战响应
ORS Operationally Responsive Space 操作响应空间
ORS Orbit Research Satellite 轨道研究卫星
ORS Orbital Refueling System 轨道燃料加注系统
ORS Orbiter Refueling System 【美国航天飞机】轨道（飞行）器燃料加注系统
ORS Orbiter Relay Simulator 【美国航天飞机】轨道（飞行）器中继模拟器

ORSC Oxygen-Rich Staged Combustion 富氧分级燃烧
ORT Operational Readiness Test 操作成熟度测试（试验）
ORT Orbit Readiness Test 轨道运行准备试验
ORT Orbit Rendezvous Technique 轨道交会技术
ORTAI Orbit-to-Air Intercept 轨道对空拦截
ORU On-Orbit Replaceable Unit 轨道上可替换部件
ORU Orbital Replacement Unit 轨道替换装置；轨道补充部件
ORV Orbital Reentry Vehicle 轨道再入飞行器
ORV Orbital Rescue Vehicle 轨道救援飞行器
ORV Orbital Return Vehicle 返回式轨道飞行器
OS On-Orbit Station 在轨站
OS Operating Software 操作软件
OS Operating System 操作系统
OS Operational Suitability 操作（作战）可适应性（能力）
OS Optics Subsystem 光学分系统
OS Orbital Servicing 轨道维修
OS Oriented System 定向系统
OSA Operational Support Area 操作保障区
OSAD Outer Space Affairs Division 【联合国】外层空间事务署
OSAR Operations Suitability Assessment Report 操作适用性评估报告
OSAS Operation Space Application System 作战空间应用系统
OSAS Orbiter Solar Array System 【美国】轨道器太阳电池阵系统
OSAS Orbiter Solar Array System 轨道器太阳电池阵系统
OSAT On-Site Acceptance Tests 现场验收测试（试验）
OSB Operations Support Building 操作保障厂房
OSB Orbital Solar Observation 轨道太阳观测
OSC Operational Systems Control 操作系统控制
OSC Operations Safety Console 操作安全控制台
OSC Operations Support Center 操作保障中心
OSC Orbital Sciences Corporation 美国轨道科学公司
OSCAR Orbital Satellite Carrying Amateur Radio 业余无线电通信用轨道卫星
OSCAR Outer Space Collector of Astronomical Radiation 外层空间大气辐射收集器
OSCARS One way Synchronous Collision Avoidance and Ranging System 单路同步避撞与测距系统
OSCARS Operational Satellite Constellation Availability and Reliability Simulation 【美国空军航天司令部】在用卫星星座可用性及可靠性模拟
OSCRS Orbital Spacecraft Consumables Resupply System 轨道航天器消耗品再补给系统
OSDH Orbiter System Definition Handbook 【美国航天飞机】轨道（飞行）器系统定义手册
OSE Operating Support Equipment 操作保障设备
OSE Orbital Sequence of Events 轨道飞行事件顺序
OSE Orbital Support Equipment 轨道保障设备

OSE Orbiter Support Equipment 【美国航天飞机】轨道（飞行）器保障设备

OSEI Operational Significant Event Imagery 提供灾害性天气事件的卫星图像

OSEIT Operations and Support Engineering Integration Tool 操作（作战）与支援（保障）工程综合工具

OSEOS Operational Synchronous Equipment Earth Observatory Satellite 运行中的同步地球观测卫星

OSF Ordnance Storage Facility 火工器存储设施

OSI Open Systems Interconnection 开放式系统互联

OSI Operator System Interface 操作员系统接口

OSIM Object Simulation 目标模拟（仿真）

OSIP Operational System Integration Plan 操作系统综合计划

OSIS Ocean Surveillance Information System 海洋监视情报资料系统

OSL Orbiting Solar Laboratory 轨道太阳实验室（美国的太阳观测卫星）

OSM Orbital Service Module 轨道服务舱

OSM Outer Space Missile 外层空间导弹

OSMP Operational Support Maintenance Plan 操作保障维护计划

OSO Orbiting Satellite Observer 轨道卫星观测器

OSO Orbiting Solar Observatory 【美国国家航空航天局】沿轨道运行的太阳观测站；奥索卫星（美国太阳观测卫星）

OSO Orbiting Solar Observatory 轨道太阳观测站

OSOB Orbiting Satellite Observer 轨道卫星观测器

OSOP Orbiter Systems Operating Procedures 【美国航天飞机】轨道（飞行）器系统操作规程

OSOSV Orbit-Surface-Orbit Space Vehicle 轨道－地面－轨道航天器

OSP Orbital Space Plane 轨道航天飞机

OSPER Ocean Space Explorer 海洋空间探测器；海洋空间探测卫星

OSR Operating Software Requirements 操作软件需求

OSR Operations Support Room 操作保障间

OSR Optimum Speed Rotor 转速最佳化旋翼

OSS Ocean Surveillance Satellite 海洋监视卫星

OSS Ocean Surveillance System 海洋监视系统

OSS Open Simulation System 开放式模拟系统

OSS Open Source Software 开放资源软件

OSS Operations Support System 操作（作战）保障系统

OSS Orbital Space Station 轨道空间站

OSSA Office of Space Science and Applications 【美国】空间科学与应用办公室

OSSE Observing Systems Simulation Experiment 观测系统模拟实验

OSSE Oriented Scintillation Spectrometer Experiment 【美国】定向闪烁光谱仪实验

OSSP Outer Solar System Probe 【美国】外太阳系探测器；外太阳系探测火箭

OSSRH Orbiter Subsystem(s) Requirement Handbook 【美国航天飞机】轨道（飞行）器分系统要求手册

OSSS Optical Space Surveillance Subsystem 航天光学监视子系统

OSSS Orbital Space Station Study 轨道空间站研究

OSSS Orbital Space Station System 轨道空间站系统

OST Operational Suitability Test 作战适用性试验

OST Orbital Solar Telescope 轨道太阳望远镜

OST Orbiter Support Trolley 【美国航天飞机】轨道（飞行）器保障运输车

OST OTV Servicing Technology 轨道转移飞行器服务技术

OST Outer Space Treaty 外层空间条约

OSTA-1 OSTA payload for Shuttle attached payload using SIR-A 空间与地球应用办公室的有效载荷用于航天飞机并搭载了附加有效载荷用于航天飞机成像雷达

OSTA-2 OSTA payload for gravity experiments 空间与地球应用办公室用于万有引力试验的有效载荷

OSTA-3 OSTA payload for photography and radar images of the Earth's surface 空间与地球应用办公室为地球摄影和雷达图像的有效载荷

OSTDS Office of Space Tracking and Data Systems 【美国】空间跟踪与数据系统办公室

OSTF Operational Support Test Facility 操作保障测试设施

OSTP Orbiting System Test Plan 【美国】轨道系统测试（试验）计划

OSTS Office of Space Transportation Systems 空间运输系统办公室

OSV Orbital Servicing Vehicle 【日本】轨道服务飞行器

OSV Orbital Space Vehicle 轨道式飞船

OSV Orbital Support Vehicle 轨道保障飞行器

OSV Oriented Space Vehicle 定向航天器

OSVS Orbiter Space Vision System 轨道（飞行）器空间观测系统

OT Operational Test 操作测试（试验）

OT Operational TIROS 【美国】业务型泰罗斯（气象）卫星

OT Operational Trajectory 移动轨迹；运行轨迹；工作轨迹

OT Optical Tracker 光学跟踪器（装置）

OT&E Operational Test and Evaluation 操作测试（试验）与鉴定

OTA Optical Telescope Assembly 光学望远镜组件

OTB Orbiting Tanker Base 轨道储箱基座

OTBV Oxidizer Turbine Bypass Valve 氧化剂涡轮旁路阀

OTDR Optical Time-Domain Reflectometer 【美国空军】（卫星控制网）光时域反射计

OTEMP Overtemperature 过热

OTES Orbiter Thermal Effects Simulater 轨道（飞行）器热效应模拟器

OTF Orbital Test Flight 轨道试飞

OTH-B Over-the-Horizon Backscatter Radar Detection System 超视距散射雷达探测系统

OTH-T Over-the-Horizon Targeting

超视距目标判定
OTHR Over the Horizon Radar 超视距雷达
OTK Oxidizer Tank 氧化剂储箱
OTL Ordnance Test Laboratory 火工品测试（试验）实验室
OTLC Orbiter Timeline Constraints 轨道（飞行）器时间线约束
OTM Orbit Transfer Module 轨道转移舱
OTM Orbit-Trim Manoeuvre 轨道调整机动
OTOS Orbit-to-Orbit Stage 轨道间芯级
OTP Operational Test Procedure 操作测试规程
OTRR Operational Test Readiness Review 作战适用性试验准备情况检查
OTS Off-the-Shelf 成品的；预制的；现用的
OTS Operations Testing and Simulations 作战试验与模拟
OTS Optical Technology Satellite 光学技术卫星
OTS Optical Test Satellite 光学试验卫星
OTS Optical Tracking System 光学跟踪系统
OTS Orbit Test Satellite 轨道试验卫星（欧洲空间局试验型通信卫星，于1978年5月10日发射。其衍生卫星包括欧洲海事通信卫星、电信1号和欧洲通信卫星）
OTS Orbital Tracking System 轨道跟踪系统
OTS/ECS Orbital Test Satellite/European Communications Satellite 轨道试验卫星/欧洲通信卫星
OTSA Off-the-Shelf Analysis 成品分析

OTSI Operating Time Since Inspection 检修后运行时间
OTSS Orbit Transfer and Servicing System 轨道转移和服务系统
OTSV Oxidizer Tank Shutoff Valve 氧化剂箱断流阀门
OTV Orbital Test Vehicle 轨道测试飞行器
OTV Orbital Transfer Vehicle 【美国】轨道转移飞行器
OTVV Oxidizer Tank Vent Valve 氧化剂箱放气阀门
OUCD Operations and Utilization Capability Development 操作和应用能力开发
OUT Orbiter Utilities Tray 【美国航天飞机】轨道（飞行）器通用垫底
OV Orbital Vehicle 轨道飞行器
OV Orbiting Vehicle (Aerospace) （航空）人造卫星；轨道飞行器
OV Oxygen Vent 氧气排放
OVA Operational Viability Assessment 操作（作战）变量评估
OVAB Orbiting Vehicle Assembly Building 轨道飞行器装配厂房
OVAM Orbital Vehicle Assembly Mode 轨道飞行器装配模式
OVCP Orbital Vehicle Checkout Procedure 轨道飞行器检查程序
OVER Optimum Vehicle for Effective Reconnaissance 能有效侦察的最佳飞行器
OVERS Orbital Vehicle Reentry Simulator 轨道飞行器再入（大气）模拟器；人造卫星重返模拟器
OVF Overfill 过量加注
OVFL Overflow 溢流
OVI Operational Validation Inspection 操作验证检查

OVLBI Orbiting Very Long Baseline Interferometry 轨道甚长基线干涉测量
OVLD Overload 过载
OVLMA Orbital Vehicle Limited Maintenance Area 轨道飞行器限定维修区
OVR Orbital Vehicle Requirement 轨道飞行器要求
OVS Operational Voice System 操作声频系统
OVS Orbital Vehicle System 轨道飞行器系统
OVSB Orbital Vehicle Support Building 轨道飞行器保障厂房
OVV Overvoltage 过电压
OW Optical Window 光学窗口
OWD One-Way Doppler 单向多普勒
OWE Optimum Working Efficiency 最佳工作效率
OWF Orbital and Weightless Flight 轨道失重飞行
OWL On-Wing Life 【主发动机】不卸下连续使用寿命
OWPS Orbiter Weather Protection System 【美国航天飞机】轨道飞行器气象防护系统
OWS Operational Weapon Satellite 作战武器卫星；实用武器卫星
OWS Operations Work Station 操作工作站
OWS Orbital Weapon System 轨道武器系统
OWS Orbital Workshop 【美国"天空实验室"】轨道工作站
OX Orbiting X-Ray Observatory 轨道 X 射线观测站；轨道 X 射线观测卫星
OYCV Optimum Yaw Control Vertical 最佳偏航控制垂直仪
OZO Orbiting Zoological Observatory 轨道动物学观测站

P

P	Pershing Missile	【美国】"潘兴"地—地导弹
P	Persistent Chemical Agent	持久性化学制剂
P	Pitch	俯仰角；纵倾角
P	Polar Distance	极距
P	Primary Frequency	主用频率
P	Priority Precedence	优先等级
P	Probability of Damage	破坏概率
P	Probability of Success	成功概率
P	Probe	探针；探测仪；探测器
P	Prohibited area	禁区
P	Soft Pad	软发射台
P&C	Performance and Control	性能与控制
P&CR	Performance and Capabilities Requirements	性能与能力的需求
P&D	Planning and Design	规划与设计
P&E	Propellant and Explosive	推进剂与炸药
P&FS	Particles and Fields Subsatellite	粒子与探测场子卫星
P&I	Performance and Interface (Specification)	性能与接口（说明书）
P&M	Performance Monitor	性能监测
P&M	Phase Modulation (Modulated)	相位调制
P&M	Planetary Mission	卫星飞行任务
P&M	Preventive Maintenance	防护性维护
P&M	Processes and Materials	工艺方法与材料
P&M	Pulse Modulation	脉冲调制
P&P	Plug and Play	即插即用
P&R	Performance and Resources	性能与资源
P&VE	Propulsion and Vehicle Engineering	推进装置与飞行器工程
P&VE-A	Propulsion and Vehicle Engineering-Administrative	推进器和飞行器实验室行政管理组
P&VE-lab	Propulsion and Vehicle Engineering Laboratory	推进器与运载器工程实验室
P-P	Peak-to-Peak (Value)	峰值之间的；正负峰间值
P-WEAR	Preliminary Work Breakdown Structure Element Audit Review	初步工作分类体系结构审查评审
P/A	Payload/Altitude (Curve)	有效载荷/飞行高度曲线（探空火箭）
P/A	Planetary Atmosphere	行星大气层
P/A	Propulsion and Vehicle Engineering-Engine Management	推进系统和飞行器工程实验室发动机管理
P/A	Propulsion/Avionics	推进系统及电子设备与控制系统
P/C	Pitch Control	变距杆；变距操纵；俯仰控制；纵向控制
P/L	Payload (Satellite, Spacecraft)	有效载荷（卫星，航天器）
P/M	Payload Module	有效载荷舱
P/P	Peak to Peak	峰间值
P/P	Polar Point	极点
P/PL	Primary Payload	主有效载荷

P/T　Prototype　原型机；样机
P/Y　Pitch/Yaw　俯仰/偏航
P^2NRTA&A　Pre-Planned Near-Real-Time Assessment and Adaptation　预先计划的近实时评估和改进
P^3　Pollution Prevention Program　污染预防计划
P^3I　Preplanned Product Improvement　预先计划产品改进
P^4A　Programmable Powdered Perform Process for Aerospace　用于航空航天的可编程粉化运行程序
PA　Approach Flight Phase　进场飞行阶段
PA　Pad Abort　发射台失事
PA　Paging and Area Warning　寻呼与区域警戒
PA　Planetary Atmosphere　行星大气
PA　Power Amplifier　功率放大器（装置）
PA　Precision Attitude　精确姿态
PA　Product Assurance　产品保证
PA　Propulsion/Avionics　推进装置与航空电子设备
PA　Pulse Amplifier　脉冲放大器
PA&E　Program Analysis and Evaluation　项目分析与鉴定
PA&E　Program Assessment and Evaluation　项目评估与鉴定
PA&R　Program Assessment and Review　项目评估与评审
PAA　Phased Adaptive Approach　相控自适应方法
PAA　Phased-Array Antenna　相控阵天线
PAAC　Program Analysis Adaptable Control　程序分析适应性控制
PAAC　Program Analysis Adaptor Control　程序分析适配器控制
PAAFB　Patrick Auxiliary Air Force Base　【美】帕特里克辅助空军基地
PAAMS　Principal Anti-Air Missile System　主要空防导弹系统
PAB　Payload Assembly Building　有效载荷组装厂房
PABLA　Problem Analysis by Logical Approach　逻辑法问题分析
PABST　Primary Adhesively Bonded Structure Technology　基本粘接结构的基本工艺
PAC　Package Attitude Control　派克卫星（美国技术试验卫星，目的是作半主动重力梯度稳定技术实验）
PAC　Patriot ATBM Capability　【美】"爱国者"反战术弹道导弹能力
PAC　Performance Analysis and Control　性能分析与控制
PAC　Pershing Airborne Computer　【美】"潘兴"导弹弹载计算机
PAC　Pneumatic Analog Computer　气动模拟计算机
PAC　Probe Aerodynamic Center　探测器空气动力中心
PAC　Program Assessment Center　项目评估中心
PAC-2　Patriot Anti-Tactical Missile Capability-2　【美】"爱国者反战术导弹能力—2"型导弹系统
PAC-3　Patriot Advanced Capability-3　【美】"爱国者先进能力—3"型导弹系统
PACAS　Personnel Access Control Accountability System　人员访问控制统计系统
PACC　Problem Action Control Center　故障措施控制中心
PACE　Performance and Cost Evaluation　性能与成本评估
PACE　Point-Ahead Compensation Ex-

periment 前沿位置补偿实验
PACE Precision Analog Computing Equipment 精确模拟计算设备
PACE Preflight Acceptance Checkout Equipment 飞行前自动检测设备
PACE Prelaunch Automatic Checkout Equipment 发射前自动检测设备
PACE/SC Preflight Acceptance Checkout Equipment for Spacecraft 航天器飞行前接收检查设备
PACEE Propulsion and Auxiliary Control Electronic Enclosure 推进装置和辅助控制电子箱
PACES Pressurized-Air-Combustor Exhaust Simulator 压缩空气燃烧室排气模拟器
PACLESS Portable Automated Communications Lightweight Expandable Search System 移动式自动化通信轻型可扩展搜索系统
PACMISRAN Pacific Missile Range 【美国】太平洋导弹靶场
PACOB Propulsion Auxiliary Control Box 推进辅助控制箱
PACOSS Passive and Active Control of Space Structure 宇航结构被动与主动控制
PACP Propulsion Auxiliary Control Panel 推进装置辅助控制板
PACPADS Portsmouth Alarm Communication Processing and Display System 朴茨茅斯报警信号传送数据处理与显示系统
PACR Performance and Compatibility Requirement 性能与兼容性需求
PACS Passive Attitude Control System 被动姿态控制系统
PACS Payload Actuation and Control System 有效载荷致动与控制系统
PACS Pitch Active Control System 俯仰主动控制系统
PACS Pitch Augmentation Control System 俯仰增稳控制系统
PACS Pitch-Axis Control System 俯仰轴控制系统
PACS Pointing and Attitude Control System 定向与姿态控制系统；瞄准与姿态控制系统
PACT Portable Automatic Calibration Tracker 便携式自动校准跟踪仪
PAD Pressure Anomaly Detection 压力异常探测
PAD Propellant Acquisition Device 推进剂获取装置
PAD Propellant Activated Device 推进剂激活装置
PAD Propellant Actuated Device 推进剂驱动装置
PADAL Pattern for Analysis, Decision, Action and Learning 分析、决策、动作与学习模式
PADE Pad Automatic Data Equipments 发射台自动数据设备
PADIL Patriot Data Information Link 【美国】"爱国者"导弹数据信息链路
PADIL Patriot Digital Information Link 【美国】"爱国者"导弹数字信息链路
PADIL Patriot Digital Interface Link 【美国】"爱国者"导弹数字接口链路
PADIL Phased-Array Tracking to Intercept of Target Digital Information Link 相控阵跟踪雷达拦截目标数字信息链路
PADIS Procedures for Analyzing the Design of Interactive Solutions 解决相互作用问题的设计分析步骤
PADLOC Passive Detection and Loca-

tion of Countermeasures (System) 无源对抗探测与定位（系统）
PADLOC Passive/Active Detection and Location (System) 无源／有源探测与定位（系统）
PADRE Portable Automatic Data Recording Equipment 便携式自动数据记录设备
PADS Passive-Active Data Simulation 主动与被动数据模拟
PADS Performance Analysis and Design Synthesis 性能分析和设计综合
PADS Performance Analysis Display System 性能分析显示系统
PADS Precision Aerial Delivery System 精确空投系统
PADS Precision Attitude Determination System 精确姿态确定系统
PADS Primary Attitude Determination System 初始姿态确定系统
PADS Propellant-Actuated Devices 推进剂驱动装置
PAE Payload Accommodations Equipment 有效载荷工作条件保证设备
PAE Payload Attach Equipment 有效载荷连接设备
PAECT Pollution Abatement and Environmental Control Technology 污染减轻与环境保护技术
PAET Planetary Atmosphere Entry Test 行星大气进入试验
PAET Planetary Atmosphere Experiments Test 行星大气实验测试
PAF Payload Attach Fitting 有效载荷连接装置
PAF Polaris Acceleration Flight (Chamber) 【美国】"北极星"导弹增速飞行（燃烧室）
PAFAM Performance and Failure Assessment Module 性能与故障评价模型
PAFAM Performance and Failure Assessment Monitor 性能与故障评价监控器
PAFB Patrick Air Force Base 【美国】帕特里克空军基地
PAFU Propulsion and Firing Unit 推进和发射装置；推进和射击装置
PAGE Programmable Aerospace Ground Equipment 可编程航空航天地面设备
PAGEOS Passive Geodetic Earth Orbiting Satellite 【美国】无源大地测量地球轨道卫星
PAGEOS Passive Geodetic Earth Orbiting Satellite 无源大地测量地球轨道卫星
PAILS Projectile Air Burst and Impact Location System 射弹空中爆炸与弹着定位系统
PAIR Precision-Approach Interferometer Radar 精确进场干涉雷达
PAL Permissive Action Link 许可行动链接
PAL Protuberance Air Load (on the ET) 【美国航天飞机】外贮箱突起部分的空气荷载
PALAPA Indonesian Communications Satellite 印度尼西亚通信卫星
PALC Point Arguello Launch Complex 【美国】阿格落角发射场
PALM Precision Altitude and Landing Monitor 精密高度与着陆监控器
PALMS Propulsion Alarm and Monitoring System 推进装置告警与监控系统
PALS Patriot Automated Logistics System 【美国】"爱国者"导弹自动后勤系统

PALS Phased Array Type L-Band Synthetic Aperture Radar L波段相控阵合成孔径雷达
PALS Photo Area and Location System 照相区与定位系统
PALS Portable Airfield Light Set 便携式机场照明设备
PALS Positioning and Location System 定位定向系统
PALS Precision Approach and Landing System 精确进场和着陆系统
PALS Program Automated Library System 【美国空间站】自动化程序库系统
PALS Protection Against Limited Strikes 针对有限攻击的保护
PAM Payload Adapter Module 有效载荷支架舱
PAM Payload Assist Module 有效载荷助推舱（从美国航天飞机上向更高轨道运送卫星的上面级火箭推进舱）
PAM Penetration Augmented Munitions 加强型穿透弹药
PAM Perigee Assist Module 近地点助推舱
PAM Perigee Assist Motor 近地点助推发动机
PAM Precision Attack Missile 精确攻击导弹
PAM Pulse Amplifier Modulation 脉冲放大器调制
PAM-A Payload Assist Module-Atlas Centaur 【美国】"宇宙神/半人马座"级有效载荷助推舱
PAM.D Payload Assist Module-Delta 【美国】"德尔它"运载火箭上的有效载荷助推舱
PAMCC Propulsion and Auxiliary Machinery Control Console 推进装置和辅助机械控制操纵台
PAMD Perimeter Antimissile Defense 反导弹环形防御
PAMIRASAT Passive Microwave Radiation Satellite 【欧洲空间局】被动微波辐射卫星
PAMISE Propulsion and Auxiliary Machinery Information System 推进装置和辅助机械信息系统
PAMS Plan Analysis and Modeling System 计划分析与建模系统
PAMS Point Antimissile System 要点反导弹系统
PAN AMSAT Pan American Satellite 泛美卫星（美国通信卫星）
PANDA Performance and Demand Analyzer 性能与要求分析器
PANFI Precision Automatic Noise Figure Indicator 自动精确噪声指数指示器
PANS Positioning and Navigation System 定位导航程序
PAP Propulseur d'Appoint à Poudre 推进剂补给火箭发动机
PAP Solid Propellant Strap-on booster/Propulseur d'Appoint à Poudre 固体推进剂捆绑式助推器
PAPI Precision Approach Path Indicators 精确进场航线指示器
PAPS Propulsion Associated Parameter Set 推进装置有关参数集
PAR Payload Accommodation Requirements 有效载荷配置要求
PAR Peak-to-Average Data Rate 从峰值到平均值的数据速率
PAR Peak-to-Average Ratio 峰值与平均值之比
PAR Perimeter Acquisition Radar 环形搜索雷达
PAR Phased-Array Radar 相控阵列

雷达

PAR Precision Approach Radar 精确进场雷达

PAR Precision Attitude Reference 精确姿态基准

PAR Preflight Acceptance Review 预飞行验收评审

PAR Product Acceptance Review 产品验收评审

PAR Propulsion and Aeroballistics Research 推进与航空弹道学研究

PAR Pulse Acquisition Radar 脉冲搜索雷达

PARCC Precision, Accuracy, Representativeness, Completeness, and Comparability 精密度、准确度、代表性、完整性和可比性

PARR Performance Analysis Reliability Reporting 性能分析可靠性报告

PARSIP Point Arguello Range Safety Impact Predictor 【美国】阿格洛角靶场安全保障弹着点预测器

PARTNER Proof of Analog Results Through a Numerical Equivalent Routine 用数字等效程序验证模拟结果

PAS Payload Analysis System 有效载荷分析系统

PAS Payload Assist System 有效载荷辅助系统

PAS Payload Attach System 有效载荷连接系统

PAS Perigee-Apogee Satellite 近地点—远地点卫星

PAS Perigee-Apogee Stage 近地点—远地点级；近地点—远地点阶段

PAS Pneumatic Actuation System 气动作动系统

PAS Pneumatic Actuator Subsystem 气动作动器分系统

PAS Power, Avionics, and Software 电力、电子设备以及软件

PAS Primary Ascent System 主上升系统

PASCOM Passive Communication Satellite 无源通信卫星

PASE Portable Avionics Support Equipment 便携式航空电子保障设备

PASI Precision Approach Slope Indicator 精确进场下滑指示器

PASIP Propulsion and Airframe Structural Integration Programme 推进装置与机体机构综合计划

PASIS Perpetually Available and Secure Information System 持续可用和可靠的信息系统

PASS Passive and Active Sensor Subsystem 无源与有源传感器子系统

PASS Planning and Scheduling System 规划与调度系统

PASS Primary Avionics Software System 主要航空电子设备软件系统

PASS Procurement Automated Source System 采购自动化来源系统

PASTA Program for Application to Stress and Thermal Analysis 热应力分析应用程序

PASTF Photovoltaic Advanced Systems Test Facility 光电先进系统试验设备

PAT Pointing, Acquisition, and Tracking 定向、获取与跟踪

PAT Production Acceptance Test 产品验收测试（试验）

PAT&E Product Acceptance Test and Evaluation 产品验收试验与鉴定

PATE Programmed Automatic Test Equipment 程序化自动测试设备

PATEC Portable Automatic Test Equip-

ment Calibrator 便携式自动测试设备校验器
PATH Postflight Attitude and Trajectory History 飞行后姿态与轨迹时间关系图
PATH Precipitation and All Weather Temperature and Humidity 降雨量与全天候温度和湿度
PATHS Precursor Above-the-Horizon Sensor 超视距传感器预报器
PATIE Pointing and Tracking Integrated Experiment 定向与跟踪集成实验
PATRES Pre-Cooled Air Turbojet/Rocket Engine with Scramjet 预冷涡轮喷气/火箭超燃冲压发动机
PATRIOT Phased Array Tracking Radar Intercept on Target (Missile) 用相控阵跟踪雷达拦截目标的导弹
PATS Precision Attack Targeting System 精确攻击瞄准系统
PATS Precision Automated Tracking Station 精确自动跟踪站
PATS Precision Automated Tracking System 精确自动跟踪系统；精密自动跟踪系统
PATS Program for Analysis of Time Series 时间序列分析程序
PATS Programmable Automatic Test System 可编程的自动测试系统
PATS Propulsion Analysis Trajectory Simulation 推进分析弹道模拟
PATS Protection Assessment Test System 防护评估测试系统
PATV Pad Abort Test Vehicle 测试发射台中止系统的测试飞船
PAV Planetary Ascent Vehicle 行星上升飞行器
PAV Pneumatic Actuated Valve 空气致动阀

PAV Pressure Actuated Valve 压力致动阀
PAVE Precision Acquisition Vehicle Entry 精确探测飞行器进入（大气层）
PAVE Precision Avionics Vectoring Equipment 精确（航空、导弹、航天）电子设备与控制系统引导设备
PAVEPAWS Precision Acquisition of Vehicle Entry and Phased Array Warning System 飞行器再入（大气层）精确搜索相控阵报警系统；"铺路爪"系统
PAVT Position and Velocity Tracking 位置与速度跟踪
PAW Physics Analysis Workstation 物理分析工作站
PAW Powered All the Way 全程动力推进的
PAWS Paging and Area Warning System 寻呼与区域警戒系统
PAWS Phased Array Warning System 相控阵报警系统
PAYCOM Payload Command 有效载荷指令
PAYCOM Payload Command Controller 有效载荷指令控制器（装置）
PAYDAT Payload Data 有效载荷数据
PAYLD Payload 有效载荷
PB Payload Bay 有效载荷舱
PB Preburner 预燃室
PB/MT/D ATD Post-Boost/Midcourse Tracking/Discrimination ATD 末助推段/中程跟踪/识别的先进技术演示验证
PBAM Performance Based Assessment Model 基于性能的评估模型
PBAN Polybutadiene Acrylonitrile

聚丁二烯—丙烯腈
PBATS Portable Battlefield Attack System 轻便式战场攻击系统（地空导弹）
PBCM PK Booster Control Module 【美国】"和平卫士"导弹助推器控制模块
PBCRAW Post-Boost Control Reaction Attitude Wafer 末助推段控制状态板
PBCS Post Boost Control System 助推段后控制系统；被动段飞行的控制系统
PBD Payload Bay Door 有效载荷舱门
PBDF Payload Bay Door FWD 有效载荷舱门前端
PBDI Position, Bearing and Distance Indicator 位置、方位与距离指示器
PBDI Postboost, Predeployment Intercept 末助推、预展开拦截
PBDM Payload Bay Door Mechanism 有效载荷舱门机械装置
PBE Piggyback Experiment 搭载实验
PBI Post-Boost Intercept 末助推段拦截
PBIC Programmable Buffer Interface Card 可编程缓冲器接口卡
PBIM Programmable Buffer Interface Module 可编程缓冲器接口模块
PBIP Pulse Beacon Impact Predictor 脉冲信标碰撞预测器
PBK Payload Bay Kit 有效载荷舱配套元件
PBL Parachute Braked Landing 降落伞制导着陆
PBL Payload Bay Liner 有效载荷低电离核区

PBL Performance Based Logistics 基于性能的后勤保障
PBL Planetary Boundary Layer 行星边界层
PBM Perigee Boost Motor 近地点（助推）发动机
PBM Power Balance Model 动力平衡模型
PBM Process Based Management 基于过程的管理
PBMA Process-Based Mission Assurance 基于过程的任务保证
PBMA-KMS Process Based Mission Assurance-Knowledge Management System 基于工艺方法的任务保证—知识管理系统
PBP Post-Boost Phase 末助推段
PBP Preburner Pump 预燃室泵
PBPS Post-Boost Propulsion System 助推段后推进系统
PBS Performance Based Specification 基于性能的规范
PBV Post-Boost Vehicle 末助推飞行器；后助推火箭
PBW Particle Beam Weapon 粒子束武器
PC Payload Container 有效载荷容器
PC Pitch Control 俯仰控制
PC Plane Change 水平面改变
PC Planetary Camera 星际照相机
PC Pressure Chamber 压力室
PC Pressurized Cabin 加压舱
PC Project Control 项目控制
PC Propellant Control 推进剂控制；火箭燃料调节
PC&IC Polaris Control and Information Center 【美国】"北极星"导弹控制与信息中心
PCA Payload Clamp Assembly 有效

载荷夹具装配
PCA Payload Controlled Area 有效载荷控制区
PCA Pitch Control Assembly 俯仰控制装置
PCA Pneumatic Control Assembly 气动控制装置
PCA Point of Closest Approach 最近的进场点
PCA Pointing and Control Assembly 定位与控制组件
PCA Power Control Assembly 功率控制组件
PCA Primer Carrier Assembly 起爆器运输装置
PCA Primer Chamber Assembly 起爆器燃烧室组装
PCA Principal Component Analysis 主要构件分析
PCA Principal Components Analysis 主要构件分析
PCA Propellant Control Assembly 推进剂控制组件
PCA Pyrotechnic Carrier Assembly 火工品运输装置
PCA Pyrotechnic Control Assembly 火工品控制装置
PCAS Pitch Control Augmentation System 俯仰控制增稳系统
PCASS Program Compliance Assurance and Status System 项目(程序)符合性保证与状态系统
PCB Power Control Box 动力控制箱
PCB Product Configuration Baseline 产品配置基线
PCB Propulsion Control Board 推进控制板
PCC Pad Control Center 发射台控制中心
PCC Payload Control and Checkout 有效载荷控制与检测
PCC Payload Control Center 有效载荷控制中心
PCC Processing Control Center 处理控制中心
PCC Production Control Centers 产品控制中心
PCCADS Panorama Cockpit Control and Display System 全景座舱控制与显示系统
PCCE Particle Cloud Combustion Experiment 粒子云燃烧实验
PCCE Payload Common Communication Equipment 有效载荷通用通信设备
PCCM Program Change Control Management 项目更改控制管理
PCCS Protected Communication Control System 防护通信控制系统
PCDA Process Control and Data Acquisition 过程控制与数据采集
PCDU Payload Command Decoder Unit 有效载荷指令译码器单元
PCF Payload Control Facility 有效载荷控制设施
PCFL Propulsion Cold Flow Lab 推进装置冷流实验室
PCG Protein Crystal Growth Experiment 【美国航天飞机】蛋白质晶体生长实验
PCGOAL Personal Computer Government Operator Aerospace Language 个人计算机政府操作员航空航天语言
PCH Program Critical Hardware 项目关键性硬件
PCI Peripheral Component Interface 外围部件接口
PCI Production Configuration Identifi-

cation 产品配置识别
PCI Program Control Input 项目控制输入
PCIL Pilot-Controlled Instrument Landing 驾驶员操纵的仪表着陆
PCIL Prime Consolidated Integration Laboratory 统一综合主实验室
PCL Precision Clean Lab 精密洁净实验室
PCL Primary Coolant Loop 主冷却回路
PCM Parallel Cluster Missile 并列集束助推器导弹
PCM Phase Change Material 相变材料
PCM Power Control Mission 动力控制任务
PCM Pressurized Cargo Module 加压货运舱
PCM Pulse Code Modulation 脉冲编码调制
PCM Pulse Code Modulator 脉冲编码调制器（装置）
PCM/FM Pulse-Coded Modulation/Frequency Modulation 脉冲编码调制/频率调制
PCMCIA Personal Computer Miniature Connector Interface Adapter 个人计算机微型连接器接口适配器
PCMMU Pulse-Code Modulation Master Units 脉冲编码调制主部件
PCMTS Pulse Code Modulation Telemetry System 脉冲编码调制遥测系统
PCMU Propellant Calibration Measuring Unit 推进剂校准测量装置
PCNS Polar Coordinates Navigation System 极坐标导航系统
PCO Postcheckout Operations 检测后的操作
PCO Program Controlled Output 项目控制输出
PCOLA Pulse Coded Optical Landing Assembly 脉冲编码光学着陆导航组件
PCOT Payload Center Operations Team 有效载荷中心操作组
PCP Payload Control Processor 有效载荷控制处理器
PCP Power Control Panel 动力控制操纵台
PCPA Pressure Control and Pump Assembly 压力控制与泵组件
PCR Payload Certification Review 有效载荷鉴定评审
PCR Payload Changeout Room 有效载荷替换间
PCR Payload Changeover Room 有效载荷转换间
PCR Payload Checkout Room 有效载荷检查间
PCR Polymerase Chain Reaction 聚合酶链式反应
PCR Preliminary Capability Review 初步能力评审
PCS Passive Communication Satellite 无源（被动）通信卫星
PCS Payload Checkout System 有效载荷检查系统
PCS Payload Containment System 有效载荷抑制系统
PCS Payload Control Supervisor 有效载荷监控员
PCS Pitch Control System 俯仰控制系统
PCS Pointing Control System 定向控制系统
PCS Power Conversion System 动力转换系统
PCS Pressure Control System 压力

控制系统
PCS Primary Coolant System 主冷却系统
PCSA Physics of Colloids in Space Apparatus 空间胶质物理学仪器
PCSA Prophylactic Cervical Shock Absorber 【航天员】颈部防冲击缓冲器
PCSPF Payload Containment System Processing Facility 有效载荷抑制系统处理设施
PCT Portable Communication Terminal 便携式通信终端
PCT Power Control Test 动力控制试验
PCTC Payload Crew Training Complex 有效载荷乘员培训大楼
PCTE Portable Commercial Test Equipment 便携式民用测试设备
PCTO Payload Cost Tradeoff Optimization 有效载荷成本均衡优化
PCTR Pad Connection Terminal Room 发射台连接终端室
PCU Parameter Control Unit 参数控制单元
PCU Payload Checkout Unit 有效载荷检测装置
PCU Pneumatic Checkout Unit 气动检测装置
PCU Power Control Unit 动力控制装置
PCU Pressure Control Unit 压力控制装置
PCU Process Control Unit 过程控制装置
PCV Precheck Verification 预检查验证
PCV Purge Control Valve 吹除控制阀
PCVB Pyro Continuity Verification Box 火工品连续性验证箱
PCVL Pilot Controlled Visual Landing 驾驶员操纵的目视着陆
PCWBS Preliminary Contract Work Breakdown Structure 初步合同工作分解结构
PCZ Physical Control Zone 物理控制区
PD Passive Detection 无源探测
PD Payload Data 有效载荷数据
PD Preliminary Design 初步设计
PD Propellant Dispersal 推进剂消散
PD&RS Payload Deployment and Retrieval Subsystem 有效载荷展开和收回子系统
PD&V Projection Definition and Validation 投放定义与验证
PD/RR Program Design and Risk Reduction 项目设计与风险降低
PDA Photon Detector Assembly 光子探测装置（太空望远镜）
PDA Power Distribution Assembly 动力分配组件
PDA Preliminary Design Audit 初步设计审查
PDA Propellant Drain Area 推进剂排放区
PDA; P.D.A. Photon Detector Assembly 光子探测器装置（太空望远镜）
PDAC Power Distribution and Control (Assembly) 功率分配与控制（装置）
PDAM Pre-Determined Deris Avoidance Maneuver 前定空间碎片规避机动
PDAR Program Description and Requirements 项目说明与要求
PDB Power Distribution Box 动力分配箱
PDB Project Data Base 项目数据库

PDC Planetary Data Center 行星数据中心
PDC Predefined Command 预先确定的指令
PDC Pre-Departure Clearance 起飞前放行许可
PDCS Power Distribution and Control System (Subsystem) 动力分配与控制系统（子系统）
PDD Point Defense Demonstration 点防御演示验证
PDE Plasma Depletion Experiment 【美国航天飞机】等离子体衰变实验
PDE Propulsion Drive Electronics 推进驱动电子设备
PDE Pulse Detonation Engine 脉冲点火发动机
PDF Precision Direction Finder 精确测向仪
PDFI Probability of Detected Fault Isolation 查明的故障隔离的概率
PDI Payload Data Interleaver 有效载荷数据交叉存取器
PDI Payload Display Indicator 有效载荷显示器
PDIP Program Development Integration Plan 项目研发一体化计划
PDL Payload Data Library 有效载荷数据库
PDL Program Design Language 项目设计语言
PDL Programmatic Design/ Development Language 纲领性设计/研发语言
PDM Predictive Maintenance 预测性维护
PDM Product Data Management 产品数据管理
PDM/FM Pulse Duration Modulation/ Frequency Modulation 脉宽调制/频率调制
PDME Peak Distortion Monitoring Equipment 峰值畸变监视设备
PDME Portable Data Module Emulator 便携式数据模块仿真器
PDMF Programmable Digital Matching Filter 可编程数字匹配滤波器
PDMR Processor Data Monitor 处理器（装置）数据监测
PDMS Point Defense Missile System 点防御导弹系统
PDP Plasma Diagnostics Package 【美国航天飞机】等离子体诊断装置
PDP Postinsertion Deorbit Preparation 入轨后脱轨准备
PDP Project Definition Phase 项目定义阶段
PDP Pulse Doppler Processor 脉冲多普勒处理器
PDR Preliminary Data Requirements 初步数据要求
PDR Preliminary Design Review 初步设计评审
PDR Process Description Report 工艺说明报告
PDR Processed Data Records 处理数据记录
PDR Processing Data Rate 数据处理速率
PDR Product Design Review 产品设计评审
PDR Program Design Review 项目设计评审
PDR Preliminary Design Review 初步设计评审
PDRC Pressure-Drop Ratio Control 压降比控制
PDRD Performance and Design Re-

quirements Document　性能与设计要求文件

PDRD　Procurement Data Requirements Document　采办数据要求文件

PDRD　Program Definition and Requirements Document　项目定义与需求文件

PDRD　Program-Level Design Requirements Document　分级程序设计需求文件

PDRE　Pulse Detonation Rocket Engine　脉冲点火火箭发动机

PDRM　Payload Deployment and Retrieval Mechanism　有效载荷展开与收回机制

PDRR　Program Definition and Risk Reduction Life Cycle Phase　"计划确定与风险处理"寿命周期段

PDRR　Program Definition (Development) and Risk Reduction　项目定义（研发）与风险降低

PDRR　Program Description, Requirements Review　项目说明、要求评审

PDRS　Payload Data and Retrieval System　有效载荷数据与恢复系统

PDRS　Payload Deployment and Retrieval System　【美国国家航空航天局】有效载荷展开与收回系统

PDRSS　Payload Development and Retrieval System Simulator　【美国国家航空航天局】有效载荷展开与收回系统模拟器

PDRSTA　Payload Deployment and Retrieval System Test Article　【美国国家航空航天局】有效载荷展开与收回系统试验件

PDS　Package Data System　数据包系统

PDS　Payload Data Segment　有效载荷数据段

PDS　Planetary Data System　行星数据系统

PDS　Power Distribution System (Subsystem)　配电系统（分系统）

PDS　Problem Data System　故障数据系统

PDSMS　Point Defense Surface Missile Systems　点目标防御舰艇导弹系统（又称"海麻雀"系统）

PDSS　Payload Data Services System　有效载荷数据服务系统

PDSS　Post-Development Software Support　研发后的软件支持（保障）

PDU　Performance Diagnostic Unit　性能诊断装置

PDU　Pilot Display Unit　驾驶显示装置

PDU　Pneumatic Drive Unit　气动装置

PDU　Power Distribution Unit　配电装置

PDU　Power Drive Unit　电源驱动装置

PDU　Pressure Distribution Unit　压力分配装置

PDU　Pulse Detection Unit　脉冲探测装置

PDUS　Primary Data-User Stations　原始数据用户站（欧洲气象卫星高分辨率图像收发站）

PDV　Pressure Disconnect Valve　压力分离阀

PDV　Program Definition and Validation　项目定义与验证

PDVF　Payload Design Verification Facility　有效载荷设计验证设施

PE　Planetary Explorer　行星探险者；行星探测器

PE Post Encounter 交会后阶段（美国"旅行者"号太空探测器与天王星交会过程中的一个阶段）

PEA Payload Encapsulation Area 有效载荷封装区

PEACSAT Pan Pacific Education and Communications Experiments by Satellite 泛太平洋卫星教育和通信实验

PEATMOS Primitive Equation and Trajectory Model Output Statistics 原始方程和弹道模型输出统计

PEB Payload Equipment Building 有效载荷设备大楼

PEBB Power Electronics Building Blocks 电子学能量模块

PEC Packet Error Control 组故障控制

PEC Preauthorized Engagement Criteria 预授权交战标准

PEC Pressure Error Correction 压力误差修正

PECS Portable Environmental Control System 移动式环境控制系统

PED Payload Element Developer 有效载荷组件研制方

PEDAS Potential Environmentally Detrimental Activities in Space 在太空进行的有潜在环境危害的活动

PEDAS Potentially Environmentally Detrimental Activities in Space 空间环境潜在放射性危害

PEDRO Pneumatic Energy Detector with Remote Optics 带遥测光学仪器的气动能量探测器

PEEL Parametric Endoatmospheric/Exoatmospheric Lethality Simulation 大气层内外杀伤力参数仿真

PEF Payload Encapsulation Facility 有效载荷封装厂房

PEFO Payload Effects Follow-on Study 有效载荷效应后续研究

PEG Powered Explicit Guidance 动力显式制导

PEGASUS NASA Unmanned Scientific Satellite 美国国家航空航天局"飞马座"不载人科研卫星

PEIR Project Equipment Inspection Record 项目设备检查记录

PEIS Programmatic Environmental Impact Statement 纲领性环境影响报告

PEL Permissible Exposure Limit 容许暴露限值

PEL Precision Elastic Limit 精确弹性限值

PEM Payload Ejection Mechanism 有效载荷弹射机构

PEM Payload Electronics Module 有效载荷电子舱

PEM Program Element Monitor 规划要素负责人；计划要素负责人

PEM Propagation Engagement Model 分布交战模型

PEM Proton Exchange Membrane 质子交换膜

PEOS Propulsion and Electrical Operating System 推进与电操作系统

PEP Performance Evaluation Profile 性能鉴定剖面

PEP Polynomial Error Protection 【美国国家航空航天局】多项式错误保护

PEPER Propellants, Explosives, Pyrotechnics Evaluation and Reapplication 推进剂、炸药、火工品鉴定与重复使用

PEPP Planetary Entry Parachute Program 星际进入降落伞计划

PEPS Protective Ensemble Perfor-

mance Standards　总体安全性能标准
PEPSY　Precision Earth Pointing System　精确瞄准地球系统；精确指向地球系统
PEQC　Production Engine Quality Control　生产发动机质量控制
PER　Perigee　近地点
PER　Preliminary Engineering Report　初步技术工艺报告
PER　Pyrotechnic Energy Release　火工品能量释出
PERD　Payload Element Requirements Document　有效载荷组件要求文件
PEREF　Propellant Engine Research Environmental Facility　研究环境对推进剂及发动机影响的设备
PERGS　Portable Earth Resources Ground Station　机动式地球资源地面站
PERME　Propellants, Explosives and Rocket Motor Establishment　推进剂、炸药和火箭发动机研究所
PERSEP　Pershing Survivability Evaluation Program　【美国】"潘兴"导弹生存能力鉴定计划
PERT　Particle Electrostatic Reverse Thruster　静电粒子反向推力器
PERT　Performance Evaluation and Record Tracking　性能鉴定与记录跟踪
PERT　Performance Evaluation and Review Technique　性能评审技术
PERT　Program Evaluation and Review Technique　计划评审技术；计划评估法；统筹法
PERTSIM　Program Evaluation and Review Technique Simulation　项目评审技术模拟
PES　Planetary Exploration Spacecraft　行星探测航天器
PES　Potential Explosion Site　潜在爆炸地点
PESHE　Programmatic Environmental Safety and Health Evaluation　纲领性环境安全与健康鉴定
PET　Parent Effectiveness Training　原效能（效力）训练
PET　Performance Evaluation Test　性能鉴定试验
PET　Periodic Evaluation Test　定期鉴定试验
PET　Production Evaluation Test(ings)　产品鉴定试验
PET　Prototype Evaluation Testing　原型鉴定试验
PETA　Performance Evaluation and Trend Analysis　性能鉴定与趋势分析
PETS　Payload Environmental Transportation System　有效载荷环境运输系统
PETS　Payload Experiment Test System　【美国肯尼迪航天中心】有效载荷实验测试系统
PETS　Polaris Engineering Technical Service　【美国】"北极星"导弹工程技术勤务
PF　Parabolic Flight　抛物线飞行
PF　Parachute Facility　降落伞设施
PF　Payload Forward　有效载荷前端
PF　Payload Function　有效载荷功能（函数）
PF　Power Factor　功率因子
PF　Powered Flight　有动力的飞行
PF　Preflight　预飞行
PF　Probability of Failure　故障（失效）概率
PF　Protoflight; Prototype Flight　样机飞行；原型机飞行

PF   Pulse Frequency   脉冲频率

PFAG   Pulse Field Alternating Gradient   脉冲场交变梯度

PFAT   Preliminary Flight Approval Test   初步试验试飞（由承制厂演示的，属于研制试验的一部分）

PFB   Payload Feedback   有效载荷反馈

PFB   Payload Forward Bus   有效载荷前部总线

PFB   Position Feedback   位置反馈

PFB   Pressure Feed Booster   压馈推器

PFBIT   Preflight Built-in Test   飞行前内部测试

PFC   Performance Flight Certification   飞行性能认证

PFC   Postflight Checkout   飞行后检测

PFC   Power Factor Corrector   功率因子修正器（装置）

PFC   Powered Flight Control (System)   飞行操作助力（系统）

PFC   Powered Flying Control (Unit)   动力飞行控制(装置)

PFC   Preflight Certification   预飞行认证

PFC   Preliminary Flight Certification   初级飞行许可证

PFC   Primary Flight Control   主飞行控制

PFC   Prototype Flight Cryocooler   样机飞行试验的低温冷却装置

PFC   Pulsed Flame Combustor   脉动式火焰燃烧室

PFCES   Primary Flight Control Electronic System   主飞行控制电子系统

PFCF   Payload Flight Control Facility   有效载荷飞行控制设施

PFCS   Primary Flight Control System   主飞行控制系统

PFCU   Payload Access Platform /Platform CU   有效载荷入口平台

PFCU   Powered Flight Control Unit   飞行操作助力器

PFCU   Powered Flying-Control Unit   动力飞行控制单元

PFD   Primary Flight Display   主飞行显示器

PFDA   Post Flight Data Analysis   飞行后数据分析

PFDP   Preliminary Flight Data Package   初级飞行数据包

PFE   Parabolic Flight Experiment   抛物线飞行实验

PFF   Precise Formation Flying   精确编队飞行

PFI   Post Flight Inspection   飞行后检查

PFJ   Payload Fairing Jettison   整流罩弹射

PFL   Propulsion Field Laboratory   推进器实验室；推进装置实验室

PFM   Multimission Platform   多任务平台

PFM   Proto Flight Model   样机飞行模型

PFM   Pulse Frequency Modulation   脉冲频率调制

PFNS   Position Fixing Navigation System   定位导航系统

PFP   Probability of Failure, Performance   性能失效概率

PFR   Part Failure Rate   零部件故障率

PFR   Post Flight Reconstruction   飞行后重建

PFR   Post Flight Review   飞行后评审

PFR   Preflight Review   起飞前检查

PFR Problem or Failure Report 问题或故障报告
PFRT Pre-Flight Reliability Test 飞行前可靠性测试
PFRT Preliminary Flight Rating Test 飞行前（发动机）额定功率测试
PFRT Preliminary Flight Readiness Test 飞行准备预先试验
PFS Planetary Fourier Spectrometer 行星傅里叶频谱仪
PFS Primary Flight System 主要飞行系统
PFS Propellant-Feed System 推进剂输送系统；燃料进给系统
PFT Transport Platform / Platform de Transport 运输平台
PFTA Payload Flight Test Article 有效载荷飞行测试部件
PFTA Payload Flight Test Article (NASA) 【美国国家航空航天局】有效载荷飞行试验件
PFTB Preflight Test Bus 飞行前测试总线
PFUO Pitch Follow up Operation 俯仰随动操作
PFUS Pitch Follow up System 俯仰随动系统
PG Propellant Grain 推进剂药柱
PGA Power Generating Assembly 动力生成组件
PGA Pressure Garment Assembly 压力服组件
PGA Programmable Gain Amplifier 可编程增益放大器（装置）
PGAS Partitioned Global Address Space 划分全球地址空间
PGCF Premixed Gas Combustion Facility 预混合气体燃烧设备
PGCP Particles and Gases Contamination Panel 微粒与气体污染控制屏
PGDCS Power Generation, Distribution, and Control Subsystem 动力生成、分配与控制分系统
PGE Purge 清洗；吹除
PGHM Payload Ground Handling Mechanism 有效载荷地面处理装置
PGM Precision Guided Munition 精确制导弹药
PGMARV Precision Guided Manoeuvring Reentry Vehicle 精确制导机动再入飞行器
PGNCS Primary Guidance, Navigation and Control System 主制导、导航与控制系统
PGOC Payload Ground Operations Contractor 有效载荷地面操作承包商
PGOR Payload Ground Operation Requirements 有效载荷地面操作要求
PGORS Payload Ground Operations Requirements Study 有效载荷地面操作要求研究
PGOWG Payload Ground Operations Working Group 有效载荷地面操作工作组
PGPE Preflight Ground Pressurization Equipment 飞行前地面增压设备
PGR Spacelab Planning and Ground Rule 【美国】天空实验室与地面规则
PGRS Precision Guided Rocket System 精确制导火箭系统
PGRTV Precisely Guided Reentry Test Vehicle 精确制导再入试验飞行器
PGRV Post-Boost Guided Reentry Vehicle 助推结束后制导再入飞行器（弹头）
PGRV Precision Guided Reentry Vehi-

cle 精确制导再入大气层飞行器；精确制导再入飞行器
PGRWG Payload Ground Requirements Working Group 有效载荷地面要求工作组
PGS Passive Geodetic Satellite 无源大地测量卫星
PGS Power Generation Subsystem 动力生成分系统
PGSC Payload and General Support Computer 【美国航天飞机】有效载荷与通用保障计算机
PGSE Payload Ground Support Equipment 有效载荷地面保障设备
PGSE Peculiar Ground Support Equipment 特殊地面保障设备
PGSM Payload Gimbals Separation Mechanism 有效载荷常平架分离机构
PGU Power Generation Unit 发电装置；动力装置
PHA Potentially Hazardous Asteroids 具有潜在危险的小行星（尤指火星和木星轨道间运行的小行星）
PHA Preliminary Hazard Analysis 初步危险分析
PHA Pulse Height Analyzer 脉冲高度分析器（装置）
PHENOS Precise Hydrid Elements for Nonlinear Operations 非线性运算精密混合元件
PHF Payload Handling Fixture 有效载荷处理装置
PHI Position and Heading Indicator 位置与航向指示器
PHIGS Programmer's Hierarchical Interactive Graphics System 程序员层次交互图形系统
PHIN Position and Homing Inertial Navigation 位置和自动寻的惯性导航
PHM Passive Homing Missile 被动式自导引导弹；被动式寻的导弹
PHM Prognostication and Health Management 故障预测与健康管理
PHMP Prognostication and Health Management Processor 故障预测与健康管理处理器
PHS&T Packaging, Handling, Storage, and Transportation 包装、装卸、储存与运输
PHSF Payload Hazardous Servicing Facility 【美国肯尼迪航天中心】有效载荷危险品勤务厂房
PI Point of Impact 弹着点；命中点
PIA Payload Interface Adapter 【美国空间站】有效载荷接口适配器
PIA Project Impact Analysis 项目影响分析
PIAPACS Psycho-Physical Information Acquisition, Processing and Control System 【航天】心理物理信息获取、处理及控制系统
PIAR Project Impact Analysis Report 项目影响分析报告
PIB Pyrotechnic Installation Building 火工品安装厂房
PIC Payload Integration Center 有效载荷组装中心
PIC Preinstallation Checkout 安装前检测
PIC Program Information Center 项目信息中心
PIC Pyro Ignition Control 火工品点火控制
PIC Pyro Initiator Capacitor 火工品起爆器电容器
PIC Pyrotechnic Initiator Controller 火工品起爆器控制器
PICA Phenolic Resin Impregnated

Carbon Ablation 酚醛树脂浸渍碳烧蚀体

PICA Phenolic Impregnated Carbon Ablator 酚碳混合烧蚀材料

PICA Pyrotechnic Initiator Control Assembly 火工品起爆器控制装置

PICC Processor, Interface Control and Communications 处理器、接口控制与通信

PICRS Program Information Control and Retrieval System 程序信息控制与检索系统

PICRS Program Information Coordination and Review Service 项目信息协调与评审服务

PICS Payload Internal Carrying Structure 有效载荷内承载结构

PICS Predefined Input Control Sequence 输入预确定控制序列

PICSS Program for Interactive Continuous System Simulation 交互式连续系统模拟程序

PICU Pyrotechnic Indicator Control Unit 火工品控制装置

PID Parameter Identification 参数识别

PID Payload Insertion Device 有效载荷注入装置

PIDA Payload Installation and Deployment Aid 有效载荷安装与展开辅助设备

PIDAS Perimeter Intrusion Detection and Assessment System 环形入侵探测与评估系统

PIDS Prime Item Development Specification 主要部件项研制说明

PIE Payload Integration Equipment 有效载荷组装设备

PIF Payload Integration Facility 有效载荷组装设施

PILOT Phased Integrated Laser Optics Technology 相位综合激光光学技术

PILS Payload Integration Library System 有效载荷综合信息库系统

PILS Portable Impact Location System 移动式碰撞定位系统

PILS Precision Instrument Landing System 精确仪表着陆系统

PIM Parameterized Ionosphere Model 参数化的电离层模型

PIM Position and Intended Movement 位置与预定的移动

PIM Pulse Interval Modulation 脉冲间隔调制

PIMS Payload Information Management System 有效载荷信息管理系统

PIMS Programmable Implantable Medication System 可编程可注入的药剂系统

PIND Particle Impact Noise Detection 微粒撞击噪声探测

PINS Portable Inertial Navigation System 轻便惯性导航系统

PINS Precise Integrated Navigation System 精确综合导航系统

PIP Payload Integration Plan 有效载荷集成计划

PIP Primary Injection Point 【人造卫星和航天器等】第一射入轨道点

PIP Polymer Infiltration and Pyrolysis 聚合物渗透和高温分解

PIPS Passivated Implanted Planar Silicon 平面硅探测器

PIPS Pneumatically-Induced Pitching System 气动导引俯仰系统

PIPS Postinjection Propulsion System 入轨后推进系统

PIR Payload Integration Review 有

效载荷综合检查
PIR Precision Instrument Runway 有精密仪表降落设施的跑道
PIR Pressure Ignition Rocket 压力点火火箭
PIRA Precision Impact Range Area 精确投弹靶场
PIRAZ Positive Identification Radar Advisory Zone 主动识别与雷达报告区
PIT Preinstallation Test 预安装测试（试验）
PIT Pulsed Inductive Thruster 脉冲感应推力器
PITG Payload Integration Task Group 有效载荷组装任务组
PITS Payload Integration Test Set 有效载荷组装测试装置
PITS Platform Integrated Transmitter System 平台综合传送系统
PIU Payload Interface Unit 有效载荷接口装置
PIU Power Interface Unit 动力接口装置
PIU Pyrotechnic Initiator Unit 火工品起爆器装置
PIX Plasma Interaction Experiment 【美国】等离子体相互作用实验卫星
PJOP Pioneer Jupiter Orbiter and Probe 【美国】"先驱者"木星轨道器与探测器
PKH Probability of Kill if Hit 命中情况下的毁伤概率
PKM Perigee Kick Motor 近地点发动机
PL Probability of Leakage 泄漏概率
PL/SNSR Payload Sensor 有效载荷感应器（装置）
PLA Parachute Location Aid 降落伞定位辅助设备
PLA Payload Accommodations 有效载荷装置
PLA Payload Adapter 有效载荷适配器
PLAADS Parachute Low Altitude Aerial Delivery System 降落伞低空空投系统
PLAAR Packaged Liquid Air Augmented Rocket 封装液态空气加力燃烧火箭
PLAC Postlaunch Analysis of Compliance 发射后符合性分析
PLAD Postlanding Attitude Determination 着陆后姿态测定
PLAID Precision Location and Identification 精确定位与识别
PLAN Payload Local Area Network 有效载荷局部区域网络
PLAN Plane Stress Analysis 平面应力分析
PLARS Position Locating and Reporting System 位置测定与报告系统
PLASI Pulse Light Approach Slope Indicator 进场下滑闪光指示器
PLAST Propellant Loading All Systems Test 推进剂加注全系统测试
PLATO Planetary Trajectory Optimization 行星际轨道最优化
PLATO Preliminary Lethality Assessment Test Object 【美国SDI计划】最初杀伤力评估试验项目
PLATO Programmed Logic for Automatic Teaching Operation 自动制导操作程序控制逻辑
PLATS Precision Location and Tracking System 精确定位与跟踪系统
PLB Payload Bay 有效载荷舱
PLBD Payload Bay Door 有效载荷舱门；（美国航天飞机的）货舱门

PLC Payload Cost 有效载荷成本
PLC Preliminary Loads Cycle 初步加载周期
PLC Pressurization Logistics Cargo 增压后勤货舱
PLC Pressurized Logistics Carrier 加压后勤舱
PLC Probe Launch Complex 探空火箭发射设施
PLC Propellant Loading Console 推进剂装填控制台
PLC Propellant Loading Control 推进剂加注控制
PLCCE Program Life Cycle Cost Estimate 计划全寿命费用估算
PLCM Propellant Loading Control Monitor 推进剂加注控制监视器
PLCS Product Life Cycle Support 产品寿命周期保障
PLCU Propellant Level Control Unit 推进剂液面控制装置
PLCU Propellant Loading Control Unit 推进剂装填控制装置
PLDM Payload Management 有效载荷管理
PLDS Payload Support 有效载荷保障
PLE Plesetsk 【俄罗斯】普列谢茨克航天发射场
PLF Payload Fairing 有效载荷整流罩
PLFPF Payload Fairing Preparation Facility 有效载荷整流罩准备厂房
PLGSS Payload Ground Support Systems 有效载荷地面保障系统
PLH Payload Handling 有效载荷处理
PLIM Post-Launch and Instrumentation Message 发射后仪表测定数据通告
PLIN Propellant Level-Sensing Indicating Unit 推进剂液面传感指示器
PLIS Portable Landing Light System 便携式着陆灯系统
PLIS Propellant Level Indicating System 推进剂水平指示系统
PLIU Propellant Level-Sensing Indicating Unit 推进剂液面敏感指示器
PLM Prelaunch Monitoring 发射前监测
PLM Propulsion Load Module 推进载荷模块
PLMS Program Logistics Master Schedule 项目后勤保障主进度
PLN Program Logic Network 程序逻辑网
PLO Pacific Launching Operations 【美国】太平洋靶场发射操作
PLODS Photogram Metric Lunar Orbital Data Processing System 摄影测量月球轨道数据处理系统
PLP Pre-Launch Phase 发射前阶段
PLPS Propellant Loading and Pressurization System 推进剂加注与增压系统
PLRR President's Launch Readiness Review 当前发射准备状态审查
PLRS Position Location Reporting System 精确定位通报系统
PLRV Payload Launch Readiness Verification 有效载荷发射准备验证
PLS Payload Support Structure 有效载荷保障结构
PLS Payload Systems 有效载荷系统
PLS Personnel Launch System 人员发射系统
PLS Post Landing and Safing 着陆后与安全
PLS Preliminary Landing Site 初级着陆场

PLS Primary Landing Site 主着陆场
PLS Propellant Loading System 推进剂装填系统
PLSL Propellants and Life Support Laboratory 推进剂与生命保障实验室
PLSP Payload Signal Processor 有效载荷信号处理器
PLSS Payload Support Structure 有效载荷支撑结构
PLSS Personal Life Support System 乘员生命保障系统
PLSS Portable Life Support Subsystem 便携式生命保障分系统
PLSS Portable Life Support System 便携式生命保障系统
PLSS Post-Landing Survival System 着陆后救生系统
PLSS Precision Location Strike System 精确定位打击系统
PLSS Prelaunch Status Simulator 发射前状态模拟器
PLSS Primary Life Support Subsystem 主生命保障分系统
PLSS Primary Life Support System 生命保障主系统
PLTC Propellant Loading Terminal Cabinet 推进剂加注终端室
PLTS Precision Laser Tracking System 精确激光跟踪系统
PLTS Propellant Loading and Transfer System 推进剂加注与输送系统
PLU Payload Unit 有效载荷单元
PLUM Payload Launch Model 有效载荷发射模型
PLUM Payload Launch Module 有效载荷发射舱
PLUM Payload Umbilical Mast 有效载荷脐带杆
PLUS Precision Loading and Utilization System 精密装载（灌注）与利用系统
PLV Payload Launch Vehicle 携带有效载荷的运载火箭
PLV Postlanding Vent 着陆后通风
PLVC Postlanding Vent Control 着陆后通风控制
PLX Propellant Loading Exercise 推进剂装填练习
PM Payload Module 有效载荷舱
PM Payload Multiplication (Factor) 有效载荷放大（系数）；有效载荷倍增（因数）
PM Pilotless Missile 飞航式导弹
PM Polymer Microgravity Experiment 【美国航天飞机】聚合物微重力实验
PM Pressurized Module 加压模块
PM Preventive Maintenance 防护性维护
PM Propulsion Module 推进舱
PM Pulse Modulation 脉冲调制
PM/GPOF Electrodynamic Plasma Motor/Generator Proof Of Function Experiment 电动等离子体发动机/发电机功能验证实验（美国绳系卫星的实验项目）
PMA Pressurization Mating Adapter 增压对接适配器
PMAD Power Management and Distribution 动力管理与分布
PMAR Preliminary Maintenance Analysis Report 初步维修分析报告
PMAR Preliminary Mission Analysis Review 初步任务分析评审
PMAS Performance Measurement Analysis System 性能测试分析系统
PMASIT PMA Software Input Tool 任务后的分析软件输入工具
PMAWS Passive Missile Approach

Warning System 被动导弹逼近警告系统
PMC　Payload Monitoring and Control　有效载荷监控
PMC　Performance Management Computer (System)　性能管理计算机（系统）
PMC　Permanently Manned Capability　长久载人能力
PMC　Polymer Matrix Composites　聚合物基复合物
PMC　Postmanufacturing Checkout　生产（制造）后检测
PMC　Propellant Monitor and Control　推进剂（火箭燃烧）监视与控制
PMCR　Partial Mission Capable Rate　能执行部分任务率
PMCS　Pre-Launch Missile Control System　预发射导弹控制系统
PMCS　Preventive Maintenance Checks and Services　预防性保养检查与维修
PMD　Propellant Management Device　推进剂管理设备
PMEL　Pacific Marine Environmental Laboratory 【美国】太平洋海洋环境实验室
PMEL　Precision Measurement Equipment Laboratory 【美国空军】精密计量设备实验室
PMF　Perigee Motor Firing　近地点发动机点火
PMF　Program Management Facility　项目管理设施
PMI　Preventive Maintenance Inspection　预防性维修检查
PMIC　Payload Mission Integration Contract　有效载荷任务综合合同
PMIS　Personnel Management Information System　人员管理信息系统
PMISS　Protein Microscope for the ISS　国际空间站使用的蛋白质显微镜
PMM　Permanent Multipurpose Module　永久性多功能舱
PMN　Program Management Network　项目管理网络
PMOC　Prototype Mission Operations Center　原型任务操作中心；试验性任务操作中心
PMOM　Performance Management Operations Manual　性能管理操作手册
PMON　Performance Management Operations Network　性能管理操作网络
PMP　Parts, Materiel and Processes　零件、材料与工艺
PMP　Payload Mounting Panels　有效载荷安装板
PMP　Premodulation Processor　预调制处理器（装置）
PMP　Pressure Measurement Package　成套压力测量设备
PMR　Pacific Missile Range 【美国】太平洋导弹靶场
PMR　Propellant Mass Ratio　推进剂质量比
PMRF　Pacific Missile Range Facility 【美国】太平洋导弹靶场设施
PMRTF　Pacific Missile Range Tracking Facility　太平洋导弹靶场跟踪设备
PMS　Payload Monitoring Subsystem　有效载荷监控分系统
PMS　Performance Management System　性能管理系统
PMS　Performance Measurement System　性能测量系统
PMS　Performance Monitoring System　性能监测系统

PMS　Permanent Measurement System　永久测量系统

PMS　Polar Meteorological Satellite　极轨气象卫星

PMS　Polaris Missile System　【美国】"北极星"导弹系统

PMS　Portable Multi-Purpose Spare　便携式多用途集装架

PMS　Probability of Mission Success　飞行任务成功概率

PMS　Project Management System　项目管理系统

PMSFN　Planetary Manned Space Flight Network　行星际载人航天飞行网络

PMT　Pre-Mission Test　任务预先试验（测试）

PMT　Production Monitoring Test　产品监控测试

PMTC　Pacific Missile Test Center　太平洋导弹试验中心

PMTS　Precision Missile Tracking System　精密导弹跟踪系统

PMU　Pressure Measuring Unit　压力测量装置

PMU　Pulse Modulation Unit　脉冲调制装置

PMUX　Propulsion Data Multiplexer　推进装置数据多路传输器

PNCS　Performance and Navigation Computer System　性能与导航计算机系统

PNGCS　Primary Navigation, Guidance and Control System　初始导航、制导与控制系统

PNT　Positioning, Navigation, and Timing　定位、导航与授时

PNVS　Pilot Night Vision Sensors　飞行员夜视传感器

PO　Parking Orbit　停泊轨道；驻留轨道

PO　Planetary Observer　【美国国家航空航天局】行星观察者

PO　Planetary Orbit　行星际轨道

POC　Payload Operations Center　有效载荷操作中心

POC　Proof of Concept　方案（概念、设计思想）论证

POC/POT　Proof of Concept/Proof of Technology　概念验证/技术验证

POCC　Payload Operations Control Center　【美国国家航空航天局】有效载荷操作控制中心

POCC　Project Operations Control Center　【美国国家航空航天局】计划实施控制中心；工程作业控制中心

POCCNET　Payload Operations Control Center Network　【美国】有效载荷操作控制中心网

POCO　Position Computer　航天器坐标位置计算机

POCT　Passive Optical Component Technology　被动光学元器件技术

POD　Pilot Ocean Data System　【美国喷气推进实验室】实验海洋数据系统

POD　Precision Orbit Determination　精确定轨

POD　Preliminary Orbit Determination　初步轨道测定

POD　Private Orientation Device　专用定向设备

POD　Probability of Detection　探测概率

POD　Proof of Design　设计验证

PODAS　Portable Data Acquisition System　便携式数据采集系统

PODIM　Project Origination Design, Implementation and Maintenance　计划项目原始设计、实施与维护

PODM　Preliminary Orbit Determination Method　初步轨道确定方法
PODS　Precise Orbit Determination System　精轨测定系统
POEMS　Positron Electron Magnet Spectrometer　正负电子磁谱仪
POES　Polar Orbiting Operational Environmental Satellite　极轨环境业务卫星（美国 NOAA 的极轨气象卫星）
POESID　Position of Earth Satellite In Digital Display　地球卫星在数字显示器上的位置
POF　Pinhole Occulted Facility　【美国空间站上研究日冕现象用的】小孔遮掩器设备
POFA　Programmed Operational Functional Appraisals　程控操作功能评估
POGO　Polar Orbiting Geophysical Observatory　【美国国家航空航天局】沿轨道航行的极地地球物理观测卫星
POGS&SSR　Polar Orbiting Geomagnetic Survey and Solid State Recorder　【美国】极轨地磁测量与固态记录器（卫星）
POI　Parking Orbit Injection　射入停泊轨道
POI　Propellant Quantity Indicator　推进剂数量指示器
POIC　Payload Operations and Integration Centre　有效载荷操作与集成中心
POITS　Payload Orientation and Instrumented Tracking System　有效载荷定向与仪表跟踪系统
POL　Petroleum, Oil, and Lubrication　石油、汽油和润滑油
POLO　Polar Orbiting Lunar Observatory　沿轨道航行的极地月球观测卫星
PON　Payload Operations Network　有效载荷操作网络
POP　Payload Optimized Program　有效载荷最佳方案
POP　Perpendicular-to-the-Orbit Plane　垂直与轨道面
POP　Polar Orbiting Platform　【日本】沿轨道航行的极地平台（地球遥感卫星）
POP　Preburner Oxidizer Pump　预燃室氧化剂泵
POP　Preflight Operations Procedures　预飞行操作规程
POP　Prelaunch Operations Plan　预发射操作计划
POP　Prise Ombilicale Pneumatique (Pneumatic Umbilical Plug)　脐带气路插头
POP　Program Operating Plan　项目操作计划
POPA　Payload Ordnance Processing Area　有效载荷武器处理区
POPS　Position and Orientation Propulsion System　推进系统的位置和方向
PORR　Preliminary Operations Requirements Review　初步操作要求评审
PORT　Full-Scale Model Test　全尺寸模型试验
PORT　Recycling test　开展着陆回收试验
PORTS　Portable Optical Radiation Testbed for Sensors　传感器便携式光辐射试验台
POS　Permanent Orbital Station　长久性轨道站
POS　Polar Orbiting Satellite　极轨

（道）卫星
POS Portable Oxygen System 便携式氧气系统
POS Primary Operating System 主操作系统
POSA Passive Optical Sample Assembly Experiment 被动光学样品装配实验
POSEIDON Positioning Ocean Solid Earth Ice Dynamics Orbiting/Orbital Navigator 海洋固体地球冰层力学定位轨道导航仪
POSI Portable Operating System Interface 便携式操作系统接口
POSIX Portable Operating System Interface 移动式操作系统接口
POSS Passive Optical Satellite Surveillance (System) 被动式光学卫星监视（系统）；无源光学卫星监视（系统）
POSS Polyhedral Oligomeric Silsesquioxane 多面体低聚倍半硅氧烷
POSSE Pluto and Outer Solar System Explorer 冥王星及外太阳系探测器
POST Passive Optical Seeker Technique 无源光学自导导弹技术；被动式光学自导导弹技术
POST Payload Operations Support Team 有效载荷操作保障组
POST Portable Optical Sensor Tester 便携式光学传感器测试装置
POST Program to Optimize Simulated Trajectories 最佳模拟轨迹程序；最佳模拟导弹程序
POT Proximity Operation Trainer 【航天员训练用的】靠近操作训练器
POTV Personnel Orbit Transfer Vehicle 乘员轨道转移飞行器

POWERCELS Spacecraft Fuel Cells That Produce Electric Power from Chemical Reactions 航天器用燃料化学电池
PP Parallel Processing 并行处理
PP Post Processing 后处理
PP Principal Polarization 主要极化
PP Propulsion Power 推进功率
PPAS Photometric Periods of Artificial Satellites 人造卫星的光度期
PPB Program Performance Baseline 项目性能基线
PPBES Planning, Programming, Budgeting, and Execution System 规划、项目、预算与执行系统
PPBS Planning, Program, and Budgeting System 规划、项目与预算系统
PPBS Positive Pressure Breathing System 增压供氧系统
PPC Preprocessing Center 预处理中心
PPC Propellant Pressurization Control 推进剂增压控制
PPD Payload Position Data 有效载荷位置数据
PPD Pitch Phase Detector 俯仰角状态探测器
PPDS Pilot Planetary Data System 【美国】实验行星际数据系统
PPDU Payload Power Distribution Unit 有效载荷配电装置
PPE Personnel Protective Equipment 人员防护设备
PPE Phase Partitioning Experiment 【美国航天飞机】液相分离实验
PPE Prototype Production Evaluation 原型机（试制样机）生产鉴定
PPEP Plasma Physics and Environmental Perturbation 等离子物理与

环境扰动

PPF Payload Processing Facility 有效载荷操作厂房；有效载荷处理设施

PPF Payload Processing Facility (USAF) 【美国空军】有效载荷处理设施

PPG Propulsion and Power Generation 推进与动力产生

PPI Pulse Position Indicator 脉冲位置指示器（装置）

PPIP Program Protection and Implementation Plan 项目保护与实施计划

PPLI Precise Position, Location and Identification 精确位置、定位与识别

PPLS Precision Position Location System 精确定位系统

PPLS Propellant and Pressurant Loading System 推进剂和增压剂加注系统

PPOD Poly Picosatellite Orbital Deployer 聚兆卫星轨道展开设备

PPP Programmable Power Processor 可编程动力处理器

PPPL Planetary Physical Processes Laboratory 【美国】行星物理过程实验室

PPQT Pre-Production Qualification Test 预先生产合格试验（测试）

PPR Payload Preparation Room 有效载荷准备室（间）

PPS Payload Planning System 有效载荷规划系统

PPS Payload Pointing System 有效载荷定点系统

PPS Pneumatic Power Subsystem 气动力分系统

PPS Precision Positioning System 精确定位系统

PPTS Prospective Piloted Transport System 新一代载人运输系统

PPU Prime Power Unit 主电源装置

PPU Propulsion Unit 推进装置

PPUs Power Processing Units 功率处理装置

PPVS Propulsion Propellant Venting System 推进器喷气燃料排气系统

PQA Procurement Quality Assurance 采办质量保证

PQGS Propellant Quantity Gages 推进剂数量检测器

PQGS Propellant Quantity Gauging System 推进剂计量系统

PQI Propellant Quantity Indicator 推进剂数量指示器（装置）

PQP Prequalification Prototype 预鉴定原型机

PR Pitch Rate 俯仰角速度

PR Pressure Ratio 压力比

PR Pressure Regulator 压力调节器（装置）

PR&R Project Review and Reporting 方案评审与报告

PR/DR Program Requirements/Design Review 项目需求/设计评审

PRA Pitch and Roll Attitude 俯仰与滚动姿态

PRA Planetary Radio Astronomy 行星际射电天文学

PRA Probabilistic Reliability Assessment 概率可靠性评估

PRA Probabilistic Risk Analysis 概率风险分析

PRA Probabilistic Risk Assessment 概率风险评估

PRACA Problem Reporting and Corrective Action 问题报告和改正措施

PRAM Preliminary Repair Level De-

cision Analysis Model　初步修理级别决定分析模型
PRAM　Producibility, Reliability, Availability, Maintainability　生产性、可靠性、可用性与维修性
PRARS　Pitch, Roll, Azimuth Reference System　俯仰、滚动、方位基准系统
PRAT　Predicted Range Against Target　打击目标的预计距离
PRAWS　Pitch/Roll Attitude Warning System　俯仰／滚动姿态报警系统
PRB　Parachute Refurbishment Building　降落伞检修厂房
PRCA　Pitch/Roll Control Assembly　俯仰／滚转操纵机构；俯仰／滚转操纵组件
PRCA　Problem Reporting and Corrective Action　故障通报与修正措施
PRCS　Pitch Reaction Control System　俯仰反应控制系统
PRCS　Primary Reaction Control System　主反作用控制系统
PRCU　Payload Rack Checkout Unit　有效载荷机架检测装置
PRD　Payload Retention Device　有效载荷留置装置
PRD　Polytechnic Research and Development　综合性研究与开发
PRD　Pressure Relief Device　压力释放装置
PRDC　Polar Research and Development Center　【美国】极地研究发展中心
PRDR　Pre-Production Reliability Design Review　预先生产可靠性设计评审
PREDICT　Prediction of Radiation Effects by Digital Computer Techniques　数字计算机技术预测的辐射效应

PREF　Propulsion Research Environment Facility　推进装置研究环境设施
PRESAIR　Pressurized Air Compressors　增压空气压缩机
PRESS　Pacific Range Electromagnetic Signature Studies　太平洋靶场电磁特征研究
PRESS　Project Review, Evaluation and Scheduling System　项目评审、鉴定与计划安排系统
PRESTO　Program Reporting and Evaluation Systems for Total Operations　总体作战规划报告和评估系统
PRF　Parachute Refurbishment Facility　【美国】降落伞整修设施
PRF　Pulse Repetition Frequency　脉冲重复频率
PRFCS　Pattern Recognition Feedback Control System　模式识别反馈控制系统
PRFECT　Prediction RF Effects Coupling Tool　预测射频效应耦合工具
PRI-PL　Primary Payload　主有效载荷
PRIME　Precision Recovery Including Maneuvering Entry　包括机动再入（大气层）的精确回收
PRIOR　Program for In-Orbit Rendezvous　轨道会合计划（反卫星系统）
PRIPG　Pseudo-Random Invertible Permutation Generator　伪随机可逆置换发生器
PRIS　Pacific Range Instrumentation Satellite　【美国】太平洋靶场测量仪器卫星
PRISM　Portable Reusable Integrated Software Modules　便携式可重用的集成软件模块
PRISM　Program Reliability Informa-

tion System for Management　管理程序可靠性信息系统

PRL　Physical Research Laboratory　【印度】物理研究实验室

PRL　Polar Research Laboratories　极地卫星（美国等离子体探测卫星）研究实验室

PRL　Propulsion Research Laboratory　推进装置研究实验室

PRM　Payload Retention Mechanism　有效载荷保持机构；有效载荷保持机械装置

PRM　Pocket Radiation Monitor　小型辐射监测器

PRM　Posigrade Rocket Motor　火箭推进发动机

PRM　Precision Runway Monitor　精密跑道监视器

PRM　Pulse Rate Modulation　脉冲速率调制

PRN　Pulse Range Navigation　脉冲范围导航

PROCAP　Protection Capability　保护（防护）能力

PROCOL　Process Oriented Control Language　面向过程的控制语言

PROCRU　Procedure Oriented Crew Model　面向程序的机组模型

PROCTOR　Procedure Routine Organizer for Computer Transfers and Operations of Registers　用于计算机转移及寄存器操作的优先程序系统

PROD　Programme for Orbit Development　轨道发展计划

PROF　Prediction and Optimization of Failure Rate　故障率预测与最佳化

PROFILE　Passive Radio Frequency Interfere Location Experiment　被动式射频干扰定位实验（卫星）

PROFILE　Passive Radio Frequency Interference Location Experiment　【美国海军】无源射频干扰定位实验（卫星）

PROFIT　Propulsion/Flight Control Integration Technology　综合推进/飞行控制技术

PROFS　Prototype Regional Observing and Forecasting Service　典型区域观测与预报服务

PROJACS　Project Analysis and Control System　项目分析与控制系统

ProSEDS　Propulsive Small Expendable Deployer System　推进式小型可扩展配置器系统

PROSEL　Process Control and Sequencing Language　过程控制与程序设计语言

PROSSS　Programming Structural Synthesis System　结构综合程序系统

PROTEC　Programmable Ordnance Technology　可编程军械技术

PROTEUS　Plate-forme Réutilisable pour l'Observation, les Télécommunications & les Usages Scientifiques　用于观测、远距离通信及科学使用的平台

PROTO　Private Rocket to Orbit Tiny Objects　使用私人企业的火箭将微小物体送入轨道

Proto　Prototype　样机

PROVIB　Propulsion System Decision and Vibration Analysis　推进系统的判定与振动分析

PRP　Personnel Reliability Program　人员可靠性项目

PRPS　Programming Requirements Process Specification　编程要求操作说明

PRR　Prelaunch Readiness Review

发射前准备状态评审

PRR　Preliminary Requirements Review　初步需求评审

PRRC　Pitch/Roll Rate Changer Assembly　俯仰/滚转率变换装置

PRS　Payload Retention Subsystem　有效载荷留置分系统

PRS　Personnel Rescue Service　人员救援服务

PRS　Personnel Rescue System　人员救援系统

PRS　Power Reactant System (Subsystem)　动力反作用系统（分系统）

PRS　Precision Ranging System　精确测距系统

PRS　Primary Recovery Site　主要回收场

PRS　Primary Rescue Site　主救援场

PRSD　Power Reactant Storage and Distribution　【美国航天飞机】动力反应剂贮存与分配

PRSDS　Power Reactant Storage and Distribution Subsystem　【美国航天飞机】（燃料电池）发电反应剂储存与分配子系统

PRSG　Pulse-Reactant Strapdown Gyro　脉冲再平衡捷联式陀螺

PRSS　Propulsion and Reboost Subsystem　推进与推升分系统

PRST　Pacific Range Support Team　【美国】太平洋导弹靶场技术支持（保障）组

PRT　Pattern Recognition Technique　【目标】模式识别技术

PRTLS　Powered Return to Launch Site　带电返回发射场

PRTSW　Pitch/Roll Trim Switch　俯仰/横滚配平开关

PRU　Power Regulator Unit　动力调节器单元

PRV　Pressure Reduction Valve　压力简化阀

PS　Parachute Subsystem　降落伞分系统

PS　Payload Shroud　有效载荷整流罩

PS　Payload Station　有效载荷站

PS　Payload Support　有效载荷支架

PS　Perigee Stage　近地点级

PS　Power Supply　动力供给

PS　Pressure Switch　压力切换

PS　Propellant Seal　推进剂封口

PSA　Parallel Staging Area　并行火箭分离区

PSA　Payload and Servicing Accommodations　有效载荷与服务条件保障

PSA　Payload Service Area　有效载荷服务区

PSA　Payload Support Avionics　有效载荷支架航空电子设备

PSA　Pilot Sensor Assembly　导航传感器组件

PSA　Pneumatic Sensor Assembly　气动传感器组件

PSA　Post-Separation Arming　分离后解除保险

PSA　Power and Servo Assembly　动力与伺服装置

PSA　Power Servo Amplifier　动力伺服放大器

PSA　Power Servo Assembly　动力伺服组件

PSA　Preferred Storage Area　优先用的存储区

PSA　Pressure Switch Assembly　压力切换组件

PSA　Probability Safety Analysis　安全概率分析

PSA　Propellant Storage Assembly　推进剂存储装置

PSAD　Precision Simulation, Adapta-

tion, Decision 精密模拟、适应与判定
PSAR Preliminary Safety Analysis Report 初步安全分析报告
PSAS Pitch Stability Augmentation System 俯仰增稳系统
PSC Protoflight Spacecraft Cryocooler 原飞行航天器低温冷却装置
PSCC Physical Security Control Center 人身安全控制中心
PSCL Propellant Systems Cleaning Laboratory 推进剂系统清洁实验室
PSCN Program Support Communications Network 项目保障通信网络
PSCRT Passive Satellite Communications Research Terminal 无源卫星通信研究终端；被动式卫星通信研究终端
PSCS Public Services Communications Satellite 公共业务通信卫星
PSCU Pupitre Sauvegarde Charge Utile (Payload Safety Console) 有效载荷安全操控台
PSD Power System Demonstrator 动力系统演示验证装置
PSDE Payload and Spacecraft Development and Experimentation 【欧洲空间局】有效载荷与航天器研制实验计划
PSDNA Propulsion Systems Development Facility 推进系统发展设施
PSDP Payload Station Distribution Panel 有效载荷站分配控制台
PSE Payload Service Equipment 有效载荷检修设备
PSE Payload Servicing Equipment 有效载荷伺服设备
PSE Payload Support Equipment 有效载荷保障设备
PSE Physiological Systems Experiment 生理系统实验
PSF Payload Servicing Fixture 有效载荷检修装置
PSF Payload-Structure-Fuel (Weight Ratio) 有效载荷、结构与燃料的重量比
PSF Processing and Staging Facility 【美国航天飞机固体燃料火箭助推器】处理与分段运输设施
PSF Processing and Storage Facility 【美国航天飞机外挂贮箱】处理与贮存设施
PSIDS Prototype Secondary Information Dissemination System 样机辅助信息识别系统
PSIV Payload Software Integration and Verification 有效载荷软件集成与验证
PSK Phase Shift Keying 移相键控
PSLV Polar Satellite Launch Vehicle 【印度】极轨卫星运载火箭
PSM Parameter Signal Measurement 参数信号测量
PSM Payload Systems Mass 有效载荷系统质量
PSM Portable Space Model 便携式太空模型
PSM Process Safety Management 操作安全管理
PSM Propellant Storage Module 推进剂贮存舱
PSM Pyro Substitute Monitor 火工品替换监控器
PSMR Pressure System Manual Regulator 压力系统手控调节器
PSOP Payload Systems Operating Procedures 有效载荷系统操作规程
PSP Payload Signal Processor 有效载荷信号处理器

PSP Payload Systems Operating Procedures 有效载荷系统操作程序
PSPA Pressure Static Probe Assembly 压力静态探测器组件
PSR Pad Safety Report 发射台安全报告
PSR Parachute Status Report 降落伞情况报告
PSR Payload Separation Ring 有效载荷分离环
PSR Pilot Supply Regulator 飞行员供氧调节器
PSR Point of Safe Return 安全返航点（出击飞机可飞抵的最远地点）
PSR Pre-Shipment Review 船运之前的评审
PSR Program Status Review 项目状态评审
PSRE Propulsion-System Rocket Engine 推进系统火箭发动机
PSRR Preliminary System Requirements Review 初步的系统需求评审
PSS Pad Safety Supervisor 发射台安全监管员
PSS Passive Surveillance Sensor 被动监视传感器
PSS Payload Separation System 有效载荷分离系统
PSS Payload Servicing Structure 有效载荷勤务结构
PSS Payload Support System 有效载荷保障系统
PSS Phase Satellite System 相控卫星系统
PSS Planetary Scan System 行星扫描系统
PSS Planetary Space Science 行星际空间科学
PSS Portable Satellite Simulator 移动式卫星模拟器
PSS Portable Simulating System 移动模拟系统
PSS Power System Stabilizer 动力系统稳定器；电力系统稳定器
PSS Power System Synthesizer 动力系统合成装置
PSS Propellant Supply Subsystem 推进剂供应分系统
PSS Propulsion Support System 推进保障系统
PSSA Payload Support Structural Assembly 有效载荷保障结构组件
PSSC Parachute Subsystem Sequence Controller 降落伞分系统序列控制器
PSSC Preliminary System Security Concept 初步的系统安全概念（方案）
PST Planetary Spectroscopy Telescope 行星光谱望远镜
PSTE Payload System Test Equipment 有效载荷系统试验设备
PSTF Payload Spin Test Facility 有效载荷自旋试验设施
PSTP Propulsion System Test Procedure 推进系统试验程序
PSU Power Sensor Unit 动力感应装置
PSU Power Switching Unit 动力切换装置
PSV Planetary Space Vehicle 行星际航天飞行器
PSV Pressure Safety Valve 压力安全阀
PSW Payload Systems Weight 有效载荷系统重量
PSWC Payload Systems Weight Capability 有效载荷系统承重量
PSWG Payload Safety Working Group

有效载荷安全工作组
PSYOP(S) Psychological Operations 心理作战
PSYWAR Psychological Warfare 心理战
PTA Post-Test Analysis 试验后分析
PTA Potential Toxic Area 潜在毒性物质区；毒性物质勘探区
PTA Propulsion Test Article 推进试验件
PTA Prototype Test Article 样机试验件
PTB Payload Timing Buffer 有效载荷计时缓冲器
PTC Passive Thermal Control 被动热控制
PTC Payload Test Conductor 有效载荷测试导体
PTC Payload Training Capability 有效载荷培训能力
PTC Payload Training Complex 有效载荷培训设施
PTC Payload Transport Canister 有效载荷运输容器
PTC Performance Technical Curves 性能技术曲线
PTC Pitch Trim Compensator 俯仰配平补偿器
PTC Pneumatic Temperature Control 气动温度控制
PTC Portable Temperature Controller 便携式温度控制器
PTC Positive Temperature Coefficient 正温度系数
PTC Power Transfer Coefficient 功率传递系数
PTC Pressure and Temperature Control 压力与温度控制
PTC Pressure Transducer Calibrator 压力传感器校准器

PTCR Pad Terminal Connection Room 发射台终端连接室
PTCS Passive Thermal Control Section 被动热控制部分
PTCS Passive Thermal Control System 被动热控制系统
PTCS Payload Module Thermal Control Subsystem 有效载荷舱热控分系统
PTCS Payload Test and Checkout System 有效载荷测试与检查系统
PTCS Planning, Training, and Checkout System 规划、培训与检测系统
PTCS Propellant Tanking Computer System 推进剂箱储计算机系统
PTDB Problem Tracking Data Base 问题跟踪数据库
PTE Performance Testing Engine 性能试验发动机
PTE Processor Test Environment 处理器试验（测试）环境
PTF Payload Test Facility 有效载荷测试设施
PTF Payload Training Facility 有效载荷培训设施
PTF Propellant Tank Flow 推进剂箱流量
PTF Propulsion Test Facility 推进装置试验设施
PTFM Pressure Testing to Failure Method 压力破坏试验方法
PTI Payload Type Indicator 有效载荷类型指示器
PTI Preliminary Test Information 初步测试（试验）信息
PTI Pre-Programmed Test Input 预编程测试（试验）输入
PTI Programmed Test Input 编程测试（试验）输入

PTIS Program Test Input System 项目测试（试验）输入系统
PTIS Propulsion Test Instrumentation System 推进装置测试系统；动力装置测试系统
PTM Polaris Tactical Missile 北极星式战术导弹
PTM Pulse Time Modulation 脉冲时间调制
PTMS Precision Torque Measuring System 扭矩精确测量系统
PTO Power Test Operations 动力测试操作
PTP Programmable Telemetry Processor 可编程的遥测处理器（装置）
PTPS Portable Telemetry Processing System 便携式遥测处理系统
PTS Payload Test Set 有效载荷试验装置
PTS Payload Transfer System 有效载荷转运系统
PTS Payload Transportation System 有效载荷运输系统
PTS Pneumatic Test Set 气动试验设备
PTS Pointing and Tracking System 瞄准跟踪系统
PTS Propellant Transfer System 推进剂输送系统
PTSS Precision Tracking and Surveillance System 精确跟踪与监视系统
PTV Pathfinder Test Vehicle 探险者实验用飞行器
PTV Pitch Thrust Vector 俯仰推力矢量
PTV Propulsion Technology Validation 推进技术验证
PTV Propulsion Test Vehicle 推进试验飞行器
PTVA Propulsion Test Vehicle Assembly 推进试验飞行器装置
PTVC Pitch Thrust Vector Control 俯仰推力矢量控制
PTVS Portable Test and Verification System 移动试验与鉴定系统
PU Power Unit 动力装置
PU Propellant Unit 推进剂装置
PU Propellant Utilization 推进剂使用
PUCS Propellant Utilization Control System 推进剂输送调节系统
PUGS Propellant Utilization and Gauging System 【美国国家航空航天局】推进剂输送调节与计量系统
PUR Payload Under Review 处于评审中的有效载荷
PUT Pop-Up Test 【弹道导弹】发射试验
PUT Propulsion Unit Test 推进单元试验
PUTT Propellant Utilization Time Trace 推进剂利用时间曲线
PUV Propellant Utilization Valve 推进剂输送阀
PV Planetary Vehicle 星际飞船
PV Pyrotechnic Valves 火工品阀门
PV Photovoltaic 光电
PV&D Purge, Vent, and Drain 吹除、通风与排放
PVA Perigee Velocity Augmentation 近地点速度增加
PVA Preburner Valve Actuator 预燃室阀门作动器
PVA Propellant Valve Actuator 推进剂阀门作动器
PVan Payload Support Van 有效载荷保障车
PVB Pressurized Vapor Burner 高压液化气燃烧器
PVC Pneumatic Vortex Control 气动

涡流控制
PVC Pressure Volume Control 压力容积控制；压力容量控制
PVD Purge, Vent, Drain System 吹除、通风与排放系统
PVDS Propelled Variable-Depth Sonar 推进式可变深度声呐
PVO Pioneer-Venus Orbiter 【美国】"先驱者—金星"轨道器
PVP Photovoltaic Power 光伏动力
PVP Prototype-Validation Phase 原型机研制阶段
PVRD Purge, Vent, Repressurize, and Drain 吹除、通风、再增压与排放
PVS Parameter Value Set 参数值集
PVS Pressure Vessels and Pressurized Systems 压力容器与增压系统
PVS Propellant Venting System 推进剂通风系统
PVT Parameter Value Table 参数数值表
PVT Preflight Verification Test 预飞行验证测试（试验）
PVT Pressure, Volume and Temperature 压力、容量与温度
PVT Pyrotechnic Verification Test 火工品鉴定试验
PVTCS Photovoltaic Thermal Control System 光电池热控系统
PVTOS Physical Vapour Transport of Organic Solids 【美国航天飞机】有机颗粒物理蒸发输运实验
PWBS Program Work Breakdown Structure 项目工作分类体系结构
PWT Portable Water Tank 便携式水箱
PWT Propulsion Wind Tunnel 推进（试验）风洞
PY Pyrotechnic Devices Storage Area 火工品装置储存区
PYC Pyrotechnic Controller 火工品控制器
PYLD Payload 有效载荷
PYRO Pyrotechnics 火工品
PZP Phase Zero Program 零相位程序；零阶段程序

Q

Q Alpha Pitch Dynamic Pressure 机翼迎角俯仰动压
Q Damping Factor 衰减系数
Q Dynamic Pressure 动压
Q Pitch Rate 俯仰角速度
Q Polaris Correction(Missiles) 【美国】"北极星"导弹校正
Q Quality Factor 质量因数；品质因数
Q Alpha Pitch Dynamic Pressure 俯仰角动态压力
Q Beta Yaw Dynamic Pressure 偏航动压
Q&R Quality and Reliability 质量与可靠性
Q&RA Quality and Reliability Assurance 质量与可靠性保证
Q/R/M Quality/Reliability/Maintainability 质量/可靠性/可维修性
QA Quality Assessment 质量评价
QA Quality Assurance 质量保证
QA Quality Audit 质量检查；品质监查（稽核）
QA & R Quality Assurance and Reliability 质量保证和可靠性
QA & R Quality Assurance and Revalidation 质量保证和重新审定
Q-D Quantity-Distance 炸药量安全距离
QA/QC Quality Assurance/Quality Control 质量保证/质量把关
QA/RM Quality Assurance and Risk Management 质量保证与风险管理
QAAR Quality Audit, Assessment, and Review 质量检查、评估与评审
QAAS Quality Assurance Acceptance Standards 质量保证验收标准
QAC Quality Assessment and Control 质量评估与控制
QACAD Quality Assurance Corrective Action Document 质量保证修正措施文件
QADS Quality Assurance Data System 质量保证数据系统
QAE Quality Assurance Evaluator 质量保证鉴定装置
QAET Quality Assurance Environment Testing 质量保证环境试验
QAET Quality Assurance Evaluation Test 质量保证鉴定试验
QAF Quality Assurance Firing 质量保证发射；质量保证射击
QAF Quality Assurance Function 质量保证功能
QAM Quadrature Amplitude Modulation 正交幅度调制
QAMSP Quality Assurance Master Surveillance Plan 质量保证主监视计划
QAP Quality Assurance Procedure 质量保证规程
QAPP Quality Assurance Program Plan 质量保证项目计划
QARI Quality Assurance Receipt Inspection 质量保证验收
QASAC Quality Assurance Spacecraft Acceptance Center 航天器质量保证接收中心
QASAR Quality Assurance Systems Analysis Review 质量保证体系的

分析评审

QASAR Quality and Safety Achievement Recognition 质量与安全成效识别

QASP Quality Assurance Surveillance Plan 质量保证监测计划

QASSOR Quality Assurance and Systems Safety Operating Review 质量保证与系统安全操作审查

QAST Quality Assurance Systems Test 质量保证体系测试

QAT Qualification Approval Test 质量合格验收试验

QAT Quality Assessment Test 质量评估测试

QAT Quality Assurance Test 质量保证测试

QATP Quality Assurance Test Procedure 质量保证测试程序

QATS Quality Assurance and Test Service 质量保证与测试业务

QATT Qualification for Acceptance Thermal Testing 热测试验收条件

QAVT Qualification Acceptance Vibration Test 振动测试验收条件

Qbar Dynamic Pressure 动态压力

QC Quality Control 质量控制

QC Quick Cleaning 快速清除

QC Quick Connect 快速连接

QC&I Quality Control and Inspection 质量控制与检查

QC&T Quality Control and Technique 质量控制与技术

QC&T Quality Control and Test 质量控制与试验

QCA Quality Control Analysis 质量控制分析

QCC Quality Control Center 质量控制中心

QCCARS Quality Control Collection Analysis and Reporting System 质量控制收集分析与报告系统

QCE Quality Control and Evaluation 质量控制与鉴定

QCI Quality Conformance Inspection 质量一致性检验

QCI Quality Control Information 质量控制信息

QCOC Quality Control Operational Chart 质量控制作业流程图

QCOP Quality Control Operating Procedure 质量控制作业程序；质量控制操作规程

QCPI Quality Conformance Preliminary Inspection（In-Plant） 质量适应性（厂内）验收初检

QCR Quality Control Reliability 质量控制可靠性

QCR Quality Control Room 质量控制室

QCR Quick Connect Relay 快速连接继电器

QCS Quality Certification System 质量认证系统

QCS Quality Control Specification 质量控制规范

QCS Quality Control Standard 质量控制标准

QCS Quality Control System 质量控制系统

QCS Quality Cost System 质量成本系统

QCSR Quality Control Service Request 质量控制业务要求

QCT Quality Conformance Test 质量一致性试验

QCT Quality Control Technology 质量控制技术

QCT Quality Control Tool 质量控制工具

QD Quick Disconnect 快速断开

QDACS Quantized Data Attitude Control System 量化数据姿态控制系统

QDC Quick Disconnect Connector 快速断开连接器

QDR Qualification Design Review 设计资格评审

QDRI Qualitative Development Requirements Information 质量改进需求信息

QDS Quality Data System 质量数据系统

QDS Quick Disconnects 快速断开

QDTA Quantitative Differential Thermal Analyzer 定量差热分析仪

QE Quantum Efficiency 量子效率

QEC Quick Engine Change 发动机快速更换

QECA Quick Engine Change Assemble 发动机快速更换装置

QECK Quick Engine Change Kit 发动机快速更换工具包

QECS Quick Engine Change Stand 发动机快速更换台

QED Quantitative Evaluation Device 定量鉴定装置

QEL Quality Evaluation Laboratory 质量鉴定实验室

QES QCSEE Engine Systems 短跑道起降（飞机）静净实验发动机系统

QEST Quality Evaluation System Test 质量鉴定系统测试

QET Quick Engine Test 发动机快速试车

QF Quality Factor 质量因子

QF Quick Fire 快速发射；快速火力

QFA Quick-Firing Ammunition 速射武器弹药

QFA Quick-Firings Alignment 速射武器调准

QFC Quantitative Flight Characteristics 定量飞行特性准则

QFC Quantum Flow Control 量子流量控制

QFD Quality Function Deployment 质量功能展开

QFD Quality Function Development 质量功能开发

QFIRC Quick Fix Interference Reduction Capacity 快速固定减轻干扰能力

QGS Quantity Gauging System 数量测量系统

QHDA Qualified Hazardous Duty Area 限定危险任务区

QHTK Quick Hard-Target Kill 迅速击毁硬目标

QIP Quality Improvement Prototype 质量改进样机

QJM Quantified Judgment Model 量化判断模型

QJMA Quantified Judgment Method of Analysis 量化分析判断法

QKD Quick Knockdown 快速拆卸

QLCS Quick Look and Checkout System 速查和检测系统

QLDA Quick Look Display Area 快视显示区

QLLC Qualified Logical Link Control 限定式逻辑链路控制

QM/DX Queen Match/Discrimination Experiment 皇后比赛（一项反导技术试验代号）识别实验

QMAC Quarter-Orbit Magnetic Attitude Control 四分之一轨道磁性姿态控制

QMCS Quality Monitoring and Con-

trol System 质量监测与控制系统
QME Quality Measurement Experiment 质量测量实验
QORGS Quasi-Optimal Rendezvous Guidance System 准最佳交会制导系统
QOS Quality of Service 服务质量
QOT&E Qualification Operational Test and Evaluation 适应性作战试验与鉴定
QQPRI Qualitative and Quantitative Personnel Requirements Information 人员数量和质量要求信息
QQPRI Qualitative and Quantitative Planing Requirements Information 定性与定量计划需求信息
QR Qualification Review 鉴定评审
QR Quantitative Restrictions 数量限制
QRA Quality Reliability Assurance 质量可靠性保证
QRA Quantitative Risk Analysis 定量风险分析
QRA Quantitative Risk Assessment 量化风险评估
QRA Quick Reaction Alert 快速反应警戒状态
QRAC Quality Review Analytical Control 质量评审分析控制
QRAS Quantitative Risk Assessment System 定量风险评估系统
QRC Quality Review Control 质量评审控制
QRC Quick Reaction Capability 快速反应能力
QRCC Quick Reaction Combat Capability 快速反应战斗能力
QRD Quick Reaction Demonstration 快速反应示范演练
QRD Quick Reaction Drill 快速反应演练
QRE Quick-Reaction Estimate 快速反应判断
QRIC Quick Reaction Installation Capacity 快速反应装配能力
QRLS Quick Reaction Launch System 快速反应发射系统
QRM Quick Response Missile 快速响应导弹
QRP Quick Response Posture 快速反应态势
QRP Radar Quick Reaction Program Radar 快速反作用项目雷达
QRR Qualification Results Review 鉴定结果评审
QRRI Qualitative Research Requirements Information 定性研究需求信息
QRS Quick Reaction Software 快速反作用软件
QRS Quick Reaction Strike 快速反应打击
QRS Quick Reaction Support 快速反应支援（保障）
QRSA Quick Reaction Satellite Antenna 快速反应卫星天线
QRSL Quick-Reaction Space Laboratory 【美国国家航空航天局】快速反应航天实验室
QRSS Quick Reaction Surveillance System 快速反应监视系统
QRT Quick Response Time 快速响应时间
QRTP Quick Response Targeting Program 快速反应瞄准程序
QSG Quasi-Stellar Galaxy 类星银河系
QSL Qualification and Standard Laboratory 鉴定与标准实验室
QSL Qualified Source List 合格来源

名单

QSL Quasi-Static Loads 准静态载荷

QSMA Quality Safety and Mission Assurance 质量安全与任务保证

QSRA Quiet Short-Haul Research Aircraft 【美国国家航空航天局】无噪声短程研究飞机

QSS Quick Service Supply 快速勤务供应

QSS Quick Supply Store 快速补给物资储存

QSS Quota Sample Survey 定额抽样检验

QSSA Quasi-Stationary Static Approximation 准静态逼近

QT Quiet Thruster 低噪声推进器

QT&E Qualification Test and Evaluation 合格性试验与鉴定

QTPT Qualification Test and Proof Test 适用性试验与验证试验

QTR Qualification Test Report 合格性检验报告；质量鉴定试验报告

Quasars Quasi-Stellar Objects 类星体

QUIC Quality Data Information and Control 质量数据信息与控制

QUICO Quality Improvement Through Cost Optimization 通过成本最佳化实现的质量改进

QUIPS Quiescent Uniform Ionosphere Plasma Simulator 静态均匀电离层等离子体模拟器

QVF Quality Verification Factor 质量验证因素

QVIS Quasi-Vertical Incidence Sounder 准垂直迎角探测器

QVT Quality Verification Testing 质量检验（鉴定）测试

R

R IMP Minimum Range to Avoid Plume Impingement 避免羽烟冲击的最小范围

R&A Reliability and Availability 可靠性与适应性

R&A Review and Approval 审查与批准

R&C Receiving and Classification 接收与分类

R&CC Recorder and Communication Control 记录仪与通信控制

R&D Requirement and Distribution 需求与分配

R&D Research and Development 研究与发展；研究与研制（开发）

R&DO Research and Development Operations 【美国马歇尔航天中心】研究与发展操作中心

R&DR Requirements and Design Review 需求与设计评审

R&E Research and Engineering 研究与工程

R&I Receiving and Inspection 接收与检查

R&I Removal and Installation 拆卸与安装

R&M Reliability and Maintainability 可靠性与可维修性

R&M Repair and Maintenance 修理与维修

R&M&S Reliability, Maintainability and Supportability 可靠性、维修性与保障性

R&PM Research and Program Management 研究与项目管理

R&QA Reliability and Quality Assurance 可靠性与质量保证

R&R Refit and Repair 改装与修理

R&R Reliability and Response 可靠性与响应性

R&R Remove and Replace 清除与替换

R&R Rendezvous and Recovery 交会与回收

R&R Repair and Return 修理与返回

R&R Reproducibility and Repeatability 重现性与重复性

R&R Resupply and Return 再补给与返回

R&S Reliability and Serviceability 可靠性与适用性

R&T Research and Technology 研究与技术

R&U Repairs and Utilities 维修与实用性

R-T Real Time 实时

R/A Radar Altimeter 雷达测高仪

R/ASR Review as Required 按要求的评审

R/C Range Clearance 靶场许可证

R/C Remote Control 远程（远距离）控制

R/D Rate of Descent 下降速度；下降速率

R/E Reentry 再进入（大气层）

R/I Receiving Inspection 接收检查

R/M Reliability/Maintainability 可靠性与维修性

R/P/Y Roll/Pitch/Yaw 滚转、俯仰

与偏转（偏航）
R/S Range Safety 靶场安全
R/S Redundant Set 冗余设备；冗余集合
R/T Real Time 实时
R/T Receive/Transmit 接收/传输
R/T Receiver/Transmitter 接收/发送
R/TCE Rotation/Translation Control Electronics 旋转/转换控制电子设备
R/UPS Recorder/Utility Processor System 【美国国家航空航天局】记录仪与公共程序处理机系统
R/W Runway 跑道
R^2D^2 Robotic Refueling Dexterous Demonstration 机器人燃料补给灵巧演示验证
R^2P^2 Rapid-Retargeting/Precision Pointing (Simulator) 快速目标再判定/精确定位（模拟器）
R^2P^2 Rapid Retargeting and Precision Pointing Simulator 【美国SDI计划】快速多次瞄准与精确定位模拟器
RA Range Area 靶场区域；范围区域
RA Reliability Analysis 可靠性分析
RA Reliability Assessment 可靠性评估
RA Risk Analysis 风险分析
RA Rocket Antenna 火箭天线
RA&D Requirement Analysis and Design 要求分析与设计
RAA Relative Azimuth Angle 相对方位角
RAA Roll-Augmentation Actuator 滚转增强作动器
RAAB Remote Amplifier and Adaptation Box 远距离放大器与适配箱
RAAB Remote Application and Advisory Box 远程应用与通报单元（组件）
RAAD Roll Attitude Anomaly Detection 滚转姿态异常检测
RAAM Roll Attitude Acquisition Mode 滚转姿态获取模式
RAAN Right Ascension of Ascending Node 升交点赤经
RAAP Rapid Application of Air Power 空中力量的快速运用
RAAWS Radar Altimeter and Altitude Warning System 雷达测高仪与高度报警系统
RAC Reliability Action Center 可靠性操作中心
RAC Reliability Analysis Center 可靠性分析中心
RAC Reliability Assessment of Components 元件可靠性评价
RAC Risk Assessment Code 风险评估码
RACC Radiation and Contamination Control 放射与污染控制
RACE Radiation Adaptive Compression Equipment 适应发射性压缩设备
RACE Rapid Automatic Checkout Equipment 快速自动检测设备
RACE Research and Applications Cooperative Experiment 研究与应用协同实验
RACEP Random Access and Correlation for Extended Performance 扩展性能的随机存取及相关性
RACER Reliability, Availability, Compatibility, Economy and Reproducibility 可靠性、可用性、兼容性、经济性与再现性
RACINE Radio Augmented and Calibrated Inertial Equipment 无线电增益与校准惯性装置

RACO Roll Attitude Cutoff 滚转姿态断开

RACP Recovery Action Command Packet 恢复动作指令包

RACS Remote Access Computing System 遥控存取计算系统

RACS Remote Automatic Calibration System 远距离自动校准系统

RACS Remote Automatic Control System 远距离自动控制系统

RACS Rotation Axis Coordinate System 旋转轴坐标系统

RACSV Reheat Acceleration Control Spill Valve 加力燃烧室加速控制溢流阀

RACU Remote Acquisition and Command Unit 远距离探测与指令装置；遥控捕获与指令装置

RAD Radar Approach Aid 进场雷达导航设备

Rad Radiation 辐射

RAD Radiation Accumulated Dose 累计辐射剂量

RAD Radiation Dosage 辐射剂量

RAD Rapid Application Development 快速应用开发

RAD Reference Attitude Display 基准姿态显示器

RAD Requirement Action Directive 需求实施指令

RAD Requirements and Analysis Division 技术要求与分析部门

RAD Research and Development 研究与开发

RAD Roll Attitude Determination 滚转姿态测定

RADAC Range Data Acquisition and Computation 靶场数据采集与计算

RADAMS Rapidly Deployed Antiaircraft Missile System 快速部署的防空导弹系统

RADAN Radar Doppler Automatic Navigator 多普勒雷达自动导航仪

RADAR Radio Detecting and Ranging 无线电探测与测距

RADAR Remote Analytical Data Acquisition and Reduction 远程分析数据获取与处理

RADC Reliability Analysis Data Center 可靠性分析数据中心

RADCAL Radiation Calibration (Satellite) 辐射校准（卫星）

RADCC Radiation Control Center 辐射控制中心

RADDAC Radar Acquisition Data Distribution and Control System 雷达测获数据分发与控制系统

RADEC Radiation Detection Capability 辐射探测能力

RADHAZ Electromagnetic Radiation Hazard 电磁辐射危害

RADHAZ Hazards Form Electromagnetic Radiation 形成电磁辐射的危害

RaDI-N Radi-N Measurements of Neutron 中子辐射测量

RADIC System Rapidly Deployable Integrated Command and Control System 快速部署的综合指挥与控制系统

RADINT Radar Intelligence 雷达情报

RADOP Radar Doppler (Missile Tracking System) 雷达多普勒效应（导弹跟踪系统）

RADOSE Radiation Dosimeter Payload 【美国】"雷多斯"卫星；辐射剂量计

RADOT Real Time Automatic Digital Optical Tracker 【美国国家航空航

天局】实时自动数字光学跟踪仪
RADOT Recording Automatic Digital Optical Tracker 记录式自动数字光学跟踪仪
RADSAT Radiation Satellite 辐射卫星（美国空间物理探测卫星）
RADU Radar Analysis and Detection Unit 雷达分析与探测装置
RADU Range and Azimuth Display Unit 距离方位显示装置
RADVS Radar Altimeter and Doppler Velocity Sensor 雷达测高仪与多普勒速度传感器
RAE Radio Astronomy Explorer 【美国】射电天文探测卫星
RAEDOT Range, Azimuth and Elevation Detection of Optical Targets 光学目标的距离、方位和标高探测
RAFS Regional Analysis and Forecast System 区域分析与预报系统
RAFT Reentry Advanced Fuzing Test 再入大气层先进引信试验
RAI Radar Altimeter Indicator 雷达测高仪指示器（装置）
RAI Reliability Assurance Instructions 可靠性保证指令
RAI Remote Attitude Indicator 遥控姿态指示仪
RAI Roll Attitude Indicator 滚转姿态指示仪
RAI Runway Alignment Indicator 跑道准线指示灯
RAIDS Marine Remote Atmospheric/Ionosphere Detection System 海洋远程大气/电离层探测系统
RAIDS Remote Atmosphere and Ionosphere Detection System 大气与电离层遥控探测系统
RAILE Retro-Assisted Imaging Laser Experiment 制动发动机辅助成像激光实验
RAIR Ram Augmented Interstellar Rocket 冲压增力星际火箭
RAIS Range Automated Information System 靶场自动化信息系统
RAL Remote Area Landing 边远地区着陆；遥控地区着陆
RAL Revue d'Aptitude au Lancement/Launch Readiness Review 发射准备状态评审
RALACS Radar Altimeter Low Altitude Control System 雷达高度表低空控制系统
RALPH Reduction and Acquisition of Lunar Pulse Heights 月球脉冲高度的简化与采集
RALS Remote Area Landing System 边远地区着陆系统；遥控地区着陆系统
RAM Redeye Air (-Launched) Missile 【美国】"红眼"（空中发射的）导弹；"红眼"空中导弹
RAM Reliability, Availability and Maintainability 可靠性、可用性与可维修性
RAM Research and Applications Module 【美国航天飞机】研究与应用舱
RAM Responsibility Assignment Matrix 责任分配矩阵
RAM Rocket Assisted Motor 火箭助推发动机
RAM Rolling Airframe Missile 弹体滚动导弹（武器系统）
RAM-D Reliability, Availability, Maintainability, and Dependability 可靠性、可用性、可维修性与可依靠性
RAM-D Reliability, Availability, Maintainability and Durability 可靠性、可用性、可维护性与耐用性

RAMA Reliability, Availability, Maintainability Analysis 可靠性、可用性、可维修性分析

RAMCAD Reliability, Availability and Maintainability Computer-Aided Design 可靠性、可用性和可维护性计算机辅助设计

RAMES Reliability, Availability, Maintainability, Engineering System 可靠性、可用性、可维修性工程系统

RAMF Revue d'Analyse de Mission Finale/Final Mission Analysis Review 最终任务分析评审

RAMIS Rapid Access Management Information System 快速存取管理信息系统

RAMIS Rapid Automatic Malfunction Isolation System 故障自动快速隔离系统

RAMIS Receive, Accept, Maintain, Issue and Store 接收、保养、发放与储存

RAMIS Receiving, Assembly, Maintenance, Inspection, Storage 接收、装配、维护、检查与储存

RAMIS Repair, Assemble, Maintain, Issue and Supply 修理、装配、维护、发放与供应

RAMMIT Reliability and Mainability Management Improvement Techniques 可靠性与可维修性管理改进技术

RAMOS Reliability, Availability, Maintainability, Operations, and Support 可靠性、可用性、可维修性、操作（作战）与保障（支援）

RAMOS Remote Automatic Meterological Observing System 遥控自动气象观测系统

RAMOS Russian-American Observation Satellite 俄罗斯－美国观测卫星

RAMP Reliability and Maintainability Program 可靠性和可维护性程序

RAMP Revue d'Analyse de Mission Préliminaire/Preliminary Mission Analysis Review 初期任务分析评审

RAMPART Radar Advance Measurement Program for Analysis of Reentry Techniques 用于分析再入技术的雷达先进测量计划

RAMS Radiation Area Monitor System 辐射区监测系统

RAMS Reliability, Availability, Maintainability, and Supportability 可靠性、适用性、维修性与保障性

RAMS Remote Activation Munitions System 远距离弹药启动系统

RAMS Remote Atmospheric Monitoring System 远距离大气监测系统

RAMS Remote Attitude Measurement System 【美国国家航空航天局】远距离姿态测定系统

RAMS Remote Automatic Multipurpose Station 远距多功能自动站

RAMTAC Reentry Analysis and Modeling of Target Characteristics 目标特性的再入分析与模拟

RAMTIP Reliability and Maintainability Technology Insertion Program 可靠性与可维修性技术引入项目

RAND Research and Development Corporation 【美国】兰德公司（研究与发展公司）

RANT Reentry Antenna Test 再入（大气层）天线试验

RAO Rocket Assisted Orbiters 火箭助推轨道飞行器

RAP Reliability Assessment Prediction

可靠性评估预测

RAP Remote Access Panel 远程访问控制板

RAPCON Radar Approach and Control 【美国帕特里克空军基地】雷达进场与控制

RAPID Real Time Acquisition Program of In-Flight Data 飞行中数据实时采集程序

RAPID Rocketdyne Automatic Processing of Integrated Data 【美国】洛克威尔公司的"洛克迪尼"火箭发动机综合数据自动处理

RAPIDS Rapid Automated Problem Identification Data System 快速自动化问题识别数据系统

RAPIDS Real Time Automated Personnel Identification System 自动化实时人员识别系统

RAPS Rate and Position System 速率与位置系统

RAPS Right AFT Propulsion System 右后部推进系统

RAPTOR Responsive Aircraft Program for Theater Operations 战区作战反应灵敏飞机计划

RARSAT Radar Ocean Reconnaissance Satellite 雷达海洋侦察卫星

RAS Radio Astronomy Satellite 射电天文卫星

RAS Reliability Analysis System 可靠性分析系统

RAS Reliability, Availability, Serviceability 可靠性、可用性与适用性

RAS Rocket Alarm System 火箭报警系统

RASA Russian Aviation and Space Agency 俄罗斯航空航天局

RASCAD Range Safety Control and Display 靶场安全控制与显示

RASER Range and Sensitivity Extending Resonator 距离和灵敏性增大谐振器

RASER Research and Seeker Emulation Radar 自动搜索引导跟踪雷达

RASIDS Range Safety Impact Display System 靶场安全弹着显示系统

RASIGMA Radar Signature Management 雷达信号控制

RASIS Reliability, Availability, Serviceability, Integrity, Security 可靠性、可用性、可维修性、完整性与安全性（性能最优化）

RASS Radio Acoustic Sounding System 无线电声学探测系统

RASS Rapid Area Supply Support 区域快速供应保障

RASSF Robotics Assembly and Servicing Simulation Facility 机器人装配与检修模拟设施

RAST Recovery Assistance and Traversing System 接应协助与转运系统

RASTA Radiation-Augmented Special Test Apparatus 辐射增强的专用试验辅助装置

RASV Reusable Advanced Space Vehicle 可重复使用的高级航天飞行器

RASV Reusable Aerospace Vehicle 可重复使用航空航天飞行器

RAT Remote Access Terminal 遥控存取终端

RAT Rock Abrasion Tool 岩石磨削工具

RAT Routing Automation Technique 路径选择自动化技术

RAT Routing-Analysis Tool 路径选择分析工具

RATCC Radar Air Traffic Control

Center　空中交通雷达控制中心
RATER　Response Analysis Tester　【美国国家航空航天局】响应分析测试仪
RATIO　Radio Telescope in Orbit　轨道射电望远镜
RATLER　Robotic All-Terrain Lunar Exploration Rover　全地形机器人月球漫游车
RATO　Rocket Assisted Takeoff　火箭辅助起飞
RATOG　Rocket Assisted Takeoff Gear　火箭助推起飞装置
RATS　Radio Theodolite System　无线电经纬仪系统
RATS　Research and Technology Studies　研究与技术研究
RAU　Remote Acquisition Unit　远程采集装置
RAUIS　Remote Acquisition Unit Interconnecting Station　远程采集装置互联站
RAV　Launch Vehicle Flight Readiness Review (Revue d'Aptitude au Vol du lanceur)　运载火箭飞行准备状态评审
RAV　Remotely Augmented Vehicle　遥控增益飞行器
RAZEL　Range, Azimuth and Elevation　距离、方位与仰角；距离、方位与高度
RB　Radar Beacon　雷达信标
RB　Reentry Body　再入（大气层）飞行器（火箭、弹头）
RBA　Radar Beacon Antenna　雷达信标天线
RBCC　Rocket Based Combined Circulation　火箭基组合循环式
RBCC　Rocket-Based Combined Cycle　火箭基组合循环（一种发动机技术）
RBDE　Radar Bright Display Equipment　雷达亮度显示设备
RBDP　Rocket Booster Development Program　火箭助推器研制计划
RBECS　Revised Battlefield Electronic CEOI System　远程战场通信、电子、情报与作战系统
RBG　Rocket Boosted Glider　火箭助推滑翔机
RBI　Rocket Balloon Instrument　火箭气球装置（膨胀球）
RBI　Rocket-Borne Instrumentation　箭载测量仪表
RBL　Range and Bearing Launch　距离与方位发射
RBS　Recoverable Booster System　可回收助推器系统
RBS　Reusable Booster System　可重复使用助推器系统
RBSC　Radar Bomb Scoring Central　雷达弹着点标定中心
RBSS　Air Force Recoverable Booster Space System Development　【美国】空军对可回收的助推器航天系统的开发
RBSS　Recoverable Booster Support System　【航天火箭】可回收助推器保障系统
RBV　Reusable Bus Vehicle　可重复使用运载飞行器；可回收运载飞行器
RC　Range Command　靶场指令；靶场指挥部
RC　Range Coordinator　靶场坐标方位仪
RC　Reaction Control　反作用控制
RC　Recovery Controller　回收控制器（装置）
RC　Recurring Cost　一次性成本
RC　Relay Center　中继中心

RC Rotation Control 旋转控制
RCA Relay Control Assembly 中断控制组件
RCA Remote Control Amplifier 远程控制放大器（装置）
RCA Root Cause Analysis 基础原因分析
RCAG Remote Control Air-Ground 空对地远距离控制
RCC Range Control Center 靶场控制中心
RCC Reinforced Carbon-Carbon 增强型碳一碳（材料）
RCC Remote Communication Centre 远程通信中心
RCC Remote Control Centre 远程控制中心
RCC Reusable Carbon-Carbon 可重复使用型碳一碳（材料）
RCC Rough Combustion Cutoff 燃烧初关机
RCCA Rough Combustion Cutoff Assembly 燃烧初关机组件
RCCC Range Control Communication Center 靶场控制通信中心
RCCP Recorder and Communications Control Panel 记录仪与通信控制操纵台
RCE Reaction Control Equipment 反作用控制设备
RCE Reliability Control Engineering 可靠性控制工程
RCE Remote Control Equipment 遥控设备
RCE Rocket Cushion Device 火箭缓冲装置
RCE Rotating-Combustion Engine 转缸式发动机
RCF Radar Correlation Function 雷达校正系数
RCF Range Correction Factor 距离修正因素
RCG Radio-Activity Concentration Guide 放射性浓度指标
RCH Reduce Cluster Handler 减装集束弹药处理装置
RCHCS Regenerable CO_2 and Humidity Control System 可再生的二氧化碳与湿度控制系统
RCJ Reaction Control Jet 反作用喷气舱；反作用控制射流（喷嘴管）
RCM Reaction Control Module 反作用控制舱
RCM Reliability-Centered Maintenance 以可靠性为中心的维修
RCM Restricted Corrosive Material 防腐蚀材料
RCMA Reliability-Centered Maintenance Analysis 以可靠性为中心的维修分析
RCMS Reconfigurable Manufacturing System 可重构制造系统
RCO Rendezvous Compatibility Orbit 交会协调轨道
RCP Radiation Constraints Panel 辐射限定控制台
RCP Refractory Concrete Panel 耐火混凝土板
RCP Right Circular Polarization 右圆极化
RCP Right Circular Polarizer 右圆极化器（装置）
RCPU Rocket Control Panel Unit 火箭控制面板装置
RCRL Reliability Critical Ranking List 可靠性临界序列表；可靠性关键序列表
RCS Radar Cross Section 雷达反射截面
RCS Range Calibration Satellite 射

程校准卫星
RCS Range Control Station 靶场控制站
RCS Rate Control System 速率控制系统
RCS Reaction Control System (Subsystem) 反作用控制系统（分系统）
RCS Reentry Control System 再入（大气层）控制系统
RCS Remote Control System 遥控系统
RCS Roll Control System 滚动控制系统
RCSAD Remote Control Safe and Arm Device 遥控安全解保装置
RCSC Reaction Control Subsystem Controller 反冲控制分系统控制器
RCSE Remote Tracking Station Control and Status Equipment 遥控跟踪站控制与状态设备
RCSR Radar Cross Section Reduction 雷达反射截面缩小
RCSS Random Communication Satellite System 随机通信卫星系统
RCSS Range Command Safety System 靶场指挥安全系统
RCT Reaction Control Thruster 反作用控制推力器
RCU Recycling Control Units 回收控制装置
RCU Remote Control Unit 远程控制装置
RCU Rocket Countermeasure Unit 火箭对抗装置
RCV Remote Control Vehicle 遥控飞行器
RCVS Remote Control Video Switch 远程控制视频切换
RD Radiation Detection 辐射探测
RD Readiness Demonstrator 成熟度演示验证
RD&A Research, Development and Acquisition 研究、发展与采办
RDA Radar Data Analysis 雷达数据分析
RDA Remote Data Access 远程数据访问
RDA Research, Development and Acquisition 研究、生成与采购
RDAISA Research, Development, and Acquisition on Information Systems Activity 研究、开发与获取信息系统活动
RDAS Rocket Data Acquisition System 火箭数据采集系统
RDAT Research and Development Acceptance Testing 研究与发展验收试验
RDAU Remote Data Acquisition Unit 远程数据采集装置
RDBMS Relational Database Management System 相关数据库管理系统
RDC Radar Data Collection 雷达数据采集
RDC Regional Dissemination Center 【美国国家航空航天局】地区传播中心
RDCS Reconfiguration Data Collection System 再配置数据采集系统
RDD Requirements Driven Design 基于要求的设计
RDD Research, Development and Demonstration 研究、开发与演示（验证）
RDDT&E Research, Development, Demonstration, Testing and Evaluation 研究、开发、演示、测试与鉴定
RDE Radar Data Exploitation 雷达数据利用

RDE Research, Development and Engineering 研究、发展与工程

RDES Requirements Determination and Execution System 需求确定与执行系统

RDF Radio Direction Finding 雷达测向

RDF Resource Data File 源数据文件

RDGT Reliability Development and Growth Test 可靠性发展和增长试验

RDI Range Doppler Imager 靶场多普勒成像仪

RDIS Research and Development Information System 研究与发展信息系统

RDL Research and Development Laboratory 研究与开发实验室

RDLS Reentry, Descent and Landing System 再入、降落与着陆系统

RDLT Reentry, Descent and Landing Technology 再入、降落与着陆技术

RDMAS Research and Development Management Analysis System 研究与发展管理分析系统

RDOC Reference Designation Overflow Code 参照设计溢流规范

RDP Receiver and Data Processor 接收器与数据处理器

RDR Raw Data Recorder 原始数据记录装置

RDS Range Destruct System 靶场自毁系统

RDS Real-Time Data Smoothing 实时数据平衡

RDS Real-Time Digital Simulator 实时数字化模拟器（装置）

RDS Regional Defense System 区域防御系统

RDS Rendezvous Docking Simulator 交会对接模拟器

RDS Russian Docking System 俄罗斯对接系统

RDSIS Radar Digital Signal Injection System 雷达数字信号射入系统

RDSS Regional Development Simulation System 区域开发模拟系统；区域发展模拟系统

RDT Reliability Demonstration Testing 可靠性演示测试

RDT&E Research, Development, Test and Engineering 研究、发展、试验与工程

RDT&E Research, Development, Test and Evaluation 研究、发展、试验与评估（美国国防科研计划的总称）

RDTA Reliability Development Test Article 可靠性发展测试项（件）

RDTU Remote Data Transmission Unit 远距离数据发送设备；远距离数据传输设备

RE Radial Error 径向偏差

RE Rocket Engine 火箭发动机

REACT Rapid Execution and Combat Targeting 快速执行与作战目标判定

REACT Reliability Evaluation and Control Technique 可靠性鉴定与控制技术

READI Rocket Engine Analyzer and Decision Instrumentation 火箭发动机分析仪与决定装置

READS Real-Time Electronic Access and Display System 实时电子存取及显示系统

READS Re-Entry Air-Data System 再入大气数据系统

READYMAIDS Ready Multipurpose

Automatic Inspection and Diagnostic System　现成多用途自动检查与诊断系统
REB　Rocket Engine Bond　火箭发动机锁键
REBA　Rechargeable EVA Battery Assembly　可再充电的舱外活动电池组件
REBOOT　Reuse Based Object-Oriented Technology　基于重复使用的面向对象技术
REC　Regional Evaluation Centers　区域鉴定中心
RECAP　Report of Evaluation, Control, Analysis and Progress　鉴定、控制、分析和进展报告
RECAS　Residual Capabilities Assessment System　剩余能力评估系统
RECEP　Relative Capability Estimating Process　相对能力估算程序
RECON　Remote Console　远距离控制台
RECON　Remote Control　远程控制
RECS　Representative Shuttle Environmental Control System　具有代表性的航天飞机环境控制系统
REDAR　Radiation and Environmental Data Acquisition and Recording　辐射和环境数据采集与记录
REDCAP　Real Time Electromagnetic Digitally Controlled Analyzer and Processor　实时电磁数字控制分析仪与处理机
REDCAP　Real-Time Digitally Controlled Analyzer and Processor　实时数字控制式分析机与处理机；实时数字控制式分析与处理程序
REE　Remote Exploration and Experimentation　【美国国家航空航天局】远距离探索与实验
REED　Rocket Exhaust Effluent Diffusion　火箭排气扩散
REEDA　Rocket Exhaust Effluent Diffusion Analysis　火箭排气扩散分析
REEP　Range Estimating and Evaluation Procedure　范围估计和鉴定程序；距离估计和鉴定程序
REFCON　Reference Configuration　参考构型
REFL　Reference Line　基准线
REGAL　Range and Elevation Guidance for Approach and Landing　进场着陆距离与仰角引导
REI　Range Elevation Indicator　距离仰角指示器
REINS　Radar-Equipped Inertial Navigation System　装备有雷达的惯性导航系统
REINS　Radio-Equipped Inertial Navigation System　装备有无线电的惯性导航系统
REIV　Rocket Engine Injector Valve　火箭发动机喷注器活门
RELACS　Radar Emission Location Attack Control System　雷达发射位置攻击控制系统
REM　Reaction Engine Modules　反作用发动机舱
REM　Reentry Module　再入（大气层）舱
REM　Release Engine Mechanism　释放发动机机构
REM　Release Engine Module　释放发动机模块
REM　Release Escape Mechanism　释放逃逸机构
REM　Release-Engage Mechanism　分离/接合机构
REM　Rocket Engine Module　火箭发动机舱

REMCA Reliability, Maintainability, Cost Analysis 可靠性、可维修性与成本分析

REMCAL Radiation Equivalent Manikin Calibration 辐射当量人体模型校准

REMCAL Relative Motion Collision Avoidance Calculator 避免相对运动时碰撞的计算装置

REMIS Reliability and Maintainability Information System 可靠性与维修性信息系统

REN RAD Rendezvous Radar 交会雷达

REND Rendezvous 交会；接合

REP Data Replay Mode (Telemetry) 数据重放模式（遥测）

REP Rendezvous Evaluation Pod 【美国】"雷普"轨道会合鉴定舱（美国航天器交会技术试验舱）

REP Rendezvous Exercise Pod 轨道会合操作舱

RERP Reliability Enhancement and Re-engine Program 可靠性增强与更换项目

RES Relay Earth Station 中继地面站

RES Reusable Booster System 可重复使用的助推器系统

RESA Radar Environment Status Assessment Algorithm 雷达环境状态评估算法

RESA Research, Evaluation, and Systems Analysis 研究、评定与系统分析

RESA Research, Evaluation, and Systems Analysis Simulation Facility 研究、鉴定与系统分析模拟设施

RESAC Regional Earth Science Application Center 【美国国家航空航天局】地区性地球科学应用中心

RESC Reconstitutable Enduring Satellite Communications 可重建的持久卫星通信系统

RESCAN Reflecting Satellite Communication Antenna 卫星通信反射天线

RESCAP Rescue Combat Air Patrol 救援作战空中巡逻

RESCU Rocket-Ejection Seat Catapult Upward 火箭助推的座椅向上弹射

RESL Radiological and Environmental Science Laboratory 辐射与环境科学实验室

REST Reentry Environment and System Technology 再入环境与系统技术

REST Reentry System Test 再入系统试验

RESTA Reconnaissance, Surveillance and Target Acquisition 侦察、监视与目标截获

RESTEC Remote Sensing Technology Center of Japan 日本遥感技术中心

RESUR-F Russian Earth Resourcing Satellite 俄罗斯地球资源卫星

RET Reliability Enhancement Testing 可靠性强化试验

RET Reliability Evaluation Test 可靠性鉴定试验

RETF Rocket-Engine Test Facility 火箭发动机测试设施

RETL Rocket-Engine Test Laboratory 火箭发动机测试实验室

RETRO Retrofit 【导弹、航天器等】改装

RETS Reconfiguration Electrical Test Stand 再配置电气测试台

RETS Remote Target System 远距离目标系统

REV Reentry Vehicle 再入飞行器
REVMAT Reentry Vehicle Materials Technology 再入飞行器材料工艺（学）
REX Radiation Experiment 辐射试验卫星
REX Radio Exploration Satellite 无线电探测卫星
REX Reentry Experiment 再入（大气层）实验
REX Rezonans Experiment on Mir Space Station 【苏联】"和平号"空间站上的共振实验
REXS Radio Exploration Satellite 无线电探测卫星
REXS Radio Wave Experiment Satellite 【日本】无限电波实验卫星
RF Radio Frequency 无线电频率
RF-ATE Radio Frequency Auto Test Equipment 无线电频率自动测试设备
RFAC Robot Flexible Assembly Cell 机器人柔性装配单元
RF-TK Radio Frequency Tracking 无线电频率跟踪
RFB Reliability Functional Block 可靠性功能块
RFC Regenerative Fuel Cell 再生型燃料电池
RFD Reentry Flight Demonstration 再入飞行演示验证
RFFEL Radio Frequency Free Electron Laser 无射频电子激光器
RFI Radio Frequency Interference 射频干扰
RFIAR Reliability Failure Items Analysis Report 可靠性故障项目分析报告；可靠性故障物品分析报告
RFID Radio Frequency Identification 射频识别
RFLINAC Radio Frequency Linear Accelerator 射频线性加速器（装置）
RFM Reliability Figure of Merit 可靠性品质因数
RFS Rover Flight Safety 【美国】"徘徊者"核火箭飞行安全
RFTA Recycle Filter Tank Assembly 回收过滤箱组件
RG Rate Gyro 速度陀螺仪
RGA Rate Gyro Assembly 速度陀螺仪组件
RGAL Rate Gyro Assembly-Left SRB 左固体火箭助推器速度陀螺组件
RGAO Rate Gyro Assembly-Orbiter 轨道器速度陀螺组件
RGAR Rate Gyro Assembly-Right SRB 右固体火箭助推器速度陀螺组件
RGI Rocket Gas Ingestion 火箭燃气注入
RGM Redundant Gyro Monitor 冗余陀螺仪监测器
RGM Remote Geophysical Monitor 远距离地球物理监测器
RGON Remote Geophysical Observing Network 远距离地球物理观测网
RGP Rate Gyro Package 速度陀螺仪装置
RGRMA Rate Gyro Redundancy Management Algorithm 速度陀螺仪冗余管理计算
RGS Remote Ground Station 远程地面站
RGS Rocket Guidance System 火箭制导系统
RGSAT Radio Guidance Surveillance and Automatic Tracking 无线电制导监视与自动跟踪
RH Radiation Hardened 辐射加固的

RH Relative Humidity 相对湿度
RH Electronics Radiation Hardened Electronics 辐射加固的电子设备
RHBD Radiation-Hardened by Design 设计加固辐射
RHBP Radiation-Hardened by Process 处理加固辐射
RHCM Relative Humidity Control/Monitor 相对湿度控制/监测
RHD Radiation Hardened Device 辐射加固的装置
RHEB Right Hand Equipment Bay 右侧设备舱
RHFEB Right Hand Forward Equipment Bay 右侧前端设备舱
RHI Risk Hazard Index 风险危害指数
RHS Rocketdyne Hybrid Simulator 【美国】洛克达因公司混合仿真器
RHT Radiant Heat Temperature 辐射热温度
RI Risk Index 风险指数
RIA Range Instrumentation Aircraft 靶场测试仪表飞机
RIACS Research Institute for Advanced Computer Science 【美国国家航空航天局】先进计算机科学研究所
RICBM Retro Intercontinental Ballistic Missile 制动发动机大陆内弹道导弹
RICC Reusable Integrated Command Center 可重复使用的综合指挥中心
RICI Remote Intercomputer Communications Interface 远程计算机互联通信接口
RICS Range Instrumentation Control System 靶场仪表控制系统
RIF Relative Importance Factor 相对重要因子
RIFCA Redundant Inertial Flight Control Assembly 冗余惯性飞行控制组件
RIFT Rocket In-Flight Test 火箭飞行试验
RIFTS Rapidly Installed Fluid Transfer System 快速安装的液体输送系统；快速安装的燃料输送系统
RIFTS Rocket In-Flight Test System 火箭飞行试验系统
RIG Rate Integration Gyro 集成式速度陀螺仪
RIGI Receiving Inspection General Instruction 验收检验总体说明
RIGS Radio Inertial Guidance System 无线电惯性制导系统
RIIP Right Instantaneous Impact Point 右侧瞬间撞击点
RIIP/LIIP Right/Left Instantaneous Impact Point 右/左侧瞬间撞击点
RILS Ranging Integrated Locating System 综合测距定位系统
RILS Rapid Integrated Logistic Support System 快速综合后勤保障系统
RIM Refrigerator Incubator Module 【美国航天飞机】恒温制冷箱
RIM Rocket Intercept Missile 火箭拦截导弹
RIM-66C Ship-Launched Medium-Range, Surface-to-Air Missile 【美国海军】舰船发射的、中程地对空导弹
RIME Radio Inertial Missile Equipment 无线电惯性制导导弹设备
RIMOS Real-Time Interactive Multiprocess Operating System 实时交互式多重处理操作系统
RIMS Roll-Stabilized Inertial Measurement System 滚转-稳定惯性测量系统

RIMU Reentry Inertial Measurement Unit 再入惯性测量设备
RINS Radio Inertial Navigation System 无线电惯性导航系统
RINT Unintentional Radiation Intelligence 无意发出情报
RINU Redundant Inertial Navigation Unit 冗余惯性导航装置
RIOMETER Relative Ionospheric Opacity Meter 相对电离层不透明度指示计；相对电离层吸收仪
RIP Remote Instrument Package 遥控仪器包
RIPE Robot Independent Programming Environment 机器人独立编程环境
RIPL Robot Independent Programming Language 机器人独立编程语言
RIPS Radar Impact Prediction System 雷达撞击预测系统
RIR Range Instrumentation Radar 靶场测量雷达
RIS Radar Integration System 雷达综合系统
RIS RADIAC (Radioactivity Detection, Indication and Computation) Instrument System 放射性探测、显示与计算仪器
RIS Ramjet Inlet System 冲压喷气发动机进气系统
RIS Range Information System 靶场信息系统；距离信息系统
RIS Range Instrumentation Ship 靶场仪表测量船
RIS Range Instrumentation Station 靶场仪表测量站
RIS Range Instrumentation System 靶场（仪表）测量系统
RIS Receipt, Inspection and Storage Facility 接收、检查及贮存厂房
RISA Rapid Imaging Spectrophotometric Array 快速成像分光光电矩阵
RISC Reduced Instruction Set Computer 简化指令集计算机
RISCAE RISC Ada Environment 简化指令集 Ada 语言环境
RISE Reliability Improvement of Selected Equipment 选择性装备的可靠性改进
RISP Recoverable Interplanetary Space Probe 可回收的行星际探测器
RISS Range Instrumentation and Support System 靶场仪表设备与保障系统
RISTA Reconnaissance, Intelligence, Surveillance, and Target Acquisition 侦察、情报、监视与目标捕获
RIT Rocket Interferometer Tracking 火箭干涉仪跟踪
RITA Recoverable Interplanetary Transport Approach 可回收的行星际运输飞行器进场
RITA Reusable Interplanetary Transport Approach Vehicle 可重复使用的星际运输飞行器
RITE Rapid Information Technique for Evaluation 鉴定用快速信息处理技术
RIU Remote Interface Unit 远程接口装置
RIV Recirculation Isolation Valve 再循环隔绝阀
RJ/EC Reaction Jet/Engine Control 反作用喷射/发动机控制
RJC Reaction Jet Control 反作用喷射控制
RJD Reaction Jet Device 反作用喷射装置

RJD Reaction Jet Driver 反作用喷射驱动器（装置）
RJDA Reaction Jet Driver Aft 反作用喷射驱动器（装置）后部
RJDF Reaction Jet Driver Forward 反作用喷射驱动器（装置）前部
RJEC Reaction Jet Engine Control 反作用喷射发动机控制
RJP Rocket Jet Plume 火箭喷气羽烟
RL Remote Launch 远程发射
RL Rocket Launcher 火箭发射器；火箭发射装置
RLAS Rocket Lunar Attitude System 登月火箭（飞行）姿态（控制）系统
RLC Remote Load Controller 远距离载荷控制器
RLCC Remote Launch and Control Center 远距离发射控制中心
RLCEU Remote Launch and Communications Enhancement Upgraded 远距离发射与通信升级
RLCS Rocket Launch Control System 火箭发射控制系统
RLD Reference Landing Distance 参考着陆距离；基准着陆距离
RLEP Robotic Lunar Exploration Program 机器人探月项目
RLG Ring Laser Gyro 环形激光陀螺仪
RLPRB Reusable Liquid Propellant Rocket Boosters 可重复使用液体（推进剂）火箭助推器
RLRIU Routing Logic Radio Interface Unit 路由选择逻辑无线电接口装置
RLS Radius of Landing Site 【月球】着陆场地半径
RLS Reusable Launch System 多次使用发射系统
RLS Rocket Launch System 火箭发射系统
RLSBO Side Looking Microwave Radar 侧视微波雷达（俄罗斯海洋卫星上的主要遥感器）
RLSS Regenerative Life Support System 可再生生命保障系统
RLST Remote Landing Site Tower 着陆场遥控塔台
RLT Return Line Tether 返回线路系留
RLV Reusable Launch Vehicle 可重复使用运载火箭
RLV Roving Lunar Vehicle 月面游动车
RM Radiation Meteoroid 【美国】辐射与流星体试验卫星
RM Reconfiguration Module 可重构模块
RM Redundancy Management 冗余管理
RM Reference Mission 参照性任务
RM Remote Manipulator 遥控机械臂
RM Rendezvous Maneuver 交会机动
RM Rescue Module 救援舱；救援模块
RM Resource Module 资源舱
RM Resupply Module 补给舱
RM Risk Management 风险管理
RM Rocket Motor 火箭发动机
RM Rolling Moment 横滚力矩
RM&A Reliability, Maintainability and Availability 【统一航天作战中心】可靠性、可维护性与可用性
RM&S Reliability, Maintainability and Supportability 可靠性、维修性与保障性

RM&T　Reliability, Maintainability and Testability　可靠性、维修性与测试性

RM/MS&C　Redundancy Management/Molding, Sequencing, and Control　冗余管理 / 成型、排序与控制

RMA　Remote Manipulator Arm　遥控机械臂

RMAS　Remote Monitoring and Alarm System　远距离监测与警报系统

RMAS　Resource Management Accounting System　资源管理统计系统

RMC　Redundancy Management Control　冗余管理控制

RMCS　Range Monitoring Control System　靶场监视控制系统

RMCT　Robotic Micro Tool　机器人微锥体工具

RME　Radiation Monitoring Equipment　辐射监测设备

RME　Relay Mirror Experiment　中继反射镜实验卫星

RMF　RCS Module Forward　反作用控制系统模块前端

RMI　Reliability Maturity Index　可靠性成熟度指数

RMI　Rocket Motor Igniter　火箭发动机点火器

RMIEP　Risk Methodology Integration and Evaluation Program　风险方法集成与评价项目

RMIS　Remote Multiplexer Instrumentation System　远距离多路调制测量系统

RMMS　Remote Maintenance Monitoring System　远程维修监控系统

RMP　Risk Management Plan　风险管理计划

RMP　Rocket Motor Propellant　火箭发动机推进剂

RMRS　Repeatable Maintenance and Recall System　可重复维修及检索系统

RMS　Radiation Monitoring Satellite　辐射监测卫星

RMS　Random Motion Simulator　随机运动模拟器（装置）

RMS　Redundancy Management System　冗余管理系统

RMS　Reentry Measurement System　再入大气层测量系统

RMS　Reference Mission Scenario　基准任务方案

RMS　Reliability, Maintainability, and Safety　可靠性，可维修性和安全性

RMS　Reliability, Maintainability, Supportability　可靠性、维修性与保障性

RMS　Remote Manipulator System　【美国航天飞机】遥控机械臂系统；远程操纵系统

RMS　Remote Manipulator System (Subsystem)　遥控机械臂系统（分系统）

RMS　Reusable Multipurpose Spacecraft　可重复使用多用途航天器

RMS　RMS Range Management Squadron　【美国】靶场可靠性、维修性与保障性管理大队

RMS　RMS Remote Manipulator System　远程可靠性、维修性与保障性操作系统

RMS　Rocket Management System　火箭管理系统

RMSA　Rocket Motor Staging Area　火箭发动机级分离段

RMSEL　Robotic Manufacturing, Science, and Engineering Laboratory

机器人制造、科学与工程实验室
RMSVP Remote Manipulator Subsystem Verification Plan 遥控机械臂子系统验证计划
RMT Radiometric Moon Tracer 辐射测量月球示踪器
RMT Reliability, Maintainability, Testability 可靠性、维修性和可测试性
RMT&S Reliability, Maintainability, Testability and Supportability 可靠性、维修性、可测试性与保障性
RMV Reentry Measurement Vehicle 再入（大气层）测量飞行器
RMV Remotely Manned Vehicle 遥控载人飞行器
RMWTS Radioactive and Mixed Waste Tracking System 放射与混合废弃物跟踪系统
RNA Rotatable Nozzle Assembly 旋转喷嘴组件
RNDS Rendezvous 交会
RNET Remote Network 远程网络
RNS Reusable Nuclear Shuttle 可重复使用核动力航天飞机
RO Range Operation 靶场操作；靶场运行
RO Recovery Operation 回收操作
ROAD Rapid Orbit Analysis and Determination 快速轨道分析与确定
ROADS Range Optimal Advanced-Development System 靶场最佳的高级开发系统
ROADS Real-Time Optical Alignment and Diagnostic System 实时光学调正与诊断系统
ROB Remote Operating Base 远程操作基地
ROBIN Rocket Balloon Instrument 火箭气球仪表装置
ROBS Rapid Optical Beam Steering (System) 快速光束操控（系统）
ROC Regional Operations Center 区域性操作中心
ROC Remote Operator's Console 远距离操作手操控台
ROC Reusable Orbital Carrier 可重复使用轨道运载器
ROCC Range Operations Control Center 靶场作战控制中心（现为Morrell作战中心）
ROCDCD Remotely Operated and Controlled Devices Countermeasures and Deception 远距离操纵与控制的（电子）干扰与诱骗设备
ROCKET Rand's Omnibus Calculator of the Kinetics of Earth Trajectories 【美国】兰德公司的地球轨道动力学计算装置
ROCKEX Rocket Exercise 火箭演练
ROCS Range Operations Control System 靶场操作控制系统
ROCS Reusable Orbital Carriers 可重复使用的轨道运载器
RoCS Roll Control system 翻滚控制系统
ROD Remote Operated Door(s) 远程控制门
RODB Remote Object Data Base 远程物体数据库
RODS Real-Time Operations, Dispatching and Scheduling System 实时操作、调度与安排系统
ROE Reflector Orbital Equipment 轨道反射器设备
ROGER Remotely Operated Geophysical Explorer 遥控地球物理探测仪
ROI Range Operating Instruction 靶场操作指令

ROI Return on Investment 投资收益率

ROLS Recoverable Orbital Launch System 【美国国家航空航天局】可回收轨道发射系统

ROMAC Range Operations Monitor Analysis Center 靶场操作监控分析中心

ROMACC Range Operations Monitoring and Control Center 靶场操作监控中心

ROMBUS Reusable Orbital Module Booster and Utility Shuttle Under Study By Douglas 【美国道格拉斯公司】可再用的轨道舱助推器与实用航天飞机

ROMOTAR Range-Only Measurement of Trajectory and Recording 弹道飞行距离只测距与记录（仪）

ROOM Real-Time Object-Oriented Methodology 实时面向目标的研究方法

ROOM Real Time Object-Oriented Modeling 实时面向对象建模

ROOST Reusable One Stage Orbital Space Truck 【美国道格拉斯公司】可重复使用的单级轨道空间运输器

ROOTS Remotely Operated Optical Tracking System 远距离操作的光学跟踪系统

ROPE Research on Orbital Plasma Electrodynamics 轨道等离子体跟踪系统研究

RORSAT Radar Ocean-Reconnaissance Satellite 雷达海洋侦察卫星

ROS Range Operations Supervisor 靶场操作监督员

ROS Real Time Operating System 实时操作系统

ROS Regulated Oxygen Supply 调节性供氧

ROS Regulated Oxygen System 调节性供氧系统

ROSA Radar Open-System Architecture 雷达开放式系统体系结构

ROSAT Roentgen Satellite 伦琴卫星（德、英、美联合的 X 射线天文卫星）

ROSCOE Remote Operating System Conventional Operating Environment 遥控系统常规操作环境；远程操作系统常规运行环境

ROSE Remote Operations Service Element 远程操作服务单元

ROSIE Reconnaissance by Orbital Ship Identified Equipment 使用轨道飞船识别设备进行侦察作业

ROSIS Reflective Optics System Imaging Spectrometer 反射光学系统成像光谱仪

ROSS Remote Orbital Servicing System 远距离轨道勤务系统

ROT Remaining Operating Time 剩余操作时间

ROT Reusable Orbital Transport 可重复使用的轨道运输机

ROTE Range Optical Tracking Equipment 靶场光学跟踪设备

ROTHR Relocatable Over the Horizon Radar 可重新部署的超视距雷达

ROTI Recording Optical Tracking Instrument 靶场光学跟踪记录仪

ROTS Reusable Orbital Transport System 可重复使用的轨道运输系统

ROTV Returnable Orbital Transfer Vehicle 【美国】返回式轨道转移飞行器

ROTV Returnable Orbital Transfer Vehicle 返回式轨道转移飞行器

ROV Remote Operated Vehicle 遥控车

ROZ Restricted Operations Zone 有限作战区域

RP Rapid Prototyping 快速试制原型机

RP Relative Pressure 相对压力

RP Release Point 投放点

RP Rocket Propellant (Hydrocarbon-Based) 火箭推进剂（碳氢基）

RP&C Resource Planning and Coordination 资源规划与协调

RP-1 Rocket Propellant-1 (Kerosene) 用于火箭和冲压式喷气发动机的煤油类燃料

RPA Record and Playback Assembly 记录与回放设备

RPAC Resource Performance Analysis Center 资源性能分析中心

RPAF Reliability Program Amplification Factor 可靠性程序放大因数

RPAM Reliability Prediction and Allocation Model 可靠性预测与分配模型

RPC Remote Position Control 远程位置控制

RPC Remote Power Controller 远程动力控制器（装置）

RPCM Remote Power Control Module 远程动力控制模块

RPCU Remote Power Control Unit 远距离动力控制装置

RPDL Rapid Prototype Development Laboratory 快速原型开发实验室

RPDP Recoverable Plasma Diagnostic Package 【美国航天飞机】可回收等离子体探测设备

RPIA Rocket Propellant Information Agency 火箭推进剂情报机构

RPL Rocket Propulsion Laboratory 【美国】火箭推进实验室

RPLV Reentry Payload Launch Vehicle 再入大气层有效载荷运载器

RPM Rendezvous Pitch Maneuver 经俯仰机动对接

RPOC Remote Payload Operations Center 远程有效载荷操作中心

RPOCC Remote Payload Operations Control Center 远距离有效载荷操作控制中心

RPP Rocket Propellant Plant 火箭推进剂工厂

RPRV Remotely Piloted Research Vehicle 遥控研究飞行器；研究用遥控飞行器

RPS Data Recording and Playback Subsystem 数据记录与回放分系统

RPS Radioisotope Power System 放射性同位素能量系统

RPSF Rotation, Processing, and Surge Facility 旋转、处理与吹除设施

RPU Remote Power Unit 远程动力装置

RPV Remotely Piloted Vehicle 远距离驾驶运载器

RQE Reliability Quality Evaluation 可靠性质量评定

RQF Rescue Flight 救援飞行

RQR Reliability Qualification Requirement 可靠性鉴定要求

RQT Reliability Qualification Tests 可靠性鉴定试验

RR Readiness Review 准备状态评审

RR Rendezvous Radar 【飞船】交会使用的雷达

RR Requirements Review 需求评审

RR Retro-Rocket 制动火箭；减速火箭；反推火箭

RRAD Roll Rate Anomaly Detection

滚转速度异常检测
RRCS Rate and Route Computer System 速率和航迹计算机系统；速率与路线计算机系统
RRCS Reentry Radar Cross Section 再入雷达截面（图）
RRCS Roll Rate Control System 滚动速度控制系统
RRDA Rendezvous, Retrieval, Docking, and Assembly 交会、收回、对接与装配
RREA Rendezvous Radar Electronics Assembly 交会雷达电子设备组件
RREB Range Radio Equipment Building 靶场无线电设备间
RREU Rendezvous Radar Electronics Unit 交会雷达电子设备
RRF Risk Reduction Flight 减少风险的飞行任务
RRFD Risk Reduction Flight Demonstration 降低风险的飞行演示验证
RRG Roll Rate Gyro 滚转角速度陀螺仪
RRG Roll Reference Gyro 【导弹】滚动基准陀螺
RRGU Redundant Rate Gyro Unit 冗余速率陀螺装置
RRI Range Rate Indicator 距离速度指示器；射程变化率指示器
RRI Rendezvous Radar Indicator 交会雷达指示器（装置）
RRIPM Rapid Response Interference Prediction Model 快速响应干扰预测模型
RRL Rocket Research Laboratory 【美国国家航空航天局】火箭研究实验室
RRM Rapid Response Manufacturing 快速响应制造
RRM Robotic Refueling Mission 机器人燃料加注演示验证
RRNS Remote Radio Navigation System 远程无线电导航系统
RRPC Rapid Response Process Improvement 快速响应处理改进
RRPI Robotic Reconnaissance Point Sensor 机器人侦察点传感器
RRS Rendezvous Radar System 【轨道】交会使用的雷达系统
RRS Repeated Retrievable Satellite 重复使用的可回收卫星
RRS Retro-Rocket System 制动火箭系统
RRS Reusable Reentry Satellite 【美国】可重复使用再入卫星
RRS Robotic Radiation Survey and Analysis System 机器人辐射监测与分析系统
RRSC Rapid Response Support Center 快速响应保障中心
RRSE Risk, Reliability, and Safety Engineering 风险、可靠性与安全工程
RRT Roll-Response Time Constant 滚转响应时间常数
RRV Reusable Reentry Vehicle 重复使用再入飞行器
RRV Rotor Reentry Vehicle 再入（大气层）旋翼飞行器
RS Radio Sport(Radio Sputnik) 无线电运动卫星（独联体业余无线电爱好者通信卫星）
RS Range Safety 靶场安全
RS Range Surveillance 靶场监视
RS Reconnaissance Satellite 侦察卫星
RS Redundancy Status 冗余状态
RS Redundant Set 冗余装置
RS Re-Entry System 再入（大气层）系统

RS　Refurbishment Spare　修整备用件
RS　Relay Satellite　中继卫星
RS　Reliability Surveillance and Control　可靠性监测与控制
RS　Remote Station　远程站
RS&S　Receiving, Shipping, and Storage　接收、船运与存储
RS/A　Range Standardization and Automation　靶场标准化与自动化
RSA　Range Safety Augmentation Program　靶场安全增强计划
RSA　Range Standardization and Automation　靶场标准化与自动化
RSA　Russian Space Agency　俄罗斯航天局
RSAP　Range Safety Augmentation Program　靶场安全增进项目
RSAP　Reentry Systems Applications Program　再进入系统应用项目
RSATS　Responsive Space Advanced Technology Study　响应型空间先进技术研究
RSC　Range Safety Command　靶场安全指令
RSC　Range Safety Console　靶场安全控制台
RSC　Range Safety Control　靶场安全控制
RSC　Rocket Space Corporation Energia　俄罗斯能源火箭航天公司
RSC　Roll Stability Control　滚转稳定性控制
RSC-S　Rate Stabilization and Control System　速率稳定与控制系统
RSCIE　Remote Station Communication Interface Equipment　远程站通信接口设备
RSCR　Range Safety Command Receiver　靶场安全指令接收机
RSCS　Range Safety Command System　靶场安全指令系统
RSCSS　Range Safety Command Shutdown System　靶场安全指令（发动机）关机系统
RSD　Range Safety Display　靶场安全显示
RSD　Recovery, Salvage and Disposal　回收、打捞与处理
RSDC　Range Safety Display Console　靶场安全显示控制台
RSDP　Remote Site Data Processor　远程场区数据处理器（装置）
RSDS　Radar and Satellite Data Store　雷达与卫星数据存储
RSDS　Range Safety Destruct System　靶场安全自毁系统
RSDS　Range Safety Display System　靶场安全显示系统
RSE　Rotational Support Equipment　旋转保障设备
RSER　Remote Sensing of Earth Resources　地球资源遥感（技术）
RSF　Refurbishment and Subassembly Facilities　整修与分组装设施
RSF　Rocket Sled Facility　火箭滑车设备
RSG　Research Satellite for Geophysics　地球物理研究卫星
RSGS　Ranges and Space Ground Support　发射场与航天地面保障
RSGS　Remote Sensing Satellite Ground Station　遥感卫星地面站
RSI　Rationalization, Standardization, Interoperability　合理化、标准化与互用性
RSI　Reusable Surface Insulation　可重复使用的表面绝缘
RSIC　Redstone Scientific Information Center　【美国】"红石"项目科学

信息中心

RSIP Radar System Improvement Program 雷达系统改进项目

RSISS Russian Segment of the International Space Station 国际空间站的俄罗斯舱段

RSL Range Safety Launch 靶场安全发射

RSL Remote Sprint Launch 【美国】"短跑"反弹道导弹遥控发射

RSLA Range Safety Launch Approval 靶场安全发射核准

RSLCC Range Safety Launch Commit Criteria 【美国】靶场安全发射放行标准

RSLP Reentry Systems Launch Program 再进入系统发射程序

RSLP Rocket Systems Launch Program 火箭系统发射程序

RSLS Redundant Set Launch Sequencer 冗余集发射程序

RSMM Redundant System Monitor Model 冗余系统监视器模型

RSMT Range Safety Modeling Toolkit 靶场安全建模工具包

RSO Resident Space Object 空间驻留物体

RSOR Range Safety Operating Requirements 靶场安全操作要求

RSP Rendezvous Station Panel 交会站控制操纵台

RSP Resupply Storage Platform 补充供给存储平台

RSR Range Safety Report 发射场安全报告；靶场安全报告

RSRB Reusable Solid Rocket Booster 【美国】可重复使用固体火箭助推器

RSRM Redesigned Solid Rocket Motor 重新设计的固体火箭发动机

RSRM Retro Solid Rocket Motor 制动固体火箭发动机

RSRM Reusable Solid Rocket Motor 可重复使用固体火箭发动机

RSRMP Range Safety Risk Management Plan 靶场安全风险管理计划

RSS Range Safety Simulator 靶场安全模拟器（装置）

RSS Range Safety Switch 靶场安全切换

RSS Range Safety System 靶场安全系统

RSS Range Scheduling System 靶场计划安排系统

RSS Reactants Supply System 反应物补给系统

RSS Received Signal Strength Indicator 接收信号强度显示器

RSS Relational Storage System 关系式存储系统

RSS Relative Signal Strength 信号相对强度

RSS Relay Station Satellite 卫星中继站

RSS Remote Sensing Satellite 【巴西】遥感卫星

RSS Resource Survey Satellite 资源勘测卫星

RSS Robotic Servicer System 机器人燃料加注系统

RSS Rotating Service Structure 旋转勤务结构（塔）

RSS Russian Space Surveillance 俄罗斯空间监视

RSSC Regional Satellite Communications (SATCOM) Support Center 地区卫星通信（SATCOM）保障中心

RSSC Regional Space Support Center 地区航天支援中心

RSSC Remote Site Simulator Console

发射场模拟器遥控台

RSSIA Regional Security Strategy Implementation Analysis 区域安全战略执行分析

RSSPS Remote Sensor System for Physical Security 物理安全措施的远距离传感器系统

RSSR Range Safety System Report 靶场安全系统报告

RSSS Reusable Space Shuttle System 可重复使用航天飞机系统

RST Radar System Technology 雷达系统技术

RST Radar System Test 雷达系统试验（测试）

RST Reheat Specified Thrust 复燃（加力）比推力

RST Remote Digital Terminal 远程数字终端

RSTA Reconnaissance, Surveillance, and Target Acquisition 侦察、监视与目标捕获

RSTCJ Remote Sensing Technology Center of Japan 日本遥感技术中心

RSTCV Reusable Space Transportation Cargo Vehicle 可重复使用的太空货物运载器

RSTER Radar Surveillance Technology Experimental Radar 雷达监视技术实验雷达

RSTKA Reconnaissance, Surveillance, Target Acquisition, Kill, and Assessment 侦察、监视、目标捕获、毁伤与评估

RSTS Range Safety and Telemetry System 靶场安全与遥测系统

RSTS Recovery System Tracking Site 回收系统跟踪站

RSTS Reusable Space Transport System 可重复使用的空间运输系统

RSTV Range Safety Test Van 靶场安全测试车

RSTWG Range Safety Trajectory Working Group 靶场安全弹道工作组

RSU Range Safety Unit 靶场安全装置

RSU Rate Sensing Unit 速度感应装置

RSU Recovery Storage Unit 回收存储装置

RSU Remote Service Unit 远距离勤务设备

RSU Remote Switching Unit 远程切换装置

RSV Resupply Vehicle 补给飞行器

RSVC Reconnaissance Satellite Vulnerability Computer 侦察卫星弱点（评估）计算机

RSVP Restartable Solid Variable Pulse Rocket 可重新启动的固体可变脉冲型火箭

RT Range Tracking 靶场跟踪

RT Reaction Time 反应时间

RT Reconfiguration Time 重构时间

RT Reference Trajectory 标准弹道

RT Remote Terminal 远程终端（设备）

RTADS Real-Time Attitude Determination System 实时姿态测定系统

RTAs Robotics, Tele-Robotics and Autonomous Systems 机器人、遥控机器人与自主系统

RTB Returned to Base 已返回（返回）基地

RTBS Real-Time Backup System 实时备份系统

RTC Real-Time Command 实时指挥

RTC Real-Time Computer 实时计算

机

RTCA Real Time Casualty Assessment 实时伤亡人员评估

RTCC Real-Time Computer Center (NASA) 【美国国家航空航天局】实时计算机中心

RTCC Real-Time Computer Command (Uplink) 实时计算机指令（上链路）

RTCC Real Time Computer Complex 实时计算机综合设备

RTCC Real-Time Command Controller 实时计算机指令控制器（装置）

RTCP Real-Time Communications Processor 实时通信处理机

RTCS Real-Time Computer System 实时计算机系统

RTD Radar Technology Demonstrator 雷达技术演示器

RTD Range Time Decoder 靶场时间译码器

RTD Real-Time Display 实时显示

RTD Receiver-Transmitter Device 接收机—发射机设备

RTD Research and Technology Development 研究与技术开发

RTD&E Research, Test, Development and Evaluation 研究、试验、发展与鉴定

RTDE Range Time Data Editor 靶场时间数据编辑程序

RTDMSA Real Time DSN (Deep Space Network) Monitor Software Assembly 实时太空网跟踪监控器软件组合

RTDP Robotics Technology Development Program 机器人技术开发项目

RTDS Real-Time Data System 实时数据系统

RTDT Real Time Data Transfer 实时数据传递

RTE Return to Earth 返回地球

RTF Return to Flight 返回飞行

RTF Rocket Test Facility 火箭试验设施

RTG Radioisotope Thermoelectric Generator 放射性同位素热电发生器

RTHS Real-Time Hybrid System 实时混合系统

RTI Real Time Interface 实时接口

RTI Real Time Interrupt 实时中断

RTIC Real-Time Information in the Cockpit 给座舱（飞行员）的实时情报

RTIF Real-Time Interface 实时接口

RTIM Radar Technology Identification Methodology 雷达技术识别方法学

RTL Ready to Launch 准备发射

RTLP Rocket Triggered Lightning Program 运载火箭触发雷电项目

RTLS Return to Launch Site (a Space Shuttle Launch Abort Mode) 【美国国家航空航天局】返回发射场（航天飞机的一种发射终止模式）

RTM Remote Telemetry Module 远距离遥测模块

RTM Remote Terminal Module 远程终端模块

RTM Requirements Traceability Matrix 需求跟踪矩阵

RTMS Rocket Thrust Measuring System 火箭推力测量系统

RTOC Real-Time Information out of the Cockpit 座舱外的实时信息

RTOP Research and Technology Objectives and Plans 研究与技术目标与计划

RTOS Real Time Operating System

实时操作系统
RTOV Real Time Operational Verification 实时操作（作战）验证
RTOVF Real Time Operational Verification Facility 实时操作（作战）验证设施
RTOW Regulated Takeoff Weight 规定起飞重量
RTP Real Time Processing 实时处理
RTPA Reaction-Time Perception Analyzer 反作用时间感知分析装置
RTR Reserve Thrust Rating 备份推力状态；备份推力额定值
RTR/WPS Real-Time Rawinsonde/Wind Processing System 实时无线电高空测风仪/风处理系统
RTS Range Time Signal 靶场时间信号
RTS Range Tracking Station 靶场跟踪站
RTS Range Tracking System (Subsystem) 靶场跟踪系统（分系统）
RTS Remote Terminal System 远程终端系统
RTS Remote Tracking Station 遥控跟踪站
RTS Requirements Tracking System 需求跟踪系统
RTS Return to Search 返回搜索
RTS Ronald Reagan Test Site, Kwajalein 【美国】位于夸贾林群礁的罗纳德·里根试验场
RTSF Real-Time Simulation Facility 实时模拟设备
RTSMP Real-Time Symmetric Multiprocessor 实时对称多处理器；实时对称多处理程序
RTSOS Real Time Secure Operating System 实时安全操作系统

RTSS Real Time Simulation System 实时模拟系统
RTTD Real Time Telemetry Data 实时遥测数据
RTTDS Real Time Telemetry Data System 实时遥测数据系统
RTTM Return to the Moon 重返月球
RTU Remote Terminal Unit 远程终端装置
RTV Reentry Test Vehicle 【美国】再入（大气层）试验飞行器
RTV Rocket Test Vehicle 火箭试验飞行器
RTV Room Temperature Vulcanizing Elastomer 室温硫化机
RU Range User 靶场用户
RU Remote Unit 远程装置
RUC Rapid Update Cycle Model 快速升级周期模型
RULER Remaining Useful Life Evaluation Routine 剩余有效期鉴定程序
RUPS Resource User ID and Password System 资源用户ID与密码系统
RUSCOM Rapid Ultrahigh Frequency Satellite Communications 快速特高频卫星通信
RUSSDPA Range User Systems Safety Data Package Approval 靶场用户系统安全数据包批准
RUT Resource Utilization Time 资源利用时间
RV Recirculation Valve 再循环阀
RV Recovery Vehicle 回收运载器
RV Recovery Vessel 回收船；回收容器
RV Reentry Vehicle 再入飞行器；再入弹头
RV Relative Velocity 相对速度

RV Relief Valve 泄放阀
RV/GC Reentry Vehicle/Ground Control 再入飞行器/地面控制
RVAO Reentry Vehicle Associated Objects 与再入飞行器相关联的目标
RVCF Remote Vehicle Checkout Facility 远距离飞行器检查设施
RVD Rendezvous and Docking 交会对接
RVFS Rendezvous Flight Software 交会飞行软件
RVI Remote Visual Inspection 远距离视觉检查
RVITS Range Visual Information Technical Services (45 Wing) 【美国第45航天联队】靶场目视信息技术服务
RVJS Reentry Vehicle Jamming Simulator 再入（大气层）飞行器干扰模拟器
RVR Reentry Vehicle Reliability 【美国第20航天联队】再入（大气层）飞行器可靠性
RVS Reentry Vehicle Separation 再入飞行器分离
RVS Reentry Vehicle Simulator 再入飞行器模拟器
RVS Reentry Vehicle System 再入飞行器系统
RVS Rendezvous Sensor 交会敏感器
RVTS Reentry Vehicle Test Set 再入飞行器试验装置
RVX Reentry Vehicle, Experimental 实验性再入飞行器
RWO Range Weather Operations 靶场气象操作
RWPD Real Time Waveform Processing Demonstration 实时波形处理演示验证
RWR Radar Warning Receiver 雷达报警接收机（装置）
RWS Rawinsonde Subsystem 无线电高空测候仪子系统
RWS Remote Workstation 远程工作站
RWS Rigid-Wall Shelter 刚性墙掩体；刚性墙体飞机掩蔽库
RWS Robotics Work Station 机器人工作站
RY Residual Yield 剩余当量
RZ Return-to-Zero 归零
RZNP Non-Polarized RZ 非极化归零制
RZP Polarized Return-to-Zero 极化归零

S

S　Scout　【美国】"侦察兵"火箭（国家航空航天局 / 小型一次运载火箭）

S&A　Safe and Arm　保险与解除保险；安全与备炸

S&A　Safety and Arming Mechanism　保险与解除保险机构；安全与备炸机构

S&A　Science and Application　科学与应用

S&A　Simulation and Analysis　仿真与分析

S&C　Stability and Control　稳定与控制

S&CS　Stabilization and Control System　稳定与控制系统

S&E　Science and Engineering　科学与工程

S&FM　Space and Flight Mission　航天飞行任务

S&IS　Space and Information System　航天与信息系统；航天与情报系统

S&M　Sequencer and Monitor　程序装置与监控器

S&M　Supply and Maintenance　供给与维修；补给与维修

S&MA　Safety and Mission Assurance　安全与任务保证

S&N　Space and Navigation　航天与导航

S&PU　Safety and Performance Upgrades　安全与性能升级

S&R　Safety and Reliability　安全与可靠性

S&R　Search and Rescue　搜索与救援

S&RE　Suspension and Release Equipment　悬挂与投放设备

S&RSAT　Search and Rescue Satellite Aided Tracking System　搜索救援卫星辅助跟踪系统

S&T　Science and Technology　科学技术

S&T　Supply and Transport　供应与运输

S&TI　Scientific and Technical Information　科学技术信息

S&TNF　Strategic and Theater Nuclear Forces　战略与战区核力量

S-IB　Saturn IB Launch Vehicle First Stage　【美国】"土星" IB 运载火箭一子级

S-IC　Saturn IC Launch Vehicle First Stage　【美国】"土星" IC 运载火箭一子级

S-IVB　Saturn IB Second Stage　【美国】"土星" IB 运载火箭二子级

S-MOWG　Super Management Operations Working Group　超级管理操作工作组

S/A　Safe/Arm　保险与解除保险

S/A　Safety/Arming　安全解保

S/A　Single Access　单个访问

S/A　Solar Array　太阳电池阵

S/A　Spacecraft Adapter　航天器适配器

S/A　Support Area　保障区

S/AC　Stabilization/Attitude Control　稳定性与姿态控制

S/AD　Sounder/Auxiliary Data　探测器 / 辅助数据

S/B	Standby 备用的；待机的；备用设备		存性/可操作性
S/C	Sensor/Controller 感应器/控制器	**S/O**	Switchover 变换；切换；转接
S/C	Signal Conditioner 信号调节器	**S/P**	Signal Processor 信号处理器（装置）
S/C	Spacecraft 航天器	**S/P**	Speed/Power Measurement Point 速度/功率测量点
S/C	Spacecraft/Capsule 航天飞机/密封座舱；航天器/密封舱	**S/R**	Send/Receive 发射与接收
S/C	Stabilization Aid Control 稳定辅助控制	**S/R**	Stimulus/Response 激励/反应
S/D	Signal-to-Distortion Ratio 信号失真比	**S/S**	Satellite/Space System 卫星/航天系统
S/D	Start Descent 开始下降	**S/S**	Signal Strength 信号强度
S/E	Single Engine 单发动机的	**S/S**	Space Shuttle【美国】航天飞机
S/E	Standardization/Evaluation 标准化/鉴定	**S/SEE**	Systems/Software Engineering Environment 系统/软件工程环境
S/EOS	Standard Earth Observation Satellite 标准地球观测卫星	**S/SI**	Simulation/System Infrastructure 模拟与系统基础设施
S/F	Safety Factor 安全系数	**S/SIA**	Shuttle/Space Lab Induced Atmosphere【美国】航天飞机－空间实验室诱导大气
S/F	Scheduled Flight 定期飞行	**S/SS**	Steering/Suspension System 转向和悬挂系统
S/L	Spacelab 天空实验室；航天实验室	**S/SU/AC**	System/System Upgrade/Advanced Concept 系统/系统更新/先进概念
S/M	Service Module 服务舱	**S/T**	Search/Track 搜索/跟踪
S/M	Service/Maintenance 勤务/维修	**S/TODS**	Strategic/Tactical Optical Disk System 战略/战术光学磁碟系统
S/M	Signal to Mean (Ratio) 信号与平均值（比）	**S/SVVT**	Software/System Verification and Validation Test 软件/系统验证与确认测试
S/M	Structural/Mechanical 结构的/机械的	**S/V**	Space Vehicle 航天飞行器
S/M	Surface to Mass Ratio 表面－质量比	**S/V**	Survivability/Vulnerability 抗毁力/易损性
S/M	Systems Management 系统管理	**S/V/L**	Survivability/Vulnerability/Lethality 抗毁能力/易损性/致命性
S/MM	Shuttle/Mir Mission 航天飞机/"和平号"空间站任务	**S/W**	Software 软件
S/N	Signal to Noise Ratio 信号噪声比；信噪比	**S³/E**	Space Based Kinetic Energy Weapon System Simulator/Emulator 天基动能武器系统模拟器/仿真器
S/N	Stress/Number of Cycles 应力/循环次数比		
S/O	Shutoff 停止，关闭；切断；断路		
S/O	Survivability/Operability 可生		

SA	Safe Area 安全区
SA	Safety Altitude 安全高度
SA	Saturn-Apollo 【美国】"土星"火箭与"阿波罗"飞船
SA	Servo Actuator 伺服传动机械；伺服作动器
SA	Signal Attenuation 信号衰减
SA	Situation Assessment 态势评估
SA	Situation Awareness 态势感知
SA	Software Assurance 软件保证
SA	Spacecraft Adapter 航天器（对接）适配器
SA	Staging Area 整装区
SA	Standard Agena 【美国国家航天局】标准型"阿森纳"火箭
SA	Standard Approach 标准进场（进近）
SA	Structural Analysis 【软件】结构分析
SA	Structural Analysis 结构（疲劳）分析
SA	Studies and Analysis 研究与分析
SA	Support Area 保障区
SA	Supportability Analysis 保障性分析
SA	System Analysis 系统分析
SA&I	System Architecture and Integration 系统体系结构与一体化
SA-CMM	Software Acquisition Capability Maturity Model 软件获得能力成熟度模型
SA/BM	Systems Analysis/Battle Management 系统分析/作战管理
SA/OR	Systems Analysis/Operational Research 系统分析/工作研究
SA/PDL	Strategic Defense Ada Process Description Language 战略防御 Ada 处理说明语言
SA/PAH	Sample Acquisition/Sample Processing and Handling 样本采集/样本加工与处理
SAA	Safety Assurance Analysis 安全保障分析
SAA	Satellite Access Authorization 卫星入网授权
SAA	Satellite Attitude Acquisition 卫星姿态探测
SAA	Saturn Apollo Applications 【美国国家航空航天局】"土星"运载火箭的"阿波罗"飞船应用
SAA	Servo Actuating Assembly/Servo Actuated Assembly 伺服机构作动装置
SAA	Space Act Agreement 【美国】航天法案协议
SAA	System Application Architecture 系统应用体系架构
SAA	System Assurance Analysis 系统保证分析
SAAHS	Stability Augmentation Altitude Hold System 稳定性增强与高度保持系统；增稳与姿态保持系统
SAAL	Signaling ATM Adaptation Layer 异步传送方式信令适配层
SAAM	Selective Availability/Anti-Spoofing Module 【全球定位系统】选择性利用/反（假脱机）欺骗组件
SAAM	Simulation Analysis and Modeling 模拟分析与建模试验
SAAP	Saturn-Apollo Application Program 【美国】"土星"火箭—"阿波罗"飞船应用计划
SAARIS	Surveys, Audits, Assessments, and Reviews Information System 勘察、审查、评估与评审信息系统
SAAS	Shuttle Aerosurface Actuator Simulation 航天飞机机面作动模拟器
SAAT	Satellite Attitude Acquisition

Technology 卫星姿态探测技术
SAAV System Approach Assistance Visit 系统分析法辅助访问
SAAWF Sector Anti-Air Warfare Facility 地段空防作战设施
SAB Satellite Assembly Building 【美国空军卫星控制网】卫星组装厂房
SAB Secondary Application Block 辅助应用程序块
SAB Shuttle Assembly Building 【美国】航天飞机组装厂房
SAB Shuttle Avionics Breadboard 【美国】航天飞机航空电子设备实验电路板
SAB Solid Assembly Building 【美国航天飞机】固体火箭发动机组装厂房
SAB Spacecraft Assembly Building 航天器组装厂房
SAB Storage and Assembly Building 储存与组装厂房
SABAR Satellite, Balloons and Rockets 卫星、气球和火箭
SABIR Satellite-Based Interceptor 卫星星载拦截器
SABLE Systematic Approach to Better Long Range Estimating 更好地进行长期估算的系统化方法
SABMIS Surface-to-Air Ballistic Missile Interception System 面对空弹道导弹拦截系统
SABMS Safeguard Anti-Ballistic Missile System 【美国】"卫兵"式反弹道导弹系统
SABR Self-Aligning Boost and Reentry 自调整助推与再入大气层
SABRS Self-Aligning Boost and Reentry System 自调整助推与再入大气层系统

SABRS Space and Atmospheric Burst Reporting System 空间与大气层爆炸通报系统
SAC Satellite De Applications Scientifics 【阿根廷】科学应用卫星
SAC Shuttle Action Center 【美国】航天飞机行动中心；航天飞机活动中心
SAC Space Applications Center 【印度】空间应用中心
SAC Strategic Air Command 【美国空军】战略空中指挥
SAC Support Action Center 保障行动中心
SACAD Stress Analysis and Computer Aided Design 应力分析和计算机辅助设计
SACC Suppression of Adversary Counterspace Capabilities 压制敌方以夺取太空优势的能力
SACCS Strategic Automated Command and Control System 战略自动化指挥与控制系统
SACE System Auxiliary Control Element 系统辅助控制单元
SACLOS Semi-Automatic Command to Line of Sight Guidance 瞄准线半自动指令制导（导弹）
SACMPS Selective Automatic Computational Matching and Positioning System 选择性自动计算匹配与定位系统
SACS Satellite Attitude Control Simulator 卫星姿态控制模拟器
SACS Software Avionics Command Support 航空电子设备软件指令支持
SACS Solar Attitude Control System 太阳基准姿态控制系统
SACS Synchronous Altitude Commu-

nication Satellite 地球同步高度通信卫星

SACS Systems Software Avionics Command Support 航空电子设备系统软件指令支持

SAD Safety, Arming and Destruct 保险、解除保险与自毁

SAD Shuttle Authorized Document 【美国】航天飞机授权文件

SAD Situational Awareness Display 态势感知显示

SAD Spacecraft Attitude Display 航天器姿态显示

SAD Systems Allocation Document 系统配置文件

SADA Solar Array Drive Assembly 太阳电池阵驱动组件

SADA Standard Advanced Dewar Assembly 标准先进"杜瓦"组件

SADAR Satellite Data Reduction 卫星数据归纳；卫星数据简化处理

SADARPS Satellite Data Reduction Processor System 卫星数据简化处理机系统

SADC Sequential Analogue-Digital Computer 时序模拟—数字计算机

SADE Solar Array Drive Electronics 太阳电池阵驱动电子设备

SADE Structural Assembly Demonstration Experiment 结构组件演示实验

SADEC Spin Axis Declination 旋转轴倾斜（偏斜）

SADIC Solid-State Analogue-Digital Computer 固态模拟—数字计算机

SADIIS Structured Analysis, Design, and Implementation of Information System 信息系统的结构分析、设计和实施

SADL Situational Awareness Data Link 态势感知数据链

SADM Solar Array Drive Mechanism 太阳电池阵驱动机械装置

SADOT Structures Assembly, Deployment and Operations Technology 结构装配、展开和操作技术

SADRG Semiactive Doppler Radar Guidance 半主动多普勒雷达制导

SADT Structured Analysis and Design Techniques 【软件】结构化分析与设计技术

SAE Site Acceptance Evaluation 发射场验收鉴定；发射场验收评估

SAE Solar Array Equipment 太阳电池阵设备

SAEB Spacecraft Assembly and Encapsulating Building 【美国】航天器装配和封装厂房

SAEF Spacecraft Assembly and Encapsulation Facility 【美国】航天器装配与封装设施

SAEF-2 Spacecraft Assembly and Encapsulation Facility 2 【美国】2号航天器装配与封装厂房

SAES Stand-Alone Engine Simulator 独立发动机模拟器

SAF Satellite Application Facilities 卫星应用设施

SAF Spacecraft Assembly Facility 航天器装配设施

SAFCS Self-Adaptive Flight Control System 自适应飞行控制系统

SAFD Systems Anomaly and Failure Detection 系统异常与故障检测

SAFE Solar Array Flight Experiment 【美国航天飞机】太阳电池阵飞行实验

SAFE System for Automated Flight Efficiency 自动飞行效能系统

SAFEA Space and Flight Equipment

Association 【美国】航天与飞行设备协会

SAFER Simplified Aid for EVA Rescue 简易舱外营救辅助装置

SAFER Simplified Aid for Extravehicular Activity Rescue 舱外活动救援的简易辅助设备

SAFIR Satellite for Information Relay 信息中继卫星

SAFIRE Spectroscopy of the Atmosphere Far Infrared Emission 大气远红外辐射光谱仪

SAGAT Situation Awareness Global Assessment Technique 态势感知全球评估技术

SAGE Semi-Automatic Ground Environment 半自动化地面环境

SAGE Sterilization Aerospace Ground Equipment 航空航天地面消毒设备

SAGE Stratospheric Aerosol and Gas Experiment Satellite 【美国】平流层气溶胶与气体实验卫星

SAGE-II Stratospheric Aerosol Gas Experiment 平流层气溶胶和气体实验

SAGGE Synchronous Altitude Gravity Gradient Experiment 同步高度重力梯度实验

SAIL Shuttle Avionics Integration Laboratory 【美国】航天飞机电子设备综合实验室

SAIL Synthetic Aperture Imaging Lidar 合成孔径成像激光雷达

SAIM System Analysis and Integration Model 系统分析与综合模型

SAINS Satellite Inspection System 卫星监视系统

SAINT Satellite Inspection and Interception 卫星监视与拦截

SAINT Satellite Inspector 卫星核查器

SAINT Satellite Interceptor 卫星拦截器

SAINT Shared Adaptive Internet Technology 共享自适应互联网技术

SAINT System Analysis of Integrated Networks of Tasks 多任务集成网络系统分析

SAIP Semi-Automated Imagery Processing 半自动化成像处理

SAIRST Situational Awareness Infrared Search and Tracking 态势感知红外搜索与跟踪

SAKT System Architecture and Key Tradeoffs 系统体系结构与关键折衷方案

SAL Satellite Applications Laboratory 卫星应用实验室

SAL Satellite Radio Line 卫星无线电线路

SAL Shuttle Avionics Laboratory 【美国】航天飞机航空电子设备实验室

SALAN Sensitivity Analysis of Linear Active Networks 线性主动网络敏感性分析

SALIN Satellite Library Information Network 卫星数据库信息网络

SALS Short Approach Light System 短距离进场照明系统

SALSA Spares Acquisition and Logistics Support Analysis 零备件采办与后勤保障分析

SAM Sensor Actuator Module 传感器执行器模块

SAM Shuttle Activation Monitor 【美国】航天飞机启动监测

SAM Shuttle Attachment Manipulator 【美国】航天飞机附加机械臂

SAM Situation Awareness Mode 态势感知模式

SAM Space Assembly and Maintenance 空间装配和维护

SAM Standard(ized) Assembly Module 标准装配模式组件

SAM Stratospheric Aerosol Measurement 平流层气溶胶测量仪

SAM System Activation and Monitoring 系统启动与监测

SAMARM Surface-to-Air Missile Routing and Maintenance System 地对空导弹航迹选择和保持系统

SAMEX Shuttle Active Microwave Experiment 【美国】航天飞机有源微波实验

SAMI Speed of Approach Measurement Indicator 进场速度测量指示器

SAMI System Acquisition Management Inspection 系统探测管理检查（系统）

SAMIDS Ship Antimissile Integrated Defense System 舰船反导弹综合防御系统

SAMIS Structural Analysis Matrix Interpretive System 结构分析矩阵解释系统

SAMM Software Acquisition Maturity Matrix 软件采购成熟度矩阵

SAMM Software Acquisition Maturity Model 软件采办成熟度模型

SAMMEX Space Active Modular Materials Experiment 空间有源模块式装备实验

SAMMI Signature Analysis Methods for Mission Identification 任务识别的特征分析法

SAMMIE Structural Analysis Model for Mission Integrated Experiments 任务综合实验结构分析模型

SAMMS Space and Missile Manpower Support 航天与导弹人力支援

SAMOPA Single Accelerator Master Oscillator-Power Amplifier 单加速器主振荡器－功率放大器

SAMOS Satellite and Missile Observation System 卫星和导弹观测系统（"萨摩斯"）

SAMOS Satellite and Missile Observatory Station 卫星和导弹观测站

SAMOS Satellite and Missile Observatory System 卫星和导弹观测系统

SAMP Shuttle Automated Mass Properties 【美国】航天飞机自动质量特性系统

SAMPEX Solar, Anomalous and Magnetosphere Particle Explorer 太阳和异常磁层粒子探测器（小型科学卫星）

SAMRO Satellite Militaried De Reconnaissance Optique 【法国】军事光学侦察卫星

SAMS Satellite Automatic Monitoring System 卫星自动监测系统

SAMS Shuttle Attached Manipulator System 【美国国家航空航天局】航天飞机附设的机械臂系统

SAMS Shuttle Automated Management System 【美国】航天飞机自动化管理系统

SAMS Software Automated Management Support 软件自动管理保障

SAMS Space Acceleration Measurement System 太空加速度测量系统

SAMS Space and Missile System 航天与导弹系统

SAMS Space Assembly, Maintenance and Servicing 空间装配、维修和

服务

SAMS Space Assembly, Maintenance Study 空间装配、维修研究

SAMS Spacecraft Assembly, Maintenance and Servicing Study 航天器装配、维修和服务研究

SAMS Stratosphere and Mesosphere Sounder 平流层与中间层探测器

SAMS-II Space Acceleration Measurement System-II 空间加速测量系统—II

SAMSAT Solar Activity Monitoring Satellites 太阳活动监视卫星

SAMSO Space and Missile Systems Organization 【美国空军】太空与导弹系统机构

SAMSOM Support Availability Multisystem Operational Model 支援可达性多系统操作模型；可用支援多系统操作模型

SAMSON System Analysis of Manned Space Operation 载人航天活动的系统分析

SAMSS Space Assembly Maintenance and Servicing Study 空间部件维修与检查研究

SAMTEC Space and Missile Test Center 【美国东/西靶场】航天与导弹试验中心

SAMTEC Space and Missile Test Evaluation Center 【美国空军】航天与导弹试验评估中心

SAMTECM Space and Missile Test Center Manual 太空与导弹测试中心手册

SAMTO Space and Missile Test Organization 【美国范登堡空军基地】太空与导弹测试机构

SAMSARS Satellite Aided Maritime Search And Rescue System 卫星辅助海上搜救系统

SANTA Systematic Analog Network Testing Approach 系统化模拟网络试验法

SAO Smithsonian Astrophysical Observatory 【美国】"史密森"天体物理观测台

SAOC Space and Astronautics Orientation Course 空间与宇宙航行定向航线

SAPIE Solar Array Plasma Interaction Experiment (ESA Study) 【欧洲空间局】太阳电池阵等离子相互作用实验

SAPL Shuttle Atmospheric Physics Laboratory 【美国】航天飞机大气物理实验室

SAPPSAC Spacecraft Attitude Precision Pointing and Slewing Adaptive Control 航天器姿态精确定位和回转自适应控制

SAR Safety Analysis Report 安全分析报告

SAR Safety Assessment Report 安全评估报告

SAR Satellite Anormaly Report 卫星异常报告

SAR S-Band Synthetic Aperture Radar S波段合成孔径雷达

SAR Single Axis Rotation 单轴旋转

SAR Site Acceptance Review 现场验收评审

SAR Software Anomaly Report 软件异常报告

SAR Spacecraft Acceptance Review 航天器验收评审

SAR Synthetic Aperture Radar 合成孔径雷达

SARA Sub-keV Volts Atomic Reflection Analyzer 亚千电子伏原子反

射分析仪

SARA System Availability and Reliability Analysis 系统可用性和可靠性分析

SARC Science Archive Research Center 科学档案研究中心

SAREX Shuttle Amateur Radio Experiment 【美国国家航空航天局】航天飞机业余无线电爱好者实验

SARL Subsonic Aerodynamic Research Laboratory 亚声速空气动力研究实验室

SARP Shuttle Astronaut Recruitment Program 【美国】航天飞机航天员招收项目

SARS Solar Array Reorientation System 太阳能电池阵重新取向系统

SARS Stellar Attitude Reference System 星体姿态参考系统

SARSS Search and Rescue Satellite System 搜索救援卫星系统

SARST Search and Rescue Satellite 搜索和救援卫星

SARV Satellite Aeromedical Research Vehicle 卫星航空医学研究运载器

SAS Satellite Automonitor System 卫星自动监控系统

SAS Scientific Application Satellite 科学应用卫星

SAS Small Astronomy Satellite 【美国】小型天文卫星

SAS Solar Acquisition Sensor 太阳能探测传感器

SAS Solar Array System 太阳能电池阵系统

SAS Space Activity Suit 航天工作服

SAS Space Adaptation Syndrome 太空适应性综合征（实验）（在航天飞机上测量长期失重情况下前庭功能、运动病易感性和空间定向能力的实验）

SAS Spacecraft Adapter Simulator 航天器适配装置模拟器

SAS Spacecraft Antenna System 航天器天线系统

SAS Stability Augmentation System (Subsystem) 稳定增强系统（子系统）

SAS Stellar Astronomy Satellite 恒星天文卫星

SAS Survivable Adaptive Systems 抗毁自适应系统

SAS System Analysis and Simulation 系统分析与模拟

SAS System Architecture Study 系统体系结构研究

SAS&R Satellite-Aided Search and Rescue 【印度】卫星辅助搜索救援系统

SAS/CSS Stability Augmentation System with Control Stick Steering 操纵杆操舵稳定性增大系统

SASE Shuttle Atmospheric Science Experiment 【美国】航天飞机大气科学实验

SASET Software Architecture Sizing and Estimating Tool 软件体系结构、规模与评估工具

SASP Science and Application Space Platform 科学和应用空间平台

SASR Satellite Aided Search and Rescue 卫星辅助搜索与救援系统

SASR Spare Available Space Required System 可用的备用航天需求系统

SASRS Satellite-Aided Search and Rescue System 卫星辅助搜索救援系统

SASS Saturn Automatic Software System 【美国】"土星"运载火箭自动软件系统

SASS Space Asset Support System 太空资源支援系统（包括天基支援平台和轨道转移飞行器）
SASTP Stand-Alone Self-Test Program 独立自测项目
SASY Satellite/SYLDA Separation System 卫星/SYLDA支架分离系统
SAT Studies, Analysis and Technology 研究、分析与技术
SAT Surveillance, Acquisition and Tracking 监视、捕获与跟踪
SAT Systems Approach to Training 进行培训的系统方法
SAT-AIS Satellite-Based Automatic Identification System 星载自动识别系统
SAT TRACGS Satellite Tracking Geolocation System 【美国】卫星跟踪地球定位系统
SATAF Shuttle Activation Task Force 【美国】航天飞机启动特遣队
SATAF Site Activation Task Force 【美国】发射场启动特遣队
SATAF Site Alteration Task Force 【美国】发射场改造特遣队
SATAN Satellite Automatic Tracking Antenna 【美国】卫星自动跟踪天线
SATAN Security Administrator Tool for Analyzing Networks 网络分析安全管理工具
SATAR Satellite for Aerospace Research 航空航天研究卫星
SATB Special Aptitude Test Battery 特殊能力倾向成套测验
SATCOM Satellite Communication System 卫星通信系统
SATCOM Satellite Communications 卫星通信
SATDAT Satellite Data Processing System 卫星数据处理系统
SATE Special Acceptance Test Equipment 专用验收测试设备
SATELAB Satellite Laboratory 卫星实验室
SATGCI Satellite Ground Controlled Interception 卫星地面控制截击
SATIC Scientific and Technical Information Center 【美国】科学技术信息中心
SATIN Satellite Inspection 卫星检测
SATKA Surveillance, Acquisition, Tracking and Kill Assessment 监视、截获、跟踪与杀伤效果评估（弹道导弹指挥、控制与通信）
SATMACS Spacecraft Assembly, Test, Monitor, and Control System 航天器组装、测试、监控与控制系统
SATMEX Satellite of Mexico 墨西哥（通信）卫星
SATNAV Satellite Navigation 卫星导航
SATO Self-Aligned Thick-Oxide (Technique) 自对准厚氧化物层（技术）
SATO Shuttle Attached Teleoperator 【美国】航天飞机上附加的遥控操纵器
SATO Space Adaptation Tests and Observation 空间适应试验与观测
SATO Supply and Transportation Operations 供应与运输操作
SATOBS Satellite Observations 卫星观测
SATODP Satellite Orbit Determination Program 卫星轨道测定程序
SATP Space Application Technology Program 航天应用技术项目
SATRAC Satellite Automatic Terminal Rendezvous and Coupling 卫星终

端自动交会与耦合

SATRAC Satellite Missile Tracking System 卫星导弹跟踪系统

SATRACK Satellite Tracking 卫星跟踪

SATRACK Satellite Tracking System 卫星跟踪系统

SATREC Satellite Technology Research Center 【韩国】卫星技术研究中心

SATS Satellite Antenna Test System 卫星天线测试系统

SATS Satellite Automated Test System 卫星自动化测试系统

SATS Shuttle Avionics Test System 航天飞机航空电子设备测试系统

SATS Small Applications Technology Satellite 小型应用技术卫星

SATSLAM Satellite-Tracked Submarinelaunched Antimissile 卫星跟踪潜艇发射的反导弹

SATVUL Satellite Vulnerability 卫星易损性

SAU Signal Acquisition Unit 信号采集装置

SAV Supportability Analysis Verification 保障性分析验证

SAVE Shuttle Avionics Verification and Evaluation 【美国】航天飞机航空电子设备验证与鉴定

SAVE Simulation Assessment Validation Environment 仿真评估确认环境

SAVE Situation Analysis and Vulnerability Estimate 形势分析与薄弱环节评估

SAVE System Analysis of Vulnerability and Effectiveness 对易损性和效能的系统分析

SAVER Stowable Aircrew Vehicle Escape Rotoseat 飞行器内乘员隐藏逃逸旋转椅

SAVES Sizing Aerospace Vehicle Structures 精整航空航天器结构

SAVI Space Active Vibration Isolation 空间主动振动隔离评估

SAW Satellite Attack Warning 卫星攻击报警

SAWAFE Satellite Attack Warning and Assessment Flight Experiment 卫星攻击报警与评估飞行实验

SAWS Satellite Attack Warning System 卫星攻击报警系统

SAWS Silent Attack Warning System 无声攻击报警系统

SAX Small X-Ray Satellite 小型X射线天文卫星；萨克斯天文卫星（意大利与荷兰合作研制）

SB Space Base 空间基地

SB Spacecraft Booster 航天器助推器

SBA Space-Based Assets 天基资产（设施）

SBA Standards Based Architecture 基于标准的体系架构

SBAEDS Satellite Based Atomic Energy Detection System 星载原子能探测系统

SBAMS Space-Based Anti-Missile System 天基反导弹系统

SBAR Space Based Infrared System 天基红外系统

SBAS S-Band Antenna Switch S波段天线切换

SBAS Space-Based Acquisition System 天基捕获系统

SBAS Space Based Architecture Study 天基系统结构体系研究

SBC Slave Bus Controller 辅助总线控制器

SBC Space Borne Computer 航天器星载计算机

SBCL Space-Based Chemical Laser 天基化学激光器

SBE Space-Based Element 天基元器件

SBE Synthetic Battlefield Environment 合成战场环境

SBEON Space Based Electro-Optical Network 天基光电网络

SBES Space-Based Experimental System 天基实验系统

SBFEL Space-Based Free Electron Laser 天基自由电子激光器

SBGCOS Space Based Global Change Observation Systems 天基全球变化观测系统

SBHE Space Based Hypervelocity Experiment 天基超高速实验

SBHEL Space Based High Energy Laser 天基高能激光器

SBHELS Space Based High Energy Laser 天基高能激光器系统

SBI Satellite Borne Instrumentation 星载仪器；星载测量设备

SBI Space-Based Interceptors 天基拦截器

SBICV Space Based Interceptor Carrier Vehicle 天基拦截弹运载器

SBILS Scanning Beam Instrument Landing System 波束扫描仪表着陆系统

SBIR Small Business Innovative Research 小型商业创新研究

SBIRS Space-Based Infrared System 天基红外系统

SBIRS GEO Space-Based Infrared System Geosynchronous Earth Orbit Satellite 天基红外系统地球同步轨道卫星

SBIRS HEO Space-Based Infrared System Infrared Sensors Hosted on Satellites in Highly Elliptical Orbits 天基红外系统大椭圆轨道卫星上装载的红外探测器

SBIRS HIGH Space-Based Infrared System High Altitude Component 天基红外系统高轨部分

SBIRS LEO Space-Based Infrared System Low Earth Orbit Satellites 天基红外系统近地轨道卫星

SBIRS Low SBIRS Low Altitude Component Consisting of SBIRS LEO Satellites 天基红外系统低轨道卫星组成的低空部分

SBIRS-High Space-Based Infrared System-High 天基红外系统－高轨

SBIS Satellite-Based Interceptor System 星载拦截系统

SBIS Space-Based Imaging Satellite 天基成像卫星

SBIS Space-Based Interceptor System 天基拦截机（导弹）系统

SBIS Sustaining Base Information Services 持续的基本信息服务；持续基地信息业务

SBKEW Space-Based Kinetic Energy Weapon 天基动能武器

SBKKV Space-Based Kinetic-Kill Vehicle 天基动能杀伤器

SBKV Space Based Kill Vehicle 天基杀伤飞行器

SBL Space-Based Laser 天基激光器

SBL/BMD Space-Based Laser/Ballistic Missile Defense 【美国】天基激光器/弹道导弹防御

SBLGS Scanning Beam Landing Guidance System 扫描波束着陆制导系统

SBLRD Space-Based Laser Readiness

Demonstrator 天基激光器成熟度演示验证装置
SBM　Space Based Mirror　天基反射镜
SBM　Strategic Ballistic Missile　战略弹道导弹
SBNPB　Space-Based Neutral Particle Beam　天基中性粒子束（武器）
SBNPBW　Space-Based Neutral Particle Beam Weapon　天基中性粒子束武器
SBP　Space Based Programmer　星载程序编制器
SBPB　Space-Based Particle Beam　天基粒子束（武器）
SBR　Signal-to-Background Ratio　信号背景噪声比
SBR　Space-Based Radar　天基雷达
SBRDC　Space-Based Range Demonstration and Certification　天基靶场的演示和认证
SBRF　Space-Based Radio Frequency　天基射频
SBRV　Small Ballistic Reentry Vehicle　小型弹道再进入飞行器
SBS　Satellite Business System　卫星商业系统
SBS　Systems Breakdown Structure　系统分类体系结构
SBSim　Space-Based Simulator　天基模拟器（装置）
SBSP　Single-Base Solid Propellant　单基固体推进剂
SBSS　Space-Based Space Surveillance　天基航天监视
SBSS　Space-Based Space Surveillance System　天基空间监视系统
SBSS　Space-Based Surveillance System　【美国】天基监视系统
SBSSO　Space-Based Space Surveillance Operation　【先进概念技术演示验证】天基空间监视作业
SBT　Simulation-Based Training　基于仿真的训练
SBV Sensor　Space-Based Visible Sensor　天基可视传感器
SBWAS　Space-Based Wide Area Surveillance　天基大面积监视卫星
SBWS　Space Based Warning System　天基预警系统
SBX　Sea-Based X-Band Radar　海基X波段雷达
SC　Simulation Center　模拟（仿真）中心
SC　Spacecraft　航天器
SC　Stress Compensated　应力补偿
SC　System Center　系统中心
SC　System Controller　系统控制器（员）
SC&CU　Signal Conditioning and Control Unit　信号调节与控制装置
SC/BM　System Concepts/Battle Management　系统概念（方案）/作战管理
SC/DoS　Space Commission/ Department of Space　【印度】航天委员会/航天部
SC/SM　Spacecraft System Monitor　航天器系统监视器
SCA　Sequence Control Assembly　序列控制组件
SCA　Shuttle Carrier Aircraft　【美国】航天飞机运输机；航天飞机驮运机
SCA　Simulation Control Area　模拟控制区
SCA　Software Communications Architecture　软件通信结构
SCA　Spacecraft Adapter　航天器适配器/航天器支架
SCA　Système de Contrôle d'Attitude/

Attitude Control System 【法语】姿态控制系统

SCA/SMS Shuttle Carrier Aircraft/Shuttle Mission Simulator 【美国】航天飞机运输机搭载航天飞机任务模拟器

SCACS Strategic Command and Control System 战略指挥与控制系统

SCADC Standard Central Air Data Computer 【美国】标准中央大气数据计算机

SCADE Signal Condition and Detection Electronics 信号处理与检测电子设备

SCADS Scanning Celestial Attitude Determination System 扫描法天体姿态测定系统

SCAI Space Command and Control architecture Infrastructure 航天指挥与控制基础结构

SCALE Space Checkout and Launch Equipment 航天测试发射设备

SCALP Suit, Contamination Avoidance and Liquid Protection 沾染规避和沾染渗透防护服

SCAMP Space Controlled Array Measurements Probe 航天控制基阵测量探测器

SCAN Antenna Scan 天线扫描

SCAN Self-Contained Automatic Navigation (System) 自主式自动导航（系统）

SCAN Shuttle Connector Analysis Network 【美国】航天飞机连接器分析网络

SCAN Space Communications and Navigation 空间通信和导航

SCAN NASA Spacecraft Communications and Navigation 美国国家航空航天局航天器通信与导航

SCANS Scheduling and Control Automation by Network System 用网络系统实现的调度与控制自动化

SCANS System Checkout Automatic Network Simulator 系统检测自动网络模拟器

SCAP Shuttle Configuration Analysis Program 【美国】航天飞机结构分析项目

SCAPE Self-Contained Atmospheric Personnel Ensemble 自备式人员大气防护服

SCAPE Self-Contained Atmospheric Protective Ensemble 自备式大气全套防护服

SCAPS Site Characterization and Analysis Penetrometer System 场地特征与分析透度系统

SCAR Satellite Capture and Retrieval 卫星捕捉与回收；卫星截获与回收

SCAR Spacecraft Assessment Report 航天器评定报告

SCARAB Scanner for Radiation Budget 【欧洲环境卫星上】放射收支扫描仪

SCARE Sensor Control Antianti-Radiation Missile Radar Evaluation 传感器控制反反辐射导弹雷达鉴定

SCARLET Solar Concentrator Arrays with Refractive Linear Element Technology 具备折射线性组件技术的太阳能聚能器阵列

SCARS Serialized Control and Record System 序列化控制与记录系统

SCARS Serialized Control and Reporting System 序列化控制与通报系统

SCAS Spacecraft Adapter Simulator 航天器适配装置模拟器

SCAS Subsystem Computer Applica-

tion Software 分系统计算机应用软件

SCAT Satellite Communications Airborne Terminal 卫星通信机载终端设备

SCAT Space Communications and Tracking 航天通信和跟踪

SCAT Spacecraft and Assemblies Transfer 卫星组装与转运设施

SCAT Speed Command of Attitude and Thrust 姿态和推力的速率指令

SCAT Storage, Checkout, and Transport 存储、检测与运输

SCAT System Calibration and Alignment Tests 系统校准与定位测试

SCATE Space Chamber Analyzer-Thermal Environment 航天舱分析仪－热环境

SCATHA Spacecraft Charging at High Altitudes 【美国军事研究卫星】高度轨道充电试验卫星

SCATS Simulation Checkout and Training System 【美国国家航空航天局】模拟检查与训练系统

SCATS Simulation Control and Training System 模拟控制与训练系统

SCATS Suspense Control and Automated Tracking System 悬浮控制与自动跟踪系统

SCBA Self-Contained Breathing Apparatus 自持式呼吸装置

SCC Safety Control Center 安全控制中心

SCC Security Consultative Committee 安全咨询委员会

SCC Security Control Center 安全控制中心

SCC Simulation Control Center 模拟控制中心

SCC Site Command Center 发射场指挥中心

SCC Space Communications Center 航天通信中心

SCC Space Computational Center 航天计算中心

SCC Space Control Center 【美国航天司令部】航天控制中心

SCC Spacecraft Control Centre 航天器控制中心

SCC Stress Corrosion Cracking 应力腐蚀裂纹

SCCC Satellite Circuit Control Center 卫星线路控制中心

SCCC Satellite Communication Control Center 【英国】卫星通信控制中心

SCCC Service Component Command Center 勤务分队指挥中心

SCCC Space Surveillance Command and Control Center 空间监视系统指挥控制中心

SCCE Satellite Configuration Control Element 卫星配置控制单元（组件）

SCCF Solar Cell Calibration Facility 太阳能电池校准设施

SCCS Signal Conditioning and Control System 信号调节与控制系统

SCCS Space Communication and Control System 航天指挥与控制系统

SCD Satellite de Coleta de Dados 【巴西】数据采集卫星

SCD Space Control Document 航天控制文件

SCDA Software Critical Design Audit 软件关键设计审查

SCDA Supervisory Control and Data Acquisition 监督控制与数据采集

SCDL Surveillance and Control Data Link 监视与控制数据链

SCDP Simulation Control Data Pack-

age 模拟控制数据包
SCDR Shuttle Critical Design Review 【美国】航天飞机关键设计评审
SCDR Software Critical Design Review 软件关键设计评审
SCDU Signal Conditioning and Display Unit 信号调节与显示装置
SCE Servo Control Electronics 伺服控制电子设备
SCE Signal Conditioning Equipment 信号调节设备
SCE Software Capability Evaluation 软件能力评估
SCE Spacecraft Command Encoder 【美国国家航空航天局】航天器指令编码装置
SCE Stabilization and Control Equipment 稳定与控制设备
SCE Station Control Electronics 空间站控制电子设备
SCE System Cost Estimate 系统成本估计
SCE/SCVM S/C Command Encoder/Shuttle Command Voice Multiplexer 航天器解码器/航天飞机指令语音多路器
SCEA Signal Conditioning Electronics Assembly 信号调节电子设备组件
SCEET Support Concept Economic Evaluation Technique 保障方案经济性评估技术
SCEPS Self-Contained Environmental Protective Suits 自主式环境防护衣
SCEPS Stored Chemical Energy Propulsion System 储存化学能推进系统
SCEPTR Suitcase Emergency Procedures Trainer 便携式应急程序练习器
SCET Satellite Control Earth Terminal 卫星控制地球终端站
SCF Satellite Control Facility 卫星控制设备；卫星辅助控制设备
SCF Satellite Control Function 卫星控制功能
SCF Seeker Characterization Flight 寻的器性能鉴定飞行
SCF Spacecraft Checkout Facility 航天器检测厂房
SCF Spacecraft Control Facility 航天器控制设备；航天器控制厂房
SCF Stack and Checkout Facility 装配和检测厂房
SCF Stress Concentration Factor 应力集中因数
SCF Sunnyvale Control Facility 【美国】森尼维尔空军基地控制中心
SCG Space Charge Grid 太空电荷栅格
SCGSS Super Critical Gas Storage System 超临界气体贮存系统
SCH Sonobuoy Cable Hold 声呐浮标电缆舱
SCHE Stowage Cargo Handling Equipment 货物配载装卸设备
SCHMOO Space Cargo Handler and Manipulator for Orbital Operation 航天货物搬运和轨道运转机械臂
SCI Scalable Coherent Interface 可变规模互连接口
SCI SES Cockpit Interface 【美国】航天飞机工程系统座舱接口
SCI Special Compartmented Information 特种分类处理信息
SCI Stress Corrosion Index 应力腐蚀指数
SCIF Sensitive Compartmented Information Facility 调度机密分类情报资料设施
SCIF Special Compartmented Infor-

mation Facility 特别分类信息处理站

SCIMITAR System for Countering Interdiction Missile and Target Radars 反遮断导弹与目标截获雷达系统

SCIRT System Control in Real Time 实时系统控制

SCIS Survivable Communications Integration System 抗毁通信综合系统

SCIT Systems Concept Integrated Technology 系统概念（方案）综合技术

SCIU Selector Control Interface Unit 选择器控制接口装置

SCL Secondary Coolant Line 辅助冷却线路

SCL Secondary Coolant Loop 辅助冷却回路

SCL Space Component Lifetime 空间元器件寿命

SCL Systems Control Language 系统控制语言

SCLIGFET Space Charge Limited Insulated-Gate Field Effect Transistor 空间电荷限制绝缘栅场效应晶体管

SCLP Snow and Cold Land Processes 雪与寒区陆面过程观测计划

SCLU Simulated Command Launch Unit 模拟指挥发射单位

SCLVA Spacecraft Launch Vehicle Adapter 航天器/运载火箭适配器

SCM Service Command Module 维护指令舱；服务命令模块

SCM Software Configuration Management 软件配置管理

SCM Spacecraft Materials 航天器材料

SCM System Control Module 系统控制舱

SCMC Satellite Communication Monitoring Center 卫星通信监控中心

SCN Satellite Control Network 卫星控制网

SCN Self-Contained Navigation 自主导航

SCN Space Communications Network 空间（航天）通信网络

SCNS Self-Contained Navigation System 自主式导航系统；自备式导航系统；独立导航系统

SCO Spacecraft Operation 航天器操作

SCO Super Critical Oxygen 超临界氧

SCOE Satellite Check-Out Equipment 卫星测试设备

SCOE Special Checkout Equipment 专检设备

SCOLE Spacecraft Control Laboratory Experiment 航天器控制实验室实验

SCOM Spacecraft Communicator 航天器通信装置

SCOMO Satellite Collection of Meteorological Observation 卫星搜集气象观测资料

SCOMP Secure Communications Processor 保密通信处理机

SCOPE Science Calibration and Onboard Performance Evaluator 科学校准和机载性能鉴定器

SCOPE Spacecraft Operational Performance Evaluation 航天器工作性能评估

SCOPE System and Component Operating Performance Evaluation 系统和部件运行性能鉴定

SCOPE System for Capability, Orders Planning and Enquiries 能力、指令计划和咨询系统

SCOPE System to Coordinate the Operation of Peripheral Equipment 外围设备操作协调系统
SCOR Self-Calibrating Omnirange 自校准全向信标
SCORE Scientific Cooperative Research Exchange 科学合作研究交流
SCORE Signal Communication by Orbiting Relay Equipment (Experiment) 【美国】轨道中继信号通信实验卫星（斯科尔卫星）
SCORE Space Communication for Orbiting Relay Equipment 轨道中继设备的空间通信
SCORE System Cost and Operational Resources Evaluation 系统成本和运行资源评估
SCORE Systematic Control of Range Effectiveness 靶场效能的系统控制
SCOS Spacecraft Control and Operations System 航天器控制与操作系统
SCOS Subsystem Computer Operating System 分系统计算机操作系统
SCOT Satellite Communications Overseas Transmission 海外卫星通信传输
SCOT Supplementary Checkout Trailer 【导弹】辅助检测拖车
SCOTDT Synchronous Continuous Orbital Three Dimensional Tracking 同步连续轨道三维跟踪
SCOTS Shuttle Compatible Orbital Transfer Subsystem 【美国】航天飞机兼容轨道转移分系统
SCOTS System Checkout Test Set 系统检测试验设备
SCP Scanner Control Power 扫描装置控制功率
SCP Spacecraft Command Processor 航天器指令处理机
SCP Spacecraft Platform 航天器平台
SCP Survey Control Point 测量控制点
SCPA Solar Cell Panel Assembly 太阳能电池板组件
SCPS Satellite Communications Protocol Standards 卫星通信协议标准
SCPS Servo-Controlled Positioning System 伺服控制定位系统
SCPS Survivable Collective Protection System 【美国空军】抗毁集体防护系统
SCR Solar Corpuscular Radiation 太阳微粒辐射
SCR Solar Cosmic Radiation 太阳宇宙辐射
SCR Solar Cosmic Ray 太阳宇宙射线
SCR Spacecraft Control Room 航天器控制间
SCR Spacecraft Receiving Time 航天器接收时间
SCR System Capability Review 系统能力评审
SCR System Concept Review 系统方案评审；系统概念评审
SCRAM Self Contained Robotic Acceleration Machine 自主式机器人加速机
SCRAM Space Capsule Regulator and Monitor 航天密封舱调节器与监测器
SCRAM Station Crew Return Alternative Module 【美国】空间站乘员返回舱
SCRAMLACE Supersonic Combustion Ramjet Liquid Air Cycle Engine 超声速燃烧冲压式液态空气循环发

动机

SCRC Satellite Communications Research Center 卫星通信研究中心

SCREWS Solar Cosmic Ray Early Warning System 太阳宇宙射线预警系统

SCRG Stationary Cosmic Ray Gas 稳态宇宙射线气体

SCRIM Supersonic Cruise Intermediate Range Missile 超声速中远程巡航导弹

SCRJ Supersonic Combustion RamJet 超声速燃烧冲压发动机

SCRR Supercircular Reentry Research 超圆形轨道速度再入（大气层）研究

SCS Satellite Communication System 卫星通信系统

SCS Satellite Control System 卫星控制系统

SCS Satellite Test Center Communications Subsystem 卫星试验中心通信分系统

SCS Secondary Coolant System 辅助冷却系统

SCS Simulation Control Subsystem 模拟控制子系统

SCS Space Cabin Simulator 航天器舱模拟器

SCS Space Collaboration System 航天协作系统；空间合作系统

SCS Space Command Station 航天指挥站

SCS Space Communication Satellite 【日本】航天通信卫星

SCS Space Communications and Sensors 航天通信与传感器系统

SCS Spacecraft Control System 航天器控制系统

SCS Spacecraft System 航天器系统

SCS Speed Control System 速度控制系统

SCS Stabilization and Control System (Subsystem) 稳定与控制系统（分系统）

SCS Stabilization Control System 稳定控制系统

SCS Standard Coordinate System 标准坐标系统

SCS Survivable Communication Satellite 抗毁通信卫星

SCS System Configuration Switch 系统配置切换

SCSC Satellite Communications System Control 卫星通信系统控制

SCSI Small Computer System Interface 小型计算机系统接口

SCSN Space Center for Satellite Navigation 卫星导航航天中心

SCSRS Shoe Cove Satellite Receiving Station 舒科夫卫星接收站

SCSS Satellite Control Simulation System 卫星控制模拟系统

SCT Satellite Control Terminal 卫星控制终端

SCTS Satellite Control and Test Station 卫星控制与试验站

SCTS Space Flight Operations Facility Communications Terminal Subsystem 航天飞行操作设施通信终端子系统

SCTV/GDHS Spacecraft TV Ground Data Handling System 航天器电视地面数据处理系统

SCU Satellite Communication Unit 卫星通信装置

SCU Secondary Control Unit 辅助控制装置

SCU Sequence Control Unit 序列控制装置

SCU Service and Cooling Umbilical

勤务和制冷脐带管线
SCU Signal Conditioning Unit 信号调节装置
SCU Signal Control Unit 信号控制装置
SCU Suit Cooling Unit 航天服冷却装置
SCU System Control Unit 系统控制装置
SCUB Satellite Central Utility Building 卫星中央动力厂房
SCUD Scud Surface-to-Surface Missile System 【苏联】"飞毛腿"地对地导弹系统
SCV Solar Constant Variations 太阳常数变化（探测仪）
SCV Space Construction Vehicle 太空建造飞行器
SCVE Spacecraft Vicinity Equipment 航天器紧接设备
SCWS Space Combat Weapon System 太空作战武器系统
SD Self-Destroying 自毁的
SD Site Defense 场区防御
SD Structured Design 【软件】结构化设计
SD System Design 系统设计
SD System Diagrams 系统图
SD System Display 系统显示
SD/FS Smoke Detector/Fire Suppression 烟探测／灭火
SDA Satellite Data Area 卫星数据区
SDA Separation Diversion Assembly 分离导流装置
SDA Strategic Defense Architecture 战略防御结构
SDAC System Data Analog Converter 系统数据模拟变换器
SDACS Solid Divert and Attitude Control System 固定转向与姿态控制系统
SDAD Satellite Digital and Analog Display 卫星数字和模拟显示
SDADS Satellite Digital and Analog Display System 卫星数字和模拟显示系统
SDADS Satellite Digital and Display System 卫星数字显示系统
SDAF Solid Rocket Booster Disassembly Facility 固体火箭助推器拆卸设施
SDAIP System Description, Analysis, and Implementation Plan 系统描述、分析和执行计划
SDAS Scientific Data Analysis System 科学数据分析系统
SDAS Source-Data Automation System 源数据自动化系统
SDAU Safety Data and Analysis Unit 安全数据与分析装置
SDB Satellite Communications Database 卫星通信数据库
SDB Space Data Base 航天数据库
SDC Satellite Data Collectors 卫星数据收集器
SDC Self-Designing Controller 自设计控制器
SDC Software Development Center 软件研发中心
SDC Software Development Computer 软件研发计算机
SDC Source Data Collection 源数据采集
SDC Space Defense Center 空间防御中心（属美国空军航空航天防御司令部）
SDC Space Shuttle Data Center 【美国】航天飞机数据中心
SDC Strategic Defense Command 战略防御司令部

SDC Structural Design Criteria 结构设计准则

SDCA Software Development Capability Assessment 软件开发能力评估

SDCC Space Defense Control Center 空间防御控制中心

SDCC Strategic Defense Command Center 战略防御指挥中心

SDCE Software Development Capability Evaluation 软件研发能力鉴定

SDCS Spaceborne Data Conditioning System 星载数据调整系统

SDCU Satellite Delay Compensation Unit 卫星延迟补偿器

SDCV Shuttle Derived Cargo Vehicle 【美国】航天飞机衍生型载货飞船

SDD Shuttle Design Directive 【美国】航天飞机设计指令

SDD Software Description Document 软件说明文件

SDD Software Design Document 软件设计文件

SDD System Design Document 系统设计文件

SDD System Development and Demonstration 系统研制与演示验证

SDDM Software Design Description Model 软件设计描述模型

SDDS Satellite Data Distribution System 卫星数据分配系统

SDDS Simulator Device Development System 【美国空军】卫星控制网模拟装置仿真系统

SDDSS Serial Distributed Decision Support System 串行分布式决策支持系统

SDE Software Development Environment 软件开发环境

SDF Saving and Deserving Facility 保存和待役设施（美国航天飞机轮休期间暂时存放处）

SDF Self-Destruct Fuse 自毁引信

SDF Software Development Facility 软件研发设施

SDF System Development Facility 系统研发设施

SDFC Space Disturbance Forecast Center 【美国】空间扰动预报中心

SDFS Spacelab Data Flow System 【美国】"空间实验室"数据流系统

SDH Software Development Handbook 软件研发手册

SDH System Definition Handbook 系统定义手册

SDH System Development Handbook 系统研发手册

SDHE Spacecraft Data Handling Equipment 航天器数据处理设备

SDHS Satellite Data Handling System 卫星数据处理系统

SDI Strategic Defense Initiative 【美国】战略防御倡议（即"星球大战"计划）

SDI Strategic Defense Initiative 战略防御倡议

SDIF Software Development and Integration Facility 软件开发与集成设施

SDIGNS Strapdown Inertial Guidance and Navigation System 捷联式惯性制导与导航系统

SDIN Space Defense Interface Network 太空防御接口网

SDL Software Development Laboratory 软件研发实验室

SDL Space Dynamics Laboratory 太空动力学实验室

SDLC Synchronous Data Link Control

同步数据链控制
SDLS Satellite Data Link Standard 卫星数据链路标准
SDLS Shuttle Derived Launch System 【美国】航天飞机衍生发射系统
SDLV Shuttle-Derived Launch Vehicle 【美国】航天飞机衍生运载器
SDM Satellite Data Management 卫星数据管理
SDM Secondary Deployment Mechanism 辅助展开机械装置
SDM Shuttle Data Management 【美国】航天飞机数据管理
SDM Software Design Method 软件设计方法
SDM Space Division Multiplexing 空间分区多路技术
SDM Standard Dynamics Model 标准动态模型
SDM System Development Multitasking 系统发展多任务处理
SDMF Space Disturbance Monitoring Facility 空间扰动监视设施
SDMS SCSI Device Management System 小型计算机系统接口设备管理系统
SDMS Software Development Maintenance System 软件研制维修系统
SDMS Spatial Data Management System 空间数据管理系统
SDMS Support Defence Missile System 支持防卫导弹系统
SDMU Serial Data Management Unit 序列数据管理装置
SDMX Space Division Matrix 太空分界线矩阵
SDN Synchronized Digital Network 同步化数字网
SDO Solar Dynamics Observatory 太阳动力学天文台；太阳动力学观测
SDO Space Defense Operation 空间防御作战
SDO Spatial Disorientation 空间方向知觉的丧失
SDOC Space Defense Operations Center 空间防御作战中心
SDOC Stopping Drift Orbit Correction 【全球定位系统】阻止漂移轨道修正
SDOF Synthesis Design Optimization Framework 综合设计优化环境
SDP Shuttle Data Processor 【美国】航天飞机数据处理机
SDP Site Data Processor 场区数据处理器（装置）
SDP Software Development Plan 软件研发计划
SDP System Development Phase 系统研制阶段
SDP System Development Plan 系统开发计划
SDP System-Definition Phase 系统定义阶段
SDPC Shuttle Data Processing Complex 【美国】航天飞机数据处理综合设备
SDPE Special Design Protective Equipment 特殊设计保护性设备
SDPF Sensor Data Processing Facility 感应器数据处理设施
SDR Signal-to-Distortion Ratio 信号失真比率
SDR Software Design Requirements 软件设计要求
SDR Software Design Review 软件设计评审
SDR Splash Detection Radar 溅落点探测雷达
SDR System Definition Review 系统

定义评审

SDR System Design Requirements 系统设计要求

SDR System Design Review 系统设计评审

SDR System Development Requirement 系统研制需求

SDRD Supplier Data Requirements Description 供应商数据要求说明

SDRD Supplier Documentation Review Data 供应商文件编制评审数据

SDRN Satellite Data and Relay Network 卫星数据与中继网络

SDRN Satellite Data Relay Network 卫星数据中继网

SDRS Satellite Data Relay System 卫星数据中继系统

SDRS Satellite Digital Radio Services 卫星数字无线电服务

SDRT Software Development Risk Taxonomy 软件开发风险分类

SDRWTD Strategic Debriefer Report Writer Training Device 战略报告写作训练器

SDS Satellite Data System 卫星数据系统

SDS Satellite Dissemination System 卫星分发系统

SDS Scientific Data System 科学数据系统

SDS Sensor Display System 感应器显示系统

SDS Servo Drive System 伺服传动系统

SDS Shuttle Dynamic Simulation 航天飞机动力模拟（装置）

SDS Simulation Data Subsystem 模拟数据子系统

SDS Software Design Specification 软件设计说明

SDS Software Development System 软件开发系统

SDS Solar Dynamical Satellite 太阳能动力卫星

SDS Spacecraft Data Simulator 航天器数据模拟器

SDS Strategic Defense System 战略防御系统

SDS Structural Dynamics Simulation 结构动力学模拟

SDS Surveillance Detection/Direction System 救援侦察或指挥系统

SDS-CC Strategic Defense System-Command Center 战略防御系统—指挥中心

SDS-OC Strategic Defense System-Operations Center 战略防御系统—操作中心

SDSC-SHAR Satish Dhawan Space Centre, Sriharikota 【印度】萨迪什·达万航天中心

SDSD Strategic Defense System Description 战略防御系统说明

SDSMS Self-Defence Surface Missile System 自卫表层导弹系统

SDSP Space Defense System Program 空间防御系统计划

SDSS Satellite Data System Spacecraft 卫星数据系统航天器

SDSS Self-Deploying Space Station 自展开空间站

SDSS Space Division Shuttle Simulator 【美国】空间部航天飞机模拟器（装置）

SDT Skylab Data Task 【美国】"天空实验室"数据任务

SDT Space Detection and Tracking 空间探测和跟踪

SDT Structural Development Test 结

构开发试验

SDT Structural Dynamic Test 结构动力测试

SDTA Structural Dynamic Test Article 结构动力测试部件

SDTI System Design and Technology Integration 系统设计与技术集成（一体化）

SDTM System Design Tradeoff Model 系统设计综合权衡模型

SDTN Space and Data Tracking Network 航天数据跟踪网

SDTN Synchronous Digital Transmission Network 同步数字传播网络

SDTS Satellite Data Transmission System 卫星数据传输系统

SDTS Space Detection and Tracking System 太空探测与跟踪系统

SDTSC Space Detection and Tracking System Center 空间探测与跟踪系统中心

SDTSI Space Detection and Tracking System Improved 改进型空间探测与跟踪系统

SDTSS Space Detection and Tracking System Sensor 空间探测与跟踪系统传感器

SDU Spacecraft Data Unit 航天器数据装置

SDU Status Display Unit 状态显示装置

SDUS Secondary Data-User Stations 二次数据用户站（欧洲气象卫星图像收发站）

SDV Shuttle Derived Vehicle 【美国】航天飞机衍生飞行器

SDV Space Delivery Vehicle 空间运输飞行器

SDVF Software Development and Verification Facilities 软件开发与验证设备

SDVS Software Design and Verification System 软件设计验证系统

SDVS Space Data Verification System 航天数据验证系统

SDVS State Delta Verification System 状态增量验证系统

SDX Satellite Data Exchange 卫星数据交换机

SE Software Engineering Environment 软件工程环境

SE Space Experiment 太空实验

SE Space Exploration 空间探索

SE Support Equipment 保障设备

SE System Effectiveness 系统有效性

SE System Element 系统元件

SE System Engineering 系统工程

SE&I Systems Engineering and Integration 系统工程与集成

SE-CMM System Engineering Capability Maturity Models 系统工程性能完备模型

SE-GAS Sustainer Engine Gimbal Actuator System 主发动机摆动作动器系统

SE/FAC Support Equipment/Facility 支撑设备/设施

SE/TA Systems Engineering Technical Assistance 系统工程技术辅助

SEA Scanning Electrostatic Analysis 扫描静电分析

SEA Sensor Electronics Assembly 感应器电子设备组件

SEA Strategic Enterprise Architecture 战略计划体系结构

SEA System Effectiveness Analyzer 系统效力分析者（器）

SEA System Engineering and Integration 系统工程与集成

SEACF Support Equipment Assembly and Checkout Facility 支撑设备组装与检测厂房

SEADS Shuttle Entry Air Data Sensor 【美国】航天飞机进入大气层数据传感器

SEADS Shuttle Entry Air Data System (Subsystem) 【美国】航天飞机进入大气层数据系统（分系统）

SEADS Survivable and Effective Air Defense System 生存能力强的有效防空系统

SEADUCER Steady-State Evaluation and Analysis of Transducers 换能器稳态评估与分析；转换器稳态评估与分析

SEALAR SEA Launch and Recovery 【多级、液体火箭】海上发射与回收（系统）

SEALS Severe Environmental Air Launch Study 恶劣环境下空中发射研究

SEAPAC Sea-Activated Parachute Automatic Crew Release 乘员释放的海上自动启动降落伞

SEARCH Standoff Detection Early Warning Agents of Biological Origin, Radiological Chemical System 对生物、辐射、化学源的远距探测和早期警报系统

SEARCHS Shuttle Engineering Approach/Rollout Control Hybrid Simulation 【美国】航天飞机工程方法/滑出控制混合模拟

SEAS System, Engineering, and Analysis Support 系统、工程和分析支持

SEAT Site-Equipment Acceptance Test 现场设备验收实验

SEAT Status Evaluation and Test 状态鉴定与实验

SEB Scientific Equipment Bay 科学设备舱

SEB Support Equipment Building 保障设备厂房；支撑设备机房

SEBO Systems Engineering Behavioral Objectives 系统工程行为目标

SEC Single Engine Centaur 【美国】单发动机型"半人马座"火箭

SEC Space Environmental Center 航天环境中心

SEC Space Environmental Chamber 太空环境室

SECAM Systems Engineering Capability Assessment Model 系统工程能力评估模型

SECM Systems Engineering Capability Model 系统工程能力模型

SECMM Systems Engineering Capability Maturity Model 系统工程能力成熟度模型

SECO Second Stage Engine Cutoff 二子级发动机关机

SECO Sustainer Engine Cutoff 主发动机关机

SECOR Sequential Collation of Range 【美国】距离按序校核卫星（西科尔卫星）

SECR Standard Embedded Computer Resource 标准嵌入式计算机资源

SECS Sequential-Events Control System 顺序事件控制系统

SECS Shuttle Events Control Subsystem 【美国】航天飞机事件控制子系统

SECS Small Experimental Communications Satellite 【美国】小型实验通信卫星

SECS Space Environmental Control System 太空环境控制系统

SECT Simulator for Electronic Combat Training 电子化战争演习模拟器

SED Scramjet Engine Demonstrator 超燃冲压发动机验证机

SEDD Systems Engineering Development Data Base 系统工程研发数据库

SEDFB Surface-Emitting Distributed Feedback 表面发射分布式反馈

SEDIS Surface-Emitter Detection Identification System 地面发射体探测识别系统（电子支援措施）

SEDS Small Expendable Deployer System 【美国"德尔它"火箭】小型一次性施放系统

SEDS Space Electronic Detection System 航天电子探测系统

SEDS Space Electronics Detection System 航天电子设备探测系统

SEDS System Effectiveness Data System 系统效能数据系统

SEE SAGE Evaluation Exercise 自动地面防空系统评估演习

SEE Software Engineering Environment 软件工程环境

SEEDS Space Exposed Experiment Developed for Students 为学生设计的空间暴露实验

SEEL Space Energy and Environment Laboratory 【日本】空间能量和环境实验室

SEER Sensor Equipment Evaluation and Review 传感器设备鉴定与评审

SEER Sensor Experimental Evaluation Review 传感器实验鉴定评审

SEER Systems Engineering Evaluation and Research 系统工程评价与研究

SEEX Systems Evaluation Experiment 【火箭】系统鉴定实验

SEF Space Education Foundation 航天教育基金会

SEF Space Environmental Facility 空间环境模拟设施

SEFC Space Environment Forecast Center 空间环境预报中心

SEFI Smart Electronic Flight Indictor 智能电子飞行指示器

SEFSP Space Experiment and Flight Support Program 航天实验与飞行保障计划

SEG Space Shuttle Entry Guidance 【美国】航天飞机进入导航

SEI Satellite Equipment Interface 卫星设备接口

SEI Space Exploration Initiative 【美国】空间探索倡议；太空探索倡议

SEI Support Equipment Installation 保障设备安装

SEI System Engineering Instrumentation 系统工程仪器仪表

SEIC System Effectiveness Information Control 系统效能信息控制

SEICO Support Equipment Installation and Checkout 支撑设备安装与检查；保障设备安装与检查

SEID Spacelab Experiment Interface Device 天空实验室实验接口装置

SEII Servoactuator Error Indication Interrupt 伺服执行器误差显示中断

SEIT Systems Engineering Integration and Test 系统工程一体化与试验（测试）

SEITE Shuttle Exhaust Ion Turbulence Experiment 【美国】航天飞机排气离子湍流实验

SEL Software Engineering Laboratory

【美国国家航空航天局】软件工程实验室

SEL Space Environment Laboratory 空间环境实验室

SEL1 Sun-Earth Lagrange: Interior to Earth's Orbit 日—地拉格朗日：地球轨道内部

SEL2 Sun-Earth Lagrange: Exterior to Earth's Orbit 日—地拉格朗日：地球轨道外部

SELDADS Space Environment Laboratory Data Acquisition and Display System 【美国】空间环境实验室数据采集和显示系统

SELSIS Space Environment Laboratory Solar Imaging System 空间环境实验室太阳成像系统

SELV Small Expendable Launch Vehicle 小型一次使用运载火箭

SELVS Small Expendable Launch Vehicle Service 小型一次使用火箭发射业务

SEM Sample Exchange Mechanism 样品交换机械装置

SEM Scanning Electron Microscope 扫描式电子显微镜

SEM Space Environment Module 空间实验模块；空间实验舱

SEM Space Environment Monitoring 空间环境监测

SEM Spacecraft Equipment Module 航天器设备舱

SEM Spacelab Engineering Model 太空实验室工程模型

SEM Standard Error of the Mean 平均值标准误差

SEM System Effectiveness Model 【全球定位系统】系统效能模型

SEM System Engineering Management 系统工程管理

SEMCAP Specification and Electromagnetic Compatibility Analysis Program 规范与电磁兼容性分析程序

SEMIS Solar Energy Monitor in Space 太空太阳能检测器

SEMMS Solar Electric Multiple Mission Spacecraft 太阳能发电推进多用途航天器

SEMMS Solar Energy Multimission Spacecraft 太阳能多用途航天器

SEMO Systems Engineering Management and Organization 系统工程管理组织

SEMOS Space Environment Monitoring System 太空环境监控系统

SEMPA Scanning Electron Microscope and Particle Analyzer 扫描电子显微镜和粒子分析器

SEMS Shuttle Environment Monitoring System 航天飞机环境监视系统（用于测量发射、飞行、着陆状态下航天飞机货舱环境的仪表）

SEO Satellite for Earth Observation 【印度】地球观测卫星

SEO Synchronous Earth Observatory 同步地球观察卫星

SEO Synchronous Equatorial Obiter 同步赤道轨道飞行器

SEOCS Sun-Earth Observatory and Climatology Satellite 【美国】日—地观测和气候学卫星

SEOS Synchronous Earth Observatory Satellite 【美国】同步轨道地球观测卫星

SEOS Synchronous Earth Orbiting Shuttle 同步地球轨道航天飞机

SEOW Solar Energy Optical Weapon 太阳能光学武器

SEP Self Elevating Platform 自动升降平台

SEP Software Engineering Process 软件工程处理过程
SEP Solar Electric Propulsion 太阳能电推进
SEP Space Electronic Package 航天电子设备组件
SEP Spherical Error Probable 球概率误差
SEP Support Equipment Package 支撑设备组件；保障设备组件
SEP System Enhancement Package 系统增进信息数据包
SEPA System Evaluation and Performance Analysis 系统评估与性能分析
SEPAC Space Experiments with Particle Accelerators 粒子加速器空间实验
SEPADS Sonar Environmental Prediction and Display System 声呐环境预测及显示系统
SEPAP Shuttle Electrical Power Analysis Program 【美国】航天飞机电源分析程序
SEPAR Shuttle Electrical Power Analysis Report 【美国】航天飞机电源分析报告
SEPAT Solar Energetic Particle Acceleration and Transport 太阳高能粒子加速和输运
SEPES Sonar Environmental Parameter Estimation System 声呐环境参数评估系统
SEPET Satellite Electrical Performance Evaluation Test 卫星电子性能评定测试
SEPIT Solar Electric Propulsion Integration Technology 太阳能电推进综合技术
SEPM Systems Engineering Program Management 系统工程项目管理
SEPRD System Element Production Readiness Demonstration 系统部件生产成熟度演示验证
SepRing Separation Ring 分离环
SEPS Solar Electric Propulsion Spacecraft 太阳能电力推进航天器
SEPS Solar Electric Propulsion Stage 太阳能供电推进级
SEPS Solar Electric Propulsion System 太阳能电力推进系统
SEPST Solar Electric Propulsion System Technology 太阳能电力推进系统技术
SER Safety Evaluation Report 安全评估报告
SER Satellite Equipment Room 卫星设备室；卫星设备间
SER Space Emergency Rescue 航天应急救生
SERAT Structurally Embedded Reconfigurable Antenna Technology 结构嵌入式可变天线技术
SERC Structural Engineering Research Center 结构工程研究中心
SERD Strategic Environment Research and Development 战略环境研究与发展
SERD Support Equipment Recommendation Data 保障装备推荐数据
SERD Support Equipment Requirement Data 保障设备需求数据
SERE Survival, Evasion, Resistance, and Escape 求生、规避、抵抗与逃脱（课程）
SERENDIP Search for Extraterrestrial Radio Emission from Nearby Developed Intelligent Populations 对地球大气圈外由临近发达智能人群发射的无线电波的搜索

SERF Space Environment Research Facility 太空环境研究设施
SERF System Engineering Research Facility 系统工程研究设施
SERJ Space Electric Ramjet 航天电冲压喷气发动机
SERM Solar and Earth Radiation Monitor 太阳与地球辐射监控器
SERPL Space Experiment Research Processing Laboratory 航天实验研究处理实验室
SERS Shuttle Equipment Record System 【美国】航天飞机设备记录系统
SERT Space Electric Rocket Test 航天电火箭试验
SERT Spinning Satellite for Electric Rocket Test 电火箭试验用自旋卫星
SERTOG Space Experiment on Relativistic Theories of Gravitation 引力相对太空实验
SERV Safety-Enhanced Reentry Vehicle 安全性增强的再入飞行器
SERV Single-Stage, Earth-Orbital Reusable Vehicle 可重复使用的单级地球轨道飞行器
SERV Space Emergency Reentry Vehicle 航天紧急重返大气层飞行器
SERVICE Space Entry Recovery Vehicle in Commercial Environments 【英国】商业航天再入回收飞行器
SES Satellite Earth Station 卫星地面站；卫星地球站
SES Separation System 分离系统
SES Ship Earth Station 飞船地球站
SES Shuttle Engineering Simulation 【美国】航天飞机工程模拟
SES Shuttle Engineering Simulator 【美国】航天飞机工程模拟器
SES Shuttle Engineering System 【美国】航天飞机工程系统
SES Single European Sky 欧洲天空一体化
SES Site Environment System 发射场环境系统
SES Solar Environment Simulator 太阳环境模拟器
SES Space Engineering Spacecraft 空间工程航天器
SES Space Environment Simulation 空间环境模拟
SES Space Erectable Structure 太空可起竖式结构
SES Systems Engineering Simulation 系统工程模拟
SESAC Space and Earth Sciences Advisory Committee 空间与地球科学顾问委员会
SESAME Severe Environmental Storms and Mesoscale Experiment 剧烈环境风暴和中等规模实验
SESC Shuttle Events Sequential Control 【美国】航天飞机事件时序控制
SESC Space Environment Service Center 【美国】空间环境服务中心
SESC Special Environmental Sample Container 环境采样专用容器
SESE Shuttle Experiment Support Equipment 【美国】航天飞机实验保障设备
SESE Software Engineering Support Environment 软件工程保障环境
SESE Space Electronics Support Equipment 航天电子支持设备
SESF Special Equipment Stowage Facility 特种设备存放设备
SESL Space Environmental Simulation Laboratory 空间环境模拟实验

室

SESS Space Environment Support System　太空环境支持系统

SESS Space Event Support Ship　【苏联】太空活动保障船

SET Satellite Earth Terminal　卫星地面终端

SET Satellite Experimental Terminal　卫星实验终端

SET Single Escape Tower　单人逃逸塔

SET Software Engineering Technology　软件工程技术

SET Spacecraft Elapsed Time　航天器累计时间

SET Split Engine Transportation　发动机分体运输

SET Stored-Energy Transmission　储存能传输

SET System Engineering Testbed　系统工程测试台

SET Space Electronics and Telemetry　航天电子设备与遥测

SETA Satellite Electrostatic Triaxial Accelerometer　卫星静电式三轴加速度计

SETA Scientific Engineering and Technical Assistance　科学工程与技术支持

SETCS Solar Energy Thermionic Conversion System　【美国国家航空航天局】太阳能热离子转换系统

SETF SNAP (Systems for Nuclear Auxiliary Power) Experimental Test Facility　辅助核动力系统实验测试设备

SETH Simulation Equipment Test Hardware　模拟设备试验硬件设备

SETOLS Surface-Effect Take-Off and Landing System　地面表面效应起飞着陆系统

SETP Solar Electric Aircraft Test Platform　太阳能电动飞行试验平台

SETS Seeker Evaluation and Test Simulation　探测器评价与试验模拟（用来代替飞行试验）

SETS Shuttle Electrodynamics Tether System　【美国】航天飞机电动系绳系统

SETS Solar Electric Test Satellite　太阳能电源试验卫星

SETS System, Environment and Threat Simulation　系统、环境与威胁模拟

SEU/SEL Single Event Upset/Single Event Latchup　单事件翻转/单事件闭锁

SEUL Support Equipment Utilization List　保障设备使用清单

SEV Space Exploration Vehicle　航天探索运输车

SEVA Stand-up Extravehicular Activity　【航天员】直立舱外活动

SEW Space Early Warning　太空预警

SEW Space Electronics Warfare　空间电子战

SEWS Satellite Early Warning System　卫星预警系统

SEWS Space-Based EWS　天基预警系统

SF Safety Factor　安全因子

SF Space Filter　太空滤波器

SF Space Forces（Russian Federation）俄罗斯航天部队

SF Static Firing　静态点火

SFA Security Fault Analysis　安全故障分析

SFA Simulated Flight-Automatic　自动飞行模拟

SFA Single Flight Azimuth　单飞行方

位角
SFA Space Force Application 航天力量应用
SFA Spaceport Florida Authority 【美国】佛罗里达航天港管理局
SFACI Software Flight Article Configuration Inspection 飞行软件项目配置检查
SFAP Spin Free Analytical Platform 自由旋转分解平台
SFAPS Space Flight Acceleration Profile System 航天加速度分布模拟器
SFC Side-Force Control 侧力控制
SFC Sideway-Force Coefficient 测向力系数
SFC Space Flight Capabilities 空间飞行能力
SFC Space Flight Center 航天飞行中心
SFC Space Forecast Center 航天预报中心
SFC Spacecraft Facility Controller 航天器设施控制台
SFC Specific Fuel Consumption 专用燃料耗用
SFCDS Space Flight Control and Data Communication 航天飞行控制与数据通信
SFCG Space Frequency Coordination Group 空间频率协调组
SFCPP Specific Fuel Consumption of Propulsion Plant 动力装置燃料消耗率
SFCS Safety Flight Control System 安全飞行控制系统
SFCS Secondary Flight Control System 辅助飞行控制系统
SFCS Survival Flight Control System 救生飞行控制系统
SFD Side Flame Deflector 侧导流器
SFDS System Functional Design Specification 系统功能设计说明
SFE Simulated Flight Environment Test 模拟飞行环境试验
SFE Solid Fuel Engine 固体燃料发动机
SFI Space Flight Instrumentation 航天飞行仪表设备
SFIR/R Solid Fuel Integral Rocket/Ramjet 整体固体燃料火箭/冲压喷气发动机
SFL Space Flight Laboratory 空间飞行实验室
SFLZ Simulated Flight Size 模拟飞行尺度
SFM Space Frequency Modulation 空间调频
SFMD Storable Fluids Management Demonstration 可存储液体管理演示验证
SFME Storable Fluid Management Experiment 可贮存液体管理实验
SFO Space Flight Operation 航天飞行操作
SFOC Space Flight Operations Center 航天飞行操作中心
SFOC Space Flight Operations Complex 航天飞行操作综合设施
SFOF Space Flight Operations Facility 【美国国家航空航天局】航天飞行操作设施
SFOM Shuttle Flight Operations Manual 【美国国家航空航天局】航天飞机飞行操作手册
SFOP Safety Operating Procedure 【美国肯尼迪航天中心】安全操作规程
SFP Single Failure Point 单一故障点
SFPA Single Failure Point Analysis

单一故障点分析

SFPS Single Failure Point Summary 单一故障点汇总

SFR Safety of Flight Requirements 飞行安全要求

SFR/R Solid Fuel Rocket/Ramjet 固体燃料火箭/冲压喷气发动机

SFRJ Solid Fuel Ramjet 固体燃料冲压喷气发动机

SFS Shuttle Flight Status 【美国】航天飞机飞行状态

SFS Space Flight System 航天飞行系统

SFSS Satellite Field Service Station 卫星场区服务站

SFSS Subminiature Flight Safety System 超小型飞行安全系统

SFT Simulated Flight Tests 模拟飞行试验；模拟试飞

SFT Static Firing Test 静态点火试验

SFT Structural Fatigue Test 结构脆性测试（试验）

SFT System Functional Test 系统功能测试

SFTA Structural Fatigue Test Article 结构疲劳试验件

SFTF Static Firing Test Facility 静态点火试验设施

SFTS Secure Flight Termination System 安全飞行终止系统

SFTS Space Flight Telecommunications System 航天远程通信系统

SFTS Space Flight Test System 航天飞行试验系统

SFTS Synthetic Flight Training System 综合飞行训练系统

SFU Space Flyer Unit 【美国国家航空航天局】空间自由飞行平台

SFU Standard Firing Unit 标准点火装置

SFU-RET Space Flyer Unit-Retrieval (Japanese Launched Free Flyer by Shuttle) 可回收的空间飞行器平台（日本通过美国航天飞机发射的自由飞行平台）

SG Signal Generator 信号生成器（装置）

SGDB Satellite Global Database 【美国】全球卫星数据库

SGEMP System/Source Generated Electromagnetic Pulse 系统/辐射源生成的电磁脉冲

SGG Superconducting Gravity Gradiometer 超导重力梯度仪

SGGM Superconducting Gravity Gradiometer Mission 超导重力梯度仪任务

SGITS Spacecraft Ground Operational Support System Interface Test System 航天器地面操作保障系统的接口测试系统

SGL Space Ground Link 空间－地面数据链

SGLS Satellite Ground Link Subsystem 卫星地面链路子系统

SGLS Space Ground Link System 空间－地面数据链系统

SGMT Simulated Greenwich Mean Time 格林尼治时间模拟

SGOS Shuttle Ground Operations Simulation 【美国】航天飞机地面操作模拟（装置）

SGP Single Ground Point 单个地面点

SGSC Strain Gauge Signal Conditioner 应变仪信号调节器（装置）

SGSN Spacelab Ground Support Network 【美国】"天空实验室"地面保障网

SGSO Space Ground Support Opera-

tions 航天地面支持活动
SGT Satellite Ground Terminal 卫星地面终端
SGTS Satellite Ground Terminal System 卫星地面终端系统
SH₂ Supercritical Hydrogen 超临界氢
SHA System Hazard Analysis 系统危险性分析
SHAG Simplified High Accuracy Guidance 【美国霍尼维尔公司】简化型高精确制导
SHAP Sample Handling and Analysis Plan 样本处理与分析计划
SHAPE Supersonic High Altitude Parachute Experiment 高超声速高空降落伞实验
SHAR Sriharikota Range 【印度】斯里哈里科塔发射场
SHARE Space Station Heat-Pipe Advanced Radiator Element 【美国国家航空航天局】空间站热管先进散热器元件
SHE Safety, Health, and Environment 安全、健康与环境
SHE Shuttle Electronic Hardware 【美国】航天飞机电子硬件
SHE Supercritical Helium 超临界氦
SHEAL Shuttle High Energy Astrophysics Laboratory 【美国国家航空航天局】航天飞机高能天文物理实验室
SHF Super High Frequency 超高频
SHFE Space Human Factors Engineering 空间人因工程
SHIELD System High Energy Laser Demonstration 系统高能激光器演示验证
SHIRAS Spaceborne High Resolution Atmospheric Spectrometer 星载高分辨率大气光谱仪
SHLB Simulator Hardware Load Boxes 模拟器（装置）硬件负载箱
SHLV Super Heavy Launch Vehicle 超重型运载火箭
SHM Structural Health Monitoring 结构健康监控
SHMAC Standard Hypersonic Missile with Attack Capability 防区外发射的具攻击能力的高超声速导弹
SHMS Structural Health Management System 结构健康管理系统
SHODOP Short-Range Doppler 近程多普勒导弹弹道测量系统
SHOOT Superfluid Helium On-Orbit Transfer 【美国航天飞机】超流动氦在轨加注实验
SHORAD Short-Range Air Defense 短程空中防御
SHORADS Short-Range All-Weather Air Defense System 近程全天候防空系统
SHORAN Short Range Navigation 短程导航
SHOTL Simulated Hot Launch 模拟热发射
SHReD Supplemental Heat Rejection Device 补充性排热装置
SHS Self-Propagating High-Temperature Synthesis 空间自蔓延高温融合实验
SHS Simulation Hardware System 模拟器（装置）硬件系统
SHS Space Head Section 上面组合体
SHSS Structure and Harness Subsystem 结构和吊带系统
SI Scientific Instrument 科学仪器
SI Solar Inertial 太阳惯性
SI Special Intelligence 特别情报

SI   System Integration   系统集成

SI&I   Systems Integration and Interoperability   系统集成与互操作性

SIA   Shuttle Induced Atmosphere   【美国】航天飞机诱发性大气层

SIA   Software Impact Assessment   软件影响评估

SIA   Solar Inertial Attitude   太阳惯性姿态

SIA   Station Interface Adapter   空间站接口连接器

SIAC   Safety Issue Assessment Center   安全事项评估中心

SIAD   Supersonic Inflatable Aerodynamic Decelerator   超声速可充气动力减速器

SIAP   System Integrity Assurance Program   系统完整性评价项目

SIAR   System Impact Assessment Report   系统影响评估报告

SIB   Satellite Integrated Buoy   卫星综合浮标

SIB   Satellite Interface Band   卫星接口基带

SIB   Satellite Ionospheric Beacon   卫星电离层无线电信标

SIB   Simulation Interface Buffer   模拟接口缓冲器

SIB   Solid Integration Building   刚性组合体

SIC   Satellite Information Center   卫星信息中心

SIC   Satellite Interface Controller   卫星接口控制器

SIC   Spacecraft Identification Code   【美国国家航空航天局】航天器识别码

SIC   Structural Influence Coefficient   结构影响系数

SIC&DH   Scientific Instrument Computer and Data Handling   科学仪器计算机和数据处理

SICBM   Small Intercontinental Ballistic Missile   【美国】小型洲际弹道导弹

SICM   Small Intercontinental Missile   小型洲际导弹

SICRAL   Sistema Italiana de Communicazione Riservente Allarmi   【意大利】西克拉尔卫星

SICS   Science Investigation Concept Studies   科学调查概念研究

SID   Shuttle Integration Device   【美国】航天飞机总装设备

SID   Simulation Interface Device   模拟接口设备

SID   Simulator Interface Device   模拟器接口装置

SID   Situation Information Display   态势信息显示器

SID   Space Intruder Detector   太空入侵探测器

SID   Standard Interface Document   标准接口文件

SID   Synchronous Identification System   同步识别系统

SID   System Identify   系统识别

SID   System Integration Demonstration   系统综合演示

SID   System Interface Document   系统接口文件

SIDAC   Single Integrated Damage Analysis Capability   统一综合损伤分析能力（模型）

SIDAT   Subsystems Integrated Design Assessment Technology   分系统综合设计评估技术

SIDC   Space Innovation and Development Center   航天创新与开发中心

SIDE   Sensor Integrated Discrimination

Experiment 【识别弹头和诱饵】传感器综合识别实验

SIDE Superthermal Ion Detector Experiment 【美国"阿波罗"飞船】超热离子探测器实验

SIDEX Signal Identification Experiment 【美国】信号识别试验卫星

SIDS Satellite Imagery Dissemination System 卫星成像传播系统

SIDS Secondary Imagery Dissemination System 辅助成像识别系统

SIDS Space Identification Device System 空间识别设备系统

SIDS Stellar Inertial Doppler System 天文惯性多普勒导航系统

SIE Satellite Imaging Experiment 卫星成像实验

SIE Shuttle Interface Equipment 【美国】航天飞机接口设备

SIEM Systems Integration Effort for MSFC 【美国】系统一体化（集成）对马歇尔航天中心的影响

SIES Supervision, Inspection, Engineering and Services 监管、检查、工程与服务

SIF Selective Identification Feature 选择性识别特点

SIF Stress Intensity Factor 应力强度因子

SIF System Integration Facility 系统集成设施

SIG Simplified Inertial Guidance 简易惯性制导

SIG Stellar/Inertial Guidance 天文惯性制导

SIG CONDR Signal Conditioner 信号调节器（装置）

SIG GEN Signal Generator 信号生成器（装置）

SIGI Space Integrated GPS Instrumentation 航天综合全球定位系统仪器

SIGINT Signals Intelligence 信号情报

SIGMA Standardized Inertial Guidance Multiple Application 标准化惯性制导的多项应用

SIGMA System for Interception, Goniometry, Monitoring and Analyzing 侦听、测角术、监控和分析系统

SIGS Stellar Inertial Guidance System 天文惯性制导系统

SIIRCM Suite of Integrated Infrared Countermeasures 【美国陆军】综合红外电子干扰成套设备

SIL Systems Integration Laboratory 系统综合实验室

SILL Strategic Illuminator Laser 战略照射激光器

SILO Signal Intercept from Low Orbit 低轨道信号截获

SILTS Shuttle Infrared Leeside Temperature Sensing 【美国】航天飞机背风面红外温度传感实验

SIM Satellite Interface Module 卫星接口模块

SIM Satellite Interpretation Message 卫星判读信息

SIM Scientific Instrument Module 科学仪器舱；科学仪表组件

SIM SEE (Software Engineering Environment) Information Model 软件工程环境信息模型

SIM Solar Interplanetary Model 太阳星系模型

SIM Solar Irradiance Monitor 太阳辐射监测仪

SIM Solar Ultraviolet Spectral Irradiance Monitor 太阳紫外光谱辐射照度监测仪

SIM Space Interceptor Missile 太空

拦截导弹
SIM Space Interferometry Mission 空间干涉测量任务
SIMA Small Incremental Motion Actuator 小型增量作用致动器（装置）
SIMALE Super Integral Microprogrammed Arithmetic Logic Expediter 高密度集成微程序运算逻辑简化器
SIMAS Shuttle Information Management Accountability System 【美国】航天飞机信息管理保障系统
SIMBAD Sensor Integrated Modeling for Biological Agent Detection 探测生物战剂的传感器集成模型
SIMBAY Scientific Instrument Module Bay 【美国"阿波罗"飞船】科学仪器舱
SIMFAC Simulation Facility 模拟（仿真）设施
SIMM Second In-Line Memory Module 单列直插式内存组件
SIMNET Simulation Network 仿真网络
SIMNET Simulator Networking 模拟器网络化
SIMPLEX Shuttle Ionospheric Modification with Pulsed Local Exhaust Experiment 【美国】航天飞机电离层改造和脉冲本地排放实验
SIMR Systems Integration Management Review 系统一体化（集成）管理评审
SIMS Security Information Management System 安全信息管理系统
SIMS Shuttle Imaging Microwave System 【美国】航天飞机成像微波系统
SIMS Shuttle Inventory Management System 【美国】航天飞机物品管理系统
SIMU Simulated Inertial Measurement Unit 模拟惯性测量设备
SIMU Stellar Inertial Measuring Unit 星体惯性测量装置
SINAD Signal-to-Noise and Distortion 信号－噪声与失真比
SINAP Satellite Input to Numerical Analysis and Prediction 数值分析与预测的卫星输入
SINBAC System for Integrated Nuclear Battle Analysis Calculus 综合核作战分析演算系统
SINCGARS Single Channel Ground and Airborne Radio Subsystem 单频道地面与机载无线电分系统
SINCGARS Single Channel Ground and Airborne Radio System 单信道地面与机载无线电系统
SIND Satellite Inertial Navigation Determination 卫星惯性导航测定
SINDA Systems Improve Numerical Differencing Analyzer 系统改进数值求差分析器（装置）
SINS Stellar Inertial Navigation System 天文惯性导航系统
SIO Satellite in Orbit 在轨卫星
SIO Serial Input/Output 序列输入／输出
SIO Staged in Orbit 在轨（级间）分离
SIOM Satellite Input/Output Module 卫星输入／输出模块
SIOSS Science Instruments, Observatories, and Sensor Systems 科学仪器、天文台与传感器系统
SIP Scientific Instrument Package 科学仪器装置
SIP Separation Instrument Package 分离仪器组件

SIP Standard Interface Panel 标准接口面板
SIP Surface Impact (Impulsion) Propulsion 表面冲击推进
SIPA Satellite Image Processing Agency 卫星图像处理机构
SIPE Scientific Instrument Protective Enclosure 科学仪器保护罩
SIPRNET Secure Internet Protocol Routing Network 安全互联网协议路由网络
SIPS Satellite Instrumentation Processor System 卫星仪表处理机系统
SIPS Small Instrument Pointing System 小型仪器定点系统
SIPS Small Integrated Propulsion System 小型综合推进系统
SIR Shuttle Imaging Radar 【美国】航天飞机成像雷达
SIR Signal-Interference Ratio 信扰比
SIR System Interface Requirements 系统接口要求
SIR Systems Integration Review 系统综合评定
SIRCS Spares Integrated Reporting and Control System 备件综合报告与控制系统
SIRE Space Infrared Experiment 空间红外实验
SIRFC Suite of Integrated Radar Frequency Countermeasures 综合雷达频率抗干扰组件
SIRIO Satellite Italiano per Ricerche Industriali Orientate 意大利工业研究卫星
SIRR Software Integration Readiness Review 软件集成成熟度评审
SIRRM Standardized Infrared Radiation Model 标准化红外辐射模型
SIRS Structural Integrity Recording System 结构完整性记录系统
SIRST System Shipboard Infrared Search and Track System 舰载红外搜索与跟踪系统
SIRTF Shuttle Infrared Telescope Facility 【美国】航天飞机红外望远镜设备
SIRTF Space Infrared Telescope Facility 空间红外望远镜设施（美国大型天文物理研究卫星）
SIS Satellite Intercept System 卫星拦截系统
SIS Scanning Imaging Spectrometer 扫描成像光谱仪
SIS Science Instruments and Sensors 科学仪器和传感器
SIS Shuttle Imaging Spectrometer 【美国】航天飞机成像光谱仪
SIS Shuttle Information System 【美国】航天飞机信息系统
SIS Shuttle Interface Simulator 【美国】航天飞机接口模拟器（装置）
SIS Simulator Interface System (Subsysterm) 模拟器（装置）接口系统（分系统）
SIS Software Implementation Specification 软件实现说明
SIS Space Imaging Satellite 空间成像卫星
SIS Space Imaging System 空间成像系统
SIS Stage Interface Simulator 级间接口模拟器
SIS Standard Interface Specification 标准接口说明
SISEX Shuttle Imaging Spectrometer Experiment 【美国】航天飞机成像光谱仪实验
SISP Surface Imaging and Sounding Package 表面成像与探测组件

SISS Scientific Instrument Support Structure 科学仪器支撑结构
SIT Satellite Interactive Terminal 卫星交互式终端
SIT Shuttle Integrated Test 【美国】航天飞机综合测试（试验）
SIT Shuttle Interface Test 【美国】航天飞机接口试验（电和机械的接口）
SIT Software Integration Test 软件集成试验
SIT Space Impact Tool 太空撞击工具
SIT Spaceborne Infrared Tracker 星载红外线跟踪仪
SIT System Integration Test 系统综合试验
SITE Spacecraft Instrumentation Test Equipment 航天器仪表测试设备
SITE Superfund Innovative Technology Evaluation 超级基金创新技术鉴定
SITVC Secondary Injection Thrust Vector Control 二次喷射推力矢量控制
SIU Servo Inverter Unit 伺服变换器机构；随动变换器机构
SIU Signal Interface Unit 信号接口装置
SIU Space Interface Unit 空间接口装置
SIVD Spacecraft Information Viewing Device 航天器信息观察设备
SIVE Shuttle Interface Verification Equipment 【美国】航天飞机接口验证设备
SIW Satellite Injection Window 卫星入轨窗口
SIX Satellite Intrusion Examination 卫星入侵检测
SIXPAC System for Inertial Experiment Pointing and Attitude Control 惯性实验定位与姿态控制系统
SJP Sun-Jupiter Probe 太阳—木星探测器
SKIRT Shuttle Kinetic Infrared Test 【美国国家航空航天局】航天飞机动能红外测试
SKIRT Spacecraft Kinetic Infrared Test 【美国国家航空航天局】航天器动能红外测试
SKKP Soviet System of Outer Space Monitoring 苏联外层空间监视系统
SL Sea Launch 海上发射公司
SL Servomechanisms Laboratory 伺服机构实验室
SL Silo Launcher 发射井
SL Soft Landing 软着陆
SL Space Lab 【欧洲空间局】空间实验室
SL Star Lab 【美国】恒星实验室
SL&I System Load and Initialization 系统载荷与初始化
SL-J Japanese Spacelab 日本天空实验室
SL-SS Spacelab Systems 【美国】天空实验室系统
SL-SSS Spacelab Subsystem(s) Segment 【美国】天空实验室系统单元
SLA Safe Launch Angle 【发射导弹时】安全横倾角
SLA Saturn Lunar Module Adapter 【美国】"土星"登月舱适配器
SLA Shuttle Laser Altimeter 【美国】航天飞机激光高度计
SLA Spacecraft Lunar Module Adapter 航天器登月舱适配器
SLA Spacecraft-Lunar Excursion Module Adapter 航天器和登月舱对接连接器
SLA Super Lightweight Ablator 【美

国航天飞机】超轻烧蚀材料
SLA Support and Logistics Areas 保障与后勤服务区
SLAA SPRINT Launch Area Antenna 固体推进剂火箭截击导弹发射区域天线
SLAC Stanford Linear Accelerator Center 【美国】斯坦福线性加速器（装置）中心
SLAC Stowage Launch Adapter Container 装载发射适配器集装箱
SLAHTS Stowage List and Hardware Tracking System 装载清单与硬件跟踪系统
SLAL Small Laser Amplifier for Ladar 用于雷达的小型激光放大器
SLALOM Space Laser Low Orbit Mission 低轨道空间激光卫星
SLAM Side Load Arrest Mechanism 侧向负载抑制机械装置
SLAM Side Load Arresting Mechanism 侧向加载制动机构
SLAM Space Launched Air Missile 太空发射的对空导弹
SLAM Stand-off Land Attack Missile 防区外发射的对地攻击导弹
SLAM-ER Standoff Land Attack Missileexpanded Response 反应增强型防区外对地攻击导弹
SLAR Side Looking Airborne Radar 侧向观察机载雷达
SLAR Sideways-Looking Airborne Radar 机载侧视雷达
SLAT Supersonic Low Altitude Target 超声速低空目标
SLBM Satellite Launched Ballistic Missile 卫星发射的弹道导弹
SLBM Space Launched Ballistic Missile 太空发射的弹道导弹
SLBMs Submarine-Launched Ballistic Missiles 潜射弹道导弹
SLC Shuttle Launch Complex 【美国范登堡空军基地】航天飞机发射场
SLC Software Life Cycle 软件寿命周期
SLC Space Launch Complex 航天发射场
SLC Space Shuttle Launch Complex 【美国】航天飞机发射场
SLCC Saturn Launch Control Computer 【美国】"土星"火箭发射控制计算机
SLCC Soft Launch Control Center 软发射控制中心
SLCM Sea-Launched Cruise Missile 海射巡航导弹
SLCS Software Lifecycle Support 软件寿命周期支持（保障）
SLCSE Software Life Cycle Support Environment 软件寿命周期支持环境
SLD Simulated Launch Demonstration 【美国国家航空航天局】模拟发射演示验证
SLD System Link Designator 系统链路标志符
SLDCOM Satellite Launch Dispenser Communications 卫星发射分配器通信
SLDPF Spacelab Data Processing Facility 【美国国家航空航天局】天空实验室数据处理设施
SLDS Spacelab Launch Data System 【美国国家航空航天局】天空实验室发射数据系统
SLED Solar Low Energy Detector 太阳低能粒子探测器
SLEDGE Simulating Large Explosive Detonable Gas Experiment 模拟大型炸药爆震气体实验

SLEMU Spacelab Engineering Model Unit 天空实验室工程模型装置
SLEP Service Life Extension Program 使用寿命延长计划
SLES Superconducting Submillimeter-wave Limb-Emission Sounds 超导亚毫米临边发射探测器
SLF Saturn Launch Facility 【美国】"土星"火箭发射设施
SLF Shuttle Landing Facility 【美国】航天飞机着陆设施；航天飞机着陆设备
SLF Silo Launch Facility 【导弹】发射井发射设施
SLG Satellite Landing Ground 卫星降落场
SLI Space Launch Initiative 【美国国家航空航天局】航天发射倡议
SLI Structural Load Indicator 结构荷载指示器（装置）
SLIC Systems and Logistic Integration Capability 系统与后勤综合能力
SLICBM Sea-Launched Intercontinental Ballistic Missile 海上发射的洲际弹道导弹
SLIP Serial Line Internet Protocol 串行线路互联网协议
SLKT Survivability, Lethality and Key Technologies 【美国星球大战计划】生存力、杀伤力与关键技术
SLLC Satellite Logical Link Control 卫星逻辑链路控制
SLM Service for Landslide Monitoring 着陆滑行监视勤务
SLM Simulated Laboratory Module 模拟实验室舱
SLM Space Laboratory Module 天空实验室舱
SLMA Spacecraft Lunar Module Adapter 飞船登月舱对接件
SLMAB Single Line Missile Assembly Building 单线导弹装配厂房
SLMAR Space Logistics, Maintenance and Rescue 航天后勤、维修和救援
SLMS Saturn Launched Meteoroid Satellite 【美国】"土星"运载火箭发射的流星体研究卫星
SLOA/A Suspended Load Operation Analysis/Approval 悬浮载荷工作分析／批准
SLOB Satellite Low Orbit Bombardment 【美国空军】卫星低轨道轰炸系统
SLOMAR Space Logistics Maintenance and Repair (Vehicle) 空间后勤维护和修理（飞行器）
SLOMAR Space Logistics Maintenance and Rescue 空间后勤、维修与救援
SLOMAR Space Logistics Maintenance and Rescue Spacecraft 空间后勤、维修与救援飞船
SLP Soft Landing Probe 软着陆探测（飞行）器
SLP Solar Lunar Planetary 太阳月球行星
SLP Surface Launch Platform 水面发射平台
SLRS Space Lift Range System 航天运载靶场系统
SLRV Shuttle Launched Research Vehicle 【美国】航天飞机发射的研究飞行器
SLRX System Lifecycle Risk Expert 系统全寿命风险专家
SLS Science Life Space Lab 【欧洲】生命科学空间实验室
SLS Space Lab Simulator 天空实验室模拟器

SLS Space Launch System 航天发射系统
SLS Spacecraft Landing Strut 航天器着陆支撑物
SLS Spacelab Life Science 生命科学天空实验室
SLS SpaceLab Simulator 天空实验室模拟器
SLS Spacelink Subnetwork 空间数据链子网
SLS Strategic Lunar System 战略月球系统
SLSA Shuttle Logistics Support Aircraft 【美国】航天飞机后期保障飞机
SLSL Space Life Sciences Laboratory 【美国国家航空航天局】空间生命科学实验室
SLSMS Spacelab Support Module Simulator 天空实验室保障舱模拟器
SLSOB Space Launch Squadron Operation Building 航天发射中队操作厂房
SLSS Supplementary Life Support System 辅助生命保障系统
SLSST Spacelab Single System Trainer 天空实验室单一系统训练机
SLST Sea-Level Static Thrust 海平面静推力;发动机地面静推力
SLSTP Space Life Sciences Training Program 空间生命科学训练项目
SLT Satellite Lourd De Telecommunication 中型通信卫星
SLT Simulated Launch Test 模拟发射试验
SLT Strategic Laser Technology 战略激光器技术
SLT Stress Limit Test 应力极限试验
SLTF Silo Launch Test Facility 地下井发射试验设施;(导弹)井下发射测试设施
SLTR Service Life Test Report 使用寿命试验报告
SLU Surface-Launched Unit 地面(水面)发射装置
SLV Satellite Launch Vehicle 【印度】卫星运载火箭
SLV Saturn Launch Vehicle 【美国】"土星"运载火箭
SLV Site de Lancement VEGA (VEGA Launch Site) 【法国】"织女星"小型火箭发射场
SLV Small Launch Vehicle 【美国】小型运载火箭
SLV Soft Landing Vehicle 软着陆飞行器
SLV Space Launch Vehicle 航天运载火箭
SLV Standard Launch Vehicle 【美国】标准运载火箭
SLWT Side Loading Warping Tug 侧载拖绞船
SLWT Super Lightweight Tank 超轻储箱
SM Satellite Simulator 卫星模拟器
SM Science Mission 科学任务
SM Service Module 【航天器】勤务舱;服务舱
SM Short Module 短舱
SM Shutter Mission 【美国】航天飞机飞行任务
SM Shuttle Management (KSC) 【美国肯尼迪航天中心】航天飞机管理
SM Space Medicine 航天医学;空间医学
SM Standard Missile 标准导弹
SM Subsystems Module 分系统舱
SM Supply Module 补给舱
SM Systems Management 系统管理

SM&R Source, Maintenance and Recoverability 来源、维护与可修复性

SM&S Systems Management and Sequencing 系统管理与程序设计

SM/PM System Management/Performance Monitor 系统管理/性能监测

SMA Safety and Mission Assurance 安全与任务保证

SMA Science Monitoring Area 科学监测区

SMA System Modeling Architecture 系统建模体系结构

SMAB Solid Motor Assembly Building 【美国】固体火箭发动机总装厂房

SMAC Space Maintenance Analysis Center 【美国】空间维护分析中心

SMAC Spacecraft Maximum Allowable Concentration 航天器最大可容许浓度

SMAC Special Missile and Astronautics Center 特种导弹与航天中心

SMACS Serialized Missile Accounting and Control System 连续发射导弹计算与控制系统；连续发射导弹计数与控制系统

SMACSS Synchronized Multiple Access Communication System by Satellite 卫星同步多址通信系统

SMALLSAT Small Satellite 小型卫星

SMAP Soil Moisture Active and Passive 土壤湿度的有源及无源观测

SMAP Systems Management Analysis Project 系统管理的分析方案

SMARF Solid Motor Assembly and Readiness Facility 【美国】固体发动机装配和准备设施

SMART Satellite Maintenance and Repair Technique 卫星维修技术

SMART Satellite Multimedia Applications Research and Trials 卫星多媒体应用研究与试验

SMART Self Monitoring, Analysis and Reporting Technology 自监视、分析与报告技术

SMART Semi-Ballistic Mission of an Atmospheric Reentry Technology 大气层再入技术的半弹道任务

SMART Shared Mobile Atmospheric Research and Teaching Radar 共享型移动大气研究与教学雷达

SMART Simulation and Modeling for Acquistion, Requirements and Training 搜索、要求和训练用模拟和建模

SMART Small Missions for Advanced Research and Technology 小型先进研究与技术任务

SMART Solid Modeling Aerospace Research Tool 航空航天固态模型研究工具

SMART Space Maintenance and Repair Techniques 空间维护与维修技术

SMART Supersonic Missile and Rocket Track 超声速导弹和火箭轨道装置

SMARTS Safety and Mission Assurance Requirements Tracking System 安全与任务保证要求跟踪系统

SMAT Satellite and Missile Analysis Tool 卫星与导弹分析工具

SMATH Space Materials Advanced Technology for Hardness 加固用的航天材料新技术

SMATS Speed Modulated Augmented Thrust System 速度调制增大推力

系统

SMATV Satellite Master Antenna Television 卫星主天线电视

SMC Space and Missile Center 太空与导弹中心

SMC Space and Missile Command 【美国】航天与导弹司令部

SMC Space and Missile Systems Center 【美国】航天与导弹系统中心

SMC Spatial Motion Compensation 空间移动补偿

SMCC Shuttle Mission Control Center 航天飞机任务控制中心

SMCC Simulation Monitor and Control Console 模拟监控器与控制台

SMCS Separation Monitor and Control System 分离监测与控制系统

SMCS Standard Monitoring and Control System 标准监视与控制系统

SMCS Structural Mode Control System 结构模态（振型）控制系统

SMCU Separation Monitoring Control Unit 分离监测控制装置

SMD Science Mission Directorate 【美国国家航空航天局】科学任务委员会

SMD Spacelab Mission Development 天空实验室任务开发

SMD Special Measuring Device 专用测量装置

SMD Special Measuring Service 专用测量服务

SMD Strategic Missile Defense 战略导弹防御

SMDC Space and Missile Defense Command 【美国陆军】航天与导弹防御司令部

SMDCOC Space and Missile Defense Command Operations Center 航天与导弹防御指挥操作中心

SMDPS Strategic Mission Data Preparation System 战略任务数据准备系统

SMEAT Skylab Medical Experiments Altitude Test 【美国】"天空实验室"医药实验的高度试验

SMEC Single Module Engine Controller 单舱发动机控制器

SMERFS Statistical Modeling and Estimation of Reliability Functions for Software 软件的统计建模与可靠性评估

SMES Shuttle Mission Engineering Simulator 【美国】航天飞机任务工程模拟器

SMES Shuttle Mission Evaluation Simulation (Simulator) 【美国】航天飞机任务评估模拟（模拟器）

SMES Superconduction Magnetic Energy Storage 超导磁能存储

SMET Simulated Mission Endurance Test 任务持续时间模拟试验

SMET Spacecraft Maneuver Engine Transients 航天器机动发动机过渡

SMEX Small Explorer 小型探测器

SMG Simulation Modeling Game System 【计算机】仿真模型演示系统

SMG Spaceflight Meteorology Group 航天飞行气象组

SMIA Serial Multiplexer Interface Adapter 序列多路调制器接口适配器

SMICBM Semi-Mobile Intercontinental Ballistic Missile 半机动发射的洲际弹道导弹

SMIDEX Spacelab Middeck Experiments 【美国】天空实验室中甲板实验

SMIIS Solar Microwave Interferometer Imaging System 太阳微波干涉

成像系统

SMILES Superconducting Submillimeterwave Limb Emission Sounder 超导亚毫米波 Limb 声探测器

SMILS Sonar Buoy Missile Locator Impact System 声呐浮标导弹落点定位系统

SMIPE Small Interplanetary Probe Experiment 小型星际探测器实验

SMIRR Shuttle Multispectral Infrared Radiometer 【美国】航天飞机多光谱红外辐射计

SML Structure Mold Line 结构模型线

SMM Shuttle-Mir Mission 【美国】航天飞机—"和平号"空间站任务

SMM Space Manufacturing Module 空间制造舱

SMM Structures, Mechanisms, and Materials 结构、机械装置与材料

SMM Subsystem Measurement Management 分系统测量管理

SMM Systems Maturity Matrix 系统成熟度矩阵

SMMD Specimen Mass Measurement Device 标本质量测量装置

SMN Scene Matching Navigation 景象匹配导航系统

SMO Small Magnetospheric Observatory (Satellite) 小型磁层观测（卫星）

SMOC Simulation Mission Operation Computer 模拟任务操作计算机

SMOS Soil, Moisture and Ocean Salinity 【欧洲空间局】土壤、湿度与海洋盐度（任务）

SMP Shape Memory Polymer 形状记忆聚合物

SMP Software Management Plan 软件管理计划

SMPAS Space Mission Payload Assessment System 航天任务有效载荷评估系统

SMPM Structural Materials Property Manual 结构材料特性手册

SMPS Scalable Multiprocessing System 可扩展多处理系统

SMPTRB Shuttle Main Propulsion Test Requirement Board 【美国】航天飞机主推进测试要求委员会

SMQ Structure Module Qualification Test 结构模块鉴定试验；结构舱鉴定测试

SMR San Marco Range 【意大利】圣马科发射场

SMR Small Missile Range 小型导弹靶场

SMR Source, Maintenance, and Recoverability 资源、维护与复原性

SMR Spaceborne Meteorological Radar 星载气象雷达

SMR Supportability, Maintainability and Reparability 保障性、维护性与修理性

SMR Code Source, Maintenance, and Recoverability Code 来源、维护与可修复性规范

SMRC Solid Motor Roll Control 固体发动机滚动控制

SMRM Solar Maximum Repair Mission 【美国国家航空航天局】太阳最大期修复任务

SMS Satellite and Missile Surveillance 卫星与导弹监视

SMS Satellite Motion Simulator 卫星运行模拟器；卫星运动仿真器

SMS Separation Mechanism Subsystem 分离机构分系统

SMS Shuttle Mission Simulator 【美国】航天飞机飞行任务模拟器

SMS Small Magnetospheric Satellite 【美国】小型磁层探测卫星
SMS Solar Maximum Satellite 最大太阳系卫星
SMS Space Mission Simulator 航天任务模拟器
SMS Space Motion Sickness 太空晕动病
SMS Synchronous Meteorological Satellite 【美国】同步气象卫星
SMSCC Shuttle Mission Simulator Computer Complex System 航天飞机任务模拟器计算机设备系统
SMSI Standard Manned Space Flight Initiator 标准载人航天飞行启动程序；标准载人航天项目发起者
SMSIMF Scaled Model Signature Intelligence Measurements Facility 几何相似模型特征情报测量设施
SMSIP Space Mission Survivability Implementation Plan 航天任务生存力执行计划
SMSR Safety and Mission Success Review 安全与任务成功评审
SMSR Small Meteorological Sounding Rocket 小型气象探空火箭
SMSRS Shipboard Meteorological Satellite Readout Station 气象卫星舰载数据接收站
SMT Saturn Missile Test 【美国】"土星"导弹试验
SMT Spacecraft Maneuvering Time 航天器机动时间
SMTA Shuttle Model Test and Analysis 【美国】航天飞机模型试验与分析
SMTAS Shuttle Model Test and Analysis System 【美国】航天飞机模型试验与分析系统
SMTC Space and Missiles Test Center 航天与导弹试验中心
SMTF Shuttle Mission Training Facility 【美国】航天飞机任务训练设施
SMTS Space and Missile Tracking System 航天和导弹跟踪系统（红外跟踪卫星）
SMU Self-Maneuvering Unit 【航天员】自主机动飞行装置
SMV Space Maneuver Vehicle 空间机动飞行器
SMVP Shuttle Master Verification Plan 【美国】航天飞机主验证计划
SMVRD Shuttle Master Verification Requirements Document 【美国】航天飞机主验证要求文件
SN Satellite Navigation 卫星导航
SN Saturn Nuclear 【美国】"土星"核动力运载火箭
SN Signal-to-Noise Ratio 信号噪声比；信噪比
SN Space Network 空间网络
SNA System Network Architecture 系统网络体系结构
SNADS System Network Architecture Distribution Services 系统网络体系结构分布服务
SNAP Space Navigation Analysis Program 空间导航分析项目
SNAPSHOT Systems for Nuclear Auxiliary Power-Shot 【美国】核辅助发电系统（试验卫星）
SNDV Strategic Nuclear Delivery Vehicle 战略核武器运载工具；战略核武器投掷工具
SNEPT Space Nuclear Electric Propulsion Test 航天核电推进试验
SNFM Serial Network Flow Monitor 串行网络流量监视器
SNIPE SDI System Network Processor Engine 战略防御倡议系统网络

处理机引擎
SNOC　Satellite Network Operation Center　卫星网操作中心
SNP　Space Nuclear Propulsion　航天核推进（装置）
SNPS　Satellite Nuclear Power Station　卫星核动力站
SNR　Signal-to-Noise Ratio　信噪比
SNS　Satellite Navigation System　卫星导航系统
SNS　Station Network Simulator　站点网络模拟器（装置）
SNS　Stellar Navigation System　星体导航系统
SNS　Sustainability and Supportability　可持续性和可支持性
SNTP　Space Nuclear Thermal Power　航天核热电源
SNTP　Space Nuclear Thermal Propulsion　航天核热电推进
SO　Stationary Orbit　静止轨道
SO　Synchronous Orbit　同步轨道
SOA　Service Oriented Architecture　面向服务的体系结构
SOA　Shuttle Orbital Application　航天飞机轨道应用
SOA　State-of-the-Art　最新型的；最优良的
SOAP　Structural Optimization and Analysis Program　结构最佳化分析程序
SOAR　Shuttle Orbital Application and Requirement　【美国】航天飞机的应用和要求
SOAR　Simulation of Apollo Reliability　【美国国家航空航天局】"阿波罗"可靠性模拟
SOARS　Satellite On-Board Attack Reporting System　【美国】卫星（星上）攻击报告系统
SOARS　Shuttle Operation Automated Reporting System　【美国】航天飞机操作自动化报告系统
SOATS　Support Operation Automated Training System　保障操作自动化训练系统
SOB　Space Order of Battle　航天作战序列
SOBs　Strap-On Boosters　捆绑式火箭助推器
SOC　Satellite Operations Center　卫星操作中心
SOC　Satellite Operations Complex　卫星操作综合设施
SOC　Satellite Orbit Control　卫星轨道控制
SOC　Science Operations Centre　科学操作中心
SOC　Separated Orbit Cyclotron　分离轨道回旋加速器
SOC　Simulation Operation Computer　模拟操作计算机
SOC　Simulation Operations Center　模拟操作中心
SOC　Single Orbit Computation　单轨道计算
SOC　Space Operations Center　航天作战中心；航天操作中心
SOC　Space Operations Controller　航天操作控制器
SOC　Spacecraft Operation Center　航天器操作中心
SOC　Special Operations Command　特种作战司令部
SOC　System Operating Concept　系统操作设计概念
SOC　System Option Controller　系统方案控制器（装置）
SOCAR　Shuttle Operational Capability Assessment Report　【美国】航天

飞机作战能力评估报告
SOCC Satellite Operations Control Center 卫星操作控制中心
SOCC Spacecraft Operations Control Center 航天器操作控制中心
SOCF Spacecraft Operations and Checkout Facility 航天器操作与检测设施；航天器操作与检测厂房
SOCH Spacelab Orbiter Common Hardware 天空实验室轨道器通用硬件
SOCOM Solar Orbital Communications 太阳轨道通信
SOCOM Special Operations Command 特种作战司令部
SOCP Satellite Orbit Control Program 卫星轨道控制程序
SOCRD Science Operations Center Requirements Document 科学操作中心要求文件
SOCS Space Operation Command System 航天操作指挥系统
SOCS Spacecraft Orientation Control System 航天器方位控制系统
SOCS Subsystem Operation and Checkout System 分系统操作与检查系统
SODAS Satellite Operation Planning and Data Analysis System 卫星操作计划与数据分析系统
SODB Shuttle Operational Data Book 【美国】航天飞机操作数据手册
SODN Shuttle Operations Data Network 【美国】航天飞机操作数据网络
SODS Saturn Operational Display System 【美国】"土星"运载火箭显示系统
SODS Shuttle Operational Data System 【美国】航天飞机操作数据系统
SODS Skylab Orbit-Deorbit System 【美国】"天空实验室"入轨－离轨系统
SODS Space Operation Data System 空间操作数据系统
SODSIM Strategic Offense/ Defense Simulator 战略进攻/防御模拟器（装置）
SOE Science Operations Element 科学操作要素
SOE Sequence of Events 事件序列
SOE Software Operating Environment 软件操作环境
SOF Safety of Flight 飞行安全
SOF Special Operations Forces 特种作战部队
SOFC Saturn Operational Flight Control 【美国】"土星"运载火箭作业飞行控制
SOFC Solid Oxide Fuel Cell 固体氧化物燃料舱
SOFIA Stratospheric Observatory for Infrared Astronomy 【美国国家航空航天局】平流层红外天文台
SOFT Safety of Flight Test 飞行安全试验
SOFT Space Operations and Flight Techniques 航天操作与飞行技术
SOGS Science Operations Ground System 科学操作地面系统
SOHO Solar and Heliospheric Observatory 日光层观测台
SOHRE Solar Hydrogen Rocket Engine 太阳－氢火箭发动机
SOI Saturn Orbit Insertion 【美国】"土星"运载火箭入轨
SOI Space Object Identification Calibration 【美国】空间目标识别校准卫星

SOIC Space Object Identification Calibration 太空物体识别校准卫星（美国的军事试验卫星）
SOIC Space Operational Intelligence Center 空间作战情报中心
SOICAS Space Object Identification Central Analysis System 太空物体识别中央分析系统
SOIF System Operation and Integration Functions 系统运行与综合功能
SOIPSA Shuttle Orbit Injection Propulsion System Analysis 【美国】航天飞机入轨推进系统分析
SOIS Space Object Identification System 空间目标识别系统
SOIS Spacelab/Orbiter Interface Simulator 天空实验室/轨道器接口模拟器
SOL Small Orbital Laboratory 小型轨道实验室
SOLACES Solar Auto-Calibrating Extreme UV/UV. Spectrophotometers 太阳能自动校准极紫外/紫外分光光度计
SOLAR HAPP Solar High Altitude Powered Platform 太阳能高空动力平台
SOLARES Solar Reflector Satellite 太阳能反射器卫星
SOLLAR Soft Lunar Landing and Return 月球软着陆和返回
SOLLO Sonic Lightning Locator 声闪电定位器
SOLOMON Simultaneous Operation Linked Orbital Modular Network 与轨道舱网络连接的同时操作
SOLRAD Solar Radiation Monitoring Satellite 【美国】太阳辐射检测卫星
SOLSPEC Solar Spectral Irradiance Measurements 太阳光谱辐照度测量
SOLSTICE Solar-Stellar Irradiance Comparison Experiment 太阳－恒星辐照度比较实验
SOLWIND Solar Wind 【美国】"太阳风"卫星
SOM Simulation Object Model 仿真对象模型
SOM Space Organization Method 空间组织方法；航天组织方法
SOM Spares Optimization Model 备件优化模型
SOM Standard Operating Manual 标准操作手册
SOM Strap-On Motor 捆绑式火箭发动机
SOM Sub-Orbital Mission 亚轨道任务
SOM System Object Model 系统目标模型
SOM&S Suite of Models and Simulations 模型与仿真设备
SOME Satellite Orbit Mission Evaluation 卫星轨道任务评估
SOMM Standoff Modular Missile 远距离模块导弹
SOMS Shuttle Orbiter Medical System 【美国】航天飞机轨道器医疗系统
SOMS Space Operations Management System 航天操作管理系统
SONAR Sound Navigation and Ranging 声导航与测距
SONET Synchronous Optical Network 同步光学网络
SONG Satellite for Orientation, Navigation and Geodesy 【美国】定向/导航和测地卫星
SOON Solar-Observatory Optical Net-

work 【美国空军】光学太阳观测网

SOP Secondary Oxygen Pack 辅助氧气单元

SOP Spacelab Opportunity Payload 【美国】天空实验室机会型有效载荷

SOP Standard Operating Procedure 标准操作规程

SOP Subsystem Operating Program 分系统操作项目

SOP Subsystem(s) Operating Procedure 分系统操作规程

SOP Systems Operation Plan 系统操作计划

SOPC Shuttle Operations and Planning Center 【美国】航天飞机操作（运营）和规划中心

SOPC Shuttle Operations Planning Complex 【美国】航天飞机操作规划设施

SOPDAS Satellite Operation Planning and Data Analysis System 卫星操作计划与数据分析系统

SOPM Standard Orbital Parameter Message 标准轨道参数信息

SOPSA Shuttle Orbit-Injection Propulsion System Analysis 【美国】航天飞机入轨推进系统分析

SOPSA Shuttle Orbit-Injection Propulsion System Analysis 航天飞机送入轨道的推进系统分析；航天飞机入轨推进系统分析

SOPSMAN Space Operations Scheduling and Management System 航天操作规划与管理系统

SOR Stable Orbit Rendezvous 稳定轨道交会

SOR Starfire Optical Range 【美国】"星火"光学试验场

SORD Systems Operations and Requirements Document 系统操作与要求文件

SORS Spacecraft Oscillograph Recording System 航天器示波器记录系统

SORT Satellite Orbital Track 卫星轨道跟踪

SORT Simulated Optical Range Tester 模拟光学距离测试器

SORT Simulation and Optimization of Rocket Trajectories 火箭轨道的模拟与优选

SORT Structures for Orbiting Radio Telescope 轨道射电望远镜结构

SORTI Satellite Orbiting Track and Intercept 卫星轨道跟踪与拦截

SORTIE Space Orbital Reentry Test Integrated Environment 空间轨道再入试验综合环境

SORTS Software Reliability Modeling and Analysis Tool Set 软件可靠性建模与分析工具集

SORTS Status of Resources and Training System 资源与训练系统状态

SOS Satellite Observation System 卫星观测系统

SOS Simulator Operating System 模拟器操作系统

SOS Source of Supply 供给源

SOS Space Operations Simulator 空间作战模拟器

SOS Space Operations Support 空间作战保障

SOS Storm Observation Satellite 【美国】风暴观测卫星

SOS Synchronous Orbit Satellite 同步轨道卫星

SOSC Space Operations Support Center 航天操作保障中心

SOSI Space Operations and Scientific Investigations 空间作战与科学调查
SOSM Strap-on Solid Motor 捆绑式固体火箭发动机
SOSO Synchronous Orbiting Solar Observatory 同步轨道太阳观测台
SOSS Satellite Optical Surveillance Station 卫星光学监视站
SOSS Satellite Optical Surveillance System 卫星光学监视系统
SOSUS Sound Surveillance System 声监视系统
SOT Solar Optical Telescope 【美国】太阳光学望远镜
SOT Strap-on Tank 捆绑式储箱
SOTA Science Operations and Test Area 科学操作与试验区
SOTC Secure Operational Telecommunications 安全操作长途通信
SOTIC Strategic Operations Target Interdiction Command 战略作战目标摧毁指令
SOTP Spacecraft Operations Test Procedure 航天器操作试验程序
SOTS Suborbital Tank Separation 亚轨道贮箱分离
SOTS Synchronous Orbiting Tracking Station 同步轨道跟踪站
SOTV Spacecraft/Orbit Transfer Vehicle 航天器/轨道转移飞行器
SOUP Solar Optical Universal Photopolarimeter 太阳光学通用偏振测光仪
SOV Shut-off Valve 关闭阀
SOV Solenoid Operated Valve 筒形操作阀
SOV Space Operation Vehicle 空间作战飞行器
SOVIM Solar Variability and Irradiance Monitor 太阳变化及辐照监测器
SOW Stand-off Weapon 防区外武器
SOX Supercritical Oxygen 超临界氧
SP Secondary Payload 二级有效载荷
SP Signal Processor 信号处理器（装置）
SP Smokeless Propellant 无烟推进剂
SP Solar Probe 太阳探测器
SP Solid Propellant 固体推进剂
SP Solid Propulsion 固体推进
SP Space Platform 空间平台
SP Space Probe 空间探测器
SP Space Propellant 航天推进剂
SP Special Processes and Sequencing 特殊处理与程序设计
SP Special Propellants 特种推进剂
SP Sustainability Plan 可持续性计划
SP Static Pressure 静态压力
SP-FGS Flight and Ground Systems Office 【美国肯尼迪航天中心】航天飞机飞行与地面系统办公室
SP-ILS Integrated Logistics Support Office 【美国肯尼迪航天中心】航天飞机一体化后勤保障办公室
SP-LMO Logistics Management Office 【美国肯尼迪航天中心】后勤保障管理办公室
SP-OPI Operations Planning and Integration Office 【美国肯尼迪航天中心】航天飞机操作、规划与一体化办公室
SP-PAI Project Assessment and Integration Staff 【美国肯尼迪航天中心】项目评估与一体化人员
SP-PCO Project Control Office 【美国肯尼迪航天中心】项目控制办公

室
SP-PMS Performance Management Systems Office 【美国肯尼迪航天中心】效能管理系统办公室
SP-SMO Site Management Office 【美国肯尼迪航天中心】航天飞机场区管理办公室
SPA S-Band Power Amplifier S波段功率放大器
SPA Servo Power Amplifier 伺服功率放大器
SPA Servo-Power Assembly 伺服动力装置
SPA Shuttle Processing Area 航天飞机处理区
SPA Signal Processor Assembly 信号处理器（装置）组件
SPA Small Payload Accommodations 小型有效载荷存放
SPA Software Product Assurance 软件产品保证
SPA Solar Panel Assembly 太阳板组件
SPA Solar-Powered Aircraft 太阳能飞机
SPA Space Processing Application 航天处理应用
SPA Spacecraft/Payload Adapter 航天器/有效载荷适配器
SPAA Spacecraft Performance Analysis Area 航天器性能分析区
SPAC Spacecraft Performance Analysis and Command 【美国国家航空航天局】航天器性能分析与指挥
SPAC Spacecraft Performance Analysis and Command 航天器性能分析和指挥
SPACC Space Command Center 航天指挥中心
SPACC Space Control Center 航天控制中心
SPACCS Space Command and Control System 航天指挥和控制系统
SPACE Spacecraft Prelaunch Automatic Checkout Equipment 航天器发射前自动检测设备
SPACECOM Space Command 航天司令部
SPACECOM Space Communications 航天通信
SPACECOM Space Communications Company 航天通信公司
SPACEHAB Space Habitat Modules 空间居住舱
SpaceX Space Exploration Technologies 【美国】太空探索技术
SpaceX Space Exploration Technology Company 美国空间探索技术公司
SPACP Steam Plant Auxiliaries Control Panel 蒸汽动力装置辅助设备控制板
SPACTS Semi-Passive Attitude Control and Trajectory Stabilization System 半被动式姿态控制和轨道稳定系统
SPAD Satellite Position Prediction and Display 卫星位置预测和显示
SPAD Shuttle Payload Accommodation Document 【美国】航天飞机有效载荷存储文件
SPAD Space Patrol Active Defense 太空巡逻主动防御
SPAD Space Patrol Antimissile Defense 太空巡逻反导弹防御
SPAD SPRINT (Solid Propellant Rocket Intercept Missile) Airdirected Defense 【美国】"短跑"固体推进剂火箭拦截弹道对空防御
SPAD Subsystem Positioning Aid Device 分系统定位辅助设备

SPADAN　Space Tracking and Data Acquisition Network　航天跟踪和数据采集网

SPADATS　Space Defense Acquisition and Tracking System　太空防御捕获与跟踪系统

SPADATS　Space Detection and Tracking System　空间探测与跟踪系统

SPADATSC　Space Detection and Tracking System Center　空间探测与跟踪系统中心

SPADATSIMP　Space Detection and Tracking System Improved　改进的空间探测与跟踪系统

SPADCCS　Space Detection Command and Control System　【战术预警/攻击效果评估】太空防御指挥与控制系统

SPADE　Space Acquisition Defense Experiment　空间捕获防御实验

SPADMS　Solar Perturbation Atmosphere Density Measurement Satellite　【美国】太阳扰动大气密度测量卫星

SPADOC　Space Defence Operation Center　【美国空军】空间防御作战中心

SPADS　Satellite Position and Display System　卫星定位和显示系统

SPADS　Shuttle Problem Action Data System　【美国】航天飞机纠错数据系统

SPADS　Shuttle Problem Analysis Data System　【美国】航天飞机问题分析数据系统

SPADTS　Space Detection and Tracking System　空间探测与跟踪系统

SPADVOS　Spaceborne Direct Viewing Optical System　航天器载直接观测光学系统

SPAF　Simulation Processor and Formatter　模拟处理器（装置）与格式化程序

SPAH　Spacelab Payload Accommodations Handbook　【美国】天空实验室有效载荷存储手册

SPAN　Solar-Particle (Proton) Alert Network　【美国国家海洋与大气管理局】太阳粒子（质子）警报网

SPAN　Space Physics Analysis Network　【美国国家航空航天局】空间物理分析网

SPAN　Space Plasma Analysis Network　空间等离子体分析网

SPAN　Spacecraft Analysis　航天器分析

SPAR　Satellite Position Adjusting Rocket　卫星位置调整火箭

SPAR　Space Processing Applications Rocket　【美国】航天处理应用火箭

SPAR　Space Project Applications Rocket　航天项目应用火箭

SPAR　Special Prelaunch Analysis Request　发射前特殊分析请求

SPARC　Space Applications and Research Center　【巴基斯坦】空间应用和研究中心

SPARC　Space Research Capsule　空间研究用的密封舱

SPARC　Stratospheric Processes and Their Role in Climate　平流层过程及其在气候中的作用

SPARCENT　Space and Atmospheric Research Centre　空间与大气研究中心

SPARCS　Solar Pointing Aerobee Rocket Control System　指向太阳的"空蜂"式火箭控制系统

SPARES　Space Radiation Evaluation

System 空间辐射鉴定系统
SPARK Solid Propellant Advanced Ramjet Kinetic Energy Missile 固体推进剂高级冲压式喷气发动机动能导弹；固体推进高级冲压式动能火箭
SPARM Solid Propellant Augmented Rocket Motor 固体推进剂改进型火箭发动机
SPARS Space Precision Attitude Reference System 航天精确姿态参考系统
SPART Space Research and Technology 空间研究和技术
SPART Special Antimissile Research Test 专用反导弹研究试验火箭
SPARTAN Shuttle Pointed Autonomous Research Tool for Astronomy 【美国国家航空航天局】"斯帕坦"卫星（由航天飞机收放的自主式天文学研究卫星）
SPAS Shuttle Pallet Satellite 【欧洲】航天飞机平台卫星
SPAS Space Pallet Satellite 【德国】空间平台卫星
SPAS Space Power Architecture Study 空间动力体系结构研究
SPASM Space Propulsion Automated Synthesis Modeling 航天推进自动化综合模拟（模型）
SPAWAR Space and Naval Warfare System Command 【美国】航天和海战系统司令部
SPB Solid Propellant Booster 【美国国家航空航天局】固体推进剂助推器
SPC Solid Propellant Combustion 固体推进剂燃烧
SPC Space Polymer Chemistry 空间聚合物化学
SPC Space Project Center 【美国】航天规划中心
SPC Standard PayLoad Computer 标准有效荷计算机
SPCAC Space Crisis Action Center 航天危机行动中心
SPCC STS Processing Control Center 【美国】空间运输系统操作控制中心
SPCS Servicing Performance Checkout System 检修效能检测系统
SPCS Supplementary Propulsion Control Set 辅助推进控制装置
SPCU Simulation Process Control Unit 模拟处理控制装置
SPDA Software Preliminary Design Audit 软件初步设计审查
SPDB Subsystem Power Distribution Box 分系统配电箱
SPDCI Standard Payload Display and Control Interface 标准有效载荷显示与控制接口
SPDM Special Purpose Dexterous Manipulator 【美国空间站】专用灵巧机械臂
SPDMS Shuttle Processing Data Management System 【美国】航天飞机处理数据管理系统
SPDR Software Preliminary Design Review 软件初步设计评审
SPDS Stabilized Payload Deployment System 稳定有效载荷配置系统（美国航天飞机释放和部署卫星的装置）
SPEAR Small Payload Ejection and Recovery 【美国航天飞机】小型有效载荷弹射和回收
SPEAR Space Power Experiments Aboard Rocket 箭载太空能源实验
SPEAR Strike Projection Evaluation

and Antiair Research 攻击（力量）投送评估与防空研究

SPEARD Satellite Photo-Electronic Analog Rectification Device 卫星光电子模拟校正装置

SPEARS Satellite Photo-Electronic Analog Rectification System 卫星光电子模拟校正系统

SPEARS Shuttle Processing Electronic Archival and Retrieval System 【美国】航天飞机处理电子存档与检索系统

SPECS Spacecraft for Environmental Control Studies 环境控制研究航天器

SPED Supersonic Planetary Entry Decelerator 超声速进入行星大气减速器

SPED Supersonic Planetary Experiment Development 超声速行星实验发展

SPEED Study and Performance Efficiency in Entry Design 进入（大气层）的设计研究与性能效能

SPEED System Planning, Engineering, and Evaluation Device 系统规划、工程和鉴定装置

SPELDA Structure Porteuse Externe Lancement Double Ariane 【欧洲空间局】"阿里安"双星发射外支承支架

SPELDA Structure Porteuse Externe Pour Lancement Double Ariane 【欧洲空间局】"阿里安"火箭双星发射系统外部支架

SPELTRA Structure Porteuse Externe Lancement Triple Ariane 【欧洲空间局】"阿里安"三星发射外支承结构

SPEMS Space Environment Monitoring System 空间环境监视系统

SPERT Schedule Performance Evaluation and Review Technique 进度性能评审技术

SPERT Scheduled [Simplified] Program Evaluation and Review Technique 定期（简化）计划评审技术

SPET Solid-Propellant Electric Thruster 固体推进剂的电控推进器

SPETC Solid Propellant Electro-Thermal-Chemical 固体推进剂电热化学

SPEX Space Plasma Experiment 空间等离子体实验

SPF Single Point Failure 单点故障；单点失效

SPF Software Production Facility 软件生产设施

SPF Space Power Facility 航天动力设施

SPF Spacecraft Processing Facility 航天器处理厂房

SPF Spacelab Processing Facility 空间实验室处理设施

SPFA Single Point Failure Analysis 单点故障分析

SPFE Special Projects Flight Experiments 特殊项目飞行实验

SPFP Single Point Failure Potential 单点失效潜能

SPG Signal Point Grounding 单点接地

SPGG Solid-Propellant Gas Generator 固体推进剂燃气发生器

SPGS Space Guidance System 航天制导系统

SPH Space Heater 空间加热器

SPHERES Synchronous Position Keeping, Orbit Reserved, Redirection Experimental Satellite 同步位置保

持、轨道预定、再定向实验卫星
SPHM Structural Prognostics and Health Management 结构预测与健康管理
SPI Single Program Initiation 单个程序启动
SPIAP Shuttle/Payload Integration Activities Plan 航天飞机/有效载荷总装操作计划
SPIBS Satellite Positiveion-Beam System 卫星正离子束系统
SPICE Smoke Point in Coflow Experiment 同向流动烟点实验
SPICE Space Integrated Controls Experiment 航天综合控制实验
SPICE Spacecraft, Planets Instrument Constants and Events 航天器、行星仪器常数与事故
SPICE Spacelab Payload Integration and Coordination in Europe 欧洲空间实验室有效载荷组装与协调机构
SPICE Superthermal Plasma Investigation of Cometary Environments 彗星环境的超热等离子体研究
SPIDER Space Inspection Device for Extravehicular Repairs 【欧洲】舱外修理空间检查装置
SPIDF Support Planning Identification File 保障规划识别文件
SPIDPO Shuttle Payload Integration and Development Program Office (JSC) 【美国约翰逊航天中心】航天飞机有效载荷集成与研发项目办公室
SPIF Shuttle Payload Integration Facility 【美国】航天飞机有效载荷装配设施
SPIF Spacecraft Processing and Integration Facility 航天器处理与组装厂房
SPIF Standard Payload Interface Facility 标准有效载荷接口设备
SPII Shuttle Program Implementation Instruction 【美国】航天飞机项目实施指南
SPILMA Structure Porteuse Intégrée pour Lancement Multiple Ariane 【欧洲空间局】"阿里安"多星发射整体式支承结构
SPIMS Shuttle Program Information Management System 【美国肯尼迪航天中心】航天飞机项目信息管理系统
SPIMS Strategic Program Information Management System 战略项目信息管理系统
SPIN Superconductive Precision Inertial Navigation 超导精密惯性导航
SPINE Shared Program Information Network 共用项目信息网络
SPIP Solid Propulsion Integrity Program 固体推进（技术）综合计划
SPIRBM Solid-Propellant Intermediate Range Ballistic Missile 固体推进剂中（远）程弹道导弹
SPIRE Space Inertial Reference Equipment 空间惯性参考（基准）装置
SPIRE Space Performance in Radiation Environments 辐射环境中的航天性能
SPIS Spacecraft Plasma Interaction System 航天器等离子体相互作用系统
SPIU System Power Interface Unit 系统电源接口装置
SPL Self-Propelled Launcher 自动推进发射装置
SPL Single Propellant Loading 单一推进剂加注
SPL Software Programming Language

软件程序设计语言
SPL　Sound Pressure Level　声压级
SPL　Space Programming Language　航天程序设计语言
SPL　Space Programs Laboratory　航天计划实验室；航天项目实验室
SPL　System Programming Language　系统程序设计语言
SPLIT　Space Program Language Implementation Tool　航天程序语言实现工具
SPLL　Self Propelled Launcher Loader　自行发射架装填器
SPLM　Space Programming Language Machine　航天程序设计语言机
SPLML　Space Programming Language Machine Language　航天程序设计语言机语言
SPM　Small Power Module　小型动力舱
SPM　Solid Propellant Motor　固体火箭发动机
SPM　Sun Probe-Mars　【美国】"火星号"太阳探测器
SPME　Solar Proton Monitoring Experiment　太阳质子监测实验
SPMS　Special Purpose Manipulator System　特殊用途机械臂系统
SPO　Sky Polarisation Observatory　【国际空间站】天空偏极观测台
SPOC　Shuttle Portable On-Board Computer　【美国】航天飞机便携式箭载计算机
SPOC　Single Point Orbit Calculator　单点轨道计算器
SPOC　Space Command Operations Center　航天指挥作战中心
SPOE　Standard Payload Outfitting Equipment　标准有效载荷装配设备
SPOF　Science Planning and Operations Facility　科学计划与操作设施
SPOM　STS (Space Transportation System) Planning and Operations Management　【美国】空间运输系统与操作管理
SPORT　Sky Polarization Observatory　天空极化天文台
SPORT　Space Probe Optical Recording Telescope　空间探测器光学记录望远镜
SPOT　Satellite Position and Track　卫星定位和跟踪
SPOT　Satellite/Système Probatoire/Pour l'Observation de la Terre　斯波特卫星（地球观测/能力测试系统/卫星）
SPOT　Speed, Position, Track　速度、位置与跟踪
SPOTS　Satellite Positioning and Tracking System　卫星定位和跟踪系统
SPP　Science and Power-Supply Platform　科学动力供应平台
SPP　Sensor Pointing Platform　遥感器定向平台
SPP　Simulation Planning Panel　模拟（仿真）规划操纵台
SPP　Solar Photometry Probe　太阳光度测定器
SPP　Solar Physics Payload　太阳物理有效载荷
SPPA　Single Pyro Released Point Attachment System　单一火工品释放点附加装置
SPPLF　Starsem Payload Processing and Launch Facilities　【俄罗斯】斯达西姆有效载荷处理和发射设施
SPPP　Spacelab Payloads Processing Project　【美国】天空实验室有效载荷处理项目
SPR　Software Problem Report　软件

故障报告

SPR Solid Propellant Rocket 固体推进剂火箭

SPRAG STS (Space Transportation System) Payloads Requirements and Analysis Group 【美国】空间运输系统有效载荷要求与分析组

SPRD Single Pallet Rotation Device 单平台旋转装置

SPRE Solid Propellant Rocket Engine 固体推进剂火箭发动机

SPREE Solid Propellant Rocket Exhaust Effect 固体推进剂火箭排气效应

SPRIAD Solid Propellant Rocket Intercept Air-Directed Defense 固体推进剂火箭拦截对空防御

SPRIM Solid-Propellant Rocket-Intercept Missile 固体推进剂火箭拦击导弹

SPRIOS Solid Propellant Rocket Intercept Operation Shelter 固体推进剂火箭拦截作战掩体

SPRN Soviet System for Missile Attack Warning 苏联导弹攻击报警系统

SPRT Sequential Probability Ratio Test 序列概率比测试

SPS Satellite Positioning System 卫星定位系统

SPS Satellite Power System 卫星电源系统

SPS Saturn Propulsion System 【美国】"土星"运载火箭推进系统

SPS Secondary Power System 辅助电源系统

SPS Secondary Propulsion System 辅助推进系统

SPS Service (Module) Propulsion System 勤务舱推进系统

SPS Shuttle Procedure Simulator 【美国】航天飞机（操作）程序模拟器

SPS Signal Processing Subsystems 信号处理分系统

SPS Solar Power Satellite 太阳动力卫星

SPS Solar Power Station 空间的太阳能发电站

SPS Solar Probe Spacecraft 太阳探测航天器

SPS Space Power System 航天动力系统

SPS Spacecraft Propulsion System 航天器推进系统

SPS Spacelab Pallet System 【美国】天空实验室平台系统

SPSA Solid Propellant Storage Area 固体推进剂贮存区

SPSS Science Planning and Scheduling System 科学规划与调度系统

SPSTP Solid Propellant Rocket Static Test Panel 固体推进剂火箭静态测试组；固体推进剂火箭静态测试板

SPTB Space Performance Test Battery 空间效能测试电池

SPU Solar Power Unit Demonstrator 太阳能动力装置演示器

SPUR Space Power Unit Reactor 航天动力装置反应堆

SPURT Spinning Unguided Rocket Trajectory 无制导自旋火箭弹道

SPVPF Shuttle Payload Vertical Processing Facility 航天飞机有效载荷垂直处理厂房

SPY-1 Naval Phased-Array Radar 【美国】海军相控阵列雷达

SQA Software Quality Assurance 软件质量保证

SQM Software Quality Management 软件质量管理

SQM Software Quality Maturity 软件质量成熟度

SQT System Qualification Testing 系统合格鉴定测试

SQUID Superconducting Quantum Interference Device 超导精密惯性导航

SQUIRT Satellite Quick Research Testbed 卫星快速研究试验台

SQV Software Quality Verification 软件质量验证

SR Solrad 【美国】太阳辐射监测卫星

SR Sounding Rocket 探空火箭

SR Space Rocket 航天火箭

SR&M Safety, Reliability and Maintainability 安全性、可靠性与维护性

SR&Q Safety, Reliability, and Quality 安全性、可靠性与质量

SR&QA Safety, Reliability and Quality Assurance 安全性、可靠性与质量保证

SRA Schedule Risk Assessment 进度风险评估

SRA Software Requirements Analysis 软件要求分析

SRA Support Requirements Analysis 保障要求分析

SRA System Requirements Analysis 系统要求分析

SRAD Shuttle Radiator Assembly Demonstration 【美国国家航空航天局】航天飞机散热器装置演示器

SRADC Space and Reconnaissance Analysis and Demonstration Center 航天与侦察分析与演示中心

SRADSPWS Short Range Air Defense Selfprotect Weapon System 近程自防护防空武器系统

SRAE Solar Radio Astronomy Experiment 太阳射电天文实验

SRAG Space Radiation Analysis Group 空间辐射分析组

SRAM Short-Range Air-to-Surface Attack Missile 近程空对地攻击导弹；近程空对舰攻击导弹

SRAM Short-Range Attack Missile 近程攻击导弹

SRARM Short-Range Anti-Radar Missile 近程反雷达导弹

SRATS Solar Radiation and Thermosphere Satellite 【日本】太阳辐射与热层卫星

SRB Solid Rocket Booster 固体火箭助推器

SRBAB Solid Rocket Booster Assembly Building 固体火箭助推器组装厂房

SRBDF Solid Rocket Booster Disassembly Facility 固体火箭助推器拆卸设施

SRBM Short Range Ballistic Missile 近程弹道导弹

SRBMD Short Range Ballistic Missile Defense 近程弹道导弹防御

SRBPF Solid Rocket Booster Processing Facility 固体火箭助推器处理厂房

SRBv Solid Rocket Booster (Fivesegment) 固体火箭助推器（五段式）

SRC Sample Return Container 样品返回容器

SRC Satellite Receiving Center 卫星接收中心

SRCC Space Research Coordination Center 【美国】航天研究协调中心

SRD Shuttle Requirements Definition 【美国】航天飞机要求定义

SRD Shuttle Requirements Document

【美国】航天飞机要求文件

SRD Systems Requirements Document 系统要求文件

SRDE Smallest Replacement Defective Element 最小可替换失效元件

SRDH System Requirements Definition Handbook 系统要求定义手册

SRDS Standard Reference Data System 标准参考数据系统

SRE Software Risk Evaluation 软件风险评估

SRE Sounding Rocket Experiment 探空火箭实验

SRE STDN Ranging Equipment 航天飞行跟踪与数据网测距设备

SREL Space Radiation Effect Laboratory 航天辐射效应实验室

SREM Software Requirements Engineering Methodology 软件需求工程研究方法

SREMP Source Region Electromagnetic Pulse 源域电磁脉冲

SRES Space Radiation Evaluation System 航天辐射鉴定系统

SRET Software Reliability Engineered Testing 软件可靠性工程测试

SRET System Requirements Evaluation Tool 系统需求评价工具

SRF Shuttle Refurbishment Facility 【美国】航天飞机整修设施

SRFCS Self-Repairing Flight Control System 自修复飞行控制系统

SRHIT Small Radar Homing Intercept Technology 小型寻的拦截技术

SRIM Short-Range Intercept Missile 短程拦截导弹

SRKKV Space-Based Rocket-Launched Kinetic Kill Vehicles 天基火箭发射动能杀伤器

SRL Shuttle Radar Laboratory 【美国】航天飞机雷达实验室

SRL System Readiness Level 系统成熟度

SRLD Small Rocket Lift Device 小型火箭升降装置

SRM Safety, Reliability and Maintainability 安全、可靠性和可维护性

SRM Sensor Response Model 传感器响应模型

SRM Small Rocket Motor 小型火箭发动机

SRM Solid Rocket Motor 固体火箭发动机

SRM Standard Reference Material 标准参照材料

SRM Super Rocket Modifications 超级火箭改装

SRM&QA Safety, Reliability, Maintainability and Quality Assurance 安全性、可靠性、可维护性和质量保证

SRMP Sounding Rocket Measurement Program 探空火箭测量项目

SRMS Shuttle Remote Manipulator System 【美国】航天飞机遥控机械臂系统

SRMU Solid Rocket Motor Upgrade 【美国】固体火箭发动机升级

SRN Satellite Radio Navigation 卫星无线电导航

SRO System Readiness Objective 系统成熟度目标

SROS Sample Return Orbiter Segment 样品返回轨道器段（火星取样返回飞行器）

SROSS Stretched Rohini Satellite Series 【印度】扩展型"罗西尼"卫星系统

SRP Sounding Rocket Program 探空火箭项目

SRP Spacecraft Readiness Panel 航天器准备状态控制板
SRP Supersonic Retro Propulsion 超声速制动火箭推进系统
SRP Supersonic Retro Propulsion 超声速制动推进
SRPSC State Research and Production Space Center (Russian Federation) 【俄罗斯】国家研究与生产航天中心
SRR Saturn Ring Rendezvous 【美国】土星光环交会任务
SRR Short-Range Recovery 近距离回收
SRR Shuttle Requirements Review 【美国】航天飞机要求评审
SRR Site Readiness Review 现场准备状态评审
SRR Software Readiness Review 软件成熟度评审
SRR Software Requirements Review 软件要求评审
SRR System Requirements Review 系统要求评审
SRRT Simultaneous Rotating and Reciprocating Technique 同时旋转运动与往复运动技术
SRS Satellite Radar Station 卫星雷达站
SRS Satellite Remote Sensing 卫星遥感
SRS Saturn Ring Surveillance 【美国】土星光环监测任务
SRS Simulated Remote Station 模拟远程站
SRS Small Research Satellite 小型研究卫星
SRS Software Requirements Specification 软件要求说明
SRS Solar Radiation Satellite 太阳辐射（测量）卫星
SRS Sounding Rocket System 探空火箭系统
SRS Space and Reentry System 航天与再入系统
SRS Space Recovery System 航天回收系统
SRS Spaceborne Reconnaissance System 天基侦察系统；星载侦察装置；航天器载侦察系统
SRS Strategic Reconnaissance Satellite 战略侦察卫星
SRS Support Requirements System 保障要求系统
SRSAT Solar Radiation Satellite 太阳辐射探测卫星
SRSB Segment Ready Storage Building 【固体发动机】分段准备贮存厂房
SRSF SRB Receiving and Subassembly Facility 固体火箭助推器接收与分装设施
SRSF SRB Refurbishment and Subassembly Facility 固体火箭助推器整修与分装设施
SRSR Schedule and Resources Status Report 进度与资源状态报告
SRSS Shuttle Range Safety System 【美国】航天飞机靶场安全系统
SRSS Soft Ride for Small Satellite 小卫星软搭载
SRSS Solar Radiation Simulator System 太阳辐射模拟器系统
SRT Satellite Regenerator Terminal 卫星再生器终端
SRT Shuttle Requirements Traceability 【美国】航天飞机要求可追溯性
SRT Software Requirements Traceability 软件要求可追溯性
SRT Supporting Research and Tech-

nology　保障研究与技术
SRTBM　Short Range Theater Ballistic Missile　短程威胁弹道导弹
SRTM　Shuttle Radar Topography Mission　【美国】航天飞机雷达地形测绘任务
SRTS　System Response Time Simulator　系统响应时间模拟器
SRU　Suspension and Release Units　悬挂与投放装置
SRUS　Solid Rocket Upper Stage　固体火箭上面级
SRV　Satellite Reentry Vehicle　卫星再入飞行器
SRV　Sensor Rendezvous　交会传感器
SRV　Single Reentry Vehicle　单个再入飞行器
SRV　Space Reentry Vehicle　空间再进入飞行器
SRV　Space Rescue Vehicle　空间救援飞行器
SRV　Surface Roving Vehicle　月面漫游车
SRV-D　Short Range Vehicle, Delta Compatible　近程飞行器("德尔它"兼容发射)
SRV-DE　Short Range Vehicle, Delta-Expandable　短程飞行器（美国用渐进型"德尔它"火箭发射的一种轨道机动飞行器）
SS　Satellite Switching　卫星交换（技术）
SS　Satellite System　卫星系统
SS　Science Simulator　科学模拟器
SS　Simulator System　模拟（仿真）系统
SS　Space Science　空间科学
SS　Space Segment　空间段
SS　Space Shuttle　航天飞机
SS　Space Station　空间站
SS　Space Suits　航天服
SS　Spaceship　太空船；宇宙飞船
SS　Stationary Satellite　地球静止轨道卫星
SS&A　Space Systems and Applications　空间系统与应用
SS-ANARS　Space Sextant, Autonomous Navigation and Attitude Reference System　航天六分仪、自主导航与姿态参考系统
SSA　Shuttle Simulation Aircraft　【美国】模拟航天飞机的飞机
SSA　Software Safety Analysis　软件安全分析
SSA　Space Situational Awareness　空间态势感知
SSA　Space Structure Assembly　空间结构装配
SSA　Space Suit Assembly　全套航天服
SSA　SPM Storage Area　固体发动机储存区
SSA　Staging and Support Area　【火箭】级间分离与支承面
SSA　System Safety Assessment　系统安全评估
SSA OPS　Space Situational Awareness Operations　空间态势感知作战
SSAA　Space Science Analysis Area　航天科学分析领域
SSAT　Shuttle Service and Access Tower　【美国】航天飞机检修与进入塔
SSAT　Space Station Airlock Trainer　国际空间站气闸室训练装置
SSAT　Space Station Assembly Technology　空间站装配技术
SSAV　Self-Sealing Aerospace Vehicle　自密封的航天飞行器；自密封的航天飞机；自密封的航空航天飞行器
SSB　Soft Service Building　无抗核加

固勤务大楼
SSB Solid Sub Booster 附属的固体助推器
SSBE Space Shuttle Booster Engine 【美国】航天飞机助推器发动机
SSBE Space Shuttle Booster Engine 航天飞机助推器发动机
SSBM Surface-to-Surface Solid-Fuel Ballistic Missile 地对地固体燃料导弹
SSBS Surface-to-Surface Ballistic System 地对地弹道导弹体系
SSBUV Shuttle Solar Back Scatter Ultraviolet Instrument 【美国】航天飞机太阳后向散射紫外仪
SSC NASA Stennis Space Center 美国国家航空航天局斯坦尼斯航天中心
SSC Satellite Situation Center 【美国】卫星状态中心
SSC Simulated Spacecraft 模拟航天器
SSC Simulation Support Center 模拟（仿真）支持（保障）中心
SSC Single-Stage Command 单级指令
SSC Site Selection Criteria 发射场选择标准
SSC Space Suit Communications 航天服通信系统
SSC Space Surface Combustion 固体面燃烧（航天飞机上的实验）
SSC Space Surveillance Center 航天监视中心
SSC Space System Center 【美国】航天系统中心
SSC Space System Console 航天系统控制台
SSC Stellar Simulator Complex 星体模拟器综合体
SSC Stennis Space Center 【美国国家航空航天局】斯坦尼斯航天中心
SSCC Space Station Control Center 空间站控制中心
SSCC Support Services Control Center 保障勤务控制中心
SSCCC Space Surveillance Command and Control Center 航天监视系统指挥控制中心
SSCE Solid Surface Combustion Experiment 固体表面燃烧实验
SSCF Space Subsystem Control Facility 航天子系统控制设备
SSCHS Space Shuttle Cargo Handling System 【美国】航天飞机货物装卸系统
SSCM Surface-to-Surface Cruise Missile 地对地巡航导弹
SSCP Small Self-Contained Payload 小型独立有效载荷（航天飞机有效载荷）
SSCS Satellite and Space Communication Systems 卫星和航天通信系统
SSCS Space Station Communication System 空间站通信系统
SSCS Space Suit Communication System 航天服内装通信系统；航天服通信系统
SSCS Synchronous Satellite Communication System 同步卫星通信系统
SSCSP Space Shuttle Crew Safety Panel 航天飞机乘员安全组
SSCTS Space Station Communication and Tracking System 空间站通信和跟踪系统
SSCV Solar Sail Cargo Vehicle 太阳帆运货飞行器（美国拟议中向火星运货的飞行器）
SSD Satellite System Development

卫星系统开发
SSD System Summary Display 系统汇总显示
SSDA Solid State Demonstration Array 固态演示验证阵列
SSDA Space System Development Agreement 空间系统发展协议
SSDC (U.S. Army) Space and Strategic Defense Command 【美国陆军】航天与战略防御司令部
SSDC Space Science Data Center 【美国】空间科学数据中心
SSDD Software System Design Document 软件系统设计文件
SSDD System/Segment Design Description 系统/分段设计说明
SSDF Space Science Development Facility 航天科学研发设施
SSDH Subsystem Data Handbook 分系统数据手册
SSDMS Space Station Data Management System 空间站数据管理系统
SSDP Satellite System Development Plan 卫星系统研发计划
SSDS Space Science Data System 空间站数据系统
SSDS Space Shuttle Display and Simulation 【美国】航天飞机显示与模拟
SSDS Space Station Data System 空间站数据系统
SSE Small Science Experiment 小型科学实验
SSE Solar System Exploration 太阳系探索
SSE Space Shuttle Engines 【美国】航天飞机发动机
SSE Space Shuttle Experiment 【美国】航天飞机实验
SSE Space Surveillance Experiment 空间监视实验
SSE Spacecraft Simulation Equipment 航天器模拟设备
SSE Subsystem Element 分系统部件
SSE Subsystem Support Equipment 分系统保障设备
SSE System Safety Engineer 系统安全工程师
SSE System Security Engineering 系统安全工程
SSE System Simulator Environment 系统模拟（仿真）器（装置）环境
SSECO Second-Stage Engine Cutoff 第二级（火箭）发动机关机
SSECS Space Station Environmental Control System 空间站环境控制系统
SSEE Security Software Engineering Environment 安全软件工程环境
SSEL Space Science and Engineering Laboratory 航天科学与工程实验室
SSEM Space System Effectiveness Model 航天系统效能模型
SSEMU Space Shuttle Extravehicular Mobility Unit 航天飞机舱外机动装置
SSEMU Space Shuttle-Extra-Vehicular Mobility Unit 【美国】航天飞机舱外机动装置
SSEOS Space Shuttle Engineering and Operations Support 航天飞机工程与操作保障
SSF Space Simulation Facility 空间模拟设备设施
SSF Space Station Freedom 【美国】"自由号"空间站
SSF SRB Storage Facility 【美国】固体火箭助推器存储设施
SSFF Space Shuttle Furnace Facility

【美国】航天飞机加热炉设备

SSFGSS Space Shuttle Flight and Ground System Specification 【美国】航天飞机飞行与地面系统说明

SSFILSS Space Station Freedom Integrated Logistics Support System 【美国】"自由号"空间站综合后勤保障系统

SSFMB Space Station Freedom Manned Base 【美国】"自由号"空间站载人基地

SSFP Space Station Freedom Program 【美国】"自由号"空间站项目

SSFS Space Shuttle Functional Simulator 【美国】航天飞机功能模拟器（装置）

SSFUF Space Station Freedom Utilization Flight 【美国】"自由号"空间站利用飞行

SSGM Strategic Scene Generation Model 战略场景生成模型

SSGM Synthetic Scene Generation Model 合成场景生成模型

SSGS Standardized Space Guidance System 标准航天制导系统

SSGW Strategic Surface-to-Surface Guided Weapon 地对地战略制导武器

SSHA Subsystem Hazard Analysis 分系统危险分析

SSHP Single Shot Hit Probability 单发命中概率

SSHPF Space Station Hazardous Processing Facility 空间站危害处理设备

SSI Second-Stage Ignition 第二级（火箭）发动机点火

SSI Space Science Instrumentation 空间科学仪器

SSI Spacecraft System Integration 航天器系统综合

SSIBD Shuttle System Interface Block Diagram 【美国】航天飞机系统接口结构图

SSIE Skylab Systems Integration Equipment 【美国】"天空实验室"系统集成设备

SSIMU Solid State Inertial Measurement Unit 固态惯性测量装置

SSIP Systems Software Interface Processing 系统软件接口处理

SSIS Space Station Information System 空间站信息系统（包括数据处理和通信）

SSIS Spacecraft System Integration Support 航天器系统综合保障

SSISS Spacecraft System Integration Support Service 航天器系统综合保障勤务

SSITP Shuttle System Integrated Test Plan 【美国】航天飞机系统综合测试计划

SSKA Spectral Sensing for Kill Assessment 毁伤评估的光谱传感

SSL Solid State Laser 固态激光器

SSL Space Science Laboratory 【美国】空间科学实验室

SSL System Software Loader 系统软件加载装置

SSLORAN Skywave Synchronized LORAN 天波同步远程导航系统

SSLS Standard Space Launch System 标准航天发射系统

SSLV Standard Small Launch Vehicle 小型标准运载火箭

SSLV Standard Space Launch Vehicle 标准航天运载火箭

SSM Sample Support Module 样品保障舱

SSM Second Stage Motor 二级子发

动机

SSM Spacecraft Systems Monitor 航天器系统监视器

SSM Standard Systems Monitor 【美国】标准系统监控器

SSM Subsystem Support Module 分系统保障舱

SSM Surface-to-Surface Missile 地对地导弹

SSM System Support Module 系统保障舱

SSM/I Special Sensor Microwave Imagery 特种传感器微波成像

SSM/T2 Special Sensor Meteorology Temperature and Vapor 特种传感器气象温度与蒸汽

SSM/TI Special Sensor Meteorology Temperature 特种传感器气象温度

SSMB Space Shuttle Maintenance Baseline 【美国】航天飞机维修基线

SSMB Space Station Manned Base 空间站载人基地

SSMCC Space Shuttle Mission Control Center 【美国】航天飞机飞行任务控制中心

SSME Satellite System Monitoring Equipment 卫星系统监控设备

SSME Space Shuttle Main Engine 【美国】航天飞机主发动机

SSMEC Space Shuttle Main Engine Controller 【美国】航天飞机主发动机控制器（装置）

SSMECA Space Shuttle Main Engine Controller Assembly 【美国】航天飞机主发动机控制器组件（组装）

SSMEPF Space Shuttle Main Engine Processing Facility 【美国】航天飞机主发动机操作厂房

SSMF Space Station Millimeter Facility 空间站毫米波设备

SSMM Space Station Mathematical Model 空间站数学模型

SSMTR Sary Shagan Missile Test Range 【美国】"萨雷沙甘"导弹试验靶场

SSN Space Surveillance Network 【美国航天司令部】航天监视网

SSO Safety Signification Operation 重大安全操作

SSO Semi-Synchronous Orbit 半同步轨道

SSO Space Ship One 【美国】"太空一号"航天器

SSO Space Shuttle Orbiter 航天飞机轨道器

SSO Subsystem Operation (in Spacelab) 【美国天空实验室】分系统操作

SSO Sun Synchronous Orbit 太阳同步轨道

SSOC Space Station Operations Center 【美国】空间站操作中心

SSOC Space Support Operations Center 航天保障操作中心

SSOC Space Surveillance Operations Center 航天监视活动中心

SSOCC Space Station Operations and Control Center 【美国】空间站操作和控制中心

SSOMC Solid State Oxygen Monitor and Controller 固态氧气监测器与控制器

SSOP Space Systems Operating Procedure(s) 航天系统操作规程

SSOPC Space Shuttle Operations and Planning Center 航天飞机操作与计划中心

SSORB Satellite Simulating Observation and Research Balloon 卫星模拟观测与研究气球

SSOS Severe Storms Observational Satellite 强风暴观测卫星
SSP Small Sortie Payload 小型一次性有效载荷
SSP Space Shuttle Program 【美国】航天飞机项目
SSP Spacesuit System Project 航天服体系方案
SSP Standard Switch Panel 标准开关板
SSP Subsatellite Point 星下点（卫星当地垂线与地球表面的交点）
SSPAR Solid State Phased Array Radar 固态相控阵列雷达
SSPC Spacelab Stored Program Command 【美国】天空实验室存储项目指挥部
SSPC Steady State Power Compensation 稳定状态功率补偿
SSPD Shuttle System Payload Data 航天飞机系统有效载荷数据
SSPD Shuttle System Payload Definition (Study) 【美国】航天飞机系统有效载荷定义（研究）
SSPD Shuttle System Payload Description 【美国】航天飞机系统有效载荷说明
SSPDA Space Shuttle Payload Data Activity 【美国】航天飞机有效载荷数据工作
SSPDB Subsystem Power Distribution Box 分系统配电箱
SSPDS Space Shuttle Payload Data Study 【美国】航天飞机有效载荷数据研究
SSPF Space Station Processing Facility 空间站加工设施
SSPGSE Space Shuttle Program Ground Support Equipment 航天飞机项目地面保障设备
SSPL Space Shuttle Payload Launcher 航天飞机有效载荷发射器
SSPM Software Standards and Procedures Manual 软件标准与规程手册
SSPN Satellite System for Precise Navigation 卫星精确导航系统
SSPO Space Segment Project Office 空间段工程局（印度卫星—1的操作管理机构）
SSPO Space Shuttle Program Office 【美国】航天飞机项目办公室
SSPO Sun-Synchronous Polar Orbit 太阳同步极轨
SSPP Strategic Sustainability Performance Plan 战略可持续性效能计划
SSPP System Safety Program Plan 系统安全项目计划
SSPPSG Space Shuttle Payload Planning Steering Group 【美国】航天飞机有效载荷规划控制组
SSPS Satellite Solar Power Station 卫星太阳能发电站
SSPS Second Stage Propulsion System 【火箭】二子级推进系统
SSPS Space Shuttle Program Schedule 【美国】航天飞机项目进度
SSPS Space Station Propulsion System 空间站推进系统
SSPS Space-Based Solar Power System 天基太阳能发电系统
SSPTS Satellite de Sensoriamento Remoto 【巴西】遥感卫星
SSR Software Specification Review 软件规范评审
SSR Spin Stabilized Rocket 旋转稳定火箭
SSR Station to Shuttle Power Transfer System 空间站－航天飞机电源转

换系统
SSR System Requirements Review 系统要求评审
SSRCAS Secondary Surveillance Radar Collision Avoidance System 辅助监视雷达避撞系统
SSRM Second Stage Rocket Motor 第二级火箭发动机
SSRMF Space Station Remote Manipulator Facility 空间站遥控机械臂设施
SSRMP Space Sounding Rocket Measurement Program 航天探空火箭测量项目
SSRMS Space Station Remote Manipulator System 空间站遥控操纵臂系统
SSRPOS Space Station Rendezvous and Proximity Operations Simulator 空间站交会和贴近操作仿真器
SSRS Start-Stop-Restart System 起动—停车—再起动系统
SSRT Single Stage Rocket Technology 单级火箭技术
SSS Satellite Server System 卫星服务器系统
SSS Satellite Surveillance System 卫星监视系统
SSS Simplified Space Station 简易空间站
SSS Small Scientific Satellite 小型科学卫星（美国空间物理探测卫星）
SSS Small Solar Satellite 小型太阳卫星
SSS Small Standard Satellite 小型标准卫星
SSS Sound Suppression System 噪声抑制系统
SSS Space Sensor System 空间传感器系统
SSS Space Shuttle Simulation 航天飞机模拟
SSS Space Shuttle System 航天飞机系统
SSS Space Station Simulator 空间站模拟器
SSS Space Station System 空间站系统
SSS Space Surveillance System 空间监视系统
SSS Spacecraft Support Systems 航天器支撑系统；航天器保障系统
SSS Spacecraft System Support 航天器系统保障
SSS Spin Stabilized Spacecraft 自旋稳定航天器
SSS Stage Separation Subsystem 芯级分离分系统
SSS Strategic Satellite System 战略卫星系统
SSS Survivable Strategic Satellite 抗毁战略卫星
SSS Synchronous Satellite System 同步卫星系统
SSS System/Segment Specification 系统/部分说明（规格）
SSSA Space Station Solar Array 空间站太阳电池板
SSSCE Space Station Science Characterization Experiment 空间站科学特性实验；空间站科学表征实验
SSSD Second Stage Separation Device 火箭二子级分离装置
SSSD Space Shuttle Simulation Display 航天飞机模拟显示器
SSSM Systems Support Service Module 系统保障服务舱
SSSS Satellite Strategic Surveillance System 卫星战略监视系统
SSSS Space Shuttle and Space Station

【美国】航天飞机和空间站

SSSS　Space Shuttle System Specification　【美国】航天飞机系统说明

SST　Satellite Servicing Technology　卫星维护技术

SST　Satellite-to-Satellite Tracking　卫星－卫星跟踪

SST　Single System Trainer　单系统培训器（装置）

SST　Space Surveillance Telescope　空间监视望远镜

SST　Spacecraft System Test　航天器系统测试

SST　Standard Serviceability Test　标准适用性试验

SST　Standard Star Tracker　标准恒星跟踪器

SST　Structural Static Test　结构静态测试（试验）

SST　Subsystem Terminal on Spacelab　天空实验室上的分系统终端

SSTA　Support System Task Analysis　保障系统工作分析

SSTB　STSS Surrogate Test Bed　天基监视与跟踪系统代用试验（测试）台

SSTB　System Simulation Test Bed　系统模拟（仿真）试验（测试）台

SSTC　Space Science and Technology Center　【印度】空间科学和技术中心

SSTC　Space Shuttle Test Conductor　【美国】航天飞机测试（试验）指挥

SSTC　Space Systems Test Capabilities　航天系统测试能力

SSTC　Space Systems Test Console　航天系统测试控制台

SSTF　Space Simulation Test Facility　空间模拟测试设施

SSTF　Space Station Training Facility　【美国】空间站训练设施

SSTG　Space Shuttle Task Group　【美国】航天飞机任务组

SSTM　Satellite Synchronous Transport Model　卫星同步传输模型

SSTO　Single Stage to Orbit　单级入轨

SSTOV　Single Stage to Orbit Vehicle　单级入轨飞行器

SSTP　System Safety Technical Plan　系统安全技术计划

SSTS　Space-Based Surveillance and Tracking System　天基监视与跟踪系统

SSTS　Space Shuttle Transportation System　【美国】航天飞机运输系统

SSTS　Space Surveillance and Tracking System　航天监视与跟踪系统

SSTS　Space-Based Surveillance and Tracking Satellite　天基监视与跟踪卫星

SSTT　Small Satellite Thermal Technologies　小型卫星热防护技术

SSU　Sea Surveillance Unit　海洋监视卫星

SSU　Space Station Utilization Office　国际空间站利用办公室

SSU　Spaceborne Subsatellite　星载子卫星

SSUPRCB　Space Shuttle Upgrades Program Requirements Control Board　【美国】航天飞机改造项目要求控制委员会

SSUS　Solid Spinning Upper Stage　固体旋转上面级

SSUS　Spinning Solid Upper Stage　自旋固体火箭上面级

SSUS-A　SSUS for Atlas-Centaur Class Spacecraft　【美国】用于"宇宙神－

半人马座"级航天器的自旋固体火箭上面级

SSUS-D　SSUS for Delta Class Spacecraft　【美国】用于"德尔它"级航天器的自旋固体火箭上面级

SSUSP　Spinning Solid Upper Stage Project　自旋固体火箭上面级项目

SSV　Space Shuttle Vehicle　【美国】航天飞机

SSV/GC&N　Space Shuttle Vehicle/Guidance, Control and Navigation　航天飞机制导、控制和导航

SSVA　Signal Susceptibility and Vulnerability Assessment　信号灵敏度与脆弱性评估

SSVDT　Sight, Stabilized, Visual Data Transmitting　瞄准、稳定、视频数据传输

ST　Shock Test　振动试验；冲击试验

ST　Shuttle Trainer　【美国】航天飞机训练器

ST　Space Telescope　空间望远镜

ST　Space Tug　空间拖船

ST　Spacelab Technology　【美国】天空实验室技术

ST　Staging Timer　级间分离定时器

ST-ECF　Space Telescope European Coordinating Facility　太空望远镜欧洲坐标设施

ST/STE　Special Tooling/Special Test Equipment　特种工具/特种试验（测试）设备

STA　Shuttle Training Aircraft　【美国】航天飞机教练机

STA　Space Technology Application　航天技术应用

STA　Static Test Article　静态测试（试验）部件

STA　Straight-in Approach　直接进场着陆

STA　Strategic Threat Assessment　战略威胁评估

STA　Structural Test Airframe　结构试验机体

STA　Structural Test Article　结构测试部件

STA　System Threat Assessment　系统威胁评估

STACS　Satellite Telemetry and Computer System　卫星遥测和计算机系统

STADA　Satellite Tracking and Data Acquisition　卫星跟踪和数据搜集

STADAC　Space Tracking And Data Acquisition Computer　空间跟踪和数据采集计算机

STADAC　Station Data Acquisition and Control　空间站数据采集与控制

STADAN　Satellite Tracking and Data Acquisition Network　【美国】卫星跟踪与数据采集网

STADAN　Space Tracking and Data Acquisition Network　空间跟踪与数据采集网络

STADSS　Strategic Transportation Analysis Decision Support System　战略运输分析决策保障系统

STADU　System Termination and Display Unit　系统终端与显示设备

STAFF　Stellar Acquisition Feasibility Flight　星体搜索可行性飞行

STAFF/S　Simulated Training and Analysis for Fixed Facilities/Sites　固定设施和场地的模拟训练和分析

STAG　Shuttle Turnaround Analysis Group　【美国】航天飞机往返飞行期分析组

STAG　Smart Tactical Autonomous Guidance　智能战术自主制导

STAGE　Simulation Toolkit and Generation Environment 模拟工具与生成环境

STAMP Satellite Telecommunication Analysis and Modeling Program 卫星电信分析与建模程序

STAMPS Stabilized Translation and Maneuvering Propulsion System 稳定传送与运行推进系统

STANS Space Target Analysis and Networking System 空间目标分析与网络系统

STAP Space-Time Adaptive Processing 空间、时间自适应处理

STAR Satellite for Telecom, Application and Research 通信、应用与研究卫星

STAR Satellite Telecommunication with Automatic Routing 自动导航卫星通信

STAR Scientific and Technical Report 科学与技术报告

STAR Shuttle Turnaround Analysis Report 【美国】航天飞机往返飞行期分析报告

STAR Space Technology and Advanced Research 航天技术与先进研究

STAR Space Telescope for Analysis of Resources 用于资源分析的太空望远镜

STAR Space Thermionic Auxiliary Reactor 空间热离子辅助反应堆

STAR Stability Analysis for Re-Entry 重新进入的稳定性分析

STAR Strategic Threat Assessment Report 战略威胁评估报告

STAR System Test Analysis Report 系统试验（测试）分析报告

STARAN Stellar Attitude Reference and Navigation 星体姿态基准与导航

STARLAB Space Technology Applications And Research Laboratory 航天技术应用与研究实验室

STARLink Satellite Telemetry and Return Link 卫星遥测与返回链路

STARR Schedule, Technical, and Resources Report 进度、技术与资源报告

STARS Satellite Telemetry Automatic Reduction System 卫星遥测数据自动简化系统

STARS Service and Technique for Advanced Real-Time System 先进实时系统维修（服务）与技术

STARS Software Technology Adaptable Reliable Systems 软件技术自适应可靠系统

STARS Software Technology for Adaptable Reliable Systems 自适应可靠系统的软件技术

STARS Space Launch Operations (SLO) Telemetry Acquisition and Reporting System 航天发射操作遥测（数据）采集与报告系统

STARS Space-Based Telemetry and Range Safety 天基遥测与靶场安全

STARS Stabilized Twin-Gyro Attitude Reference System 稳定双陀螺姿态基准系统

STARS Stellar Tracking Attitude Reference System 星体跟踪姿态基准系统

STARS Strategic Tactical Airborne Range System 战略战术机载测距系统

STARS Strategic Target System 战略目标系统

STARS Surveillance and Target Attack

Radar System 监视与目标攻击雷达系统
STARS System Test and Astronaut Requirements Simulation 系统试验和航天员条件模拟
START NASA Standards and Technical Assistance Resource Tool 美国国家航天航空局标准与技术资源协助工具
START Space Technology and Re-Entry Tests 航天技术与再入试验
START Space Test and Reentry Technology 航天试验和再入技术
START Space Transport and Re-Entry Test 空间运输和再入试验
START Spacecraft Technology and Advanced Re-Entry Tests 【美国空军】航天器技术和先进再入试验
START System of Transportation Applying Rendezvous Technology 采用交会技术的运输系统
STAS Space Transportation Architecture Study 航天运输体系结构研究
STASS Space Transportation Architecture System Study 航天（空间）运输体系结构系统研究
STB Satellite Testbed 卫星试验台
STB Surveillance Test Bed 监视试验（测试）台
STB System Test Bed 系统试验（测试）台
STBE Space Transportation Booster Engine 【美国】航天运输助推器发动机
STC Satellite Test Center 【美国空军】卫星试验中心
STC Space Technology Center 航天技术中心
STC Space Test Center 航天（空间）试验中心
STC Standard Test Configuration 标准测试（试验）配置
STC Systems Test Complex 系统测试综合设施
STC Systems Test Control 系统试验（测试）控制
STCC Spacecraft Technical Control Center 航天器技术控制中心
STCDHS Spacecraft Telemetry Command Data Handling System 航天器遥测指令数据处理系统
STCS Service Module Thermal Control Subsystem 勤务舱热控制子系统
STD Software Test Description 软件试验（测试）说明
STD System Technology Demonstration 系统技术验证
STDCE Surface Tension Driven Convection Experiment 表面张力驱动对流实验
STDCF Space Telescope Data Capture Facility 太空望远镜数据采集设备
STDN Secure Tactical Data Network 保密战术数据网络
STDN Space Tracking and Data Network 航天跟踪与数据网点
STDN Space Tracking Data Network 空间跟踪数据网络
STDN Spacecraft Tracking and Data Network 航天器跟踪与数据网络
STDN Spaceflight Tracking and Data Network 【美国国家航空航天局】航天飞行跟踪和数据网
STE Satellite Test Equipment 卫星测试设备
STE Software Test Environment 软件测试环境
STE Special Test Equipment 专用试验设备

STE Suitability Test Evaluation 适用性试验鉴定

STEAMS Study Towards European Autonomous Manned Space-Flight 【欧洲空间局】欧洲自主载人航天研究

STEAP Simulated Trajectories Error Analysis Program 模拟弹道误差分析程序

STEAP Space Trajectories Error Analysis Program 航天轨道误差分析程序

STEDI Space Thrust Evolution and Disposal Investigation 空间推力发展与处理方法调查研究

STEM Science, Technology, Engineering, and Mathematics 科学、技术、工程与数学

STEM Spaceflight Trainer and Exercise Module 航天训练器与练习舱

STEM System Trainer and Exercise Module 系统训练器与练习舱

STEP Satellite Telecommunications Experiments Project 【印度】卫星通信试验计划

STEP Simulation, Test and Evaluation Process 模拟、试验与评估过程

STEP Software Test and Evaluation Project 软件试验与评审计划

STEP Space Technology Experiments Platform 航天技术实验平台

STEP Space Test Experimental Platform 【美国空军】空间试验实验平台

STEP Structural Technology Evaluation Program 结构工艺评审程序

STEP Surveillance Tracking and Experiment Program 【美国】监视跟踪和试验计划

STEPS Solar Thermionic Electric Power System 太阳能热离子电源系统

STEPS Solar Thermionic Electric(al) Propulsion System 太阳能热离子电推进系统

STEREO Solar Terrestrial Relations Observatory 太阳地球生物关系观测

STEREO Solar-Terrestrial Relations Observatory 【美国】太阳地球关系观测站

STEVE Space Tool for Extravehicular Emergencies 舱外应急用航天工具

STEX Sensor Technology Experiment 感应器技术实验

STF Spacecraft Test Facility 航天器测试设施

STF Spin Test Facility 旋转测试设施

STF Static Test Facility 静态试验（测试）设施

STG Satellite Terminal Guidance 卫星末端制导

STI Scientific and Technical Information 科学技术信息

STIFC Space Track Interim Fire Control 太空跟踪间歇射击控制

STIL Software Test and Integration Laboratory 软件测试与集成实验室

STILAS Scientific and Technical Information Library Automation System 科学与技术信息库自动化系统

STIM Slew Thruster Impulse Monitor 回转推进器脉冲监视器

STIMS Shuttle Thermal Infrared Multi-Spectral Scanner 航天飞机热红外多光谱扫描仪

STINFO Science and Technical Information 科学与技术信息

STINGS Stellar Inertial Guidance Sys-

tem　星体惯性制导系统
STIS　Space Telescope Imaging Spectrograph　空间望远镜成像光谱仪
STL　Space Technology Laboratory　【美国】空间技术实验室
STL　System Test Laboratory　系统测试实验室
STM　Signal Termination Module　信号终止模块
STM　Silo Test Missile　发射井试验导弹
STM　Software Test Management　软件测试管理
STM　Special Test Missile　特种实验导弹
STM　Static Test Model　静态测试（试验）模型
STM　Structural Test Model　结构试验模型
STM　Supersonic Tactical Missile　超声速战术导弹
STME　Space Transportation Main Engine　【美国】航天运输主发动机
STMS　Shuttle Thermal Infrared Multispectral Scanner　航天飞机热红外多光谱扫描仪
STN　Satellite Tracking Network　卫星跟踪网
STNORAOS　Space Track Network of Radar and Optical Sensors　雷达与光传感器太空跟踪网
STO　Science and Technology Objective　科学与技术目标
STO　Solar-Terrestrial Observatory　日地观测台
STO　Special Technical Operations　特种技术作战（操作）
STOBAR　Short Take-Off But Arrested Recovery　短距起飞拦阻索伴机回收

STOCC　Space Telescope Operations Control Center　空间望远镜操作控制中心
STOL　Saturn Test Oriented Language　【美国】"土星"火箭试验专用语言
STOL　Systems Test and Operations Language　系统测试与操作语言
STOL/MTD　Short Take-Off and Landing/Maneuvering Technology Demonstrator　短距起飞与着陆/机动（交会）技术演示验证装置
STOM　System Test Object Model　系统试验（测试）目标模型
STORMSAT　Storm Satellite　【美国】"风暴"卫星
STOVL　Short Take-Off and Vertical Landing　短距起飞与垂直着陆
STP　Satellite Tracking Program　卫星跟踪系统
STP　Self-Test Program　自测试项目
STP　Shuttle Technology Panel　【美国】航天飞机工艺控制台
STP　Solar Terrestrial Probe　日地探测器
STP　Solar Terrestrial Program　【欧洲】日地探测计划
STP　Solar Thermal Propulsion　太阳热能推进
STP　Space Technology Payload　空间技术有效载荷
STP　Space Technology Program Satellite　空间技术试验卫星
STP　Space Test Program　空间测试（试验）项目
STP　Subsystem Test Plan　分系统测试计划
STP-S　Space Test Program Satellite　【美国】空间试验计划卫星
STPH　Static Phase Error　静态相错误

STPS Space Test Program Satellite 空间实验计划卫星

STR System Test Review 系统测试评审

STRAMST S&T Reliance Assessment for Modeling & Simulation Technology 建模与模拟（仿真）技术的科技可靠性评估

STRAP Star Tracking Rocket Attitude Positioning (System) 恒星跟踪火箭姿态定位（系统）

STRAP Star Tracking Rocket Attitude Positioning System 恒星跟踪火箭姿态定位系统

STRAT Special Telemetry Research and Tracking 特殊遥测研究与跟踪

STRAT Stratospheric Tracers of Atmospheric Transport 大气传输的遗迹超常上升

STRATCOM US Strategic Command 美国战略司令部

STRD Space Track Research and Development Facility 航天跟踪研究发展设施

STREP Space Trajectory Radiation Exposure Procedure 空间弹道放射性照射程序

STRICOM Simulation, Training, and Instrumentation Command 模拟、训练与仪表指令

STROBE Satellite Tracking of Balloons and Emergencies 气球和紧急事件卫星跟踪

STRV Space Technology Research Vehicle 航天技术研究飞行器

STRV Space Test Research Vehicle 空间试验研究飞行器

STS Satellite Tracking Station 卫星跟踪站

STS Satellite Transit System 卫星转运系统

STS Shuttle Test Station 【美国】航天飞机测试站

STS Shuttle Transportation System 【美国】航天飞机运输系统

STS Simulation Training System 模拟训练系统

STS Skylab Terminal System 【美国】"天空实验室"终端系统

STS Space Technology Satellite 空间技术试验卫星

STS Space Transportation System 航天运输系统（即航天飞机）

STS Spacecraft Telecommunications System 航天器远距离通信系统

STS Stockpile to Target Sequence 从库存场到目标的顺序

STS Structural Transition Section 结构过渡段

STSC Software Technology Support Center 软件技术保障中心

STSLA Satellite-Tracked Submarine-launched Antimissile 卫星跟踪潜艇发射的反导弹

STSOPO Shuttle Transportation Systems Operations Program Office 【美国约翰逊航天中心】航天飞机运输系统操作项目办公室

STSP Solar-Terrestrial Science Program 【欧洲空间局】日地科学计划

STSR System Test Summary Report 系统测试汇总报告

STSS Space Tracking and Surveillance System 空间跟踪与监视系统

STT Small Tactical Terminal 小型战术终端

STT Spacecraft Terminal Trust 航天器末端推力

STT Spacelab Transfer Tunnel 【美国】

天空实验室转移通道

STTACS Space-Based Telemetry, Tracking, and Command Subsystem 天基遥测、跟踪与指挥分系统

STTP Space Technology Training Program 【美国国家航空航天局】航天技术训练项目

STTR Small Business Technology Transfer Research 小企业技术转移研究

STU Shuttle Test Unit 【美国】航天飞机测试装置

STU Special Test Unit 专用测试装置

STV Satellite Test Vehicle 卫星试验运载器

STV Separation Test Vehicle 【火箭级】分离试验飞行器

STV Space Transfer Vehicle 空间转移飞行器

STV Spacecraft Transport Vehicle 航天器运输车

STV Special Test Vehicle 特种试验飞行器

STV Standard Test Vehicle 标准试验飞行器

STV Supersonic Test Vehicle 超声速试验飞行器

STV System Test Vehicle 系统试验飞行器

STW Satellite Threat Warning 卫星威胁预警

STWAR Satellite Threat Warning and Attack Reporting 卫星威胁预警和攻击报告

SUDM Small User-Dedicated Military Communication Satellite 【美国】小型用户专用军事通信卫星

SUE LAN System Utilization Enhancement Local Area Network 【美国国家航空航天局】系统利用增强局域网

SUMC Space Ultra Reliable Modular Computer 航天超级可靠模块计算机

SUMMITS Scenario Unrestricted Mobility Model of Intra-Theater Simulation 战区内无限制机动模拟想定模型

SUMS Shuttle Upper-Atmosphere Mass Spectrometer 【美国】航天飞机上面大气层质量质谱仪

SUMTS Satellite-Universal Mobile Telecommunication System 卫星—宇宙间的移动通信系统

Super Survivable Solar Power Subsystem Demonstrator 耐用的太阳能电源分系统演示验证装置

SUR/VIAC Survivability/Vulnerability Information Analysis Center 抗毁能力/弱点信息分析中心

SURCAL Surveillance Calibration 【美国】监视校准卫星（瑟克尔卫星）

SURE Shuttle Users Review and Evaluation 【美国】航天飞机用户评审与鉴定

SURE Space Ultraviolet Radiation Environment 空间紫外线辐射环境

SURF DMSP Launch and Checkout Facility at VAFB 【美国】范登堡空军基地的国防气象卫星发射和检测设备

SURF Space Ultra-Vacuum Research Facility 航天超真空研究设施

SURFRR Standford University Radio Frequency Radiation Receiver 【美国】斯坦福大学射频辐射接收机（技术试验卫星）

SURS Standard Umbilical Retraction System 标准脐带回收系统

SURSEM Surveillance Radar Systems

Evaluation Model 监视雷达系统评估模型

SURTASS Surveillance Towed Array Sonar System 拖曳式阵列监视传感器系统

SURVSAT Survival Satellite Communication System 抗毁卫星通信系统；抗毁通信卫星

SURVSATCOMS Survivable Satellite Communications System 抗毁损卫星通信系统

SUSS Shuttle Upper Stage System 【美国】航天飞机上面级系统

SUT Satellite Under Test 处于测试中的卫星

SUTAGS Shuttle Uplink Text and Graphics Scanner 【美国】航天飞机上链文本与图形扫描仪

SUTARS Search Unit Tracking and Recording System 搜索装置跟踪与记录系统

SV Safety Valve 安全阀

SV Shuttle Vehicle 【美国】航天飞机运载器

SV Simulation Validation 模拟鉴定

SV Software Validation 软件认证

SV Space Vehicle 空间飞行器；航天器

SVA&C Shuttle Vehicle Assembly and Checkout 【美国】航天飞机组装与检查

SVAB Shuttle Vehicle Assembly Building 【美国】航天飞机组装厂房

SVAC Shuttle Vehicle Assembly and Checkout 【美国】航天飞机飞行器装配和检测

SVACS Shuttle Vehicle Assembly and Checkout Station 【美国】航天飞机飞行器装配和检测站

SVAFB South Vandenberg Air Force Base 【美国】范登堡空军基地南区

SVB Shuttle Vehicle Booster 【美国】航天飞机助推器

SVB Space Vehicle Booster 航天飞行器助推器

SVDS Space Vehicle Dynamic Simulator 航天飞行器动力模拟器

SVDSS Space Vehicle Data Systems Synthesizer 航天运载器数据系统合成器

SVE Software Validation Equipment 软件验证设备

SVER Spatial Visual Evoked Response 空间视觉诱发响应

SVEX Switching Program Verification Expert System 交换程序验证专家系统

SVF Software Validation Facility 软件验证设施

SVLA Steered Vertical Line Array 易操纵的垂直线性阵列（系统）

SVM Sensor Verification Mechanism 感应器验证机械装置

SVM State Variable Model 状态变量模型

SVMA Space Vehicle Mission Analysis 航天运载器任务分析

SVMF Shuttle Vehicle Mockup Facility 【美国】航天飞机模型设备

SVMF Space Vehicle Mockup Facility 飞船模拟实验室

SVMS Space Vehicle Motion Simulator 航天运载器运动模拟器

SVO Space Vehicle Operations 【美国肯尼迪航天中心】航天运载器操作

SVOD Soviet Aircraft Navigation and Landing System 苏联飞机导航和着陆系统

SVPF Space Vehicle Processing Facility 航天飞行器操作设施

SVR System Verification Review 系统验证评审

SVRR Software Verification Readiness Review 软件验证成熟度评审

SVRR System Verification Readiness Review 系统验证完备性评审

SVS Space Vehicle Simulator 航天器模拟装置

SVS Spinning Vehicle Simulator 自旋飞行器模拟器

SVS Suit Ventilation System 航天服通风系统

SVS Synthetic Vision System 【美国国家航空航天局】合成视觉系统

SVT Spacecraft Validation Test 航天器验证测试

SVTS Space Vehicle Test Supervisor 航天运载器试验监控装置

SW Space Wing 【美国】航天联队

SW/PM System Management/Performance Monitor 系统管理/性能监测

SWA Support Work Authorization 保障工作授权

SWAA Spacelab Window Adapter Assembly 【美国】天空实验室窗口适配器组件

SWACS Space Warning and Control System 航天告警与控制系统

SWAMR Small Warhead and Maneuverable Reentry 小型弹头与机动再入

SWAMRS Small Warhead and Maneuverable Reentry System 小型弹头与机动再入系统

SWAP Size, Weight and Power 大小、重量和功率

SWAS Submillimeter Wave Astronomy Satellite 【美国】亚毫米波天文卫星

SWAT Sidewinder Angle Tracking 【美国】"响尾蛇"导弹角度跟踪

SWAT Split Waveform Analysis Technique 分裂式波型分析技术

SWAT Stress Wave Analysis Technique 应力波分析技术

SWC Space Warfare Center 空间战中心

SWCE Solar Wind Composition Experiment 【美国"阿波罗"飞船登月航天员在月面上】太阳风成分实验装置

SWEAT Severe Weather Threat Index 极端天气威胁指数

SWFT Shuttle Wide-Field Telescope 【美国】航天飞机宽视场望远镜

SWFT Shuttle Wide-Field Telescope 航天飞机宽视场望远镜

SWG Software Working Group 软件工作组

SWI Space Wing Instruction 【美国】航天联队指南

SWIL Software-in-the-Loop 回路中的软件

SWIM Solar Wind Interplanetary Measurements 太阳风星际测量

SWIPE Simulated Weapon Impact Predicting Equipment 预测武器弹着点模拟设备

SWIS Satellite Weather Information System 卫星天气信息系统

SWL Strategic Weapon Launcher 战略武器发射装置

SWNT Single-Walled Carbon Nanotubes 单壁碳纳米管

SWOT Surface Water and Ocean Topography 地表水与海洋地形学

SWP Safe Working Pressure 安全工作压力

SWR Space Wing Regulation 【美国】

航天联队规定
SWS Saturn Workshop 【美国】"土星"运载火箭工厂
SWS Software Simulator 软件仿真器
SWS Space Weapon System 空间武器系统
SWS Space Weapon System Analysis 空间武器系统分析
SWS Strategic Warning Satellite 【美国】战略预警卫星
SWSA Spatial Weapons System Analysis 太空武器系统分析
SWSC Space and Warning System Center 空间与预警系统中心
SWT Supersonic Wind Tunnel 超声速风洞
SXTF Satellite X-Ray Test Facility 卫星X射线实验设备
SYCOM Synchronous Altitude Communications Satellite 地球同步高度通信卫星
SYDAS Synchro Data Acquisition System 同步机数据采集系统
SYLDA SYstème de Lancement Double Ariane 【欧洲空间局】"阿里安"火箭双星发射系统
SYLDSO SYstème de Lancement Double Soyuz/Payload Internal Carrying Structure 有效载荷内部支承结构
SYNCOM Hughes Spin Stabilized Spacecraft 【美国】休斯公司
SYNCOM Synchronous Communications Satellite 【美国】同步通信卫星（辛康卫星）
SYRACUSE Systeme de Radio Communication Utilisant un Satellite 【法国】锡拉库斯军用通信卫星系统
Sys C/O System Check Out 系统检测
Sys T&E System Test and Evaluation 系统试验（测试）与鉴定
SYSCOM Systems Command 系统指令
SYSML Systems Modeling Language 系统建模语言
SYSTIM Systematic Interaction Model 系统互作用模型
SYSTRAN Systems Analysis Translator 系统分析
SZ Surface Zone 地（水）面爆炸中心投影点
SZA Sur-Zone Array (Explosive System) 地面爆炸中心投影点排列（爆炸系统）

T

T　Temperature Rate　温度率
T　Throttle Command　扼流指令
T&A　Temperature and Altitude　温度与高度
T&A　Test and Adjust　测试和调整
T&A　Transcription and Analysis　记录与分析
T&BI　Turn and Bank Indicator　转弯与倾斜指示器
T&C　Target and Control　目标与控制
T&C　Telemetry and Command　遥测和指令
T&C　Test and Control　试验（测试）与控制
T&C　Tracking and Control　跟踪与控制
T&C/O　Test and Checkout　测试与检查
T&CD　Timing and Countdown　计时与倒计时
T&CP　Test and Checkout Procedure　测试与检查程序
T&D　Telemetry and Data　遥测技术（装置）与数据
T&D　Test and Diagnostic　测试与诊断
T&D　Transmission and Distribution　传输与分配；发射与分配
T&D　Transport and Diffusion　输送与扩散
T&D　Transportation and Docking　运送与入坞（库）检修（航天器对接）
T&DA　Tracking and Data Acquisition　跟踪与数据采集
T&DE　Test and Diagnostic Equipment　测试与诊断设备
T&DR　Tracking and Data Relay　跟踪与数据中继
T&DRE　Tracking and Data Relay Experiment　跟踪和数据中继实验
T&DS　Tracking and Data System　跟踪与数据系统
T&E　Test and Evaluation　测试与评定
T&E-EW　Test and Evaluation-Electronic Warfare　电子战试验与评定
T&EL　Test and Evaluation Laboratory　测试与评定实验室
T&G　Tracking and Guidance　跟踪和制导
T&O　Test and Operations　测试与操作
T&O　Training and Operations　训练与作战；训练与操作
T&P　Target and Penetration　目标与穿透
T&QA　Test and Quality Assurance　试验与质量保证
T&T　Transportation and Transportability　运输与可运输性
T&V　Test and Verification　测试与验证
T-　Time Prior to Launch　发射前时间
T-A　Thrust-Augmented　推力增大的；推力加强的
T-AGM　Missile Range Instruction Ship　导弹靶场测量船
T-O　Takeoff　起飞
T-UAV　Tactical Unmanned Aerial Vehicle　战术无人机

T.A.E	Time, Azimuth, Elevation	时间、射向和高度	
T.B.	Toggle Buffer	触发缓冲装置	
T.P.	Transition Period	过渡期	
T/A	Throat Area	喷口面积	
T/A	Turnaround	工作（检修）周期；往返飞行	
T/C	Technical Control	技术控制	
T/C	Telecommunications	远程通信	
T/C	Temperature Compensating	温度补偿	
T/C	Termination Check	终止检查	
T/C	Thrust Chamber	推力室；燃烧室	
T/D	Telemetry Data	遥测数据	
T/D	Time Delay	时间延误	
T/D	Touchdown	着地；降落	
T/E	Transporter/Erector	运输车/起竖车	
T/ESM	Targeting/Effects Synchronization Matrix	攻击（目标）效果同步矩阵	
T/F	Time of Fail	故障时间	
T/F	Time of Fall	降落时间	
T/F	Training Flight	飞行训练	
T/L	Timeline	时间线	
T/L	Total Loss	全部损失	
T/L	Transport/Launcher	运输/发射装置	
T/M	Telemetry	遥测	
T/M	Torque Meter Telemetering	扭矩遥测	
T/MGS	Transportable/Mobile Ground Station	移动式/机动地面站	
T/MPATCH	Target-to-Missile Patch Unit	目标至导弹碎片装置	
T/R	Transmitter/Receiver	发射机/接收机	
T/R	Turnaround Requirements	往返周期要求	
T/REA	Transmit/Receive Element Assembly (of A Radar)	雷达传输/接收要素组件	
T/RIA	Telemetry/Range Instrumented Aircraft	靶场遥测仪器飞机	
T/T	Terminal Timing	终端计时	
T/T	Timing/Telemetry	计时/遥测	
T/TCA	Thrust/Translation Control Assembly	推力/移位控制组件	
T/V	Thermal/Vacuum	热真空	
T/W	Thrust to Weight (Ratio)	推重（比）	
T^2	Technology Transfer	技术转让	
T^2E	Technical Training Equipment	技术训练设备	
T=	Time Equivalent (Used for Duration of Event)	时间当量	
TA	Target Acquisition	目标截获	
TA	Target Analysis	目标分析	
TA	Target Angle	瞄准角；进入角	
TA	Target Area	目标地域	
TA	Task Analysis	任务分析	
TA	Technical Architecture	技术体系结构	
TA	Technical Area	技术领域	
TA	Technical Assembly	技术组装；技术装配	
TA	Technology Assessment	工艺评估	
TA	Terminal Adaptation	终端适配	
TA	Terminal Adapter	终端适配器	
TA	Test Analyzer	测试分析器	
TA	Test Article	测试（试验）部件	
TA	Threat Assessment	威胁评估	
TA	Thruster Assembly	推进器装配（组件）	
TA	Type Availability	型号可用性	
TA&R	Test Analysis & Reporting	试验（测试）分析与通报	
TA&TP	Temporary Assembly and Test	

Procedure 临时装配与测试程序
TA(DRS) Target Acquisition, Designation and Reconnaissance System 目标截获、指示和侦察系统
TA/CE Technical Analysis and Cost Estimate 技术分析和成本估算
TA1 ROCKOT Low Rate Telemetry Device 【俄罗斯】"轰鸣"号运载火箭低速率遥测装置
TAA Tactical Assembly Area 战术装配区
TAA Temporary Access Authorization 临时访问授权
TAA Temporary Area Access 临时区域访问（进入）
TAADF TACAN Air-to-Air Direction Finding 战术空中控制与导航系统空对空测向
TAADM TACAN Air-to-Air Distance Measuring 战术空中控制与导航系统空对空测距
TAAF Test, Analyze and Fix 试验、分析与改进
TAAM Testing Air-to-Air Missile 试验性空-空导弹
TAAM Tomahawk Airfield Attack Missile 【美国】"战斧"式机场攻击导弹
TAAM Training Air-to-Air Missile 训练用空-空导弹
TAB Tactical Atomic Bomb 战术原子炸弹
TAB Trunk Test Adaptation Board 中继测试适配器
TABCASS Tactical Air Beacon Command and Surveillance System 战术防空信标指挥与监视系统
TABMS Tactical Air Battle Management System 战术空中作战管理系统
TABMS Tactical Antiballistic Missile System 战术反弹道导弹系统
TABS Tactical Airborne Beacon System 战术机载信标系统
TABS Technology Area Breakdown Schedule 技术领域分类体系结构
TABSAC Targets and Backgrounds Signature Analysis Center 目标与背景特征分析中心
TABSDA Tactical Airborne Sonar Decision Aid 战术机载声呐目标判断辅助装置
TABSTONE Target Against Background Signal-to-Noise Evaluation 目标与背景信号噪声比评估
TAC Tactical Air Command 战术空中指挥部
TAC Target Acquisition Capability 目标截获能力
TAC Target Acquisition Center 目标截获中心
TAC Technical Application Center 技术应用中心
TAC Technical Assistance Center 技术协助中心
TAC Telemetry and Command Computer 遥测与指挥计算机
TAC Terminal Access Control 终端进入控制
TAC Terminal Attack Control 终端攻击控制
TAC Threat Adaptive Countermeasures 威胁自适应对抗措施
TAC Total Average Cost 总平均成本
TAC Tracking Accuracy Control 跟踪精度控制
TAC/RA TACAN Radar Altimeter 塔康雷达测高仪
TACAN Tactical Air Command and Navigation System 战术空中指挥

与导航系统
TACAN Tactical Air Control and Navigation 战术空中控制与导航
TACAN Tactical Air Navigation 战术空中导航
TACAN Tactical Control and Navigation 战术控制与导航
TACBE Tactical Beacon 战术信标
TACC Tactical Air Command Center 战术空中指挥中心
TACC Tactical Air Control Center 战术空中控制中心
TACC Tracking and Control Center 跟踪与控制中心
TACCIMS Theater Automated Command and Control Information Management System 战区自动化指挥与控制信息管理系统
TACCOPS Tactical Air Control Center Operations 战术空中指挥作战中心
TACCS Tactical Air Command and Control System 战术空中指挥与控制系统
TACCSF Theater Air Command and Control Simulation Facility 战区空中指挥与控制模拟设施
TACDACS Target Acquisition and Data Collection System 目标截获与数据采集系统
TACDAR Tactical Detection and Reporting 战术探测与通报
TACDEW Tactical Advanced Combat Direction and Electronics Warfare Complex 先进战术战斗指挥与电子战综合设施
TACDS Threat Adaptive Countermeasure Dispenser System 威胁自适应干扰投放系统
TACELIS Tactical Automated Communication Emitter Location and Identification System 战术自动通信辐射源定位识别系统
TACEVAL Tactical Evaluation 【美国空军】战术使用评估
TACFAM Tactical Frequency Assignment Model 战术频率分配模型
TACINTEL Tactical Intelligence Information 战术情报信息
TACIT Time Authenticated Cryptographic Identity Transmission 时间鉴别的密码识别传输
TACLAN Tactical Landing System 战术着陆系统
TACLAN Tactical Local Area Network 战术局域网
TACMS Tactical Missile System 战术导弹系统
TACNAVSAT Tactical Navigation Satellite 战术导航卫星
TACO Task Control and Location System 任务控制与定位系统
TACO Test and Checkout Operations 测试与检查操作
TACODA Target Coordinated Data 目标坐标数据
TACOM Tactical Area Communications 战术区域通信
TACOMSAT Tactical Communications Satellite 战术通信卫星
TACON Tactical Control 战术控制
TACOS Tactical Air Combat Simulation 战术空中作战模拟
TACS Theater Air Control System 战区空中控制系统
TACS Thruster Attitude Control System 推进器姿态控制系统
TACS Tracking and Command Station 跟踪控制站
TACSAT Tactical Communications Satellite 【美国】战术通信卫星

TACSATCOM Tactical Satellite Communication 战术卫星通信

TACSI Tactical Air Control System Improvements 战术空中控制系统改进

TACSIM Tactical Simulation 战术模拟（仿真）

TACT (NASA) Transonic Aircraft Technology 【美国国家航空航天局】跨声速飞机技术

TACTASS Tactical Towed Array Sonar System 战术拖曳声呐阵列系统

TACTERM Tactical Terminal 战术终端

TACTIFS Tactical Integrated Flight System 战术综合飞行指示系统

TACTS Tactical Aircrew Combat Training System 战术空中人员作战训练系统

TACTS Telemetry, Acquisition, and Command Transmission System 遥测、截获、指令传送系统

TACWARS Tactical Air Warfare Simulation 战术空战模拟

TAD Tactical Air Defense 战术空中防御

TAD Technical Acceptance Demonstration 技术验收演示（验证）

TAD Theater Air Defense 战区空中防御

TAD C² Theater Air Defense Command and Control 战区空中防御指挥与控制

TADAR Target Acquisition, Designation and Aerial Reconnaissance 目标截获、标示与空中侦察

TADARS TROPO Automated Data Analysis Recorder System 对流层散射通信系统自动数据分析记录系统

TADAS Tactical Air Defense Alerting System 战术防空警报系统

TADC Tactical Air Direction Center 战术空中定向中心

TADCOM Theater Air Defense Command 战区空中防御司令部

TADIL Tactical Digital Information Link 战术数字信息链

TADIX Tactical Data Information Exchange 战术数据信息交换

TADIXS Tactical Data Information Exchange System 战术数据信息交换系统

TADL Tactical Data Link 战术数据链

TADS Tactical Air Defense Systems 战术防空系统

TADS Target Acquisition and Designation Sight 目标截获与标识瞄准器

TADS Target Acquisition and Designation System 目标截获与标识系统

TADS Target Acquisition and Detection System 目标截获与探测系统

TADS Target and Activity Display System 目标活动显示系统

TADS Tracking and Display System 跟踪与显示系统

TADSIM Theater Air Defense Simulation 战区空中防御模拟（仿真）

TADSS Training Aids and Devices, Simulators, and Simulations 训练辅助设备与装置、模拟器与仿真

TAER Telemetry Analog Equipment Room 遥测模拟设备间

TAER Time, Azimuth, Elevation, Range 时间、方位、仰角、距离

TAERS Telecommunications and Earth Resources Stations 通信和地球资源站（大型空间结构）

TAFCO Time and Frequency Control

时间与频率控制

TAFCOS Total Automatic Flight Control System 全自动飞行控制系统

TAFI Turn-Around Fault Isolation 往返飞行故障隔离

TAFIIS Tactical Air Force Integrated Information System 空军战术综合信息系统

TAFIM Technical Architecture Framework for Information Management 信息管理的技术体系框架结构

TAFT Test, Analyze, Fix and Test 试验、分析、改正与再试验

TAG Technical Air-to-Ground 技术性空对地

TAG Time Automated Grid 时间自动格栅

TAGIS Tracking and Ground Instrumentation System 跟踪与地面仪表测量系统

TAGIU Tracking and Ground Instrumentation Unit 跟踪与地面测量装置

TAGS Technology for Automated Generation of Systems 新一代自动化系统技术

TAGS Text and Graphics System 【美国航天飞机】电文图像系统

TAGW Take-off Gross Weight 发射总重量；起飞全重

TAIARA Two-Axis Inertial Attitude Reference Assembly 双轴惯性姿态参考（基准）组件

TAID Thrust Augmented Improved Delta 【美国】加大推力改进型"德尔它"运载火箭

TAIP Terminal Air impact Point 终点区弹着点

TAIR Test Assembly Inspection Record 测试装配检查记录

TAIS Tactical Airspace Integration System 战术空域一体化系统

TAIS Technology Applications Information System 技术应用信息系统

TAL Transatlantic Abort Landing 大西洋彼岸中止着陆（美国航天飞机发射故障处理模式之一）

TAL Transoceanic Abort Landing 跨大洋中止着陆

TALAR Tactical Approach and Landing Radar 战术进场与着陆雷达

TALBE Talk and Listen Beacon Equipment 小型导航设备；无线电应答装置

TALCM Tactical Air-Launched Cruise Missile 空（中发）射（的）战术巡航导弹

TALM Tactical Air-Launched Missile 空射战术导弹

TALON TACSIM Analysis and Operations Node 战术模拟分析与操作节点

TALRNS Tactical Airborne Long-Range Navigation System 机载战术远距离导航系统

TALTT Thrust Augmented Long Task Thor 【美国】加大推力长舱"雷神"导弹

TAM Task Analysis Methodology 任务分析方法

TAM Technical Area Management 技术区（技术阵地）管理

TAM Theater Analysis Model 战区分析模型

TAM Theater Attack Model 战区攻击模型

TAM Thermal Analytical Model 热分析模型

TAM Thrust-augmented Missile 加大推力导弹

TAM Trajectory Application Method 弹道应用法

TAMD Theater Air and Missile Defense 战区空中与导弹防御

TAME Tactical Missile Encounter 战术导弹对抗

TAMF Tactical Automated Maintenance Facility 战术自动化维修设施

TAMS Tactical Avionics Maintenance Simulation 战术航空电子维修模拟

TAMS Test and Monitoring System 测试与监控系统

TAMS Transportation Analysis, Modeling and Simulation 运输分析、建模与模拟

TAN Tactical Air Navigation 战术空中导航

TAN Transonic Aerodynamic Nozzle 跨声速气动喷嘴

TAOC Tactical Air Operations Center 战术空中作战（操作）中心

TAOM Tactical Air Operations Module 战术空中操作模块

TAOS Technology for Autonomous Operation of Satellites 卫星自主操作技术

TAOS Technology for Autonomous Operational Survivability 【美国】自主作战生存技术卫星

TAOS Thrust-Assistant Orbit Shuttle 助推轨道航天飞机

TAOS Thrust-Assistant Orbiter System 带助推的轨道飞行系统

TAP Technical Area Planning 技术区（技术阵地）规划

TAP Telemetry Acceptance Pattern 遥测验收模式

TAP Temporary Assembly Procedure 临时装配程序

TAP Total Air Pressure 总大气压力

TAPES Toxicologic Agent Protective Ensemble, Self-Contained 独立式毒剂防护装置

TAPR Toxic Altitude Propulsion Research 高空毒气推进研究

TAPS Target Acquisition Processing System 目标截获处理系统

TAPS Target Analysis and Planning System 目标分析与规划系统

TAPS Telemetry Automatic Processing System 遥测自动处理系统

TAPS Terminal Application Processing System 终端应用处理系统

TAPS Trajectory Accuracy Prediction System 弹道精度预测系统

TAPS Two-Axis Pointing System 两轴定位系统

TAR Target Acquisition Radar 目标捕获雷达

TAR Technical Analysis Request 技术分析请求

TAR Technical Application Reference 技术应用参考

TAR Test Action Requirement 测试措施要求

TAR Trajectory Analysis Room 轨迹分析室

TARA Technology Area Reviews and Assessments 技术领域评审与评估

TARA Track Analyzing and Recording Apparatus 跟踪分析与记录装置

TARAN Tactical Radar and Navigation 战术雷达与导航

TARAN Test and Repair as Necessary 按需测试与修理

TARC Through Axis Rotational Control 过轴旋转控制

TARCAP Target Combat Air Patrol 目标区作战空中巡逻

TARGET Theater Analysis and Replanning Graphical Execution Toolkit 战区分析与再规划图形实施工具

TARM Tactical Antiradiation Missile 战术反辐射导弹

TARM TIROS Atmospheric Radiance Module 【美国】泰罗斯卫星大气辐射组件

TARN Turn Around 回车道（场）；往返周转；有效半径

TARPS (DI) Tactical Airborne Reconnaissance Pod System (Digital Imagery) 战术机载侦察吊舱系统（数字成像）

TARS Tethered Aerostat Radar System 系留高空气球雷达系统

TAS Target-Acquisition System 目标截获系统

TAS Technological Alarming System 技术预警系统

TAS Telemetry Acquisition System 遥测数据采集系统

TAS Telemetry Antenna Subsystem 遥测天线分系统

TAS Test Article Specification 测试（试验）部件规格

TAS Tether Applications in Space 空间绳系应用

TAS Thermal Analysis for Space 空间热分析

TASA Task and Skills Analysis 任务与技能分析

TASE Tactical Support Equipment 战术保障装备

TASE Thrust Assessment Support Environment 推力评估保障环境

TASM Tactical Air-to-Surface Missile 战术空对地导弹

TASR Tactical Automated Situational Receiver 战术状态自动接收器

TASR Technology Area Strategic Roadmap 【美国】技术领域战略路线图

TASS Technical Analytical Study Support 技术分析研究保障

TASS Towed Array Sonar System 拖曳式阵列声呐系统

TAT Thrust-Augmented Thor 【美国】增大推力的"雷神"火箭

TAT Total Air Temperature 总大气温度

TAT Turn Around Time 往返飞行时间

TATI Total Air Temperature Indicator 总大气温度指示器（装置）

TATI Trim and Tailplane Incidence (Indicator) 配平与水平尾翼倾角指示器

TAUL Test and Upgrade Link 试验（测试）与更新链路

TAV Trans Atmospheric Vehicle 跨大气层飞行器

TAVERNS Test and Verification Environment for Remotely Networked Systems 远程网络系统试验与鉴定环境

TAW Tactical Assault Weapon 战术攻击武器

TAWDS Tactical Automated Weather Distribution System 战术自动化气象分布系统

TB Time Duration of Burn 燃烧持续时间

TB Top Boost 最高助推力

TBC Terminal Buffer Controller 终端缓冲控制器（装置）

TBC Thermal Barrier Coating 热防护涂层

TBCC Turbine Based Combined Circulation 涡轮基组合循环式

TBCC Turbine Based Combined Cycle

涡轮基组合循环
TBF Time Between Failures 故障间隔时间；失效间隔时间
TBL Terminal Ballistic Laboratory 末弹道实验室
TBM Tactical Ballistic Missile 战术弹道导弹
TBM Theater Ballistic Missile 战区弹道导弹
TBMCS Theater Battle Management Core Systems 战区作战管理核心系统
TBMD Tactical Ballistic Missile Defense 战术弹道导弹防御
TBMD Terminal Ballistic Missile Defense 弹道导弹末端防御
TBMD Theater Ballistic Missile Defense 战区弹道导弹防御
TBMDSE Theater Ballistic Missile Defense System Exercise 战区弹道导弹防御系统演习
TBMEWS Tactical Ballistic Missile Early Warning System 战术弹道导弹预警系统
TBMPGIP Tactical Ballistic Missile Predicted Ground Impact Point 战术弹道导弹预计地面弹着点
TBMW Theater Ballistic Missile Warning 战区弹道导弹预警
TBMX Theater Ballistic Missile Experiment 战术弹道导弹试验
TBO Time Between Overhaul 大修间隔期
TBONE Theater Battle Operations Network Centric Environment 战区作战操作网络化环境
TBS Tactical Broadcast System 战术广播系统
TBS Task Breakdown Structure 任务分类体系架构
TBT Terminal Ballistic Track 终端弹道轨迹
TBX Tactical Ballistic Missile, Experimental 实验型战术弹道导弹
Tc Adiabatic Flame Temperature of Rocket 火箭发动机绝热火焰温度
TC Technical Characteristics 技术特性（参数）
TC Technical Complex 技术设施
TC Technical Control 技术控制
TC Telecommand 遥控
TC Temperature Compensating 温度补偿
TC Thermal Control 热控制
TC Thrust Chamber 推进舱（室）
TC Titan/Centaur 【美国】"大力神/半人马座"运载火箭
TC&R Tracking, Command and Ranging 跟踪、指令与测距
TC^{2}S Tactical Command and Control System 战术指挥与控制系统
TCA Thrust Chamber Assembly （火箭发动机）推进舱组件
TCA Time of Closest Approach 最近进场时间
TCAMS Technical Control and Monitoring System 技术控制与监控系统
TCAS Technical Control and Analysis System 技术控制与分析系统
TCATS Target Cueing and Tracking System 目标提示与跟踪系统
TCB Task Control Block 任务控制块
TCC Tactical Command Center 战术指挥中心
TCC Test Control Center 测试（试验）控制中心
TCC Thermal Control Coating 热控制涂层
TCC Thrust Control Computer 推力

控制计算机

TCC Tracking and Control Center 跟踪与控制中心

TCCCS Tactical Command, Control, and Communications System 战术指挥、控制与通信系统

TCCF Tactical Communications Control Facility 战术通信控制设施

TCCS Tactical Command and Control System 战术指挥与控制系统

TCCS Thrust Control Computer System 推力控制计算机系统

TCD Terminal Countdown Demonstration 终端倒计时演示

TCDT Terminal Count Demonstration Test 终端计数验证试验

TCE Thermal Canister Experiment 热容器实验

TCF Tank Checkout Facility 储箱检测设施

TCFP Thrust Chamber Fuel Purge 推进室燃料吹除

TCG Time Code Generator 时间编码生成器(装置)

TCG Timing and Control Generator 定时与控制发生器

TCI Technical Critical Item 技术关键项

TCIM Tactical Communications Interface Module 战术通信接口模块

TCIS Tactical Communications Interface Software 战术通信接口软件

TCL Troposcatter Communications Link 对流层散射通信链路

TCLSM Total Cycle Life System Management 全寿命周期系统管理

TCM Trajectory Correction Maneuver 弹道修正机动

TCM Trauma Control Model 损伤控制模型

TCMD Theater Cruise Missile Defense 战区巡航导弹防御

TCMS Test and Checkout Management System 测试与检测管理系统

TCMS Test, Checkout and Monitoring System 测试、检测与监控系统

TCMS Test, Control and Monitor System 测试、控制与监控系统

TCN Tactical Component Network 战术部门网络

TCO Tactical Combat Operations 战术作战行动

TCO Test and Checkout 测试与检测

TCO Thrust Cutoff 推力关停

TCOPS Trajectory Computation & Orbital Products System 弹道计算与轨道产品系统

TCOS Tactical Combat Operations System 战术战操作系统

TCP Test Checkout Procedure 测试检测规程

TCP Test Control Package 测试控制包

TCP Thrust Center Position 推力中心位置

TCP Thrust Chamber Pressure 推进室压力

TCP Transmission Control Protocol 传输控制协议

TCPS Thrust Chamber Pressure Switch 【火箭发动机】推进室压力开关

TCR Telemetry, Command and Ranging 遥测、指令与测距

TCR Test Compare Results 测试对比结果

TCR Test Constraints Review 试验限制条件评审

TCR Thermal Concept Review 热设计评审

TCS Tactical Control System 战术

控制系统
TCS Tanking Control System 【燃料】加注控制系统
TCS Telemetry and Command Simulator 遥测与指挥模拟器
TCS Telemetry and Command System 【卫星通信】遥测与指挥系统
TCS Telemetry Control System 遥测控制系统
TCS Test Control System 测试控制系统
TCS Theater Combat System 战区作战系统
TCS Thermal Control System (Subsystem) 热控系统（分系统）
TCS Tracking and Communications Subsystem 跟踪与通信分系统
TCS Tracking Control Sensors 跟踪控制敏感器
TCS Trajectory Control Sensor 弹道控制传感器
TCS Trajectory Control System 轨迹控制系统
TCS Trajectory Correction System 弹道修正系统
TCSE Thermal Control Surfaces Experiment 热控制表面实验
TCSSS Thermal Control Subsystem Segment 热控制分系统段
TCV Temperature Control Voltage 温度控制电压
TCV Thrust Chamber Valve 推力室阀门
TD&E Tactics Development and Evaluation 战术研发与评估
TD&E Test, Demonstrations and Experiments 测试、验证与实验
TD&E Transposition, Docking and Lunar Module Ejection 换位、对接和登月舱弹射
TD-2 Taepo'o-dong-2 Missile 【朝鲜】"大浦洞"2 导弹
TD/CMS Technical Data/Configuration Management System 技术数据/配置管理系统
TDA Target Damage Assessment 目标损伤评估
TDA Technical Data Analysis 数据分析工具
TDA Technology Development Approach 技术研发方法
TDA Telecommunications and Data Acquisition 远程通信与数据采集
TDACS Throttleable Divert and Attitude Control System 可调转向与姿态控制系统
TDADT Total Distribution Advanced Technology Demonstration 全分布式先进技术演示验证
TDARDS Truth Data Acquisition, Recording, and Display System 真值数据采集、记录与显示系统
TDAS Test Data Acquisition System 试验数据采集系统
TDAS Test Data Analysis System 试验数据分析系统
TDAS Theater Defense Architecture Study 战区防御体系结构研究
TDAS Tracking and Data Acquisition Satellite 跟踪与数据采集卫星
TDAS Tracking and Data Acquisition System 跟踪与数据采集系统
TDASS Theater Defense Architecture Scoping Study 战区防御体系结构目标研究
TDASS Tracking and Data Acquisition Satellite System 跟踪和数据采集卫星系统
TDAT Teleprocessing Diagnostic Analysis Tester 远程处理诊断分析测试

程序

TDATD Total Distribution Advanced Technology Demonstration 总分配先进技术验证

TDB Target Designation Box 目标指示框

TDB Time Distribution Bus 时间分配总线

TDBMS Tactical Database Management System 战术数据库管理系统

TDC Tactical Display Console 战术显示控制台

TDC Target Designation Control 目标指示控制

TDC Target Designator Control 目标指示器控制

TDC Technology Demonstration Center 技术演示验证中心

TDC Telemetry Data Center 【美国】遥测数据中心

TDC Telemetry Data Controller 【美国国家航空航天局】遥测数据控制器

TDC Theater Deployable Communications 战区可部署通信

TDC Type Design Change 型号设计更改

TDCC Test Data Collection Center 试验（测试）数据采集中心

TDD Tactical Display Device 战术显示装置

TDD Target Detecting (Detection) Device 目标探测装置

TDD Top-Down Design 自顶向下设计

TDDS Tactical Related Applications (TRAP) Data Dissemination System 战术相关应用数据传播系统

TDDS Telemetry Data Display System 遥测数据显示系统

TDDS Telemetry Decommutation and Display Systems 遥测信息译码与显示系统

TDDS TRAP Data Dissemination System 战术应用数据分发系统；跟踪分析程序数据分发系统

TDI Technical Data Interchange 技术数据交换

TDM Technology Demonstration Mission 技术演示验证任务

TDMS Tactical Decision Making Simulation 战术决策模拟

TDMS Technical Data Management System 技术数据管理系统

TDP Technology Development Project 技术发展计划

TDP Time-Space-Position-Information Data Processor 时空位置数据处理器

TDP Tracking Data Processor 跟踪数据处理器（装置）

TDQRC Technology Demonstration, Quick-Reaction Capability 技术演示验证快速反作用能力

TDR Technical Design Review 技术设计评审

TDR Terminal Defense Radar 终端防御雷达

TDRR Test Data Recording and Retrieval 测试数据记录与查询

TDRS Tracking and Data Relay Satellite 【美国】跟踪与数据中继卫星

TDRS Tracking and Data Relay Station 跟踪与数据中继站

TDRSS Tracking and Data Relay Satellite System 跟踪与数据中继卫星系统

TDS Technology Demonstration Satellite 技术演示验证卫星

TDS Technology Demonstration Sys-

tem　技术演示验证系统
TDS　Terminal Defense Segment　【弹道导弹防御】末端防御部分
TDS　Test Data System　测试数据系统
TDSSPA　Technical Development for Solid State Phased Arrays　固态相阵技术研发
TDT　Target Development Test　目标研制试验（测试）
TDTC　Test, Development and Training Center　试验（测试）、研发与训练中心
TDTS　Telemetry Data Transmitting Station　遥测数据传输站
TDTS　Telemetry Data Transmitting System　遥测数据传输系统
TDU　Technical Demonstration Unit　技术演示验证单元
TDU　Techniques Development Unit　技术研发单元
TDU　Time Display Unit　时间显示装置
TDUGS　Target Data Update Ground Station　目标数据更新地面站
TDUGS　Target Data Uplink Ground Station　目标数据上链路地面站
TDWR　Terminal Doppler Weather Radar　终端多普勒气象雷达
TE　Test Equipment　测试（试验）设备
TE　Timing Electronics　计时电子设备
TEA　Technical Exchange Agreement　技术交换协议
TEAM　Technology Exchange Assessment Methodology　技术交换评估方法
TEAM　Telemetry Evaluation and Analysis Monitor　遥测评估与分析监控装置
TEAM　Test, Evaluation, Analysis, and Modeling　试验、评估、分析与建模
TEAMS　Technology Experiments for Advancing Missions in Space　前沿航天任务技术实验
TEAMS　Test Evaluation and Monitoring System　试验评估与监控系统
TEAMS　Trend and Error Analysis Methodology System　趋向与误差分析法系统
TEAS　Thrust Evaluation and Action Selection　推力估计与行动选择
TEASE　Tracking Errors and Simulation Evaluation　跟踪误差与模拟评估
TEC　Test Equipment Center　测试（试验）设备中心
TEC　Test Execution Control　试验（测试）执行控制
TEC　Thermal Electric Cooler　热电制冷器
TEC　TOS Evaluation Center　【美国】泰罗斯业务卫星计算中心
TECC　Test, Evaluation and Calibration Center　试验、评估与校准中心
TECCS　Tactical Engineer Command and Control System　战术工程（兵）指挥与控制系统
TECHCON　Technical Control　技术控制
TECHEVAL　Technical Evaluation　技术鉴定
TECOM　Test and Evaluation Command　试验（测试）与鉴定指令
TECS　Thermal Environmental Control Subsystem　热环境控制分系统
TED　Technology Exploitation Demonstration　技术利用演示验证

TED Theater Exploitation Demonstration 战区开发演示验证

TED Total Energy Detector 总能量探测器

TEDAC Test & Evaluation Data Analysis Capability 试验与鉴定数据分析能力

TEFF Turbine Engine Fatigue Facility 涡轮发动机疲劳（试验）设施

TEG Thermo-Electric Generator 热电发电机

TEI Trans-Earth Insertion 【航天器从月球返回时】返回地球导入轨道

TEI&R Tactical Engagement, Instrumentation and Ranges 战术对抗、仪表与范围

TEL Transporter-Erector-Launcher 运输－起竖－发射装置

TEL-4 Eastern Range Central Telemetry Facility 【美国】东部靶场中央遥测设施

TELAC Technology Laboratory for Advanced Composites 先进复合材料工艺实验室

TELAR Transporter/Erector/Launcher/Radar 带有升降架、发射架与雷达的运输车

TELESAT TELESAT (Canadian Payload, Communication Satellite) 加拿大有效载荷通信卫星

TELINT Telemetry Intelligence 遥测情报

TELS Turbine Engine Loads Simulator 涡轮发动机载荷模拟器

TEMISAT Telespazio Micro Satellite 【意大利】空间通信公司微型卫星（泰米卫星）

TEMO Training Exercises and Military Operations 训练、演习与军事行动

TEMP Test and Evaluation Master Plan 试验（测试）与鉴定总计划

TEMPEST Transient Electromagnetic Pulse Emanation Standard 瞬态电磁脉冲发射标准

TEMS Transport Environment Monitoring System 运输环境监测系统

TENA Test and Training Enabling Architecture 试验与训练使能体系结构

TENCAP Tactical Exploitation of National Space Capabilities 【美国】国家空间能力的战术利用

TEPC Tissue Equivalent Proportional Radiation Counter 组织等效比例辐射计数器

TEPOP Tracking Error Propagation and Orbit Prediction Program 跟踪误差传播和轨道预测程序

TEPOP Tracking-Error Propagation and Orbit Prediction 跟踪误差传播与轨道预测

TER Triple Ejector Rack 三角型起竖架

TERA Terminal Effects Research and Analysis 终端效应研究与分析

TERC Turbine Engine Research Center 涡轮发动机研究中心

TERLS Thumba Equatorial Rocket Launching Station 【印度】顿巴赤道火箭发射站

TERS Tactical Event Reporting System 战术事件通报系统

TERS Tropical Earth Resources Satellite 【印度尼西亚】热带地球资源卫星

TERSE Tunable Etalon Remote Sounder of Earth 可调基准地球遥测器

TES Tactical Event System 战术事件系统

TES　Theater Event System　战区事件系统

TES　Transition Edge Sensors　转换边界传感器

TES　Tropospheric Emission Spectrometer　对流层辐射摄谱仪

TESS　Tactical Environment Support System　战术环境保障系统

TESSE　Test Environment Support System Enhancement　增强的试验环境保障系统

TestPAES　Test Planning, Analysis and Evaluation System　试验（测试）规划、分析与鉴定系统

TET　Test Equipment Tester　试验设备检验装置

TETAM　Tactical Effectiveness Testing of Antitank Missiles　反坦克导弹战术效能检验

TETR　Test and Training Satellite　【美国】试验与训练卫星

TEV　Test, Evaluation and Verification　试验（测试）、鉴定与确认

TEVS　Test Environment Support Systems　试验（测试）环境保障系统

TEVS　Test Environment System　试验（测试）环境系统

TEX　Transceiver Experiment　【美国】无线电收发机试验（卫星）

TEXCOM　Test and Experimentation Command　试验（测试）与实验指令

TF　Test Facility　测试设施

TF　Test Fixture　测试装置

TFCC　THAAD Fire Control and Communications　战区高空区域防御发射控制与通信

TFCS　Triplex Flight Control System (Subsystem)　三重飞行控制系统（分系统）

TFD　Technical Feasibility Decision　技术可行性决策

TFDE　Tank Fluid Dynamics Experiments　储箱流体动力实验

TFG　Thrust Floated Gyroscope　推力悬浮陀螺仪

TFIM　Technical (Architecture) Framework for Information Management　信息管理的技术（体系结构）框架

TFR　Terrain Following Radar　地形跟踪雷达

TFR　Trouble and Failure Report　故障通报

TFRAMES　Tools to Facilitate the Rapid Assembly of Missile Engagement Simulations　快速建立导弹交战模拟的工具

TFS　Thermal and Fluid Systems　【美国国家航空航天局】散热与流体系统

TFU　Test Facility Utilization　测试设施利用

TGA　Thermal Gravimetric Analysis　热重力分析

TGG　Triple Graph Grammar　三相图原理

TGIF　Tactical Ground Intercept Facility　战术地面拦截设施

TGIF　Transportable Ground Intercept Facility　移动式地面拦截设施

TGLV　Terminal Guidance for Lunar Vehicle　月球飞行器末端制导

TGPSG　Tactical Global Positioning System Guidance　战术全球定位系统制导

TGS　Telemetry Ground Station　遥测地面站

TGS　Track Generation System　跟踪生成系统

TGS　Transportable Ground Station

移动式地面站
TGSE Telemetry Ground Support Equipment 遥测地面保障设备
TGSE Test Ground Support Equipment 地面试验保障设备
TGSS Thermal Gradient Sensor System 热梯度感应系统
THAAD Terminal (formerly theater) High Altitude Area Defense 终端（原称为"战区"）高空区域防御
THAAD Theater High Altitude Area Defense 【美国】战场高空区域防御
THAAD-GBR Theater High-Altitude Area Defence Ground-Based Radar 战区高空区域防御地基雷达
THAADS Theater High Altitude Area Defense System 【美国】战区高空区域防御体系
THAD Terminal-Homing Accuracy Demonstrator 末段寻的精度演示（导）弹
THC Toxic Hazard Corridor 【美国东靶场】有毒危险性通道
THCU Thermal Hydraulic Control Unit 液压温控系统组件
THE Test Equipment Hookup 试验设备连接装置
THE Thrust Engine 大推力发动机
THE Turbojet Engine 涡轮喷气发动机
THEL Tactical High-Energy Laser 战术高能激光器
THEMIS Time History of Events and Macroscale Interactions during Substorms 亚暴中的事件时间关系曲线图与宏观交互作用
THEO Theoretical Trajectory 理论弹道
THERMO Thermal and Hydrodynamic Experiment Research Module in Orbit 热力学和流体动力学实验研究轨道舱
THM Thermal Health Monitoring 热健康状态监控
THMT Tactical High Mobility Terminal 战术高机动性终端
THOR Tiered Hierarchy Overlayed Research 多层次重叠研究（美国系列天基动能杀伤武器试验）
THREADS Technology for Human/Robotic Exploration and Development of Space 载人/机器人空间探索和发展技术
THS Hughes Satellite 【美国】休斯公司卫星
THZ Toxic Hazard Zone 【美国西靶场】有毒危险性区
TI Technical Integration 技术集成（一体化）
TI Thermal Phase Initiator 热相引发器（装置）
TIARA Tactical Intelligence and Related Activities 战术情报及相关活动
TIBS Tactical Information Broadcast System 战术信息广播系统
TIBS Theater Intelligence Broadcast System 战区情报广播系统
TIC Technical Information Center 技术信息中心
TICM Test Interface and Control Module 测试接口控制模块
TICS Terminal Interface Control System 终端接口控制系统
TID Total Ionizing Dose 总电离剂量
TIDE Thermal Ion Dynamics Experiment 热电离动态实验
TIE Technical Independent Evaluation 技术独立鉴定

TIE Technology Integration Experiments 技术集成实验
TIES Technology Integration Equipment System 技术集成设备系统
TIFS Total In-Fight Simulator 全飞行过程模拟器
TIG Time of Ignition 点火时间
TIGER Tactical Intelligence and Evaluation Relay 战术情报与评估中继
TIGER Tactical Interactive Ground Equipment Repair 战术交互式地面设备维修
TIGER Terminal Guided and Extended Range (Missile) 终端制导与增程（导弹）
TIIAP Telecommunications and Information Infrastructure Assistance Program 远程通信与信息基础设施辅助项目
TIL Technical Insertion Laboratory 技术引入实验室
TILL Tracking Illuminator Laser 跟踪照明激光器
TIMS Training Integration Management System 训练综合管理系统
TIN Theater Intelligence Networks 战区情报网络
TINA Thermal Imaging Navigation Aid 热成像导航设备
TINS Thermal-Imaging Navigation System (Set) 热成像导航系统（仪）
TINS Train Inertial Navigation System 序列惯性导航系统
TIP Tracking and Impact Predication 跟踪与弹着点预测
TIP Transit Improvement Program 【美国】子午仪改型卫星
TIPRS Tomahawk Inflight Position Reporting System 【美国】"战斧"巡航导弹飞行中定位报告系统
TIPS Telemetry Integrated Processing System 遥测综合处理系统
TIR Terminal Imaging Radar 终端成像雷达
TIR Total Internal Reflection 全内反（映）射
TIROS Television Infrared Observational Satellite 【美国航天飞机】红外电视观测卫星；泰罗斯卫星
TIRS Telemetry, Instrumentation and Range Safety 遥测、测量与靶场安全
TIRS Tether Initiated Recovery System 系绳回收系统（美国系绳卫星实验项目）
TIS Technical Information System 技术信息系统
TISRS Tether Initiated Space Recovery System 系留引发空间回收系统
TIU Test and Maintenance Bus Interface Unit 试验与维修总线接口单元
TJP Turbojet Propulsion 涡轮喷气推进装置
TKSC Tsukuba Space Center 【日本】筑波空间研究中心
TLA Time Line Analysis 时间线分析
TLAM Theater Land Attack Missile 战区地面攻击导弹
TLAM Tomahawk Land Attack Missile 【美国】"战斧"对地攻击导弹
TLC Transport and Launch Container 运输与发射容器
TLC Transport Launching Container 发射运输容器
TLCE Transmission Line Conditioning Equipment 传输线调节设备

TLCF Technical Laboratory Computer Facility 技术实验室计算机设施
TLE Target Location Error 目标定位误差
TLI Trans-Lunar Insertion 进入飞向月球轨道
TLIR Time Limited Impulse Response 时限脉冲响应
TLOS Target Location and Observation System 目标定位与观测系统
TLP Transient Lunar Phenomena 月球瞬变现象
TLS Tandem Launch System 一箭多星发射系统；串联（多星）发射系统
TLS Translunar Shuttle 跨月球的航天飞机
TLSA Torso Limb Suit Assembly 躯干肢体套装（航天服）组件
TLT Thrust Line Translation 推力线平移
TLV Target Launch Vehicle 目标运载火箭
TM Tactical Missile 战术导弹
TM Technological Model 工艺模型
TM Test Maintenance 测试与维修
TM Theater Missile 战区导弹
TM Training Model 训练模型
TM/ACS True-Motion Anti-Collision System 真实运动防撞系统
TMA Target Motion Analysis 目标运动分析
TMC Test Monitoring Console 测试（试验）监视控制操纵台
TMCC Test Monitor and Control Center 试验（测试）监视与控制中心
TMD Theater Missile Defense 战区导弹防御
TMD C² Theater Missile Defense Command and Control 战区导弹防御指挥与控制
TMD ESM Theater Missile Defense Existing System(s) Modification 战区导弹防御现有系统改造
TMD GBR Theater Missile Defense Ground-Based Radar 战区导弹防御地基雷达
TMD IA Theater Missile Defense Interoperability Architecture 战区导弹防御拦截体系结构
TMDAS Theater Missile Defense Architecture Study 战区导弹防御体系结构研究
TMDE Test, Measurement and Diagnostic Equipment 试验、测量与诊断设备
TMDS Test, Measurement, and Diagnostic System 试验、测量和诊断系统
TMECO Time of Main Engine Cutoff 主发动机停车时间
TMF Transporter Maintenance Facility 运输装置维护设施
TMG Thermal/Meteoroid Garment 加温防宇宙尘（流星体）服
TMI TRMM Microwave Imager 热带降雨测量卫星微波成像仪
TMIA Trans-Mars Injection Assembly 穿越火星飞行发射装置（组件）
TMIG Telemetry Inertial Guidance Data 遥测惯性制导数据
TML Total Mass Loss 总质量损失
TMM Thermal Math Model 热数模型
TMOS Transformational Satellite Mission Operations System 转型卫星任务作战系统
TMOS-EP Total Ozone Mapping Spectrometer-Earth Probe 地球探测器的臭氧总量绘图分光计

TMPC Tomahawk Theater Mission Planning Center 【美国】"战斧"导弹战区任务规划中心

TMPCU Tomahawk Theater Mission Planning Center Upgrade 【美国】"战斧"导弹战区任务规划中心升级

TMT Theater Missile Tracker 战区导弹跟踪装置

TMU Temperature Measurement Unit 温度测量装置

TMV Thermal Margin Verification 热系数（限度，裕度）验证

TNAR Telemetry Doppler Nominal Acceleration 遥测多普勒额定加速

TO Transfer Orbit 转移轨道

TOB Takeoff Boost 起飞助推（器）；发射加速器

TOC Tactical Operations Center 战术作战（操作）中心

TOC Telemetry Operations Control 遥测操作控制

TOC Test Operations Center 测试（试验）操作中心

TOC Test Operations Change 测试（试验）操作更改

TOC Total Organic Carbon 总有机碳量

TOCA Total Organic Carbon Analyzer 总有机碳分析仪

TOCC Tactical Operations Control Center 战术作战控制中心

TOCC Test Operations Control Center 试验作战控制中心；测试操作控制中心

TOCS Terminal Operations Control System 终端操作控制系统

TOF Test Operations Facility 测试（试验）操作设施

TOF Time of Flight 飞行时间

TOG Technical Objectives and Goals 技术目的与目标

TOI Tethered Orbiting Interferometer 系留环轨干涉仪

TOI Transfer Orbit Insertion 转移轨道射入

TOL Time of Launching 发射时间

TOLIP Trajectory Optimization and Linearized Pitch 轨道最佳化与线性化俯仰

TOM Target Object Map 目标图

TOMS Total Ozone Mapping Spectrometer (NASA) 【美国国家航天局】总臭氧含量测绘光谱仪（卫星）

TOMUIS Total Ozone Mapping with UV Imaging Spectrometer 臭氧总量紫外线成像测绘光谱仪

TOO Target of Opportunity 临时目标

TOO Test of Opportunity 临时试验（测试）

TOOL Target of Opportunity Launch 临时目标发射；意外目标发射

TOP Technical Operating Procedure 技术操作规程

TOPAS Transport Operation of Micro-G Payloads Assembled on Scout 【美国】"侦察兵"火箭发射微重力有效载荷舱

TOPEX Topography Experiment Satellite 海洋地貌实验卫星；托佩克斯卫星

TOPO Topographical Position Satellite 【美国】地形测量卫星

TOPS Thermoelectric Outer-Planet Spacecraft 热电外层行星航天器

TOR Time of Release 投弹时间

TORF Time of Retrofire 制动发动机点火时间

TOS Tactical Operations Station 战

术作战（操作）站

TOS Television Infrared Operational Satellite (TIROS) Operational System 【美国】电视红外应用卫星（泰罗斯）应用系统

TOS Test Operating System 测试(试验）操作系统

TOS TIROS Operational Satellite 【美国】泰罗斯业务卫星

TOS Transfer Orbit Stage 【火箭】转移轨道级

TOS Type of Shipment 运输方式

TOSSA Transient Or Steady State Analysis 瞬态或稳态分析

TOT Time On Track 跟踪时间；在轨时间

TOT Time-Over-Target 穿越目标时间；飞临目标时间

TOT Transfer Of Technology 技术转让

TOTS Target Oriented Tracking System 面向目标的跟踪系统

TOTS Transportable Orbital Tracking System 移动式轨道跟踪系统

TOVS TIROS Operational Vertial Sounder 泰罗斯卫星实用垂直分布探测器

TOW Tank and Orbiter Weight 储箱与轨道器重量

TOW Tube Launched, Optically Tracked, Wire Guided 【美国】导管发射、光学跟踪与有线制导（导弹）；陶式导弹

TP Test Pad 测试台；试验台

TP Test Point 测试（试验）点

TP Test Procedure 测试（试验）规程

TP Testing Program 测试程序

TP Time Pulse 时间脉冲

TP Time to Perigee 近地点时间

TP&C Thermal Protection and Control 热防护与控制

TPA Test Preparation Area 测试（试验）准备区

TPALS Theater Protection Against Limited Strikes 防御有限攻击的战区防护系统

TPALS Transportable Protection Against Accidental Launch System 移动式防止意外发射系统

TPC Telemetry Preprocessor Computer 遥测预处理器（装置）计算机

TPD Mobile Tactical Radar 移动式战术雷达

TPDM Three-Point Docking Mechanism 三点停泊机械装置

TPDR Total Processing Data Rate 总处理数据率

TPDRS Time-Phased Downgrading and Reclassification System 分阶段降低密级和重新分类系统

TPDS Test Procedures Development System 测试（试验）规程研发系统

TPEC THAAD Performance Evaluation Center 战区高空防御区域防御性能鉴定中心

TPF-C Terrestrial Planet Finder-Coronagraph 类地行星探测器－日冕仪

TPFDD Timed Phased Force Deployment Data 分阶段部队部署数据

TPFEL Transportation Feasibility Estimate 运输可行性评估

TPI Terminal Phase, Initiation 弹道末段开始阶段

TPITS Two-Phased Integrated Thermal System 二相集成热系统

TPL Trans-Pacific Landing 跨太平洋着陆（美国航天发动故障处理方式之一）

TPM Technical Performance Measurement (System) 技术性能测量（系统）

TPMT Total Preventative Maintenance Time 总预防性维修时间

TPOCC Transportable Payload Operations Control Center 可运输的有效载荷操作与控制中心

TPR Terminal Phase Radar 终端相雷达

TPS Thermal Protection System (Subsystem) 热保护系统（分系统）

TPS Thrust Chamber Pressure Switch 推力室压力开关

TPS Tracking Antenna Pedestal System 跟踪天线基座系统

TPSA Trace Parameter Status Area 追踪参数状态区

TPSE Thermal Protection Subsystem Experiments 热防护子系统实验

TPST Thermal Protection Structure 热防护结构

TPTA Transient Pressure Test Article 瞬态压力测试部件

TPU Transverse Propulsion Unit 横向推进装置

TPY-2 Forward-Based X-Band Radar 前沿配置的X波段雷达

TQC Total Quality Control 全面质量控制

TQCS Time, Quality, Cost, Service 周期、质量、成本与服务

TQE Technical Quality Evaluation 技术质量鉴定

TQM Total Quality Management 总体质量管理

TQVS Training, Qualification and Validation Subsystem 训练、鉴定和检验子系统

TR Technical Report 技术报告

TR Technical Review 技术评审

TR Thrust Reverser 推进转向器（装置）

TR Time to Retrofire 制动发动机点火前时间

TR&D Training Research and Development 训练研究与发展

TR/SBS Teleoperator Retrieval/Skylab Boost System 遥控操纵器回收的/"天空实验室"助推系统

TR/TV Thrust Reverser/Thrust Vectoring 反推力装置/推力矢量

TRAAMS Time Reference Angle of Arrival Measurement System 到达测量系统的时间基准角

TRAAV Test Range Facility for Advanced Aerospace Vulnerability 先进航空航天易损性测试的试验靶场设施

TRAC Trials Recording and Analysis Console 试射记录与分析操作台

TRACE Test-Equipment for Rapid Automatic Checkout and Evaluation 快速自动检测与评估使用的测试设备

TRACE Time-Shared Routines for Analysis, Classification and Evaluation 分析、分类和鉴定的分时程序

TRACE Transition Region and Coronal Explorer 过渡区与日冕探测器

TRACE-P Total Risk Assessing Cost Estimate for Production 生产成本核算综合风险评估

TRACE-P Transport and Chemical Evolution over the Pacific 【美国国家航空航天局试验任务】太平洋上的运输与化学演化试验

TRACS Transportation Requirements and Capabilities Simulation model 运输要求与能力模拟模型

TRAD Tile Repair Ablator Dispenser

热防瓦修理工具

TRADAC Trajectory Determination and Acquisition Computation 弹道测定与目标探测计算

TRADAT Trajectory Data System 弹道数据系统

TRADEX Target Resolution and Discrimination Experiment 目标分辨与识别实验

TRAE Time, Range, Azimuth, and Elevation 时间、范围、方位角和仰角

TRAJEX Trajectory Executive 轨迹执行程序

TRAM Test, Reliability and Maintainability 测试、可靠性与可维修性

TRAMP Testing, Reporting, and Maintenance Program 试验（测试）、通报与维护项目

TRANSATEL Transportable Satellite Telecommunications 移动式卫星通信

TRAP Tactical Receiver and Related Applications 战术接收装置与相关应用

TRAP Tactical Receive Applications Program 战术接收应用项目

TRAP Threat Risk Assessment Program 威胁风险评估项目

TRAP TRE and Related Applications 战术接收设备与相关应用

TRAPS Telemetry and Radar Acquisition Processing System 遥测与雷达获取处理系统

TRAVEL Transportable Vertical Erectable Launcher 移动式垂直竖立发射装置

TRC Thrust Rating Computer 推力额定值计算机

TRCCC Tracking Radar Central Control Console 跟踪雷达中央控制台

TRE Tactical Receive Equipment 战术接收设备

TRE Transmitter/Receiver Equipment 发射/接收设备

TRE Transponder Equipment 应答机设备

TREA Transmit/Receive Element Array 传输/接收要素阵列

TREE Transient Radiation Effects on Electronics 对电子设备的瞬时辐射效应

TREM Total Radiation Environment Model 总辐射环境模型

TRESIM Tactical Receive Equipment Simulator 战术接收设备模拟器（装置）

TRF Turbine Research Facility 涡轮研究设施

TRG Target Area 目标区域

TRI Tactical Reconnaissance/Intelligence 战术侦察/情报

TRI-TAC Tri-Service Tactical Digital Communications System 三军战术数字通信系统

TRIM Targets, Receivers, Impacts and Methods 目标、接收器、弹着和方式

TRIMM Transmit/Receive Integrated Microwave Modules 发射/接收综合微波模式

TRL Thermodynamics Research Laboratory 【美国】热力学研究实验室

TRL Technology Readiness Level 技术成熟度等级

TRM Technical Reference Model 技术参照模型

TRM Tension Release Mechanism 张力释放装置

TRM Time Release Mechanism (Ejection Seat System) 【弹射座椅系统】定时释放机构

TRM Transmit/Receive Modules 传输/接收模块

TRMM Tropical Rainfall Measuring Mission 【日本、美国合作】热带降雨测量卫星

TRMP Test Resource Master Plan 试验（测试）源总计划

TRN Terrain Relative Navigation 地形相对导航

TRO Temporary Release Order 临时释放指令

TRP Technical Requirements Package 技术要求数据包

TRR Technology Readiness Review 技术准备状态评审

TRR Technology Risk Reduction 技术风险降低

TRR Test Readiness Review 试验完备性评审

TRR Transfer Readiness Review 转运准备状态评审

TRS Teleoperator Retrieval System 遥控机构回收系统

TRS Tetrahedral Research Satellite 【美国】四面体研究卫星

TRS Time Reference System 时间参照系统

TS Tensile Strength 张力强度

TS Test Site 测试（试验）场

TS Test Stand 测试（试验）台

TS Test Station 测试（试验）站

TS Timing System 计时系统

TSA Technology Security Analysis 技术安全分析

TSAC Tracking System Analytical Calibration 跟踪系统分析校准

TSAD Trajectory and Signature Data 弹道与特征数据

TSAS Tactile Situational Awareness System 触觉态势感知系统

TSAT Technology Satellite 【英国】技术卫星

TSAT Transformational Communications Satellite 转型通信卫星

TSAT Transformational Satellite Communications System 转型卫星通信系统

TSB Technical Support Building 技术保障厂房

TSB Thrust Section Blower 推力舱增压器

TSC Theater Surface Combatant 战区地面作战

TSC Thruster System Control 推力器系统控制

TSC Transportable Satellite Communication 移动卫星通信

TSC Two-Stage Command 二级指令

TSCLT Transportable Satellite Communication Link Terminal 移动式卫星通信数据链终端

TSCM Tomahawk Strike Coordination Module 【美国】"战斧"导弹打击协调模块

TSCP Training Simulator Control Panel 培训模拟器（装置）控制操纵台

TSCS Transportable Satellite Communications System 移动式卫星通信系统

TSCT Transportable Satellite Communications Terminal 移动式卫星通信终端（站）

TSD Tactical Surveillance Demonstration 战术监视演示验证

TSDE Tactical Surveillance Demonstration Enhancement 增强的战术监视演示验证

TSE Transportation Support Equipment 运输保障设备
TSES Transportable Satellite Earth Station 移动式卫星地面站
TSEU Technology Seeker Evaluation Unit 技术寻的鉴定装置
TSF Time to System Failure 系统出现故障前的时间
TSF Tower Shielding Facility 塔式屏蔽设备
TSF Tracking and Simulation Facility 跟踪与模拟设施
TSFP Time to System Failure Period 系统失效周期时间
TSGMS Test Set Guided Missile System 导弹系统测试装置
TSGP Test Sequence Generator Program 测试（试验）序列生成器（装置）程序
TSL Terrestrial Science Laboratory 地球科学实验室
TSM Tail Service Mast 尾部勤务塔
TSM TIROS Stratospheric Mapper 【美国】电视红外观测卫星平流层探测资料映射模块；"泰罗斯"卫星平流层探测资料映射模块
TSMA Theater of Strategic Military Action 战略军事行动战区
TSO Time Sharing Operation 分时操作
TSO Time Sharing Option 分时选择
TSP Test Software Program 测试（试验）软件程序
TSP Track Sensor Payload 跟踪传感器有效载荷
TSPI Time-Space-Position Information 时间/空间/位置信息
TSPM Target Strength Predictive Model 目标强度预测模型
TSR Technical Status Review 技术状态评审
TSR Test Status Report 测试（试验）状态报告
TSRA Total System Requirements Analysis 系统总体要求分析
TSRM Third-Stage Rocket Motor 三级火箭发动机
TSRP Time Stress Measurement Device 时间应力测量装置
TSS Telemetric Test and Simulation System 遥测试验和模拟系统
TSS Tethered Satellite System 系留卫星系统；绳系卫星系统
TSS Thrust Sensitive Signal 推力敏感信号
TSS Time Sharing System 时间共享系统
TSS Tromso Satellite Solids 【挪威】特罗姆瑟卫星站
TSS-1 Tether in Space 空间绳
TSSAM Tri-Service Standoff Attack Missile 三军防区外攻击导弹
TSSLS Titan Standardized Space Launch System 【美国】"大力神"标准化航天发射系统
TSSM Titan Saturn System Mission 【美国】"土卫六"系统任务
TSSP Tactical Satellite Signal Processor 战术卫星信号处理器
TSSU Test Signal Switching Unit 测试（试验）信号切换装置
TSTO Two Stage to Orbit 两级入轨
TSTOL The System Test and Operations Language 系统测试与操作语言
TSTS Thrust Structure Test Stand 推力结构试验台
TSTS Tracking System Test Set 跟踪系统测试装置
TSTV Tracked and Semi Tracked Ve-

hicle 履带式及半履带式车辆
TSU Thermal Structural Unit 热结构装置
TSW Test Software 测试（试验）软件
TT&C Telemetry, Tracking and Commanding 遥测、跟踪与指挥
TT&C Tracking, Telemetry and Control 跟踪、遥测与控制
TT&E Targets, Test, and Evaluation 目标、测试与评估
TT&E Technical Test and Evaluation 【美国陆军】技术试验（测试）与鉴定
TTA Technology Targeting Assessment 技术瞄准评估
TTA Time to Apogee 远地点时间
TTAC Telemetry, Tracking and Command 遥测、跟踪与指挥
TTB Technology Test Bed 技术测试台
TTC Time-to-Criticality 时间临界
TTC Tracking, Telemetry and Command 跟踪、遥测与指挥
TTC Translunar Trajectory Characteristics 月球轨道外的轨道特性
TTC&M Tracking, Telemetry, Command and Monitoring 跟踪、遥测、指挥与监控
TTCA Thrust Translation Controller Assembly 推进变换控制器（装置）组件
TTCA Translation Thrust Control Assembly 变换推力控制装置
TTCC Telemetering, Timing, Command and Control 遥测、定时、指挥与控制
TTCS Target Tracking and Control System 目标跟踪与控制系统
TTI Transfer Trajectory Insertion 射入转移轨道
TTIM Total Thruster Impulse Monitor 总推力脉冲监控
TTL Total Time of Launch 发射前总时间
TTL Tracking/Targeting or Launching 跟踪/瞄准或发射
TTP Tactics, Techniques, and Procedures 战术、技术与规程
TTS Test and Training Satellite 试验与训练卫星（美国"阿波罗"飞船跟踪网的试验卫星）
TTS Thrust Termination System 推力终止系统
TTS Transportable Telemetry Systems 移动式遥测系统
TTT Test Technology Transfer 试验（测试）技术转让
TTU Thrust Termination Unit 推力终止装置
TTU Tracer Test Unit 跟踪试验装置
TTV Teleoperator Transfer Vehicle 遥控操作器转移飞行器
TTV Termination, Test, and Verification 终止、测试与鉴定
TV Thermal Vacuum 热真空
TV Thrust Vectoring 推力矢量
TVA Thrust Vector Alignment 推力矢量校准
TVAC Thrust Vector Activation Control 推力矢量启动控制
TVBS Television Broadcasting Satellite 电视广播卫星
TVC Thermal Vacuum Chamber 热真空室（舱）
TVC Thrust Vector Control 推力矢量控制
TVC Trapped Vortex Combustor 汽阀涡流喷射引擎燃烧室
TVCA Thrust Vector Control Actuator

推力矢量控制作动器
TVCD Thrust Vector Control Driver 推力矢量控制驱动器
TVCS Thrust Vector Control System 推力矢量控制系统
TVDS Toxic Vapor Detection System 毒气雾探测系统
TVE Technology Validation Experiment 技术验证试验
TVE Test Vehicle Engine 试验飞行器发动机
TVM Target Via the Missile 导弹跟踪
TVM Thrust Vector Maneuver 推力矢量动机
TVN Test Verification Network 测试（试验）验证网络
TVSA Thrust Vector Control Servoamplifier 推力矢量控制伺服放大器
TVTA Thermal Vacuum Test Article 热真空测试（试验）部件
TVV Technology Validation Vehicle 技术验证飞行器
TW Tactical Warning 战术报警
TW/AR Threat Warning/Attack Reporting 威胁报警/攻击通报
TW/SD Tactical Warning and Space Defense 战术报警与空间防御
TWAA Tactical Warning and Attack Assessment 战术预警与攻击判定
TWL Total Weight Loss 总重量损失
TWR Thrust-Weight Ratio 推重比
TWS TOMAHAWK Weapons System 【美国】"战斧"导弹武器系统

U

U&S　Utilities and Subsystem　通用系统与分系统
U/L　Uplink　上行链路
U/M　Unmanned　不载人、无人（驾驶的）
U/R　Up Range　上靶场
U/S　Upper Stage　上面级
U/V　Under Voltage　电压不足
UA　Uncontrolled Airspace　非管制空域
UACTE　Universal Automatic Control and Test Equipment　通用自动控制与试验设备
UADPS　Uniform Automatic Data Processing System　统一自动数据处理系统
UAG　Upper Atmosphere Geophysics　高层大气地球物理学
UAI　Universal Azimuth Indicator　通用方位指示器
UALA　Universal Adaptive Logic Arrays　通用自适应逻辑阵列
UAM　Underwater-to-Air Missile　潜－空导弹
UAO　Upper-Air Observation　高空观测
UAOS　Unmanned Aerial Observation System　空中无人观测系统
UAP　Unidentified Atmosphere Phenomena　尚未探明的大气现象
UAP　Upper Atmosphere Phenomena　高层大气现象
UAR　Upper Atmosphere Research　高层大气研究
UARS　Upper Atmosphere Research Satellite　【美国】高层大气研究卫星
UARV　Unmanned Airborne Reconnaissance Vehicle　无人驾驶机载侦察飞行器
UAS　Unmanned Aerial System　无人航空系统
UAS　Unmanned Aerospace Surveillance　无人空间监视
UAS　Upper Atmosphere Sounder　高层大气探测器
UASN　Upper Air Sounding Network　高空探测网
UASs　Uninhabited Aerial Systems　无人机系统
UAUM　Underwater-to-Air-to-Underwater Missile　潜－空－潜导弹
UAV　Unmanned Aerial Vehicle　无人驾驶飞行器
UAV-CR　Unmanned Aerial Vehicle-Close Range　近距无人驾驶飞行器
UAV-E　Unmanned Aerial Vehicle-Endurance　持久飞行无人机
UAV-GCS　Unmanned Aerial Vehicle-Ground Control Station　无人驾驶飞行器－地面控制站
UAV-MR　Unmanned Aerial Vehicle-Medium Range　无人驾驶飞行器－中距
UAV-SR　Unmanned Aerial Vehicle-Short Range　无人驾驶飞行器－短距离
UBIC　Universal Bus Interface Controller　通用总线接口控制器（装置）
UBM　Underground Ballistic Missile　地下发射的弹道导弹

UBM Unified Battlescale Model 统一的战场规模模型
UCAS Unmanned Combat Air System 无人操纵空战系统
UCAV Unmanned Combat Aerial Vehicle 无人驾驶作战飞行器
UCC Universal Checkout Console 通用检测控制台
UCIF Upper Composite Integration Facility 上面组合体组装厂房
UCN Uniform Control Number 统一控制编号
UCNI Unified Communication/Navigation/Identification 通信/导航/识别综合系统
UCOS Up-Range Computer Output System 上靶场计算机输出系统
UCS Uniform Coding System 统一编码系统
UCS Universal Control System 通用控制系统
UCS Unmanned Control System 无人控制系统
UCS Utilities Control System 公用设施控制系统
UDAM Universal Digital Avionics Module 通用数字式航空电子设备模块
UDLC Universal Data Link Control 通用数据链路控制
UDMH Unsymmetrical Dimethy Hydrazine 偏二甲肼
UDOP Ultra High Frequency Doppler Velocity and Position 特高频多普勒速度与位置测量系统
UDS Universal Documentation System 通用文件编制系统
UDSM User Data Summary Message 用户数据汇编信息
UEET Ultra-Efficient Engine Technology 超高效发动机技术
UEL Upper Explosive Limit 爆炸上限
UELS Unintentional Emitter Locating System 随机发射器定位系统
UER Unique Equipment Register 设备专用注册器
UERA Umbilical Ejection Relay Assembly 脐带式管缆弹出中继电器组件
UET Underground Explosion Test 地下爆炸试验
UETS Universal Emulating Terminal System 通用仿真终端系统
UEWR Upgraded Early Warning Radar 升级后的预警雷达
UF Utilization Flight 【航天飞机】利用率飞行
UFD Universal Firing Device 通用发射装置
UFO UHF Follow-On 超高频后继星（美国军事通信卫星）
UFTAS Uniform Flight Test Analysis System 【美国空军飞行试验中心】统一试飞分析系统
UGML Universal Guided Missile Launcher 通用制导导弹发射装置
UGS SR TBM Unattended Ground Sensors Support to Special Reconnaissance of Theater Ballistic Missile 对战区弹道导弹实施特殊侦察保障所用无人值守的地面传感器
UGSADCE Unattended Ground Sensor and Associated Data Communication Equipment 无人值守的地面传感器和相关的数据通信设备
UGSS Unattended Ground Sensor System 无人监控的地面传感器系统
UGT Underground Test 地下试验
UHF Ultra High Frequency 超高频

UHTC Ultra High Temperature Ceramic 超高温陶瓷
UHTS Universal Hydraulic Test Stand 通用液压测试台
UI User Interface 用户接口
UIC User Identification Code 用户识别码
UIM Unique Interface Module 单一接口模块；专用接口模块
UIP Utility Interface Panel 通用接口面板
UIS Upper Information System 高空信息系统
UIT Ultraviolet Imaging Telescope 紫外线成像望远镜
UKS United Kingdom Satellite 英国卫星
ULA United Launch Alliance 联合发射联盟
ULAIDS Universal Locator Airborne Integrated Data System 通用定位器机载综合数据系统
ULC Unpressurized Logistics Carrier 非增压后勤货舱
ULC Unpressurized Logistics Center 非增压后勤保障中心
ULF Utilization & Logistics Flight 【航天飞机】用途和后勤飞行
ULI Unmanned Lunar Impact 无人探测器月球硬着陆
ULNC Unit Logic Network Control 单元逻辑网络控制
ULO Unmanned Launch Operation 无人发射操作
ULPR Ultralow Pressure Rocket 超低压火箭（发动机）
ULRGW Ultra Long-Range Guided Weapon 超远程制导武器
ULSV Unmanned Launch Space Vehicles 无人驾驶航天器

ULT Ultimate Load Test 最大有效载荷测试（试验）
UM Upper Module 上面舱
Umb Umbilical 地面缆线和管道，脐带式管缆；脱落插头；地面缆线及管道
UMBC Umbilical Cord Cable 脐带芯电缆
UMC Uniform Mechanical Code 统一机械编码
UML Unified Modeling Language 统一建模语言
UML Universal Mission Load 多用任务载荷
UMLS Universal Microwave Landing System 通用微波着陆系统
UMO Unmanned Orbital 无人轨道
UMS Unmanned Multifunction Satellite 不载人多功能卫星
UMS Urine Monitoring System 【美国航天飞机】尿液监测系统
UMS Utilization Mission Systems 运行任务系统
UMTE Unmanned Threat Emitters 无人值守威胁辐射源
UMV Unrestricted Motion Vector 无限制的运动向量
UMVF Unmanned Vertical Flight 无人垂直飞行
UN Unipolar Navigation 单极导航
UNICOM Universal Integrated Communications System 通用综合通信系统
UNISPACE United Nations Conference on the Exploration and Peaceful Uses of Outer Space 联合国探索与和平利用外层空间大会
UNIVAC Universal Automatic Computer 通用自动计算机
UNOOSA United Nations Office for

Outer Space Affairs 联合国外层空间事务办公室
UOC Ultimate Operational Capability 最大作战能力
UOF User Operations Facility 用户操作设施
UOMS Unmanned Orbital Multifunction Satellite 无人轨道多功能卫星
UOSAT University Of Surrey Satellite 【英国】萨里大学卫星
UOSC Unmanned Orbital Satellite Communication 无人轨道卫星通信
UOSC Unmanned Orbital Satellite Controller 无人轨道卫星控制器
UOSS Unmanned Orbital Satellite System 无人轨道卫星系统
UPA Urine Processing Assembly 尿处理装置
UPACS Universal Performance Assessment and Control System 通用性能评价与控制系统
UPC Unmanned Payload Carrier 无人有效载荷运输工具
UPG Usine de Propergol Guyane 【法语】圭亚那推进剂工厂
UPLF Universal Panel Fairing 通用仪表舱整流罩
UPM Universal Processor Module 通用处理器模块
UPP Universal Plug and Play 通用型即插即用
UPP User Parameter Processor 用户参数处理器
UPS Uninterruptible Power Supply 不间断供电
UPS Uninterruptible Power System 不间断供电系统
UPS Upright Perigee Stage 立式近地点芯级
UPSTARS Universal Propulsion Stabilization, Retardation and Separation 通用推进系统稳定、延迟和分离
UPTLM Uplink Telemetry 上链路遥测
UR Ullage Rocket 气垫增压火箭
UR Universal Rockets 通用火箭
URA Uniformly Redundant Array 统一冗余阵列
URA User Range Accuracy 【卫星导航系统】用户测距精度
URBM Ultimate Range Ballistic Missile 超远程弹道导弹
URCU Upper-Stage Remote Control Unit 上面级遥控装置
UREST Universal Range, Endurance, Speed and Time 通用航程、续航能力、速度与时间
URM Universal Rocket Module 通用火箭模块
URSF Universal Remote Support Facility 通用远程支持设施
US Upper Stage 上面级
US Upper Stratosphere 高空同温层
USA Upper Stage Adapter 上面级适配器
USAFSMC US Air Force Space and Mission Center 美国空军基地航天与导弹中心
USAKA U.S. Army Kwajalein Atoll 美国陆军夸贾林基地
USAM Unified Space Application Mission 【美国国家航空航天局】统一航天应用任务
USB Upper-Stage Booster 上面级助推器
USC Ultrasonic Cleaning 超声波清洗
USCG United States Coast Guard 美国海岸警卫队
USE User Support Equipment 用户

保障设备
USE Upper Stage Engine 上面级发动机
USEF Unmanned Space Experiments Free Flyer【日本】无人空间实验自由飞行器
USERS Unmanned Space Experiment Recovery System 非载人航天实验恢复系统
USET Upper Stage Engine Technology 上面级发动机技术
USFE Upper Stage Flight Experiment 上面级飞行实验
USL US Laboratory【美国空间站】美国实验室
USML United States Microgravity Laboratory【航天飞机】美国微重力实验室
USMP United States Microgravity Payload 美国微重力有效载荷
USNTC United States National Training Centre【美国】国家训练中心
USOC User Support and Operations Center 用户支持和操作中心
USOS United States on-Orbit Segment 美国在轨段
USS Unmanned Scientific Satellite【美国国家航空航天局】无人操纵科学卫星
USS Upper Stage Simulator 上面级模拟器
USSS Unmanned Sensing Satellite System 无人传感卫星系统
UST Universal Servicing Tool 通用检修工具
USV Upper Stage Vehicle 上面级运载器
UT Umbilical Tower 脐带塔
UT Upper Troposphere 高空对流层
UT&GS Uplink Text and Graphics System 上链路文本与图形系统
UTC Universal Test Console 通用测试（试验）控制台
UTDF Universal Tracking Data Format 通用跟踪数据格式
UTE Universal Test Equipment 通用测试（试验）设备
UTRC United Technologies Research Center 联合技术研究中心
UV Ultra-Violet 紫外线
UVAS Ultraviolet Astronomical Satellite 紫外天文卫星
UVCS Ultraviolet Coronagraph Spectrometer 紫外日冕光谱仪
UVF Unmanned Vertical Flight 无人驾驶的垂直飞行
UVOIR UV-Optical-Near IR-Visible 紫外—光学—近红外—可见光
UVPI Ultraviolet Plume Instrument【美国 SDI 计划】紫外羽焰探测仪
UVS Unmanned Vehicle System 无人机系统
UVX Ultraviolet Cosmic Background Experiment 宇宙紫外背景实验
UVX Ultraviolet Experiment 紫外线实验
UWB Ultra Wide Band 超宽波段
UWS User Work Station 用户工作站

V

V Velocity 速度；速率
V Virtual System Architecture 虚拟系统体系结构
V&DA Video and Data Acquisition 视频与数据采集
V&DA Video and Data (Processing) Assembly 视频与数据（处理）组件
V&V Verification and Validation 验证与审定（尤指软件）
V-CITE Vertical-Cargo Integration Test Equipment 货物垂直组装测试设备
V-RTIF Vandenberg Real Time Interface 【美国】范登堡空军基地实时接口
V(INT)^2 Vehicle Integrated Intelligence 飞行器综合情报
V/C Vector Control 矢量控制
V/C Velocity Counter 速度（率）计数器
V/CTOL Vertical/Conventional Takeoff and Landing 垂直或常规起落
V/H Velocity/Height (Ratio) 速度与高度比
V/H Velocity-to-Height 速度－高度
V/M Velocity Meter 速度计量器
V/STOL Vertical/Short Takeoff and Landing 垂直／短距起落
V/STOL Very Short Takeoff and Landing 甚短起飞与着陆
V/TS Viewfinder/Tracking System 探视器／跟踪系统
V/V Vent Valve 排放阀
VA Value Analysis 值分析

VA Variable Area 可变区域
VA Vehicle Accommodations 运载器居住舱
VA Velocity at Apogee 远地点的速度
VA Verification Analysis 核查分析；验证分析
VA Vibroacoustic Test 振动声测试（试验）
VA Virtual Address 虚拟地址
VA/T/VTA Vibro-Acoustic/Thermal/Vacuum Test Article 振动－声音／热／真空测试件
VAA Vehicle Assembly Area 【美国国家航空航天局】运载器组装车间（区）
VAA Viewpoint Adapter Assembly 视点适配器组件
VAAC Vectored Thrust Aircraft Advanced Flight Control 推力矢量飞机先进飞行控制
VAATE Versatile, Affordable, Advanced Turbine Engine 多用途、可承受性高级涡轮发动机
VAB Vehicle Assembly Building 运载火箭总装大楼
VAC Vector Analogue Computer 矢量模拟计算机
VAC Vehicle Assembly and Checkout 运载器组装与检查；飞行器组装与检查；车辆组装与检测
VAD Velocity/Azimuth Display 速度／方位显示器
VAD Vertical Azimuth Display 垂直方位显示

VADS Vulcan Air Defense System 【美国】"火神"防空系统
VAFB Vandenberg Air Force Base 【美国】范登堡空军基地
VAHIRR Volume Averaged Height Integrated Radar Reflectivity 体积均高综合雷达发射率
VAIS Visually Artificial Intelligent Surveillance 可视人工智能监视
VAK Vertical Access Kit 垂直检修工具箱
VAK Vertical Assembly Kit 垂直装配工具箱
VAL Variable-Angle Launcher 可变方向角发射器
VAL Virtual Assembly Language 虚拟汇编语言
VAL Vulnerability Assessment Laboratory 易损性评估实验室
VALID Virtual Assembly Language Identification 虚拟汇编语言识别
VALT VTOL Approach and Landing Technology 垂直起降进场与降落技术
VAM Vulnerability Assessment and Management 易损性评估与管理
VAMOSC Visibility and Management of Operations and Support Cost 运行和保障成本可视性与管理
VAMS Vector Airspeed Measuring System 空速矢量测量系统
VANA Vector Automatic Network Analysis (or Analyzer) 矢量自动网络分析（或分析仪）
VANVIS Visual and Near-Visual Intercept System 可见与近可见截击系统
VAPI Vertical Approach Path Indicator 垂直进场航迹指示器
VAPI Visual Approach Path Indicator 目视进场航迹指示器
VAR Verification Analysis Report 验证分析报告
VAR Vertical Aircraft Rocket 垂直机载火箭
VARR Volume Averaged Radar Reflectivity 体积平均雷达反射率
VARS Vertical and Azimuth Reference System 垂直与方位参考（或基准）系统
VAS Vertical Atmospheric Sounder 大气垂直探测器
VAS VISSR Atmospheric Sounder 可见光—红外线自旋扫描辐射计大气探测器
VASI Visual Approach Slope Indicator 目视进场下滑道指示器
VASIMR Variable Specific Impulse Magnetoplasma Rocket 可变比冲磁等离子体火箭
VASN Virtual Analog Switching Node 虚拟模拟交换节点
VAST Versatile Avionics System Test 多用途电子设备与控制系统测试
VAST Vibration Analysis System Technique 振动分析系统技术
VAT Vacuum Arc Thruster 真空电弧推力器
VAT Vehicle Acceptance Test 运载器验收测试（试验）
VAT Vertical Access Tower 垂直发射勤务塔
VATA Vertical Assembly and Test Area 垂直装配与测试区
VATA Vibroacoustic Test Article 振动声测试（试验）部件
VATE Versatile Automatic Test Equipment 多用途自动测试设备
VATE Vertical Anisotropic Etch 垂直各向异性腐蚀

VATF Vibration and Acoustic Test Facility 振动与声学测试设施
VATLS Versatile Automatic Target Locating System 多用途自动目标定位系统
VATOL Vertical Attitude Takeoff and Landing 垂直起飞与着陆
VBL Voyager Biological Laboratory "旅行者"号太空探测器生物实验室
VBS Verification by Simulation 模拟验证
VBSP Video Baseband Signal Processor 视频基带信号处理器
VBW Vertical Ballistic Weapon 【德国】垂直弹道武器
VBWBN Vandenberg Base Wide Broadband Network 【美国】范登堡基地宽带网
VC Vector Control 矢量控制
VC Virtual Channel 虚拟信道
VC Vortex Combustor 涡流燃烧室
VCAM Vehicle Cabin Air Monitor 航天器舱内空气监测仪
VCATS Visually Coupled Acquisition and Targeting System 目视耦合捕捉与瞄准系统
VCB Vertical Location of the Center of Buoyancy 浮力中心的垂直位置
VCC Vandenberg Control Center 【美国】范登堡控制中心
VCC Vehicle Control Center 飞行器管制中心
VCC Verification Code Counter 验证码计数器
VCCT Variable Confinement Cookoff Test 可变限制火烧试验
VCD Vapor Compression Distillation 蒸汽压缩蒸馏
VCD Vertical Climb and Descend 垂直爬升与下降
VCD(S) Vapor Compression Distillation (Subsystem) 蒸汽压缩蒸馏(分系统)
VCE Variable Cycle Engine 变循环发动机
VCE Vehicle Cycle Engine 运载器循环发动机
VCEE Variable Cycle Experimental Engine 可变循环实验发动机
VCG Vapor Crystal Growth 蒸汽晶体生长
VCG Vertical Location of the Center of Gravity 重力中心的垂直位置
VCGS Vapor Crystal Growth System 蒸汽晶体生长系统
VCI Velocity Change Indicator 速度改变指示器
VCI Volatile Corrosion Inhibitor 挥发性防腐蚀剂
VCK Verification Check 验证检查；审核检查
VCMS Vehicle Control and Management System 飞行器控制与管理系统
VCOS Vehicle Control and Operating System 运载器控制和操作系统；车辆控制与操作系统
VCPE Vehicle Charging and Potential Experiment 运载器放电与电势实验
VCPI Virtual Control Program Interface 虚拟控制程序接口
VCRS Video Compression and Reconstruction System 视频压缩与再现系统
VCS Vehicle Control System 飞行器控制系统
VCSFA Virginia Commercial Space Flight Authority 【美国】弗吉尼亚

航天飞行管理局
VCWS Vector Control Wheel Steering 矢量（推力）控制盘操纵
VDA Valve Driver Assembly 阀门驱动组件
VDA Vapor Diffusion Apparatus 蒸汽弥散装置
VDA Variable Data Area 可变数据区
VDB Verification Data Base 验证数据库
VDEP Vehicle Design Evaluation Program 飞行器设计评审程序
VDG Vertical and Direction Gyro 垂直与航向陀螺仪
VDI Vertical Displacement Indicator 垂直位移指示器
VDI Vertical Display Indicator 垂直显示指示器
VDL Vehicle Dynamics and Control Laboratory 运载器动力学与控制实验室
VDM Venus Descent Module 【美国设想中的】金星降落舱
VDP Validation Demonstration Phase 确认/演示验证阶段（装备设计的第二阶段）
VDPI Vehicle Direction and Position Indicator 飞行器方向与位置指示器
VDR Vehicle Directional Response 飞行器方向响应
VDS Vehicle Dynamics Simulator 运载器动力模拟器；车辆动力模拟器
VDU Visual Display Unit 可视显示装置
VE Equivalent Velocity 等效速度
VE Virtual Environment 虚拟环境
VE-GAS Vernier Engine Gimbal Actuator System 游动发动机摆动作动器系统

VEB Vehicle Equipment Bay 运载火箭仪器舱
VEBAL Vertical Ballistic 垂直发射的弹道导弹
VECO Vehicle Checkout Set 火箭测试装置；车辆测试装置
VECO Vernier Engine Cutoff 游标发动机关机
VEDS Vehicle Emergency Detection System 飞行器应急探测系统
VEEGA Venus-Earth-Earth Gravity Assist 金星－地球－地球重力飞行
VEEI Vehicle Electrical Engine Interface 运载器电发动机接口
VEFCO Vertical Functional Checkout 垂直功能性检查
VEL Virtual Environment Laboratory 虚拟环境实验室
VEL Visual Electrodiagnostic Laboratory 可视电诊断实验室
VEN Variable Exhaust Nozzle 可调截面排气喷管
VEP Vehicle Evaluation Payload 日本火箭性能鉴定卫星
VER Vertical Ejector Rack 垂直弹射架
VERAS Vehicle for Experimental Research in Aerodynamics and Structures 空气动力学与结构学实验研究用飞行器
VERLORT Very Long Range Tracking Radar 非常长距离跟踪雷达
VERT VEL Vertical Velocity 垂直速度
VERTOL Vertical Takeoff and Landing 垂直起落
VES Vehicle Ecological System 运载器（或飞行器、车辆）生态系统
VESTA Vehicle for Engineering and Scientific Test Activity 工程与科学

试验用飞行器
VETS Vehicle Electrical Test System 飞行器电气试验系统
VFA Variable Flight Azimuth 可变飞行方位角
VFCS Vehicle Flight Control System 飞行器飞行控制系统
VFDR Variable Flow Ducted Rocket 可变流量冲压火箭复合发动机
VFI Verification Flight Instrumentation 验证飞行仪表
VFR Visual Flight Rules 可视飞行规则
VFT Verification Flight Test 验证飞行测试（试验）
VGH Velocity, Acceleration, Height 速度、加速度与高度
VGOR Vandenberg Ground Operations Requirement 【美国】范登堡空军基地地面操作要求
VGOR Vehicle Ground Operation Requirements 运载器地面操作要求
VGP Vehicle Ground Point 运载器地面点
VGT Vehicle Ground Test 运载器地面测试（试验）
VGVT Vertical Ground Vibration Test 垂直地面振动试验
VHAA Very High Altitude Abort 甚高度姿态中止
VHDL Very High Level Design Language 甚高层设计语言
VHDL VHSIC Hardware Description Language 超高速集成电路硬件描述语言
VHF Very High Frequency 甚高频
VHF/AM Very High Frequency Amplitude Modulator 甚高频调幅器
VHI Vehicle Hit Indicator 运载器命中指示器

VHM Vehicle Health Monitor 运载器健康监控
VHSIC Very High Speed Integrated Circuit 甚高速集成回路
VI Design Limit Speed 设计极限速度
VIA Virtual Interface Architecture 虚拟接口体系架构
VIA Voice-Interactive Avionics 人机对话式航空电子设备
VIAGRA Vertical Installation Automated Group Receiver Antenna 垂直安装自动化组接收机天线
VIB Vertical Integration Building 【美国】垂直装配厂房
VIC Vaporized Corrosion Inhibitor 汽化腐蚀抑制剂
VIC Vehicle Identification Code 飞行器识别码
VIDD Vertical Interval Data Detector 垂直间隔数据探测器
VIDS Vehicle Integrated Defense System 运载器（或车辆）综合防御系统
VIEWS Vibration Imbalance Early Warning System 振动不平衡预警系统
VIEWS Vibration Indicator Early Warning System 振动指示器预警系统
VIF Vehicle Management System Integration Facility 飞行器管理系统综合（试验）设备
VIF Vertical Integration Facility 垂直组装设施
VIICS Vehicular Inter/Intra Communications System 运载器内部通信系统；车载内部通信系统
VIL Vertical Integrated Liquid 垂直集成液体
VIM Voyager Interstellar Mission 【美国】"旅行者"号空间探测器星

际飞行任务

VIP　Verification Integration Plan　验证综合计划

VIR　Visible Infrared Radar　可视红外雷达

VIRD　Verification Implementation Requirements Document　验证实施要求文件

VIS　Verification Information System　验证信息系统

VISTA　Variable-Stability in-Flight Simulation Test Aircraft　可变稳定性飞行模拟试验机

VISTA　Very Intelligent Surveillance and Target Acquisition　高智能监视与目标截获

VITT　Vehicle Integration Test Team　运载器一体化测试（试验）组

VJ　Vacuum Jacketed　真空护套

VL　Vertical Landing　垂直降落（着陆）

VL　Vertical Lift　垂直升力

VLA　Very Large Array　甚大阵列

VLAAS　Vertical Launch Autonomous Attack System　垂直发射自主攻击系统

VLBA　Very Long Base Antenna　甚长基天线

VLBA　Very Long Base Array　甚长基阵列

VLBI　Very Long Baseline Interferometer　甚长基线干涉仪

VLBI　Very Long Baseline Interferometry　甚长基线干涉测量法（术）

VLBI　Very Long Baseline Intervention　【日本】甚长基线干涉卫星

VLF　Very Low Frequency　甚低频

VLM　Vertical Lift Module　垂直起升舱

VLM　Virtual Loadable Module　虚拟可装载模块

VLPS　Vandenberg Launch Processing System　【美国】范登堡空军基地发射处理系统

VLS　Vandenberg Launch and Landing Site　【美国】范登堡空军基地发射与着陆系统

VLS　Vandenberg Launch Site　【美国】范登堡空军基地发射场

VLS　Vehiculo Lanzador de Satellite　【巴西】卫星运载器

VLS　Vertical Launch(ing) System　垂直发射系统

VLSIC　Very Large Scale Integrated Circuit　甚大规模集成回路

VLSID　Very Large Scale Integrated Device　甚大规模集成装置

VLSM　Vertically Launched Standard Missile　垂直发射的标准导弹

VLT　Vertical Launching Test　垂直发射试验

VLV　Vanguard Launch Vehicle　【美国】"先锋"号运载火箭

VMC　Vehicle Mission Computer　运载器任务计算机

VMDI　Vector Miss-Distance Indicator　矢量脱靶距离指示器

VMF　Vertical Maintenance Facility　垂直维护设施

VMHIRR　Volume Maximum Height Integrated Radar Reflectivity　体积最大高度综合雷达反射率

VMPB　Vertical Missile Packaging Building　导弹垂直总装厂房

VMRR　Volume Maximum Radar Reflectivity　体积最大雷达反射率

VMS　Vehicle Motion Sensor　飞行器运动传感器

VMS　Velocity Measuring System　速度测量系统

VMS Vertical Motion Simulator 垂直运动模拟器

VMSEL Virtual Manufacturing and Synthetic Environment Laboratory (SNL) 虚拟制造与综合环境实验室

VNAV Vertical Navigation 垂直导航

VNIM Voyager-Neptune Interstellar Mission 【美国】"旅行者"号海王星星际飞行任务

VO Vehicle Operations 运载器操作

VOC Volatile Organic Compound 挥发性有机化合物

VOIR Venus Orbiting Imaging Radar 【美国】金星轨道成像雷达

VOV Venus Orbiting Vehicle 【美国设想中的】金星轨道飞行器

VP Vacuum Pump 真空泵

VP Verification Polarization 验证极化

VP Vertical Polarization 垂直极化

VPC Vapor Phase Compression 蒸汽段压缩

VPF Vertical Processing Facility 垂直处理（操作）厂房

VPHD Vertical Payload Handling Device 有效载荷垂直装卸设备

VPI Valve Position Indicator 阀门位置指示器

VPI Vertical Position Indicator 垂直位置指示器

VPL Virtual Processing Laboratory 虚拟（操作）处理实验室

VPLCC Vehicle Propellant Loading Control Center 飞行器推进剂加注控制中心

VPM Virtual Product Modeling 虚拟产品建模

VPQ-1 Tactical Range Threat Generator 战术靶场战区生成器

VPS Vernier Propulsion System 微调推进系统

VR Vehicle Recovery 飞行器回收

VR Virtual Reality 虚拟实境

VRBM Variable Range Ballistic Missile 可变射程弹道导弹

VRCCC Vandenberg Range Communications Control Center 【美国】范登堡靶场通信控制中心

VRCS Vector Reaction Control System 矢量反作用控制系统

VRM Venus Radar Mapper 【美国】金星雷达测绘器

VRMAT Virtual Reality Maintenance Trainer 虚拟现实维修训练器

VRML Virtual Reality Modeling Language 虚拟现实建模语言

VS Staging Velocity 芯级速度

VSD Vertical Situation Display 垂直状态显示

VSE Vehicle System Engineering 飞行器系统工程

VSE Virtual System Environment 虚拟系统环境

VSER Vertical Speed and Energy Rate 垂直速度与能量变化率

VSFC Virginia Space Flight Center 【美国】弗吉尼亚州航天飞行中心

VSI Vertical Speed Indicator 垂直速度指示器

VSI Video Simulation Interface 视频模拟接口

VSIL Virtual System Integration Laboratory 虚拟系统综合实验室

VSM Vehicle Simulation Model 飞行器模拟模型

VSRA V/STOL Research Aircraft 【美国国家航空航天局】垂直/短距起落研究机

VSS Vehicle Systems Simulator 飞行器系统模拟器

VSS Vertical Separation System 垂直分离系统

VSS Vertical Sounding System 垂直探测系统

VSS Virtual Space Simulator 虚拟空间模拟器

VSSC Vikram Sarabhai Space Center 【印度】维克兰·沙拉拜空间中心

VSTAA Vehicle Systems Technology and Architecture Analysis 飞行器系统技术与结构分析

VSTOL Vertical/ Short Takeoff and Landing 垂直/短距离起飞与着陆

VT/FT Verification Test/Flight Test 验证试验/飞行试验

VTA Vehicle Test Area 运载器测试（试验）区

VTA Vertical Test Area 垂直测试（试验）区

VTC Vectored (Vectoring) Thrust Control 矢量推力控制

VTDP Vectored Thrust Ducted Propeller 矢量推力涵道推进器

VTE Variable Thrust Engine 可变推力发动机

VTF Vertical Test Facility 垂直试验设施

VTF Vertical Test Flight 垂直试验飞行

VTL Virtual Test Laboratory 虚拟测试实验室

VTM Vertical Thrust Margin 垂直推力余量

VTM Vibration Test Module 振动测试（试验）模块

VTN Verification Test Network 验证测试（试验）网络

VTO Vertical Takeoff 垂直起飞；垂直发射

VTOGW Vertical Takeoff Gross Weight 垂直起飞总重量

VTOHL Vertical Take Off, Horizontal Landing 垂直起飞水平着陆

VTOL Vertical Take Off and Landing 垂直起飞和垂直着陆

VTRS Vandenberg Telemetry Receiving Site 【美国】范登堡遥测接收站

VTS Vandenberg Tracking Station 【美国】范登堡跟踪站

VTS Vertical Test Site 垂直试验场；垂直测试场

VTS Vertical Test Stand 垂直试验台

VTS Vertical Test System 垂直测试系统；垂直试验系统

VUIM Voyager-Uranus/Interstellar Mission 【美国】"旅行者"号太空探测器—天文星际飞行任务

VV&A Verification, Validation and Accreditation 验证、确认与认可

VV&A/C Verification, Validation and Accreditation/Certification 验证、确认与认可/认证

VV&C Verification, Validation and Certification 验证、确认与认证

VV&T Verification, Validation, and Testing 验证、确认与测试

VVCS Vertical Velocity Correction System 垂直速度校正系统

VVI Verification, Validation, and Inspection 验证、确认与检查

VVI Vertical Velocity Indicator 垂直速度指示器

VVRM Vortex Valve Rocket Motor 涡流阀火箭发动机

VVSA Velocity Vector Senor Assemblies 速度矢量传感器组件

VVT Verification, Validation, and Testing 验证、确认与测试

VVV Vertical Velocity Vector 垂直

速度矢量
VX Velocity Along the X-Axis 沿 X 轴速度
VY Velocity Along the Y-Axis 沿 Y 轴速度

VZ Velocity Along the Z-Axis 沿 Z 轴速度

W

W　Window　【航天】（火箭等的）最佳发射时限（=Launch～）；发射窗口

W&A　Warning and Assessment　警告与评估

W&B　Weight and Balance　重量与平衡

W&C　Wire and Cable　电线与电缆

W-G　Water-Glycolal　水—乙醇醛

W/ATO　With Assisted Takeoff　助推起飞

W/B　Wideband　宽（频）带的，宽波段

W/D　Wind Direction　风向

W/E　Wing Elevon　升降舵补助翼

W/E&SP　With Equipment and Spare Parts　带有装备与备件

W/L　Weapon Loading　武器悬挂；装弹；挂弹；武器安装

W/O ATO　Without Assisted Takeoff　不用助推器起飞

W/S　Work Station　工作站

W/SP　Warheads and Special Projects Laboratory　弹头与特殊项目实验室

W/T　Wind Tunnel　风洞

W/V　Wind Vector　风矢量

W/V　Wind Velocity　风速

W/W　Wheel Well　滑轮井；齿轮井

W/W　With Weapons　带（配）有武器

WA　Wide Angle　广角

WAAM　Wide Area Anti-Armor Munition　大面积穿甲弹药

WAAS　Warning and Attack Assessment　警告与袭击评估

WAAS　Wide Area Active Surveillance (Radar)　广域有源监视（雷达）

WAAT　Wide Area Applications Testbed　广域应用试验台

WAATS　Weights and Analysis for Advanced Transportation Systems　重量与先进运输系统分析

WAC　Weapon-Aiming Computer　武器瞄准计算机

WACS　Weapon-Aiming Computer System　武器瞄准计算系统

WACV　Wake Analysis and Control Vehicle　尾流分析与控制飞行器

WAD　Work Authorization Document　工作审定文件

WAD/SO　Work Authorization Document/Shop Order　工作审定文件/工作〔加工〕单

WAM　Wide Area Missile　大面积毁伤导弹

WAM　Workload Assessment Monitor　工作负荷评估监视

WAMDII　Wide-Angle Michelson-Doppler Imaging Interferometer　广角迈克逊—多普勒成像干涉仪

WAN　Work Authorization Number　工作审定号

WAR　Work Authorization Report　工作审定报告

WARM　Wartime Reserve Mode　战备模式

WARM　Weapon Assignment Resource Management　武器分配资源管理

WARN　Warning　警告；告警

WARS War Analysis and Research System 战争分析与研究系统
WARS Worldwide Ammunition Reporting System 全球弹药报告系统
WAS Weapons Alerting System 武器警戒系统
WASAAMM Wide Area Search Autonomous Attack Miniature Munition 广域搜索自主攻击小型弹药
WASP Wide Area Special Projectile 大面积特种投射武器
WASP Wide-Body Airborne Sensor Platform 宽体机载传感器平台
WAT Wide-Angle Track 大角度跟踪
WATCH Wide Angle Telescope for Cosmic and Hard X-Ray 宇宙和硬X射线的广角望远镜
WATR Western Aeronautical Test Range （美国）西部航空试验靶场
WAVES Waveform and Vector Exchange Specifications 波形与矢量交换说明
WB Wet Bulb（Thermometer） 湿球（温度计）
WB Wideband 宽带
WBB Wide-Body Booster 宽体助推器
WBDI Wideband Data Interleaver 宽带数据交错（关联）器
WBDS Wide Band Data Separator 宽带数据分离器
WBMOD Wide-Band Scintillation Model 宽波段调制载频模型
WBOSS Wide-Band Oxygen Sensor System 宽带氧传感器系统
WBR Work Bench Rack 工作台支架
WBRS Wide Band Ranging System 宽带测距系统
WBS Work Breakdown Structure 工作分解结构
WBSC Wide-Band Signal Conditioner 宽带信号调节器
WBSE Web-Based Simulation Environment 基于网络的模拟环境
WBT Wide-Band Terminal 宽带终端
WBTS Wideband Transmission System 宽带传输系统
WCCS Window Cavity Conditioning System 窗口空腔调节系统
WCCS Window Contamination Control Number 窗口污染控制数
WCCU Wireless Communications Control Unit 无线电通信控制装置
WCDB Work Control Data Base 工作控制数据库
WCDDT Wet Countdown Demonstration Test 加燃料的倒计时演示测试
WCL Warfighting Capability Level 作战能力水平
WCL Water Coolant Line 冷却水输送管线
WCL Water Coolant Loop 冷却水回路
WCL Wet Chemistry Laboratory 湿化学实验室
WCP Wing Chord Plane 翼弦平面
WCRP World Climate Research Programme 世界气候研究程序
WCS Waste Collection System 废物收集系统
WCS Waste Control System 废物控制系统
WCS Writeable Control Storage 可写控制存储器
WCSC Western Commercial Space Center 【美国】西部商业航天中心
WCTRV Winged Cargo Transport and

Return Vehicle　带翼的货物输送与返回飞行器
WD　Width　宽度
WD　Wired Discrete　有线离散；有线分立系统
WD　Work Days　工作日
WDCS　Weapon Digital Control System　武器数字控制系统
WDEL　Weapons Development and Engineering Laboratories　【美国】武器发展与工程实验室
WDNS　Weapon Delivery and Navigation System　武器投放与导航系统
WDSS-II　Warning Decision Support System Integrated Information　告警判定支持系统综合信息
WEA　Weather　天气
WEDS　Weapon Effect Display System　武器效应显示系统
WEES　Weapon Engagement Evaluation System　武器交战评估系统
WEFAX　Weather Facsimile Format　气象传真格式（气象卫星与地面站之间传输数据的一种方式）
WES　Water Electrolysis System　电解水系统
WES　Weapon Effect Simulation　武器效应（效能）模拟
WESP　Weapons Evaluation System Program　武器鉴定系统项目
WESTE　Weapons Effectiveness and System Test Environment　武器效能与系统试验环境
WET　Weapons Effectiveness Testing　武器效能测试
WETF　Weightless Environment Test Facility　失重环境试验设施
WETF　Weightless Environment Training Facility　失重环境训练设施
WETL　Weapons Evaluation Test Laboratory　武器鉴定试验实验室
WETS　Weightless Environment Training System　失重环境训练系统
WF　Wide Field　广视野；宽视场
WF/PC　Wide Field/Planetary Camera　广视野行星摄像机
WFC　Wide Field Camera　广视野照相机
WFE　Wall Plug Efficiency　光电转换效率
WFF　Wallops Flight Facility　【美国国家航空航天局】沃洛普斯飞行设施
WFIRST　Wide-Field Infrared Survey Telescope　广角红外巡天望远镜
WFO　Weather Forecast Office　天气预报办公室
WFOP/MO　Windows Flight Operations Procedures/Mission Operations Information System　【美国】微软公司的飞行操作程序/任务操作信息系统
WFOV　Wide Field of View　宽视场
WFSC　Wavefront Sensing and Control　波前传感与控制
WG　Wave Guide　波导管；导波器
WG　Wing　（机，弹）翼；翼形物
WGL　Weapon Guidance Laboratory　武器制导实验室
WGO　Wallops Geophysical Observatory　【美国国家航空航天局】沃洛普斯航天中心地球物理观测台
WGS　Water Glycolal Service Unit　水—乙醇醛供给（加注）装置
WGS　Wideband Global SATCOM　宽波段全球卫星通信
WGT　Weapon Guidance and Tracking　武器制导与跟踪
WHC　Waste and Hygiene Compartment　废水及卫生间

WHDE Weapon Handling and Discharge Equipment 武器装卸设备
WHDEVAL Warhead Evaluation 弹头鉴定
WHDS Warhead Section 弹头段；弹头截面
WHL Wheel 轮，盘；驾驶盘；起落架
WIDB Weather Information Database 气象信息数据库
WIDT Weapon Integration and Design Technology 武器集成与设计技术
WIF Water Immersion Facility 水浸装置
WIM Weapon Launch Module 武器发射模块
WIN WWMCCS (Worldwide Military Command and Control System) Intercomputer Network 全球军事指挥与控制系统部内计算机网络
WIN-T Warfighter Information Network-Tactical 作战人员信息网－战术
WINCS Wind Ion-Drift Neutral-Ion Composition 风离子漂移中性离子合成物
WINDS Weather Information Network Display System 气象信息网显示系统
WINDSAT Wind Satellite 【美国】测风卫星
WISE Women International Spacesimulation for Exploration 国际女性空间探索模拟
WISI World Index of Space Imagery 全球空间图像索引
WISP Waves in Space Plasma 空间等离子体波
WITS Weather Information Telemetry System 气象信息遥测系统

WL Wavelength 波长
WLA Weapon Lethality Analysis 武器致命性分析
WLE Wing Leading Edge 翼前缘
WM Waste Management 废物处理
WMC Waste Management Compartment 废物处理舱
WMCCS Worldwide Military Command and Control System 【美国】全球军事指挥与控制系统
WMD Weapons of Mass Destruction 大规模杀伤性武器
WME Water Membrane Evaporator 膜式水蒸发器
WME Weapon of Mass Effects 大规模效应武器
WMO World Meteorological Organization 世界气象组织
WMS Weapon Monitoring System 武器监控系统
WNC Weapons/Navigation Computer 武器／导航计算机
WO ATO Without Assisted Takeoff 不用助推器起飞
WORF Window Observation Research Facility 舷窗观测研究设施
WOTS Wallops Orbital Tracking Station 【美国】沃洛普斯轨道跟踪站
wp Argument of Perigee 近地点角距；近地点幅角
WPDS Western Pacific Data System 【美国】西太平洋数据系统
WPHM Weibull Proportional Hazards Models 威布尔比例危险模型
WPP Water Pump Package 水泵
WPP Work Package Plan 工作任务计划
WPPS Work Package Planning Sheet 工作任务计划表
WPS Words Per Second 每秒字符数

WPSS Work Preparation Support System 准备工作保障系统
WPT Wireless Power Transmission 无线电力传输
WQT Weapon Quality Track 武器质量跟踪
WR Weather Radar 气象雷达
WR Western Range 【美国】西部试验靶场
WR Wrist Roll 肘节辊
WRATS Wing and Rotor Aeroelastic Testing System 翼与旋翼气动弹性试验系统
WRCC Western Range Control Center 【美国】西部试验靶场控制中心
WRCS Weapon Release Control System 武器投放控制系统
WREBUS Weapons Research Establishment Breakup System 【澳大利亚】武器研究所爆破（自毁）系统
WRESAT Weapons Research Establishment Satellite 【澳大利亚】武器研究所卫星
WRG Wiring 线路，电路；导线
WRL Wing Reference Line 机翼基准线
WRM Water Recovery and Management 水回收与管理
WRNSC Western Range Network Segment Control 【美国】西部试验靶场网络段控制
WRO Work Release Order 工作释放指令
WRR Weapon Response Range 武器反应范围；武器投放距离
WRR Western Range Regulation 【美国】西部试验靶场条例
WRS Water Regeneration System 水再生系统
WRSS Weather Radar Simulation System 气象雷达模拟系统
WRTTM Warhead Replacement Tactical Telemetry Module 弹头替换战术遥测舱
WRWS Western Range Weather System 【美国】西部试验靶场气象系统
WS Weapon System 武器系统
WS Work Statement 布置任务，安排工作
WSAC Weapon System Analysis Capability 武器系统分析能力
WSAT Weapon System Accuracy Trial 武器系统精度试验
WSC White Sands Complex 【美国】白沙综合发射场（地）
WSC Wide-Band Signal Conditioner 宽带信号调节器
WSCC Weapon System Configuration Control 武器系统型号控制
WSCC Work Station Control Center 工作站控制中心
WSCT Weapon System Compatibility Test 武器系统兼容性试验
WSD Wide-Band Data 宽带数据
WSDA Weapon System Data Acquisition (Unit) 武器系统数据采集（装置）
WSDC Weapon System Design Criteria 武器系统设计准则
WSDD Weapon System Development and Demonstration 武器系统开发和演示
WSEF Weapon System Effectiveness Factors 武器系统效能系数
WSEM Weapon System Evaluation Missile 导弹武器系统评估
WSEP Weapon System Evaluation Program 武器系统评估程序
WSGS White Sands Ground Station

【美国】白沙地面站
WSGT　White Sands Ground Terminal　【美国】白沙靶场地面终端
WSII　Weapon Systems Intelligence Integration　武器系统智能组装
WSIL　Weapon System Integration Laboratory　武器系统综合实验室
WSM　Waste Storage Module　废物储存舱
WSMAS　Weapon Separation Motion Analysis System　武器分离运动分析系统
WSMC　Western Space and Missile Center　【美国空军】西部航天与导弹中心
WSMCR　Western Space and Missile Center Regulation　西部航天与导弹中心条例
WSMR　White Sands Missile Range　【美国】白沙导弹靶场
WSO　Water Servicer Operator　水加注车操作员（驾驶员）
WSO　World Space Organization　世界航天组织
WSOA　Weapon Systems Open Architecture　武器系统开放式结构
WSOC　Weapon System Operation Concept　武器系统作战概念
WSOC　Wideband Satellite Operations Center　宽带卫星运作中心
WSOE　Worldwide Satellite Observing Equipment　全球卫星观测设备
WSON　Worldwide Satellite Observing Network　全球卫星观测网
WSPACS　Weapon System Planning and Control System　武器系统计划与控制系统
WSR　Weapon System Reliability　武器系统可靠性
WSR　Weather Surveillance Radar　气象监视（警戒）雷达
WSRN　Western Satellite Research Network　【美国】西部卫星研究网
WSRT　Weapon System Readiness Test　武器系统完备性试验
WSS　Weapon Separation Simulation　武器分离仿真
WSS　Weapon Systems Software　武器系统软件
WSSA　Weapons System Support Activity　武器系统保障活动
WSSC　Weapon System Support Center　武器系统保障中心
WSSG　Warning System Signal Generator　告警系统信号发生器
WSSH　White Sands Space Harbor　【美国】白沙航天港；白沙空间港
WSSS　Weapon Storage and Security System　武器储存与安全系统
WST　Weapon System Trainer　武器系统训练装置
WST　Weapons Systems Test　武器系统试验
WST　World Satellite Terminal　世界卫星终端
WSTA　Waste Water Storage Tank Assembly　废水储箱装置
WSTF　White Sands Test Facility　【美国】白沙导弹靶场试验设施
WSTT　Weapon System Tactical Test　武器系统战术测试
WSU　Water Servicing Unit　水维修设备
WSW　Weapon Status Windows　武器状态窗口
WSWL　Warhead and Special Weapon Lab　弹头和特种武器实验室
WSWR　Variable Standing Wave Ratio (Rate)　可变驻（定）波系数（率）
WT　Watchdog Timer　监视计时器，

监视时钟；程序控制定时器

WT Weight 重量；质量

WTA Wire Traceability and Accountability 有线跟踪能力和可衡算（计量）性

WTF Wind Tunnel Facility 风洞设施

WTM Wind Tunnel Model 风洞试验模型

WTN Wind Tunnel Note 风洞记录

WTR Western Test Range 【美国】西部试验靶场（美国航天发射场）

WTR SYS Water System 水系统

WTSC Wet Tantalum Slug Capacitor 湿钽棒电容器

WTT Wind Tunnel Test 风洞试验

WUC Work Unit Code 工作单位代码

WUCF Work Unit Code File 工作单位代码文件

WUPPE Wisconsin Ultraviolet Photo Polarimeter Experiment 威斯康星紫外线光电偏振计实验（室）

WUSCI Western Union Space Communications 西部联盟空间通信

WVCF Western Vehicle Checkout Facility 【美国】西部靶场火箭检测设施

WVRAAM Within Visual Range Air-to-Air Missile 视距内有效的空（对）空导弹

WW Water Waste 废水

WW Wire Wrap 绕（线连）接

WWMS Waste Water Management System 废水管理系统

WWRP World Weather Research Programme 世界天气研究程序

WZTD Wet Zenith Tropospheric Delay 对流层最高湿气点延迟

X

X X-Coordinate X 坐标
X X-Windows X—窗口
X&D Experiment and Development 试验与研制
XA Auxiliary Amplifier 辅助放大器
XA Auxiliary Antenna 辅助天线
XA Extended Architecture 扩展构型，扩展构造；扩展的体系结构
XAA Extended Attribute Area 扩展属性区
XAAM Experimental Air-to-Air Missile 实验型空对空导弹
XAIM Experimental Air Intercept Missile 实验型空中拦截导弹
XAS Experimental Air Specification Weapons 实验型航空武器
XASM Experimental Air-to-Surface Missile 实验型空对水面/地面导弹
XB Experimental Bomber Prototype 实验型轰炸机样机
XBC External Block Controller 外部程序块控制器
XBM Experimental Ballistic Missile 试验性弹道导弹
XBSA X-Band Satellite Antenna X 波段卫星天线
XBSTS X-Band Satellite Tracking System X 波段卫星跟踪系统
XC Auxiliary Channel 辅助信道
XCRV Experimental Crew Return Vehicle 实验型乘员返回飞行器
XCVR Transceiver 无线电收发（两用）机；收发报机
XDA X-Ray Detector Assembly X 射线探测器组件
XDB External Data Bus 外部数据总线
XDC Control Center for Nike-X System 【美国】"奈基—X"导弹系统控制中心
XDCR Transducer 变（转）换器；传感器
XDM/X Model Experimental Development Model/ Exploratory Development Model 实验开发模型/应用开发模型
XDP X-Ray Density Probe X 射线密度探测器
XDR External Data Representation 外部数据表示
XDS Exoatmospheric Defense System 外大气层防御系统
XE Experimental Engine 实验型发动机
XECF Experimental Engine Cold Flow 实验性冷流发动机
XEPF Extended EPF 超长型整流罩
XEPF Extra Extended Payload Fairing 外延的整流罩
XEUS X-Ray Evolving Universe Spectroscopy Mission X 射线演变宇宙光谱任务
XEUS X-Ray Universe Evolution Spectrum Detection Plan 宇宙演化 X 射线光谱探测计划
XFD Crossfeed 交叉馈（供）电；横向送进
XFD X-Ray Flaw Detection X 射线探伤

XFER Transfer 转移；输送
XFT Extremely Fast Tracker 极速跟踪装置
XFV Exoskeleton Flying Vehicle 无骨架式飞行器
XGAM Experimental Guided Air Missile 实验型空中制导导弹
XIES Xenon Ion Engine System 氙离子发动机系统
XIMER Expendable Intelligent Multiple Ejector Rack 一次性智能多弹射装置滑轨
XISP Xenon-Ion Propulsion System 氙离子推进系统
XL X-Axis of Spacelab 【美国】"天空实验室空间站"的 X 轴
XLD Experimental Laser Device 实验性激光装置
XLDB Cross-Linked Composite-Modified Double-Base 交联改性复合双基（推进剂）
XLR Experimental Liquid Rocket 实验型液体火箭
XLT Experimental Lunar Tyres 实验性登月轮胎
XLTN Translation 转换；移位
XM Experimental Missile 实验型导弹
XMAS Extended Mission Apollo Simulation 【美国】"阿波罗"延长（飞行）任务模拟
XMGM Experimental Mobilelaunched Surface Attack Guided Missile 实验型机动发射的对地攻击导弹
XML Extensible Markup Language 可扩展的标识语言
XMM X-Ray Multi-Mirror Satellite X 射线多镜卫星
XMM X-Ray Multi-Mirror Mission X 射线多镜面任务
XMM X-Ray Multi-Mirror Observatory X 射线多镜观测卫星
XMM X-Ray Multimission 多任务 X 射线
XMR CCAFS Balloon Facility 3-Letter Identifier 【美国卡纳维拉尔角空军基地】气球探测装置字母识别器
XMR CCAFS Rawinsonde 3-Letter Identifier 【美国卡纳维拉尔角空军基地】无线电高空测风仪字母识别器 / 标志符
XMS Experimental Manned Spacecraft 载人试验航天器
XMT Transmit 发射
XMTR Transmitter 发射机
XNAV X-Ray Navigation X 射线导航
XNS Xerox Network Systems 静电复印网系统
XO Orbiter Structural Body Reference (X-Axis) 轨道飞行器结构体基准点 / 参考（X 轴）
XO X-Axis of Orbiter 【美国航天飞机】轨道器的 X 轴
XOS Experimental Operating System 实验型操作系统
XP Payload Structural Body Reference (X-Axis) 有效载荷结构体基准点 / 参考（X 轴）
XP X-Axis of Payload 有效载荷的 X 轴
XPAS External Payload Analysis System 外部有效载荷分析系统
XPC External Payload Carrier 外部整流罩托架
XPNDR Transponder 应答机；应答器；脉冲转发器
XPOP X-Axis Perpendicular to Orbital Plane X 轴与轨道面垂直线
xPRM Exploration Robotic Precursor Mission 探索机器人先驱任务

XRC Experimental Research Center 【美国】试验研究中心

XRC Extraterrestrial Research Center 【美国】宇宙研究中心

XRCF X-Ray Calibration Facility X 射线校准装置

XRD X-Ray Diffraction X 射线衍射

XRD X-Ray Diffractometer X 射线衍射仪

XRF X-Ray Fluorescence X 射线荧光

XRL Extended Range Lance 增程"长矛"导弹

XRP X-Ray Polychromator X 射线多色仪

XRS X-Ray Spectrometer X 射线分光仪；X 射线光谱仪

XRT X-Ray Telescope X 射线望远镜

XS X-Axis of Solid Rocket Booster 固体火箭助推器的 X 轴

XSCC China Xi'an Satellite Control Center 中国西安卫星测控中心

XSLC China Xichang Satellite Launch Center 中国西昌卫星发射中心

XSM Experimental Strategic Missile 实验战略导弹

XSM X-Ray Solar Monitor 太阳 X 射线监测仪

XSPV Experimental Solid-Propellant Vehicle 实验固体推进剂运载器

XSS Experimental Space Station 实验空间站

XSS Experimental Spacecraft System 实验飞船系统

XSSM Experimental Surface-to-Surface Missile 面对面实验导弹

XSST X-Ray Spectrometer/Spectrograph Telescope X 射线分光仪/摄谱仪望远镜

XST Experimental Stealth Technology 隐身技术实验机

XST Experimental Survivable Testbed 实验可生存性试验台

XSTA X-Band Satellite Tracking Antenna X 波段卫星跟踪天线

XT X-Axis of External Tank 【美国航天飞机】外贮箱的 X 轴

XTA X-Ray Telescope Assembly X 射线望远镜组件

XTASI Exchange of Technical Apollo Simulation Information 【美国】"阿波罗"飞船模拟器技术情报交换

XTE X-Ray Timing Explorer 【美国】X 射线定时探测器

XTM Experimental Test Model 试验测试模型

XTX Experimental Test Vehicle 实验性试验运载器（飞行器）

XUV Experimental Unmanned Vehicle 实验性无人机

XUV Extreme Ultraviolet Radiation (also EUV) 极远紫外线辐射

XV-15 Experimental Tilt-Rotor Aircraft 试验偏转旋翼飞机

XW Experimental Warhead 实验型弹头

XWS Experimental Weapons System 实验武器系统

XXI XXU/X-Ray Imager (Orbiting Solar Laboratory) 极远紫外线/X 射线成像（轨道太阳实验室）

Y

Y　Yaw　偏航；偏航（转）角
Y　Y-Axis　Y 轴
Y　Y-Axis Rate of Change　Y 轴的变化率；对 Y 轴的导数
Y　Y-Axis, Horizontal-Width of Vehicle/Structure　运载器/结构件的 Y 轴、水平宽
Y　Y-Coordinate　Y 坐标
Y/D　Yaw Damper　偏航阻尼器
YAP　Yaw and Pitch　偏航与俯仰
YAPS　Yaw and Pitch Sensing　偏航与俯仰传感（器）
YCP　Yaw Coupling Parameter　偏航耦合参数
yd　Yard　码（长度单位）
YDA　Yaw Damper Actuator　偏航阻尼作动器
YDC　Yaw Damper Computer　偏航阻尼计算机
YDMA　Yaw Damper Monitor Accelerometer　偏航阻尼监控加速计
YGC　Yaw Gyrocompass　偏航旋转罗盘
YL　Y-Axis of Spacelab　【美国】"天空实验室"空间站的 Y 轴
YM　Yawing Moment　偏航力矩
YMS　Yaw Microwave Sensor　偏航微波传感器
YO　Y-Axis of Orbiter　【美国航天飞机】轨道器的 Y 轴
YOS　Year of Service　服役年限
YP　Y-Axis of Payload　【美国航天飞机】有效载荷的 Y 轴
YP　Yield Point　屈服点
YP　Yield Pointer　屈服指示器
YP　Yield Pressure　屈服压力
YS　Y-Axis of Solid Rocket Booster　【美国航天飞机】固体火箭助推器的 Y 轴
YS　Yield Strength　屈服强度
YS　Yield Stress　屈服应力
YSE　Yaw Steering Error　偏航操纵误差
YSF　Yield Safety Factor　屈服安全因数
YSLF　Yield Strength Load Factor　屈服强度载荷因数
YSM　Prototype Strategic Missile　原型战略导弹
YSM　Yaw Steering Mode　偏航操纵模式
YT　Y Axis of External Tank　【美国航天飞机】外贮箱的 Y 轴

Z

Z	Z-Axis	Z 轴；垂直轴
Z	Z-Axis Direction	Z 轴方向
Z	Z-Axis Rate of Change	Z 轴的变化率；对 Z 轴的导数
Z	Zulu (Greenwich Mean Time-GMT)	格林尼治平均时
Z	Z-Coordinate	Z 坐标
Z/L	Zero Line	零位线；基准线
ZBB	Zero Base Budget	零基本预算
ZBO	Zero Boil Off	零蒸发损耗
ZC	Zero Calibration	零校准
ZCA	Zone Control Alarm	区域控制报警
ZCS	Zone Communication System	区域通信系统
ZCSC	Zone Communication System Control	区域通信系统控制
ZD	Zero Defects	无缺点；无故障；无缺陷；无瑕疵
ZDC	Zeus Defense Center	【美国】"宙斯"导弹防御中心
ZEL	Zero Launching	无导轨发射
ZELL	Zero Length Launching	零长发射；无导轨发射
ZERKT	Rocket Operations Plant	【俄罗斯】火箭操作间
ZERO	Prohibition of Transit of Contaminated Area	禁止穿越污染区
ZF	Zero Force	零力（分离）；无外力（级间分离）
ZFC	Zero Failure Criteria	零故障准则；零失事准则
ZFT	Zero Flight Time	无飞行时间
ZFW	Zero Fuel Weight	无燃料重量
ZG	Zero Gravity	失重
ZGE	Zero Gravity Effect	失重效应；零重效应
ZGF	(NASA) Zero Gravity Facility	【美国国家航空航天局】失重设施
ZGS	Zero Gravity Simulator	失重模拟器
ZGT	Zero Gravity Trainer	失重训练设备（员）
ZIF	Zero Insertion Force	无插拔力，零插入力
ZL	Launch Pad/Zone de Lancement	【法语】发射台；发射区
ZL	Z-Axis of Spacelab	【美国】"空间实验室"的 Z 轴
ZL	Zero Lift	零升力
ZL	Zero Line	零线；基线
ZLD	Zero-Liquid Discharge	废液零排放
ZLIS	Zero Lift Inertial System	零升力惯性系统
ZLS	Zero Landing System	零位指示着陆系统
ZLV	Zero Local Vertical (Payload Bay Toward Earth)	零局部垂直（朝向地球的有效载荷舱）
ZO	Z-Axis of Orbiter	【美国航天飞机】轨道器的 Z 轴
ZOC	Zero Operation Control	零操作控制
ZODIAC	Zone Defense Integrated Active Capability	区域防御综合有效能力
ZODIAC	Zone Digital Automatic Communication	区域数字式自动通信
ZOE	Zone of Exclusion	禁（止）区

（域）
ZOND USSR Interplanetary Probe 【苏联】星际探测器
ZOP Zero Operational Adjoint Code 零运算共轭码（伴随码）
Zp Perigee Altitude 近地点高度
ZP Preparation Zone/Zone de Préparation 【法语】准备区
ZP Z-Axis of Payload 【美国航天飞机】有效载荷的Z轴
ZP Zone of Protection （避雷针）保护区
ZPM Zero Propulsion Motor 零推进机动
ZPS Zero Pre-Breathe Suit 【美国】零预吸氧航天服
ZS Zero Shift 零偏移
ZSI Z Solar Inertial (Payload Bay Facing Away From Sun) 零太阳惰性（远离太阳的有效载荷舱）
ZSP Pyrotechnics Storage Facility/Zone de Stockage de Pyrotechnique 【法语】火工品贮存设施
ZSR Zero-G Stowage Rack 零G装载导轨
ZSRB Z-Axis of Solid Rocket Booster 【美国航天飞机】固体火箭助推器的Z轴
ZT Z-Axis of External Tank 【美国航天飞机】外贮箱的Z轴
ZTDL Zero Thrust Descent and Landing 零推力下降与着陆
ZTV Zone Trim Valve 【飞机或宇宙飞船】环境控制系统分区调节阀
ZZV Zero-Zero Visibility 能见度极差